# Biochemistry

# Biochemistry

## Mary K. Campbell

*Mount Holyoke College*

Illustrations by J/B Woolsey Associates
Chapter Introductions and art contributions by Irving Geis

SAUNDERS COLLEGE PUBLISHING
Philadelphia Fort Worth Chicago
San Francisco Montreal Toronto
London Sydney Tokyo

Text Typeface: Baskerville
Compositor: Black Dot Graphics
Acquisitions Editor: John J. Vondeling
Developmental Editors: Sandi Kiselica, Janet Nuciforo
Managing Editor: Carol Field
Project Editor: Margaret Mary Anderson
Copy Editors; Jay Freedman, Bonnie Boehm
Manager of Art and Design: Carol Bleistine
Art and Design Coordinator: Doris Bruey
Text Designer: Edward A. Butler
Cover Designer: Lawrence R. Didona
Text Artwork: J/B Woolsey Associates and Irving Geis
Layout Artist: Dorothy Chattin
Director of EDP: Tim Frelick
Production Manager: Bob Butler
Marketing Manager: Marjorie Waldron

About the cover: DNA Fibers © Phillip A.
Harrington from Fran Heyl Associates
(Inset): Computer graphic representation of
the hemoglobin molecule from a red blood
cell (erythrocyte). The cylindrical shapes are
the polypeptide chains (protein), of which
there are four in two identical pairs. The
protein combines with four bead-like discs
which contain the iron pigment, heme.
Hemoglobin is the medium by which
oxygen is transported within the body.
Laboratory of Molecular Biology,
MRC/Science Photo Library © Photo
Researchers, Inc.

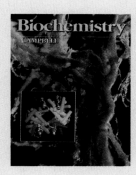

Figures p. 96, p. 164, p. 213, p. 534, 9.9, 9.10a, 9.16, 9.19, 9.21, 9.22ab, 9.25,
20.10ac, 20.12a, 20.18, p. 113, p. 114, p. 161, p. 167, p. 179, p. 180, p. 619 copyright
© Irving Geis.

Printed in the United States of America

Biochemistry

0-03-052213-7

Library of Congress Catalog Card Number: 90-052775

1234 032 987654321

## Dedication

*To everyone, especially my students, who made this book possible in every way.*

### Author
## Mary K. Campbell

Mary K. Campbell is a professor of Chemistry at Mount Holyoke College, where she frequently teaches the one-semester biochemistry course and advises undergraduates working on biochemical research projects. She received her Ph.D. from Indiana University and did postdoctoral work in biophysical chemistry at Johns Hopkins University. Professor Campbell's research interests are in the area of the physical chemistry of biomolecules, specifically, spectroscopic studies of protein–nucleic acid interactions.

Mary Campbell can be found frequently hiking the Appalachian trail with her Bernese mountain dog, Lolly.

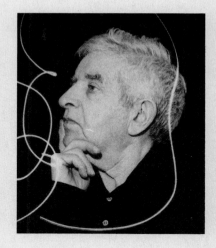

### Contributor
## Irving Geis

Irving Geis is well known for his lucid visualizations of molecular structures, particularly proteins and nucleic acids. These have appeared in *Scientific American* for the past thirty years and in major chemistry, biology, and biochemistry textbooks. He is a co-author with R.E. Dickerson, Director of the Molecular Biology Institute of UCLA, of *Chemistry, Matter and the Universe; The Structure and Action of Proteins* and *Hemoglobin: Structure, Function, Evolution and Pathology.*

In addition to drawing, painting, and writing, Irving Geis is a frequent lecturer at universities and medical schools on protein structure and function.

A recent Guggenheim fellowship made possible the assembly and cataloging of his drawings and paintings into the The Geis Archives of molecular structure.

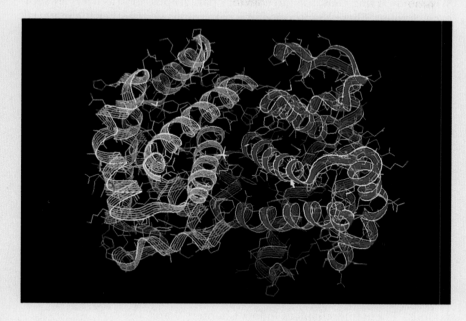

α-helixes of hemoglobin.

This text is intended for students in any field of science or engineering who want a one-semester introduction to biochemistry but who do not intend to be biochemistry majors. My main goal in writing this book is to make biochemistry as clear and interesting as possible and to familiarize all science students with the major aspects of biochemistry. There are many reasons why students in the pure and applied sciences should be acquainted with biochemistry. For students of biology, chemistry, physics, geology, and nutrition, biochemistry impacts greatly on the content and approach of their fields, especially in the areas of medicine and the pharmaceutical industry. For engineers, studying biochemistry is especially important for those who hope to enter a career in biomedical engineering.

Students who will use this text are at an intermediate level in their studies. A beginning biology course, general chemistry, and at least one semester of organic chemistry are assumed in preparation.

## SPECIAL FEATURES

### • Visual Impact

One of the most distinctive features of this text is its visual impact. Its extensive full-color art program includes artwork by the well-known biochemical illustrator, Irving Geis, and by professional artists, John and Bette Woolsey. The illustrations convey meaning so powerfully that many have become, or are certain to become, standard presentations in the field.

- **Chapter Overviews**

  Written by Irving Geis, the introductory paragraphs serve as overviews for each chapter. These chapter-opening paragraphs tie together the material from previous chapters with the topics to be discussed. They serve as building blocks for new ideas.

- **Interviews**

  Each of the five parts opens with an interview with a research biochemist, one a Nobel laureate. These outstanding scientists talk about both their research interests and teaching biochemistry. The interviews include a brief look at both their professional and personal lives and are included to encourage students to look closely at a career in science.

- **Boxes**

  Boxes on special topics, such as lactose intolerance, sickle cell anemia, and artificial sweeteners, highlight points of particular interest to students.

- **Interchapters**

  Two flexible interchapters explain the experimental methods for determining protein structure and the anabolism of nitrogen-containing compounds. These interchapters follow Chapters 9 and 18.

- **Summaries**

  Each chapter closes with a summary, a broad selection of exercises, and an annotated bibliography of up-to-date references.

- **Glossary and Answers**

  The book ends with an answer section, a glossary of important terms and concepts, and a detailed index.

## ORGANIZATION

Because biochemistry is a multidisciplinary science, the first task in presenting it to students of widely different backgrounds is to put it in context. Part I provides the necessary background and connects biochemistry to the other sciences. Part II focuses on biomolecules, while Part III discusses biochemical reactions. Metabolism is covered in Part IV. The final part of the book is devoted to the flow of genetic information.

In Part I, The Position of Biochemistry in the Sciences, three chapters relate biochemistry to other fields of science. Chapter 1 deals with the connections of biochemistry to physics, astronomy, and geology, mostly in the context of the origins of life. Chapter 2 discusses the more readily apparent linkage of biochemistry with biology, especially with respect to the distinction between prokaryotes and eukaryotes, as well as the role of organelles in eukaryotic cells. Chapter 3 builds on material familiar from general chemistry, such as buffers and the solvent properties of water, but emphasizes the biochemical point of view toward such material.

Part II, The Molecular Nature of Cellular Components, focuses on the comparison and contrast that arises from the juxtaposition of organic chemistry and biochemistry. An introduction to the section explicitly addresses the differences in emphasis and approach between the two disciplines. Chapters 5, 6, and 7 deal with amino acids and peptides, carbohydrates, and lipids. Students may have some familiarity with the first two topics from organic chemistry, but detailed discussions of lipid structure are likely to be new.

In Part III, the dynamic aspects of biochemical reactions, rather than the structure of biomolecules, become the center of attention. Thermodynamic concepts learned in general chemistry are applied specifically to biochemical topics such as coupled reactions and hydrophobic interactions. Chapters 9 and 10 cover the structure and action of proteins, one emphasizing structure and the other the behavior of proteins as enzymes. Finally, there is a discussion of the dynamics of membrane structure in Chapter 11.

In Part IV, metabolism is discussed. In Chapter 12, the connection is made between metabolism and electron transfer (oxidation-reduction) reactions. Coenzymes are introduced in this chapter and are discussed in later chapters in the context of the reactions in which they play a role. Glycolysis and the citric acid cycle are treated in Chapters 13 and 14, followed by the electron transport chain (Chapter 15) and oxidative phosphorylation. A discussion of photosynthesis rounds out the discussion of carbohydrate metabolism (Chapter 17). The catabolic and anabolic aspects of lipid metabolism are dealt with in a single chapter (Chapter 16). The metabolism of nitrogen-containing compounds such as amino acids, porphyrins, and nucleobases is covered in Chapter 18. Chapter 19 summarizes metabolism and gives an integrated look at metabolism, including a treatment of hormones and second messengers. A brief discussion of nutrition, specifically showing the relationship of nutritional requirements to metabolic pathways, is also included in this chapter.

The final section, Chapters 20–22 (Part V), deals with molecular biology. The structure of nucleic acids (Chapter 21) is presented in detail as a preparation for discussing the replication of DNA, the translation of the genetic message in RNA (Chapter 21), and the ultimate translation of that message in the synthesis of proteins (Chapter 22).

This text attempts to give an overview of important topics of interest to biochemists and to show how the remarkable recent progress of biochemistry impinges on other scientists. The length is intended to allow instructors to choose favorite topics but not to be so long as to be overwhelming for the limited amount of time available in one semester.

## ALTERNATIVE TEACHING OPTIONS

The order in which individual chapters are covered can be changed to suit the needs of specific groups of students. With students who have a strong background in organic chemistry it should be possible to omit Chapter 4 and the early sections of Chapters 5 and 6. It is possible to delay coverage of the material on thermodynamics until the start of the metabolism section or to move thermodynamics to the beginning of the course. Some instructors may prefer to cover molecular biology before intermediary metabolism, and these sections are flexible enough to be moved.

## SUPPLEMENTS

The text is accompanied by the following supplements:

- CAMPBELL'S COMPANION AND PROBLEMS BOOK by William M. Scovell (Bowling Green State University) accompanies and complements the text, with the objective of helping students gain a more comprehensive understanding of biochemistry.

  1. Each chapter begins with an introductory paragraph outlining the major topics discussed in the text.
  2. Each chapter contains Learning Objectives to help focus attention on important concepts.
  3. The heart of the book is the additional problems with detailed explanations and answers. These are intended to develop a fuller understanding of general concepts, and in some cases, focus on important details of a structure, reaction, mechanism, metabolic cycle, or pathway. In addition, some problems go beyond the text material to provide a glimpse of rapidly evolving areas.
  4. A section is included which reviews many of the more important organic reactions underlying the anabolic and catabolic reactions in biochemistry. This section provides a foundation for clearly realizing that biochemistry is simply "chemistry in living systems."
  5. A number of important topics, which are currently experiencing rapid if not explosive development, are "SPOTLIGHTED" to point out their role and impact on today's burgeoning understanding of the molecular basis of biology.

- INSTRUCTOR'S MANUAL by Mary Campbell. Includes chapter summaries, lecture outlines, answers to all exercises, and a bank of 25 multiple-choice questions for each chapter.
- OVERHEAD TRANSPARENCIES. One hundred full-color figures from the text.
- COMPUTERIZED TEST BANK. Compatible with IBM and Macintosh computers, it contains 25 multiple-choice questions for each chapter.

## ACKNOWLEDGMENTS

The help of many others made this book possible. A grant from the Dreyfus Foundation made possible the experimental introductory course that was the genesis of many of the ideas for this text. My colleagues Edwin Weaver and Francis DeToma gave much of their time and energy in initiating that course. Many others at Mount Holyoke were generous with their support, encouragement, and good ideas, especially Anna Harrison, George Hall, Lilian Hsu, Sheila Browne, Susan Hixson, Janice Smith, Jeffrey Knight, Sue Ellen Frederick Gruber, Peter Gruber, Carolyn Quarles, and Sue Rusiecki. I am deeply indebted to my colleagues at the University of Arizona, where I did much of the writing while on sabbatical. Particular thanks go to Professor Leslie Forster, in whose lab I worked both on this text and on protein fluorescence.

Many biochemistry students have used and commented on early versions of this text, with outstanding contributions by Heidi Behforouz and Clelia Biamonti at Mount Holyoke and Douglas Murrow and Stephanie Williams at the University of Arizona. The thought of my own mentors has been an inspiration to me: Walter Moore and Henry Mahler, whose untimely death was such a loss to us all, at Indiana University and Paul O. P. Ts'o at The Johns Hopkins University.

I would like to acknowledge my colleagues who contributed their ideas and critiques of my manuscript:

Robert Armstrong, Michigan State University

Bruce Banks, University of North Carolina-Greensboro

Charles H. Brueske, Mt. Union College

Clyde Denis, University of New Hampshire

Lester R. Drewes, University of Minnesota-Duluth

John R. Edwards, Villanova University

C. Dan Foote, Eastern Illinois University

Charles Grisham, University of Virginia

Richard Hewitson, Eureka College

Charles F. Hosler, Jr. University of Wisconsin-LaCrosse

Albert Light, Purdue University

Sabeeha Merchant, University of California, Los Angeles

Patrick W. Mobley, California State Polytechnic University-Pomona

Lawrence J. Tirri, University of Nevada, Las Vegas

Chen-Pei D. Tu, The Pennsylvania State University

Barbara D. Wells, University of Wisconsin-Milwaukee

The efforts of Richard Morel, Tom Thompson, Jeff Holtmeier, and Kathy Walker at Harcourt Brace Jovanovich contributed greatly to the development of the manuscript. John Vondeling, publisher, Janet Nuciforo, Sandi Kiselica, and Margaret Mary Anderson at Saunders College Publishing took over the herculean task of turning manuscript into a book. Carol Bleistine directed the art and design efforts that have had such magnificent results. I feel privileged that Irving Geis has contributed not only some of his classic illustrations but the introductory overviews to each chapter. John and Bette Woolsey created the illustrations and turned crude sketches into works of art. Computer ray trace space-filling molecular models, which set a new standard of excellence for computer art, were produced by Leonard Lessin, F.B.P.A., in conjunction with Hans Dijkman, Ph.D., and Waldo Feng. I extend my most sincere gratitude to those listed here and to all others to whom I owe the opportunity to do this book. Finally, I thank my family, whose moral support has meant so much to me in the course of my work.

**Mary K. Campbell**
November 1990

# CONTENTS OVERVIEW

Ascorbic acid.

# CONTENTS

Mimosa blossom and foliage, Louisiana.

XV

A dose of the peptide hormone melanotropin evident on the black frog.

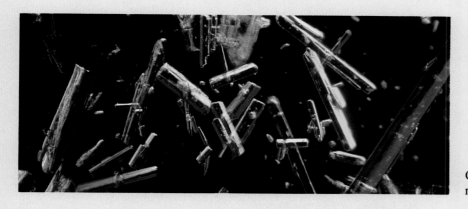

Glycine: an amino acid and possibly a neurotransmitter.

Citric acid.

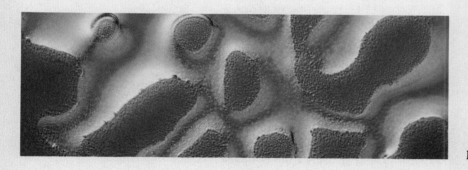

Lecithin.

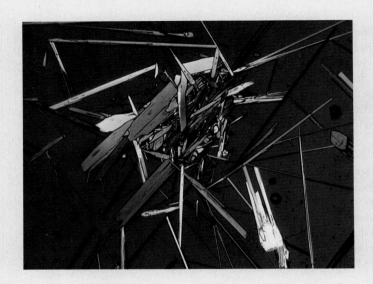

Cholesterol.

L-tryptophan.

DNA fibers.

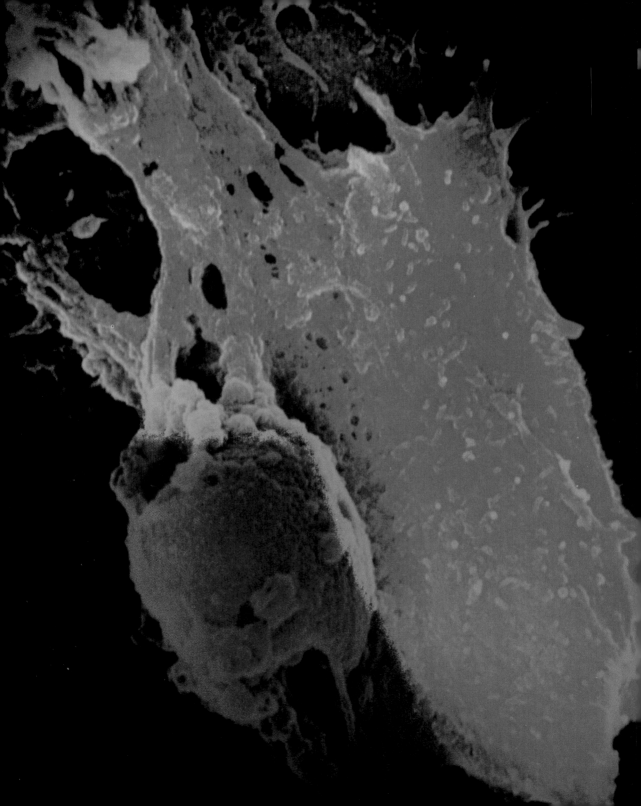

# The Position of Biochemistry in the Sciences

# INTERVIEW

# John R. Riordan

John R. Riordan, who is a Professor of Biochemistry at the University of Toronto and Director of the Cystic Fibrosis Research Program at Toronto's Hospital for Sick Children, has done research in the field of membrane biochemistry since he was a graduate student at the University of Toronto in the late 1960s. His goal is to understand how the protein molecules, embedded in the membrane that surrounds cells, accomplish both the segregation of their interior and exterior parts and the regulated communication between these compartments. One particular focus of his attention has been the common childhood disease cystic fibrosis (CF). In patients with this fatal disorder, there is defective control of the transport of inorganic ions across the membranes of epithelial cells lining the digestive and respiratory tracts. As a result, food is poorly digested and absorbed, and the lungs become severely congested and infected. In 1989, the gene that is mutated in CF was discovered in the laboratories of Dr. Riordan and his collaborators, Lap-Chee Tsui and Francis Collins. This gene, which is a member of a superfamily of membrane transporter genes, codes for a protein named CFTR (cystic fibrosis transmembrane conductance regulator). Identification of the mutations in this gene enables the screening of carriers. Knowledge of the characteristics of the gene product also provides a new basis for the development of therapies. In addition to these practical benefits, further insight is gained into the molecular mechanisms whereby the transcellular movement of salt and water is controlled. This work is an example of the constructive interplay that often takes place between applied research and that which is motivated primarily by curiosity.

**At what point did your interest in biochemistry lead you to pursue a research career?**

As an undergraduate, I was in a special program intended to prepare students for biomedical research. It was an intensive 4-year science course that included the first year of medical school. At the end of the program, I had the option of continuing in medical school or going directly into a graduate program in basic science. I had a great deal of difficulty deciding but took the latter route. At that time, basic science technology was not nearly as directly applicable to problems of human health as it is now.

Fortunately, during the past 15 years the power of the direct application of tools such as recombinant DNA technology to medical problems has grown almost exponentially and continues to do so.

**Upon completion of your Ph.D., you became a postdoctoral fellow in the laboratories of Herman Passow at the Max-Planck Institute for Biophysics in Germany. Can you briefly describe the research you pursued in this laboratory?**

My Ph.D. research dealt with enzymes at the surfaces of cells. For my postdoctoral training, I chose to go to a biophysics laboratory with the aim of learning a more quantitative, mathematically based approach to understanding ion permeation across membranes. However, I learned that my strengths were not especially in this discipline. In establishing my own laboratory, I returned to a more qualitative approach, utilizing cells in culture, their isolated membranes, and the molecules contained within the membranes. However, the main subject of research in Professor Passow's laboratory was anion transport, the area in which, 20 years later, we have found the basic CF defect to reside.

**In 1989, you published your findings on the identification of the CF gene. How long have you been working on this problem?**

The identification and characterization of the CF gene were the culmination of many years of work directed at establishing the basic defect that causes the myriad of symptoms in the disease. Certainly most of the 1980s was intensively devoted to this problem.

I was greatly influenced by the demonstrations of an epithelial cell ion conductance defect by Dr. Paul Quinton at the University of California, Riverside, and Dr. Richard Boucher of the University of North Carolina and their co-workers. Thus, in my own laboratory, we shifted our attention from the more readily available skin fibroblasts and white blood cells, which had proved useful in the study of other genetic diseases, to the epithelial cells of the sweat gland. We chose that tissue because increased salt in sweat is the most constant diagnostic feature of the disease. Unfortunately, the glands are too small to permit the types of experiments on isolated membranes and molecules that we wished to perform. However, it was possible to overcome this limitation by developing methods to prepare and expand the glandular epithelial cell populations in culture. These cultures retained the characteristics of CF, and hence we were assured that they must have the CF gene. Therefore, using RNA from cultures derived from normal individuals and patients, we could obtain copies of DNA that contained all the genes present, including the one mutated in CF. Simultaneously, Lap-Chee Tsui's laboratory, in an almost herculean effort over several years, had established and progressively refined the chromosomal localization of the CF gene. During 1988 and 1989, we were able to combine forces and isolate genes from the appropriate chromosomal localization that were expressed in sweat gland cells. One of these was found to carry a unique mutation in about 70% of CF chromosomes and had other properties that convinced us—and the rest of the world—that it was indeed the CF gene.

**How many people are affected by CF in the world? What are the symptoms of this disease? What is the mortality rate among people with CF?**

In Caucasian populations, in which CF is far more common than in others, approximately 1 in 20 individuals is a carrier of one copy of the defective gene. When two of these carriers mate, there is a 1 in 4 chance that their offspring will have CF. In North America, more than 2000 new cases are diagnosed each year. Similar numbers might be expected in Latin America, although detection and recording are incomplete. Births of those affected by CF should be more numerous in Europe than in North America. Since there are about 30,000 living patients in North America, a very crude estimate of the number worldwide might be on the order of 100,000. For most patients, the major symptoms are what amounts to malnutrition, because of an inability to digest food, and grossly impaired lung function, because of clogging of the airways and colonization by antibiotic-resistant bacteria. CF is eventually fatal for all patients, although the time of death ranges from infancy to the fourth decade. In parts of the world where health care is optimal and records are well documented, the median life expectancy is approximately 25 years.

**Will your discovery of the CF gene help others to find a cure for this disease? Will we see this cure in our lifetime?**

Knowledge of the gene and its products has served to focus all aspects of CF research on the development of new therapies.

3

Previously, all the research efforts were based on symptoms of the disease, which are complex and varied. Now that we know that dysfunction of the CFTR protein is the fundamental cause of the disease, a precise target for the action of drugs, for example, is available. A great deal more fundamental knowledge about the structure-function relationships of this protein molecule will need to be gained before the design of pharmaceuticals can be accomplished. However, progress has already been made in the delivery of the normal CFTR gene or protein. Since enzyme supplements can adequately counteract the problem of digestion and absorption of nutrients, the lung is clearly the target of new therapies. Delivery of good copies of the gene or of its protein product, and a drug that will restore function to airways are goals of further research and development. While it is impossible to say if there will be a cure in our lifetime, I am confident that major advances in control of the disease will result from these kinds of undertakings.

*Professor Riordan working in the lab with one of his colleagues, Olga Augustinas.*

### Where do you see your research going from here?

Our own CF research will focus on understanding how the CFTR protein works. For this purpose, we shall need to express it in large quantities using the techniques of molecular biology. Then we will attempt to evaluate features of its shape using the techniques of physical biochemistry and how these features change when the protein is activated by binding ATP or being phosphorylated, for example. It will also be necessary to reconstitute it into pure synthetic membranes so that its function can be analyzed in isolation, free from the influences of the many other proteins present in epithelial cells and their membranes.

### What advice can you give to students who plan to become research scientists?

New knowledge from research in the life sciences will have a tremendous impact on our world in the next few decades. Medicine, agriculture, and management of the environment will all be greatly influenced. In contrast to physics and chemistry, in which very great strides have already been made, molecular biology is a very young science in which essentially everything remains to be done.

*Jack Riordan analyzing DNA sequences in his labs at the Hospital for Sick Children in Toronto.*

Opportunities will be as great as applications of the new basic knowledge are increasingly exploited by industry. Furthermore, demographic studies indicate that researchers in the sciences will be in short supply as the new century approaches. Students considering a research career should get a substantial grounding in physical sciences and mathematics.

Although intuitive reasoning is extremely important to success in research in the life sciences, the sophisticated techniques demand a good undergraduate training in physics and chemistry. Molecular science is clearly the strongest area of current and future research in biology. In addition, it is important to maintain and nurture one's natural curiosity and inclination to question everything. Once an in-depth knowledge of one's chosen area has been achieved, research is largely a guessing game. One repeatedly has an opportunity to say, "I wonder if it may work this way," and do experiments to test the possibility. More often than not, it does not work the way you reasoned, but occasionally it does and that provides immense satisfaction. In this sense, scientific research enables one to make a career of playing with ideas, some of which can turn out to be of real significance.

# 1

# The Context of Biochemistry

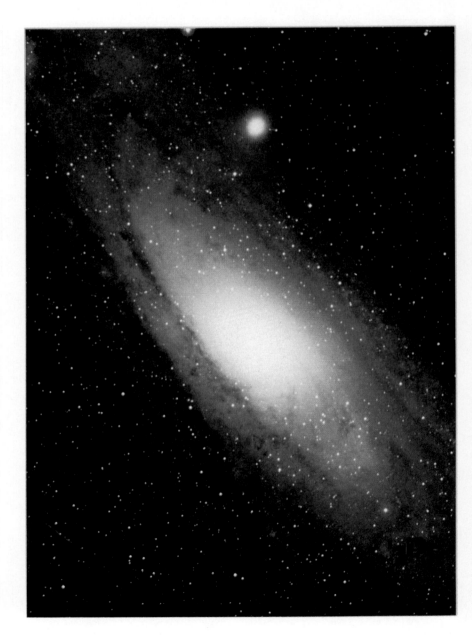

*The Andromeda Galaxy, also known as M31 and NGC 224, the nearest galaxy to the Milky Way.*

*Complex living organisms originate with simple light elements—chiefly carbon, hydrogen, oxygen, and nitrogen (C, H, O, N), along with sulfur and phosphorus. Combinations of carbon, hydrogen, and oxygen make carbohydrate molecules, from which we derive our energy. Adding nitrogen makes possible the amino acids that combine to form large protein molecules, which are essential to the machinery of life. The addition of phosphorus provides the ingredients for making DNA, the double-helical molecule containing coded instructions for synthesizing proteins. Thus living beings arise from a "building up" process going from atoms to small molecular units to large biomolecules such as proteins and the nucleic acids DNA and RNA. Collections of interacting molecules, encased in a suitable membrane, become a cell—the basic unit of life. Even the most primitive living cell has the capacity to take in and process nutrients from the environment as well as to reproduce itself and to change with changing conditions. Although living organisms show fantastic variation in size, shape, and temperament, their components and processes at the molecular level are remarkably alike.*

## 1.1
## SOME BASIC THEMES

Living organisms, and even the individual cells of which they are composed, are enormously complex. Despite the diverse complexity of the living world, certain unifying features are common to all things that live. All make use of the same types of *biomolecules* and all use energy. As a result, they can be studied by the methods of chemistry and physics. The belief in "vital forces" (forces supposedly existing only in living organisms) held by 19th century biologists has long since given way to the understanding that there is an underlying unity throughout the natural world. Other disciplines that appear to be unrelated can provide answers to important biochemical questions. The discovery in physics, for example, that X-rays can be diffracted by crystals led to the elucidation of the three-dimensional structure of molecules as complex as proteins and nucleic acids. Biochemistry is a field that draws on many disciplines, and its multidisciplinary nature allows it to use results from many sciences to answer questions about the *molecular nature of life processes.*

As cells evolved from comparatively simple unicellular organisms (similar to bacteria) to more complex ones such as those in multicelled plants and animals, they developed a variety of *metabolic pathways* as adaptations to their environments. Certain organisms, both single-celled and multicellular, are capable of *photosynthesis;* some organisms have metabolic pathways for the *aerobic oxidation* of foodstuffs. While several pathways are specific to restricted groups of organisms, many metabolic pathways are common to all organisms. The glycolytic pathway by which foodstuffs are oxidized anaerobically is an example. Many important processes such as protein synthesis can be studied in bacteria and the results applied with little change to·more complex organisms. *Escherichia coli,* for example, is an organism widely used in laboratory studies. Like all bacteria, it is easily grown, has a short generation time, and is reasonably easy to deal with in large quantities. The basic features of many important processes such as protein synthesis were elucidated in *E. coli,* and the general results have been applied to all organisms. It is helpful to use *model systems* that are comparatively simple but that still have the required specificity of the system of interest to provide a useful approach to understanding complex processes.

The metabolic relationship between *E. coli* and humans, which undeniably exists in spite of the differences between these two species, makes it interesting to speculate on the origins of life. Such speculations can be quite illuminating. Even comparatively small biomolecules such as amino acids and nucleotides have structures that consist of several parts. Large biomolecules such as proteins and nucleic acids have highly complex structures, and living cells are enormously more complex. However, *both molecules and cells must have arisen ultimately from very simple molecules* such as water, methane, ammonia, nitrogen, and hydrogen. These simple molecules must have arisen in turn from atoms. The way in which the universe itself, and the atoms of which it is composed, came to be is a topic of great interest to astrophysicists as well as other scientists. We have raised a number of questions here, and they touch on many scientific disciplines. However, we can narrow them down to specific points that are useful for the purposes of this text. A brief look at some theories of the origins of life will point out some central questions in biochemistry and the connections between biochemistry and other sciences.

## 1.2
## ORIGINS OF LIFE

### The Earth and Its Age

To date, we are aware of only one planet that supports life: our own earth. The earth and its waters are universally understood to be a source and mainstay of life, and a natural first question is how the earth came to be, along with the universe of which it is a part. At the present time, the most widely accepted cosmological theory for the origin of the universe is the *big bang,* a cataclysmic explosion that marked the beginning of the universe.

According to big bang cosmology, all the matter in the universe was originally confined to a comparatively small volume of space. As a result of a tremendous explosion, this "primordial fireball" started to expand with great force. Immediately after the big bang the universe was extremely hot, on the order of 15 billion ($15 \times 10^9$) K. (Kelvin temperatures are written without a degree symbol.) The average temperature of the universe has been decreasing ever since as a result of expansion, and the lower temperatures have permitted the formation of stars and planets.

The composition of the universe in its earliest stages was fairly simple. Hydrogen, helium, and some lithium were present, having been formed in

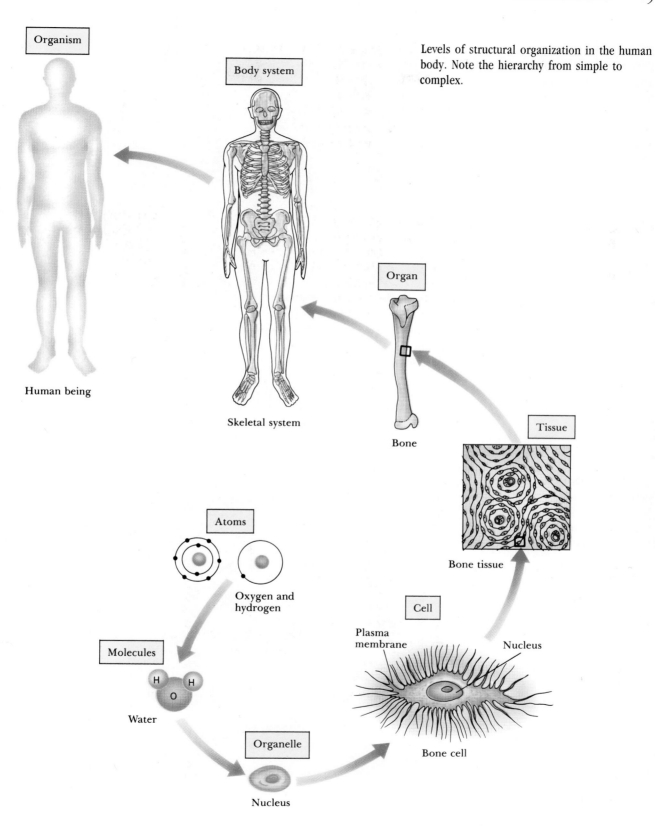

Organism

Human being

Body system

Skeletal system

Organ

Bone

Tissue

Bone tissue

Cell

Plasma membrane

Nucleus

Bone cell

Organelle

Nucleus

Molecules

H    H
  O

Water

Atoms

Oxygen and hydrogen

Levels of structural organization in the human body. Note the hierarchy from simple to complex.

The 67-m radiotelescope. Radioastronomy has provided strong evidence for the big bang theory of the origin of the universe and has detected important precursors of biomolecules, such as formaldehyde and hydrogen cyanide, in interstellar space.

the original big bang explosion. The rest of the chemical elements are thought to have been formed in three ways: first, by the thermonuclear reactions that normally take place in stars; second, in the explosions of stars; and third, by the action of cosmic rays outside the stars since the formation of the galaxy. The process by which the elements are formed in stars is a topic of interest to chemists as well as to astrophysicists. For our purposes it is noteworthy that the most abundant isotopes of biologically important elements such as *carbon, oxygen, nitrogen, phosphorus, and sulfur*

The Hubble space telescope was designed to provide images about seven times clearer than can be obtained with ground-based telescopes.

**TABLE 1.1     Relative Abundance of Important Elements***

| ELEMENT | ABUNDANCE IN ORGANISMS | ABUNDANCE IN UNIVERSE |
|---|---|---|
| Hydrogen | 80–250 | 10,000,000 |
| Carbon | 1,000 | 1,000 |
| Nitrogen | 60–300 | 1,600 |
| Oxygen | 500–800 | 5,000 |
| Sodium | 10–20 | 12 |
| Magnesium | 2–8 | 200 |
| Phosphorus | 8–50 | 3 |
| Sulfur | 4–20 | 80 |
| Potassium | 6–40 | 0.6 |
| Calcium | 25–50 | 10 |
| Manganese | 0.25–0.8 | 1.6 |
| Iron | 0.25–0.8 | 100 |
| Zinc | 0.1–0.4 | 0.12 |

*Abundance is given as the number of atoms relative to a thousand atoms of carbon.

*have particularly stable nuclei.* These elements were produced by nuclear reactions in first-generation stars, the original stars produced after the beginning of the universe (Table 1.1). Many of these stars were destroyed by explosions called supernovas, and their stellar material was recycled to produce second-generation stars such as our own sun, along with our solar system.

The age of the earth (and the rest of the solar system) can be determined by radioactive dating methods, which make use of the decay of unstable nuclei. The results indicate that the earth is 4 to 5 billion ($4-5 \times 10^9$) years old. The atmosphere of the early earth was very different from the present one and probably went through several stages before reaching its present composition. The most important difference is that, according to most theories of the origins of the earth, very little or no free oxygen ($O_2$) existed in the early stages (Figure 1.1). The early earth was constantly irradiated with ultraviolet light from the sun, since there was no ozone ($O_3$) layer to block it. Under these conditions, the chemical reactions that produced simple biomolecules took place.

The gases usually postulated to have been present in the atmosphere of the early earth include $NH_3$, $H_2S$, $CO$, $CO_2$, $CH_4$, $N_2$, $H_2$, and $H_2O$ (present in both liquid and vapor forms). However, there is no universal agreement on the relative amounts of these components from which biomolecules ultimately arose. Many of the earlier theories of the origin of life postulated $CH_4$ as the carbon source, but more recent studies have shown that appreciable amounts of $CO_2$ must have existed in the atmosphere at least 3.8 billion ($3.8 \times 10^9$) years ago. This point is based on geological evidence about the age of the earliest known rocks. These rocks are 3.8 billion years old, and they are carbonates, which arise from $CO_2$. Any $NH_3$ originally present must have dissolved in the oceans, leaving $N_2$ in the atmosphere as the nitrogen source required for formation of proteins and nucleic acids.

**FIGURE 1.1**    Conditions on earth would have been inhospitable for most of today's life forms. Very little or no free oxygen ($O_2$) existed. Volcanoes erupted, spewing gases, and violent thunderstorms produced torrential rainfall that covered the earth.

## Biomolecules

Experiments have been done in which the simple compounds of the early atmosphere were allowed to react under the various sets of conditions that might have been present on the early earth. The variations have included using $CH_4$, $CO_2$, and CO as the carbon source and $NH_3$ and $N_2$ as the nitrogen source. Liquid water and hydrogen gas have usually been present in the reaction mixture. Energy is supplied either as ultraviolet light or, more frequently, as an electrical discharge that simulates lightning. The components are allowed to react for 7 to 10 days. The results of such experiments indicate that these simple compounds react *abiotically* (*a,* "not" and *bios,* "life"), or as the word indicates, in the absence of life, to

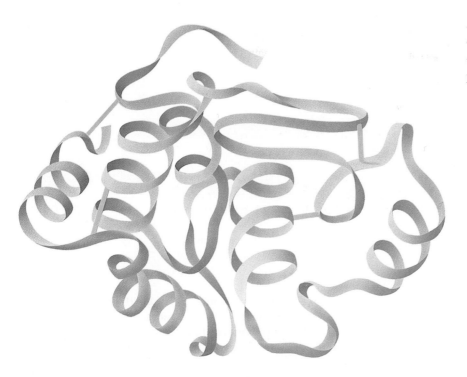

The ribbon structure of a protein. The order of amino acids determines the three-dimensional folding of the protein.

give rise to biologically important compounds such as *amino acids, sugars, purines, and pyrimidines.* According to one theory, reactions such as these took place in the earth's early oceans; however, other researchers postulate that such reactions took place on the surface of clay particles present on the early earth. It is certainly true that mineral substances similar to clay can serve as catalysts in many types of reactions. Both theories have their proponents, and more research will be needed to answer the many questions that remain.

Other important reactions can take place under abiotic conditions. For example, purines and pyrimidines can be linked with sugars such as ribose or deoxyribose to produce **nucleosides.** Nucleosides in turn can be linked to a phosphate group or groups to give rise to **nucleotides,** mimicking an abiotic process that could have taken place in the presence of phosphate ions of mineral origin in the earth's early oceans. All the compounds produced this way are significant not only because they are biologically important themselves but because of the large molecules that are produced when they combine further.

Living cells are assemblages that include very large molecules, such as proteins, nucleic acids, and polysaccharides, which are larger by many powers of ten than the amino acids, nucleotides, and simple sugars from which they are built. Hundreds or thousands of these smaller molecules or **monomers** can be linked to produce macromolecules, which are also called **polymers.** In present day cells, amino acids combine by polymerization to give **proteins,** and nucleotides combine to give **nucleic acids;** the polymerization of sugars produces polysaccharides. Polymerization experi-

**FIGURE 1.2**  The order in which monomers are linked together can produce different molecules in heteropolymers. (a) A homopolymer: all monomers are the same. (b) Two heteropolymers with monomers linked in different order. It is possible for two heteropolymers to have the same number and kind of monomer units and to have different properties because the order in which the monomers are linked is different.

(a)

$$\text{X—X—X—X—X—X—X—X—X—X} \cdots \text{X}$$
$$1 \quad 2 \quad 3 \quad 4 \quad 5 \quad 6 \quad 7 \quad 8 \quad 9 \quad 10 \cdots n$$

**X** represents a monomer

(b)

$$\text{X—Z—V—Y—W—Z—W—X—Y—V} \cdots \text{Z}$$
$$1 \quad 2 \quad 3 \quad 4 \quad 5 \quad 6 \quad 7 \quad 8 \quad 9 \quad 10 \cdots n$$

$$\text{V—W—X—Y—Z—X—V—Y—W—Z} \cdots \text{X}$$
$$1 \quad 2 \quad 3 \quad 4 \quad 5 \quad 6 \quad 7 \quad 8 \quad 9 \quad 10 \cdots n$$

**V, W, X, Y, Z** represent different monomers

ments with amino acids carried out under early earth conditions have produced *proteinoids,* which are proteinlike polymers. Similar experiments have been done on the abiotic polymerization of nucleotides and sugars.

A polymer in which all the monomers have the same chemical identity is called a *homopolymer.* Since all the monomers are the same, the order in which they are linked makes no difference in the nature of the polymer. In contrast, different types of monomers can be linked together to form a *heteropolymer,* in which the order of monomer units is not necessarily the same as another polymer of the same size and overall composition. Thus, the order in which the monomers are linked together can make an important difference in the properties of the molecule (Figure 1.2). For example, there are several different types of amino acids and nucleotides, and they can easily be distinguished from one another. The sequence of amino acids determines the properties of proteins. The genetic code lies in the sequence of purines and pyrimidines in nucleic acids. In polysaccharides the order of monomers does not have an extremely important effect on the properties of the polymer, nor does the order of the monomers carry any genetic information.

The effect of monomer sequence on the properties of polymers can be illustrated by another example. Proteinoids are artificially synthesized polymers of amino acids, and their properties can be compared with those of true proteins. There is some evidence that the order of amino acids in artificially synthesized proteinoids is not completely random. There is a preferred order, but not a definite amino acid sequence. However, *a well-established, unique amino acid sequence exists for each protein produced by present day cells.* Proteinoids, like the class of proteins called *enzymes,* display **catalytic activity,** which means that both proteins and proteinoids increase the rates of chemical reactions in comparison to uncatalyzed reactions. As might be expected, enzymes are far more effective than proteinoids at increasing the rates of reactions.

The role of lightning in the synthesis of biomolecules on the early earth has been simulated by electric discharges in laboratory experiments on abiotic synthesis of amino acids.

The specific amino acids present and their sequence ultimately determine the properties of all types of proteins, including enzymes. The order in which individual amino acids are linked determines the three-dimensional structure of the protein, also called its *conformation,* which in turn determines its function and mode of action. One of the most important functions of proteins is catalysis, and the catalytic effectiveness of a given enzyme depends on its amino acid sequence.

The sequence of amino acids in proteins is determined in present day cells by the sequence of purines and pyrimidines in nucleic acids, and the process by which the transfer of this genetic information takes place is very complex. *DNA,* one of the nucleic acids, serves as the genetic coding material. The **genetic code** is the means by which the information for the structure and function of all living things is passed from one generation to the next. The workings of the genetic code are no longer completely mysterious, but they are far from completely understood. Theories of the origins of life have the greatest difficulty explaining the rise of a coding system, and new insights in this area could throw some light on the present-day genetic code.

## Molecules to Cells

One of the most puzzling questions about the origin of life is which came first: catalytic activity (associated with proteins) or coding (associated with nucleic acids)? Another way of saying this is which came first: proteins or nucleic acids? There are proponents for both viewpoints, but both will require modification in light of the recent discovery that *RNA,* another nucleic acid, is capable of catalyzing its own further processing. Until this discovery, catalytic activity was associated exclusively with proteins. RNA, rather than DNA, is considered by many scientists to have been the original coding material, and it still serves this function in some viruses. The idea of catalysis and coding both occurring in one molecule will certainly provide a point of departure for more research on the origins of life. (See the article by Cech in the bibliography at the end of this chapter.)

In the theory that proteins came first, it is thought that aggregates of proteinoids formed on the early earth, probably in the oceans or at their edges. Other abiotically produced precursors of biomolecules were taken up into these aggregates. Model systems for *protocells,* the aggregates that led to true cells, have been devised by several workers. In one type of model, artificially synthesized proteinoids are induced to aggregate, forming structures called **microspheres.** Proteinoid microspheres are spherical in shape, as the name implies, and in a given sample they are approximately uniform in diameter (Figure 1.3). Such microspheres are certainly not cells, but they do provide a model for protocells.

Microspheres prepared from proteinoids with catalytic activity will exhibit the same catalytic activity as the proteinoids. Furthermore, it is possible to construct such aggregates with more than one type of catalytic activity. In some cases enzymes isolated from living cells are used instead of

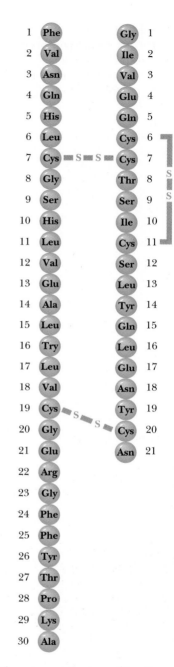

The primary structure of insulin, showing the amino acid sequence. A protein with the same number and kinds of amino acids in a different order is a different protein.

**FIGURE 1.3** Proteinoid microspheres, a type of protobiont, are tiny spheres (1-2 $\mu$m in diameter) that exhibit some of the properties of life.

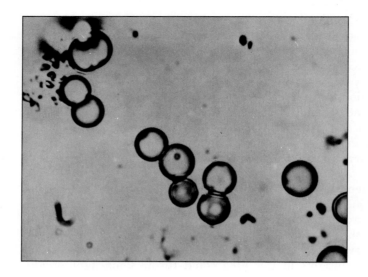

artificially synthesized proteinoids. A collection of enzymes is far from being a living cell even though the enzymes were produced by a cell. A microsphere incorporating a suitable group of enzymes can mimic a metabolic pathway. For example, a microsphere can be prepared containing an enzyme that catalyzes the formation of starch by polymerization of the simple sugar glucose. Another microsphere can be formed under the same conditions, differing from the first only by the addition of another enzyme that catalyzes the breakdown of starch to maltose, another carbohydrate. This microsphere is a model for a carbohydrate metabolic pathway involving formation and breakdown of a polymer.

In the theory that proteins were first, the rise of a coding system is a crucial point but the one with the most tenuous support from experimental models. It is more difficult to produce models for nucleic acids, especially for DNA, under abiotic conditions that it is to produce proteinoids. Some protonucleic acids have been produced, however, under abiotic conditions. The incorporation of protonucleic acids into microspheres to produce protocell models with an effective coding system presents a major challenge for researchers in this field.

According to the theory that nucleic acids arose before proteins, the appearance of a form of RNA capable of coding for its own replication was the central point in the origin of life. Experiments have been done under various sets of conditions to test whether RNA can be polymerized from monomers in the absence of the factors usually required in the reaction mixture. Such factors include a preexisting RNA to be copied or an enzyme to catalyze the process. (These experiments were done before the discovery, as previously mentioned, that some currently existing RNAs can catalyze their own processing once they have been formed by other means.) It has been shown that RNA can be produced in the absence of *one* of these factors, but replication in the absence of *both* factors has yet to be demonstrated. "Naked" RNA, that is, RNA not associated with any other substance, does not exist independently. Even viruses, the simplest entities capable of replicating themselves, consist of a nucleic acid core, which may be RNA or DNA, and a protein coat. Many unanswered questions about the role of RNA in the origin of life still exist, but it is clear that RNA must play an important role in the process.

Attempts have been made recently to combine these two approaches into a *double-origin theory.* According to this idea, the development of metabolic pathways and the development of a coding system came about separately, and the combination of the two produced life as we know it. The rise of aggregates of molecules capable of catalyzing metabolic reactions was one origin of life, and the rise of a nucleic acid–based coding system was another origin. A double-origin theory acknowledges the importance of coding and also provides an answer to the objection based on instability of unprotected RNA.

The theory that life began on clay particles, which was mentioned earlier, is a form of double-origin theory. According to this point of view, coding arose first, but the coding material was the surface of naturally occurring clays. The pattern of ions on the clay surface is thought to have served as the code, and the process of crystal growth was responsible for replication. Simple molecules, and then protein enzymes, arose on the clay surface, eventually giving rise to aggregates similar to the microspheres already described. At some later date, the rise of RNA provided a far more efficient coding system than clay, and RNA-based cells replaced clay-based cells.

At this writing none of the theories of the origin of life are definitely established and none are definitely disproved. The topic is still under active investigation.

## 1.3
## DEVELOPMENT OF CELLS

Once true cells, no matter how primitive, came into existence, they started to develop in response to their environment. First, they required energy to maintain their cellular organization. Organisms need energy to carry out cellular functions, to grow, and to reproduce. The very earliest cells could have met their energy needs by taking in molecules that could be broken down in reactions that release energy. The process of breakdown of nutrient molecules to provide energy is called **catabolism.** The molecule **ATP** (*a*denosine *tri*phosphate), an energy carrier universally found in organisms, must have served the function of providing energy from the earliest times. ATP has been produced under abiotic conditions, but the earliest cells would have depleted the supply of ATP in the environment very quickly. In response to this need for energy, cells evolved the process of *fermentation,* a metabolic pathway in which sugars are broken down to produce simpler organic molecules. Another feature of the fermentation process is that, in addition to breakdown of sugars, ATP is produced by the reaction of a similar molecule, **ADP** (*a*denosine *di*phosphate), with phosphate ion, designated $P_i$.

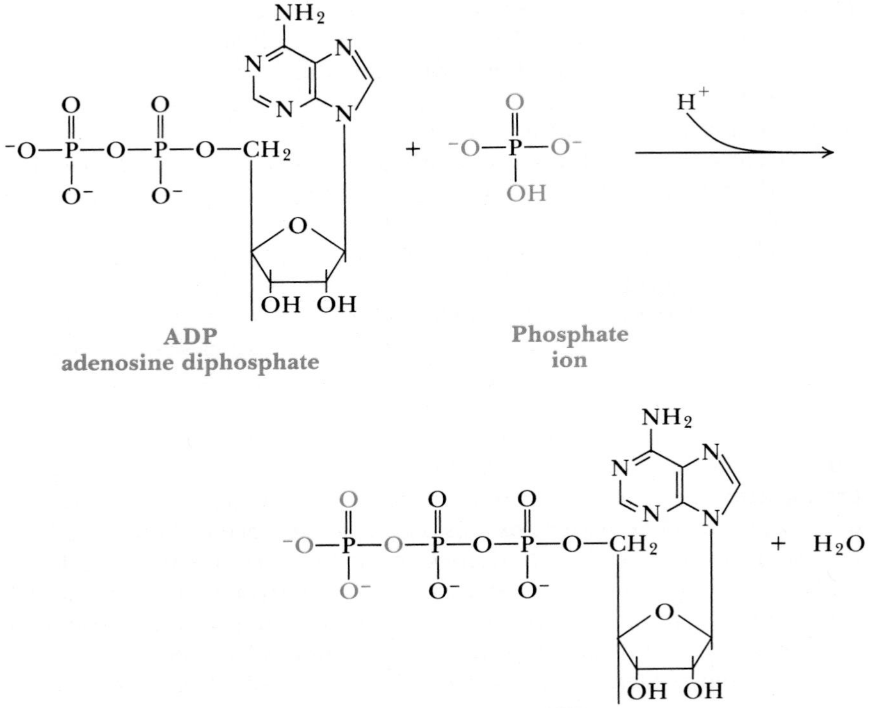

ADP
adenosine diphosphate

Phosphate
ion

ATP
adenosine triphosphate

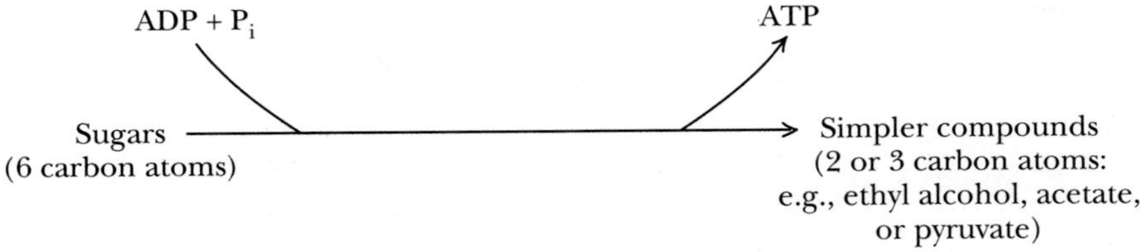

ADP + P$_i$ → ATP

Sugars
(6 carbon atoms)

Simpler compounds
(2 or 3 carbon atoms:
e.g., ethyl alcohol, acetate,
or pyruvate)

**Basic equation of fermentation**

The process of fermentation is *anaerobic,* that is, it operates in the absence of oxygen. However, in the breakdown of sugars in fermentation, oxidation takes place; the term *oxidation* as used by scientists today means *loss of electrons,* not just reaction with oxygen. The oxidation of sugars produces energy, which is used by the cells to form ATP from ADP. The ATP that is produced in this way is subsequently broken down to release energy. Cells trap energy from the oxidation of sugars in the chemical energy of ATP, and this chemical energy is used as needed.

The evolution of fermentation gave rise to other problems for the first organisms. As early cells produced ATP by fermentation, they depleted the supply of abiotically produced foodstuffs such as sugars, amino acids, and nucleotides. At that point the growth of cells was limited by the abiotic production of new food molecules in the environment (resulting from the action of ultraviolet light from the sun on simpler molecules). The stage was set for the development of *autotrophic organisms,* organisms that could

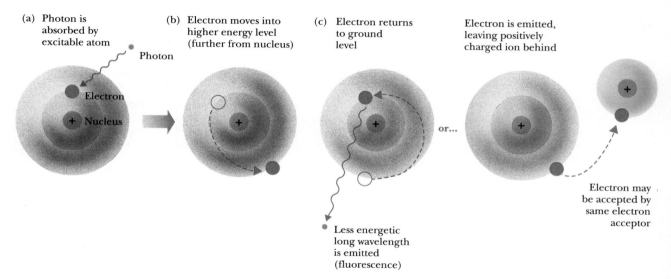

(a) Photon is absorbed by excitable atom

Photon

Electron

+ Nucleus

(b) Electron moves into higher energy level (further from nucleus)

(c) Electron returns to ground level

or...

Less energetic long wavelength is emitted (fluorescence)

Electron is emitted, leaving positively charged ion behind

Electron may be accepted by same electron acceptor

**FIGURE 1.4** (a) A photon of light energy strikes an atom or the molecule of which the atom is a part. (b) The energy of the photon may push the electron to an orbit farther from the nucleus. (c) If the electron "falls" back to the next lower energy level, a less energetic photon is re-emitted. Alternatively, if the appropriate electron acceptors are available, the electron may leave the atom. In photosynthesis a chain of such acceptors captures the energy of the electron.

produce their own food from simpler molecules. Some organisms developed pathways for the synthesis of important biomolecules from simpler molecules. This process is called **anabolism.** Some types of organisms developed a particularly efficient way of extracting energy from the environment by using light energy from the sun to drive the synthesis of carbohydrates, the process of **photosynthesis.** The trapping of light energy is a separate reaction from the synthesis of carbohydrates (a reaction that can and does take place in the dark), and the two processes combine in photosynthesis. The two processes, light trapping and synthesis of carbohydrates, are called the *light* and *dark reactions,* respectively, and they are treated separately in discussions of photosynthesis.

Anabolic processes such as the formation of sugars from simpler molecules require energy and take place as a result of *reduction* reactions, ones in which *the reacting substance gains electrons.* This anabolic process is the dark reaction of photosynthesis. Six molecules of carbon dioxide are reduced to one molecule of a six-carbon sugar ($C_6H_{12}O_6$); in the process of reduction there is a transfer of hydrogen atoms as well as electrons.

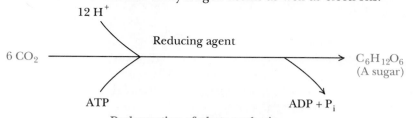

$12\ H^+$

Reducing agent

$6\ CO_2$ ⟶ $C_6H_{12}O_6$ (A sugar)

ATP

$ADP + P_i$

**Dark reaction of photosynthesis**

The ultimate source of energy is the sun, and the other reaction that is a part of photosynthesis requires light. ATP is produced as a result of the light reaction, which is also a reduction reaction. We will discuss the details of this reaction and photosynthesis in general in Chapter 17. For now it is enough to say that a source of electrons (a reducing agent) must be present in the environment in large enough quantities to serve the needs of organisms as they carry out the light reaction of photosynthesis (Figure 1.4). Another point is that the reducing agent must be able to give up its electrons fairly easily. The earliest examples of photosynthesis were not as

19

efficient as are modern organisms, so they needed even better reducing agents.

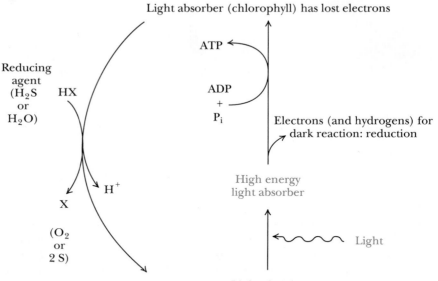

**Light reaction of photosynthesis**

One of the common reducing agents for the earliest forms of photosynthesis must have been hydrogen sulfide ($H_2S$), which gives up electrons easily; it was also present in quantity on the early earth as a result of volcanic activity. The first photosynthetic organisms were probably similar to the present-day green sulfur bacteria. These organisms use $H_2S$ as the reducing agent for photosynthesis, producing food for the organism

These stromatolites at Shark Bay in Western Australia are approximately 2000 years old. The formations are composed of mats of cyanobacteria and minerals like calcium carbonate. Some fossil stromatolites are 3.5 billion years old.

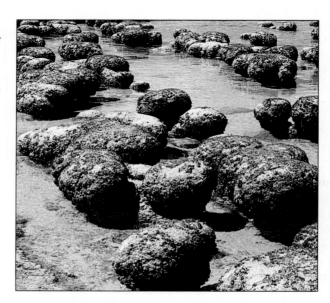

as well as elemental sulfur (S) as products of the reaction. The whole process is still anaerobic since no molecular oxygen ($O_2$) is produced.

If water ($H_2O$) is the reducing agent in photosynthesis, molecular oxygen ($O_2$) is produced. There is an advantage for organisms that use water rather than hydrogen sulfide as a reducing agent: much more $H_2O$ is now present in the environment than $H_2S$. The drawback is that it is much more difficult for water to give up its electrons than for hydrogen sulfide to do so. Some organisms, probably similar to present-day *Cyanobacteria*, eventually developed the form of photosynthesis that uses water as the reducing agent and produces oxygen as a by-product. Their presence made a drastic change in the earth's atmosphere by markedly increasing the amount of free oxygen available.

Atmospheric oxygen in large quantities presented a new challenge to the organisms that existed at the time. Many of them could not tolerate this gas, which was toxic to them, and the descendants of these organisms became the obligate anaerobes that today occupy specialized ecological niches in the absence of oxygen. However, the fact that oxygen is highly reactive made possible the eventual rise of *aerobic metabolism,* as opposed to anaerobic metabolism.

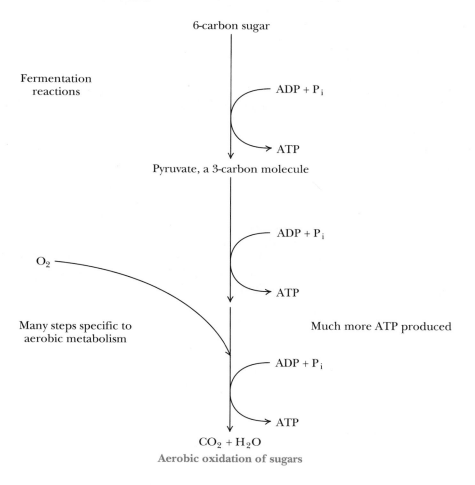

Aerobic oxidation of sugars

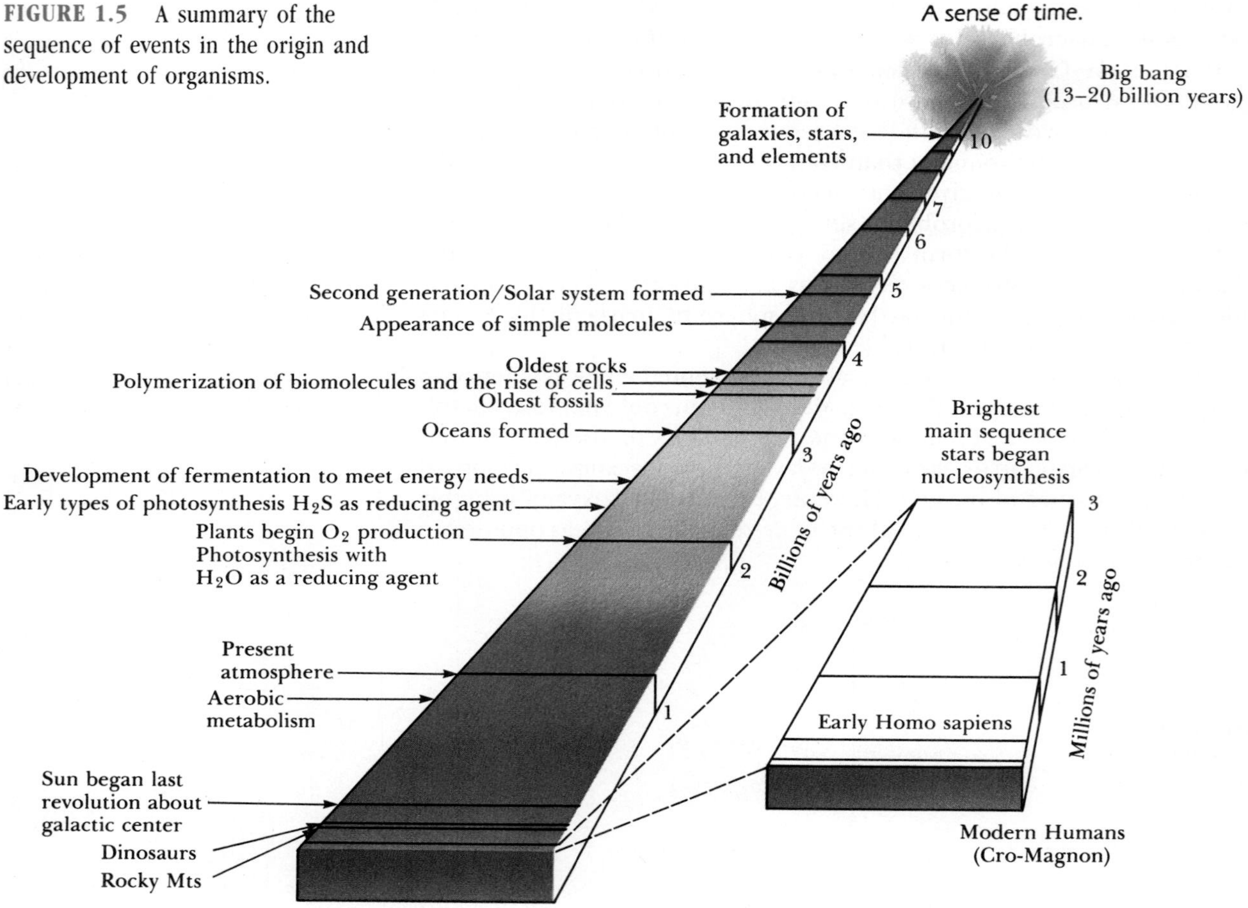

**FIGURE 1.5**    A summary of the sequence of events in the origin and development of organisms.

This more efficient pathway for extracting energy from carbohydrates in turn made possible the appearance of more complex organisms, especially animals, which cannot exist without oxygen. The series of events leading to the origin and development of living organisms can be summarized in a simple flow chart, although the events themselves were extremely complex and spanned billions of years (Figure 1.5).

## SUMMARY

Biochemistry is a multidisciplinary field that addresses questions about the molecular nature of life processes. In the cell, biomolecules undergo the many series of reactions that constitute metabolic pathways. Speculations on the origins of life have arisen from the fundamental biochemical similarities observed in all living organisms.

It has been shown that important biomolecules such as amino acids, sugars, purines, and pyrimidines can be produced under abiotic conditions (in the absence of life) from simple compounds postulated to have been present in the atmosphere of the early earth. These simple biomolecules can polymerize, also under abiotic conditions, to give rise to compounds that resemble proteins and nucleic acids.

All cellular activity depends on the presence of catalysts, which increase the rates of chemical reactions, and on the genetic code, which directs the synthesis of the catalysts. In present-day cells, catalytic activity is associated with proteins, and transmission of the genetic code is associated with nucleic acids, particularly with DNA. Both these functions may once have been carried out by a single biomolecule, RNA. It has been postulated that RNA was the original coding material, and it has recently been shown to have catalytic activity as well.

Cells capable of catalysis and coding are continually evolving and developing metabolic pathways in response to their environment. They can meet their energy needs by catabolism, which is the breakdown of nutrients to release energy. Catabolic pathways can be aerobic (requiring oxygen) or anaerobic (operating in the absence of oxygen). The synthesis of biomolecules of importance to organisms takes place by the process of anabolism. A particularly important form of anabolism is photosynthesis, in which light energy from the sun drives the synthesis of carbohydrates. The development of photosynthesis linked to oxygen made possible the rise of aerobic metabolism and the organisms that possess this pathway.

## EXERCISES

1. Define each of the following terms: polymer, protein, nucleic acid, catalysis, genetic code, anabolism, and catabolism.
2. Compare and contrast theories of the origin of life.

Pay special attention to theories that proteins arose first and to theories that nucleic acids arose first.
3. Comment on the role of RNA in catalysis and coding in theories of the origin of life.

## ANNOTATED BIBLIOGRAPHY

Research progress is very rapid in biochemistry, and the literature in the field is vast and growing. Many books appear each year, and a large number of primary research journals and review journals present original research. References to this literature are given at the end of each chapter. Some useful references are *Annual Review of Biochemistry* and *Scientific American*, both of which have articles giving a general overview of the topic discussed. *Trends in Biochemical Sciences* and *Science* (a journal published weekly by the American Association for the Advancement of Science), *Nature*, and *Cell* can be particularly useful as the first place to find information about a given topic.

Adams, R. L. P., ed. *J. N. Davidson's The Biochemistry of the Nucleic Acids.* 10th ed. New York: Academic Press, 1986. [A classic introduction to the subject.]

Cairns-Smith, A. G. The First Organisms. *Sci. Amer.* **252** (6), 90–100 (1985). [A presentation of the point of view that the earliest life processes took place in clay rather than in the "primordial soup" of the early oceans.]

Cairns-Smith, A. G. *Genetic Takeover and the Mineral Origins of Life.* Cambridge: Cambridge Univ. Press, 1982. [A book length presentation of the idea that life began in clay.]

Cech, T. R. RNA as an Enzyme. *Sci. Amer.* **255** (5), 64–75 (1986). [A discussion of the ways in which RNA can cut and splice itself.]

Dickerson, R. E., and I. Geis. *Hemoglobin: Structure, Function, Evolution and Pathology.* Menlo Park: Benjamin-Cummings, 1983. [An important protein used as a case study of biochemical approaches to understanding life processes; particularly good illustrations.]

Dyson, F. *Origins of Life.* Cambridge: Cambridge Univ. Press, 1985. [A comparison of various theories of the origin of life.]

Eigen, M., W. Gardiner, P. Schuster, and R. Winkler-

Oswatitsch. The Origin of Genetic Information. *Sci. Amer.* **244** (4), 88–118 (1981). [A presentation of the case for RNA as the original coding material.]

Lewin, R. RNA Catalysis Gives Fresh Perspective on the Origin of Life. *Science* **231,** 545–546 (1986). [A consideration of RNA splicing in terms of catalysis and coding.]

Margulis, L. *Early Life.* Boston: Science Books Intl., 1982. [A clearly written discussion of the origin and development of cells.]

Mather, K. F. *The Permissive Universe.* Albuquerque: Univ. of New Mexico Press, 1986. [A philosophical as well as scientific discussion of the nature of evolution.]

# The Organization of Cells

2

## OUTLINE

*Lichens, shown here, are a classic example of symbiosis. A form of symbiosis in which one cell completely envelops another plays an important role in theories of the origin of eukaryotes.*

*Every cell has a central core of the hereditary material DNA, which contains the information to make the complete organism. In primitive one-celled prokaryotes, such as bacteria, the nuclear material is loosely organized. Plant and animal cells (called eukaryotes) are more highly organized; the nucleus is enclosed in a separate membrane. Specialized compartments for particular functions are characteristic of eukaryotic cells. In plants, photosynthesis takes place in chloroplasts where energy is captured from light and stored in carbohydrate molecules. In the mitochondria of animal cells, the stored energy of carbohydrates is recovered in a process called respiration where carbohydrates are oxidized to carbon dioxide and water. Cells vary in form according to their function. The red blood cell transports oxygen in the blood stream. It is a highly flexible biconcave disk that can squeeze through tiny capillaries. By contrast, nerve cells are long, with branches from a central nucleus. Their function is to connect with other nerve cells, often at long distances.*

## 2.1
## PROKARYOTES AND EUKARYOTES: DIFFERENCES IN LEVELS OF ORGANIZATION

The earliest true cells that evolved must have been very simple, having the minimum apparatus necessary for life processes. The types of organisms living today that probably most resemble the earliest cells are the **prokaryotes.** This word is of Greek derivation and literally means "before the nucleus." Examples of prokaryotes include *bacteria* and *cyanobacteria*. (Cyanobacteria are the organisms formerly called blue-green algae; they are more closely related to bacteria, as the newer name indicates.) Prokaryotes are single-celled organisms, but groups of single-celled prokaryotes can exist in association, forming colonies. Some differentiation of cellular functions exists in such colonies.

   **Eukaryotes** (which means "true nucleus") are more complex organisms and can be multicellular as well as single-celled. A well-defined nucleus, set off from the rest of the cell by a membrane, is one of the chief features distinguishing eukaryotes from prokaryotes. There is a growing body of fossil evidence indicating that eukaryotes evolved from prokaryotes about 1.5 billion ($1.5 \times 10^9$) years ago, about 2 billion years after life first appeared on earth. Examples of single-celled eukaryotes include yeasts and *Paramecium* (an organism frequently discussed in beginning biology courses); all multicellular organisms such as animals and plants are eukaryotes. As might be expected, eukaryotic cells are more complex and usually much larger than prokaryotic cells. The diameter of a typical prokaryotic cell is on the order of 1 to 3 $\mu$m (1 to $3 \times 10^{-6}$ m), while that of a typical eukaryotic cell is about 10 to 100 $\mu$m. The distinction between prokaryotes and eukaryotes is such a basic one that this difference is now a key point in the classification of living organisms, far more important than the distinction between plants and animals.

*The main difference between prokaryotic and eukaryotic cells is the existence of organelles, especially the nucleus, in eukaryotes.* An **organelle** is a part of the cell that has a distinct function; membranes mark the portion of a cell occupied by an organelle. In a prokaryotic cell the cell structure is relatively simple, lacking membrane-bounded organelles (particularly a nucleus). A prokaryotic cell, like a eukaryotic cell, has a **cell membrane,** or plasma membrane, separating it from the outside world, but unlike the eukaryotes, the cell membrane is the only membrane found in the prokaryotic cell. The general composition of cell membranes is similar in prokaryotes and eukaryotes; in both cases the membranes consist of a double layer (bilayer) of lipid molecules with a variety of proteins embedded in it.

The organelles in eukaryotic cells have specific functions. A typical cell of a eukaryotic organism has a *nucleus* with a nuclear membrane, and it shows a more complex structure than a prokaryotic cell. *Mitochondria,* which are the respiratory organelles, and an internal membrane system known as the *endoplasmic reticulum* are common to all eukaryotic cells. Oxidation reactions take place in eukaryotic mitochondria, and similar reactions occur on the plasma membrane in prokaryotes. *Ribosomes* (particles consisting of RNA and protein), which are the site of protein synthesis in all living organisms, are frequently bound to the endoplasmic reticulum in eukaryotes. In prokaryotes, ribosomes are found free in the cytosol. A distinction can be made between the cytoplasm and the cytosol. The *cytoplasm* refers to the portion of the cell outside the nucleus, while the *cytosol* is the soluble portion of the cell that lies outside the membrane-bounded organelles.

*Chloroplasts,* which are photosynthetic organelles, are found in plant cells in which photosynthesis takes place. In prokaryotes that are capable of photosynthesis, the various reactions take place in laminar arrays called *chromatophores* rather than in chloroplasts. Both prokaryotes and eukaryotes have exterior structures called *flagella* that play a role in the

**TABLE 2.1    A Comparison of Prokaryotes and Eukaryotes**

| ORGANELLE | PROKARYOTES | EUKARYOTES |
|---|---|---|
| Nucleus | No definite nucleus; DNA present but not bound to anything | Present |
| Cell membrane (plasma membrane) | Present | Present |
| Mitochondria | None; enzymes for oxidation reactions located on plasma membrane | Present |
| Endoplasmic reticulum | None | Present |
| Ribosomes | Present | Present |
| Chloroplasts | None; photosynthesis localized in chromatophores | Present in green plants |

motion of the cell. Flagella are similar in appearance in prokaryotes and eukaryotes, but the molecular structures of flagella differ drastically in the two types of cells. This difference is great enough that it has been suggested that the term *flagella* should be confined to prokaryotes, and that the eukaryotic structures be called *undulipodia*. *Pili* and *cilia* are shorter exterior projections in prokaryotes and eukaryotes, respectively. Pili play a role in intercellular communication, while cilia aid in locomotion. Table 2.1 summarizes the basic differences between the two cell types.

## 2.2
## PROKARYOTIC CELLS

Although no well-defined nucleus is present in prokaryotes, the DNA of the cell is concentrated in one region called the **nuclear region.** This part of the cell directs the workings of the cell, very much like the eukaryotic nucleus. The DNA of prokaryotes is not complexed with proteins, as is the case with the DNA of eukaryotes. In general there is only one chromosome in a bacterial cell—a single circular molecule of DNA. This closed circle of DNA, also called the **genome,** is attached to the cell membrane. Before a prokaryotic cell divides, the DNA replicates itself and both DNA circles are bound to the plasma membrane. The cell then divides, and each of the two daughter cells receives one copy of the DNA (Figure 2.1).

The *cytosol* (the portion of the cell outside the nuclear region) of a prokaryotic cell frequently has a slightly granular appearance because of the presence of **ribosomes.** These consist of RNA and protein and thus are also called *ribonucleoprotein particles;* they are the site of protein synthesis in all organisms. The presence of ribosomes is the main feature of prokaryotic cytosol. Membrane-bounded organelles, characteristic of eukaryotes, are not found in prokaryotes.

All cells are separated from the outside world by a **cell membrane,** also called the **plasma membrane,** which consists of an assemblage of lipid molecules and proteins. In addition to the cell membrane and external to it, prokaryotic bacterial cells have a **cell wall,** made up mostly of polysaccharide material, a feature they share with eukaryotic plant cells. Some differences exist in the chemical nature of prokaryotic and eukaryotic cell walls, but a common feature is that the polymerization of sugars produces the polysaccharides found in both. Since the cell wall is made up of rigid material, it presumably serves as protection for the cell.

The exterior features of prokaryotes are **pili** and **flagella.** The pili (derived from the Latin *pilus,* meaning "hair") are short projections from the cell wall. Although comparatively few genera of bacteria can form pili, the role of pili in sexual conjugation in these bacteria is important in the transfer of genetic material between the cells involved. The flagella (derived from the Latin term meaning "whip") are longer tubular projections. Prokaryotic flagella are made up of the protein flagellin and have a circular structure at their base. This circular structure serves as a swivel from which flagella can rotate. Their rotatory motion aids in locomotion of the cells as they swim through the medium in which they live.

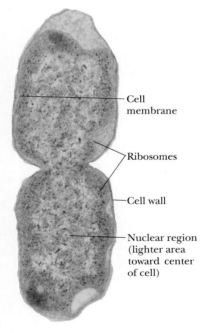

Cell membrane

Ribosomes

Cell wall

Nuclear region (lighter area toward center of cell)

**FIGURE 2.1**  A typical prokaryote: the bacterium *Escherichia coli* (magnified 100,000 ×). Division into two cells is nearly complete.

## 2.3
## EUKARYOTIC CELLS

Multicelluar plants and animals are both examples of eukaryotes, but obvious differences between plants and animals exist that are reflected on the cellular level. Plant cells, like bacteria, have cell walls. In the case of plants, the cell wall material is mostly cellulose, giving the plant cell its shape and mechanical stability. Chloroplasts, the photosynthetic organelles, are found in green plants. Animal cells have neither cell walls nor chloroplasts. Figure 2.2 shows some of the important differences between a typical plant cell, a typical animal cell, and a prokaryote.

**FIGURE 2.2**   A comparison of a typical plant cell, a typical animal cell, and a prokaryotic cell. (a) Animal. (b) Plant. (c) Prokaryote.

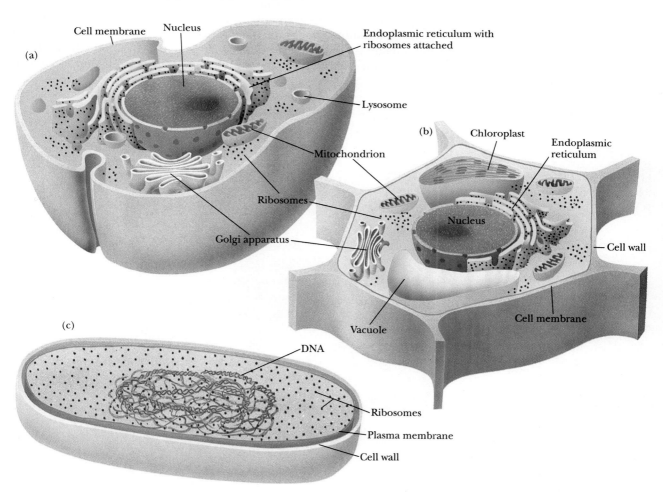

## Important Organelles

The **nucleus** is perhaps the most important eukaryotic organelle. A typical nucleus exhibits several important structural features (Figure 2.3). It is bounded by a *nuclear double membrane.* A prominent feature of the nucleus is the **nucleolus,** which is rich in RNA. The RNA of a cell (with the exception of the small amount produced in such organelles as mitochondria and chloroplasts) is produced on a DNA template in the nucleus. This RNA, ultimately destined for the ribosomes, is prepared in the nucleolus for export to the cytoplasm through pores in the nuclear membrane. **Chromatin** is an aggregate of DNA and protein that is also visible in the nucleus, frequently near the nuclear membrane. The cellular DNA is duplicated before cell division takes place, as occurs in prokaryotes. In eukaryotes both copies of DNA, which are to be distributed between the daughter cells, are associated with protein. When a cell is about to divide, the loosely organized strands of chromatin become tightly coiled and the **chromosomes** can be seen under the electron or light microscope. The **genes,** responsible for the transmission of inherited traits, are part of the DNA found in each chromosome.

A second very important eukaryotic organelle is the **mitochondrion,** which has a double membrane like that of the nucleus (Figure 2.4). The outer membrane has a fairly smooth surface, but the inner membrane exhibits many folds called *cristae.* Oxidation processes that occur in

**FIGURE 2.3**   The nucleus of a tobacco leaf cell (magnified 15,000 ×).

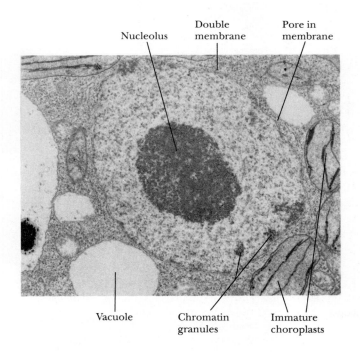

Nucleolus    Double membrane    Pore in membrane

Vacuole    Chromatin granules    Immature choroplasts

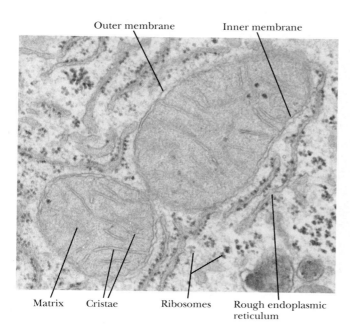

Outer membrane     Inner membrane

Matrix    Cristae      Ribosomes    Rough endoplasmic
                                    reticulum

**FIGURE 2.4**  Mouse liver mitochondria (magnified 50,000 ×).

mitochondria yield energy for the cell. Most of the enzymes responsible for these important reactions are located on the inner mitochondral membrane. The internal mitochondrial *matrix* is the location of other enzymes needed for oxidation reactions. The matrix also contains DNA that differs from that found in the nucleus, a point to which we shall return in Section 2.6. A similar point of interest is that mitochondria contain ribosomes similar to those found in bacteria. Mitochondria are approximately the same size as many bacteria, typically about 1 $\mu$m in diameter and 2 to 8 $\mu$m in length, and it is theorized that they may have arisen from the absorption of aerobic bacteria by larger host cells.

The **endoplasmic reticulum** (ER) is part of a continuous single membrane system throughout the cell; the membrane doubles back on itself to give the appearance of a double membrane in electron micrographs. The endoplasmic reticulum is attached to the cell membrane and to the nuclear membrane. It occurs in two forms, rough and smooth. The *rough endoplasmic reticulum* is studded with ribosomes bound to the membrane, although ribosomes can also be found free in the cytosol (Figure 2.5). Ribosomes are the site of protein synthesis in all organisms. The *smooth endoplasmic reticulum* does not have ribosomes bound to it.

**FIGURE 2.5**   Rough endoplasmic reticulum from mouse liver cells (magnified 50,000 ×).

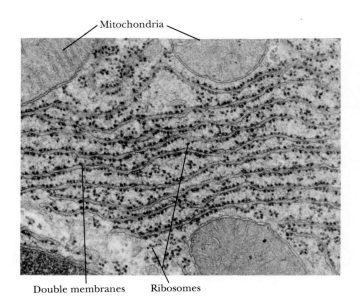

**Chloroplasts** are important organelles found only in green plants. They have a double membrane and are relatively large, typically up to 2 $\mu$m in diameter and 5 to 10 $\mu$m in length. The photosynthetic apparatus is located in a specialized structure within the chloroplast. These stacked membranous bodies within the chloroplast, called *grana,* are easily seen in the electron microscope (Figure 2.6). Chloroplasts, like mitochondria, contain a characteristic DNA which is different from that found in the nucleus. Chloroplasts also contain ribosomes similar to those found in bacteria.

## Other Organelles and Cellular Constituents

Membranes are important in the structure of some less well understood organelles, such as the Golgi apparatus. The **Golgi apparatus** is a membrane-bounded organelle (with a single membrane) that is separate

**FIGURE 2.6**   Chloroplasts from tobacco mesophyll cell (magnified 25,000 ×).

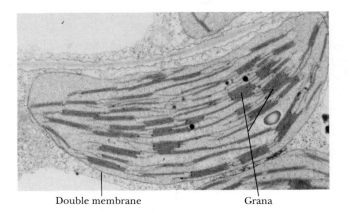

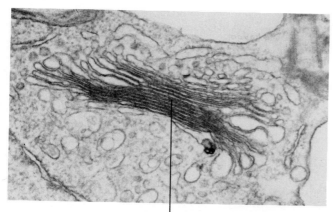

**FIGURE 2.7**  Golgi apparatus from the green alga *Dunaliella tertiolecta* (magnified 59,000 ×).

Stack of flattened
membranous vesicles

from the endoplasmic reticulum, but frequently found close to the smooth endoplasmic reticulum. It is an assemblage of flattened vesicles or sacs (Figure 2.7). It is involved in secretion of proteins from the cell, but it also occurs in cells in which the primary function is not secreting proteins. The Golgi apparatus also appears to be involved in the metabolism of sugars. The function of this organelle is still a subject of research.

There are other organelles in eukaryotes similar to the Golgi apparatus in that they are bounded by single, smooth membranes and have specialized functions. **Lysosomes** are an example. These membrane-bounded sacs contain enzymes that could cause considerable damage to the cell if they were not separated from the lipids, proteins, or nucleic acids that they attack. These enzymes are released as needed to break down their target molecules, usually from outside sources, to provide for the needs of the cell. **Peroxisomes** are similar to lysosomes; their principal characteristic is that they contain enzymes involved in the metabolism of hydrogen peroxide ($H_2O_2$), which is toxic. The enzyme *catalase,* which occurs in peroxisomes, catalyzes the conversion of $H_2O_2$ to $H_2O$ and $O_2$. **Glyoxysomes** are found in plant cells only. They contain the enzymes that catalyze the *glyoxylate cycle,* a pathway that converts lipids to carbohydrate with glyoxylic acid as an intermediate (see Section 15.5 in Chapter 15).

The **cytosol** was long considered to be nothing more than a viscous liquid, but recent studies by electron microscopy have revealed that there is some internal organization in this part of the cell. The various organelles are held in place by a lattice of fine strands that seem to consist mostly of protein. This **microtrabecular lattice** is connected to all organelles (Figure 2.8). Many questions remain about the nature of this **cytoskeleton,** as it is also called.

(a)                                                    (b)

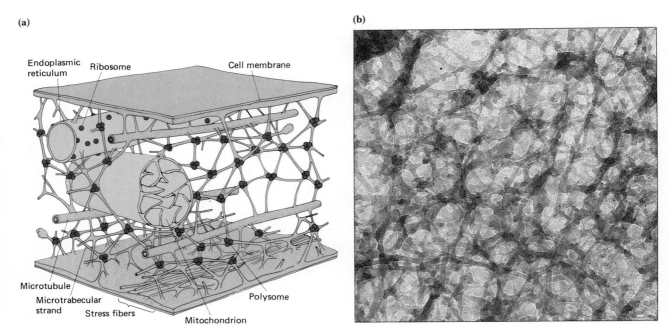

Endoplasmic reticulum   Ribosome              Cell membrane

Microtubule
Microtrabecular strand
Stress fibers          Polysome
Mitochondrion

**FIGURE 2.8**   The microtrabecular lattice. (a) This network of filaments, also called the cytoskeleton, pervades the cytosol. Organelles such as mitochondria are attached to the filaments. (b) An electron micrograph of the microtrabecular lattice (magnification 87,450 ×)

The **cell membrane** of eukaryotes serves to separate the cell from the outside world. It consists of a double layer of lipids, with several types of proteins embedded in the lipid matrix. Some of the proteins transport specific substances across the membrane barrier. Transport can take place in both directions, with substances useful to the cell being taken in and ones to be exported by the cell going out.

Plant cells, but not animals cells, have **cell walls** external to the plasma membrane. The cellulose that makes up plant cell walls is a major component of plant material; wood, cotton, linen, and most types of paper are mainly cellulose. Another feature of plant cells not often seen in animal cells is the presence of **vacuoles,** sacs in the cell surrounded by a single membrane. They tend to increase in number and size as plant cells age. An important function of vacuoles is to isolate waste substances that are toxic to the plant and are produced in a greater amount than the plant can secrete to the environment. These waste products may be unpalatable or even poisonous enough to discourage herbivores (plant-eating organisms) and may thus provide some protection for the plant.

Eukaryotic flagella are called **undulipodia** and consist of microtubules, which are long, hollow cylinders made up of the protein tubulin. Many molecules of this protein are arranged in a repeating pattern similar to a coil of rope to produce the cylindrical structure of microtubules (Figure 2.9). Microtubules in turn are arranged in a characteristic pattern in undulipodia. There are nine pairs of microtubules on the outside of each undulipodium and two unpaired microtubules in the interior. Undulipodia

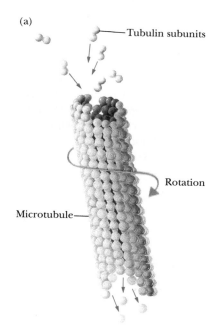

(a)

Tubulin subunits

Rotation

Microtubule

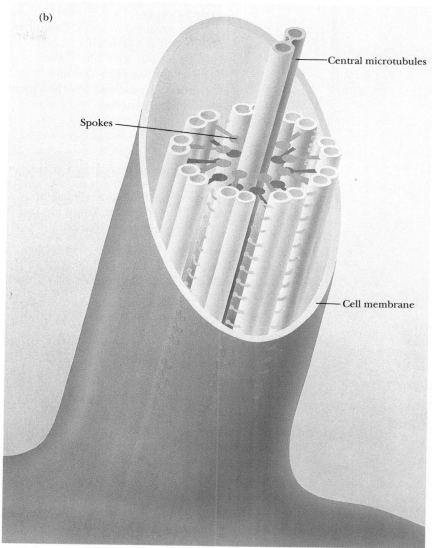

(b)

Central microtubules

Spokes

Cell membrane

**FIGURE 2.9** Microtubules, cilia, and undulipodia. (a) Schematic diagram of the subunit structure of a microtubule. Tubulin (a protein consisting of two subunits) is constantly added at one end of the microtubule, while subunits are removed by breakdown at the other end. (b) and (c) Cross-section and electron micrograph of a cilium. The arrangement of undulipodia is the same, with two central microtubules (shown in light blue) connected to nine pairs of microtubules by spokes (shown in light purple). The peripheral paired microtubules are attached to each other by extensions of the protein subunits, shown here as spokes.

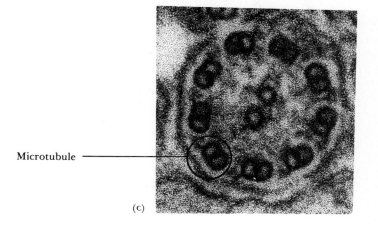

Microtubule

(c)

occur alone or in pairs; other appendages, called **cilia,** are shorter and occur in greater numbers. Undulipodia and cilia do not have the ringlike structure at the base characteristic of prokaryotic flagella, only a straight connection. As a result, the rotatory motion of prokaryotic flagella is replaced by wavelike motion in cilia and undulipodia. The sliding of microtubules past one another gives rise to the beating motion of the eukaryotic organelles.

## 2.4
## THE FIVE-KINGDOM CLASSIFICATION SYSTEM

The original biological classification scheme divided all organisms into two categories: the plant kingdom and the animal kingdom. In this scheme, plants are organisms that obtain food directly from the sun and animals are

**FIGURE 2.10**    The five-kingdom classification scheme.

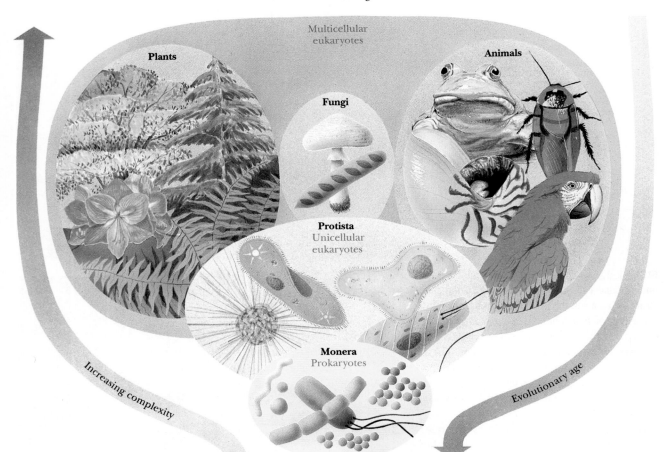

organisms that move about to search for food. Some organisms, bacteria in particular, do not have an obvious relationship to either kingdom. It has also become clear that the fundamental difference in living organisms is actually not between plants and animals, but rather between prokaryotes and eukaryotes. Classification schemes have now been introduced that divide living organisms into more than the two traditional kingdoms. The five-kingdom system takes into account the differences between prokaryotes and eukaryotes and also provides classifications for eukaryotes that are neither plants nor animals.

There is one kingdom that consists of only prokaryotic organisms—the Monera. Bacteria and cyanobacteria are members of this kingdom. The other four kingdoms are made up of eukaryotic organisms. The kingdom Protista includes unicellular organisms such as *Euglena, Volvox, Amoeba,* and *Paramecium.* Some protists, such as *Volvox,* form colonies. There is some question among biologists about whether multicellular organisms should be included in this kingdom. Most biologists do not classify multicellular organisms as protists, but the question continues to be discussed. The three kingdoms that consist of mainly multicellular (with a few unicellular) eukaryotes are Fungi, Plantae, and Animalia. Fungi include molds and mushrooms. Fungi, plants, and animals must have evolved from simpler eukaryotic ancestors; however, the major evolutionary change was the development of eukaryotes from prokaryotes. (Figure 2.10).

## 2.5
## ARE THERE LIFE FORMS OTHER THAN PROKARYOTES AND EUKARYOTES?

A group of organisms exists that can be classified as bacteria on the basis of not having a well-defined nucleus but that differ from both prokaryotes and eukaryotes in several important ways. They are called *Archaebacteria* (early bacteria) to distinguish them from *Eubacteria* (true bacteria), since archaebacteria are very primitive organisms. Most of the differences between archaebacteria and other organisms are biochemical in nature, such as the molecular structure of the cell walls, membranes, and some types of RNA. The article by Woese listed in the bibliography at the end of this chapter gives a comparison between archaebacteria and other life forms based on the biochemical differences.

There are three groups of archaebacteria: methanogens, halophiles, and thermacidophiles, all of which live in extreme environments. *Methanogens* are strict anaerobes that produce methane ($CH_4$) from carbon dioxide ($CO_2$) and hydrogen ($H_2$). *Halophiles* require very high salt concentrations for growth, such as the conditions found in the Dead Sea. *Thermacidophiles* require high temperatures and acid conditions for growth; typical conditions are 80 to 90°C and pH 2. These growth patterns may have resulted from adaptations to harsh conditions on the early earth. Some of these points will be useful to keep in mind as we discuss pathways by which eukaryotes may have evolved from prokaryotes.

## 2.6
## ENDOSYMBIOTIC THEORY OF THE ORIGIN OF EUKARYOTES

The complexity of eukaryotes presents many questions about how such cells arose from simpler progenitors. Symbiosis plays a large role in current theories of the rise of eukaryotes; the symbiotic association between two organisms is seen as giving rise to a new organism that combines characteristics of both the original ones. The type of symbiosis called *mutualism* is a relationship that is mutually beneficial to both species involved, as opposed to a *parasitic* symbiosis in which one gains and the other is harmed. A classic example of mutualism, although it has been questioned from time to time, is found in lichens, which consist of fungi and algae. The fungus provides water and protection for the alga; the alga is photosynthetic and provides food for both partners. Another example is the root nodule system formed by leguminous plants such as alfalfa or beans and anaerobic nitrogen-fixing bacteria. The plant gains useful compounds of nitrogen, and the bacteria are protected from oxygen that is harmful to them. Still another example of mutualistic symbiosis of great practical interest is that between humans and such bacteria as *Escherichia coli* that live in the intestinal tract. The bacteria receive nutrients and protection from the environment; in return, they are an important aid in our digestive process. Without intestinal bacteria, we would soon develop dysentery and other intestinal disorders.

Examples of hereditary symbiosis are known in which a larger host cell contains a genetically determined number of smaller organisms. An example is the protist *Cyanophora paradoxa,* a eukaryotic host that contains a genetically determined number of cyanobacteria (blue-green algae). This situation is an example of **endosymbiosis,** since the cyanobacteria are contained within the host organism. The cyanobacteria are aerobic prokaryotes and are capable of photosynthesis. The host cell gains the products of photosynthesis; in return, the cyanobacteria are protected from the environment and still have access to oxygen and sunlight because of the host's small size. This arrangement can be considered a model for the origin of chloroplasts. In this model, over the passage of many generations the cyanobacteria would gradually lose the ability to exist independently and would become organelles within a new and more complex type of cell. Such a situation in the past may well have given rise to chloroplasts, which are not capable of independent existence. Their autonomous DNA and ribosomal protein-synthesizing apparatus can no longer provide all their needs, but the very fact that these organelles have their own DNA and are capable of protein synthesis suggests that they may have existed as independent organisms in the distant past.

A similar model can be proposed for the origin of mitochondria. Consider this scenario. A large anaerobic host cell assimilates a number of smaller aerobic bacteria. The larger cell protects the smaller ones and serves as a source of nutrients for them. As in the example we used for the development of chloroplasts, the smaller cells still have access to oxygen. Since the larger cell is not capable of aerobic oxidation of nutrients, some of the end-products of anaerobic oxidation can be oxidized further by the more efficient aerobic metabolism of the smaller cell. As a result, the larger cell gets more energy out of a given amount of food than would otherwise be possible. In time the two associated organisms evolve to form a new aerobic organism containing mitochondria derived from the original aerobic bacteria.

The fact that both mitochondria and chloroplasts have their own DNA, which is different from the DNA found in the nucleus of the cell, is an important point of biochemical evidence in favor of this model. Additional support comes from the fact that both mitochondria and chloroplasts have their own apparatus for synthesis of RNA and proteins. The remains of these systems for synthesis of RNA and protein could reflect their former existence as free-living cells. It is reasonable to conclude that large unicellular organisms that assimilated aerobic bacteria went on to evolve mitochondria from the bacteria and eventually gave rise to animals. Other types of unicellular organisms assimilated both aerobic bacteria and cyanobacteria and evolved both mitochondria and chloroplasts, eventually giving rise to green plants.

A particularly striking example of endosymbiosis, one that may provide a model for the origin of undulipodia, is the microorganism *Mixotricha paradoxa,* which lives in the gut of Australian termites and aids in the efficient digestion of wood pulp. This microorganism, which is a protist, has a mutualistic relationship with three species of bacteria: one found internally whose function appears to be digestion of wood pulp and two species of spirochete (corkscrew-shaped bacteria) attached externally. *Mixotricha* appears to have many undulipodia, but it actually has only a few true undulipodia of its own. The attached spirochetes have the appearance of undulipodia. This example provides a model for the origin of undulipodia as a result of incorporation of spirochetes as an integral part of the organism.

The possible rise of undulipodia from external attachment of spirochetes leads to questions about possible consequences of internal assimilation of spirochetes. These questions have a bearing on the important topic of mitosis. The mitotic process provides an important distinction between prokaryotes and eukaryotes besides the well-defined nucleus. When eukaryotic cells divide by mitosis, there is a thoroughly organized pattern of chromosome division and separation. The separation stages are mediated by spindle fibers and centrioles that provide support for the dividing chromosomes. Spindle fibers and centrioles are composed of microtubules, as are undulipodia. It is possible that internally assimilated spirochetes may have had a part in the development of mitosis. This idea is not definitely established, but it does provide an interesting point of departure for further research.

The endosymbiotic theory is far from complete, and many questions remain to be answered. One of the biggest questions is that of the origin of the nuclear double membrane in eukaryotes. Did it arise from extensions of the plasma membrane that folded around the DNA of the cell, or did it have some other origin? No clear answer to this question is currently available.

Another question about the distribution of DNA in the cell concerns the fate of many of the original genes of those bacteria that eventually became mitochondria and chloroplasts. The protein-synthesizing apparatus of these organelles does not produce all the proteins the organelles need. Many of the oxidative enzymes of mitochondria, for example, are products of nuclear genes. There must have been some wholesale loss of organellar DNA and transfer to the nucleus, but details of such a process are not clear. (For a discussion of this point, see the article by Lewin cited in the bibliography at the end of the chapter.) The origin of eukaryotes is a question that still provides many opportunities for research.

## SUMMARY

Two main cell types occur in organisms. In prokaryotes the cell does not exhibit a well-defined nucleus. Another distinguishing feature of prokaryotes is that they have no internal membranes, only the cell membrane that separates the cell from the outside world. In contrast, eukaryotic cells have a well-defined nucleus, internal membranes as well as a cell membrane, and a considerably more complex internal structure than prokaryotes.

Prokaryotes have a nuclear region, the portion of the cell that contains DNA. The other principal feature of a prokaryotic cell's interior is the presence of ribosomes, the site of protein synthesis.

In eukaryotes the nucleus is separated from the rest of the cell by a double membrane. Eukaryotic DNA in the nucleus is associated with proteins, which does not occur in prokaryotes. There is a continuous membrane system, called the endoplasmic reticulum, throughout the cell. Eukaryotic ribosomes are frequently bound to the endoplasmic reticulum, but ribosomes free in the cytosol also occur. Membrane-bounded organelles are characteristic of eukaryotic cells. Two of the most important are mitochondria, the sites of energy-yielding reactions, and chloroplasts, the sites of photosynthesis.

The endosymbiotic theory of the origin of eukaryotes from prokaryotes postulates that small prokaryotic cells were absorbed by larger ones and eventually became incorporated into the larger cell as organelles. The combined organisms evolved to become eukaryotes.

## EXERCISES

1. List five differences between prokaryotes and eukaryotes.
2. Draw an idealized animal cell and identify the parts by name and function.
3. Draw an idealized plant cell and identify the parts by name and function.
4. What are the differences between the photosynthetic apparatus of green plants and photosynthetic bacteria?
5. Which organelles are bounded by a double membrane?
6. Which organelles contain DNA?
7. Which organelles are the sites of ATP synthesis?
8. Does the site of protein synthesis differ in prokaryotes and eukaryotes?
9. Briefly describe the role of microtubules in mitosis.
10. Compare and contrast the following organelles in terms of structure and function: Golgi apparatus, lysosomes, peroxisomes, and glyoxysomes.

## ANNOTATED BIBLIOGRAPHY

Alberts, B., D. Bray, J. Lewis, M. Raff, K. Roberts, and J. D. Watson. *Molecular Biology of the Cell.* 2nd ed. New York: Garland Publishing, Inc., 1989. [A particularly well-written and well-illustrated textbook of cell biology.]

Allen, R. D. The Microtubule as an Intracellular Engine. *Sci. Amer.* **256** (2), 42–49 (1987). [The role of the microtrabecular lattice and microtubules in the motion of organelles is discussed in this article.]

Hinkle, P. C., and R. E. McCarty. How Cells Make ATP. *Sci. Amer.* **238** (3), 104–123 (1978). [An article dealing with the role of chloroplasts and mitochondria in generating energy for the cell. An old but particularly good article.]

Lewin, R. No Genome Barriers to Promiscuous DNA. *Science* **224,** 970–971 (1984). [A discussion of the movement of DNA between mitochondrial, chloroplast, and nuclear genomes.]

Lipsky, N. G., and R. E. Pagano. A Vital Stain for the Golgi Apparatus. *Science* **228,** 745–747 (1985). [A description of the first staining method for the Golgi apparatus in living cells.]

Margulis, L. *Early Life.* Boston: Science Books Intl., 1982. [A clearly written discussion of the origin and development of cells.]

Margulis, L. *Symbiosis in Cell Evolution.* San Francisco: W. H. Freeman, 1981. [A description of the endosymbiotic theory of the rise of eukaryotes, presented by one of its leading proponents.]

Pool, R. Pushing the Envelope of Life. *Science* **247,** 158–160 (1990). [A Research News article describing the extreme conditions under which archaebacteria can flourish.]

Rothman, J. E. The Compartmental Organization of the Golgi Apparatus. *Sci. Amer.* **253** (3), 74–89 (1985). [A description of the various functions of the Golgi apparatus.]

Vidal, G. The Oldest Eukaryotic Cells. *Sci. Amer.* **250** (2), 48–57 (1984). [Describes fossil evidence for the rise of eukaryotes.]

Weber, K., and M. Osborn. The Molecules of the Cell Matrix. *Sci. Amer.* **253** (4), 100–120 (1985). [An extensive description of the cytoskeleton.]

Woese, C. R. Archaebacteria. *Sci. Amer.* **244** (6), 98–122 (1981). [A detailed description of the biochemical differences between archaebacteria and other types of organisms.]

# 3

# Water, Hydrogen Bonding, pH, and Buffers

## OUTLINE

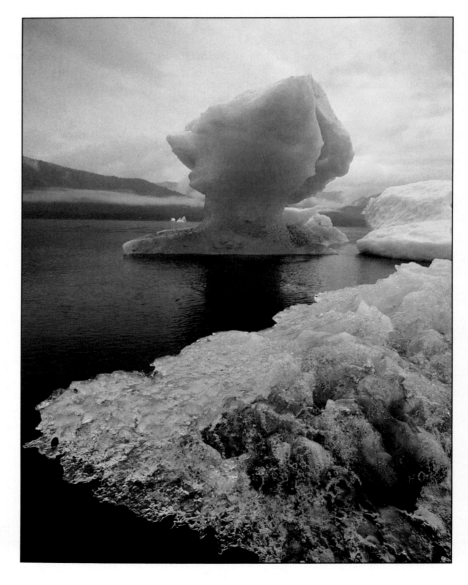

*A striking example of water in both the solid and liquid states. Life as we know it depends on the properties of this simple molecule.*

*Virtually all of the chemical reactions of the cell involve water. Life has evolved around its special properties. Water plays a crucial role in photosynthesis, where $H_2O$ is split, and in respiration when hydrogen is rejoined with oxygen. Important structural considerations follow from the nature of the water molecule, where there is a partial positive charge on each of its hydrogen atoms and a partial negative charge on its oxygen atom. This allows a water molecule to associate with four others of its kind. Four hydrogen bonds can be formed, pointing to the corners of a tetrahedron. The three-dimensional conformation of large protein molecules is determined by the interactions of the protein chain with water. Hydrophobic (water hating) side chains of amino acids aggregate in the interior of the folding protein. At the same time, hydrophilic side chains are attracted to water. The final result of these activities is a specific three-dimensional folding pattern for each type of protein molecule. Hydrogen bonds are important everywhere in biomolecular structures—stitching together parts of protein chains of enzymes as well as holding together the two complementary chains of the DNA double helix.*

## 3.1
## THE POLAR NATURE OF THE WATER MOLECULE

Water is the principal component of most cells. The geometry of the water molecule and its properties as a solvent play a major role in determining the properties of living systems.

When electrons are shared between atoms in a chemical bond, they need not be shared equally; bonds with unequal sharing of electrons are referred to as **polar.** The tendency of an atom to attract electrons to itself in a chemical bond (i.e., to become negative) is called **electronegativity.** Atoms of the same element will of course share electrons equally in a bond, but different elements do not necessarily have the same electronegativity. Fluorine, oxygen, and nitrogen are all highly electronegative; carbon and hydrogen have about the same electronegativity, which is less than the values for fluorine, oxygen, and nitrogen (Table 3.1).

The difference in electronegativity between oxygen and hydrogen in the O—H bond in water is such that the oxygen, the more electronegative element, has a larger share of the electrons. The difference in electronegativity between oxygen and hydrogen gives rise to *partial* positive and negative charges, usually pictured as $\delta^+$ and $\delta^-$, respectively (Figure 3.1). The O—H bond is thus a polar bond. In situations where the electronegativity difference is quite small, such as in the C—H bond in methane, $CH_4$, the sharing of electrons in the bond is very nearly equal, and the bond is essentially **nonpolar.** The C—O bond is polar because of the electronegativity difference between carbon and oxygen.

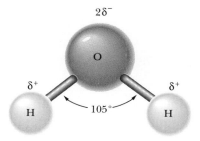

**FIGURE 3.1** The structure of the water molecule, showing the polar bonds.

**TABLE 3.1   Electronegativities of Selected Elements**

| ELEMENT | ELECTRONEGATIVITY* |
|---|---|
| Fluorine | 4.0 |
| Oxygen | 3.5 |
| Nitrogen | 3.0 |
| Carbon | 2.5 |
| Hydrogen | 2.1 |

*The values of electronegativity are relative ones, and are chosen to be positive numbers ranging from less than 1 for some metals to 4 for fluorine.

It is possible for a molecule to have polar bonds but still be nonpolar because of its geometry. Carbon dioxide is an example. The two C—O bonds are polar, but because the $CO_2$ molecule is linear, the attraction of the oxygen for the electrons in one bond is cancelled out by the equal and opposite attraction for the electrons by the oxygen on the other side of the molecule.

$$\delta^- \quad 2\delta^+ \quad \delta^-$$
$$O=C=O$$

Water is a bent molecule with a bond angle of 105°, and the uneven sharing of electrons in the two bonds is not cancelled out as in $CO_2$ (Figure 3.1). The result is that the bonding electrons are more likely to be found at the oxygen end of the molecule than at the hydrogen end. Molecules with positive and negative ends are called **dipoles.**

## Solvent Properties of Water

The polar nature of water largely determines its solvent properties. *Ionic* compounds with full charges such as potassium chloride (KCl) and *polar* compounds with partial charges such as ethyl alcohol ($C_2H_5OH$) or acetone (($CH_3)_2C=O$) tend to dissolve in water (Figure 3.2). The underlying physical principle is that of electrostatic attraction between unlike charges. The negative end of a water dipole will attract a positive ion or the positive end of another dipole. The positive end of a water molecule will attract a negative ion or the negative end of another dipole. The aggregate of unlike charges, held in close proximity to one another because of electrostatic attraction, has a lower energy than the two charges when they are widely separated from one another. The lowering of energy makes the system more stable and more likely to exist. These *ion–dipole* and *dipole–dipole* interactions are similar to the interactions between water molecules themselves in terms of the quantities of energy involved. Examples of polar compounds that dissolve easily in water are small organic molecules containing one or more electronegative atoms (such as oxygen or nitrogen), including alcohols, amines, and carboxylic acids. The attraction between the dipoles of these molecules and the water dipole makes them tend to dissolve. Ionic and polar substances are referred to as **hydrophilic** ("water-loving," from the Greek) because of this tendency.

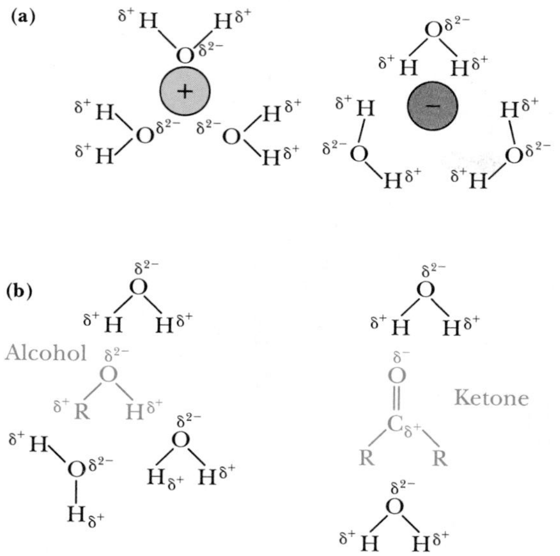

**(a)**

**(b)**

Alcohol

Ketone

FIGURE 3.2  Ion–dipole and dipole–dipole interactions help ionic and polar compounds dissolve in water. (a) Ion–dipole interactions with water. (b) Dipole–dipole interactions of polar compounds with water. The examples shown here are an alcohol (ROH) and a ketone ($R_2C=O$) (both shown in red).

Hydrocarbons (compounds that contain only carbon and hydrogen) are nonpolar. The favorable ion–dipole and dipole–dipole interactions responsible for the solubility of ionic and polar compounds do not exist for nonpolar compounds, which tend not to dissolve in water. The interactions between nonpolar molecules and water molecules are weaker than dipolar interactions. The permanent dipole of the water molecule can induce a temporary dipole in the nonpolar molecule by distorting the spatial arrangements of the electrons in its bonds. Electrostatic attraction is possible between the induced dipole of the nonpolar molecule and the permanent dipole of the water molecule, but the attraction is not as strong as that between permanent dipoles. This electrostatic attraction and the consequent lowering of energy by the *dipole–induced dipole* interaction is less than that produced by attraction of the water molecules for one another. The association of nonpolar molecules with water is far less likely to occur than the association of water molecules with themselves.

A full discussion of why nonpolar substances are insoluble in water requires the thermodynamic arguments that we will develop in Chapter 8. However, the points that we make here about intermolecular interactions will be useful background information for that discussion. For the moment it is enough to know that it is less favorable thermodynamically for water molecules to be associated with nonpolar molecules than with other water molecules. As a result, nonpolar molecules do not dissolve in water and are referred to as **hydrophobic** ("water-hating"). Hydrocarbons in particular tend to sequester themselves from an aqueous environment. Nonpolar solids will not dissolve in water, leaving undissolved material. Nonpolar liquids and water will form a two-layer system, an example of which is an oil slick on water.

It is possible for a molecule to have both polar (hydrophilic) and nonpolar (hydrophobic) portions. Substances of this type are called **amphi-**

**FIGURE 3.3** (a) A sodium salt of a fatty acid with an ionized polar head and a nonpolar tail. (b) Formation of a micelle, with the ionized polar groups in contact with the water and the nonpolar parts of the molecule protected from contact with water.

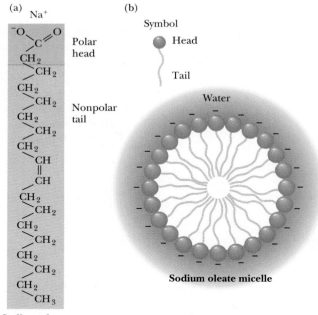

**philic** (from the Greek *amphi*, "on both ends," and *philic*, "loving"), because one part of the molecule tends to dissolve in water and another part in a nonpolar environment. A long-chain fatty acid having a polar carboxylic acid group and a long nonpolar hydrocarbon portion is a prime example of an amphiphilic substance. The carboxylic acid group, the "head" group, contains two oxygen atoms in addition to carbon and hydrogen; it is very polar and can form a carboxylate anion. The rest of the molecule, the "tail," contains only carbon and hydrogen and is thus nonpolar. A compound such as this in the presence of water tends to form structures called **micelles,** in which the polar head groups are in contact with the aqueous environment and the nonpolar tails are sequestered from the water (Figure 3.3).

Interactions between nonpolar molecules themselves are very weak and depend on the attraction between transient dipoles and induced dipoles. In a large sample of nonpolar molecules there will always be some molecules with temporary or transient dipoles, which are caused by a momentary clumping of bonding electrons at one end of the molecule. This transient dipole can induce another dipole in a neighboring molecule (Figure 3.4). These associations are short-lived, and the interaction energy is small. They are called **van der Waals bonds.** [A caveat about terminology is in order here. Different nomenclature is often used by physical chemists to describe some noncovalent interactions, and that usage appears in many current textbooks of general chemistry and other books of that nature. In the terminology of physical chemistry the interactions we have called van der Waals bonds (those between transient dipoles and induced dipoles) are called London forces. Usually the terminology is reasonably clear from the context and does not cause confusion.]

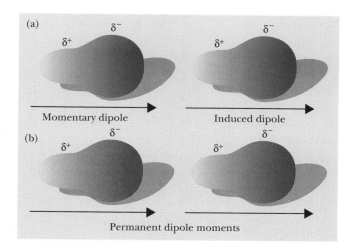

(a)

δ⁺          δ⁻

Momentary dipole          Induced dipole

(b)

δ⁺          δ⁻

Permanent dipole moments

**FIGURE 3.4** A comparison of van der Waals bonds with dipole–dipole interactions. (a) van der Waals bonds. (b) Dipole–dipole attraction.

## 3.2
## THE HYDROGEN BOND

There is another important type of noncovalent interaction, **hydrogen bonding.** Hydrogen bonding is of electrostatic origin and can be considered a special case of dipole–dipole interaction. When hydrogen is covalently bonded to an electronegative atom such as oxygen, nitrogen, or fluorine, it has a partial positive charge due to the polar bond. This partial positive charge can interact with an unshared (nonbonding) pair of electrons (a source of negative charge) on another electronegative atom. All three atoms lie in a straight line, forming a hydrogen bond. This arrangement allows for the greatest possible partial positive charge on the hydrogen and consequently for the strongest possible interaction with the unshared pair of electrons on the second electronegative atom (Figure 3.5). The electronegative atom to which the hydrogen is covalently bonded is called the *hydrogen-bond donor,* while the electronegative atom that contributes the unshared pair of electrons to the interaction is the *hydrogen-bond acceptor.* The hydrogen is not covalently bonded to the acceptor.

A consideration of the hydrogen bonding situation in HF, $H_2O$, and $NH_3$ can yield some useful insights. It is apparent from Figure 3.6 that water represents an optimum situation in terms of the number of hydrogen bonds that each molecule can form. Water has two hydrogens to enter into hydrogen bonds, and two unshared pairs of electrons on the oxygen to which other water molecules can be hydrogen bonded. Each water molecule is involved in four hydrogen bonds, two as donor and two as acceptor. Hydrogen fluoride has only one hydrogen to enter into a hydrogen bond as a donor, but it has three unshared pairs of electrons on the fluorine that could bond to other hydrogens. Ammonia has three hydrogens to donate to a hydrogen bond but only one unshared pair of electrons on the nitrogen.

The geometric arrangement of hydrogen-bonded water molecules has important implications for the properties of water as a solvent. The bond angle in water is 105°, as was shown in Figure 3.1, and the angle

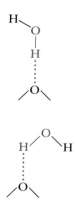

**FIGURE 3.5** A comparison of linear and nonlinear hydrogen bonds. Nonlinear bonds are weaker than ones in which all three atoms lie in a straight line.

**FIGURE 3.6**    A comparison of the numbers of hydrogen-bonding sites in HF, $H_2O$, and $NH_3$. (Actual geometries are not shown.)

$$H—\ddot{\overset{\displaystyle ..}{F}}: \cdots H—\ddot{\overset{\displaystyle ..}{F}}: \cdots H—\ddot{\overset{\displaystyle ..}{F}}:$$

$$
\begin{array}{cccccccc}
H—\overset{..}{\underset{|}{O}}: & \cdots & H—\overset{..}{\underset{|}{O}}: & \cdots & H—\overset{..}{\underset{|}{O}}: & \cdots & H—\overset{..}{\underset{|}{O}}: \\
H & & H & & H & & H \\
\vdots & & \vdots & & \vdots & & \vdots \\
H—\overset{..}{\underset{|}{O}}: & \cdots & H—\overset{..}{\underset{|}{O}}: & \cdots & H—\overset{..}{\underset{|}{O}}: & \cdots & H—\overset{..}{\underset{|}{O}}: \\
H & & H & & H & & H
\end{array}
$$

$$
\begin{array}{cccccccc}
& & H & & H & & H \\
& & | & & | & & | \\
H—\overset{..}{N}—H & \cdots & :N—H & \cdots & :N—H & \cdots & :N—H \\
| & & | & & | & & | \\
H & & H & & H & & H
\end{array}
$$

between the unshared pairs of electrons is similar. The result is a tetrahedral arrangement of water molecules. Liquid water consists of hydrogen-bonded arrays containing up to a hundred water molecules. These arrangements resemble ice crystals, the structure of which is shown in Figure 3.7. There are several differences between hydrogen-bonded arrays of this type in liquid water and the structure of ice crystals. In liquid water hydrogen bonds are constantly breaking and new ones forming, with some molecules breaking off and others joining the cluster. A cluster can

**FIGURE 3.7**    Tetrahedral hydrogen bonding in $H_2O$. An array of $H_2O$ molecules in an ice crystal.

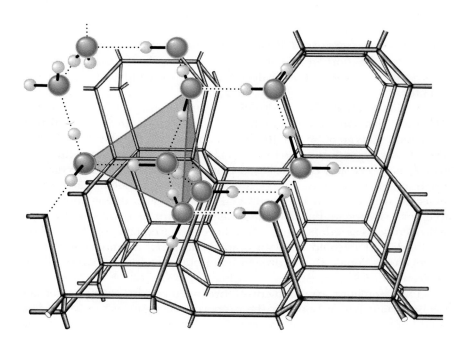

**TABLE 3.2    Some Bond Energies**

| | TYPE OF BOND | ENERGY | |
|---|---|---|---|
| | | *(kJ mol⁻¹)* | *(kcal mol⁻¹)* |
| Covalent Bonds (Strong) | O—H | 460 | 110 |
| | H—H | 416 | 100 |
| Noncovalent Bonds (Weaker) | Hydrogen bond | 20 | 5 |
| | Ion–dipole interaction | 20 | 5 |
| | Hydrophobic interaction | 4–12 | 1–3 |
| | van der Waals bonds | 4 | 1 |

*Note that two units of energy are used throughout this text. The kilocalorie (kcal) is a commonly used unit in the biochemical literature. The kilojoule (kJ) is an SI unit and will come into wider use as time goes on.

break up and re-form in $10^{-10}$ to $10^{-11}$ seconds in liquid water at 25°C. An ice crystal has a more or less stable arrangement of hydrogen bonds, and of course contains many, many more than a hundred molecules, exceeding that figure by numerous orders of magnitude.

Hydrogen bonds are much weaker than normal covalent bonds. The energy required to break the O—H covalent bond is 460 kJ mol⁻¹ (110 kcal mol⁻¹), while the energy of hydrogen bonds in water is about 20 kJ mol⁻¹ (5 kcal mol⁻¹) (Table 3.2). Even this comparatively small amount of energy is enough to affect the properties of water drastically, especially its melting point, its boiling point, and the density of liquid water relative to that of ice. Both the melting point and boiling point of water are significantly higher than would be predicted for a molecule of its size. Other substances of about the same molecular weight, such as methane and ammonia, have much lower melting and boiling points. The forces of attraction between the molecules of these substances are weaker than the forces between water molecules. The energy of attraction due to the hydrogen bonds in water has to be overcome to melt ice or boil water.

Ice has a lower density than liquid water because the fully hydrogen-bonded array in an ice crystal is very open, with a lot of empty space between the molecules. Liquid water is less extensively hydrogen-bonded, and thus has a greater density than ice. Most substances contract when they freeze, but the opposite is true for water. Thus ice cubes and icebergs float, and the replacement of liquid water by ice crystals can cause damage to surrounding materials as a result of expansion. Cars require antifreeze in cold climates to avoid a cracked engine block, which can occur as a result of freezing and expansion of the water in the cooling system. The same principle can be used in a method for disrupting cells by several cycles of freezing and thawing.

Hydrogen bonding also plays a role in the behavior of water as a solvent. If a polar solute can serve as a donor or an acceptor of hydrogen bonds, it can form hydrogen bonds with water in addition to being involved in nonspecific dipole–dipole interactions. Some examples are shown in Figure 3.8. Alcohols, amines, carboxylic acids, and esters, as well as aldehydes and ketones, can all form hydrogen bonds with water, so they are

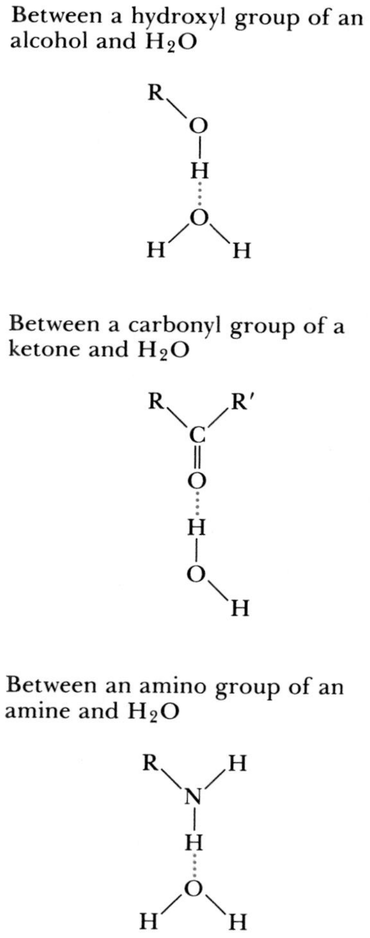

Between a hydroxyl group of an alcohol and $H_2O$

Between a carbonyl group of a ketone and $H_2O$

Between an amino group of an amine and $H_2O$

**FIGURE 3.8**  Examples of hydrogen bonding between polar groups and water.

soluble in water. It is difficult to overstate the importance of water to the existence of life on earth, and difficult to imagine life based on another solvent.

### Biologically Important Hydrogen Bonds Other Than to Water

Before ending the discussion of hydrogen bonds, it is useful to point out their vital involvement in stabilizing the three-dimensional structures of biologically important molecules including DNA, RNA, and proteins. The two strands of the double helix in DNA are held together by hydrogen bonds between complementary bases (Section 20.2). Transfer RNA also has a complex three-dimensional structure stabilized by hydrogen bonds (Section 20.3). Hydrogen bonding in proteins gives rise to two important structures, the $\alpha$-helix and $\beta$-pleated sheet conformations. Both types of conformation are widely encountered in proteins (Section 9.5).

## 3.3
## ACIDS AND BASES

The biochemical behavior of many important compounds depends on their acid–base properties. Acids are defined as proton (hydrogen ion) donors and bases as proton acceptors. How readily acids or bases lose or gain protons depends on the chemical nature of the compounds involved. The degree of dissociation of acids in water, for example, ranges from essentially complete dissociation for a strong acid to practically no dissociation for a very weak acid, and any intermediate value is possible.

It is useful to derive a numerical measure of the **strength** of an acid, which is the amount of hydrogen ion released when a given amount of acid is dissolved in water. Such an expression, called the **acid dissociation constant** or $K_a'$, can be written for any acid, HA, that reacts according to the equation

$$HA \rightleftharpoons H^+ + A^-$$
**Acid**        **Conjugate base**

$$K_a' = \frac{[H^+][A^-]}{[HA]}$$

In this expression the square brackets refer to molar concentration, that is, the concentration in moles per liter. For each acid, the quantity $K_a'$ has a fixed numerical value at a given temperature. The numerical value of the $K_a'$ will be larger when the acid is more completely dissociated; **the larger the $K_a'$, the stronger the acid.**

Strictly speaking, the acid–base reaction we have just written is a proton-transfer reaction in which water acts as a base as well as the solvent.

$$HA(aq) + H_2O(l) \rightleftharpoons H_3O^+(aq) + A^-(aq)$$
**Acid**     **Base**     **Conjugate**     **Conjugate**
                              **acid to $H_2O$**    **base to HA**

The notation (aq) refers to solutes in aqueous solution, while (l) refers to water in the liquid state. It is well established that there are no "naked protons" (free hydrogen ions) in solution; even the hydronium ion ($H_3O^+$) is an underestimate of the degree of hydration of hydrogen ion in aqueous solution. All solutes are extensively hydrated in aqueous solution. We will write the short form of equations for acid dissociation in the interests of simplicity, but the role of water should be kept in mind throughout our discussion.

## 3.4
## THE SELF-DISSOCIATION OF WATER AND THE pH SCALE

The acid–base properties of water play an important part in biological processes because of the central role of water as a solvent. The extent of self-dissociation of water to hydrogen ion and hydroxide ion

$$H_2O \rightleftharpoons H^+ + OH^-$$

is small, but the fact that it takes place determines important properties of many solutes. Both the hydrogen ion ($H^+$) and the hydroxide ion ($OH^-$) are associated with several water molecules, as is true of all ions in aqueous solution, and the water molecule in the equation is itself part of a cluster of such molecules.

It is especially important to have a quantitative estimate of the degree of dissociation of water. We can write the expression

$$K_a' =' \frac{[H^+][OH^-]}{[H_2O]}$$

The molar concentration of pure water, $[H_2O]$, is quite large compared to any possible concentrations of solutes and can be considered a constant. (The numerical value is 55.5 M, which can be obtained by dividing the number of grams of water in one liter, 1000 grams, by the molecular weight of water, 18 grams/mole; $1000/18 = 55.5$ M.) Thus,

$$K_a' = \frac{[H^+][OH^-]}{55.5}$$

$$K_a' \times 55.5 = [H^+][OH^-] = K_w$$

A new constant, $K_w$, the **ion product constant for water,** has just been defined, where the concentration of water has been included in the value of $K_w$.

The numerical value of $K_w$ can be determined experimentally by measuring the hydrogen ion concentration of pure water. The hydrogen ion concentration is also equal by definition to the hydroxide ion concentration, because water is a monoprotic acid (one that releases a single proton per molecule). At 25°C in pure water

$$[H^+] = 10^{-7} \text{ M} = [OH^-]$$

Thus at 25°C the numerical value of $K_w$ is given by the expression

$$K_w = [H^+][OH^-] = (10^{-7})(10^{-7}) = 10^{-14}$$

**TABLE 3.3   Dissociation Constants of Some Acids**

| ACID | HA | A⁻ | $K'_a$ | $pK'_a$ |
|---|---|---|---|---|
| Pyruvic acid | $CH_3COCOOH$ | $CH_3COCOO^-$ | $3.16 \times 10^{-3}$ | 2.50 |
| Formic acid | $HCOOH$ | $HCOO^-$ | $1.44 \times 10^{-4}$ | 3.75 |
| Lactic acid | $CH_3CHOHCOOH$ | $CH_3CHOHCOO^-$ | $1.38 \times 10^{-4}$ | 3.86 |
| Benzoic acid | $C_6H_5COOH$ | $C_6H_5COO^-$ | $6.46 \times 10^{-5}$ | 4.19 |
| Acetic acid | $CH_3COOH$ | $CH_3COO^-$ | $1.76 \times 10^{-5}$ | 4.76 |
| Ammonium ion | $NH_4^+$ | $NH_3$ | $5.6 \times 10^{-10}$ | 9.25 |
| Oxalic acid (1) | $HOOC—COOH$ | $HOOC—COO^-$ | $5.9 \times 10^{-2}$ | 1.23 |
| Oxalic acid (2) | $HOOC—COO^-$ | $^-OOC—COO^-$ | $6.4 \times 10^{-5}$ | 4.19 |
| Malonic acid (1) | $HOOC—CH_2—COOH$ | $HOOC—CH_2—COO^-$ | $1.49 \times 10^{-3}$ | 2.83 |
| Malonic acid (2) | $HOOC—CH_2—COO^-$ | $^-OOC—CH_2—COO^-$ | $2.03 \times 10^{-6}$ | 5.69 |
| Malic acid (1) | $HOOC—CH_2—CHOH—COOH$ | $HOOC—CH_2—CHOH—COO^-$ | $3.98 \times 10^{-4}$ | 3.40 |
| Malic acid (2) | $HOOC—CH_2—CHOH—COO^-$ | $^-OOC—CH_2—CHOH—COO^-$ | $5.5 \times 10^{-6}$ | 5.26 |
| Succinic acid (1) | $HOOC—CH_2—CH_2—COOH$ | $HOOC—CH_2—CH_2—COO^-$ | $6.17 \times 10^{-5}$ | 4.21 |
| Succinic acid (2) | $HOOC—CH_2—CH_2—COO^-$ | $^-OOC—CH_2—CH_2—COO^-$ | $2.3 \times 10^{-6}$ | 5.63 |
| Carbonic acid (1) | $H_2CO_3$ | $HCO_3^-$ | $4.3 \times 10^{-7}$ | 6.37 |
| Carbonic acid (2) | $HCO_3^-$ | $CO_3^-$ | $5.6 \times 10^{-11}$ | 10.2 |
| Citric acid (1) | $HOOC—CH_2—C(OH)$ $(COOH)—CH_2—COOH$ | $HOOC—CH_2—C(OH)$ $COOH—CH_2—COO^-$ | $8.14 \times 10^{-4}$ | 3.09 |
| Citric acid (2) | $HOOC—CH_2—C(OH)$ $(COOH)—CH_2—COO^-$ | $^-OOC—CH_2—C(OH)$ $(COOH)—CH_2—COO^-$ | $1.78 \times 10^{-5}$ | 4.75 |
| Citric acid (3) | $^-OOC—CH_2—C(OH)$ $(COOH)—CH_2—COO^-$ | $^-OOC—CH_2—C(OH)$ $(COO^-)—CH_2—COO^-$ | $3.9 \times 10^{-6}$ | 5.41 |
| Phosphoric acid (1) | $H_3PO_4$ | $H_2PO_4^-$ | $7.25 \times 10^{-3}$ | 2.14 |
| Phosphoric acid (2) | $H_2PO_4^-$ | $HPO_4^{2-}$ | $6.31 \times 10^{-8}$ | 7.20 |
| Phosphoric acid (3) | $HPO_4^{2-}$ | $PO_4^{3-}$ | $3.98 \times 10^{-13}$ | 12.4 |

This relationship, which we have derived for pure water, is valid for *any* aqueous solution, whether neutral, acidic, or basic.

The wide range of possible hydrogen ion and hydroxide ion concentrations in aqueous solution makes it desirable to define a quantity for expressing these concentrations more conveniently than by exponential notation. This quantity is called **pH** and is defined by

$$pH = -\log_{10} [H^+]$$

with the logarithm taken to the base 10. The pH values of some typical aqueous samples can be determined by a simple calculation:

pure water, $[H^+] = 10^{-7}$ M; pH = 7

$10^{-3}$ M HCl, $[H^+] = 10^{-3}$ M; pH = 3

$10^{-4}$ M NaOH, $[OH^-] = 10^{-4}$ M so $[H^+] = 10^{-10}$ M; pH = 10

## BOX 3.1
## THE HENDERSON-HASSELBALCH EQUATION

There is an equation that connects the $K_a'$ of any weak acid with the pH of a solution containing both that acid and its conjugate base. This relationship has wide use in biochemical practice, especially where it is necessary to control pH for optimum reaction conditions. To derive this expression, it is first necessary to solve the $K_a'$ equation for the hydrogen ion concentration caused by dissociation of the acid HA.

$$K_a' = \frac{[H^+][A^-]}{[HA]}$$

Solving for $[H^+]$ gives

$$[H^+] = \frac{K_a'\,[HA]}{[A^-]}$$

Taking the negative logarithm of both sides and using the definition of pH,

$$pH = -\log\left[K_a'\,\frac{[HA]}{[A^-]}\right]$$

Recalling that logarithms are added in multiplication and subtracted in division,

$$pH = -\left[\log K_a' + \log\frac{[HA]}{[A^-]}\right]$$

$$pH = -\log K_a' - \log\frac{[HA]}{[A^-]}$$

Recalling that $pK_a' = -\log K_a'$,

$$pH = pK_a' - \log\frac{[HA]}{[A^-]}$$

By making use of the properties of logarithms, this equation can be rewritten as

$$pH = pK_a' + \log\frac{[A^-]}{[HA]}$$

This relationship is known as the **Henderson-Hasselbalch equation,** and it is of use in predicting the properties of buffer solutions used to control the pH of reaction mixtures. When buffers are discussed in Section 3.6, we will be interested in the situation where the concentration of acid, [HA], and the concentration of the conjugate base, $[A^-]$, are equal ($[HA] = [A^-]$). The ratio $[A^-]/[HA]$ is then equal to 1, and the logarithm of 1 to any base is equal to zero. Therefore, when a solution contains equal concentrations of weak acid and its conjugate base, the pH of that solution is equal to the $pK_a'$ value of the weak acid.

For the strong base NaOH, the $[OH^-]$ is the same as the solution concentration; then $[H^+]$ is found from $[H^+] = K_w/[OH^-]$. Pure water with a pH of 7 is neutral; acidic solutions have pH values lower than 7; and basic solutions have pH values higher than 7.

A similar quantity, $pK_a'$, can be defined by analogy with the definition of pH:

$$pK_a' = -\log_{10}K_a'$$

The $pK_a'$ is another numerical measure of acid strength; the smaller the value of $pK_a'$, the stronger the acid. This is the reverse of the situation with $K_a'$, where larger $K_a'$ values imply stronger acids (Table 3.3).

As we will discuss in Section 3.6, a buffer solution is a mixture of a weak acid and its conjugate base; its most useful property is that its pH remains almost constant when a small amount of strong acid or base is added. Box 3.1 shows how to derive an equation for calculating the pH of a buffer solution if you know the $pK_a'$ of the weak acid and the concentrations of the acid and its conjugate base.

## 3.5
## TITRATION CURVES

When base is added to a sample of acid, the pH of the solution changes. A **titration** is an experiment in which measured amounts of base are added to a measured amount of acid. It is convenient and straightforward to follow the course of the reaction with a pH meter. The amount of base required for complete reaction with the acid present is referred to as one **equivalent.** The point in the titration at which the acid is exactly neutralized is called the **equivalence point.** In order to discuss a given type of titration in a general sense without reference to the number of moles of acid or base in a specific sample or to the volume of a specific sample, it is useful to refer to the number of equivalents of base added to a sample of acid in a titration.

If the pH is monitored as base is added to a sample of acetic acid in the course of a titration, an inflection point in the titration curve is reached when the pH equals the $pK'_a$ of acetic acid (Figure 3.9a). As we saw in Box 3.1 on the Henderson-Hasselbalch equation, a value of pH equal to the $pK'_a$ corresponds to a mixture with equal concentrations of the weak acid and its conjugate base, in this case acetic acid and acetate ion. The pH at the inflection point is 4.76, equal to the $pK'_a$ of acetic acid. The inflection point occurs when 0.5 equivalent of base has been added. Near the inflection point the pH changes very slowly as more base is added.

When 1 equivalent of base has been added, the equivalence point is reached. At the equivalence point essentially all the acetic acid has been converted to acetate ion. (See Exercise 9.) Another graphical representation (Figure 3.9b) shows the relative abundance of acetic acid and acetate ion with increasing amounts of NaOH added. Notice that the percentage of acetic acid plus the percentage of acetate ion adds up to 100%. The acid (acetic acid) is progressively converted to its conjugate base (acetate ion) as more NaOH is added and the titration proceeds. It can be helpful to keep

**FIGURE 3.9** (a) Titration of acetic acid with NaOH. (b) Relative abundance of acetic acid and acetate ion during a titration.

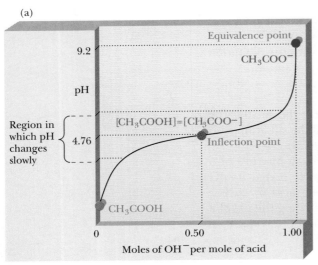

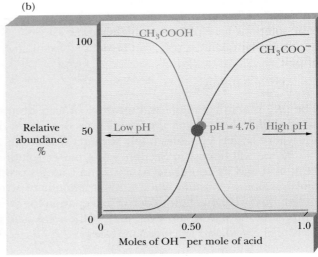

track of the percentages of a conjugate acid and base in this way to understand the full significance of the reaction taking place in a titration. The form of these curves represents the behavior of any monoprotic weak acid, but the value of the $pK_a'$ for each individual acid determines the pH values at the inflection point and at the equivalence point.

## 3.6
## BUFFERS

A **buffer solution** consists of a mixture of a weak acid and its conjugate base. Buffer solutions tend to resist a change in pH on addition of moderate amounts of strong acid or base. Let us compare the changes in pH that occur on adding equal amounts of strong acid or strong base to pure water at pH 7 or to a buffer solution at pH 7. If 1.0 mL of 0.1 M HCl is added to 99.0 mL of pure water, the resulting solution will be 0.001 M = $10^{-3}$ M HCl and have a pH of 3. If the same experiment is done with 0.1 M NaOH in place of 0.1 M HCl, the solution will be 0.001 M = $10^{-3}$ M NaOH, and $[OH^-] = 10^{-3}$ M; thus $[H^+] = 10^{-11}$ M and pH = 11.

The results are different when 99.0 mL of buffer solution is used instead of pure water. A solution that contains the monohydrogen phosphate and dihydrogen phosphate ions, $HPO_4{}^{2-}$ and $H_2PO_4{}^-$, in suitable proportions can serve as such a buffer. The Henderson-Hasselbalch equation can be used to calculate the $[HPO_4{}^{2-}]/[H_2PO_4{}^-]$ ratio that corresponds to pH 7.0. It is left as an exercise to show that the proper ratio for pH 7.0 is 0.6 parts $HPO_4{}^{2-}$ to 1 part $H_2PO_4{}^-$. For purposes of illustration we shall consider a solution in which the concentrations are $[HPO_4{}^{2-}] = 0.06$ M and $[H_2PO_4{}^-] = 0.10$ M. If 1.0 mL of 0.1 M HCl is added to 99.0 mL of the buffer, the reaction

$$HPO_4{}^{2-} + H^+ \rightleftharpoons H_2PO_4{}^-$$

will take place, and almost all the added $H^+$ will be used up. The concentrations of $HPO_4{}^{2-}$ and $H_2PO_4{}^-$ will change, and the new concentrations can be calculated. The new pH can then be calculated using the Henderson-Hasselbalch equation and the phosphate ion concentrations. The new pH will be 6.99, a much smaller change than in the unbuffered pure water. Similarly, if 1.0 mL of 0.1 M NaOH is used, the same reaction will take place as in a titration:

$$H_2PO_4{}^- + OH^- \rightleftharpoons HPO_4{}^{2-} + H_2O$$

Almost all the added $OH^-$ is used up, but a small amount remains. Since this buffer is an aqueous solution, it is still true that $K_w = [H^+][OH^-]$. The increase in hydroxide ion concentration implies that the hydrogen ion concentration decreases and that the pH increases. The new pH can be calculated, and the result is pH = 7.01, again a much smaller change in pH than in the pure water. Many biological reactions will not take place unless the pH remains within fairly narrow limits, and as a result buffers have great practical importance.

A consideration of titration curves can give insight into the origin of this property (Figure 3.10a). The pH of a sample being titrated changes

(a)

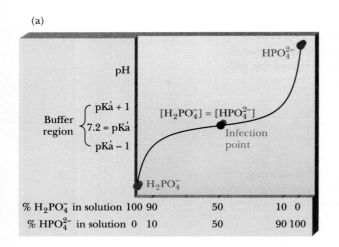

(b)

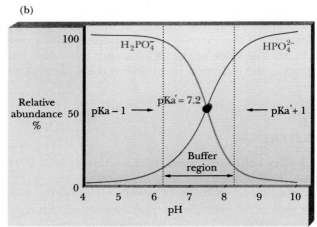

H$_2$PO$_4^-$ in excess ←——|——→ HPO$_4^{2-}$ in excess

**FIGURE 3.10**    The relationship between the titration curve and buffering action in H$_2$PO$_4^-$. (a) The titration curve of H$_2$PO$_4^-$ showing the buffer region for the H$_2$PO$_4^-$/HPO$_4^{2-}$ pair. (b) Relative abundances of H$_2$PO$_4^-$ and HPO$_4^{2-}$.

very slowly in the vicinity of the inflection point of a titration curve. Also, at the inflection point half the amount of acid originally present has been converted to the conjugate base. The second stage of ionization of phosphoric acid, H$_2$PO$_4^- \rightleftarrows$ H$^+$ + HPO$_4^{2-}$, was the basis of the buffer just used as an example. The pH at the inflection point of the titration is 7.2, a value numerically equal to the p$K_a'$ of the dihydrogen phosphate ion. At this pH the solution contains equal concentrations of the dihydrogen phosphate and monohydrogen phosphate ions, the acid and base forms.

A buffer solution is able to maintain the pH at a relatively constant value because of the presence of appreciable amounts of both the acid and its conjugate base. This condition is met at pH values at or near the p$K_a'$ of the acid. If OH$^-$ is added, there is an appreciable amount of the acid form of the buffer present in solution to react with the added base. If H$^+$ is added, there is also an appreciable amount of the basic form of the buffer to react with the added acid.

The H$_2$PO$_4^-$/HPO$_4^{2-}$ pair is suitable as a buffer near pH 7.2 and the CH$_3$COOH/CH$_3$COO$^-$ pair near pH 4.76. At pH values below the p$K_a'$ the acid form predominates, and at pH values above the p$K_a'$ the basic form predominates. A ratio of acid to base of 10:1 corresponds to a pH lower than the p$K_a'$ by about 1 pH unit, because each unit change in the pH scale corresponds to a tenfold change in the hydrogen ion concentration. A ratio of acid to base of 1:10 corresponds to a pH about 1 unit higher than the p$K_a'$. The plateau region in a titration curve, the part in which the pH does not change rapidly, covers a pH range approximately 1 pH unit on either side of the p$K_a'$. Thus, there is a range of about 2 pH units in which the buffer is effective (Figure 3.10b). The condition that a buffer contain appreciable amounts of both a weak acid and its conjugate base refers both to the ratio of the two forms and to the absolute amount of each present in

## BOX 3.2
## SOME PHYSIOLOGICAL CONSEQUENCES OF BLOOD BUFFERING

The process of respiration plays an important role in the buffering of blood. In particular, an increase in $H^+$ concentration can be dealt with by increasing the rate of respiration. The added hydrogen ion binds to bicarbonate ion, forming carbonic acid.

$$H^+(aq) + HCO_3^-(aq) \rightleftharpoons H_2CO_3(aq)$$

An increased level of carbonic acid gives rise to more dissolved carbon dioxide and ultimately to more gaseous carbon dioxide in the lungs.

$$H_2CO_3(aq) \rightleftharpoons CO_2(aq) + H_2O(l)$$

$$CO_2(aq) \rightleftharpoons CO_2(g)$$

An increased rate of respiration removes more carbon dioxide from the lungs, starting a shift in the equilibrium positions of all the preceding reactions. Removal of gaseous $CO_2$ decreases the amount of dissolved $CO_2$. To restore the equilibrium, $H^+$ reacts with $HCO_3^-$ and in the process lowers the $H^+$ concentration of blood, keeping the pH constant in the presence of added $H^+$.

*Hyperventilation* (excessively deep and rapid breathing) removes large amounts of carbon dioxide from the lungs and raises the pH of blood, sometimes to dangerously high levels, bringing on weakness and fainting. Sprinters, however, have learned how to make use of the increase in blood pH caused by hyperventilation. Short bursts of strenuous exercise produce high levels of lactic acid in the blood as a result of the breakdown of glucose. The presence of so much lactic acid tends to lower the pH of the blood, but a short (30-second) period of hyperventilation before a 100-meter dash counteracts the effects of the added lactic acid and maintains the pH balance.

An increase in the level of $H^+$ in blood can be caused by large amounts of any acid entering the bloodstream. Aspirin, like lactic acid, is an acid, and *aspirin poisoning* can be caused by the increased acidity due to ingesting large doses of aspirin. The effect of exposure to *high altitudes* is similar to hyperventilation at sea level. The increase in the rate of respiration in response to the tenuous atmosphere increases the pH of the blood. As is the case with hyperventilation, more carbon dioxide is expired from the lungs, ultimately lowering the $H^+$ level in blood and raising the pH. When a person who normally lives at sea level is suddenly placed at a high elevation, there is a temporary elevation of the blood pH until the person becomes acclimated.

---

a given solution. If a buffer solution contained a suitable ratio of acid to base, but very low concentrations of both, it would take very little added acid to use up all of the base form or vice versa. A buffer solution with low concentrations of the acid and base forms is said to have a low **buffering capacity,** while a buffer that contains larger amounts of both acid and base has a higher buffering capacity.

Buffer systems in living organisms and in the laboratory are based on many types of compounds. Since physiological pH in most organisms is around 7, it might be expected that the phosphate buffer system would be widely used in living organisms. This is the case where phosphate ion concentrations are high enough for the buffer to be effective, as in most intracellular fluids. Phosphate ion levels in blood are inadequate for buffering, however, and a different system is operative there.

The buffering system in blood is based on the dissociation of carbonic acid ($H_2CO_3$)

$$H_2CO_3 \rightleftharpoons H^+ + HCO_3^-$$

where the $pK_a'$ of $H_2CO_3$ is 6.37. The pH of human blood, 7.4, is near the

**TABLE 3.4    Acid and Base Forms of Some Useful Biochemical Buffers**

| ACID FORM | | BASE FORM | $pK_a$ |
|---|---|---|---|
| N—Tris[hydroxymethyl]aminomethane (TRIS) | | | 8.3 |
| TRIS—H$^+$ (protonated form) $(HOCH_2)_3CNH_3^+$ | $\rightleftharpoons$ | TRIS (free amine) $(HOCH_2)_3CNH_2$ | |
| N—Tris[hydroxymethyl]methyl-2-aminoethane sulfonate (TES) | | | 7.55 |
| $^-$TES—H$^+$ (zwitterionic form) $(HOCH_2)_3C\overset{+}{N}H_2CH_2CH_2SO_3^-$ | $\rightleftharpoons$ | $^-$TES (anionic form) $(HOCH_2)_3CNHCH_2CH_2SO_3^-$ | |
| N—2—Hydroxyethylpiperazine-N′-2-ethane sulfonate (HEPES) | | | 7.55 |
| $^-$HEPES—H$^+$ (zwitterionic form) 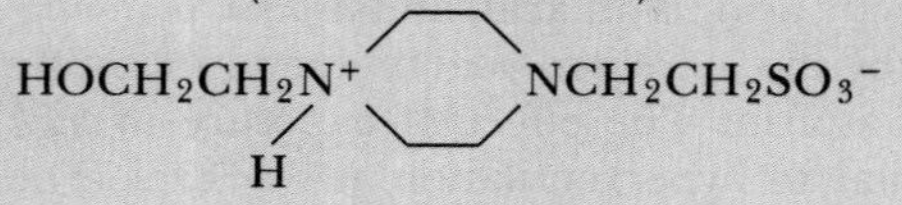 $HOCH_2CH_2\overset{+}{N}\!\!\!\!\!\overset{\displaystyle\phantom{x}}{\underset{H}{}}\!\!\!NCH_2CH_2SO_3^-$ | $\rightleftharpoons$ | $^-$HEPES (anionic form) 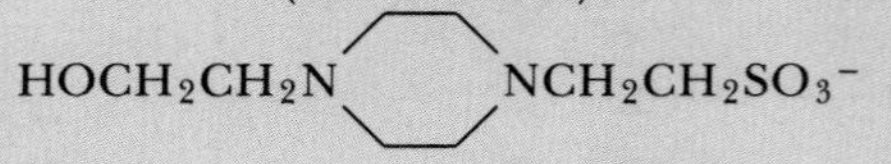 $HOCH_2CH_2N\!\!\!\!\!\phantom{x}\!\!\!NCH_2CH_2SO_3^-$ | |
| 3—[N—Morpholino]propane-sulfonic acid (MOPS) | | | 7.2 |
| $^-$MOPS—H$^+$ (zwitterionic form) 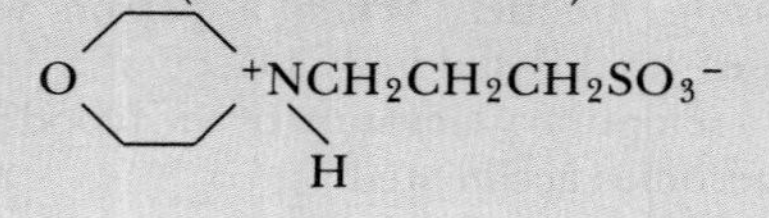 $O\phantom{xxx}\overset{+}{N}HCH_2CH_2CH_2SO_3^-$ | $\rightleftharpoons$ | $^-$MOPS (anionic form) 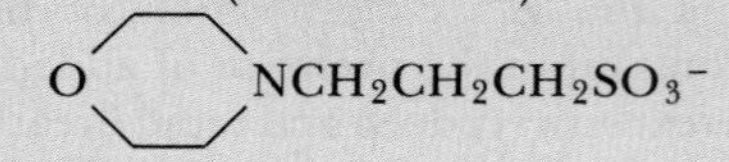 $O\phantom{xxx}NCH_2CH_2CH_2SO_3^-$ | |
| Piperazine—N,N′-bis[2-ethanesulfonic acid] (PIPES) | | | 6.8 |
| $^{2-}$PIPES—H$^+$ (protonated dianion) 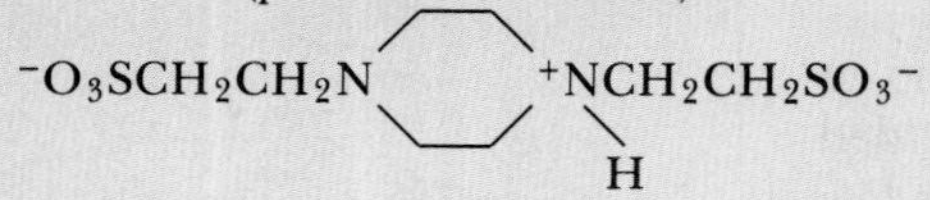 $^-O_3SCH_2CH_2N\phantom{x}\overset{+}{N}HCH_2CH_2SO_3^-$ | $\rightleftharpoons$ | $^{2-}$PIPES (dianion) 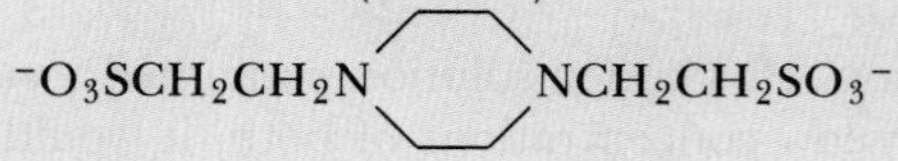 $^-O_3SCH_2CH_2N\phantom{x}NCH_2CH_2SO_3^-$ | |

end of the buffer range of this system, but another factor enters into the situation. The protein hemoglobin is a major constituent of red blood cells, and its function is to carry oxygen from the lungs to the tissues and carbon dioxide from the tissues to the lungs. Carbon dioxide can dissolve in water and in water-based fluids such as blood. The dissolved carbon dioxide forms carbonic acid, which in turn reacts to produce bicarbonate ion, $HCO_3^-$.

$$CO_2(g) \rightleftharpoons CO_2(aq)$$

$$CO_2(aq) + H_2O(l) \rightleftharpoons H_2CO_3(aq)$$

$$H_2CO_3(aq) \rightleftharpoons H^+(aq) + HCO_3^-(aq)$$

Net equation: $CO_2(g) + H_2O(l) \rightleftharpoons H^+(aq) + HCO_3^-(aq)$

At the pH of blood, which is about 1 unit higher than the $pK'_a$ of carbonic acid, most of the dissolved $CO_2$ will be present as $HCO_3{}^-$. The $CO_2$ being transported to the lungs to be expired takes the form of bicarbonate ion, which is bound to the hemoglobin molecule by ionic attraction to a charged group on the protein. There is a direct relationship between the pH of the blood and the pressure of carbon dioxide gas in the lungs.

The phosphate buffer system is a common one in the laboratory (*in vitro*) as well as in living organisms (*in vivo*). The buffer system based on tris(hydroxymethyl)aminomethane (called TRIS) is also widely used *in vitro*. Other buffers that have come into wide use more recently are **zwitterions,** compounds that have both a positive and a negative charge. They are usually considered to be less likely to interfere with biochemical reactions than some of the earlier buffers (Table 3.4).

Most living systems operate at pH values close to 7. The $pK'_a$ values of many functional groups, such as the carboxyl and amino groups, are well above or well below this value. As a result, many important biomolecules exist as charged species to a greater or lesser extent under physiological conditions. The practical consequences of this fact are explored in Box 3.2.

## SUMMARY

The properties of the water molecule have a direct effect on the behavior of biomolecules. Water is a polar molecule, with partial charges of opposite signs at opposite ends of the molecule. There are forces of attraction between the unlike partial charges. In addition, in both the liquid and solid states water molecules are extensively hydrogen bonded to one another. Polar substances tend to dissolve in water, while nonpolar substances do not. Hydrogen bonding between water and polar solutes takes place in aqueous solutions. The three-dimensional structures of many important biomolecules, including proteins and nucleic acids, are stabilized by hydrogen bonds.

The degree of dissociation of acids in water can be characterized by an acid dissociation constant, $K'_a$, which gives a numerical indication of the strength of the acid. The self-dissociation of water can be characterized by a similar constant, $K_w$. Since the hydrogen-ion concentration of aqueous solutions can vary by many orders of magnitude, it is desirable to define a quantity, pH, which expresses the concentration of hydrogen ions in convenient fashion. A similar quantity, $pK'_a$, can be used as an alternate expression for the strength of any acid. The pH of a solution of a weak acid and its conjugate base can be related to the $pK'_a$ of that acid by the Henderson-Hasselbalch equation.

In aqueous solution, the relative concentrations of a weak acid and its conjugate base can be related to the titration curve of that acid. In the region of the titration curve in which the pH changes slowly on addition of acid or base, the acid/base concentration ratio varies within a fairly narrow range (10:1 at one extreme, and 1:10 at the other). The tendency to resist a change in pH on addition of relatively small amounts of acid or base is characteristic of buffer solutions. The control of pH by buffers depends on the fact that their composition reflects the acid/base concentration ratio in the region of the titration curve where there is little change in pH.

## EXERCISES

1. Rationalize the fact that hydrogen bonding has not been observed between $CH_4$ molecules.
2. Many properties of acetic acid can be rationalized in terms of a hydrogen-bonded dimer. Propose a structure for such a dimer.
3. Identify the conjugate acids and bases in the following pairs of substances.
   $(CH_3)_3NH^+ / (CH_3)_3N$
   $^+H_3N—CH_2COOH / ^+H_3N—CH_2—COO^-$
   $^+H_3N—CH_2—COO^-/H_2N—CH_2—COO^-$
   $^-OOC—CH_2—COOH/^-OOC—CH_2—COO^-$
   $^-OOC—CH_2—COOH/HOOC—CH_2—COOH$
4. Calculate the hydrogen ion concentration, $[H^+]$, for each of the following materials.
   Blood plasma, pH 7.4
   Orange juice, pH 3.5
   Human urine, pH 6.2
   Household ammonia, pH 11.5
   Gastric juice, pH 1.8
5. Suggest a suitable buffer range for each of the following substances.
   Lactic acid ($pK_a' = 3.86$) and its sodium salt
   Acetic acid ($pK_a' = 4.76$) and its sodium salt
   TRIS (see Table 3.4, $pK_a' = 8.3$) in its protonated form and its free amine form
   HEPES (see Table 3.4, $pK_a' = 7.55$) in its zwitterionic form and its anionic form
6. What is the $[CH_3COO^-]/[CH_3COOH]$ ratio in an acetate buffer at pH 5.00?
7. How would you prepare one liter of a 0.05 M phosphate buffer at pH 7.5 using crystalline $K_2HPO_4$ and a solution of 1 M HCl?
8. The buffer needed for Exercise 7 can also be prepared using crystalline $NaH_2PO_4$ and a solution of 1 M NaOH. How would you prepare this buffer using these materials?
9. In Section 3.5 we say that at the equivalence point of a titration of acetic acid *essentially all* the acid has been converted to acetate ion. Why do we not say that *all* the acetic acid has been converted to acetate ion?
10. Define buffering capacity. How do the following buffers differ in capacity? How do they differ in pH?
    0.01 M $Na_2HPO_4$ and 0.01 M $NaH_2PO_4$
    0.1 M $Na_2HPO_4$ and 0.1 M $NaH_2PO_4$
    1.0 M $Na_2HPO_4$ and 1.0 M $NaH_2PO_4$
11. Identify the zwitterions in the list of substances in Exercise 3.
12. If you mixed equal volumes of 0.1 M HCl and 0.2 M TRIS (free amine form), is the resulting solution a buffer? Give the reason for your answer.
13. The measured pH of a sample of lemon juice is 2.1. Calculate the hydrogen ion concentration.
14. What is the ratio of concentrations of acetate ion and undissociated acetic acid in a solution which has a pH of 5.12?
15. You need to carry out an enzymatic reaction at pH 7.5. A friend suggests a weak acid with a $pK_a'$ of 3.9 as the basis of a buffer. Will this substance and its conjugate base make a suitable buffer? Give the reason for your answer.
16. A frequently recommended treatment for hiccups is to hold one's breath. This condition of hypoventilation causes a buildup of carbon dioxide in the lungs. Predict the effect on the pH of blood.
17. Aspirin is an acid with a $pK_a'$ of 3.5; its structure includes a carboxyl group. To be absorbed into the bloodstream it must pass through the membrane lining the stomach and the small intestine. Electrically neutral molecules can pass through a membrane more easily than can charged molecules. Would you expect more aspirin to be absorbed in the stomach, where the pH of gastric juice is about 1, or in the small intestine, where the pH is about 6? Give the reason for your answer.

## ANNOTATED BIBLIOGRAPHY

Barrow, G. M. *Physical Chemistry for the Life Sciences.* 2nd ed. New York: McGraw-Hill, 1981. [Acid–base reactions are discussed in Chapter 4, with titration curves treated in great detail.]

Fasman, G. D., ed. *Handbook of Biochemistry and Molecular Biology: Physical and Chemical Data Section.* 2 vols. 3rd ed. Cleveland: The Chemical Rubber Company, 1976. [Includes a section on buffers and directions for preparation of buffer solutions (volume 1, pp. 353–378). Other sections cover all important types of biomolecules.]

Ferguson, W. J., and N. E. Good. Hydrogen Ion Buffers. *Anal. Biochem.* **104,** 300–310 (1980). [A description of useful zwitterionic buffers.]

Franks, F. *Polywater.* Cambridge, MA: MIT Press, 1981. [An amusing account of a scientific controversy about the existence of a polymeric form of water.]

Pauling, L. *The Nature of the Chemical Bond.* 3rd ed. Ithaca, NY: Cornell University Press, 1960. [A classic. Chapter 12 is devoted to hydrogen bonding.]

[Some works of fiction touch on the role of water in living systems: *Stranger in a Strange Land* by Robert A. Heinlein, *Dune* by Frank Herbert, and *Cat's Cradle* by Kurt Vonnegut. Numerous paperback editions exist.]

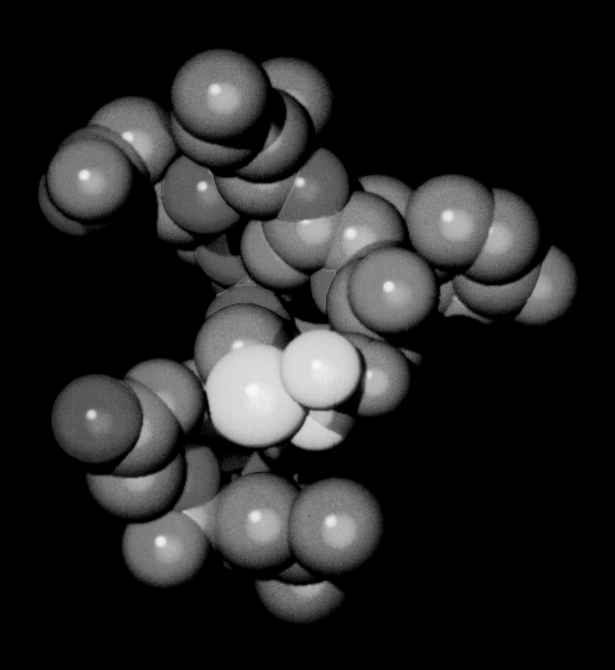

PART

II

# The Molecular Nature of Cellular Components

# Sidney Altman

Sidney Altman was born in Montreal in 1939, the son of immigrants. His mother worked in a textile mill, and his father ran a grocery store in the Montreal district of Notre-Dame-de-Grace.

He studied physics at the Massachusetts Institute of Technology, where he was also co-founder and editor of a literary magazine. After graduating in 1960, he got a job as a science and poetry editor for a New York publisher, where he edited some Robert Frost books.

A chance meeting with physicist George Gamow at a physics institute in Colorado led him to become a graduate student in biophysics at the University of Colorado Medical Center under Leonard Lerman. He completed his Ph.D. in biophysics in 1967. Altman did postdoctoral work at Harvard and at Cambridge, England, in the laboratories of Sydney Brenner and Francis Crick, the co-discoverer of the structure of DNA. At Cambridge he began research into the enzymatic properties of RNA.

In 1971, Altman joined the faculty of Yale. He served as dean of Yale College from 1985 to 1989. It was at Yale that Altman conducted the groundbreaking and controversial research that demonstrated that RNA, in addition to its role in the translation of a cell's genetic material into proteins, also serves as an enzyme. This discovery, which contradicted the then prevalent scientific dogma that RNA was simply DNA's messenger in cellular functions, began a revolution in the scientific understanding of how the body's genetic function works. Altman and Thomas R. Cech of the University of Colorado, who independently made the same discovery, shared the Nobel Prize for Chemistry in 1989.

## What made you want to be a scientist?

I think one of the reasons I went into science was that I thought if you did something in science it would be judged objectively. I grew up in an immigrant family, and I wanted my achievements to be judged as objectively as possible. I didn't want to have to spend a lot of time becoming a perfectly assimilated person in order to establish myself professionally. I just wanted to get out there and do what I could do and have it judged objectively.

I got interested in science at an early age. When I was about 12 or 13 years old, I was given a book, *Explaining the Atom*, on the periodic table of the elements. It was a description of how Mendeleev constructed the periodic table and predicted the properties of the elements that had not yet been found. It gave me a great sense of the beauty and elegance of science. I had never had that appreciation before.

Scientific education was very limited in schools in those days. In fact, compared with what my kids get these days, we had virtually no science aside from mathematics. It really wasn't until tenth grade that we were taught anything that approached science, real science. And it wasn't until my senior year at MIT that I had a sense of what really working in science could be like.

## What happened at MIT?

In my first years there, I had been soaking up a lot of knowledge. It was interesting, but I must admit that sitting in large lecture halls with lots of kids around wasn't that exciting. Then, in my senior year, I had to do a senior thesis, and I was able to spend a great deal of time working on my own project under the close supervision of a faculty member. I was fortunate enough to hook up with a mentor, my thesis supervisor, who was very sympathetic. He was a new, young faculty member. He was enthusiastic about his work, and he set me on to an interesting problem.

I felt that I was really making an individual contribution to the problem, that I wasn't just working out things that he told me to do. I spent a lot of time working on it—gladly, willingly—and I was getting some interesting results. It was all great. It was what science should be. I think there is nothing like a little bit of success to give you confidence at any stage of your development. If you go into a laboratory and you do an experiment and it works, you really feel that you have control of things, that you can handle things.

## You got involved in biophysics almost by accident?

There were a whole series of unrelated and fortuitous events that got me there. After MIT, I started graduate school, then I quit after a year. I was attending a summer institute in physics in Boulder, Colorado, and I met George Gamow, who was in many ways one of the great physicists of the century and was also very involved in biological problems. He contributed some important ideas in the early days of the coding problem. I met him at a student party. Gamow was very gregarious and quite boyish in person, so a lot of students enjoyed speaking to him. We got to talking about molecular biology and the genetic code. He told me there were people at the University of Colorado Medical School who were doing interesting things in molecular biology and that if I was at all interested I should speak with them. So I did, and I was accepted as a Ph.D. student, although I first had to spend about a year learning all the chemistry I never knew.

I started working on my Ph.D. with Leonard Lerman, who had discovered the interpolation of acridines into DNA. Those are the flat molecules that slide between the bases of the DNA and produce deletion and addition mutations. My work was on the effect of acridines on the replication of bacteriophage T4 DNA.

I had a postdoctoral appointment at Harvard, where I worked on enzymes that cut up bacteriophage T4 DNA, and then I went to Cambridge University, where I worked for Crick and Brenner. This was one of the greatest opportunities for a young scientist. It was the premier lab in the world. Earlier researchers discovered that some RNA molecules undergo a chemical cleavage that enables cells to perform various functions essential to their growth. These researchers recognized that this was a fundamental process leading to protein synthesis, but little was understood about how the process worked. While I was at Crick's lab, I isolated an enzyme that governed this cleaving of RNA.

## In the 1970s, after you came to Yale, your RNA research became controversial. What sort of problems did you face?

It took approximately 2 years before we were able to get a key finding published in the mid-70s because it was not believed by many people, or not understood, or resented, or whatever you want to say. We were studying how the genetic code of the DNA was transcribed into RNA. This process requires, apart from actual transcription, a cleaving of the RNA molecule. Like all chemical reactions in the cell, this RNA cleavage requires enzymes. It was during this research for the enzymatic properties of these reactions that we made a surprising discovery—the enzymes were not proteins but nucleic acids.

In 1978, we studied the RNA-cutting enzyme from the bacterium *Escherichia coli*. This enzyme, named RNase P, is a component of a complex between one protein and one RNA molecule. When we chemically split the RNase P and separated the protein from the nucleic acid, the enzyme was no longer functional. However, we could restore the enzymatic activity by remixing the two different components. This was the first time that an RNA molecule had been shown to be necessary for a catalytic reaction. However, not until 5 years later did we show that the RNA molecule itself could carry out RNA-shearing activity, that RNA was a biocatalyst.

There was a great deal of resistance to the idea at first. Certainly, some people didn't believe it. I think it was around 1980 that my funding was cut off for 3 or 4 months because NIH

*The King of Sweden presents Sidney Altman with the 1989 Nobel prize for chemistry.*

didn't believe that what I was doing was interesting. I shouldn't really say NIH; I should say my peers who reviewed my grant applications. After a while, the opposition broke down because other people started working on the system and repeated our results, so there was no question anymore. I think some people may have tried to repeat them originally in an attempt to show the idea was wrong, but it didn't turn out that way. Once other people successfully repeat your experiments, there is no question that the conclusions are true.

Tom Cech's work on the splicing of RNA also supported our findings. Cech was studying the splicing of RNA in a unicellular organism called *Tetrahymena thermaphilia*. He discovered, much

to his surprise, that when he put an unprocessed RNA molecule into a test tube in the absence of protein, it started to splice itself. In other words, the RNA molecule could cut itself and join the important RNA fragments together again. Through the discovery of this chemically very complicated self-splicing reaction, Cech in 1982 became the first to show that RNA molecules can have a catalytic function. Subsequent development has been rapid, and many RNA enzymes are now known.

**Is your current research still moving along the same lines?**

Yes, because we don't fully understand yet just how this RNA enzyme works. We want to understand its mechanism in great detail so that we can make

comparisons with protein enzymes and see if it is significantly different in any way. We want to understand all its functions inside the cell. We have to see if our ideas are really valid or not. So there are a lot of things to do. But I guess the main thing is I am not dropping out of this and going into a completely different problem area. I am staying with this kind of problem.

**Let's talk about our future scientists. What kind of qualities should a student bring to science?**

A student who studies science is interested in nature, in understanding things, and gets some pleasure from fooling around in the laboratory. In general, I think you have to gain some satisfaction from solving problems. In addition, you have to recognize

that you don't really get anywhere without working hard and taking whatever you are doing seriously. None of these things, none of these endeavors of scholarship succeed unless you are persistent. I think those are important qualities for all of us, essential to any trade, or any discipline, in order to make it successful. The world can be unforgiving unless you take it seriously. Nature is not going to reveal itself to you unless you make some effort. Anything that is worth knowing is difficult to find out.

### Beyond the need to work hard, what else would you emphasize?

If possible, students should start doing independent research in the second semester in their junior year and then work all through their senior year on a research project. That is a good preparation for a career in science. Probably in all areas of scholarship, that is really the best kind of experience a student can have. By the time they are second-semester juniors or seniors, students have matured enough so they can really get some sense of excitement about doing some scholarly research on their own. That will really give them the inspiration and the confidence to study on their own. I am not saying necessarily that they have to get Ph.D.'s, but they simply must have the confidence to go out and research a particular project, whether it just be in the stacks of the library or in the laboratory. They need the sense of what it is to learn new things.

### What advice would you give to future scientists?

One message that I would want to give is that there seldom is a direct path from where you start to where you end up. In pursuing a career in research, one ought to be prepared for all kinds of twists and turns. Don't be discouraged, for example, if you start college wanting to be a neurosurgeon, but it turns out you don't become a neurosurgeon, but rather do something else. There is nothing necessarily disappointing about that. Fate takes you along the road, and you just do the best you can and work seriously on whatever you do. There are some people who have very firm, set ideas once they are in high school about what they want to do, but I think relatively few of them ever get the chance to realize those set goals. You have to be prepared for all kinds of different events in your life.

### Has that been true in your life?

I am impressed by the different, almost random events that occurred in my life that have pushed me one way or the other. I couldn't say that I would do any particular thing differently because I don't know what the consequences of making an alternative choice would have been. I feel so unexpectedly blessed with success, how could I possibly complain about where I am and what I have done? I don't think I have a right to do that. I have been very privileged in many ways.

# CHAPTER

## 4

# The Molecules of Life

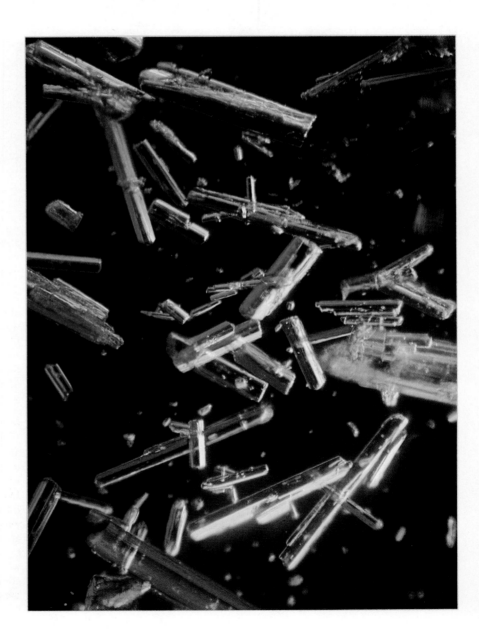

*Crystals of glycine, an amino acid, viewed under polarized light.*

*Molecules of life—carbohydrates, lipids, proteins, and the nucleic acids—are all built from simple precursors. Their structure (how the building blocks are put together) is the key to their activity in life processes. Carbohydrates are polymers built up from repeating sugar monomers. Glucose, the most common sugar, when assembled one way can form starch, our chief energy source. Assembled in another way, glucose units form cellulose, the chief ingredient of wood and the most plentiful organic substance on earth. Lipids, likewise, are built from the simple precursors, mainly the long hydrocarbon chains of fatty acids. The principal component of biological membranes is a phospholipid bilayer. The fatty acid portion of lipids contains a rich source of stored energy. Protein molecules are constructed from 20 different kinds of amino acids, giving proteins their great structural as well as functional variability. The structural protein collagen, for example, is a major component of tendons, skin, and bones. Hemoglobin, a transport protein, ferries oxygen in the bloodstream from the lungs to working tissues. Enzymes are protein catalysts that speed up chemical reactions in all living organisms.*

## 4.1
## THE IMPORTANCE OF FUNCTIONAL GROUPS IN BIOCHEMISTRY

**Organic chemistry** is the study of compounds of carbon, specifically compounds of carbon and hydrogen and their derivatives. Since the cellular apparatus of living organisms is made up of carbon compounds, biomolecules are part, but not all, of the subject matter of organic chemistry. There are many carbon compounds that are not found in any organism, and many topics of importance to organic chemistry have little connection with living things.

Until the early part of the 19th century, there was a widely held belief in "vital forces," forces unique to living things. A part of this theory was the idea that the compounds found in living organisms could not be produced in the laboratory. The critical experiment that disproved this belief was performed by the German chemist Freidrich Wöhler in 1828. Wöhler synthesized urea, a well-known waste product of animal metabolism, from ammonium cyanate, a compound obtained from mineral (*i.e.*, non-living) sources.

$$NH_4OCN \longrightarrow H_2NCONH_2$$

It has since been shown that any compound that occurs in a living organism can be synthesized in the laboratory, although in many cases the synthesis represents a considerable challenge to even the most talented organic chemist.

**TABLE 4.1    Functional Groups of Biochemical Importance**

| CLASS OF COMPOUND | GENERAL STRUCTURE | CHARACTERISTIC FUNCTIONAL GROUP | NAME OF FUNCTIONAL GROUP | EXAMPLE |
|---|---|---|---|---|
| Alkenes | $RCH\text{=}CH_2$ <br> $RCH\text{=}CHR$ <br> $R_2C\text{=}CHR$ <br> $R_2C\text{=}CR_2$ | $C\text{=}C$ | double bond | $CH_2\text{=}CH_2$ |
| Alcohols | $ROH$ | $-OH$ | hydroxyl group | $CH_3CH_2OH$ |
| Ethers | $ROR$ | $-O-$ | ether linkage | $CH_3OCH_3$ |
| Amines | $RNH_2$ <br> $R_2NH$ <br> $R_3N$ | $-N\big\langle$ | amino group | $CH_3NH_2$ |
| Thiols | $RSH$ | $-SH$ | sulfhydryl group | $CH_3SH$ |
| Aldehydes | $R-\overset{\displaystyle O}{\overset{\|}{C}}-H$ | $-\overset{\displaystyle O}{\overset{\|}{C}}-$ | carbonyl group | $CH_3\overset{\displaystyle O}{\overset{\|}{C}}H$ |
| Ketones | $R-\overset{\displaystyle O}{\overset{\|}{C}}-R$ | $-\overset{\displaystyle O}{\overset{\|}{C}}-$ | carbonyl group | $CH_3\overset{\displaystyle O}{\overset{\|}{C}}CH_3$ |
| Carboxylic acids | $R-\overset{\displaystyle O}{\overset{\|}{C}}-OH$ | $-\overset{\displaystyle O}{\overset{\|}{C}}-OH$ | carboxyl group | $CH_3\overset{\displaystyle O}{\overset{\|}{C}}OH$ |
| Esters | $R-\overset{\displaystyle O}{\overset{\|}{C}}-OR$ | $-\overset{\displaystyle O}{\overset{\|}{C}}-OR$ | ester group | $CH_3\overset{\displaystyle O}{\overset{\|}{C}}OCH_3$ |
| Amides | $R-\overset{\displaystyle O}{\overset{\|}{C}}-NR_2$ <br> $R-\overset{\displaystyle O}{\overset{\|}{C}}-NHR$ <br> $R-\overset{\displaystyle O}{\overset{\|}{C}}-NH_2$ | $-\overset{\displaystyle O}{\overset{\|}{C}}-N\big\langle$ | amide group | $CH_3\overset{\displaystyle O}{\overset{\|}{C}}N(CH_3)_2$ |

*The symbol R refers to any carbon-containing group. When there are several R groups in the same molecule, they may be different groups or they may be the same.

The reactions of biomolecules can be described by the methods of organic chemistry, and one of the most useful tools of organic chemistry is to classify compounds according to **functional groups.** *The reactions of molecules are the reactions of the functional groups.* Table 4.1 lists some biologically important functional groups. Some groups of vital importance to organic chemists are missing from this table. Molecules containing these groups, such as alkyl halides and acyl chlorides, do not have any particular applicability in biochemistry. Conversely, carbon-containing derivatives of phosphoric acid are little mentioned in beginning courses on organic chemistry, but esters and anhydrides of phosphoric acid (Figure 4.1) are of vital importance in biochemistry. ATP (adenosine *tri*phosphate, introduced in Section 1.4) contains both ester and anhydride linkages involving phosphoric acid.

**(a.1)**

$$HO-\overset{\displaystyle O}{\underset{\displaystyle OH}{\overset{\|}{P}}}-OH + HO-R \xrightarrow{\quad\quad} HO-\overset{\displaystyle O}{\underset{\displaystyle OH}{\overset{\|}{P}}}-O-R$$

$H_2O$

**Phosphoric acid    Alcohol**                                    **Phosphoric acid ester**

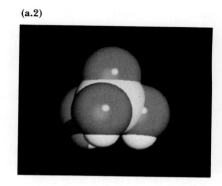

**(a.2)**

**(b.1)**

$$HO-\overset{\displaystyle O}{\underset{\displaystyle OH}{\overset{\|}{P}}}-OH + HO-\overset{\displaystyle O}{\underset{\displaystyle OH}{\overset{\|}{P}}}-OH \xrightarrow{\quad\quad} HO-\overset{\displaystyle O}{\underset{\displaystyle OH}{\overset{\|}{P}}}-O-\overset{\displaystyle O}{\underset{\displaystyle OH}{\overset{\|}{P}}}-OH$$

$H_2O$

**Anhydride**

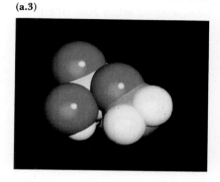

**(a.3)**

**(c)**

NH₂

Ester

$$HO-\overset{O}{\overset{\|}{P}}-O-\overset{O}{\overset{\|}{P}}-O-\overset{O}{\overset{\|}{P}}-O$$

OH    OH    OH    CH₂  O

**Anhydride**

ATP

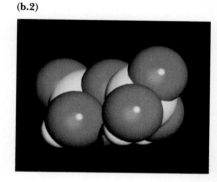

**(b.2)**

**FIGURE 4.1**    Structures of some important monosaccharides. (a.1) Reaction of phosphoric acid with a hydroxyl group to form an ester, which contains a P—O—R linkage. (Phosphoric acid is shown in its nonionized form in this figure.) (a.2) Space-filling model of phosphoric acid. (a.3) Space-filling model of the methyl ester of phosphoric acid. (b.1) Reaction of two molecules of phosphoric acid to form an anhydride, which contains a P—O—P linkage. (b.2) Space-filling model of the anhydride of phosphoric acid. (c) The structure of ATP (*a*denosine *tri*phosphate), showing two anhydride linkages and one ester linkage.

We shall discuss reactions of the various functional groups when we consider compounds in which they occur. In the remainder of this chapter we shall review the structure of some compounds you have probably encountered in earlier courses, including amino acids and sugars. Structures of lipids and of nucleotides (the monomers of nucleic acids) are also included for reference. These important classes of biomolecules have characteristic functional groups, which determine the reactions of the molecules. In the next three chapters we shall go on to some aspects of the biochemistry of peptides, carbohydrates, and lipids. The structure of nucleic acids is discussed in detail in Chapter 20.

## 4.2
## AMINO ACIDS

There are twenty amino acids usually found in proteins. They have a general structure in common, but variations in side chains that distinguish the compounds from one another. The general structure of amino acids, as

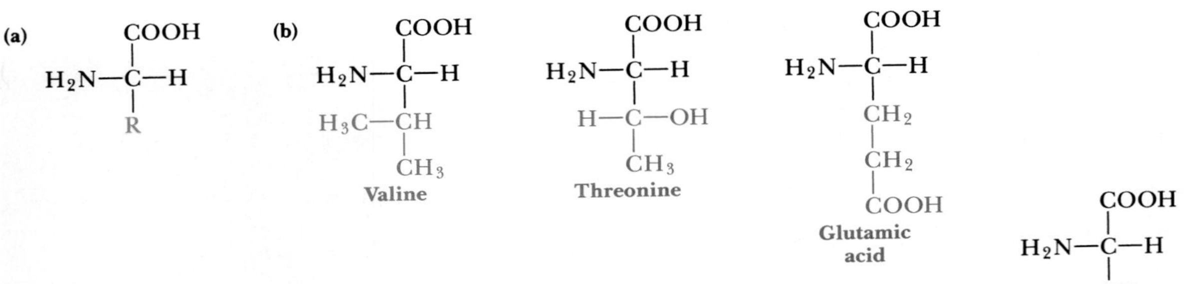

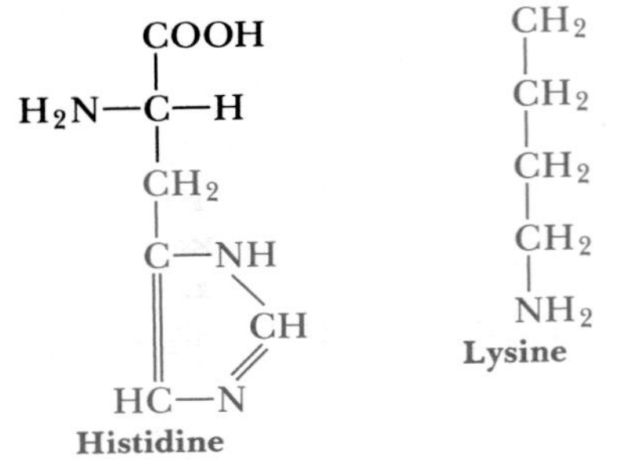

(a)

$$COOH$$
$$H_2N-C-H$$
$$R$$

(b)

$$COOH$$
$$H_2N-C-H$$
$$H_3C-CH$$
$$CH_3$$
**Valine**

$$COOH$$
$$H_2N-C-H$$
$$H-C-OH$$
$$CH_3$$
**Threonine**

$$COOH$$
$$H_2N-C-H$$
$$CH_2$$
$$CH_2$$
$$COOH$$
**Glutamic acid**

$$COOH$$
$$H_2N-C-H$$
$$CH_2$$
$$C-NH$$
$$\diagdown CH$$
$$HC-N$$
**Histidine**

$$COOH$$
$$H_2N-C-H$$
$$CH_2$$
$$CH_2$$
$$CH_2$$
$$CH_2$$
$$NH_2$$
**Lysine**

**FIGURE 4.2**   Structures of amino acids. (a) The general formula of an amino acid. (b) Structures of some representative amino acids. The R groups are highlighted. (Figure 5.4 in Chapter 5 shows 20 amino acids.)

the name implies, involves an **amino group** and a **carboxyl group.** Both groups are bonded to the same carbon atom, the $\alpha$ carbon, which is the one next to the carboxyl group. The carbon atoms in amino acids are designated by the letters of the Greek alphabet [alpha ($\alpha$), beta ($\beta$), gamma ($\gamma$), delta ($\delta$), epsilon ($\epsilon$), and so forth] in the order in which they are bonded, starting with the carbon adjacent to the carboxyl group. The $\alpha$ carbon is also bonded to a hydrogen atom and to the side chain group, designated as R. The identity of the particular amino acid depends on the nature of the **R group** (Figure 4.2).

## 4.3
## MONOSACCHARIDES

The word "carbohydrate" was coined in the late 19th century to refer to a class of compounds with the general formula $C_n(H_2O)_n$. This general formula implies that the structure is a hydrate of carbon, but strictly speaking this applies only to some carbohydrates, such as simple sugars (called monosaccharides). The **monosaccharides** are compounds that contain a single carbonyl group and two or more hydroxyl groups. In other words, monosaccharides can be polyhydroxy aldehydes or polyhydroxy ketones, depending on the isomeric form of the carbonyl compound. The ones that contain an aldehyde group are called **aldoses,** while those that contain a ketone group are called **ketoses.** Six-carbon sugars are most abundant in nature, but two five-carbon sugars, ribose and deoxyribose, occur in the structures of RNA and DNA, respectively (Figure 4.3).

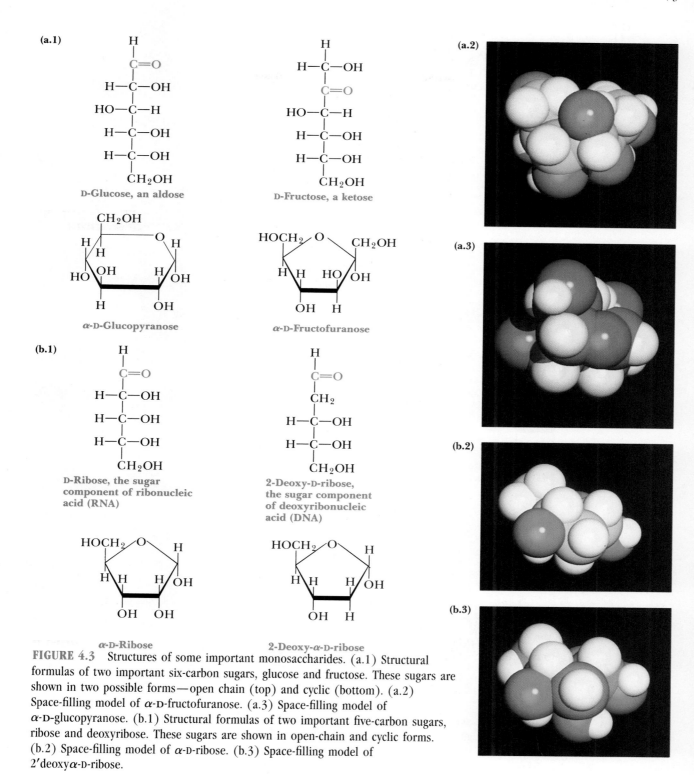

**(a.1)**

D-Glucose, an aldose

α-D-Glucopyranose

D-Fructose, a ketose

α-D-Fructofuranose

**(b.1)**

D-Ribose, the sugar component of ribonucleic acid (RNA)

2-Deoxy-D-ribose, the sugar component of deoxyribonucleic acid (DNA)

α-D-Ribose

2-Deoxy-α-D-ribose

**FIGURE 4.3**  Structures of some important monosaccharides. (a.1) Structural formulas of two important six-carbon sugars, glucose and fructose. These sugars are shown in two possible forms—open chain (top) and cyclic (bottom). (a.2) Space-filling model of α-D-fructofuranose. (a.3) Space-filling model of α-D-glucopyranose. (b.1) Structural formulas of two important five-carbon sugars, ribose and deoxyribose. These sugars are shown in open-chain and cyclic forms. (b.2) Space-filling model of α-D-ribose. (b.3) Space-filling model of 2′deoxyα-D-ribose.

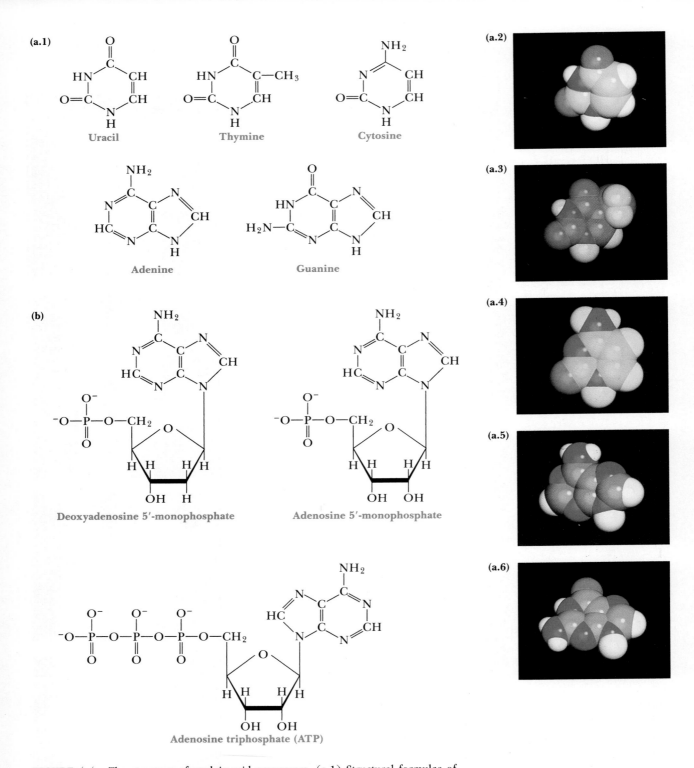

**FIGURE 4.4** The structure of nucleic acid monomers. (a.1) Structural formulas of uracil, thymine, cytosine, adenine, and guanine. Space-filling models of uracil (a.2), thymine (a.3), cytosine (a.4), adenine (a.5), and guanine (a.6). (b) Adenine nucleotides: a deoxyribonucleotide, a ribonucleotide, and a nucleoside triphosphate (ATP) are shown.

74

## 4.4
## THE STRUCTURE OF NUCLEIC ACID MONOMERS

The two kinds of nucleic acids are DNA and RNA. Although there are differences between DNA and RNA, they share important similarities. Nucleic acids are very large molecules (macromolecules) formed by the polymerization of monomer units. The monomers of nucleic acids are **nucleotides,** each of which consists of a base, a sugar, and a phosphoric acid residue covalently bonded together. Nucleic acids can be hydrolyzed to their constituent nucleotides by acids, bases, or enzymes. The main distinction between the nucleotide monomers of RNA and DNA lies in the sugar portion; in RNA the sugar is ribose, while in DNA the sugar is deoxyribose, which differs from ribose by lacking an oxygen atom in a specific position (Figure 4.3 and Section 6.2). These sugars are reflected in the names RNA (*ribo*nucleic *a*cid) and DNA (*d*eoxyribo*n*ucleic *a*cid).

The **nucleic-acid bases** (also called nucleobases) are nitrogen-containing aromatic compounds that fall into two classes, **pyrimidines** and **purines** (Figure 4.4). Three pyrimidine bases commonly occur in nucleic acids: **cytosine, thymine,** and **uracil.** Cytosine is found in both RNA and DNA, while uracil occurs only in RNA. In DNA thymine is substituted for uracil; thymine is also found to a small extent in some forms of RNA. The common purine bases are **adenine** and **guanine,** both of which are found in both RNA and DNA. A **nucleoside** consists of a base and a sugar covalently linked together. When phosphoric acid is esterified to one of the hydroxyl groups of the sugar portion of a nucleoside, a **nucleotide** is formed. Nucleoside di- and triphosphates are particularly important kinds of nucleotides.

## 4.5
## LIPIDS

The name **lipid** refers to a class of biomolecules that are insoluble in water and soluble in organic solvents. The glycerol esters of fatty acids (long-chain carboxylic acids) and derivatives of these esters are important examples of lipids. Steroids, which have a characteristic fused ring structure, are another important group of lipids (Figure 4.5).

**(a.1)**

CH₂—OH          HOOC—(CH₂)ₙ—CH₃
|                          A saturated fatty acid
CH—OH
|
CH₂—OH          HOOC—(CH₂)₇—CH=CH—(CH₂)ₘ—CH₃
Glycerol                 An unsaturated fatty acid

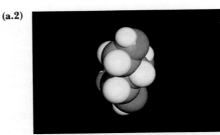

**(a.2)**

$$CH_2-O-\overset{\overset{O}{\parallel}}{C}-R_1$$

$$CH-O-\overset{\overset{O}{\parallel}}{C}-R_2$$

$$CH_2-O-\overset{\overset{O}{\parallel}}{C}-R_3$$

Glycerol esterified to
three different fatty acids

$$CH_2-O-\overset{\overset{O}{\parallel}}{C}-R_4$$

$$CH-O-\overset{\overset{O}{\parallel}}{C}-R_5$$

$$CH_2-O-\overset{\overset{O}{\parallel}}{P}-O-R_6$$
$$OH$$

Glycerol esterified to two different
fatty acids and to phosphoric acid.
Phosphoric acid is esterified in
turn to another alcohol (R₆—OH).

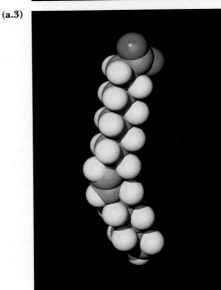

**(a.3)**

**(b)**

Cholesterol

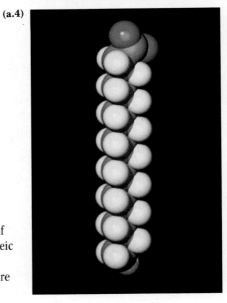

**(a.4)**

FIGURE 4.5   Some lipid structures. (a.1) Structural formulas of glycerol esters of fatty acids and related compounds. Space-filling models of glycerol (a.2), palmitoleic acid (an unsaturated fatty acid) (a.3), and palmitic acid (a saturated fatty acid) (a.4). (b) The structure of cholesterol, a typical steroid. The characteristic rings are designated A through D.

## SUMMARY

Both organic chemistry and biochemistry deal with the reactions of carbon-containing molecules. Both disciplines base their approach on the behavior of functional groups, but their emphases differ because some functional groups important to organic chemistry do not play a role in biochemistry and vice versa. Functional groups of importance in biochemistry include carbonyl groups, hydroxyl groups, carboxyl groups, amines, amides, and esters; derivatives of phosphoric acid such as esters and anhydrides are also important.

Amino acids, the monomer units of proteins, have a general structure in common, with an amino group and a carboxyl group bonded to the same carbon atom. The side chains, referred to as R groups, are different in the various amino acids.

The simplest examples of carbohydrates are monosaccharides, sugars that contain a single carbonyl group and two or more hydroxyl groups. Monosaccharides frequently encountered in biochemistry contain from three to seven carbon atoms.

The nucleic acids, DNA and RNA, are large polymers formed by the covalent linking of monomers. The monomers of nucleic acids are nucleotides. An individual nucleotide consists in turn of three parts—a base, a sugar, and a phosphoric acid residue—all of which are covalently bonded together. There are two classes of bases, pyrimidines and purines. Pyrimidine bases have six-membered rings, while purine bases consist of a six-membered ring fused to a five-membered ring. The pyrimidine bases cytosine, thymine, and uracil are commonly found in nucleic acids (thymine is found mostly, but not exclusively, in DNA, and uracil occurs only in RNA). The purine bases found in nucleic acids are adenine and guanine. The sugar ribose is found in RNA, giving rise to the name *ribo*nucleic *acid*, while the sugar found in DNA is deoxyribose, giving rise to the name *deoxyribo*nucleic *acid*. The formation of a covalent bond between a sugar and a base produces a nucleoside. When phosphoric acid is esterified to one of the hydroxyl groups of the sugar portion of a nucleoside, a nucleotide is formed.

## EXERCISES

**1.** Match each entry in column a with one in column b; column a shows the names of some important functional groups, and column b shows their structures.

| Column a | Column b |
|---|---|
| Amino group | $CH_3SH$ |
| Carbonyl group (ketone) | $CH_3CH{=}CHCH_3$ |
| Hydroxyl group | $CH_3CH_2\overset{\displaystyle O}{\overset{\|}{C}}H$ |
| Carboxyl group | $CH_3CH_2NH_2$ |
| Carbonyl group (aldehyde) | $CH_3\overset{\displaystyle O}{\overset{\|}{C}}OCH_2CH_3$ |
| Thiol group | $CH_3CH_2OCH_2CH_3$ |
| Ester linkage | $CH_3\overset{\displaystyle O}{\overset{\|}{C}}CH_3$ |
| Double bond | $CH_3\overset{\displaystyle O}{\overset{\|}{C}}OH$ |
| Amide linkage | $CH_3OH$ |
| Ether | $CH_3\overset{\displaystyle O}{\overset{\|}{C}}N(CH_3)_2$ |

**2.** Identify the functional groups in the compounds shown below.

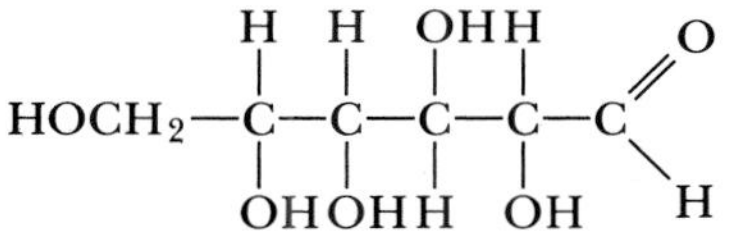

$$HOCH_2-\overset{\displaystyle H}{\underset{\displaystyle OH}{C}}-\overset{\displaystyle H}{\underset{\displaystyle OH}{C}}-\overset{\displaystyle OH}{\underset{\displaystyle H}{C}}-\overset{\displaystyle H}{\underset{\displaystyle OH}{C}}-\overset{\displaystyle O}{\underset{\displaystyle H}{C}}$$

**glucose**

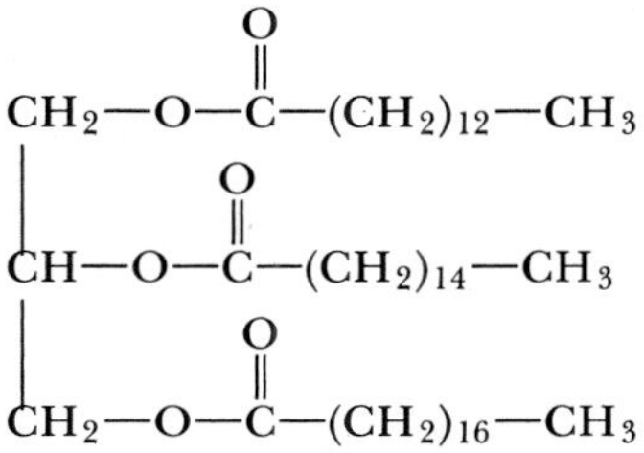

$$CH_2-O-\overset{\displaystyle O}{\overset{\|}{C}}-(CH_2)_{12}-CH_3$$
$$CH-O-\overset{\displaystyle O}{\overset{\|}{C}}-(CH_2)_{14}-CH_3$$
$$CH_2-O-\overset{\displaystyle O}{\overset{\|}{C}}-(CH_2)_{16}-CH_3$$

**a triglyceride**

$$H_2N-CH_2-\overset{\displaystyle O}{\overset{\|}{C}}-\underset{\displaystyle H}{N}-CH_2-\overset{\displaystyle O}{\overset{\|}{C}}-\underset{\displaystyle H}{N}-\overset{\displaystyle CH_3}{\underset{\displaystyle H}{C}}-\overset{\displaystyle O}{\overset{\|}{C}}-OH$$

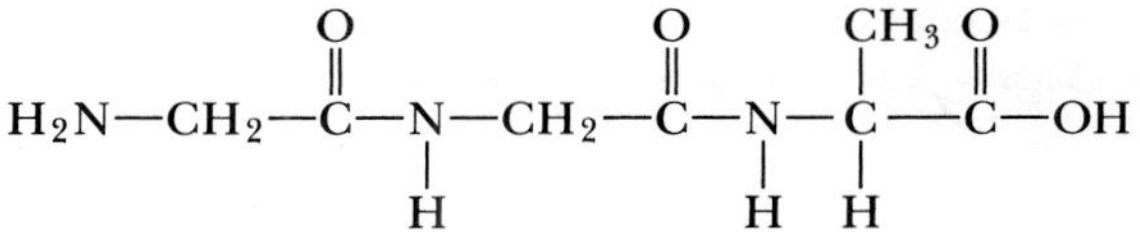

**a peptide**

(Continued)

**vitamin A**

3. The following equations show the hydrolysis of an amide and of an ester. Identify the functional groups in the products and name the type of compound formed.

$$RCNR'_2 + H_2O \rightarrow RCOH + R'_2NH$$

$$RCOR' + H_2O \rightarrow RCOH + R'OH$$

4. A friend who is enthusiastic about health foods and organic gardening asks you whether urea is "organic" or "chemical." How do you reply to this question?

5. The synthetic textile fiber polyester is formed by polymerization of terephthalic acid

**terephthalic acid**

and ethylene glycol ($HO-CH_2-CH_2-OH$). Suggest the reaction by which such a polymerization might take place and how such a reaction might be related to the formation of biomolecules.

6. Nylon is formed by the polymerization reaction of adipic acid ($HOOC-CH_2CH_2CH_2CH_2-COOH$) and hexamethylenediamine ($H_2N-CH_2CH_2CH_2CH_2CH_2CH_2-NH_2$). Suggest the reaction by which such a polymerization might take place and how such a reaction might be related to the formation of biomolecules.

7. Compounds such as ethyl acetate

($CH_3-C-O-C_2H_5$) are responsible for the characteristic odors of many kinds of fruit. To what class of compound does ethyl acetate belong? Identify the functional group(s) in this molecule.

8. How is the structure of caffeine related to that of nucleobases?

**caffeine**

9. Distinguish between the structures of a phosphoric acid ester and a carboxylic acid ester.

10. Write an equation for the reaction responsible for the formation of vinegar from wine.

## ANNOTATED BIBLIOGRAPHY

Adams, R. L. P., ed. *J. N. Davidson's The Biochemistry of the Nucleic Acids.* 10th ed. New York: Academic Press, 1986. [A classic introduction to the subject.]

Barrett, G.C., ed. *Chemistry and Biochemistry of the Amino Acids.* New York: Chapman and Hall, 1985. [Wide coverage of many aspects of reactions of amino acids.]

Sharon, N. Carbohydrates. *Sci. Amer.* **243** (5), 90–102 (1980). [A good overview of structures.]

Westheimer, F. H. Why Nature Chose Phosphates. *Science* **235**, 1173–1178 (1987). [A discussion of the importance of phosphate groups in biochemistry, particularly in the backbone of nucleic acids. The author of this article is an eminent organic chemist.]

Any current textbook of organic chemistry can be expected to have a discussion of functional groups and their reactions. Most organic chemistry texts have several chapters which discuss carbohydrates, lipids, proteins, and nucleic acids from the point of view of organic chemists.

# Amino Acids and Peptides

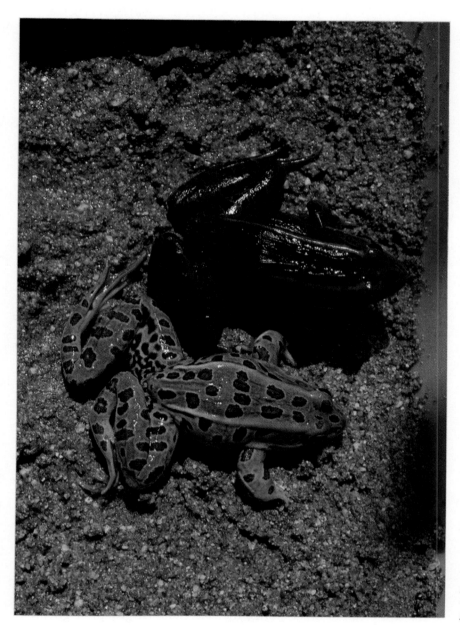

The black frog received a dose of a peptide hormone called melanotropin. Its skin color was originally the same green as that of the other frog.

*Proteins are long chains of amino acids linked together by peptide bonds between a positively charged nitrogen (amino) group at one end and a negatively charged carboxyl group at the other end. Along the chain is a series of side chains that are different for each of the 20 amino acids. A dipeptide is a linkage of two amino acids; three form a tripeptide. The sequence of the amino acids is of the utmost importance. Glycine-lysine-alanine is a different peptide, with a different chemical significance than alanine-lysine-glycine. In the same manner, the motto, "Talk little, do much," has a different meaning from "Do little, talk much." There are 8,000 ways that the 20 amino acids can be sequenced; in a protein chain of 100 amino acids there are more ways than there are atoms in the Universe! Literally, the sequence is the message. It determines exactly how the protein will fold up in a three-dimensional conformation to perform its precise biochemical function.*

**(a)**

$$\underset{R}{\overset{COO^-}{H_3\overset{+}{N}-C-H}}$$

**(b)**

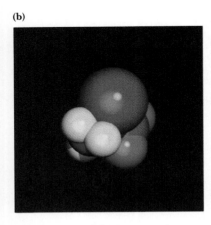

**FIGURE 5.1** (a) The general formula of amino acids showing the ionic forms predominant at pH 7. (b) Space-filling model of an amino acid. The large green sphere represents the R group.

## 5.1
## AMINO ACIDS, THEIR GENERAL FORMULA AND THREE-DIMENSIONAL STRUCTURE

Among all the possible amino acids, only twenty are usually found in proteins. As discussed in Section 4.2, the general structure of amino acids involves an **amino group** and a **carboxyl group,** both of which are bonded to the $\alpha$ carbon (the one next to the carboxyl group). The $\alpha$ carbon is also bonded to a hydrogen and to the **side chain group,** which is represented as R. The R group determines identity of the particular amino acid (Figure 5.1).

The two-dimensional formula shown here gives only a partial representation of the common structure of amino acids, because one of their most important properties is their three-dimensional shape, or **stereochemistry.**

Every object has a mirror image. Many objects that are mirror images of each other can be superimposed on each other; coffee mugs (ignoring any lettering) are examples of mirror-image objects that can be superimposed on each other. In other cases the mirror-image objects cannot be superimposed on one another but are related to each other as the right hand is to the left. Such nonsuperimposable mirror images are **chiral** (from the Greek *cheir*, "hand"); many important biomolecules are chiral (Figure 5.2).

A frequently encountered chiral center in biomolecules is a carbon atom with four different groups bonded to it. Such a center occurs in all amino acids except glycine, so the mirror images of these molecules cannot be superimposed on each other. Glycine has two hydrogen atoms bonded to the $\alpha$ carbon; in other words, the side chain (R group) of glycine is

(a)

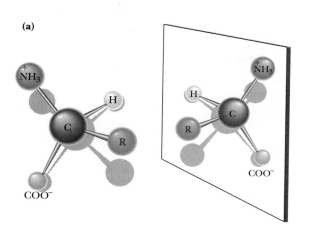

(b)

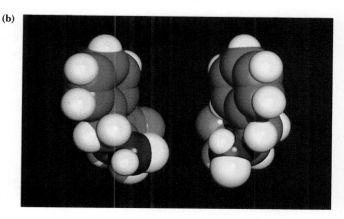

**FIGURE 5.2** (a) A pair of mirror image amino acids. Amino acids found in proteins have the configuration shown on the right side of the mirror. (b) Space-filling models of the mirror image forms of the amino acid phenylalanine.

hydrogen. Glycine is not chiral (or, alternatively, is achiral) because of its symmetry. In all the other commonly occurring amino acids the $\alpha$ carbon has four different groups bonded to it, and as a result all amino acids except glycine can have two nonsuperimposable mirror-image forms. The two possibilities or **stereoisomers** are shown in perspective drawings of the two forms of alanine, where the R group is —CH$_3$ (Figure 5.3). The dashed wedges represent bonds directed down, behind the plane of the paper, and the solid triangles represent bonds directed up, out of the plane of the paper in the direction of the observer.

The two possible stereoisomers of another chiral compound, L- and D-glyceraldehyde, are shown for comparison with the corresponding forms of alanine. The two forms of glyceraldehyde are the basis of the classification of amino acids into L and D forms. The two different stereoisomers of each amino acid are designated as **L and D amino acids** on the basis of their similarity to the glyceraldehyde standard. The terminology comes from the Latin *laevus* and *dexter,* meaning "left" and "right," respectively. In the L form of glyceraldehyde the hydroxyl group is on the left side of the molecule; in the D form it is on the right side, as shown in the perspective drawing. In amino acids, the position of the amino group on the left or right side of the $\alpha$ carbon determines the choice of the L or D designation. The amino acids that occur in proteins are all of the L form; although D

**FIGURE 5.3** Stereochemistry of alanine and glyceraldehyde. The amino acids found in proteins have the same chirality as L-glyceraldehyde and opposite to that of D-glyceraldehyde. (b) Space-filling model of L-alanine.

(b)

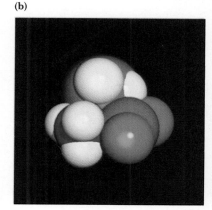

(a)

| | | | |
|---|---|---|---|
| H O \\ / C | ⁻O O \\ / C | O O⁻ \\ / C | O ‖ C—H |
| HO—C◄H | H$_3$N⁺—C◄H | H—C◄N⁺H$_3$ | H◄C—OH |
| CH$_2$OH | CH$_3$ | CH$_3$ | CH$_2$OH |
| **L-Glyceraldehyde** | **L-Alanine** | **D-Alanine** | **D-Glyceraldehyde** |

amino acids occur in nature, most often in bacterial cell walls and in some antibiotics, they are not found in proteins.

There are other systems for assignment of stereochemistry; one of these, the R,S system, is the one most frequently used in organic chemistry. We do not need it in this text and shall not mention it again. The L and D classification persists in biochemical usage because it is sufficient for most applications.

## 5.2
## THE STRUCTURES AND PROPERTIES OF THE INDIVIDUAL AMINO ACIDS

The R groups, and thus the individual amino acids, are classified according to several criteria, two of which are particularly important. The first of these is the **polar or nonpolar nature** of the side chain. The second depends on the presence of an **acidic or basic group** in the side chain. Other useful criteria include the presence of functional groups in the side chains and the nature of those groups.

In the simplest amino acid, glycine, the side "chain" is a hydrogen atom, and in this case alone there are two hydrogen atoms bonded to the $\alpha$ carbon. In all other amino acids the side chain is larger and more complex (Figure 5.4). We shall frequently refer to amino acids by three-letter or one-letter abbreviations for their names; these abbreviations are listed in Table 5.1.

### Group 1—Amino Acids with Nonpolar Side Chains

One group of amino acids has **nonpolar side chains.** In several members of this group (namely alanine, valine, leucine, and isoleucine), each side chain is an aliphatic hydrocarbon group. (In organic chemistry, the term "aliphatic" refers to the absence of a benzene ring or related structure.) Proline has an aliphatic cyclic structure and strictly speaking is an *imino acid,* since the nitrogen is bonded to two carbon atoms. In phenylalanine the hydrocarbon group is aromatic (contains a cyclic group similar to a benzene ring) rather than aliphatic. In tryptophan the side chain contains an indole ring, which is also aromatic. In methionine the side chain contains a sulfur atom in addition to aliphatic hydrocarbon groupings. (See Figure 5.4.)

### Group 2—Amino Acids with Electrically Neutral Polar Side Chains

Another group of amino acids has **polar side chains that are electrically neutral at neutral pH.** In serine and threonine the polar group is a hydroxyl (—OH) bonded to aliphatic hydrocarbon groups. The hydroxyl

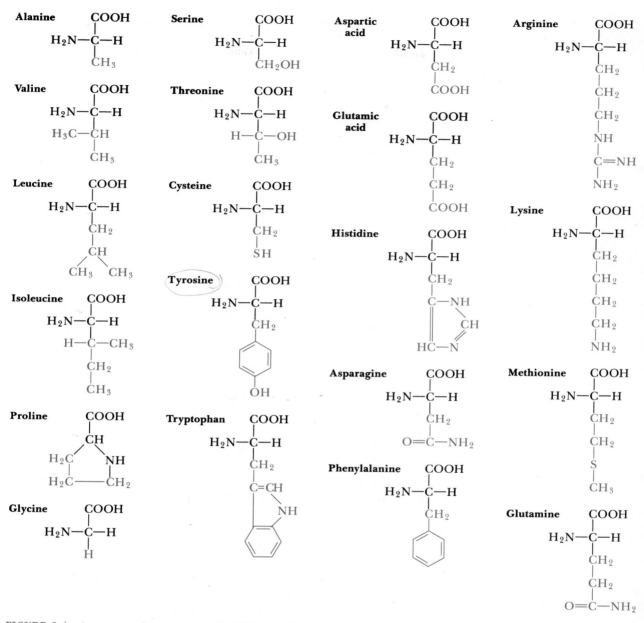

**FIGURE 5.4**  Structures of the amino acids. The 20 amino acids found in proteins. The R groups are shown in green.

group in tyrosine is bonded to an aromatic hydrocarbon grouping, which does eventually lose a proton at higher pH. In cysteine the polar side chain consists of an —SH group, which can react with other cysteine —SH groups to form disulfide (—S—S—) bridges in proteins. (See Figure 5.4.)

**TABLE 5.1   Names and Abbreviations of the Common Amino acids**

| AMINO ACID | THREE-LETTER ABBREVIATION | ONE-LETTER ABBREVIATION |
|---|---|---|
| Alanine | Ala | A |
| Arginine | Arg | R |
| Asparagine | Asn | N |
| Aspartic acid | Asp | D |
| Cysteine | Cys | C |
| Glutamine | Gln | Q |
| Glutamic acid | Glu | E |
| Glycine | Gly | G |
| Histidine | His | H |
| Isoleucine | Ile | I |
| Leucine | Leu | L |
| Lysine | Lys | K |
| Methionine | Met | M |
| Phenylalanine | Phe | F |
| Proline | Pro | P |
| Serine | Ser | S |
| Threonine | Thr | T |
| Tryptophan | Trp | W |
| Tyrosine | Tyr | Y |
| Valine | Val | V |

## Group 3—Amino Acids with Carboxyl Groups in Their Side Chains

Two amino acids have **carboxyl groups in their side chains** in addition to the one present in all amino acids. A carboxyl group can lose a proton, forming the corresponding carboxylate anion (Section 3.3). Because of the presence of the carboxylate, the side chains of these two amino acids are **negatively charged at neutral pH.** The amino acids in this group, glutamic acid and aspartic acid, differ only by a —CH$_2$— group in the side chain (Figure 5.4). They are frequently referred to as glutamate and aspartate, respectively. These side-chain carboxyl groups frequently bond to —NH$_2$ to form **side chain amide groups,** yielding the analogous amino acids glutamine and asparagine (Figure 5.4).

## Group 4—Amino Acids with Basic Side Chains

There are three amino acids with **basic side chains,** and in all three the side chain is **positively charged at or near neutral pH.** In histidine the pK$_a'$ of the side-chain imidazole group is very close to physiological pH, and the properties of many proteins depend on whether individual histidine residues are or are not charged. In lysine the side chain amino group is attached to an aliphatic hydrocarbon tail. In arginine the side chain basic group, the guanidino group, is more complex in structure than the amino group, but it is also bonded to an aliphatic hydrocarbon tail (Figure 5.4).

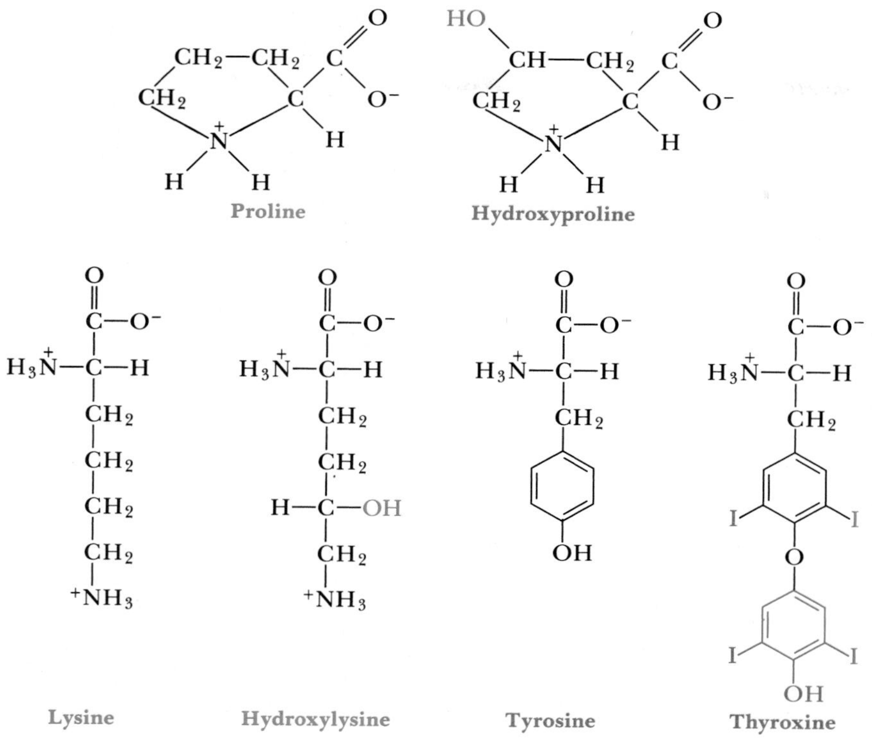

**FIGURE 5.5** Structures of hydroxyproline, hydroxylysine, and thyroxine. The structures of the parent amino acids—proline for hydroxyproline, lysine for hydroxylysine, and tyrosine for thyroxine—are shown for comparison. All amino acids are shown in the predominant ionic form at pH 7.

Several other amino acids are known to occur in some, but by no means all, proteins (Figure 5.5). They are derived from the common amino acids and are produced by modification of the parent amino acid after the protein is synthesized by the organism. Hydroxyproline and hydroxylysine differ from the parent amino acids in having hydroxyl groups on the side chain; they are found only in a few connective tissue proteins such as collagen. Thyroxine differs from tyrosine in having an extra iodine-containing aromatic group on the side chain; it is found only in thyroglobulin (a protein that occurs in the thyroid).

## 5.3
## TITRATION CURVES OF THE AMINO ACIDS

In free amino acids the carboxyl group and amino group of the general structure are charged at neutral pH; the carboxylate portion is negatively charged and the amino group has a positive charge. Amino acids without charged groups on their side chains exist in neutral solution as **zwitterions** with no net charge. When an amino acid is titrated, its titration curve indicates the reaction of each functional group that is capable of reacting with hydrogen ion.

In alanine, there are two titratable groups, the carboxyl and amino groups. At very low pH alanine has a protonated (and thus uncharged)

**(a)**

$$\underset{\text{+1 net charge}}{\text{H}_3\text{C}-\underset{\overset{|}{\text{NH}_3{}^+}}{\text{CH}}-\text{COOH}} \underset{2.3}{\overset{\mathbf{p}K'_\text{a}}{\rightleftharpoons}} \underset{\substack{\text{0 net charge}\\ \text{Isoelectric zwitterion}}}{\text{H}_3\text{C}-\underset{\overset{|}{\text{NH}_3{}^+}}{\text{CH}}-\text{COO}^-} \underset{9.7}{\overset{\mathbf{p}K'_\text{a}}{\rightleftharpoons}} \underset{\text{-1 net charge}}{\text{H}_3\text{C}-\underset{\overset{|}{\text{NH}_2}}{\text{CH}}-\text{COO}^-}$$

**(b)**

$$\underset{\text{+2 net charge}}{\text{structure}} \underset{1.8}{\overset{\mathbf{p}K'_\text{a}}{\rightleftharpoons}} \underset{\text{+1 net charge}}{\text{structure}} \underset{6.0}{\overset{\mathbf{p}K'_\text{a}}{\rightleftharpoons}} \underset{\substack{\text{0 net charge}\\ \text{Isoelectric species}}}{\text{structure}} \underset{9.2}{\overset{\mathbf{p}K'_\text{a}}{\rightleftharpoons}} \underset{\text{-1 net charge}}{\text{structure}}$$

**FIGURE 5.6**   The ionization of amino acids. (a) The ionization of alanine (a neutral amino acid). (b) The ionization of histidine (an amino acid with a titratable side chain).

**FIGURE 5.7**   Titration curves of amino acids. (a) The titration curve of alanine. (b) The titration curve of histidine.

carboxyl group and a positively charged amino group that is also protonated. Under these conditions the alanine has a net positive charge of 1. As base is added, the carboxyl group loses its proton to become a negatively charged carboxylate group (Figure 5.6(a)), and the pH of the solution increases. Alanine now has no net charge. As the pH increases still further with addition of more base, the protonated amino group (a weak acid) loses its proton; the alanine molecule now has a negative charge of 1. The titration curve of alanine is that of a dibasic acid (Figure 5.7(a)).

In histidine, the imidazole side chain also contributes a titratable group. At very low pH values, the histidine molecule has a net positive charge of 2, because both the imidazole and amino groups have a positive charge. As base is added and the pH increases, the carboxyl group loses a proton to become a carboxylate as before, and the histidine now has a

(a)

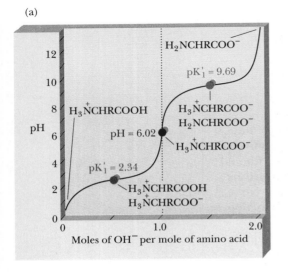

(b)

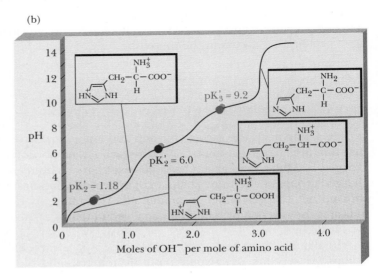

**TABLE 5.2    p$K'_a$ Values of Common Amino Acids**

| ACID | $\alpha$ COOH | $\alpha$ NH$_3^+$ | RH or RH$^+$ |
|------|------|------|------|
| Gly | 2.34 | 9.60 | |
| Ala | 2.34 | 9.69 | |
| Val | 2.32 | 9.62 | |
| Leu | 2.36 | 9.68 | |
| Ile | 2.36 | 9.68 | |
| Ser | 2.21 | 9.15 | |
| Thr | 2.63 | 10.43 | |
| Met | 2.28 | 9.21 | |
| Phe | 1.83 | 9.13 | |
| Trp | 2.38 | 9.39 | |
| Asn | 2.02 | 8.80 | |
| Gln | 2.17 | 9.13 | |
| Pro | 1.99 | 10.6 | |
| Asp | 2.09 | 9.82 | 3.86* |
| Glu | 2.19 | 9.67 | 4.25* |
| His | 1.82 | 9.17 | 6.0* |
| Cys | 1.71 | 10.78 | 8.33* |
| Tyr | 2.20 | 10.07 | 9.11* |
| Lys | 2.18 | 8.95 | 10.53 |
| Arg | 2.17 | 9.04 | 12.48 |

*For these amino acids the R group ionization occurs before the $\alpha$-NH$_3^+$ ionization.

positive charge of 1 (Figure 5.6(b)). As still more base is added, the charged imidazole group loses its proton, and this is the point at which the histidine has no net charge. At still higher values of pH the amino group loses its proton as was the case with alanine, and the histidine molecule now has a negative charge of 1. The titration curve of histidine is that of a tribasic acid (Figure 5.7(b)).

Like the acids we discussed in Chapter 3, the amino acids have characteristic values for the $K'_a$s and p$K'_a$s of their titratable groups. The p$K'_a$s of $\alpha$-carboxyl groups are fairly low, around 2. The p$K'_a$s of amino groups are reasonably high, with values ranging from 9 to 10.5. The p$K'_a$s of side chain groups, including side chain carboxyl and amino groups, depend on the chemical nature of the group (Table 5.2). The classification of amino acids as acidic or basic depends on the p$K'_a$ of the side chain. These groups can still be titrated after the amino acid is incorporated into a peptide or protein, but the p$K'_a$ of the titratable group on the side chain is not necessarily the same in a protein as it is in free amino acids.

The fact that amino acids, peptides, and proteins have different p$K'_a$s gives rise to the possibility that they can have different charges at a given pH. Alanine and histidine, for example, both have a net charge of $-1$ at high pH, above 10; the only charged group is the carboxylate anion. At lower pH, around 5, alanine is a zwitterion with no net charge, but histidine has a net charge of $+1$ at this pH because the imidiazole group is protonated. This property is the basis of **electrophoresis,** a method for separating proteins and protein fragments on the basis of charge. This method is extremely useful in determining the amino acid sequence of proteins (Interchapter A, Section 2).

**(a)**

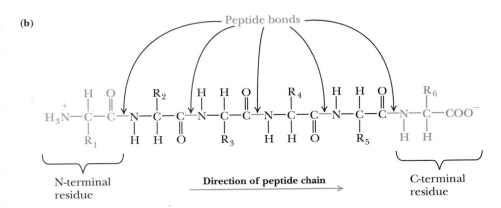

Dipeptide

**(b)**

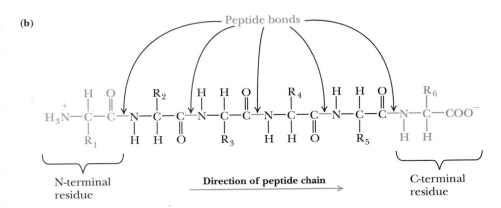

Peptide bonds

N-terminal residue

**Direction of peptide chain**

C-terminal residue

**(c)**

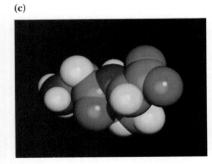

FIGURE 5.8    The peptide bond. (a) Formation of the peptide bond. (b) A small peptide showing the direction of the peptide chain (N-terminal ⟶ C-terminal). (c) Space-filling model of the dipeptide formed by reacting the carboxyl group of glycine with the amino group of alanine.

## 5.4
## THE PEPTIDE BOND

Individual amino acids can be linked together by formation of covalent bonds. The bond is formed between the $\alpha$-carboxyl group of one amino acid and the $\alpha$-amino group of the next one. Water is eliminated in the process, and the linked amino acid **residues** are what remain after the elimination of water (Figure 5.8(a)). A bond formed in this way is called a **peptide bond.** In a protein many amino acids (usually more than a hundred) are linked by peptide bonds to form a **polypeptide chain** (Figure 5.8(b)). **Peptides** are compounds formed by linking together smaller numbers of amino acids, ranging from two to several dozen. Another name for a compound formed by the reaction between an amino group and a carboxyl group is an *amide,* and the term "amide bond" is a synonym for "peptide bond."

The carbon-nitrogen bond formed when two amino acids are linked in a peptide bond is usually written as a single bond, with one pair of electrons shared between the two atoms. It is quite possible to write this bond as a double bond, with a simple shift in the position of a pair of electrons. This shifting of electrons is well known in organic chemistry and results in **resonance structures,** structures that differ only in the position of electrons. The positions of double and single bonds are frequently different in the resonance structures of a compound. No single resonance structure actually represents the bonding in a compound for which resonance structures can be written, but all resonance structures contribute to the actual bonding situation. The peptide bond can be written as a resonance hybrid of two structures (Figure 5.9), one with a single bond between the carbon and nitrogen and the other with a double bond

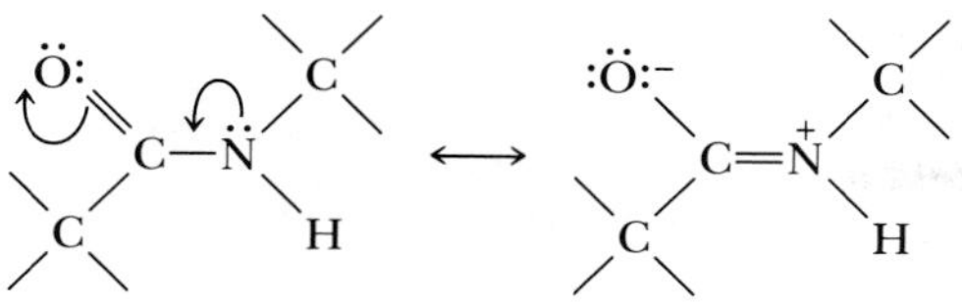

**FIGURE 5.9** Resonance structures of the peptide group.

between the carbon and nitrogen. The peptide bond has **partial double bond character,** and as a result the peptide group that forms the link between the two amino acids is planar.

This structural feature has important implications for the three-dimensional conformations of peptides and proteins. There is free rotation around the bonds between the $\alpha$-carbon of a given amino acid residue and the amino nitrogen and the carboxyl carbon of that residue, but there is no significant rotation around the peptide bond. This stereochemical constraint plays an important role in determining how the protein backbone can fold.

## 5.5
## SOME SMALL PEPTIDES OF PHYSIOLOGICAL INTEREST

The simplest possible covalently bonded combination of amino acids is a dipeptide, in which two amino acid residues are linked by a peptide bond. An example of a naturally occurring dipeptide is carnosine, which is found in muscle tissue. This compound, which has the alternate name $\beta$-alanyl-L-histidine, has an interesting structural feature. [In the systematic nomenclature of peptides, the **N-terminal** amino acid residue (the one with the free amino group) is mentioned first; then other residues are given as they occur in sequence, and the **C-terminal** amino acid residue (the one with the free carboxyl group) is given last.] The N-terminal amino acid residue, $\beta$-alanine, is structurally different from the $\alpha$ amino acids we have seen up to now. As the name implies, the amino group is bonded to the second or $\beta$ carbon of the alanine (Figure 5.10). The peptide bond in this dipeptide is formed between the carboxyl group of the $\beta$-alanine and the amino group of the histidine, which is the C-terminal amino acid. Box 5.1 discusses another dipeptide of some interest.

Glutathione is a commonly occurring tripeptide; it has considerable physiological importance because it is a scavenger for oxidizing agents. (It is

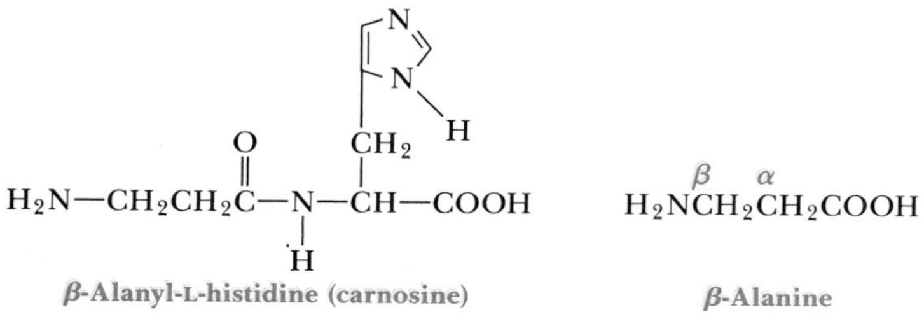

**FIGURE 5.10** Structures of carnosine and $\beta$-alanine.

$\beta$-Alanyl-L-histidine (carnosine)

$\beta$-Alanine

## BOX 5.1
## ASPARTAME, THE SWEET PEPTIDE

The dipeptide L-Aspartyl-L-phenylalanine is of considerable commercial importance. The aspartyl residue has a free $\alpha$ amino group (it is the N-terminal end of the molecule), and the phenylalanyl residue has a free carboxyl group, the C-terminal end. This dipeptide is about 200 times sweeter than sugar, as is a derivative of even greater commercial importance than the dipeptide itself. The derivative has a methyl group at the C-terminal end in an ester linkage to the carboxyl group. The methyl ester derivative is called *aspartame* and is marketed as a sugar substitute under the trade name Nutra-Sweet.

The consumption of common table sugar in the United States is about 100 pounds per person per year. Many people want to curtail their sugar intake in the interests of fighting obesity. Others must limit their sugar intake because of diabetes. One of the commonest ways of doing so is by drinking diet soft drinks. The soft drink industry is one of the largest markets for aspartame. The use of this sweetener was approved by the U.S. Food and Drug Administration in 1981 after extensive testing. Note that both amino acids have the L configuration. If a D-amino acid is substituted for either amino acid or for both of them, the resulting derivative is bitter rather than sweet.

**(a)**

L-Aspartyl-L-phenylalanine (methyl ester)

**(b)**

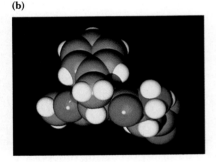

**UN 5.a**   (a) Structure of aspartame. (b) Space-filling model of aspartame.

thought that some oxidizing agents are harmful to organisms and play a role in the development of cancer.) In terms of its amino acid composition and bonding order, it is $\gamma$-glutamyl-L-cysteinylglycine (Figure 5.11(a)). Once again the N-terminal amino acid is mentioned first. In this case the $\gamma$-carboxyl group of the glutamic acid is involved in the peptide bond; the amino group of the cysteine is bonded to it. The carboxyl group of the cysteine is bonded in turn to the amino group of the glycine. The carboxyl group of the glycine forms the other end of the molecule, the C-terminal end. The glutathione molecule shown here is the reduced form. It scavenges oxidizing agents by reacting with them. The oxidized form of glutathione is formed from two molecules of the reduced peptide by formation of a disulfide bond between the —SH groups of the two cysteine residues (Figure 5.11(b)).

Two pentapeptides found in the brain are enkephalins, naturally occurring analgesics. Abbreviations for the amino acids are more convenient than structural formulas for molecules of this size. The same notation is used for the amino-acid sequence, with the N-terminal amino acid listed

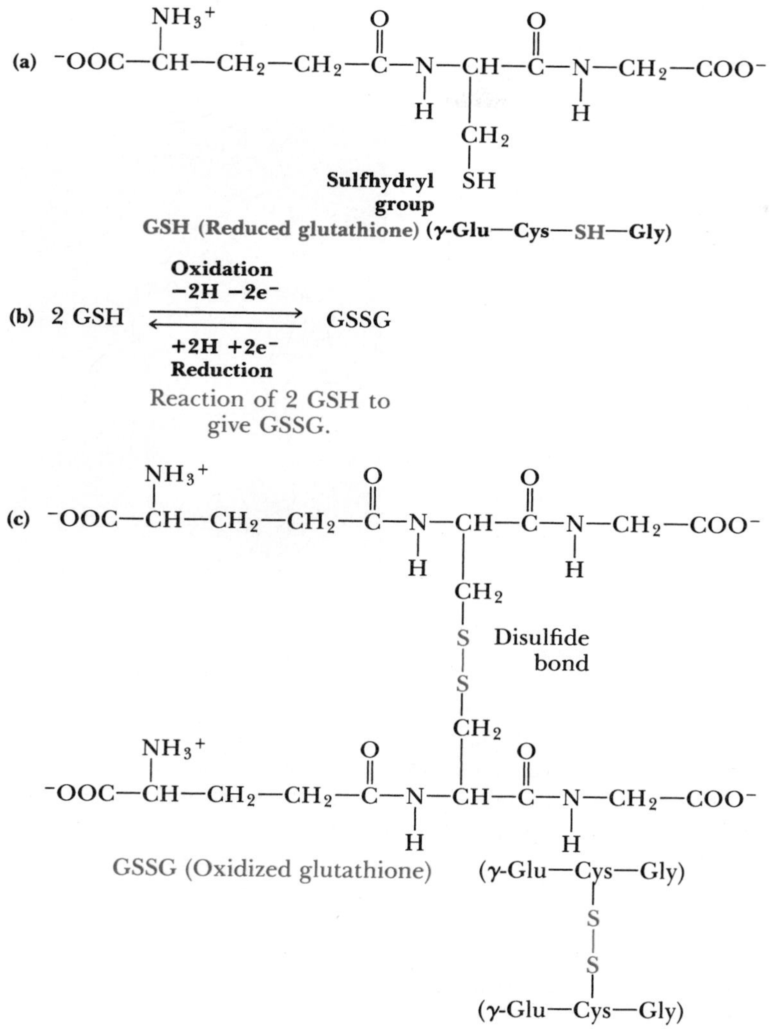

FIGURE 5.11  The oxidation and reduction of glutathione.

(a)
$$\overset{\overset{NH_3^+}{|}}{^-OOC-CH-CH_2-CH_2-\overset{\overset{O}{\|}}{C}-\underset{\underset{H}{|}}{N}-\underset{\underset{\underset{\underset{SH}{|}}{CH_2}}{|}}{CH}-\overset{\overset{O}{\|}}{C}-\underset{\underset{H}{|}}{N}-CH_2-COO^-}$$

Sulfhydryl group    SH

GSH (Reduced glutathione) (γ-Glu—Cys—SH—Gly)

(b)  2 GSH $\underset{\underset{\text{Reduction}}{+2H\ +2e^-}}{\overset{\overset{\text{Oxidation}}{-2H\ -2e^-}}{\rightleftarrows}}$ GSSG

Reaction of 2 GSH to give GSSG.

(c)
$$\overset{\overset{NH_3^+}{|}}{^-OOC-CH-CH_2-CH_2-\overset{\overset{O}{\|}}{C}-\underset{\underset{H}{|}}{N}-\underset{\underset{\underset{\underset{S}{|}}{CH_2}}{|}}{CH}-\overset{\overset{O}{\|}}{C}-\underset{\underset{H}{|}}{N}-CH_2-COO^-}$$

S    Disulfide bond
S
CH_2

$$\overset{\overset{NH_3^+}{|}}{^-OOC-CH-CH_2-CH_2-\overset{\overset{O}{\|}}{C}-\underset{\underset{H}{|}}{N}-CH-\overset{\overset{O}{\|}}{C}-\underset{\underset{H}{|}}{N}-CH_2-COO^-}$$

GSSG (Oxidized glutathione)    (γ-Glu—Cys—Gly)

S
S

(γ-Glu—Cys—Gly)

first and the C-terminal listed last. The two peptides in question, leucine and methionine enkephalin, differ only in their C-terminal amino acids.

Tyr-Gly-Gly-Phe-Leu

**Leucine Enkephalin**

Tyr-Gly-Gly-Phe-Met

**Methionine Enkephalin**

It is thought that the aromatic side chains of tyrosine and phenylalanine in these peptides play a role in their activity. It is also thought that there are similarities between the three-dimensional structures of opiates such as morphine and those of the enkephalins. As a result of these

**FIGURE 5.12** (a) Structures of oxytocin and vasopressin. (b) Space-filling model of oxytocin.

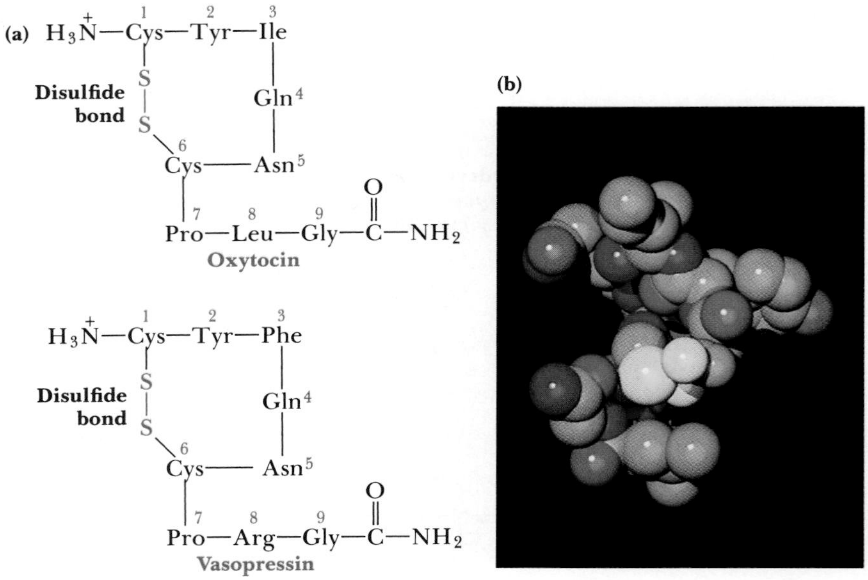

(a)

$H_3\overset{+}{N}$—$\underset{1}{Cys}$—$\underset{2}{Tyr}$—$\underset{3}{Ile}$

Disulfide bond

S | S

$\underset{6}{Cys}$——$\underset{5}{Asn}$

$Gln^4$

$\underset{7}{Pro}$—$\underset{8}{Leu}$—$\underset{9}{Gly}$—$\overset{O}{\underset{\parallel}{C}}$—$NH_2$

**Oxytocin**

$H_3\overset{+}{N}$—$\underset{1}{Cys}$—$\underset{2}{Tyr}$—$\underset{3}{Phe}$

Disulfide bond

S | S

$\underset{6}{Cys}$——$\underset{5}{Asn}$

$Gln^4$

$\underset{7}{Pro}$—$\underset{8}{Arg}$—$\underset{9}{Gly}$—$\overset{O}{\underset{\parallel}{C}}$—$NH_2$

**Vasopressin**

(b)

structural similarities, opiates bind to the receptors in the brain intended for the enkephalins and thus produce their physiological activity.

Some important peptides have cyclic structures. Two well known examples with many structural features in common are oxytocin and vasopressin (Figure 5.12). In both peptides there is an —S—S— bond similar to that in the oxidized form of glutathione; the disulfide bond is responsible for the cyclic structure. Both peptides contain nine amino acid residues, both have an amide group at the C-terminal end rather than a free carboxyl group, and both have a disulfide link between cysteine residues at positions 1 and 6. The difference between these two peptides is that oxytocin has an isoleucine residue at position 3 and a leucine residue at position 8, while vasopressin has a phenylalanine residue at position 3 and an arginine residue at position 8. Both these peptides have considerable physiological importance as hormones (see Box 5.2).

**FIGURE 5.13** Structures of ornithine, gramicidin S, and tyrocidine A.

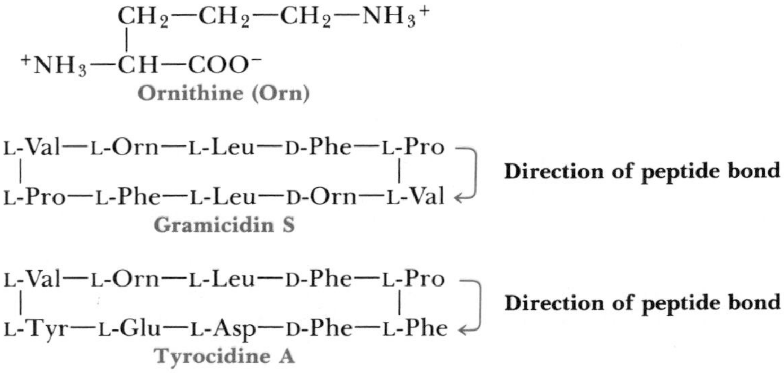

$$CH_2—CH_2—CH_2—NH_3{}^+$$
$$^+NH_3—\underset{|}{CH}—COO^-$$
**Ornithine (Orn)**

L-Val—L-Orn—L-Leu—D-Phe—L-Pro ⌐

L-Pro—L-Phe—L-Leu—D-Orn—L-Val ⌐

**Direction of peptide bond**

**Gramicidin S**

L-Val—L-Orn—L-Leu—D-Phe—L-Pro ⌐

L-Tyr—L-Glu—L-Asp—D-Phe—L-Phe ⌐

**Direction of peptide bond**

**Tyrocidine A**

## BOX 5.2
## PEPTIDE HORMONES

Oxytocin induces labor in pregnant women and controls contraction of uterine muscle. During pregnancy the number of receptors for oxytocin in the uterine wall increases. At term the number of receptors for oxytocin is large enough to cause contraction of the smooth muscle of the uterus in the presence of small amounts of oxytocin. The fetus moves toward the cervix of the uterus because of the strength and frequency of the uterine contractions. The cervix stretches, sending nerve impulses to the hypothalamus. When the impulse reaches this part of the brain, positive feedback leads to the release of still more oxytocin by the posterior pituitary gland. The presence of more oxytocin leads to stronger contractions of the uterus so that the fetus is forced through the cervix and the baby is born. Oxytocin also plays a role in stimulating the flow of milk by a nursing mother. The process of suckling sends nerve signals to the hypothalamus of the mother's brain. Oxytocin is released and carried by the blood to the mammary glands. The presence of oxytocin causes contraction of smooth muscle in the mammary glands, forcing out the milk that is in them. As suckling continues, more hormone is released, producing still more milk.

Vasopressin plays a role in control of blood pressure by regulating contraction of smooth muscle. Like oxytocin, vasopressin is released by the action of the hypothalamus on the posterior pituitary and transported by the blood to specific receptors. Vasopressin stimulates reabsorption of water by the kidney, thus having an antidiuretic effect. More water is retained and the blood pressure increases.

In some other peptides the cyclic structure is formed by the peptide bonds themselves. Two cyclic decapeptides (peptides containing ten amino-acid residues) produced by the bacterium *Bacillus brevis* are interesting examples. Both of these peptides, Gramicidin S and Tyrocidine A, are antibiotics, and both contain D amino acids as well as the more usual L amino acids (Figure 5.13). In addition, both contain the amino acid ornithine (Orn), which does not occur in proteins, but which does play a role as a metabolic intermediate in several common pathways (Section 18.7).

## SUMMARY

Amino acids, the monomer units of proteins, have a general structure in common, with an amino group and a carboxyl group bonded to the same carbon atom. The nature of the side chains, referred to as R groups, is the basis of the difference between the various amino acids.

Except for glycine, amino acids can exist in two forms, designated L and D. These two stereoisomers are nonsuperimposable mirror images. The amino acids found in proteins are all of the L form, but some D amino acids occur in nature. A classification scheme for amino acids can be based on the nature of the side chain. Two particularly important criteria are the polar or nonpolar nature of the side chain and the presence of an acidic or basic group in the side chain.

In free amino acids at neutral pH, the carboxylate group is negatively charged and the amino group has a positive charge. Amino acids without charged groups on their side chains exist in neutral solution as zwitterions with no net charge. Titration curves of amino acids indicate the pH ranges in which titratable groups gain or lose a proton. Side chains of amino acids can also contribute titratable groups; the charge (if any) on the side chain must be taken into consideration in determining the net charge on the amino acid.

Peptides are formed by linking the carboxyl group of one amino acid to the amino group of another amino acid in a covalent (amide) bond. Proteins consist of polypeptide chains; the number of amino acids in a protein is usually 100 or more. The peptide group is planar; this stereochemical constraint plays an important rôle in determining the three-dimensional structures of peptides and proteins. Small peptides, containing two to several dozen amino acid residues, can have marked physiological effects in organisms.

## EXERCISES

1. Write equations to show the ionic dissociation reactions of the following amino acids: aspartic acid, valine, histidine, serine, and lysine.
2. Predict the ionized form of the following amino acids at pH 7: glutamic acid, leucine, threonine, histidine, and arginine.
3. Based on the information in Table 5.2, is there any amino acid that could serve as a buffer at pH 8? If so, which?
4. Given a peptide with the following amino-acid sequence:

   Val-Met-Ser-Ile-Phe-Arg-Cys-Tyr-Leu

   identify the polar amino acids, identify the aromatic amino acids, and identify the sulfur-containing amino acids.
5. Identify the charged groups in the peptide shown in Exercise 4 at pH 1 and at pH 7.
6. What are the sequences of all the possible tripeptides that contain the amino acids serine, leucine, and phenylalanine? Use three-letter abbreviations to express your answer.
7. Draw structures of the following amino acids, indicating the charged form that exists at pH 4: histidine, asparagine, tryptophan, proline, and tyrosine.
8. Consider the following peptides:

   Phe-Glu-Ser-Met    and    Val-Trp-Cys-Leu

   Do these peptides have different net charges at pH 1? At pH 7? Indicate the charges at both pH values.
9. Sketch a titration curve for the amino acid cysteine; indicate the $pK'_a$ values for all titratable groups. Also indicate the pH at which this amino acid will have no net charge.
10. Sketch a titration curve for aspartic acid; indicate the $pK'_a$ values of all titratable groups. Also indicate the pH range in which the conjugate acid-base pair +1 asp and 0 asp will act as a buffer.
11. Consider the peptides Ser-Glu-Gly-His-Ala and Gly-His-Ala-Glu-Ser. How do these two peptides differ? Would you expect the titration curves for these peptides to differ from one another? Give the reason for your answer.
12. What are the structural differences between the peptide hormones oxytocin and vasopressin? How do they differ in function?
13. How do the oxidized and reduced forms of glutathione differ from one another?
14. Give the amino acid sequence of a peptide that contains one or more D amino acids. What is the biological role of this peptide?
15. What is the stereochemical basis of the observation that D-aspartyl-D-phenylalanine has a bitter taste while L-aspartyl-L-phenylalanine is significantly sweeter than sugar?
16. Name and give the structure of an amino acid produced by modification of one of the usual twenty amino acids after protein synthesis. List a specific protein in which such an amino acid can be found.
17. Sketch resonance structures for the peptide group and indicate how they contribute to the planar arrangement of this group of atoms.

## ANNOTATED BIBLIOGRAPHY

Barrett, G.C., ed. *Chemistry and Biochemistry of the Amino Acids.* New York: Chapman and Hall, 1985. [Wide coverage of many aspects of reactions of amino acids.]

Larsson, A., ed. *Functions of Glutathione: Biochemical, Physiological, Toxicological and Chemical Aspects.* New York: Raven Press, 1983. [A collection of articles on the many roles of a ubiquitous peptide.]

McKenna, K.W., and V. Pantic, eds. *Hormonally Active Brain Peptides: Structure and Function*. New York: Plenum Press, 1986. [A discussion of the chemistry of enkephalins and related peptides.]

Stegink, L.D., and L.J. Filer, Jr. *Aspartame—Physiology and Biochemistry*. New York: Marcel Dekker, 1984. [A comprehensive treatment covering metabolism, sensory and dietary aspects, preclinical studies and issues relating to human consumption, including ingestion by phenyketonurics and during pregnancy.]

Wold, F. *In vivo* Chemical Modification of Proteins (Post-translational Modification). *Ann. Rev. Biochem.* **50,** 788–814 (1981). [A review article on the modified amino acids found in proteins.]

# CHAPTER

6

# Carbohydrates

## OUTLINE

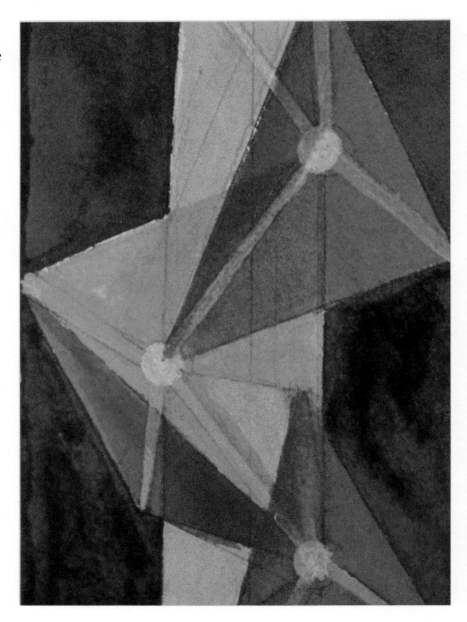

*A schematic representation of the carbon chain in carbohydrates.*

*More than half of all the organic carbon on planet Earth is stored in just two carbohydrate molecules—starch and cellulose. Both are polymers of the sugar monomer,* glucose. *The only difference between them is the manner in which the glucose units are joined together. Glucose is made by green plants and stored in starch as the plant's energy reserve. Animals (including humans) have an enzyme that recognizes the helical conformation of starch and can degrade it into its glucose units. Glucose, oxidized to carbon dioxide and water, is our primary energy source. Cellulose, a major component of plant cell walls as well as of cotton and wood, is a polymer of glucose monomers that all lie in the same plane. We don't possess the enzyme cellulase to break down cellulose, but termites do. A protozoan in their intestines contains the necessary enzyme, cellulase, to digest wood. We learn this the hard way when the house falls down. Modified carbohydrates are constituents of bacterial cell walls where they are cross-linked by short polypeptide chains. By combining different sugar monomers with amino acids,* glycoproteins *can be formed and act as cell surface markers to be recognized by other biomolecules.*

---

As noted in Section 4.3, the name "carbohydrate" originally referred to compounds of the general formula $C_n(H_2O)_n$. However, only the simple sugars or **monosaccharides** fit this formula exactly. The other types of carbohydrates, oligosaccharides and polysaccharides, are based on the monosaccharide units and have a slightly different general formula. **Oligosaccharides** are formed when a few ("oligo" in Greek) monosaccharides are linked together; **polysaccharides** are formed when many ("poly" in Greek) monosaccharides are bonded together. The reaction that adds monosaccharide units to a growing carbohydrate molecule involves loss of one $H_2O$ for each new link formed, accounting for the difference in the general formula.

Many commonly encountered carbohydrates are polysaccharides, including **glycogen,** which is found in animals, and **starch** and **cellulose,** which occur in plants. Carbohydrates play two important roles in biochemistry. First, they are major energy sources, and we shall devote Chapters 13 through 15 to carbohydrate metabolism. Second, polysaccharides are essential structural compounds in plants; cellulose is a major component of grass and trees.

## 6.1
## MONOSACCHARIDES: STRUCTURE AND STEREOCHEMISTRY

Recall from Chapter 4 that a monosaccharide can be a polyhydroxy aldehyde (**aldose**) or a polyhydroxy ketone (**ketose**). The simplest monosaccharides contain three carbon atoms and are called trioses—"tri,"

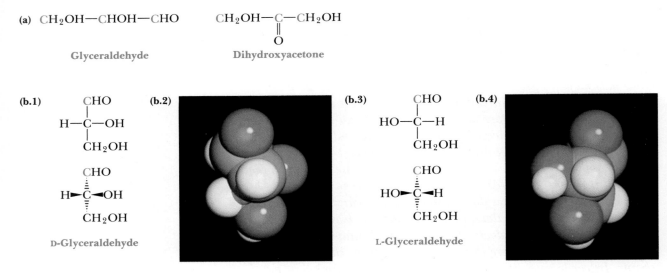

**(a)**  $CH_2OH$—$CHOH$—$CHO$         $CH_2OH$—$\underset{\overset{\|}{O}}{C}$—$CH_2OH$

Glyceraldehyde                    Dihydroxyacetone

**(b.1)**

$$CHO$$
$$H—\underset{|}{\overset{|}{C}}—OH$$
$$CH_2OH$$

$$CHO$$
$$H—\overset{\vdots}{C}◀OH$$
$$CH_2OH$$

D-Glyceraldehyde

**(b.3)**

$$CHO$$
$$HO—\underset{|}{\overset{|}{C}}—H$$
$$CH_2OH$$

$$CHO$$
$$HO▶\overset{\vdots}{C}◀H$$
$$CH_2OH$$

L-Glyceraldehyde

**FIGURE 6.1**  The structure of the simplest carbohydrates, the trioses. (a) A comparison of glyceraldehyde (an aldotriose) and dihydroxyacetone (a ketotriose). (b.1) Structure of D-glyceraldehyde. (b.2) Space-filling model of D-glyceraldehyde. (b.3) Structure of L-glyceraldehyde. (b.4) Space-filling model of L-glyceraldehyde.

three. **Glyceraldehyde** is the aldose with three carbons (an aldotriose), and **dihydroxyacetone** is the ketose with three carbon atoms (a ketotriose). Figure 6.1 shows these molecules. Aldoses with four, five, six, and seven carbon atoms are called aldotetroses, aldopentoses, aldohexoses, and aldoheptoses. The corresponding ketoses are ketotetroses, ketopentoses, ketohexoses, and ketoheptoses. Six-carbon sugars are the most abundant in nature, but two five-carbon sugars, ribose and deoxyribose, occur in the structures of RNA and DNA, respectively. Four-carbon and seven-carbon sugars play a role in photosynthesis.

We have already seen (Section 5.1) that some molecules are not superimposable on their mirror images and that these mirror images are **optical isomers (stereoisomers)** of each other. A chiral (asymmetric) carbon atom is the usual source of optical isomerism, as was the case with amino acids. The simplest carbohydrate that contains a chiral carbon is glyceraldehyde, which can exist in two isomeric forms that are mirror images of each other (Figure 6.1b). (Dihydroxyacetone does not contain a chiral carbon atom and does not exist in nonsuperimposable mirror-image forms.) The two forms of glyceraldehyde are designated D-glyceraldehyde and L-glyceraldehyde. Mirror-image stereoisomers are also called **enantiomers,** and D-glyceraldehyde and L-glyceraldehyde are enantiomers of each other. There is a convention for drawing a two-dimensional picture of the three-dimensional structures of stereoisomers. The dashed wedges represent bonds directed down, below the plane of the paper, and the solid triangles represent bonds directed up, out of the plane of the paper in the direction of the observer. The term **configuration** refers to the three-dimensional arrangement of groups around a chiral carbon atom, and stereoisomers differ from each other in their configuration.

The two enantiomers of glyceraldehyde are the only possible stereoisomers of three-carbon sugars, but the possibilities for stereoisomerism increase as the number of carbon atoms increases. Let us see what happens

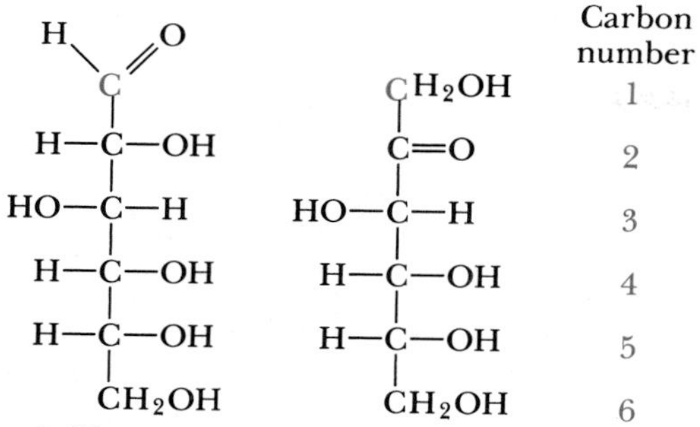

FIGURE 6.2  Examples of an aldose (D-glucose) and a ketose (D-fructose) showing numbering of carbon atoms.

as a —CH$_2$OH group is added to glyceraldehyde. In other words, what are the possible stereoisomers for an aldotetrose? In order to show the structures of the resulting molecules, we need to say more about the convention for a two-dimensional perspective of the molecular structure. This convention is called the **Fischer projection** method after the German chemist Emil Fischer, who established the structures of many sugars. In the Fischer projection, bonds written vertically in the two dimensions on the paper represent bonds behind the paper in three dimensions, while bonds written horizontally in two dimensions represent bonds directed in front of the paper in three dimensions. The most highly oxidized carbon, in this case the one involved in the aldehyde group, is written at the "top" and is designated C-1 (Figure 6.2). (In the ketose we have shown the ketone group becomes C-2, the carbon atom next to the "top." Most common sugars are aldoses rather than ketoses, so our discussion will focus mainly on aldoses.) The other carbon atoms are numbered in sequence from the "top." The designation of the configuration as L or D depends on the arrangement at the chiral carbon with the highest number. In the Fischer projection of the D configuration, the hydroxyl group is on the right of the highest numbered chiral carbon; in the L configuration the hydroxyl group is on the left of the highest numbered chiral carbon.

The aldotetroses (Figure 6.3) have two chiral carbons, numbers 2 and 3, and there are four possible stereoisomers. Two of the isomers have the D configuration, and two have the L configuration. The two D isomers have the same configuration at C-3, but they differ from each other in the configuration (arrangement of the —OH group) at the other chiral carbon (C-2). These two isomers are called D-erythrose and D-threose. They are not superimposable on each other, but neither are they mirror images of each other. Such nonsuperimposable, non-mirror-image stereoisomers are called **diastereomers.** The two L isomers are L-erythrose and L-threose. L-Erythrose is the enantiomer (mirror image) of D-erythrose, and L-threose is the enantiomer of D-threose. L-Threose is a diastereomer of both D- and L-erythrose, and L-erythrose is a diastereomer of both D- and L-threose. Diastereomers that differ from each other in the configuration at only one chiral carbon are called **epimers;** D-erythrose and D-threose are epimers.

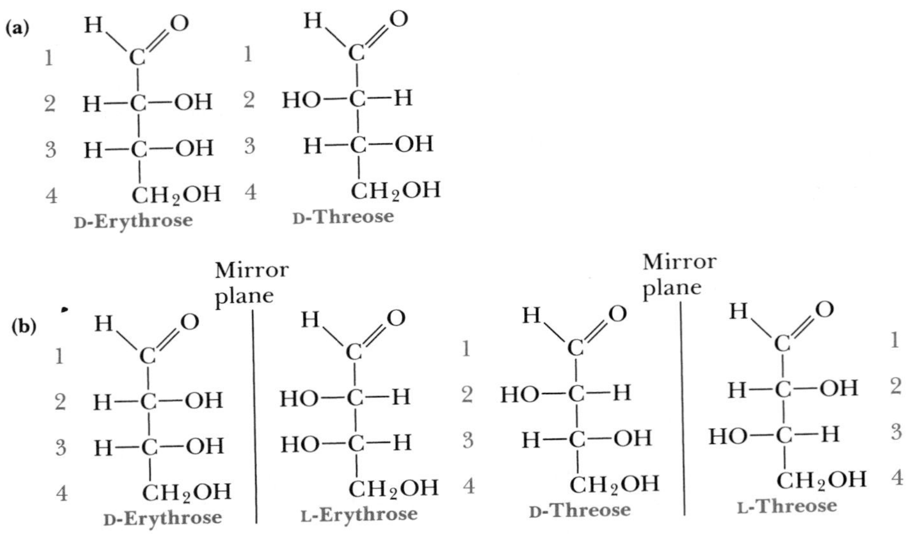

**FIGURE 6.3** Stereoisomers of an aldotetrose. (a) Diastereomers D-erythrose and D-threose. (b) Enantiomers D- and L-erythrose and D- and L-threose. Carbons are numbered.

(a)

D-Erythrose

D-Threose

(b)

Mirror plane

D-Erythrose    L-Erythrose

Mirror plane

D-Threose    L-Threose

Aldopentoses have three chiral carbons, and there are eight possible stereoisomers, four D forms and four L forms. Aldohexoses have four chiral carbons and sixteen stereoisomers, eight D forms and eight L forms (Figure 6.4). Some of the possible stereoisomers are much more commonly encountered in nature than others, and most biochemical discussion centers on the common naturally occurring sugars. For example, D sugars predominate in nature rather than L sugars. Most of the sugars that play an important role in metabolism contain either five or six carbon atoms. We shall discuss D-glucose (an aldohexose) and D-ribose (an aldopentose) far more than many other sugars. Glucose is a ubiquitous energy source, and ribose plays an important role in the structure of nucleic acids.

## Cyclic Structures: Anomers

Sugars, especially those with five and six carbon atoms, normally exist as cyclic molecules rather than the open-chain forms we have shown so far. The cyclization takes place as a result of interaction between the functional groups on distant carbons such as C-1 and C-5 to form a cyclic **hemiacetal** (in aldohexoses). Another possibility is interaction between C-2 and C-5 to form a cyclic **hemiketal** (in ketohexoses). See Figure 6.5. In either case the carbonyl carbon becomes a new chiral center called the **anomeric carbon.** The cyclic sugar can take either of two different forms, designated $\alpha$ and $\beta$, and these two forms are called **anomers** of each other.

The Fischer projection of the $\alpha$ anomer of a D sugar has the anomeric hydroxyl group to the right of the anomeric carbon (C—OH), and the $\beta$ anomer of a D sugar has the anomeric hydroxyl group to the left of the anomeric carbon (Figure 6.6). The free carbonyl species can readily form either the $\alpha$ or $\beta$ anomer, and the anomers can be converted from

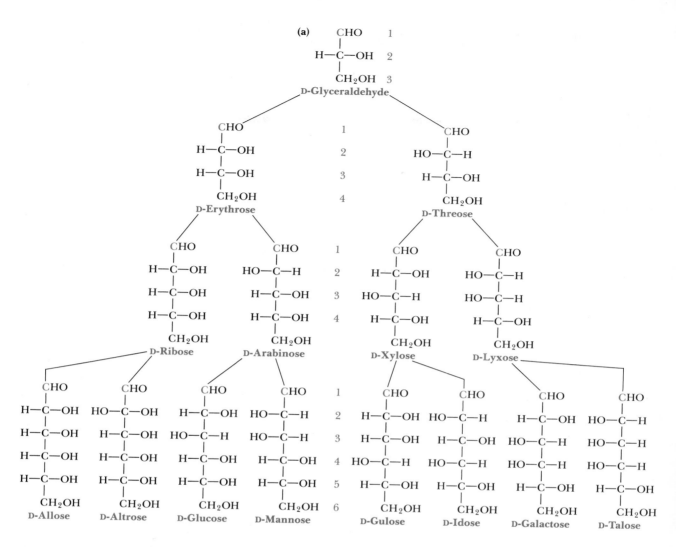

**(a)**

**D-Glyceraldehyde**

**D-Erythrose**  **D-Threose**

**D-Ribose**  **D-Arabinose**  **D-Xylose**  **D-Lyxose**

**D-Allose**  **D-Altrose**  **D-Glucose**  **D-Mannose**  **D-Gulose**  **D-Idose**  **D-Galactose**  **D-Talose**

**FIGURE 6.4** Stereochemical relationships among monosaccharides. (a) Aldoses containing from three to six carbon atoms, with the numbering of the carbon atoms shown. Note that the figure shows only half the possible isomers. For each isomer shown there is an enantiomer which is not shown, the L series. (b) The relationship between mirror images is of interest to mathematicians as well as to chemists. Lewis Carroll (C. L. Dodgson) was a contemporary of Emil Fischer.

**(b)**

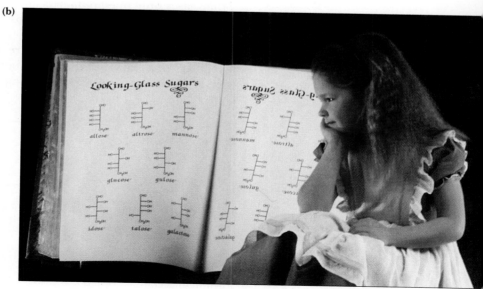

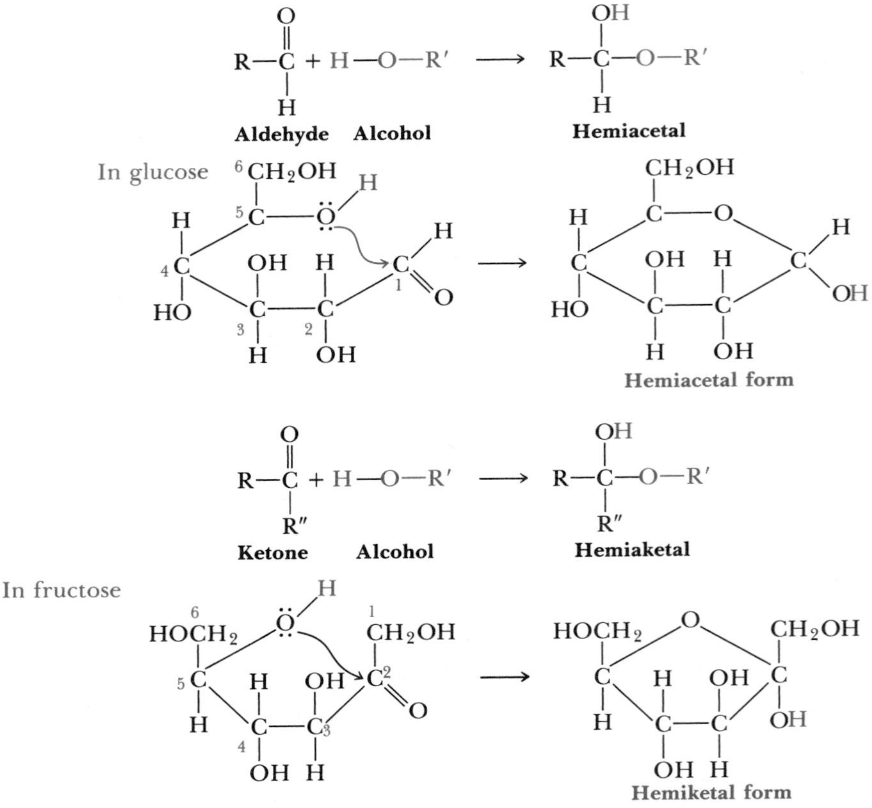

FIGURE 6.5    Formation of hemiacetals and hemiketals. An alcohol reacts with a carbonyl to give rise to a chiral center. In the six-carbon sugars, the hydroxyl group on carbon-5 reacts with the carbonyl group. In the cyclic form there are two possible configurations at the carbon atom which had been the carbonyl carbon in the open-chain form.

one form to another through the free carbonyl species. In some biochemical reactions any anomer of a given sugar can be used, but in other cases only one anomer occurs. For example, in living organisms only $\beta$-D-ribose and $\beta$-D-deoxyribose are found in RNA and DNA, respectively.

Fischer projection formulas are useful in describing the stereochemistry of sugars, but their long bonds and right angle bends do not give a realistic picture of the bonding situation in the cyclic forms, nor

FIGURE 6.6    Fischer projection formulas of three forms of glucose. Note that the $\alpha$ and $\beta$ forms can be converted to one another through the open chain form.

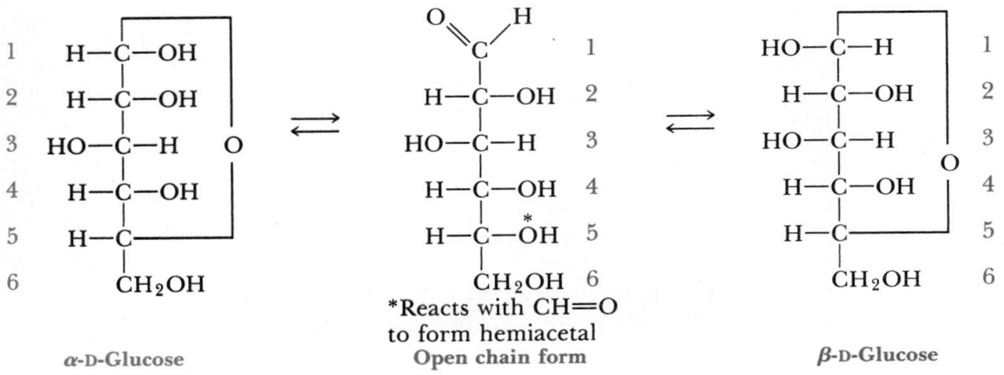

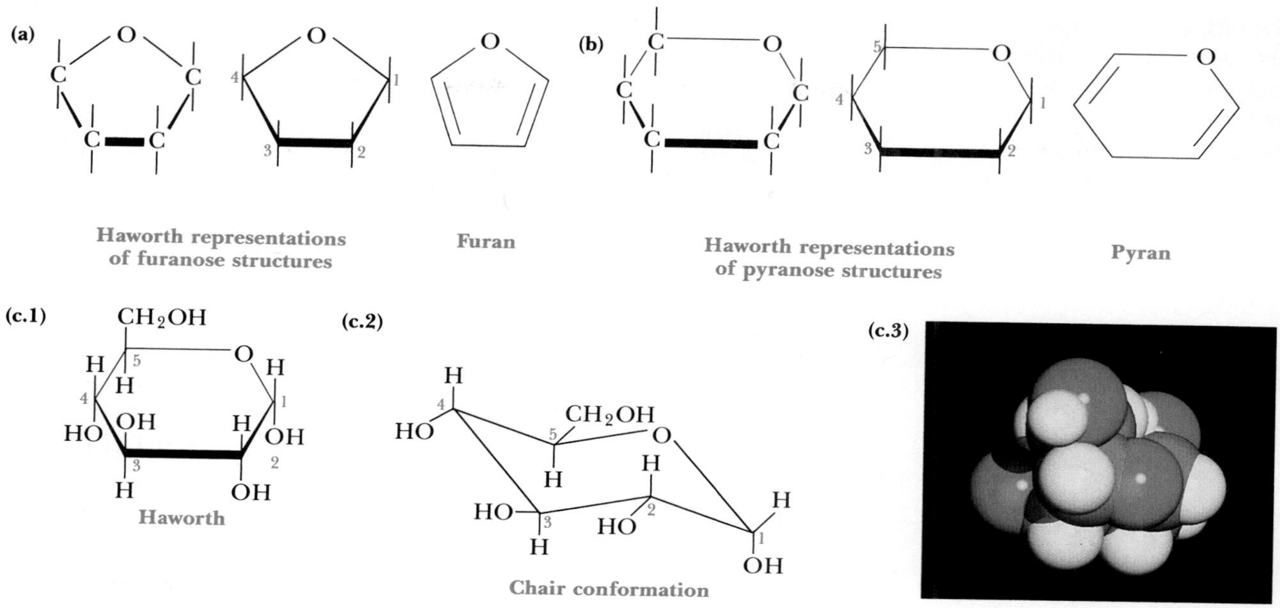


FIGURE 6.7 Haworth representations of sugar structures. (a) A comparison of the structure of furan with Haworth representations of furanoses. (b) A comparison of the structure of pyran with Haworth representations of pyranoses. (c.1) α-D-glucopyranose in the Haworth representation (1), in the chair conformation (c.2), and as a space-filling model (c.3).

do they represent the overall shape of the molecules accurately. **Haworth projection formulas** are more useful for those purposes. In Haworth projections the cyclic structures of sugars are shown in perspective drawings as planar five- or six-membered rings viewed nearly edge on. A five-membered ring is called a **furanose** because of its resemblance to furan; a six-membered ring is called a **pyranose** because of its resemblance to pyran (Figure 6.7a and b). These cyclic formulas are a better approximation to the shapes of the actual molecules for furanoses than for pyranoses. The five-membered rings of furanoses are in reality very nearly planar, but the six-membered rings of pyranoses actually exist in solution in the chair conformation (Figure 6.7c). Even though the Haworth formulas are approximations, they are very useful ones because they are easy to draw. The Haworth projections give a more realistic representation of the stereochemistry of sugars than do the Fischer projections, and the Haworth scheme is adequate for our purposes. We shall continue to use Haworth projections in our discussion of sugars.

For a D sugar, any group that is written to the right of the carbon in a Fischer projection has a downward direction in a Haworth projection; any group that is written to the left in a Fischer projection has an upward direction in a Haworth projection. The terminal —$CH_2OH$ group, which contains the carbon atom with the highest number in the numbering scheme, is shown in an upward direction. The structures of

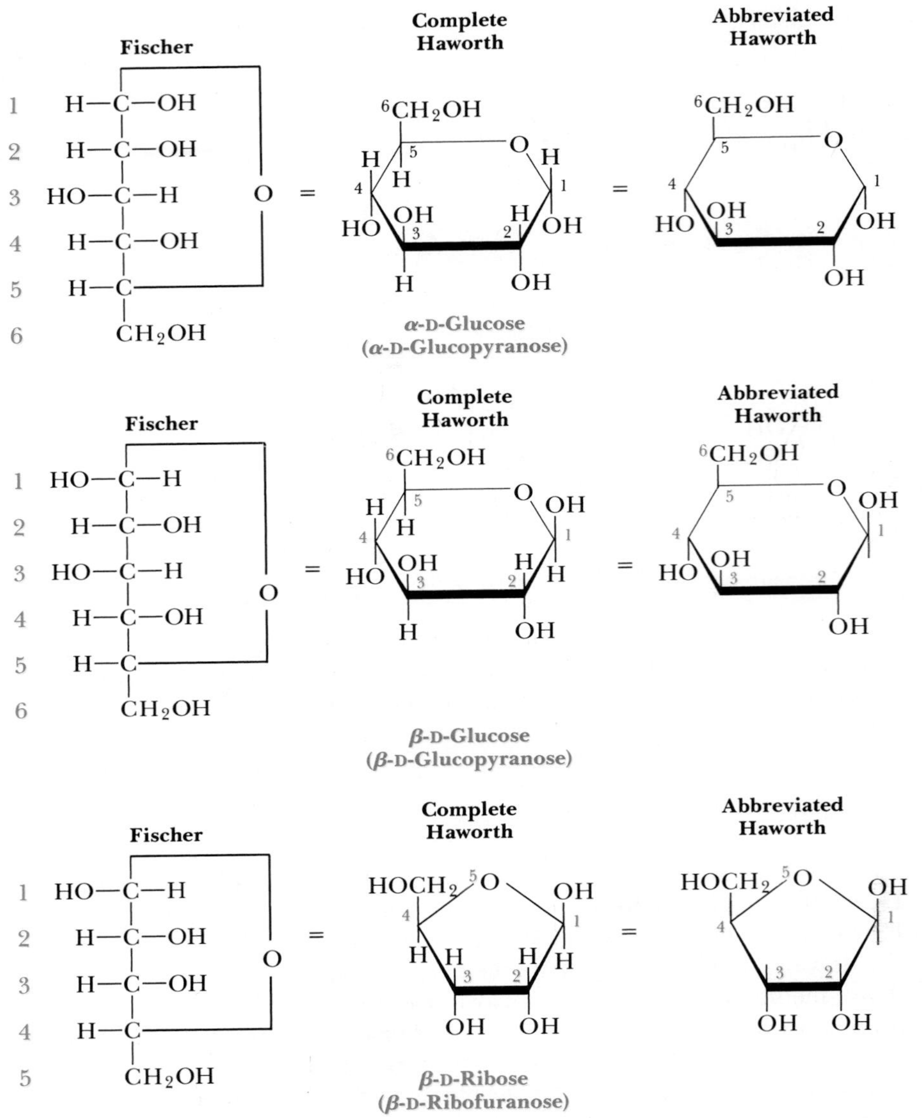

**FIGURE 6.8** A comparison of the Fischer, complete Haworth, and abbreviated Haworth representations of $\alpha$- and $\beta$-D-glucose (glucopyranose) and $\beta$-D-ribose (ribofuranose).

$\alpha$- and $\beta$-D-glucose, which are both pyranoses, and of $\beta$-D-ribose, which is a furanose, illustrate this point (Figure 6.8).

## 6.2
## REACTIONS OF MONOSACCHARIDES

### Oxidation–Reduction Reactions

Oxidation and reduction reactions of sugars play a key role in biochemistry. Oxidation of sugars provides energy for organisms to carry out their life processes; the highest yield of energy from carbohydrates occurs when

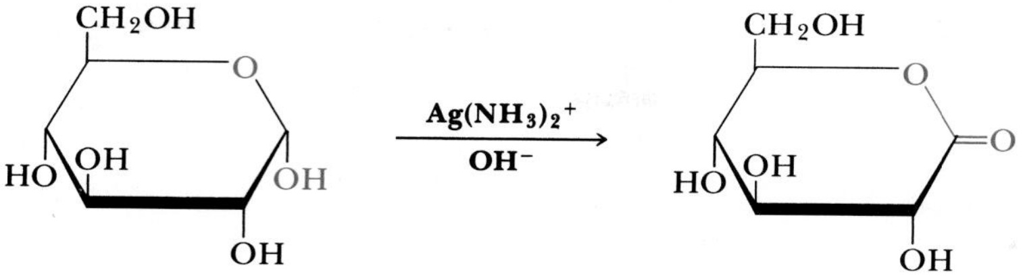

FIGURE 6.9   An example of an oxidation reaction of sugars. Oxidation of $\alpha$-D-glucose hemiacetal to give a lactone. Deposition of free silver as a silver mirror indicates that the reaction has taken place.

sugars are completely oxidized to $CO_2$ and $H_2O$ in aerobic processes. The reverse of complete oxidation of sugars is the reduction of $CO_2$ and $H_2O$ to form sugars, a process that takes place in photosynthesis.

Several oxidation reactions of sugars are of some importance in laboratory practice, because they can be used for identification of sugars. Aldehyde groups can be oxidized to give the carboxyl group characteristic of acids, and this reaction forms the basis of a test for the presence of aldoses. When the aldehyde is oxidized, some oxidizing agent must be reduced. Aldoses are called **reducing sugars** because of this type of reaction; ketoses can also be reducing sugars because they isomerize to aldoses. In the cyclic form the compound produced by oxidation of an aldose is a **lactone** (a cyclic ester linking the carboxyl group and one of the sugar alcohols, as shown in Figure 6.9). Two types of reagent are used in the laboratory to detect the presence of reducing sugars. The first of these is Tollens' reagent, which uses the silver ammonia complex ion, $Ag(NH_3)_2{}^+$, as the oxidizing agent. A silver mirror is deposited on the wall of the test tube if a reducing sugar is present, as shown by the equation for the reaction:

$$RCHO + 2\ Ag(NH_3)_2{}^+ + 2\ OH^- \longrightarrow$$
$$RCOO^- + 2\ Ag + 3\ NH_3 + NH_4{}^+ + H_2O$$

The use of other types of oxidizing agents is based on the reaction of copper ion to produce a red precipitate of cuprous oxide, $Cu_2O$:

$$RCHO + 2\ Cu^{2+} + 5\ OH^- \longrightarrow RCOO^- + Cu_2O + 3\ H_2O$$

There are two common reagents that use copper ion as the oxidizing agent: in Fehling's solution the copper ion is dissolved in a buffer containing tartrate ion, and in Benedict's solution the copper ion is used with sodium carbonate and citrate buffer. In both cases the solution is quite basic. Ions such as tartrate and citrate can form complexes with the $Cu^{2+}$, which would otherwise form the insoluble hydroxide. Solubility considerations can make either Fehling's solution or Benedict's solution the reagent of choice for a particular test. A more recent method for detection of glucose, but not other reducing sugars, is based on the use of the enzyme glucose oxidase, which is specific for glucose.

In addition to oxidized sugars, there are some important reduced sugars that we should mention at this point. In **deoxy sugars** a hydrogen atom is substituted for one of the hydroxyl groups of the sugar. One of these deoxy sugars is L-fucose (L-6-deoxygalactose), which is found in the carbohydrate portion of some glycoproteins. The name ''glycoprotein'' indicates that these substances are conjugated proteins that contain some carbohydrate group (*glykos,* Greek for ''sweet'') in addition to the

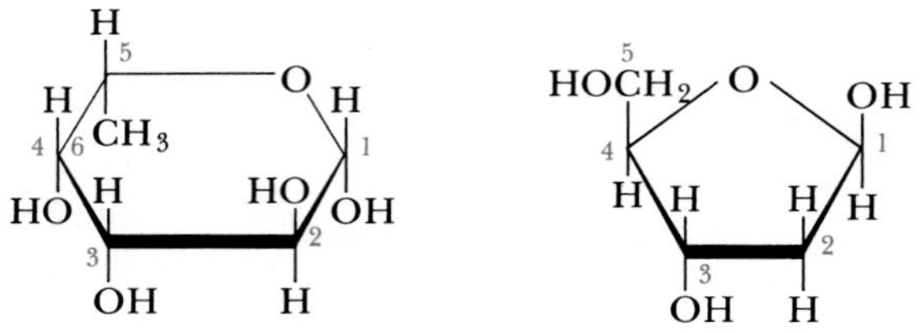

**FIGURE 6.10**    Structures of two deoxy sugars.

β-L-Fucose
(6-Deoxy-β-L-galactose)

β-D-Deoxribose
(2-Deoxy-β-D-ribose)

polypeptide chain. An even more important example of a deoxy sugar is D-2-deoxyribose, the sugar found in DNA (Figure 6.10).

## Esterification Reactions

The hydroxyl groups of sugars behave exactly like all other alcohols in the sense that they can react with acids and derivatives of acids to form esters. The phosphate esters are particularly important ones because they are the usual intermediates in the breakdown of carbohydrates, not simply the breakdown of oligo- and polysaccharides to monosaccharides but the further metabolism of sugars to provide energy. Phosphate esters are frequently formed by transfer of a phosphate group from ATP (*adenosine triphosphate*) to give the phosphorylated sugar and ADP (*adenosine diphosphate*), as shown in Figure 6.11. Such reactions play an important role in the metabolism of sugars (Section 13.2).

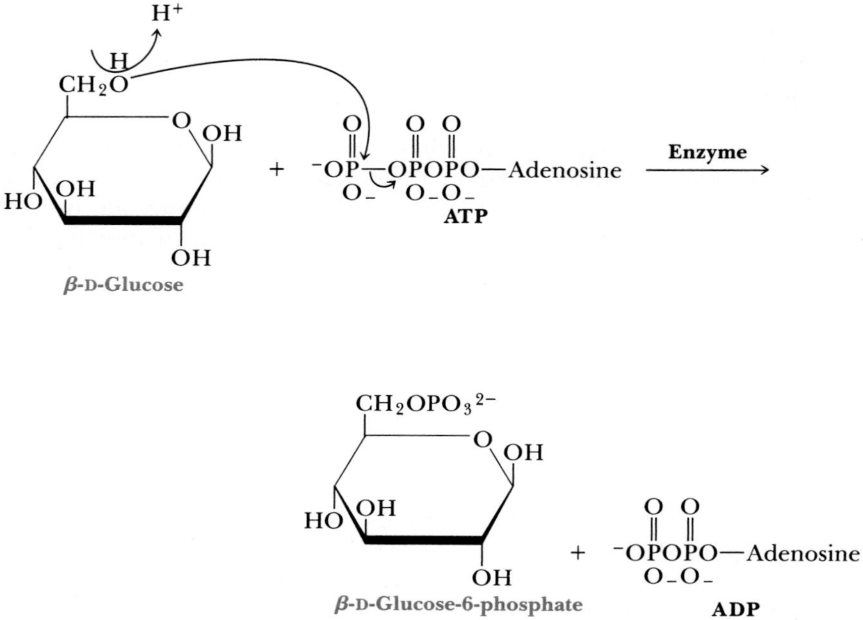

**FIGURE 6.11**    The formation of a phosphate ester of glucose. ATP is the phosphate group donor. The enzyme specifies the interaction with —$C^6H_2OH$.

β-D-Glucose

ATP

Enzyme

β-D-Glucose-6-phosphate

ADP

## The Formation of Glycosides

It is possible for a sugar hydroxyl group (ROH) to react with another hydroxyl (R′OH) to form an ether linkage (R′—O—R). This type of reaction frequently involves the —OH group bonded to the anomeric carbon of a sugar in its cyclic form. (Recall that the anomeric carbon is the carbonyl carbon of the open chain form of the sugar, and is the one that becomes a chiral center in the cyclic form.) To say the same thing in a slightly different way, a hemiacetal can react with an alcohol such as methyl alcohol to give a **full acetal** or **glycoside** (Figure 6.12). The newly formed bond is called a **glycosidic bond.** Glycosides derived from furanoses are called **furanosides,** and the ones derived from pyranoses are called **pyranosides.** (Recall that we saw earlier in this chapter that furanoses are cyclic sugars with a five-membered ring, while pyranoses are analogous compounds with a six-membered ring.)

Glycosidic bonds between monosaccharide units are the basis for the formation of oligosaccharides and polysaccharides. Glycosidic linkages can take various forms; either the $\alpha$ or $\beta$ anomer of one sugar can be bonded to any one of the —OH groups on the other sugar. Many different combinations are found in nature. The —OH groups are numbered to distinguish them, and the numbering scheme follows that of the carbon atoms. The notation for the glycosidic linkage between the two sugars specifies which anomeric form of the sugar is involved in the bond and also specifies which carbon atoms of the two sugars are linked together. Two ways that two $\alpha$-D-glucose molecules could be linked together are $\alpha(1 \longrightarrow 4)$ and $\alpha(1 \longrightarrow 6)$. In the first example, the $\alpha$ anomeric carbon (C-1) of the first glucose molecule is joined in a glycosidic bond to the fourth carbon atom, C-4, of the second glucose molecule; the C-1 carbon of the first glucose molecule is linked to the C-6 carbon of the second glucose molecule in the second example (Figure 6.13). Another possibility of a glycosidic bond, this time between two $\beta$-D-glucose molecules, is a $\beta,\beta(1 \longrightarrow 1)$ linkage. The anomeric form at both C-1 carbons must be specified because the linkage is between the two anomeric carbons, each of which is C-1 (Figure 6.14).

When oligosaccharides and polysaccharides are formed as a result of glycosidic bonding, their chemical nature depends on which monosaccharides are linked together and also on the particular glycosidic bond formed (*i.e.*, which anomers and which carbon atoms are linked

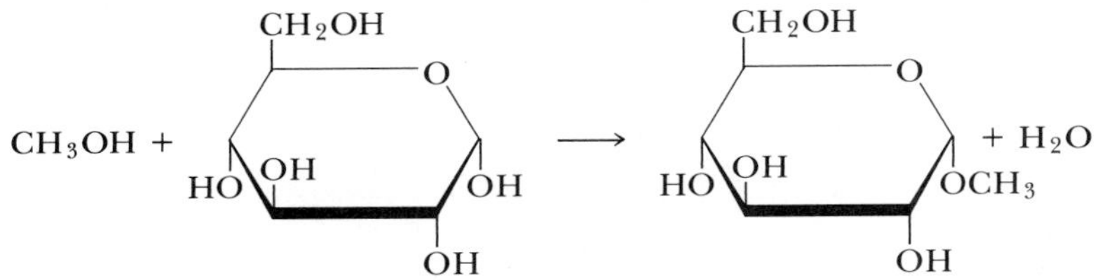


**Methyl alcohol** **α-D-Glucopyranose (hemiacetal)** **Methyl-α-D-glucopyranoside, a glycoside (full acetal)**

**FIGURE 6.12** An example of the formation of a glycoside. Methyl alcohol ($CH_3OH$) and $\alpha$-D-glucopyranose react to form the corresponding glycoside.

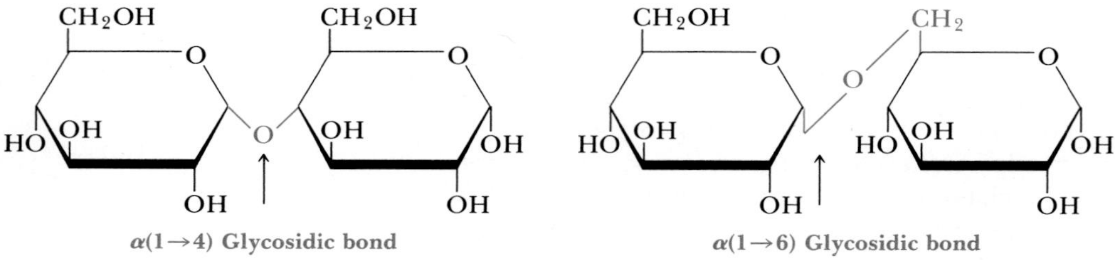

α(1→4) Glycosidic bond                    α(1→6) Glycosidic bond

**FIGURE 6.13**    Two different disaccharides of α-D-glucose. These two chemical compounds have different properties because one has an α(1——→4) linkage and the other has an α(1——→6) linkage.

together). The difference between cellulose and starch depends on the glycosidic bond formed between glucose monomers. Because of the variation in glycosidic linkages, both linear and branched chain polymers can be formed. If the internal monosaccharide residues that are incorporated in a polysaccharide form two glycoside bonds, the polymer will be linear. (Of course, the end residues will be involved in only one glycosidic linkage.) Some internal residues can form three glycosidic bonds, leading to the formation of branched chain structures (Figure 6.15).

There is another point worth mentioning about glycosides before we leave the topic. We have already seen that the anomeric carbon is frequently involved in the glycosidic linkage, and also that the test for the presence of sugars—specifically for reducing sugars—requires a reaction of the group at the anomeric carbon. The internal anomeric carbons in oligosaccharides are not free to give the test for reducing sugars. Only if the end residue is a free hemiacetal rather than a glycoside will there be a positive test for a reducing sugar (Figure 6.16). The level of detection can be important for such a test. A sample that contains only a few molecules of a large polysaccharide, each molecule with a single reducing end, might well give a negative test because there are not enough reducing ends to detect.

**FIGURE 6.14**    A disaccharide of β-D-glucose. Both anomeric carbons (C-1) are involved in the glycosidic linkage.

β,β(1→1) Glycosidic bonds

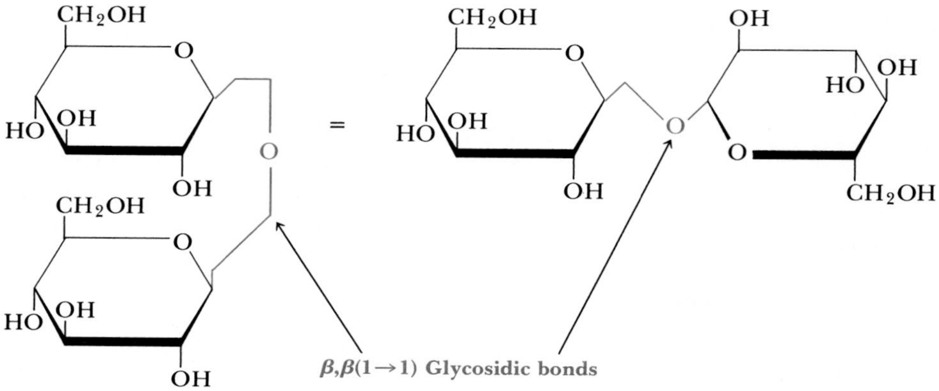

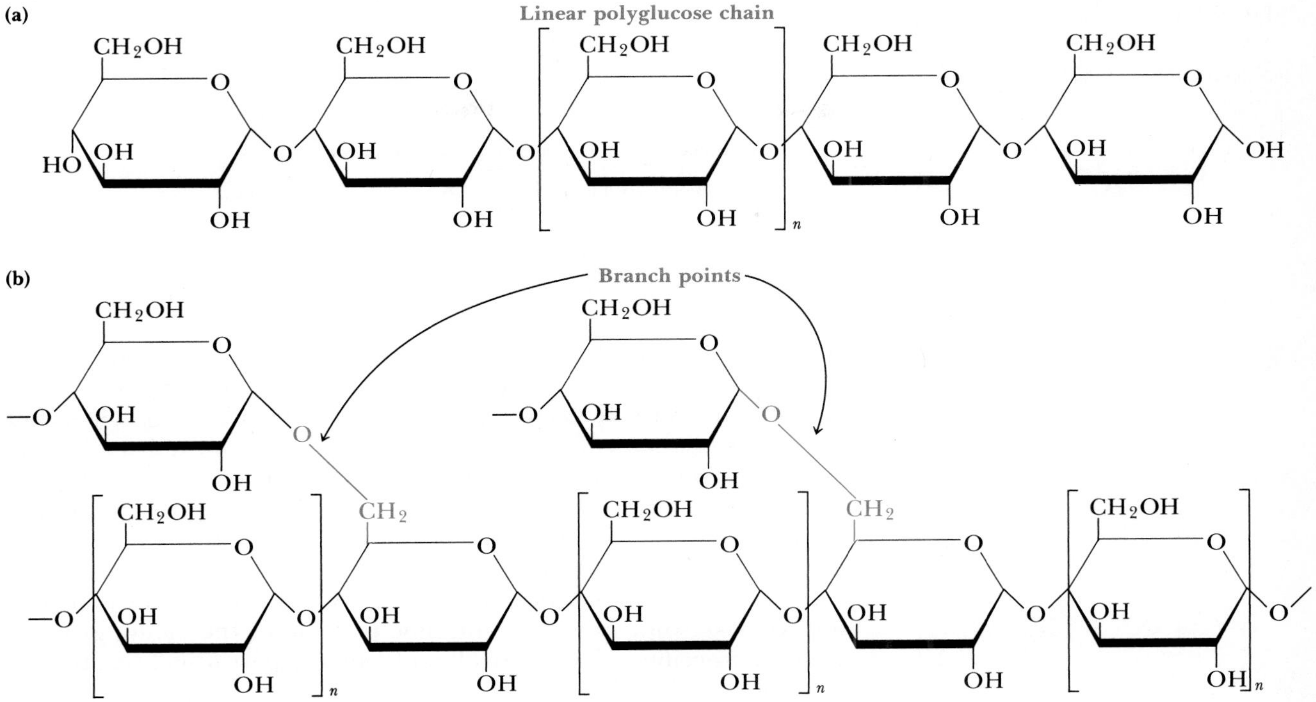

**(a)**

Linear polyglucose chain

**(b)**

Branch points

FIGURE 6.15 Linear and branched chain polymers of $\alpha$-D-glucose. (a) The linear polyglucose chain occurs in amylose, while the branched chain polymer occurs in amylopectin and glycogen (Section 6.4). All glycoside bonds are $\alpha(1 \longrightarrow 4)$. (b) Branched polyglucose chain glycoside bonds are $\alpha(1 \longrightarrow 6)$ at branch points. Again, all glycoside bonds along the chain are $\alpha(1 \longrightarrow 4)$.

## Other Derivatives of Sugars

**Amino sugars** are an interesting class of compounds related to the monosaccharides. We shall not go into the chemistry of their formation, but it will be useful to have some acquaintance with them when we discuss polysaccharides. In sugars of this type an amino group ($-NH_2$) or one of its derivatives is substituted for the hydroxyl group of the parent sugar. In **N-acetyl amino sugars** the amino group itself carries an acetyl group

FIGURE 6.16 A disaccharide with a free hemiacetal end is a reducing sugar because of the presence of a free anomeric aldehyde carbonyl or potential aldehyde group.

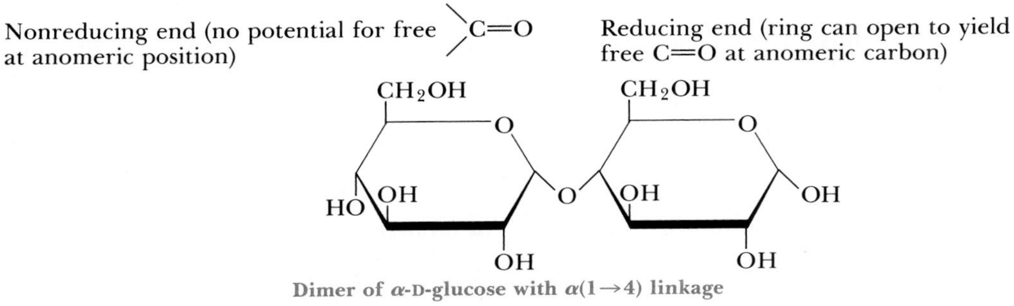

Nonreducing end (no potential for free $>C=O$ at anomeric position)

Reducing end (ring can open to yield free $C=O$ at anomeric carbon)

Dimer of $\alpha$-D-glucose with $\alpha(1 \rightarrow 4)$ linkage

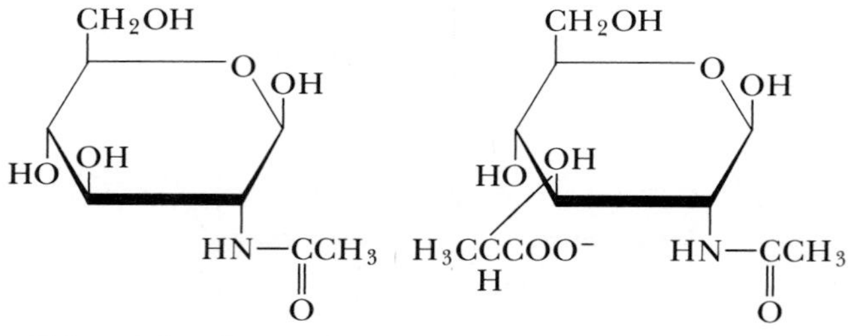

**FIGURE 6.17**  The structures of *N*-acetyl-β-D-glucosamine and *N*-acetylmuramic acid.

*N*-Acetyl-β-D-glucosamine          *N*-Acetylmuramic acid

($CH_3$—CO—) as a substituent. Two particularly important examples are *N*-acetyl-β-D-glucosamine and its derivative *N*-acetyl-β-muramic acid, which has an added carboxylic acid side chain (Figure 6.17). These two compounds are components of bacterial cell walls. We did not specify whether *N*-acetylmuramic acid belongs to the L or the D series of configurations, and when we mentioned it in this sentence we did not specify the α or β anomer. This type of shorthand is usual practice with β-D-glucose and its derivatives, since the D configuration and β anomeric form are so common that we need not specify them all the time unless we want to make some specific point. The position of the amino group is also left unspecified, since discussion of amino sugars usually centers on a few compounds whose structure is well known.

## 6.3
## OLIGOSACCHARIDES

Oligomers of sugars frequently occur as **disaccharides,** formed by linking two monosaccharide units by glycosidic bonds. Three of the most important examples of oligosaccharides are disaccharides. They are sucrose, lactose, and maltose.

**Sucrose** is the common table sugar extracted from sugar cane and sugar beets. The monosaccharide units that make up sucrose are α-D-glucose and β-D-fructose. Glucose (an aldohexose) is a pyranose, and fructose (a ketohexose) is a furanose. The α C-1 carbon of the glucose is linked to the β C-2 carbon of the fructose (Figure 6.18a) in a glycosidic linkage that has the notation α,β(1⟶2). Sucrose is not a reducing sugar because both anomeric groups are involved in the glycosidic linkage. Free glucose is a reducing sugar, and free fructose can also give a positive test even though it is a ketone rather than an aldehyde in the open-chain form. Fructose and ketoses in general can act as reducing sugars because they can isomerize to aldoses in a rather complex rearrangement reaction. We need not concern ourselves with the details of this isomerization.

When sucrose is consumed by animals, it is hydrolyzed to glucose and fructose, which are then degraded by metabolic processes to provide energy. Humans consume large quantities of sucrose, and excess consumption can contribute to health problems. This last point has led to a search

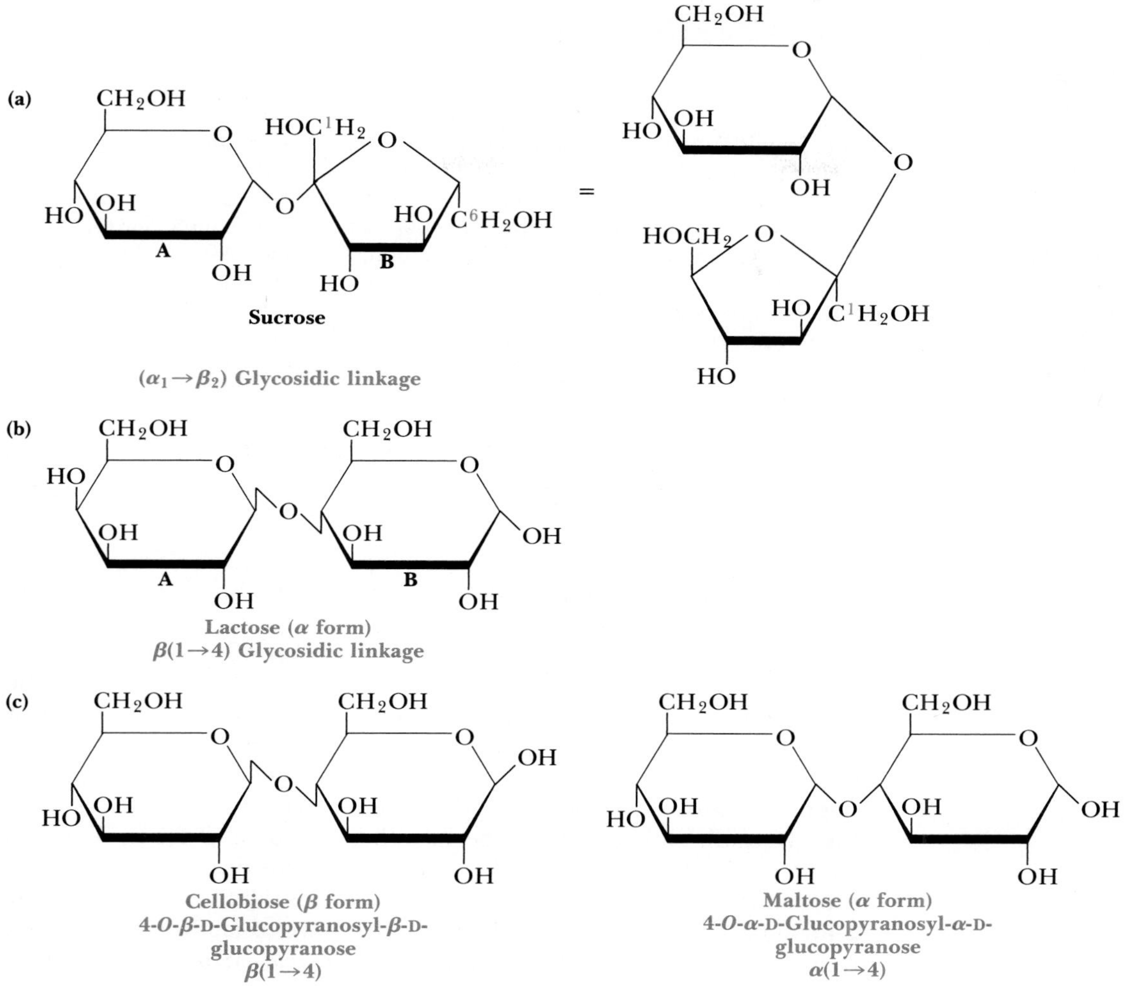

**FIGURE 6.18** Structures of some disaccharides. (a) In sucrose, both anomeric carbons are involved in the glycosidic linkage; no free carbonyl (C=O) group can be obtained from ring opening. As a result, sucrose is not a reducing sugar. (b) In lactose, the anomeric carbon of glucose is not involved in the glycosidic bond. (α form refers to this free anomeric end.) A free carbonyl group can be obtained from ring opening. Lactose is a reducing sugar. (c) Cellobiose and maltose. (β form and α form refer to the free anomeric end.)

for other sweetening agents. One such sweetener that has been proposed is fructose itself, which is sweeter than sucrose. Because of the difference, a smaller amount (by weight) of fructose than sucrose can produce the same sweetening effect with fewer "calories" and none of the apparent harmful side effects that are suspected with artificial sweeteners. Saccharin, for example, has been found to cause cancer in laboratory animals; aspartame (NutraSweet, Section 5.5) has been suspected of causing neurological

## BOX 6.1
## LACTOSE INTOLERANCE

Lactose is sometimes referred to as milk sugar because it occurs in milk. Humans can be intolerant of milk and milk products for several reasons. In some adults a deficiency of the enzyme lactase (which degrades lactose to galactose and glucose) causes a buildup in the level of the disaccharide. The enzyme normally occurs in the intestinal villi; without it, the disaccharide is not broken down to the component monosaccharides and absorbed into the bloodstream. Lactose can be acted on by the lactase of intestinal bacteria, producing hydrogen gas, carbon dioxide, and organic acids. The products of the bacterial

lactase reaction lead to digestive problems such as bloating and diarrhea, as does the presence of undegraded lactose. This disorder affects only about one-tenth of the white population of the United States, but it occurs more commonly among African-Americans, Asians, Native Americans, and Hispanics. Even if lactose can be broken down by the body, other problems can occur, since the galactose must be isomerized to glucose if it is to enter the usual metabolic pathways. If there is a buildup of galactose, a condition known as galactosemia can result; this problem is especially serious in infants and can lead to mental retardation.

problems, especially in individuals whose metabolism cannot tolerate phenylalanine (Section 18.8).

**Lactose** (Box 6.1) is a disaccharide made up of $\beta$-D-galactose and D-glucose. Galactose is an epimer of glucose. In other words, the difference between glucose and galactose is inversion of configuration at carbon C-4. The glycosidic linkage is $\beta(1\longrightarrow4)$, between the anomeric carbon C-1 of the $\beta$ form of galactose and the C-4 carbon of glucose (Figure 6.18b). Since the anomeric carbon of glucose is not involved in the glycosidic linkage, it can be in either the $\alpha$ or $\beta$ form. The two anomeric forms of lactose can be specified, and the designation refers to the glucose residue; galactose must be present as the $\beta$ anomer, since the $\beta$ form of galactose is required by the structure of lactose. Lactose is a reducing sugar because the group at the anomeric carbon of the glucose portion is not involved in glycosidic linkage, so it is free to react with oxidizing agents.

**Maltose** is a disaccharide obtained from the hydrolysis of starch. It consists of two residues of D-glucose in an $\alpha(1\longrightarrow4)$ linkage. Maltose differs from **cellobiose,** a disaccharide that is obtained from the hydrolysis of cellulose, only in the glycosidic linkage. In cellobiose the two residues of D-glucose are bonded together in a $\beta(1\longrightarrow4)$ linkage (Figure 6.18c). Maltose can be digested by mammals, but cellobiose is indigestible to them.

### 6.4
### POLYSACCHARIDES

Polysaccharides that occur in organisms are usually composed of a very few types of monosaccharide components. A polymer that consists of only one type of monosaccharide is a **homopolysaccharide;** a polymer which

CH₂OH ... CH₂OH ... CH₂OH ... CH₂OH

β(1→4)   β(1→4)

Repeating disaccharide
in cellulose
(β-cellobiose)

FIGURE 6.19   The polymeric structure of cellulose. β-Cellobiose is the repeating disaccharide.

consists of more than one type of monosaccharide is a **heteropolysaccharide.** Glucose is the most common monomer; when there is more than one type of monomer, there are frequently only two types of molecules in a repeating sequence. A complete characterization of a polysaccharide includes specifying which monomers are present, and if necessary the sequence of monomers. It also requires that the type of glycosidic linkage be specified. We shall see the importance of the type of glycosidic linkage as we discuss different polysaccharides.

## Cellulose

Cellulose is the major structural component of plants, especially of wood and plant fibers. It is a linear homopolysaccharide of β-D-glucose, and all residues are linked in β(1⟶4) glycosidic bonds (Figure 6.19). Individual polysaccharide chains are hydrogen bonded together, leading to the mechanical strength of plant fibers. Animals do not have enzymes, called cellulases, that hydrolyze cellulose to glucose. Such enzymes

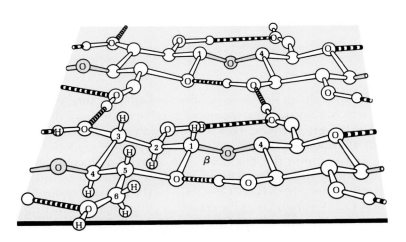

The monomer of cellulose is the β anomer of glucose, giving rise to long chains that can hydrogen bond to one another.

The monomer of starch is the $\alpha$ anomer of glucose, giving rise to a chain that folds into a helical form.

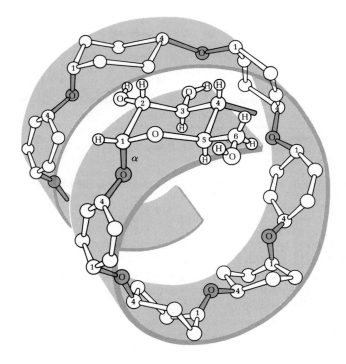

attack the $\beta$-linkage, which is common to structural polymers; the $\alpha$-linkage, which animals can digest, is characteristic of energy-storage polymers such as starch. Cellulases are found in bacteria, including the bacteria that inhabit the digestive tracts of insects such as termites and of grazing animals such as cattle and horses. The presence of these bacteria explains why cows and horses can live on grass and hay while humans cannot do so. The damage that termites do to the wooden parts of buildings arises from their ability to use cellulose as a nutrient; the presence of suitable bacteria in the digestive tract of the termite allows the insect to use wood as a food source.

## Chitin

A polysaccharide that is similar to cellulose in both structure and function is chitin, which is also a linear homopolysaccharide with all the residues linked in $\beta(1\longrightarrow4)$ glycosidic bonds. Chitin differs from cellulose in the nature of the monosaccharide unit; in cellulose the monomer is $\beta$-D-glucose, and in chitin the monomer is $N$-acetyl-$\beta$-D-glucosamine. The latter compound differs from glucose only by the substitution of the $N$-acetylamino group (—NH—CO—CH$_3$) for the hydroxyl group (—OH) on carbon C-2 (Figure 6.20). Like cellulose, chitin plays a structural role and has a fair amount of mechanical strength because the individual strands are held together by hydrogen bonds. It is a major structural component of the exoskeletons of invertebrates such as insects and

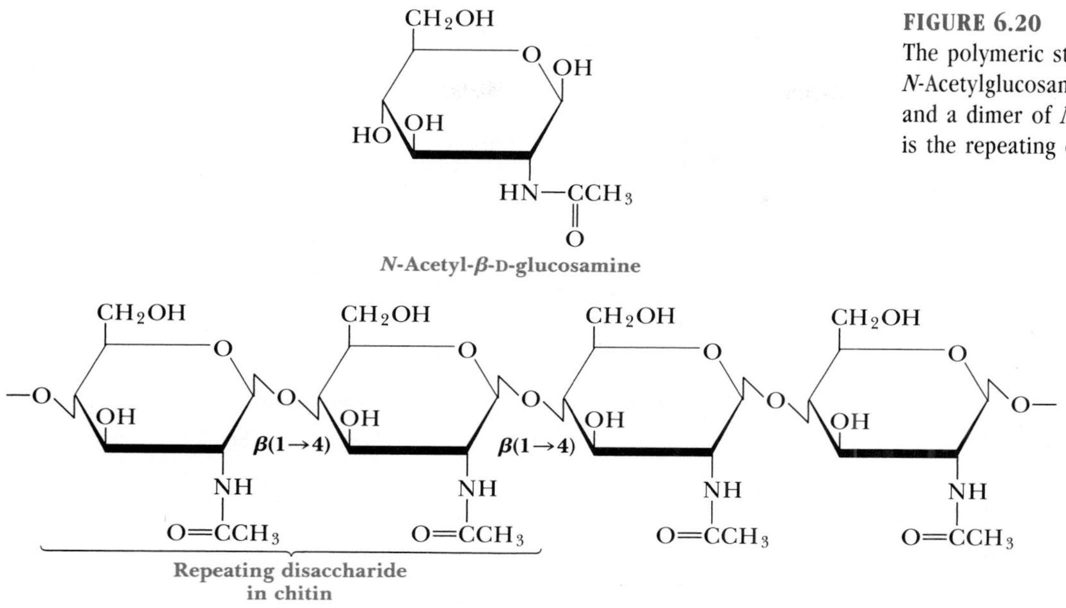

**FIGURE 6.20**
The polymeric structure of chitin. *N*-Acetylglucosamine is the monomer, and a dimer of *N*-acetylglucosamine is the repeating disaccharide.

*N*-Acetyl-*β*-D-glucosamine

*β*(1→4)    *β*(1→4)

Repeating disaccharide in chitin

crustaceans (a group that includes lobsters and shrimp), and it also occurs in cell walls of algae, fungi, and yeasts.

## Other Polysaccharides That Play a Structural Role in Organisms

Several heteropolysaccharides occur in the connective tissue of animals; some of the most important are hyaluronic acid, chondroitin sulfate, and heparin. Many of these polymers have a highly viscous, gelatinous consistency in aqueous solution, and as a result they are given the name **mucopolysaccharides.** In many cases the mucopolysaccharides serve a lubricating function in connective tissue.

**Hyaluronic acid** is a polymer that consists of repeating units of two derivatives of glucose, *N*-acetylglucosamine and glucuronic acid. The second type of monomer unit, glucuronic acid, is derived from glucose by oxidation of the hydroxyl group at the C-6 carbon of glucose (Figure 6.21). The monomer units are linked together by alternating *β*(1⟶3) and *β*(1⟶4) glycosidic bonds. Hyaluronic acid occurs in the lubricating fluid of joints.

**Chondroitin sulfate,** another mucopolysaccharide that occurs in connective tissue, is similar to hyaluronic acid in some respects. Both have alternating *β*(1⟶3) and *β*(1⟶4) glycosidic bonds within a repeating unit that consists of two monomers. Also one of the monomers, glucuronic acid, is the same in both polysaccharides. The second monomer in chondroitin sulfate is a derivative not of glucose but of galactose. Recall that galactose is an epimer of glucose, where the difference in configuration is at carbon C-4. The derivative of galactose in the repeating unit of chondroitin sulfate is *N*-acetyl-*β*-D-galactosamine with a sulfate group

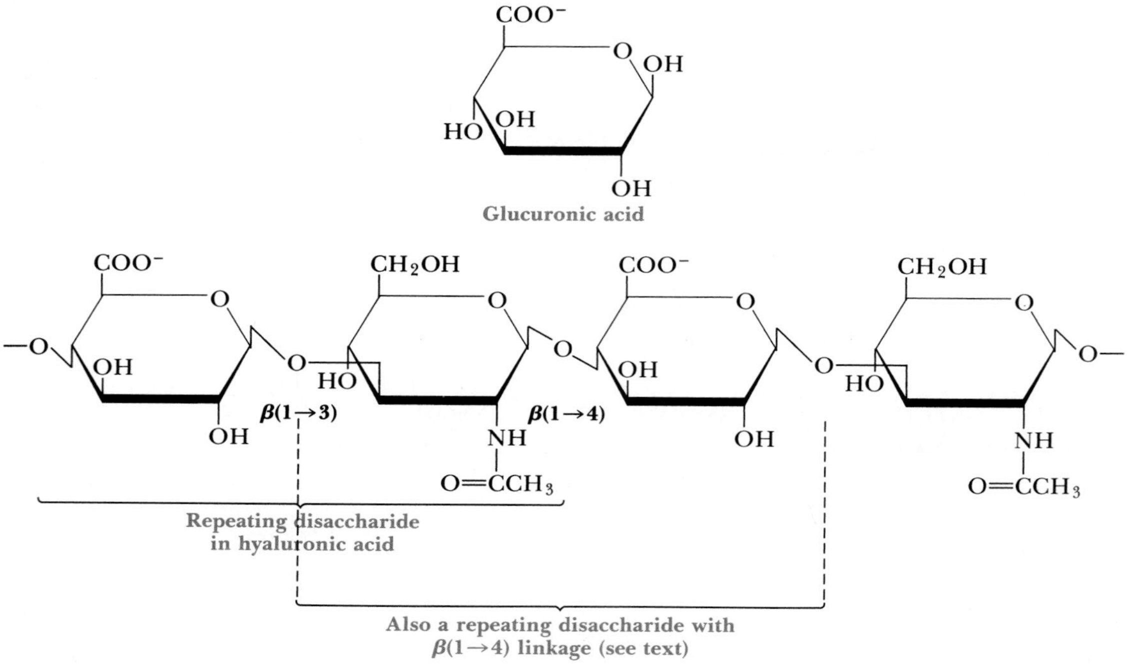

**FIGURE 6.21**   Hyaluronic acid is an alternating copolymer. The repeating disaccharide consists of glucuronic acid and N-acetylglucosamine joined by a β(1——➔3) linkage.

bonded to it (Figure 6.22). The sulfate group forms an ester bond to one of the hydroxyl groups of the N-acetylgalactosamine, usually at the C-6 carbon, but esterification at the C-4 and C-5 carbons is also known; the position and number of ester bonds depends on the source of the polysaccharide.

**Heparin** is best known for its anticoagulant properties, and it finds wide use in medical practice when it is necessary to prevent blood clotting. The structure of heparin is not completely understood, and it is probably a mixture of polysaccharides. Some possible structures of repeating units are shown in Figure 6.23.

All the mucopolysaccharides we have discussed are highly ionized at the essentially neutral pH values that occur in organisms. These ionic groups, all of which are negatively charged, are extensively hydrated and hydrogen bonded, resulting in a gelatinous texture. These interactions in turn lead to the occurrence of these polymers as cement in connective tissue. Their gelatinous nature can hold components together and can also serve the function of lubrication when they occur in the "ground structure" of connective tissue.

## The Role of Polysaccharides in the Structure of Cell Walls

In organisms that have cell walls, such as bacteria and plants, this feature of cellular structure consists largely of polysaccharides. There are some differences in the biochemistry of the cell walls of bacteria and plants, and

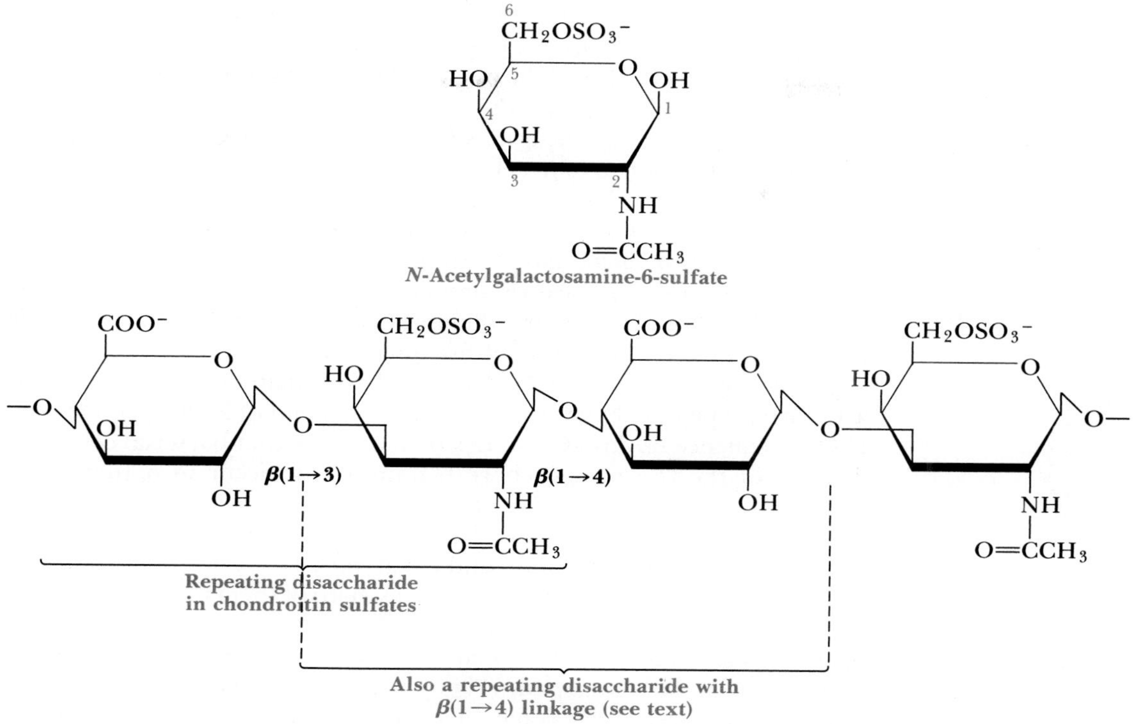

$N$-Acetylgalactosamine-6-sulfate

FIGURE 6.22   Chondroitin sulfate is an alternating copolymer. The repeating disaccharide consists of glucuronic acid and $N$-acetylgalactosamine sulfate joined by a $\beta(1\longrightarrow3)$ linkage.

these differences serve to emphasize the differences between prokaryotes and eukaryotes.

Heteropolysaccharides are major components of bacterial cell walls. A distinguishing feature of prokaryotic cell walls is that the polysaccharides are cross-linked by peptides. The repeating unit of the polysaccharide consists of two residues held together by $\beta(1\longrightarrow4)$ glycosidic links, as was the case in cellulose and chitin. One of the two monomers is $N$-acetyl-D-

FIGURE 6.23   Some possible repeating units found in heparin.

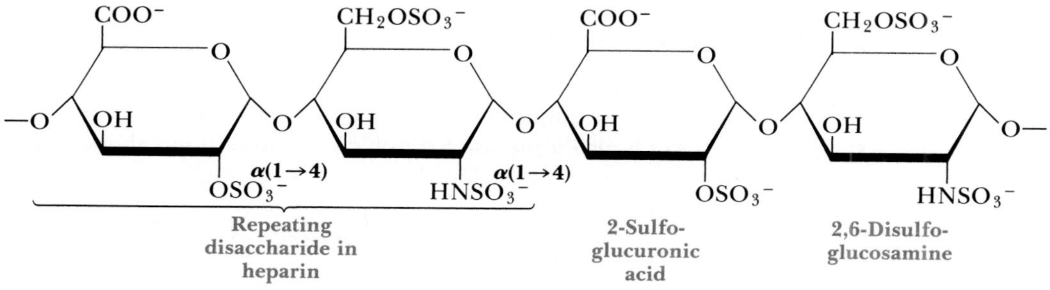

glucosamine, which occurs in chitin, and the other monomer is *N*-acetylmuramic acid (Figure 6.24a). The structure of *N*-acetylmuramic acid differs from that of *N*-acetylglucosamine by the substitution of a lactic acid side chain (—O—CH(CH$_3$)—COOH) for the hydroxyl group (—OH) on carbon 3. *N*-Acetylmuramic acid is found only in prokaryotic cell walls; it does not occur in eukaryotic cell walls.

The crosslinks in bacterial cell walls consist of small peptides. We shall use one of the best known examples as an illustration. In the cell wall of the bacterium *Staphylococcus aureus* an oligomer of four amino acids (a tetramer) is bonded to *N*-acetylmuramic acid, forming a side chain (Figure 6.24b). The tetrapeptides are themselves crosslinked by another small peptide, in this case consisting of five amino acids.

The carboxyl group of the lactic acid side chain of *N*-acetylmuramic acid forms a peptide bond with the N-terminal end of a tetrapeptide that has the sequence L-Ala-D-Gln-L-Lys-D-Ala. Recall that bacterial cell walls are one of the few places where D-amino acids occur in nature. The occurrence of D-amino acids and *N*-acetylmuramic acid in bacterial cell walls but not in plant cell walls shows a biochemical as well as structural difference between prokaryotes and eukaryotes.

The tetrapeptide forms two crosslinks, both of them to a pentapeptide that consists of five glycine residues, (Gly)$_5$. The glycine pentamers form peptide bonds to the C-terminal end and to the side chain $\epsilon$-amino group of the lysine in the tetrapeptide (Figure 6.24c). This extensive crosslinking produces a three-dimensional network of considerable mechanical strength, which is why bacterial cell walls are extremely difficult to disrupt. The material resulting from the crosslinking of polysaccharides by peptides is called a **peptidoglycan** (Figure 6.24d), so named because it has both peptide and carbohydrate components.

Plant cell walls consist largely, but not exclusively, of **cellulose.** The other important polysaccharide component found in plant cell walls is **pectin,** a polymer made up mostly of D-galacturonic acid, a derivative of galactose that has the same relationship to the parent sugar as glucuronic acid has to glucose (Figure 6.25). Pectin is extracted from plants because it has commercial importance in the food processing industry as a gelling agent in fruit preserves, jams, and jellies. The major nonpolysaccharide component in plant cell walls, especially in woody plants, is **lignin** (Latin, *lignum,* "wood"). Lignin is a polymer of coniferyl alcohol, and it is a very tough and durable material (Figure 6.26). Unlike bacterial cell walls, plant cell walls contain comparatively little peptide or protein.

## The Various Forms of Starch

All the various forms of polysaccharides we have discussed so far have a structural role, but the importance of carbohydrates as energy sources suggests that there is also some use for polysaccharides in metabolism. We shall now discuss some polysaccharides, such as starches, that serve as vehicles for storage of glucose.

Starches are polymers of $\alpha$-D-glucose that occur in plant cells, usually as starch granules in the cytosol. Note that there is an $\alpha$-linkage in

**(a)**

CH$_2$OH     CH$_2$OH

H$_3$C
CH
COO$^-$     CH$_3$     CH$_3$

NAM
(*N*-Acetylmuramic acid)     NAG
(*N*-Acetylglucosamine)

**(b)**

CH$_2$OH     CH$_2$OH

H$_3$C
CH
C=O
H—N
L-Ala
D-Gln
L-Lys—$\epsilon$-NH$_3{}^+$
D-Ala
C=O
O$^-$

**(c)**

CH$_2$OH     CH$_2$OH

H$_3$C
CH
C=O
H—N
L-Ala
D-Gln
L-Lys—$\epsilon$-NH—C—(Gly)$_5$—NH—   To tetrapeptide side chain
D-Ala
C=O
H—N
(Gly)$_5$
C=O

To tetrapeptide
side chain

**(d)**

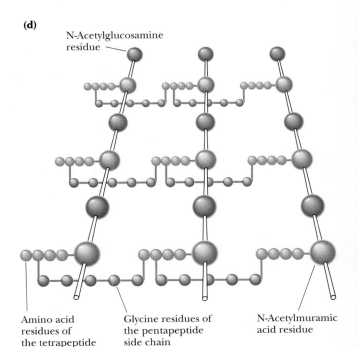

N-Acetylglucosamine residue

Amino acid residues of the tetrapeptide side chain

Glycine residues of the pentapeptide side chain

N-Acetylmuramic acid residue

**FIGURE 6.24** The structure of the peptidoglycan of the *Staphylococcus aureus* cell wall. (a) The repeating disaccharide. (b) The repeating disaccharide with the tetrapeptide side chain (shown in color). (c) Adding the pentaglycine cross-links (shown in color). (d) Schematic diagram of the peptidoglycan. The sugars are the larger circles. The triangles are the amino acid residues of the tetrapeptide, and the smaller circles are the glycine residues of the pentapeptide.

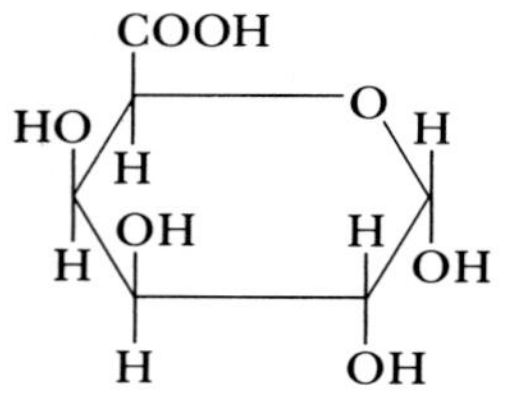

FIGURE 6.25   The structure of
D-galacturonic acid.

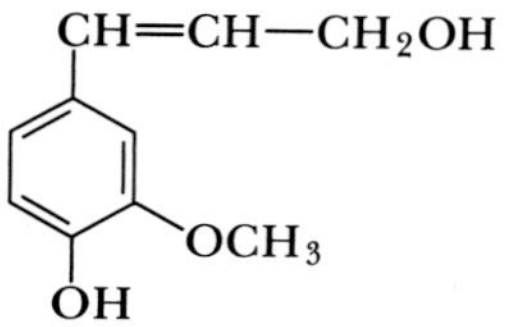

Coniferyl alcohol

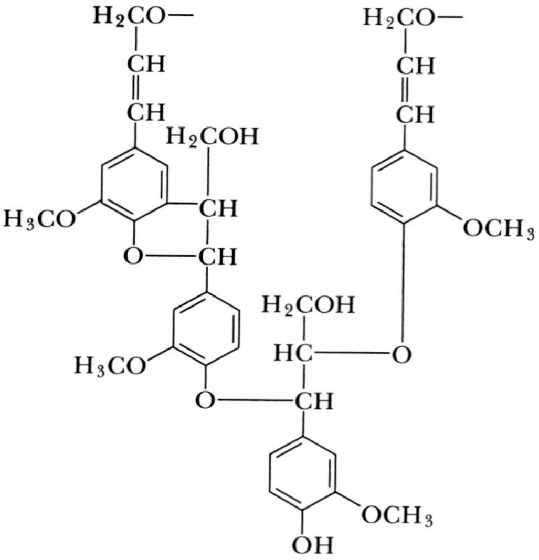

Lignin

FIGURE 6.26   The structure of lignin. Lignin is a polymer of coniferyl alcohol.

starch, in contrast to the $\beta$-linkage in cellulose. The different types of starches can be distinguished from one another by their degree of chain branching. Amylose is a linear polymer of glucose, with all the residues linked together by $\alpha(1\longrightarrow4)$ bonds. Amylopectin is a branched chain polymer, with the branches starting at $\alpha(1\longrightarrow6)$ linkages along the chain of $\alpha(1\longrightarrow4)$ linkages. Refer to Figure 6.15 for the detailed structures involving these two kinds of glycosidic linkages. The most usual conformation of amylose is a helix with six residues per turn. Iodine molecules can fit inside the helix to form a starch-iodine complex, which has a characteristic dark-blue color (Figure 6.27). The formation of this complex is a well-known test for the presence of starch. If there is a preferred conformation for amylopectin, it is not known at this time.

Since starches are storage molecules, there must be a mechanism for releasing glucose from starch when the organism needs energy. Both plants and animals contain enzymes that hydrolyze starches. Two of these enzymes, known as $\alpha$- and $\beta$-amylase (the $\alpha$ and $\beta$ designations do not refer to anomeric forms in this case), attack $\alpha(1\longrightarrow4)$ linkages. One of these enzymes, $\beta$-amylase, is an **exoglycosidase** that cleaves from the nonreducing end of the polymer. Maltose, a dimer of glucose, is the product of reaction. The other enzyme, $\alpha$-amylase, is an **endoglycosidase,** which can hydrolyze a glycosidic linkage anywhere along the chain to produce glucose and maltose as products. Amylose can be completely degraded to glucose and maltose by the two amylases, but amylopectin is not completely degraded because the branching linkages are not attacked. There are, however, **debranching enzymes** that occur in both plants and animals; they degrade the $\alpha(1\longrightarrow6)$ linkages. When these enzymes are

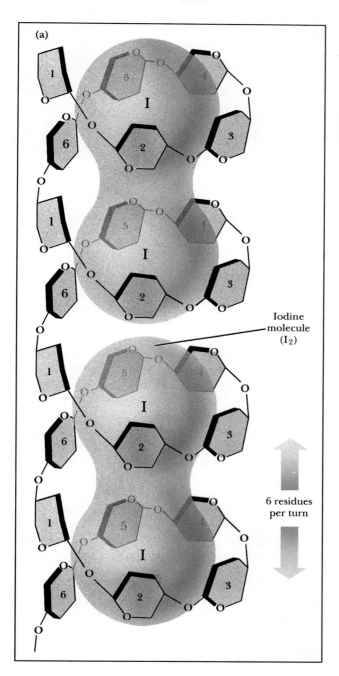

**(a)**

Iodine molecule (I₂)

6 residues per turn

FIGURE 6.27 The starch–iodine complex. Amylose occurs as a helix with six residues per turn. In the starch–iodine complex, the iodine molecules are parallel to the long axis of the helix. Four turns of the helix are shown here. Six turns of the helix, containing 36 glycosyl residues, are required to give the characteristic blue color of the complex.

**FIGURE 6.28**  A comparison of the degree of branching in starch amylopectin and glycogen amylopectin.

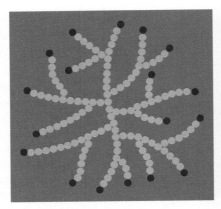

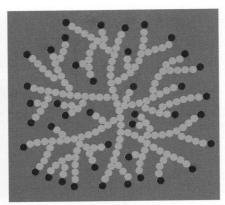

Starch amylopectin                    Glycogen amylopectin

combined with the amylases, they contribute to the complete degradation of both forms of starch.

## Glycogen

Although starches occur only in plants, in animals there is a similar carbohydrate storage polymer called glycogen. Glycogen is a branched-chain polymer of $\alpha$-D-glucose, and in this respect it is similar to the amylopectin fraction of starch. Like amylopectin, glycogen consists of a chain of $\alpha(1\longrightarrow4)$ linkages with $\alpha(1\longrightarrow6)$ linkages at the branch points. The main difference between glycogen and amylopectin is that glycogen is more highly branched (Figure 6.28). The branch points in glycogen occur about every 10 residues; in amylopectin the branch points occur about every 25 residues. Glycogen is found in animal cells in granules similar to the starch granules in plant cells. When the organism needs energy, various degradative enzymes remove glucose units. **Glycogen phosphorylase** is one such enzyme; it cleaves one glucose at a time from the nonreducing end of a branch to produce glucose-1-phosphate, which then enters the metabolic pathways of carbohydrate breakdown. De-branching enzymes also play a role in the complete breakdown of glycogen. Some athletes, particularly long-distance runners, try to build up their glycogen reserves before a race by eating large amounts of carbohydrates.

## 6.5
## GLYCOPROTEINS

Glycoproteins contain carbohydrate residues in addition to the polypeptide chain. Some of the most important examples of glycoproteins are involved in the immune response; **antibodies,** which bind to and immobilize antigens (the substances attacking the organism), are glycoproteins. Carbo-

---

## BOX 6.2
## GLYCOPROTEINS, BLOOD TRANSFUSIONS, AND CELL-CELL RECOGNITION

If a transfusion is attempted with incompatible blood types, such as blood from a type A donor given to a type B recipient, an antigen–antibody reaction takes place because the type B recipient has antibodies to the type A blood. The characteristic oligosaccharide residues of type A blood cells serve as the antigen. A crosslinking reaction takes place between antigens and antibodies, and the blood cells clump together. Antibodies to type B blood produce the same result in the case of a transfusion of type B blood to a type A recipient. Type O blood does not have either antigenic determinant, and so people with type O blood are considered universal donors. Type O persons, however, have antibodies to both type A and type B blood,

and they are not universal acceptors. Type AB persons have both antigenic determinants, and, as a result, they do not produce either type of antibody; the universal acceptors are the type AB persons.

The behavior of blood-group determinants is not the only example of the role of glycoproteins in cell-surface recognition and interaction. Cell surfaces in general are coated with glycoproteins. In normal cells, growth stops when cells touch each other. This phenomenon is known as **contact inhibition.** In cancer cells there is no contact inhibition, and cancer cells proliferate wildly. It is also known that the cell-surface glycoproteins of cancer cells differ from those of normal cells, and this point is the subject of active investigation.

---

hydrates also play an important role as **antigenic determinants,** which are the portions of an antigenic molecule that antibodies recognize and to which they bind.

An example of the role of the oligosaccharide portion of glycoproteins as antigenic determinants is found in human blood groups. There are four human blood groups, A, B, AB, and O (see Box 6.2). The distinction between the four groups depends on the oligosaccharide portion of the glycoproteins on the surface of the blood cells called erythrocytes. In all blood types the oligosaccharide portion of the molecule contains the sugar L-fucose, which we saw earlier in this chapter as an example of a deoxy sugar. N-Acetylgalactosamine is found at the nonreducing end of the oligosaccharide in the type A blood-group antigen. In type B blood, α-D-galactose takes the place of N-acetylgalactosamine. In type O blood neither of these terminal residues is present, and in type AB blood both kinds of oligosaccharide are present (Figure 6.29).

**β-N-Acetylgalactosamine** (1→3) **β-Galactose** (1→3) **β-N-Acetylgalactosamine**

$$\uparrow^2_1$$

**Nonreducing end**        α-L-Fucose

**A blood group antigen**

**α-Galactose·**(1→3) **β-Galactose** (1→3) **β-N-Acetylgalactosamine**

$$\uparrow^2_1$$

**Nonreducing end**    α-L-Fucose

**B blood group antigen**

**FIGURE 6.29** The structures of the blood group antigenic determinants.

## SUMMARY

The simplest examples of carbohydrates are monosaccharides, compounds that contain a single carbonyl group and two or more hydroxyl groups. Monosaccharides frequently encountered in biochemistry are sugars that contain from three to seven carbon atoms. Sugars contain one or more chiral centers; the configurations of the possible stereoisomers can be shown by Fischer projection formulas. Sugars frequently exist as cyclic molecules rather than in open-chain form. Haworth projection formulas give a more realistic representation of the cyclic forms of sugars than do Fischer projection formulas. Many stereoisomers are possible for five- and six-carbon sugars, but only a few of the possibilities are frequently encountered in nature. Monosaccharides can undergo various reactions, including oxidation and esterification, but the most important reaction by far is the formation of glycosidic linkages, which give rise to oligosaccharides and polysaccharides.

Three important examples of oligosaccharides are the disaccharides sucrose, lactose, and maltose. Sucrose is common table sugar, lactose occurs in milk, and maltose is obtained by the hydrolysis of starch. In polysaccharides the repeating unit of the polymer is frequently limited to one or two kinds of monomer. Cellulose and chitin are polymers based on one kind of monomer unit, glucose and N-acetylglucosamine, respectively; both polymers play a structural role in organisms. Structural polysaccharides that contain more than one kind of monomer unit include hyaluronic acid and chondroitin sulfate. Starch, found in plants, and glycogen, which occurs in animals, are energy-storage polymers of glucose. They differ from each other in the degree of branching in the polymer structure, and from cellulose in the stereochemistry of the glycosidic linkage between monomers. In glycoproteins, carbohydrate residues are covalently linked to the polypeptide chain; glycoproteins play a role in the recognition site of antigens.

## EXERCISES

1. Pectin, which occurs in plant-cell walls, exists in nature as a polymer of D-galacturonic acid methylated at carbon atom number 6 of the monomer. Draw a Haworth projection for a repeating disaccharide unit of pectin with one methylated and one unmethylated monomer unit in $\alpha(1\longrightarrow4)$ linkage. Is the methyl group attached by an ester or an ether linkage?
2. Draw a Haworth projection for the disaccharide gentibiose, given the following information:
   (a) it is a dimer of glucose,
   (b) the glycosidic linkage is $\beta(1\longrightarrow6)$,
   (c) the anomeric carbon not involved in the glycosidic linkage is in the $\alpha$ configuration.
3. An amylose chain is 5000 glucose units long. At how many places must it be cleaved to reduce the average length to 2500 units? To 1000? To 200? What percentage of the glycosidic links is hydrolyzed in each case? (Even partial hydrolysis can drastically alter the physical properties of polysaccharides and thus affect their structural role in organisms.)
4. Suppose that a polymer of glucose with alternating $\alpha(1\longrightarrow4)$ and $\beta(1\longrightarrow4)$ glycosidic linkages has just been discovered. Draw a Haworth projection for a repeating tetramer (two repeating dimers) of such a polysaccharide. Would you expect this polymer to have primarily a structural role or an energy storage role in organisms? What sort of organisms, if any, could use this polysaccharide as a food source?
5. Define the following terms: polysaccharide, furanose, pyranose, aldose, ketose, glycosidic bond, oligosaccharide, glycoprotein.
6. Which of the following, if any, are epimers of D-glucose: D-mannose, D-galactose, D-ribose?
7. Is any of the following groups NOT an aldose-ketose pair: D-ribose and D-ribulose, D-glucose and D-fructose, D-glyceraldehyde and dihydroxyacetone?
8. What is the metabolic basis for the observation that many adults cannot ingest large quantities of milk without gastric difficulties?
9. Fischer projections are shown on p.125 for a group of five-carbon sugars, all of which are aldopentoses. Identify the pairs that are enantiomers and the pairs that are epimers. (The sugars shown here are not all the possible five-carbon sugars.)

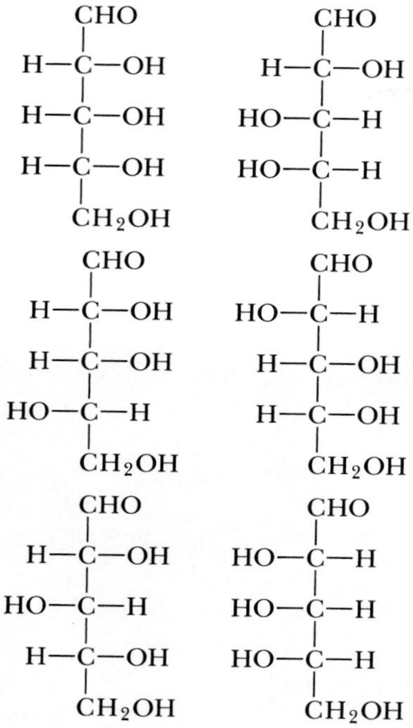

```
      CHO                CHO
  H—C—OH            H—C—OH
  H—C—OH           HO—C—H
  H—C—OH           HO—C—H
    CH₂OH              CH₂OH

      CHO                CHO
  H—C—OH           HO—C—H
  H—C—OH            H—C—OH
 HO—C—H             H—C—OH
    CH₂OH              CH₂OH

      CHO                CHO
  H—C—OH           HO—C—H
 HO—C—H            HO—C—H
  H—C—OH           HO—C—H
    CH₂OH              CH₂OH
```

10. Draw Haworth projection formulas for dimers of glucose with the following types of glycosidic linkages:
    (a) $\beta(1\longrightarrow4)$ linkage (both molecules of glucose in the $\beta$ form)
    (b) an $\alpha,\alpha(1\longrightarrow1)$ linkage
    (c) a $\beta(1\longrightarrow6)$ linkage (both molecules of glucose in the $\beta$ form)

11. What are some of the main differences between the cell walls of plants and of bacteria?

12. How does chitin differ from cellulose in structure and function?

13. How does glycogen differ from starch in structure and function?

14. How do the enzymes $\alpha$-amylase and $\beta$-amylase differ from one another?

15. Briefly indicate the role of glycoproteins as antigenic determinants for blood groups.

## ANNOTATED BIBLIOGRAPHY

Aspinall, G.O., ed. *The Polysaccharides.* 2 vols. New York: Academic Press, 1982. [Good coverage of topics of current interest in carbohydrate chemistry.]

Sharon, N. Carbohydrates. *Sci. Amer.* **243** (5), 90–102 (1980). [A good overview of structures.]

Most organic chemistry textbooks have one or more chapters on the structure and reactions of carbohydrates.

# CHAPTER

## 7

# Lipids

## OUTLINE

*A cholesterol rainbow formed by adding napthalene to cholesterol and heating the mixture.*

*The most striking features of lipids are their long oily hydrocarbon chains, which are insoluble in water. As fatty acids, lipids contain a carboxyl head group attached to a hydrocarbon "tail." With three long-chain fatty acids, the triacylglycerols (also referred to as fats) are ideal reservoirs for energy storage in the cell. Some lipids have large charged polar heads in addition to their uncharged hydrocarbon tails. The chief ingredients of biological membranes are the phospholipids. In water, they spontaneously form lipid bilayers with their flexible tails in the hydrophobic interior of the membrane and their polar heads on exterior surfaces in contact with water. About half of the membrane consists of protein molecules associated with the lipid bilayer. As fat, the lipids can store about six times the energy of carbohydrates. While we can survive less than a day on the body's reserve of carbohydrates in the form of glycogen and glucose, we might survive without food for two or three months on our reserve store of fat. W.C. Fields, the late great comedian (and dedicated alcoholic) once complained, "I have nothing to live on but food and water." Even without food, the corpulent actor might have lasted several months.*

## 7.1
## THE DEFINITION OF A LIPID

The definition of a lipid is based on solubility. Lipids are insoluble in water and soluble in nonpolar organic solvents such as chloroform or acetone. Fats and oils are typical lipids in terms of their solubility, but that statement does not really define their chemical nature. In terms of chemistry, lipids are a mixed bag of compounds that share some properties on the basis of structural similarities, mainly a preponderance of nonpolar groups.

There are various types of lipids, and the classification depends on their chemical nature. According to this classification scheme, the lipids fall into two main groups. One group consists of open-chain compounds with a polar head group and a long nonpolar tail; this group includes **fatty acids, triacylglycerols, sphingolipids, phosphoacylglycerols,** and **glycolipids.** The second major group consists of fused-ring compounds, the **steroids;** an important representative of this group is cholesterol.

## 7.2
## THE CHEMICAL NATURE OF VARIOUS TYPES OF LIPIDS

### Fatty Acids

A fatty acid has a carboxyl group at the polar end, while the nonpolar tail of the molecule consists of a hydrocarbon chain. Fatty acids are amphiphilic compounds; in other words, each molecule has a hydrophilic, polar part

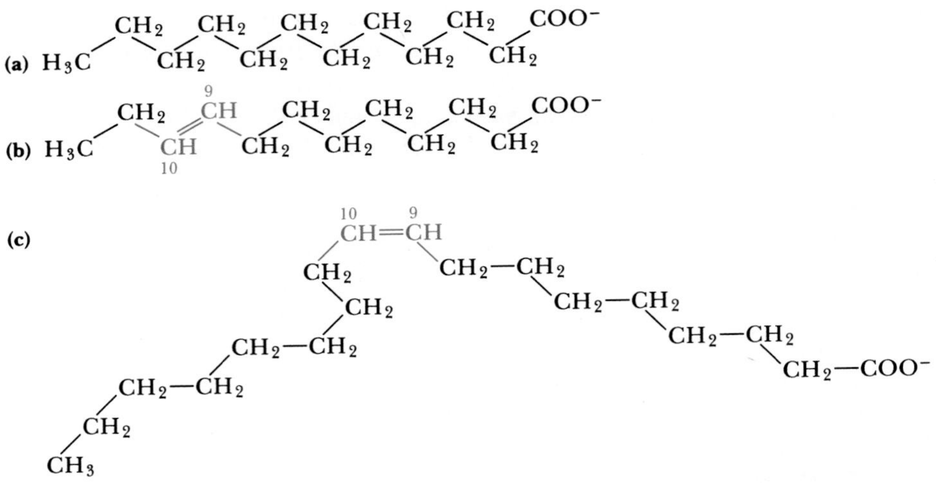

**FIGURE 7.1** Structures of representative fatty acids. (a) Dodecanoate, the ionized form of the saturated fatty acid dodecanoic acid (orlauric acid). (b) Dodecenoate, the ionized form of the unsaturated fatty acid, dodecenoic acid, showing the effect of a *trans* double bond. (c) Oleate, the ionized form of oleic acid, another unsaturated fatty acid. Note that the *cis* double bond introduces a kink in the hydrocarbon chain. The double bonds in both unsaturated fatty acids are at the ninth carbon atom from the carboxyl end.

(the carboxyl group) and a hydrophobic, nonpolar part (the hydrocarbon tail). The carboxyl group can ionize under the proper conditions.

Fatty acids that occur in living systems normally contain an even number of carbon atoms, and the hydrocarbon chain is usually unbranched (Figure 7.1). If there are carbon–carbon double bonds in the chain, the fatty acid is **unsaturated;** if there are only single bonds, the fatty acid is **saturated.** A few examples of the two classes are listed in Tables 7.1 and 7.2. In unsaturated fatty acids the stereochemistry at the double bond is usually *cis* rather than *trans* (Figure 7.1c). Note that the double bonds are isolated from one another by several singly-bonded carbons; fatty acids do not have conjugated double bond systems. The notation used for fatty acids indicates the number of carbon atoms and the number of double bonds. In this system the notation 18:0 refers to an 18-carbon saturated fatty acid (no

**TABLE 7.1  Typical Naturally Occurring Saturated Fatty Acids**

| ACID | NUMBER OF CARBON ATOMS | FORMULA |
|---|---|---|
| Lauric | 12 | $CH_3(CH_2)_{10}CO_2H$ |
| Myristic | 14 | $CH_3(CH_2)_{12}CO_2H$ |
| Palmitic | 16 | $CH_3(CH_2)_{14}CO_2H$ |
| Stearic | 18 | $CH_3(CH_2)_{16}CO_2H$ |
| Arachidic | 20 | $CH_3(CH_2)_{18}CO_2H$ |

**TABLE 7.2    Typical Naturally Occurring Unsaturated Fatty Acids**

| ACID | NUMBER OF CARBON ATOMS | DEGREE OF UNSATURATION* | FORMULA |
|------|------------------------|--------------------------|---------|
| Palmitoleic | 16 | 16:1 | $CH_3(CH_2)_5CH{=}CH(CH_2)_7CO_2H$ |
| Oleic | 18 | 18:1 | $CH_3(CH_2)_7CH{=}CH(CH_2)_7CO_2H$ |
| Linoleic | 18 | 18:2 | $CH_3(CH_2)_4CH{=}CHCH_2CH{=}CH(CH_2)_7CO_2H$ |
| Linolenic | 18 | 18:3 | $CH_3(CH_2CH{=}CH)_3(CH_2)_7CO_2H$ |
| Arachidonic | 20 | 20:4 | $CH_3(CH{=}CHCH_2)_4(CH_2)_2CO_2H$ |

*Degree of unsaturation refers to the number of double bonds.

double bonds), while the notation 18:1 refers to an 18-carbon fatty acid with one double bond. Note that in the unsaturated fatty acids in Table 7.2 (except arachidonic acid) there is a double bond at the ninth carbon atom from the carboxyl end.

Fatty acids are rarely found free in nature, but they form a part of many commonly occurring lipids.

## Triacylglycerols

**Glycerol** is a simple compound that contains three hydroxyl groups (Figure 7.2). When all three of the alcohol groups form ester linkages with fatty acids, the resulting compound is a **triacylglycerol;** an older name for this type of compound is **triglyceride.** It is usual for three different fatty acids to be esterified to the alcohol groups of the same glycerol molecule. Triacylglycerols do not occur as components of membranes (as do other types of lipids), but they accumulate in adipose tissue (primarily fat cells) and provide a means of storing fatty acids, particularly in animals. Triacylglycerols do not play an important role in plants.

When fatty acids are put to use by an organism, the ester linkages of triacylglycerols are hydrolyzed by enzymes called **lipases.** The same hydrolysis reaction can take place outside organisms, using acids or bases as catalysts. When a base such as sodium or potassium hydroxide is used, the products of the reaction are glycerol and the sodium or potassium salts of the fatty acids. These salts are soaps, and the reaction is called **saponification** (Figure 7.3).

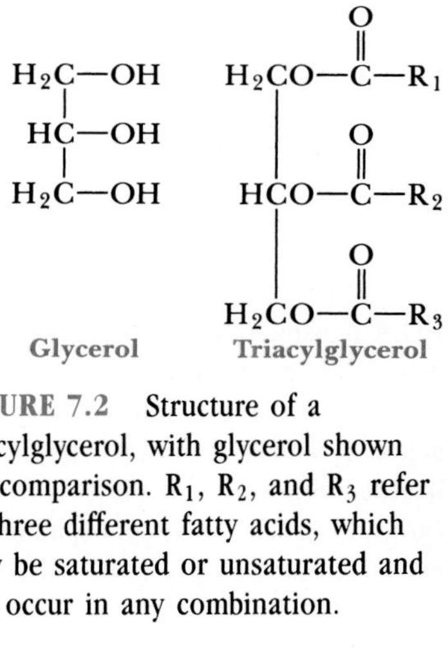

**FIGURE 7.2**  Structure of a triacylglycerol, with glycerol shown for comparison. $R_1$, $R_2$, and $R_3$ refer to three different fatty acids, which may be saturated or unsaturated and can occur in any combination.

## Phosphoacylglycerols (Phosphoglycerides)

It is possible for one of the alcohol groups of glycerol to be esterified by a phosphoric acid molecule rather than by a carboxylic acid. In such lipid molecules, there are also two fatty acids esterified to the glycerol molecule. The resulting compound is called a **phosphatidic acid** (Figure 7.4a). Fatty acids are usually monobasic acids with only one carboxyl group able to form an ester bond, but phosphoric acid is tribasic and thus able to form more than one ester linkage. One molecule of phosphoric acid can form

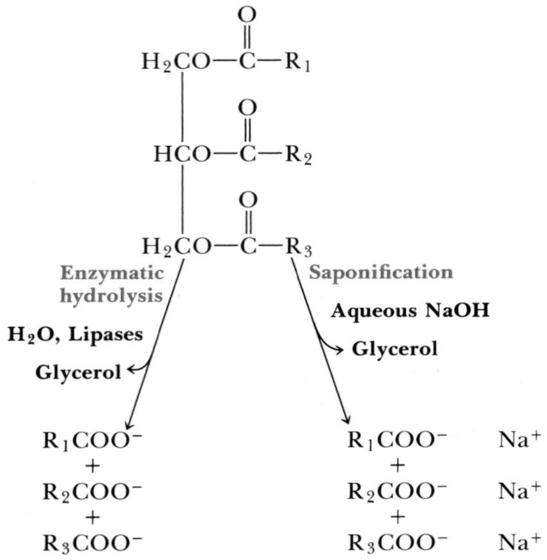

FIGURE 7.3 Hydrolysis of triacyglycerols. The term *saponification* refers to the reaction of a glyceryl ester with sodium or potassium hydroxide to produce a soap, which is the corresponding salt of the long-chain fatty acid.

ester bonds both to glycerol and to some other alcohol, creating a **phosphatidyl ester** (Figure 7.4b). Both phosphatidic acids and phosphatidyl esters are classed as **phosphoacylglycerols.** The nature of the fatty acids varies widely, as is the case with triacylglycerols; as a result, the names of the types of lipids (such as triacylglycerols and phosphoacylglycerols) that contain fatty acids must be considered generic names.

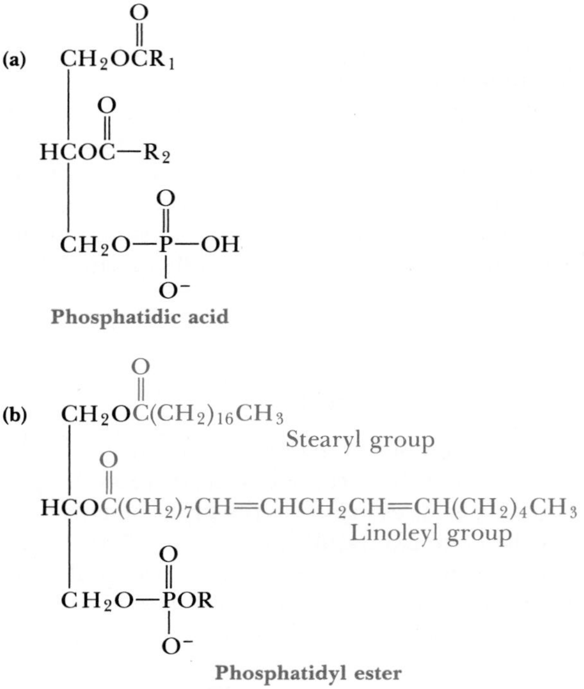

FIGURE 7.4 The molecular architecture of phosphoacylglycerols. (a) A phosphatidic acid, in which glycerol is esterified to phosphoric acid and to two different carboxylic acids. $R_1$ and $R_2$ represent the hydrocarbon chains of the two carboxylic acids. (b) A phosphatidyl ester (phosphoacylglycerol). Glycerol is esterified to two carboxylic acids, stearic acid and linoleic acid, as well as to phosphoric acid. Phosphoric acid in turn is esterified to a second alcohol, ROH.

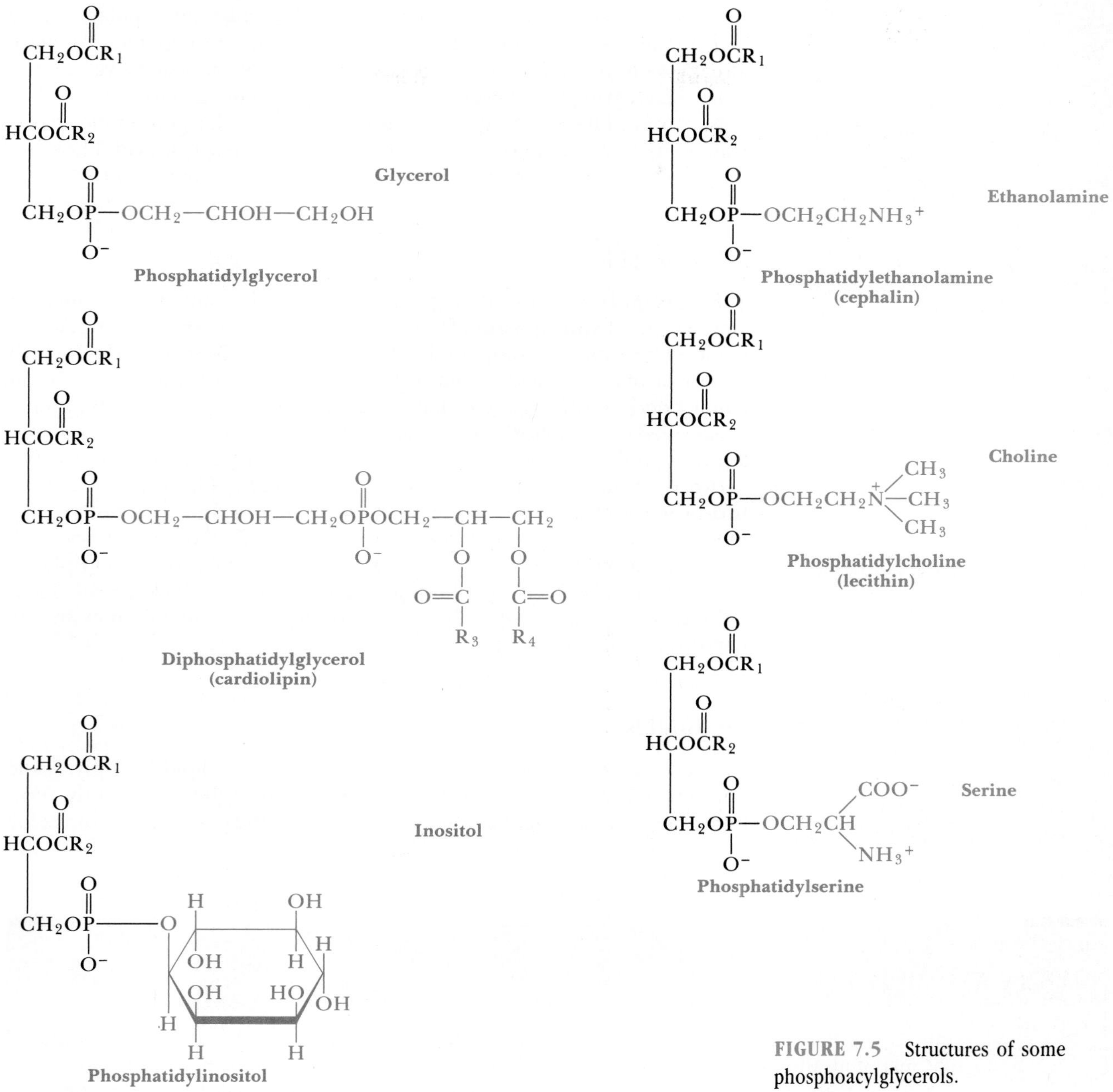

FIGURE 7.5   Structures of some phosphoacylglycerols.

The classification of a phosphatidyl ester depends on the nature of the second alcohol esterified to the phosphoric acid. Some of the most important lipids in this class are **phosphatidyl ethanolamine** (cephalin), **phosphatidyl serine, phosphatidyl choline** (lecithin), **phosphatidyl inositol, phosphatidyl glycerol,** and **diphosphatidyl glycerol** (cardiolipin) (Figure 7.5). In all these types of compounds there can be wide variation in the nature of the fatty acids in the molecule. All these compounds have long, nonpolar, hydrophobic tails and a polar, highly hydrophilic head group and thus are markedly amphiphilic. We have already seen this characteristic for fatty acids. In phosphoacylglycerols the polar head group is charged, since the phosphate group is ionized at neutral pH. There is frequently also a positively charged amino group contributed by an amino alcohol esterified to the phosporic acid. Phospho-acylglycerols are important components of biological membranes.

## Sphingolipids

**Sphingolipids** do not contain glycerol, but they do contain the long-chain amino alcohol **sphingosine,** from which this class of compounds takes its name. The simplest compounds of this class are the **ceramides,** which consist of one fatty acid linked to the amino group of sphingosine by an amide bond. In **sphingomyelins** the primary alcohol group of sphingosine is esterified to phosphoric acid, which in turn is esterified to another amino alcohol, choline (Figure 7.6). Note that there are structural similarities between sphingomyelin and other phospholipids. There are two long hydrocarbon chains attached to a backbone that contains alcohol groups. One of the alcohol groups of the backbone is esterified to phosphoric acid. A second alcohol, choline in this case, is also esterified to the phosphoric acid. Choline also occurs in phosphoacylglycerols, as we have already seen. Sphingomyelins are amphiphilic; they occur in cell membranes in the nervous system (Box 7.1).

## Glycolipids

If a carbohydrate is bound to an alcohol group of a lipid by a glycosidic linkage, the resulting compound is a **glycolipid.** Quite frequently cera-mides (a type of sphingolipid; see Figure 7.6) are the parent compounds for

## BOX 7.1
## MYELIN AND MULTIPLE SCLEROSIS

Myelin is the lipid-rich membrane sheath that surrounds the axons of nerve cells. It consists of many layers of plasma membrane that have been wrapped around the nerve cell. Unlike many other types of membranes (Section 11.3), myelin is essentially all lipid bilayer with only a small amount of embedded protein. The structure of myelin, consisting of segments with nodes separating them, promotes rapid transmission of nerve impulses from node to node. Loss of myelin leads to the slowing and eventual cessation of the nerve impulse.

The myelin sheath is progressively destroyed in **multiple sclerosis,** a crippling and eventually fatal disease. The brain and spinal cord are affected by **sclerotic plaques** that destroy the myelin sheath. The cause of these plaques is unknown, but there is some evidence for an autoimmune origin. The progress of the disease is marked by periods of active destruction of myelin interspersed with periods in which no destruction of myelin takes place. Persons affected by multiple sclerosis suffer from weakness, lack of coordination, and speech and vision problems.

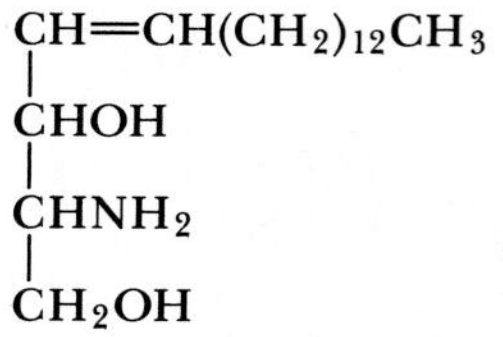

$$CH=CH(CH_2)_{12}CH_3$$
$$|$$
$$CHOH$$
$$|$$
$$CHNH_2$$
$$|$$
$$CH_2OH$$

**Sphingosine**

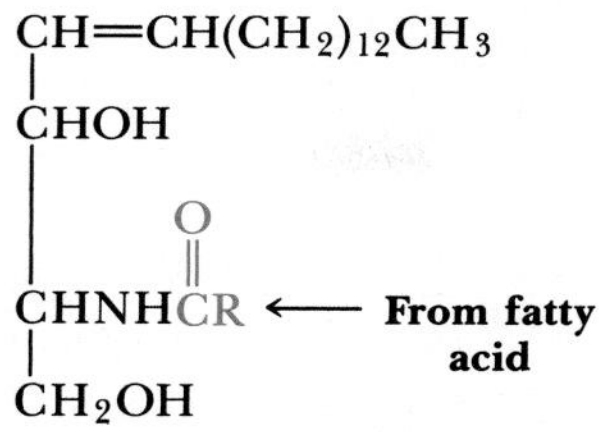

$$CH=CH(CH_2)_{12}CH_3$$
$$|$$
$$CHOH$$
$$|$$
$$\overset{O}{\overset{\|}{CHNHCR}} \longleftarrow \text{**From fatty acid**}$$
$$|$$
$$CH_2OH$$

**A ceramide (N-acylsphingosine)**

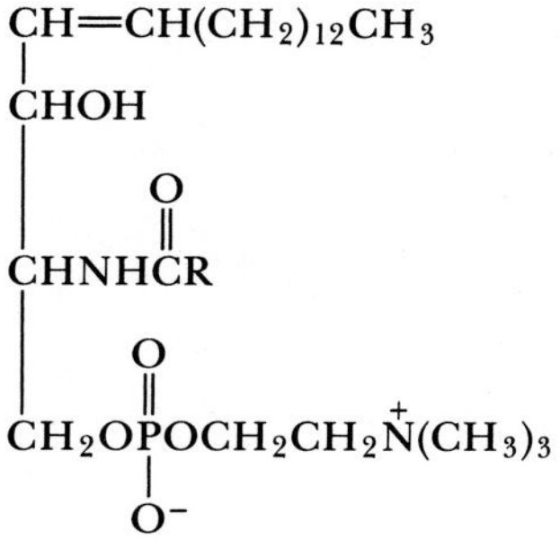

**FIGURE 7.6**   Structures of some sphingolipids.

$$CH=CH(CH_2)_{12}CH_3$$
$$|$$
$$CHOH$$
$$|$$
$$\overset{O}{\overset{\|}{CHNHCR}}$$
$$|$$
$$\overset{O}{\overset{\|}{CH_2OPOCH_2CH_2\overset{+}{N}(CH_3)_3}}$$
$$|$$
$$O^-$$

**A sphingomyelin**

glycolipids, and the glycosidic bond is formed between the primary alcohol group of the ceramide and a sugar residue. The resulting compound is called a **cerebroside.** In most cases the sugar is glucose or galactose; for example, a glucocerebroside is a cerebroside that contains glucose.

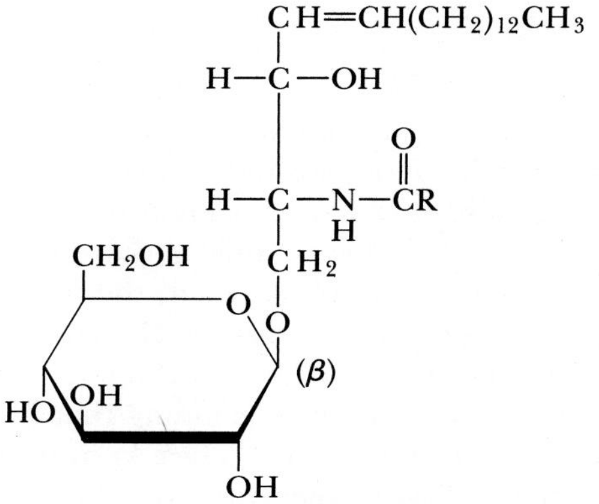

$$CH=CH(CH_2)_{12}CH_3$$
$$|$$
$$H-C-OH$$
$$|$$
$$\overset{O}{\overset{\|}{H-C-N-CR}}$$
$$\underset{H}{}$$
$$CH_2OH \qquad CH_2$$
$$|$$
$$O \quad O$$

$(\beta)$

$$HO \quad OH$$

$$OH$$

**STRUCTURE OF A GLUCOCEREBROSIDE**

As the name indicates, cerebrosides are found in nerve and brain cells, primarily in cell membranes.

## Steroids

Many compounds of widely differing function are classified as **steroids** on the basis of the same general structure: a fused-ring system consisting of three six-membered rings (the A, B, and C rings) and one five-membered

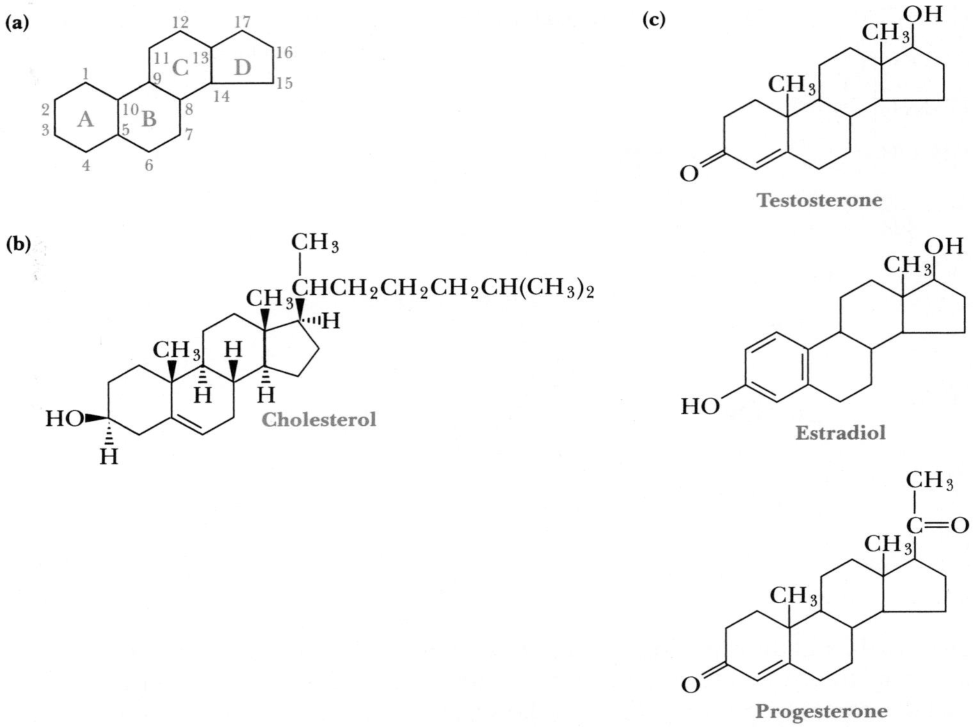

(a)

(b)

Cholesterol

(c)

Testosterone

Estradiol

Progesterone

**FIGURE 7.7**    Structures of some steroids. (a) The fused-ring structure of steroids. (b) Cholesterol. (c) Some steroid sex hormones.

ring (the D ring). There are many important steroids, including sex hormones. (See Section 16.8 for more steroids of biological importance.) The steroid that will be of most interest in our discussion of membranes is **cholesterol** (Figure 7.7). The only hydrophilic group in the cholesterol structure is the single hydroxyl group. As a result, the molecule is highly hydrophobic. Cholesterol is widespread in biological membranes, especially in animals, but it does not occur in prokaryotic cell membranes. In spite of its many important biological functions, including that of precursor of other steroids and of vitamin $D_3$, cholesterol is best known for its harmful effects on health. Cholesterol plays a role in the development of **athero-sclerosis,** a condition in which lipid deposits block the blood vessels and lead to heart disease.

## 7.3
## LIPID-SOLUBLE VITAMINS

There is a group of vitamins, having a variety of functions, that are of interest in this chapter because they are soluble in lipids. The lipid-soluble vitamins are themselves hydrophobic in nature, accounting for their solubility (Table 7.3).

**TABLE 7.3   Lipid-Soluble Vitamins and Their Functions**

| VITAMIN | FUNCTION |
| --- | --- |
| Vitamin A | Serves as the site of the primary photochemical reaction in vision |
| Vitamin D | Regulates calcium (and phosphorus) metabolism |
| Vitamin E | Serves as an antioxidant—necessary for reproduction in rats, may be necessary for reproduction in humans |
| Vitamin K | Has a regulatory function in blood clotting |

## Vitamin A

The extensively unsaturated hydrocarbon **β-carotene** is the precursor of **vitamin A,** also known as **retinol.** As the name suggests, β-carotene is abundant in carrots, and it also occurs in other vegetables, particularly the yellow ones. When an organism requires vitamin A, β-carotene is converted to the vitamin (Figure 7.8a). A derivative of vitamin A plays a crucial role in vision when it is bound to a protein called **opsin.**

The cone cells in the retina of the eye are responsible for vision in bright light and for color vision; cone cells contain several types of opsin. The rod cells in the retina are responsible for vision in dim light, and they contain only one type of opsin. The chemistry of vision has been more extensively studied in rod cells than in cone cells, and we shall discuss the events that take place in rod cells.

Vitamin A has an alcohol group that is enzymatically oxidized to an aldehyde group, forming **retinal** (Figure 7.8b). Two isomeric forms of retinal, involving *cis-trans* isomerization around one of the double bonds, are important in the behavior of this compound *in vivo*. The aldehyde group of retinal forms an imine (also called a Schiff base) with the side-chain amino group of a lysine residue in rod-cell opsin (Figure 7.9). The product of the reaction between retinal and opsin is **rhodopsin.** The outer segment of rod cells contains flat membrane-bounded discs; the membrane consists of about 60% rhodopsin and 40% lipid. For more details about rhodopsin, see Box 7.2 on p. 138.

## Vitamin D

There are several forms of **vitamin D;** these compounds play a major role in regulation of calcium and phosphorus metabolism. One of the most important, vitamin $D_3$ (cholecalciferol), is formed from cholesterol by the action of ultraviolet radiation from the sun. Vitamin $D_3$ is further processed in the body to form hydroxylated derivatives, which are the metabolically active form of this vitamin (Figure 7.10).

A deficiency of vitamin D can lead to **rickets,** a condition in which the bones of growing children become soft, resulting in various skeletal deformities. Children, especially infants, have higher requirements for vitamin D than do adults. Vitamin D supplements are available for children

FIGURE 7.8    Reactions of vitamin A. (a) The conversion of β-carotene to vitamin A. (b) The conversion of vitamin A to 11-*cis*-retinal.

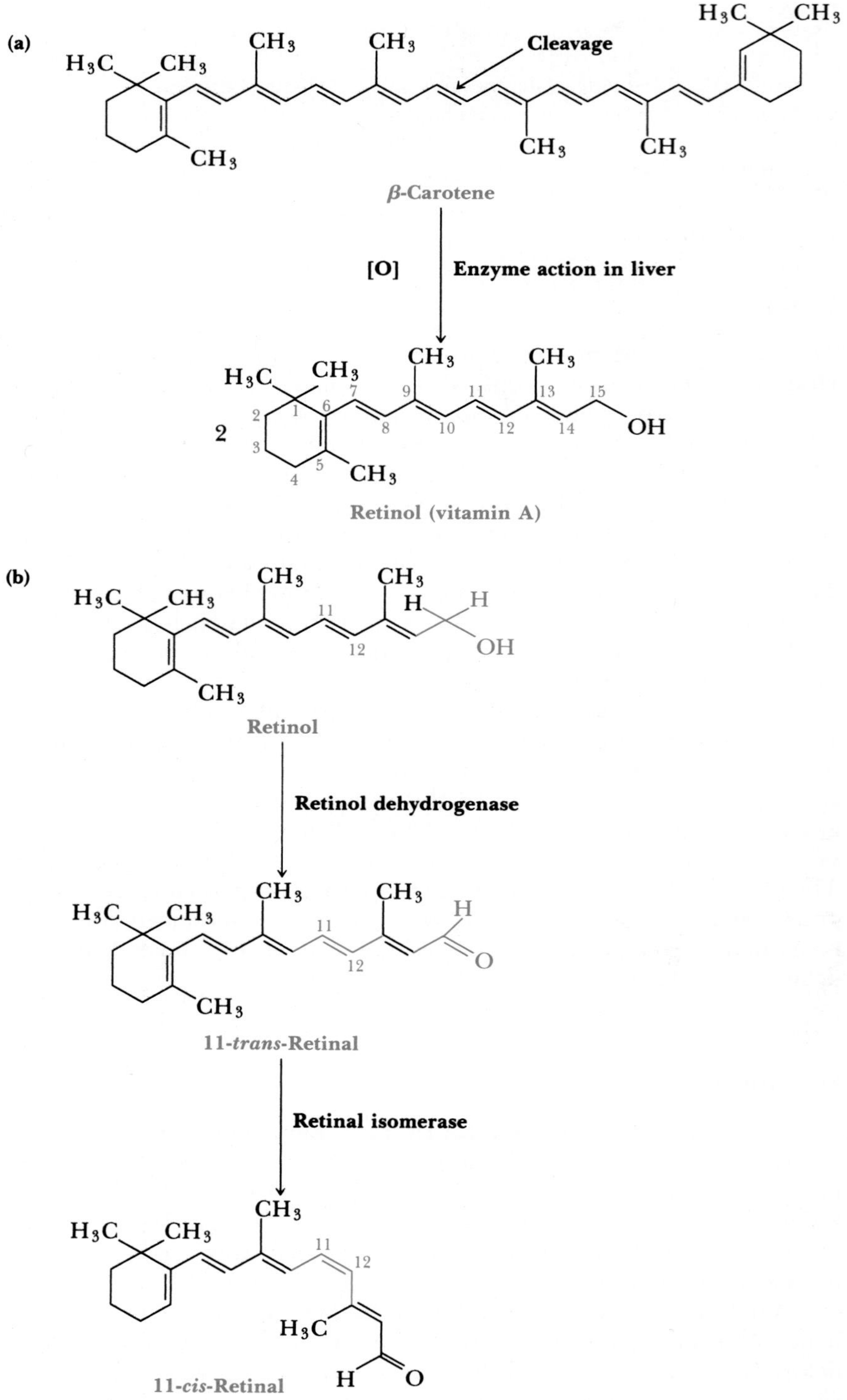

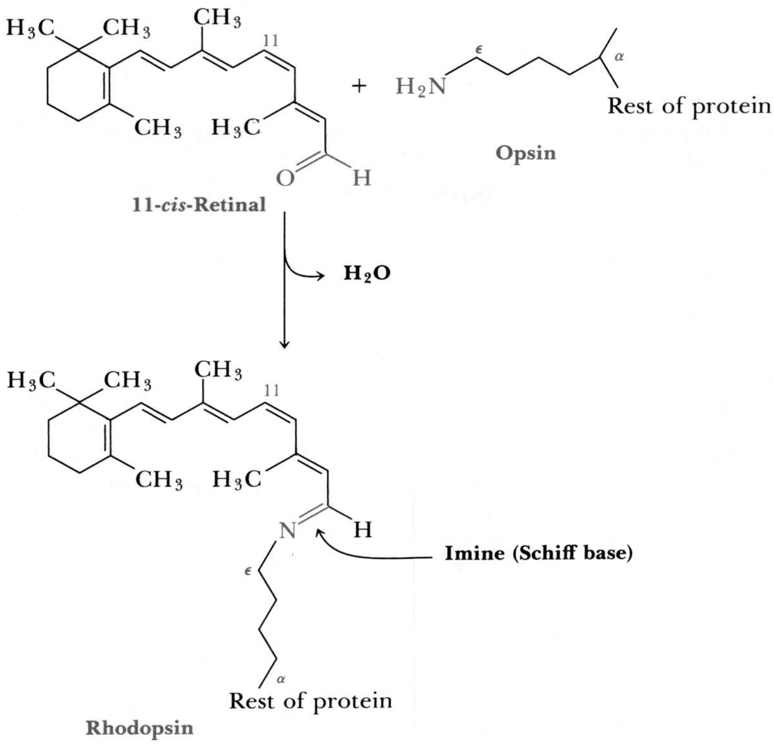

**FIGURE 7.9** The formation of rhodopsin from 11-*cis*-retinal and opsin.

**FIGURE 7.10** Reactions of vitamin D. The photochemical cleavage occurs at the bond shown by the red arrow; electron rearrangements after the cleavage yield the product vitamin D₃. The final product, 1,25-dihydroxycholecalciferol, is the form of the vitamin most active in stimulating the intestinal absorption of calcium and phosphate and in mobilizing calcium for bone development.

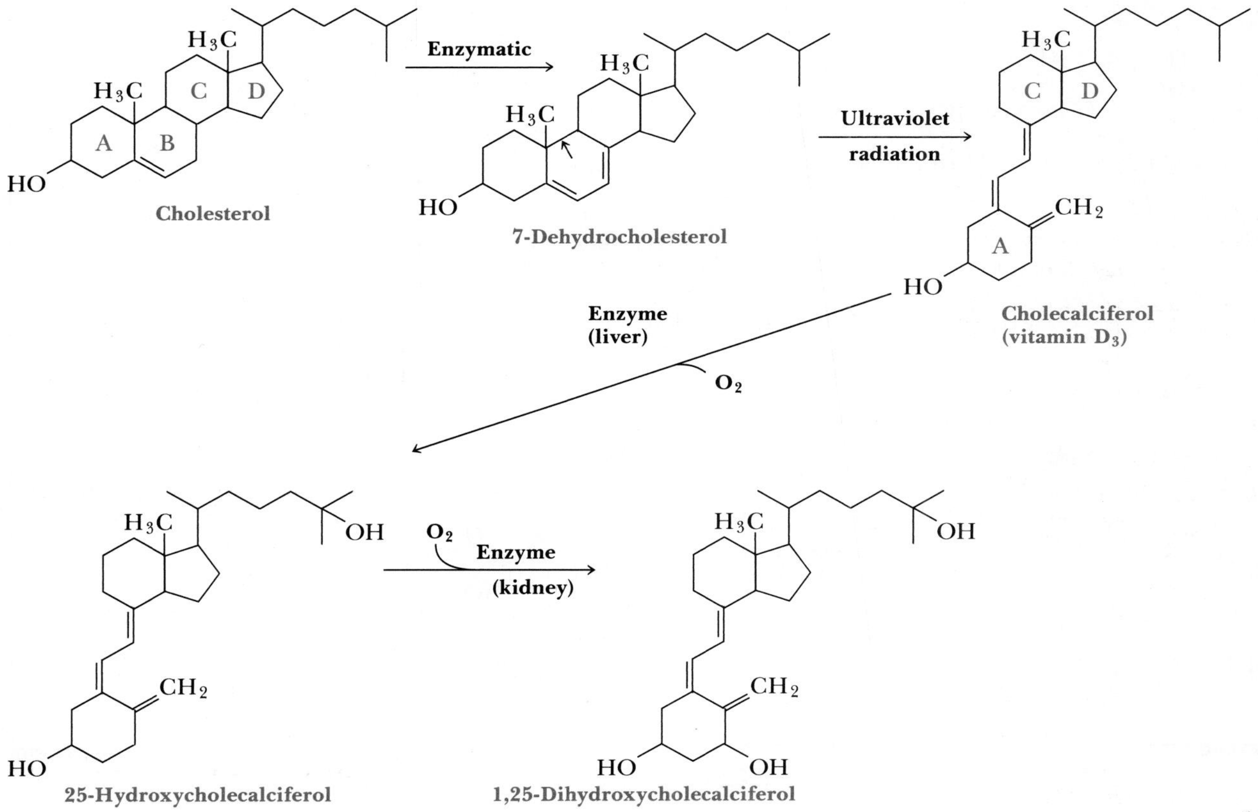

137

## BOX 7.2
## THE CHEMISTRY OF VISION

The primary chemical reaction in vision, the one responsible for generating an impulse in the optic nerve, involves *cis-trans* isomerization around one of the double bonds in the retinal portion of rhodopsin. When rhodopsin is active (that is, when it can respond to visible light), the double bond between carbon atoms 11 and 12 of the retinal (11-*cis*-retinal) has the *cis* orientation. Under the influence of light an isomerization reaction occurs at this double bond, producing all-*trans*-retinal. The all-*trans* form of retinal cannot bind to opsin, and the imine hydrolyzes to release opsin and all-*trans*-retinal. As a result of this reaction, an electrical impulse is generated in the optic nerve and transmitted to the brain to be processed as a visual event. The active form of rhodopsin is regenerated by enzymatic isomerization of the all-*trans*-retinal back to the 11-*cis* form and subsequent re-formation of the imine.

Vitamin A deficiency can have drastic consequences, as can be predicted from its importance in vision. Night blindness and even total blindness can result, especially in children. On the other hand, an excess of vitamin A can cause harmful effects such as fragile bones. Lipid-soluble compounds are not excreted as readily as water-soluble substances, and it is possible for excessive amounts of lipid-soluble vitamins to accumulate in adipose tissue.

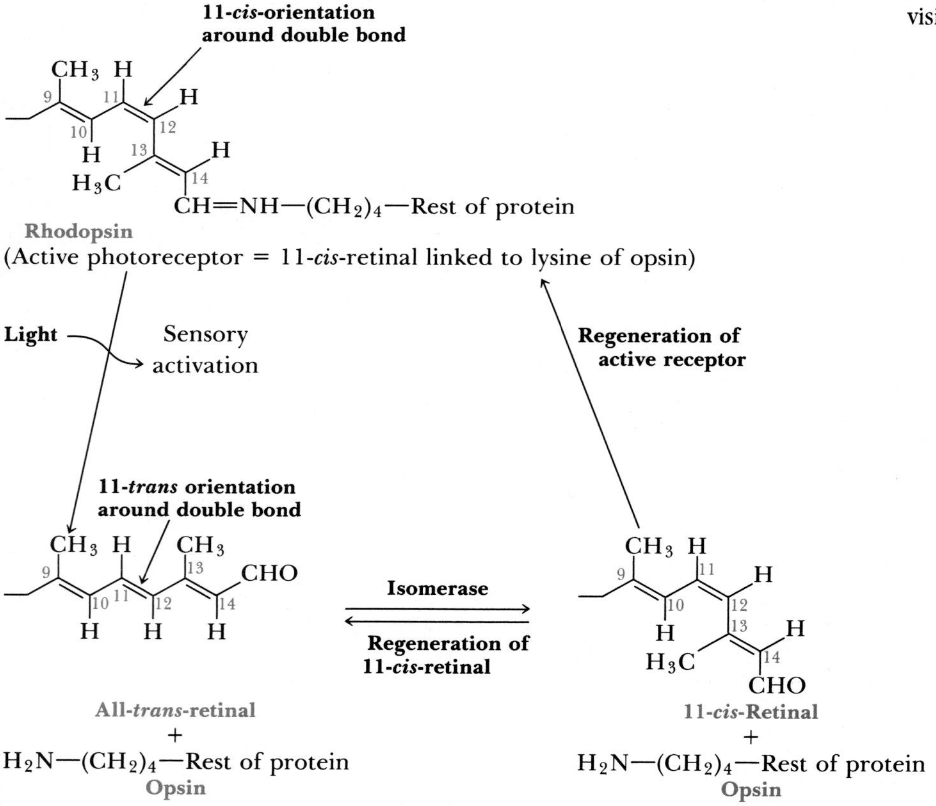

The primary chemical reaction of vision.

in milk and in cod-liver oil. Adults who are exposed to normal amounts of sunlight do not usually require vitamin D supplements.

## Vitamin E

The most active form of **vitamin E** is *α*-**tocopherol.**

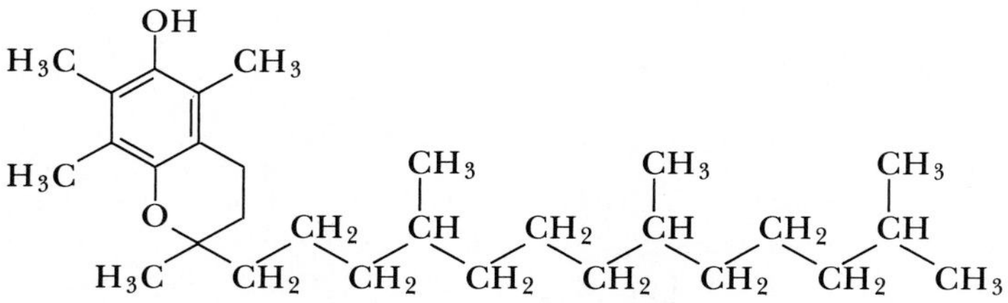

Vitamin E (*α*-tocopherol)

In rats vitamin E is required for reproduction and for prevention of the disease **muscular dystrophy.** It is not definitely known whether this requirement exists in humans. A well established chemical property of vitamin E is that it is an **antioxidant.** That is, it is a good reducing agent, so it reacts with oxidizing agents before they can attack other biomolecules. It has been shown in the laboratory that the antioxidant action of vitamin E protects important compounds, including vitamin A, from degradation; vitamin E probably also serves this function in organisms. Another function of antioxidants such as vitamin E is to react with and thus remove the very reactive and highly dangerous substances known as **free radicals.** A free radical has at least one unpaired electron, which accounts for the high degree of reactivity. Free radicals may play a part in the development of cancer and in the aging process.

## Vitamin K

The name of **vitamin K** comes from the German "Koagulation," because this vitamin is an important factor in the blood clotting process. The bicyclic ring system contains two carbonyl groups, the only polar groups on the molecule (Figure 7.11). There is a long unsaturated hydrocarbon side chain that consists of repeating **isoprene** units, the number of which determines the exact form of vitamin K. Several forms of this vitamin can be found in the same organism, and the reason for this variation is not well understood. Vitamin K is not the first vitamin we have encountered that contains isoprene units, but it is the first one in which the number of isoprene units makes a difference. (See whether you can pick out isoprene-derived portions of the structures of vitamins A and E.) It is also known that the steroids are biosynthetically derived from isoprene units, but the structural relationship is not immediately obvious (Section 16.8).

The presence of vitamin K is required in the complex process of blood clotting, which involves many steps and many proteins; there are numerous

**FIGURE 7.11** The structure of vitamin K. (a) The general structure of vitamin K, which is required for blood clotting. The value of $n$ is variable but usually $<10$. (b) Vitamin $K_1$ has four isoprene units; vitamin $K_2$ has five.

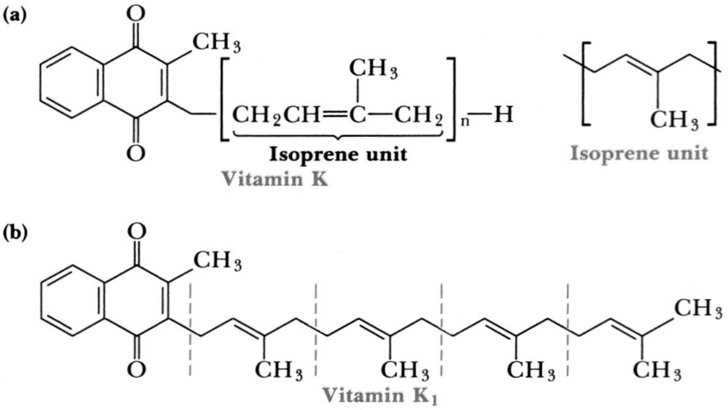

**(a)**

Vitamin K

Isoprene unit

Isoprene unit

**(b)**

Vitamin $K_1$

unanswered questions about the process. It is known definitely that vitamin K is required to modify the protein prothrombin. This process is the modification of the side chains of several glutamate residues of prothrombin by the addition of another carboxyl group. The modification of glutamate produces $\gamma$-carboxyglutamate residues (Figure 7.12). The two carboxyl groups in close proximity form a **bidentate** ("two teeth") **ligand,** which can bind calcium ion ($Ca^{2+}$). If prothrombin is not modified in this way, it does not bind $Ca^{2+}$. Even though there is a lot more to be learned about blood clotting and the role of vitamin K in the process, this point at least is well established, because $Ca^{2+}$ is required for blood clotting.

**FIGURE 7.12**   The role of vitamin K in the modification of prothrombin. The detailed structure of the $\gamma$-carboxyglutamate at the calcium complexation site is shown below.

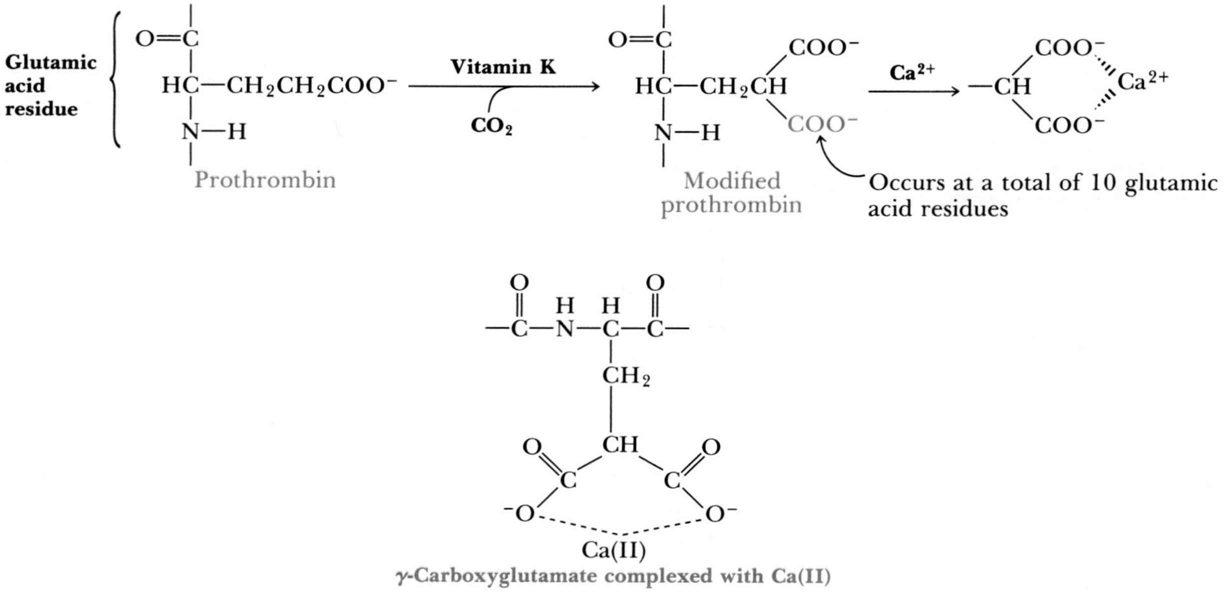

Glutamic acid residue

Prothrombin

Vitamin K

$CO_2$

Modified prothrombin

$Ca^{2+}$

Occurs at a total of 10 glutamic acid residues

Ca(II)

$\gamma$-Carboxyglutamate complexed with Ca(II)

## 7.4
## PROSTAGLANDINS AND LEUKOTRIENES

A group of substances derived from fatty acids has a wide range of physiological activities; these compounds are called **prostaglandins** because they were first detected in seminal fluid, which is produced by the prostate gland. It has since been shown that they are widely distributed in a variety of tissues. The metabolic precursor of all prostaglandins is **arachidonic acid,** a fatty acid that contains 20 carbon atoms and four double bonds. The double bonds are not conjugated. The production of the prostaglandins from arachidonic acid takes place in several steps, which are catalyzed by enzymes. There is a five-membered ring in the prostaglandins themselves; the compounds of this group differ from one another in the numbers and positions of double bonds and oxygen-containing functional groups (Figure 7.13).

Elucidation of the structures of prostaglandins and their laboratory synthesis have been topics of great interest to organic chemists, largely because of the many physiological effects of these compounds and their possible usefulness in the pharmaceutical industry. Some of the functions of prostaglandins include control of blood pressure, stimulation of smooth-muscle contraction, and induction of inflammation. Aspirin inhibits the synthesis of prostaglandins, a property that accounts for its anti-inflammatory and fever-reducing properties. Cortisone and other steroids also show anti-inflammatory properties, which arise from their inhibition of prostaglandin synthesis.

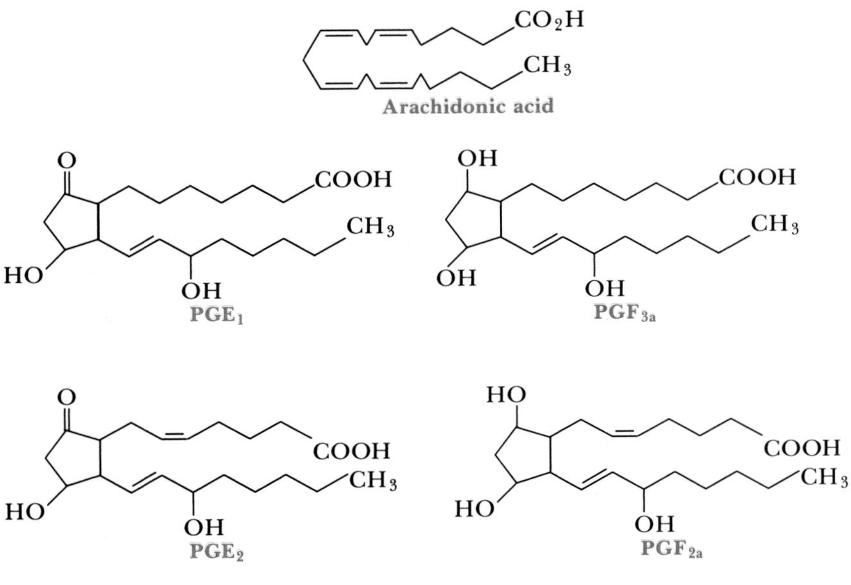

**FIGURE 7.13**  Arachidonic acid and some prostaglandins.

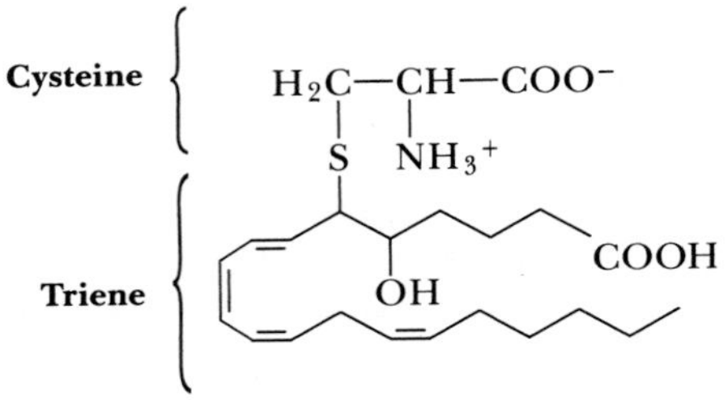

FIGURE 7.14    Leukotriene C.

Prostaglandins are known to inhibit the aggregation of blood plate-lets. As a result they may be of therapeutic value in preventing the types of strokes and heart attacks that arise from blood clots which cut off the blood supply to the brain or the heart. Even if this behavior were the only useful property of prostaglandins, it would justify considerable research effort for therapeutic purposes.

**Leukotrienes** are compounds that, like prostaglandins, are derived from arachidonic acid. They are found in leukocytes (white blood cells) and have three conjugated double bonds; these two facts account for their name. (Fatty acids and their derivatives do not normally contain conjugated double bonds.) Leukotriene C (Figure 7.14) is a typical member of this group; note the 20 carbon atoms in the carboxylic acid backbone, a feature structurally related to arachidonic acid. An important property of leukotri-enes is their constriction of smooth muscle, especially in the lungs. Asthma attacks may result from this constricting action, since the synthesis of leukotriene C appears to be facilitated by allergic reactions, such as a reaction to pollen. Drugs that inhibit the synthesis of leukotriene C show promise for the treatment of asthma. Leukotrienes may also have inflam-matory properties and may be involved in rheumatoid arthritis.

## SUMMARY

Lipids are compounds that are insoluble in water and soluble in nonpolar organic solvents. One group of lipids consists of open-chain compounds with a polar head group and a long nonpolar tail; this group includes fatty acids, triacylglycerols, phospho-acylglycerols, sphingolipids, and glycolipids. A sec-ond major group consists of fused-ring compounds, the steroids.

Fatty acids, saturated or unsaturated, have a car-boxyl group at the polar end and a hydrocarbon chain as the nonpolar tail. Fatty acids do not normal-ly occur free in nature but are incorporated into triacylglycerols and phosphoacylglycerols. In triacyl-

glycerols, three fatty acids are esterified to the three hydroxyl groups of glycerol. Phosphoacylglycerols differ from triacylglycerols in that one of the three hydroxyls of glycerol is esterified to phosphoric acid; the phosphoric acid in turn is esterified to a second alcohol. The nature of the second alcohol deter-mines the identity of the phosphoacylglycerol. Tri-acylglycerols are the storage form of fatty acids, and phosphoacylglycerols are important components of biological membranes. Sphingolipids contain the long-chain amino alcohol sphingosine in place of glycerol; otherwise their structure is similar to that of the lipids already discussed. In glycolipids, a carbohydrate is bound to an alcohol group of a lipid by a glycosidic linkage. Steroids have a characteristic fused-ring structure.

Lipid-soluble vitamins are hydrophobic, accounting for their solubility properties. A derivative of vitamin A plays a crucial role in vision. Vitamin D controls calcium and phosphorus metabolism, affecting the structural integrity of bones. Vitamin·E is known to be an antioxidant; its other metabolic functions are not definitely established. The presence of vitamin K is required in the blood clotting process.

The unsaturated fatty acid arachidonic acid is the precursor of prostaglandins and leukotrienes, compounds that have a wide range of physiological activities. Stimulation of smooth-muscle contraction and induction of inflammation are common to both classes of compounds. Prostaglandins are also involved in control of blood pressure and inhibition of blood platelet aggregation.

## EXERCISES

1. What structural features do a triacylglycerol and a phosphatidyl ethanolamine have in common? How do the structures of these two types of lipids differ?
2. What structural features do a sphingomyelin and a phosphatidyl choline have in common? How do the structures of these two types of lipids differ?
3. Which of the following lipids are *not* found in animal membranes?
   (a) phosphoglycerides      (d) glycolipids
   (b) cholesterol            (e) sphingolipids
   (c) triacylglycerols
4. Draw the structure of a phosphoacylglycerol that contains glycerol, oleic acid, stearic acid, and choline.
5. You have just isolated a pure lipid that contains only sphingosine and a fatty acid. To what class of lipid does it belong?
6. Write the structural formula for a triacylglycerol and name the component parts.
7. Write an equation, with structural formulas, for the saponification of the triacylglycerol in Exercise 6.
8. Briefly discuss the structure of myelin and its role in the nervous system.
9. How does the structure of steroids differ from that of the other lipids discussed in this chapter?
10. What is the role in vision of the *cis-trans* isomerization of retinal?
11. Give a reason for the toxic effects noted in cases of overdoses of lipid-soluble vitamins.
12. What is the structural relationship between vitamin $D_3$ and cholesterol?
13. List an important chemical property of vitamin E.
14. List two classes of compounds derived from arachidonic acid; suggest some reasons for the amount of biomedical research devoted to these compounds.

## ANNOTATED BIBLIOGRAPHY

Chakrin, L.W., and D.M. Bailey. *The Leukotrienes—Chemistry and Biology.* Orlando, FL: Academic Press, 1984. [A collection of articles on the structure and action of leukotrienes.]

DeLuca, H.F., and H.K. Schneos. Vitamin D: Recent Advances. *Ann. Rev. Biochem.* **52**, 411–439 (1983). [A review article on the physiological function of vitamin D.]

Hakomori, S. Glycosphingolipids. *Sci. Amer.* **254** (5), 44–53 (1986). [A possible role for this class of cell membrane components in diagnosis and treatment of cancer.]

Hammarstrom, S. Leukotrienes. *Ann Rev. Biochem.* **52,** 355–377 (1983). [A review article on the chemistry of substances of possible therapeutic value.]

Keuhl, F.A., and R.W. Egan. Prostaglandins, Arachidonic Acid and Inflammation. *Science* **210**, 978–984 (1980). [An article on both the chemistry of these compounds and their physiological effects.]

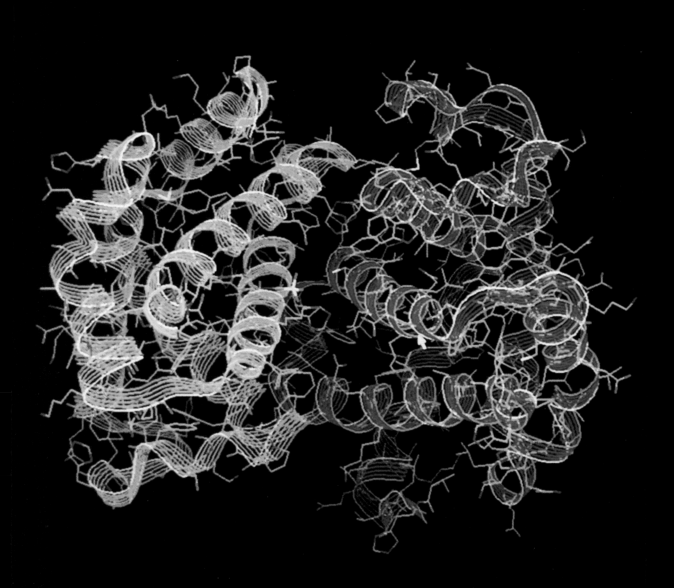

# The Dynamic Aspects of Biochemical Reactions

# INTERVIEW

## JoAnne Stubbe

JoAnne Stubbe, currently Ellen Swallow Richards Professor in the chemistry and biology department at MIT, has had a productive and successful career in biochemistry.

She received her B.A. in chemistry from the University of Pennsylvania, graduating in 1968 with cum laude honors. From there Ms. Stubbe went on to recieve her doctorate in organic chemistry at the University of California, Berkeley.

JoAnne Stubbe has published a number of works on mechanism-based inhibitors and mechanism of antitumor antibiotics that cleave DNA. Ms. Stubbe's major research interests include developing methods to understand enzyme mechanisms involved in nucleotide metabolism, the design of suicide inhibitors, and the study of the mechanism of drug-mediated DNA cleavage.

She belongs to several professional organizations. Among these are the American Chemical Society and the American Society of Biological Chemistry. In addition, she has served on many committees and editorial boards and has been a guest lecturer often throughout her career.

In 1986, JoAnne Stubbe received the ACS-Pfizer Award in Enzyme Chemistry. In 1989, she received the ICI–Stuart Pharmaceutical Award for Excellence in Chemistry.

### How did you get interested in science?

I've always been interested in the biological end of things. As a child, I always played in the backwoods, building camps and trout fishing with my grandfather. By the time I

was 8 or 9 years old, I had formed a nature club and I used to make people write reports for me.

My interest in education results from my family life. Both of my parents were educators, and everyone in the family was expected to do well in school. Since we didn't have a television, we read all the time, and we read anything that we were interested in.

I was really excited about chemistry in high school and I started to work in a chemistry lab as a freshman in college. Then, during the summer of my first year in college, I had an NSF Fellowship and I worked all of my subsequent summers in labs.

I went to graduate school in chemistry, not biochemistry, although eventually my interests changed.

## What led you to shift to biochemistry?

My first job was teaching at Williams College, and even though I was trained as an organic chemist, I had to teach biochemistry and physical chemistry. That typically happens when you teach—you don't end up teaching in your area of expertise. Teaching biochemistry got me more and more interested in it and I decided that while I liked teaching, I really wanted to do research. You can't do research on the forefront of science at a small liberal arts school; I had to get back to the university atmosphere. I took a sabbatical and worked in Robert Abeles lab at Brandeis.

## What drew you to laboratory research?

Science is the scientific method. It's asking a question and then designing experiments to figure out how you are going to prove or disprove your theory. Or, you get a result from an experiment and then change your hypothesis and begin again. I think the only way you can actually do science is in a laboratory setting.

What drives me is that I'm excited about solving a problem that I'm interested in. And that's it! I can get excited about something that wouldn't interest most people at all!

## Was your chemical training a help or an obstacle when you decided to do research in biochemistry?

I think it's easier to go into biochemistry with a chemical background than to start on the biological side and go into chemistry. The problems that really excite me now are biological problems, but I approach them from a chemical point of view.

I think the people who are going into molecular biology, the really hot area of research right now, don't know enough chemistry. We're in the Age of the Gene Hunters—everybody is finding new genes. But, ultimately, to understand things from a biochemical perspective, you need to study all of the proteins that these genes produce and the ways that they interact with each other. That's what a biochemist does. Even though biology is incredibly powerful, I think students need to know something about structure in order to think about what's really going on.

## Your research is in the area of enzymes. How did you become interested in that?

Enzymes are amazingly efficient catalysts. They work in aqueous solutions at room temperature, to give you 100% yield of a product, and they do this at rates $10^6$ to $10^{10}$ that of nonenzymatic reactions.

What causes these tremendous rate accelerations? What are the basic chemical principles involved? These questions intrigue me.

You can study enzymes from an organic chemist's point of view because the basic principles of organic chemistry are directly applicable to enzyme systems. Enzymes are involved in catalyzing the reaction from "A" to some product "P," but there are some enzyme systems in which there is no chemical precedent for the reactions. These are the systems that I have become interested in because nobody knows how they really work.

## Some of your research is done in conjunction with the search for new drugs. How basic and how practical is your work?

I work in the area of drug design, and the information that we obtain is utilized by the pharmaceutical companies. They use it to do the kind of science that ultimately leads to something that might have an applied effect and help lots of people.

But my research is not directly applicable to something that practical. Research is the heart of everything but it's not, tomorrow, going to have direct consequences in terms of the marketplace the way that the work of pharmaceutical companies will.

Our approach is to rationally design inhibitors of a specific enzyme, required for cell viability, by understanding the detailed mechanism by which you convert the substrate "A" into the product "P." The inhibitor looks like the substrate "A" so that it can bind at the site where chemistry occurs. Then using the normal catalytic mechanism, the inhibitor is chemically transformed by the enzyme into a highly reactive molecule within the enzyme's active site and inactivates itself. These are called mechanism-based inhibitors.

An example of how this can be applied to human health is a class of drugs that control blood pressure. Both Merck and Squibb have designed potent mechanism-based inhibitors of a key enzyme, ACE (angiotensin-converting enzyme), involved in regulation of blood pressure; these are now available and used by thousands of people.

## You don't see your task as creating new drugs?

What we do is design the basic principles of how the target enzyme works, and the drug companies can take it from there. I think the research for designing the actual

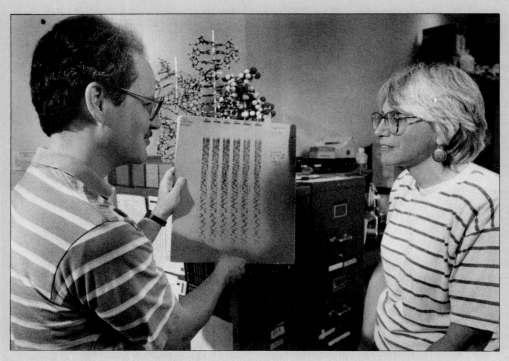

*JoAnne Stubbe and post-doctoral fellow, Tod Holler, analyze the autoradiogram of a DNA sequencing gel.*

drug—which will eventually cure somebody—is best done in a place like Merck or Squibb. We design mechanism-based inhibitors and then let the drug company go and make the 10,000 compounds they need to find the active one, *in vivo.*

Some of the drug companies, depending on how much money they have, are even trying to do a lot of their own basic research. However, they don't do it in anywhere near the depth that it would be done in a university.

Lots of times, though, there's collaboration. I collaborate with a drug company in England on an enzyme targeted for an antiviral design. The drug company provides me with money to support a postdoctoral fellow in my

laboratory, and we use each other's information in order to solve the problem faster. That's the bottom line.

**Why the emphasis on collaboration rather than competition?**

Competition is still there, but the problem is that there are so many new techniques, making it very difficult to be good at all of them. It used to be that you would do everything yourself. Now, different groups of people with different kinds of expertise end up collaborating because there's so much involved. In addition, the problem is solved in a much more efficient way than it would have been 10 years ago. It's exciting, especially in a place like MIT,

where it's really easy to have collaborative interactions. People in biology and chemistry have pretty good interactions here because the subjects are contiguous and there are a lot of common interests.

**Has the technology changed everything?**

The technology is absolutely phenomenal. The things that you can do now, you couldn't have dreamed of doing 5 years ago. For example, I used to go into the laboratory when we were isolating enzymes, and it would take 3½ months to isolate a microgram of protein. Nowadays, you can go into a laboratory using molecular biological methods and isolate a gram of protein (that's $10^6$ as much

148

material) in a day. Molecular biology is an incredibly powerful tool for someone interested in biochemistry or enzymology, which is my area.

When you look at what's happened to the whole field, it's mind-boggling! There are dramatic changes that allow us to solve problems in a way I don't think I could have ever imagined when I was in graduate school. I think that's part of the challenge.

You're continually learning something new so that you can use the best methods to solve the problems you want to solve. And the bottom line is that if you don't continue to study, you're left behind. Things are changing so fast that methods people routinely used 7 years ago are archaic today.

### It is expensive to do science. Does that affect how you work and what you work on?

Once you get to a certain stage in your career, you spend all of your time trying to get money to support the people in your lab by writing papers and selling your science. Research is expensive. I have a lab of ten people which costs about $500,000 a year.

The question is, can you get money to work on basic research, and where do you get that money from? That changes with the times. You have to change your outlook on what your research projects are going to be. You have to go with the flow; otherwise you lose your funding.

For example, I haven't had any problems getting funding because I work in biochemistry, or the biology/organic chemistry interface, where there is a fair amount of money right now. However, I have colleagues who work in organometallic chemistry,

which is not as "trendy," who do experience difficulty with funding. I think it depends on the area that you're in.

### Beyond the perennial funding problem, the technology, and the new emphasis on collaboration, what else characterizes your science?

There's competition. That's also what drives science. You want to be there first, right? You don't want someone else to beat you out with that idea. I think, in general, that competition turns out to be really healthy if somebody makes an important discovery. The result will be reproduced in 2 or 3 weeks, and if there's anything wrong with it, the problem will be immediately known. Scientists end up policing themselves because there are overlaps of interests. Nobody approaches the experiments in exactly the same way, and if it's something important, it's not going to stay in the literature for very long before it's challenged.

### How would you advise a student who was thinking about a career in biochemistry?

I think right now is a wonderful time to be in biochemistry. Every time I come to work I learn something new. There are so many changes going on so rapidly and the methods are so powerful—who knows what we'll be able to do 5 years from now?

I'd also tell a student to get a good chemical background. I think the future of biochemistry is looking at macromolecular interactions, DNA-protein interactions, and protein-sugar interactions. You can do that only with a knowledge of structure, which is the chemical side of

biochemistry. In the long run, it will help you solve problems in a unique way that people with more biological backgrounds can't solve because they don't have that kind of perspective and insight.

Finally, you need to be highly motivated and interested. If what makes you happy is making discoveries in the laboratory, then you're in the right field, and if you don't get incredibly excited about that, you're in the wrong field. It really takes a certain kind of personality to deal with a lot of the failures that you have in science on the day-to-day level. Only one out of ten experiments might work, so you have to get really excited about the one that does. You have to have the kind of personality that gets incredibly high from very little result because that's what keeps you going to ask the next 20 questions, which eventually leads to the answer.

I think the way you find out as an undergraduate if you have the right personality to do science is to do research in somebody's laboratory, where not all the recipes work. You do something and it fails, and it fails again, and it fails again. How do you figure out how to make it work? Are you excited by that challenge?

149

# CHAPTER

## 8

# Energy Requirements for Life: Thermodynamics

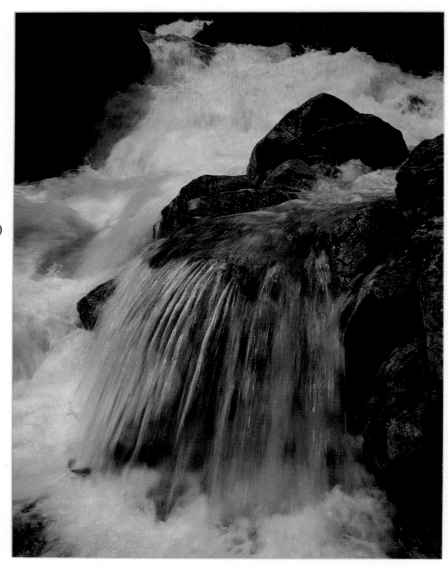

*The potential energy of the water at the top of the waterfall is transformed into kinetic energy in spectacular fashion.*

150

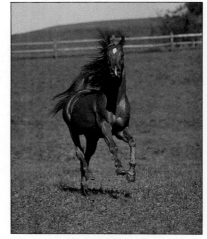

*Everyone knows that order in the dining room begets disorder in the kitchen. Cleaning up and getting ready for the next meal—restoring order—requires energy. So it is with life processes on the molecular level. Molecules taken in as nutrients are torn apart to extract energy and also to provide the building blocks to create new molecules. To maintain a steady state, a living organism needs a constant supply of energy from without to bring order to the constant turmoil within. Speed is essential. Large protein molecules provide the machinery to make rapid chemical changes possible. The enzyme carbonic anhydrase attaches a water molecule to carbon dioxide 600,000 times a second so that $CO_2$ can be transported to the lungs and expelled during respiration, making way for the next inhalation of oxygen. The shapes, and thus the activities, of enzymes themselves are determined by favorable interactions with water. Hydrophobic residues of proteins are sequestered in the interior of the protein chain. At the same time, hydrophilic residues rise to the surface of the protein where they contact the surrounding water. This arrangement lowers the energy level so that the protein encounters minimum resistance folding up into its native conformation and can carry out its necessary biological function.*

## 8.1
## ENERGY AND CHANGE

All living organisms require and use energy. However, energy can take several forms and can be converted from one form to another. Organisms require energy in different forms; for example, motion involves mechanical energy, and maintaining body temperature uses thermal energy. Photosynthesis requires light energy from the sun. Some organisms, such as several species of fish and eels, are striking examples of the use of chemical energy to produce electrical energy. The formation and breakdown of biomolecules involve changes in chemical energy.

Any process that will actually take place is **spontaneous** in the specialized sense used in thermodynamics. The laws of thermodynamics can be used to predict whether any change involving transformations of energy will take place. An example of such a change is a chemical reaction in which covalent bonds are broken and new ones formed. Another example is the formation of noncovalent interactions, such as hydrogen

Three examples of transformations of energy in biological systems. (a) The horse represents conversion of chemical energy into mechanical energy. (b) The electric fish (Torpedinidae) converts chemical energy into electrical energy, while the phosphorescent bacteria (c) converts chemical energy into light energy.

J. Willard Gibbs (1839–1903). The symbol $G$ is given to free energy in his honor. His work is the basis of biochemical thermodynamics, and he is considered by some to be the greatest scientist born in the U.S. (The Bettmann Archive)

bonds or hydrophobic interactions when a protein folds to produce its characteristic three-dimensional structure. The tendency of polar and nonpolar substances to exist in separate phases is a reflection of the energies of interaction between the individual molecules, in other words, the thermodynamics of the interaction.

## 8.2
## THE CRITERION FOR SPONTANEITY

The most useful criterion for predicting the spontaneity of a process is the **free energy,** which has the symbol $G$. (Strictly speaking, the use of this criterion requires conditions of constant temperature and pressure, but these conditions are usual in biochemical thermodynamics.) It is not possible to measure absolute values of energy, only the *changes* in energy that occur during a process. The value of the change in free energy ($\Delta G$, where the symbol $\Delta$ indicates change) gives the needed information about the spontaneity of the process under consideration.

The free energy of a system decreases in a spontaneous process, so $\Delta G$ is negative ($\Delta G < 0$). Such a process is called **exergonic,** meaning that energy is given off. When the free energy change is positive ($\Delta G > 0$), the process is nonspontaneous; the reverse of such a process is spontaneous. Energy must be supplied to make a nonspontaneous process occur. Nonspontaneous processes are also called **endergonic,** meaning that energy is absorbed. For a process at **equilibrium,** with no net change in either direction, the free energy change is zero ($\Delta G = 0$).

| | | |
|---|---|---|
| $\Delta G < 0$ | spontaneous | exergonic—energy released |
| $\Delta G = 0$ | equilibrium | |
| $\Delta G > 0$ | nonspontaneous | endergonic—energy required |

An example of a spontaneous process is the aerobic metabolism of glucose, in which glucose reacts with oxygen to produce carbon dioxide, water, and energy for the organism. An example of a nonspontaneous process is the phosphorylation of ADP (adenosine *di*phosphate) to give ATP (adenosine *tri*phosphate). This reaction takes place in living organisms because energy is supplied as a result of metabolic processes.

ADP
adenosine diphosphate

phosphate
ion

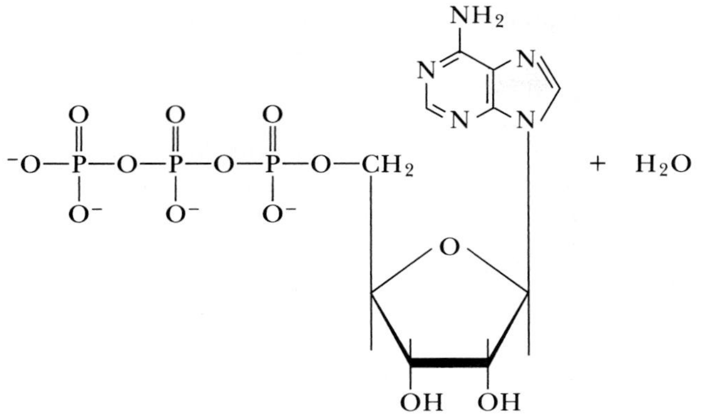

**ATP**
**adenosine triphosphate**

8.1

## 8.3
## STANDARD STATES AND THE STANDARD FREE ENERGY CHANGE

We can define **standard conditions** for any process, and then we can use these standard conditions as the basis for comparing reactions. The choice of standard conditions is arbitrary. For a process under standard conditions, all substances involved in the reaction are in their **standard states,** at which they are said to be at **unit activity.** For pure solids and pure liquids, the standard state is the pure substance itself. For gases the standard state is usually taken as a pressure of 1 atmosphere of that gas. For solutes, the standard state is usually taken as 1 molar concentration. Strictly speaking, these definitions for gases and for solutes are approximations, but they are valid for all but the most exacting work.

For any general reaction

$$a\text{A} + b\text{B} \rightleftharpoons c\text{C} + d\text{D}$$

we can write an equation that relates the free energy change ($\Delta G$) for the reaction under *any* conditions to the free energy change under *standard* conditions ($\Delta G°$); the superscript ° refers to standard conditions. This equation is

$$\Delta G = \Delta G° + 2.303\ RT \log \frac{[\text{C}]^c[\text{D}]^d}{[\text{A}]^a[\text{B}]^b}$$

In this equation, the square brackets indicate molar concentrations, $R$ is the gas constant (8.31 J mol$^{-1}$ K$^{-1}$), and $T$ is the absolute temperature. This equation holds under all circumstances; the reaction does not have to be at equilibrium.

When the reaction is at equilibrium, $\Delta G = 0$, and thus

$$0 = \Delta G° + 2.303\ RT \log \frac{[\text{C}]^c[\text{D}]^d}{[\text{A}]^a[\text{B}]^b}$$

or

$$\Delta G° = -2.303\ RT \log \frac{[\text{C}]^c[\text{D}]^d}{[\text{A}]^a[\text{B}]^b}$$

The concentrations are now equilibrium concentrations, and this equation can be rewritten

$$\Delta G^\circ = -2.303\ RT \log K_{eq}$$

where $K_{eq}$ is the equilibrium constant for the reaction. We have here a relationship between the equilibrium concentrations of reactants and products and the standard free energy change. Once we have determined the equilibrium concentrations of reactants by any convenient method, we can calculate the equilibrium constant, $K_{eq}$. We can then calculate the standard free energy change, $\Delta G^\circ$, from the equilibrium constant.

## 8.4
## A MODIFIED STANDARD STATE FOR BIOCHEMICAL APPLICATIONS

We have just seen that the calculation of standard free energy changes includes the stipulation that all substances are at unit activity, which for solutes can be approximated as 1 molar concentration. Many important biochemical reactions include hydrogen ion as a product or reactant. If the hydrogen ion concentration of a solution is 1 molar, the pH is zero. (Recall that the logarithm of 1 to any base is zero.) The interior of a living cell is in many respects an aqueous solution of the cellular components, and the pH of such a system is normally in the neutral range. Biochemical reactions in the laboratory are usually carried out in buffers that are also at or near neutral pH. For this reason it is convenient to define a modified standard state for biochemical practice, one that differs from the original standard state only by the change in hydrogen ion concentration from 1 M to $1 \times 10^{-7}$ M, implying a pH of 7. When free energy changes are calculated on the basis of this modified standard state, they are designated by the symbol $\Delta G^{\circ\prime}$. Note that the standard state for water (pure water, approximately 55.5 M) remains unchanged.

Here is an example of the use of equilibrium constants to determine the $\Delta G^{\circ\prime}$. Let us assume that the relative concentrations of reactants have been determined for a reaction carried out at pH 7 and 25°C (298 K). Such concentrations can be used to calculate an equilibrium constant, $K'_{eq}$, which in turn can be used to determine the standard free energy change, $\Delta G^{\circ\prime}$, for the reaction. A typical reaction to which this kind of calculation can be applied is the reverse of a reaction we met earlier in this chapter. The reaction we have in mind here is the hydrolysis of ATP at pH 7, yielding ADP, monohydrogen phosphate ion (written as $P_i$), and $H^+$.

$$ATP + H_2O \rightleftharpoons ADP + P_i + H^+$$

$$K'_{eq} = \frac{[ADP][P_i][H^+]}{[ATP]} \qquad \text{pH 7, 25°C}$$

The concentrations of the solutes are used to approximate their activities, and the activity of the water is one. The experimentally determined value for $K'_{eq}$ is $2.23 \times 10^5$. Substituting $R = 8.31$ J mol$^{-1}$ K$^{-1}$, $T = 298$ K, and $\log K'_{eq} = 5.35$,

$$\Delta G^{\circ\prime} = -2.303 \; RT \log K'_{eq}$$

$$\Delta G^{\circ\prime} = -(2.303)(8.31 \text{ J mol}^{-1} \text{ K}^{-1})(298 \text{ K})(5.35)$$

$$\Delta G^{\circ\prime} = -30.5 \text{ kJ mol}^{-1} = -7.3 \text{ kcal mol}^{-1}$$

In addition to illustrating the point about the usefulness of a modified standard state for biochemical work, this reaction is an example of one in which energy is released.

## 8.5
## THERMODYNAMICS AND LIFE

From time to time one encounters the statement that the existence of living things is a violation of the laws of thermodynamics, specifically of the second law. A look at the laws of thermodynamics will answer the question about whether life is thermodynamically possible, and more discussion of thermodynamics will increase our understanding of this important topic.

The laws of thermodynamics can be stated in several ways. According to one formulation, the first law is "you can't win," and the second is "you can't break even." Less flippantly, the first law states that it is impossible to convert energy from one form to another at greater than 100% efficiency. In other words, the first law of thermodynamics is the law of conservation of energy. The second law states that even 100% efficiency in energy transfer is impossible.

It is possible to relate the two laws of thermodynamics to the free energy by means of a well-known equation:

$$\Delta G = \Delta H - T\Delta S$$

In this equation $G$ is the free energy, as before; $H$ stands for the **enthalpy** and $S$ for the **entropy.** Discussions of the first law focus on the change in enthalpy, which is the **heat of a reaction at constant pressure.** This quantity is relatively easy to measure. Enthalpy changes for many important reactions have been determined and are widely available, with tables given in textbooks of general chemistry. Discussions of the second law focus on changes in entropy, a concept that is less easily described and measured than enthalpy. Entropy changes are particularly important in biochemistry.

One of the most useful definitions of entropy arises from statistical considerations. From the statistical point of view, an increase in the entropy of a system (the substance or substances under consideration) represents an increase in its disorder or randomness. Books scattered around the reading room of a library have a higher entropy than when they are in their proper places on the shelves. The scattered books are clearly in a more random state than the ones on the shelves. The natural tendency of the universe lies in the direction of greater disorder. Another statement of the second law is this: **in any spontaneous process the entropy of the universe increases** ($\Delta S_{universe} > 0$). (This statement is a general one that applies to any set of conditions; it is not confined to the special case of constant temperature and pressure, as is the statement that the free energy decreases in a spontaneous process.)

Ludwig Boltzmann (1844–1906). His equation for entropy in terms of the disorder of the universe was one of his supreme achievements; this equation is carved on his tombstone. (The Bettmann Archive)

The argument that life is a violation of the second law is based on the observation that any living organism is a highly specific, organized, thoroughly nonrandom arrangement of matter, not at all a reflection of increasing disorder. However, the metabolic processes of organisms involve the breakdown of complex molecules in foodstuffs, and the production of wastes and heat. Carbon dioxide and water vapor are given off as gases, with the totally random arrangement characteristic of gases. Living things definitely represent local decreases in entropy, with the high degree of order found in complex molecules such as proteins and nucleic acids, and the still higher degree of order found in organelles such as nuclei, mitochondria, and chloroplasts. The incorporation of organelles into cells, cells into tissues, and tissues into organisms represents a high degree of order, a low entropy. As we have just seen, though, this high degree of organization is achieved by greater increases in entropy elsewhere in the universe. The local decrease in the entropy of a small part of the universe (*i.e.,* the lower entropy of the system) associated with a living organism is more than offset by the increase in the entropy of the rest of the universe, the surroundings of the system. The entropy of the universe as a whole increases, as must be true for a spontaneous process.

**OPTIONAL**

## 8.6
## HYDROPHOBIC INTERACTIONS: A CASE STUDY IN THERMODYNAMICS

Nonpolar substances do not dissolve in water. The fact that polar and nonpolar substances do not mix is ascribed to hydrophobic interactions. Unfavorable entropy changes that accompany attempts to mix polar and nonpolar substances are in turn the basis of hydrophobic interactions.

Hydrophobic interactions actually occur and thus are spontaneous processes. The entropy of the universe increases when hydrophobic interactions are formed.

$$\Delta S_{universe} > 0$$

We are specifically interested in the polar and nonpolar substances that do not dissolve, rather than the whole universe. We can define the system as these two substances (polar and nonpolar) in which we are interested, and the surroundings as the rest of the universe. The entropy change for the whole universe can be broken up into the entropy change for the system and the entropy change for the surroundings.

$$\Delta S_{universe} = \Delta S_{system} + \Delta S_{surroundings}$$

The entropy change of the surroundings can be very important, as we saw in our discussion about living things and the second law of thermodynamics. We can also relate $\Delta S_{surroundings}$ to other, easily measured thermodynamic parameters.

We have already seen one definition of entropy, one that emphasizes its relation to randomness or disorder. There is another definition of entropy that emphasizes the relation between the entropy and other quantities, such as heat and temperature, encountered in thermodynamics. This second definition will be useful for our current discussion.

Heat is a reflection of the random motion of molecules; the more heat that flows into a system, the greater is the randomizing influence. There is a direct relationship between heat and the entropy of a system. The molecules of a substance are less randomly arranged at low temperatures than high ones; as a result, there is an inverse relationship between entropy and temperature. The thermodynamic definition of entropy, $S$, involves the absolute temperature, $T$, and the heat flow, $q$; the **change in entropy** is

$$\Delta S = \frac{q_{rev}}{T}$$

The subscript rev refers to *reversible* heat flow, which takes place so slowly that the direction of heat flow can easily change direction. Recall that we defined enthalpy changes in terms of heat flow as well. (See Section 8.5. Enthalpy is the heat of reaction at constant pressure.) This relationship to heat flow is an important connection between enthalpy and entropy.

It is known (although we shall not derive the equation here) that at constant temperature and pressure

$$\Delta S_{surroundings} = - \frac{\Delta H_{system}}{T}$$

We can now write an expression for the entropy change of the universe in terms of changes in the system.

$$\Delta S_{system} - \frac{\Delta H_{system}}{T} = \Delta S_{universe}$$

We multiply this expression by $T$,

$$T\Delta S_{system} - \Delta H_{system} = T\Delta S_{universe}$$

If we then change the signs of all quantities involved in this equation, we obtain

$$\Delta H_{system} - T\Delta S_{system} = -T\Delta S_{universe}$$

Note that the left side of this last equation is equal to $\Delta G_{system}$ (since $\Delta G = \Delta H - T\Delta S$). We have established the important connection that at constant temperature and pressure

$$\Delta G_{system} = -T\Delta S_{universe}$$

and we have also seen that the enthalpy change involved in the formation of hydrophobic interactions is intimately associated with the entropy change ($-\Delta H_{sys}/T = \Delta S_{surr}$). In fact, the enthalpy change is a reflection of the unfavorable entropy change.

As an example, let us assume that we have tried to mix the liquid hydrocarbon hexane ($C_6H_{14}$) with water, but we have obtained not a solution but a two-layer system, one layer of hexane and one of water. The process is nonspontaneous and

$$\Delta S_{sys} + \Delta S_{surr} = \Delta S_{univ} < 0$$

As we have just seen, $\Delta S_{surr} = -\Delta H_{sys}/T$; this enthalpy change is a quantity known as the heat of solution. Some substances either give off heat when they dissolve (the solution is hot to the touch) or take up heat (the solution is cold to the touch). Such substances are said to have a large heat of solution. Hexane dissolves in ethanol ($C_2H_5OH$), a polar liquid but less polar than water. When a solution of hexane in ethanol is prepared, it is cold to the touch because heat from the surroundings is needed to supply energy for hexane to dissolve in ethanol. In the case of hexane in water, the surroundings cannot supply enough energy to allow hexane to dissolve. Before we go into detail on the relationship between the enthalpy change and the unfavorable entropy change, we should point out that $\Delta S_{sys}$, as opposed to $\Delta S_{surr}$, favors the dissolving of hexane in water, since a mixture of two substances has more possible random arrangements than do the separate pure substances.

The heat of solution of nonpolar substances in water is a reflection of the unfavorable entropy change in such a situation.

$$\Delta H_{soln} = \Delta H_{sys}$$

$$-\frac{\Delta H_{sys}}{T} = \Delta S_{surr}$$

The heat of solution in turn can be broken down into three parts, all of which are related to the intermolecular interactions we discussed in Chapter 3 (Section 3.1). The first part ($\Delta H_1$) is the enthalpy required to separate the *solute* molecules from one another by overcoming the interactions between them, and the second part ($\Delta H_2$) is the enthalpy required to separate the *solvent* molecules by overcoming the interactions between them (Figure 8.1). The third part ($\Delta H_3$) is the enthalpy released as a result of the interaction between solute and solvent molecules. The relative magnitudes of these three enthalpy changes determine the overall heat of solution.

$$\Delta H_{soln} = \Delta H_1 + \Delta H_2 + \Delta H_3$$

In our example of an attempt to dissolve hexane in water, we can see that it takes very little energy to separate hexane molecules. They are held

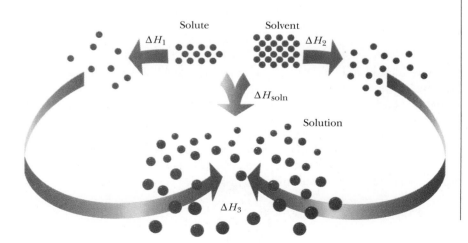

**FIGURE 8.1** Enthalpy changes in solution formation. $\Delta H_{soln} = \Delta H_1 + \Delta H_2 + \Delta H_3$; where $\Delta H_1$ is the enthalpy required to overcome interactions between solute molecules; $\Delta H_2$ is the enthalpy required to overcome interactions between solvent molecules (green circles); and $\Delta H_3$ is the enthalpy of interaction between solvent and solute molecules.

together by van der Waals interactions, which are weak. Considerably more energy is needed to overcome the dipole–dipole interactions and hydrogen bonds between water molecules. The energy released as a result of interaction between molecules of water and of hexane is not enough to overcome the interaction between water molecules themselves (Figure 8.2). In contrast, the heat of solution for dissolving polar solutes in water is favorable.

The only way to get a lowering of energy (a favorable enthalpy change) in this situation is to take advantage of the open hydrogen-bonded structure of water. It is possible for a nonpolar molecule to interact with a water molecule that already has four hydrogen bonds and lower the energy, albeit by a small amount. If the water molecule involved in the interaction has fewer than four hydrogen bonds, the nonpolar molecule adjacent to that water molecule will take the place that could be occupied by a water molecule. In effect, a dipole–dipole interaction (relatively strong) has been replaced by a weaker interaction, raising the energy.

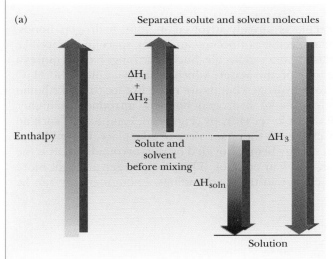

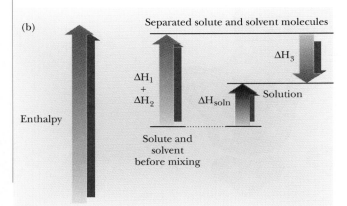

**FIGURE 8.2** Enthalpies of solution ($\Delta H_{soln}$). (a) A favorable $\Delta H_{soln}$ ($\Delta H$ is negative). Exothermic reaction favors solution formation. (b) An unfavorable $\Delta H_{soln}$ ($\Delta H$ is positive). Endothermic reaction does not favor solution formation.

**FIGURE 8.3**   Enthalpies of interaction between water molecules. The energy of water molecules in pure water depends on the number of hydrogen bonds. Each added hydrogen bond gives a more stable state, lowering the energy by a favorable enthalpy of interaction. It is energetically unfavorable to replace a hydrogen bond by a weaker interaction. Replacing a hydrogen bond by a dipole–induced dipole interaction with a nonpolar molecule is particularly unfavorable. The only arrangement that is favorable from the standpoint of enthalpy is to add the stabilization energy of a dipole–induced dipole interaction to that of four hydrogen bonds around a given water molecule. Such an arrangement, however, is highly unfavorable because of the drastic decrease in the entropy of the water.

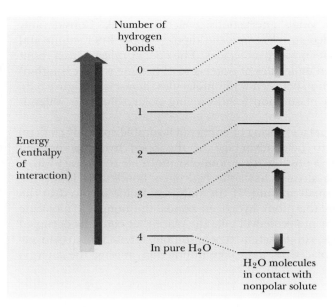

When a water molecule is already involved in four hydrogen bonds, however, a nonpolar molecule can occupy an empty space in the ice-like lattice without replacing a water molecule. Such an arrangement requires a lowering of the entropy of the water, which is unfavorable. The $H_2O$ molecules that have four hydrogen bonds are more like ice than like liquid water. The greater degree of hydrogen bonding introduces a higher degree of order and thus a lower entropy (Figures 8.3 and 8.4). If such an unfavorable entropy change in the water were to take place, it would have to do so as a result of energy being provided by the surroundings, as is the case when hexane dissolves in ethanol. There is a larger and still more unfavorable entropy change in the surroundings. Enough energy can be

**FIGURE 8.4**   A nonpolar molecule in the presence of a partial "cage" of hydrogen-bonded water molecules. The water molecules in the interior of the cluster are involved in four hydrogen bonds, which are not shown in this two-dimensional representation. The enthalpy change ($\Delta H$) for placing a nonpolar solute in water is unfavorable except for a small favorable contribution from water molecules with four hydrogen bonds. The entropy change ($\Delta S$) is highly unfavorable, requiring more extensive hydrogen bonding of water (greater degree of order).

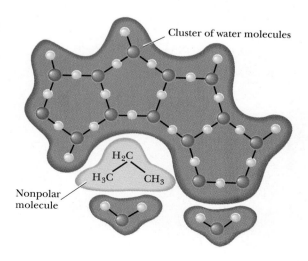

supplied from the surroundings, decreasing the entropy of the surroundings, when hexane dissolves in ethanol but not when we try to dissolve hexane in water. The required entropy decrease in the surroundings is too large for the process to take place. Therefore, nonpolar substances do not dissolve in water; rather, nonpolar molecules associate with one another by hydrophobic interactions and sequester themselves from water.

Hydrophobic interactions have important consequences in biochemistry. Large arrays of molecules can take on a definite structure as a result of hydrophobic interactions. Phospholipid bilayers are an example of such an array. We have seen (Section 3.1) that phospholipids are molecules that have a polar head group and a long nonpolar tail of hydrocarbon chains. Under suitable conditions, a double-layer arrangement is formed such that the polar head groups of a large number of molecules face the aqueous environment; the nonpolar tails are in contact with each other and are kept away from the aqueous environment. These bilayers form three-dimensional structures called **liposomes.** Such structures are useful model systems for biological membranes, which consist of similar bilayers with proteins embedded in them. The very existence of membranes depends on hydrophobic interactions.

Hydrophobic interactions are a major factor in determining the folding of proteins into the specific three-dimensional structures required for them to function as enzymes, oxygen carriers, or structural elements. The order of amino acids (*i.e.*, the nature of the side chains) automatically determines the three-dimensional structure of the protein. It is known experimentally that proteins tend to be folded so that the nonpolar side chains are sequestered from water in the interior of the protein, while the polar side chains lie on the exterior of the molecule and are accessible to the aqueous environment.

**(a)**    **(b)**

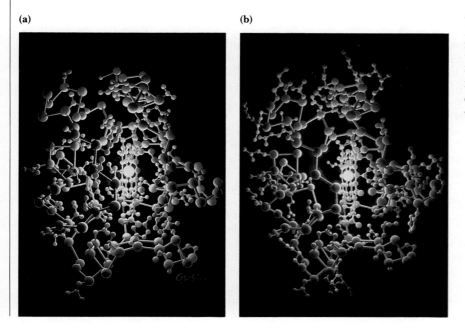

The three-dimensional structure of the protein cytochrome c. (a) The hydrophobic side chains (shown in red) are found in the interior of the molecule. (b) The hydrophilic side chains (shown in green) are found on the exterior of the molecule.

## SUMMARY

Thermodynamics deals with the changes in energy that determine whether or not a process will take place. In a spontaneous process (one that will take place), the free energy decreases. In a nonspontaneous process (one that will not take place), the free energy increases. The symbol for change in free energy is $\Delta G$. Changes in free energy under any set of conditions can be compared to the free energy change under standard conditions ($\Delta G°$). The standard conditions are the pure substance for solids and liquids, 1 atmosphere pressure for gases, and 1 molar concentration for solutes. For biochemical usage, a modified standard state has been defined, in which the standard concentration for hydrogen ion is $1 \times 10^{-7}$ M, implying a pH of 7. Free energy changes under standard conditions can be related to the equilibrium constant of a reaction by the equation

$$\Delta G° = -2.303\, RT \log K_{eq}$$

In addition to the free energy, entropy is an important quantity in thermodynamics. The entropy of the universe increases in any spontaneous process. Local decreases in entropy can take place within a larger overall increase in entropy. Living organisms represent a local decrease in entropy. Unfavorable entropy changes accompany attempts to mix polar and nonpolar substances. The underlying cause of the insolubility of nonpolar substances in water is the decrease in the entropy of the water. These thermodynamic considerations are the basis of hydrophobic interactions, which play an important role in the structure of biological membranes and in maintaining the characteristic three-dimensional structure of proteins.

## EXERCISES

1. For the process

   Nonpolar solute + $H_2O \longrightarrow$ solution

   what are the signs of $\Delta S_{univ}$, $\Delta S_{sys}$, and $\Delta S_{surr}$? What is the reason for your answer to each part of this question?
2. Which of the following are spontaneous processes? Give the reason for your answer in each case.
   (a) The hydrolysis of ATP to ADP and $P_i$
   (b) The oxidation of glucose to $CO_2$ and $H_2O$ by an organism
   (c) The phosphorylation of ADP to ATP
   (d) The production of glucose and $O_2$ from $CO_2$ and $H_2O$ in photosynthesis
3. In which of the following processes does the entropy increase? Give the reason for your answer in each case.
   (a) A bottle of ammonia is opened. The odor of ammonia is soon apparent throughout the room.
   (b) Sodium chloride dissolves in water.
   (c) A protein is completely hydrolyzed to the component amino acids.
4. Which statements are true about the modified standard state for biochemistry? What is the reason for your answer?

$[H^+] = 10^{-7}$ M, not 1 M

$[H_2O] = 1$ M

5. The standard free energy of hydrolysis ($\Delta G°'$) of creatine phosphate at pH 7 and 37°C is $-43.0$ kJ mol$^{-1}$ ($-10.3$ kcal mol$^{-1}$). Calculate the equilibrium constant for the reaction

   creatine phosphate + $H_2O \longrightarrow$ creatine + $P_i$

6. For the hydrolysis of ATP at 25°C (298 K) and pH 7

   ATP + $H_2O \longrightarrow$ ADP + $P_i$ + $H^+$

   the standard free energy of hydrolysis ($\Delta G°'$) is $-30.5$ kJ mol$^{-1}$ ($-7.3$ kcal mol$^{-1}$), and the standard enthalpy change ($\Delta H°'$) is $-20.1$ kJ mol$^{-1}$ ($-4.8$ kcal mol$^{-1}$). Calculate the standard entropy change ($\Delta S°'$) for the reaction in both joules and calories. Why is the positive sign of the answer to be expected in view of the nature of the reaction?
7. Which of the following statements are true? Give the reason for your answer.
   (a) An unfavorable entropy change for the water is the most important factor in the insolubility of nonpolar substances in water.
   (b) The entropy of the universe decreases in a spontaneous process.

(c) The enthalpy change for a reaction is the heat of reaction measured at constant pressure.
(d) Heat is a reflection of the random motion of molecules.
(e) An endergonic reaction occurs spontaneously.

**8.** In most proteins, the majority of the nonpolar side chains are found in the interior of the molecule, while the polar residues tend to be found on the surface. Comment on this observation in light of the material in this chapter.

**9.** The standard free energy change ($\Delta G^{\circ\prime}$) for the reaction

$$CO_2 + \text{pyruvate} + ATP \longrightarrow$$
$$\text{oxaloacetate} + ADP + P_i$$

is 0.26 kJ mol$^{-1}$. Calculate the equilibrium constant for this reaction.

**10.** Comment on the statement that the existence of life is a violation of the second law of thermodynamics.

**11.** Comment on the role of hydrophobic interactions in maintaining the structure of biological membranes.

**12.** Would you expect the biosynthesis of a protein from the constituent amino acids in an organism to be an exergonic or an endergonic process? Give the reason for your answer.

**13.** Adult humans synthesize large amounts of ATP in the course of a day, but their body weight does not change significantly. The structure and composition of their bodies does not change appreciably in the same time period. Explain this apparent contradiction.

**14.** Would you expect an increase or decrease of entropy to accompany the hydrolysis of phosphatidylcholine to the constituent parts (glycerol, two fatty acids, phosphoric acid, and choline)? Give the reason for your answer.

# ANNOTATED BIBLIOGRAPHY

Atkins, P.W. *The Second Law*. San Francisco: W.H. Freeman, 1984. [A highly readable nonmathematical discussion of thermodynamics.]

Campbell, J.A. Reversibility and Returnability, or When Can You Return Again? *J. Chem. Ed.* **57**, 345–348 (1980). [A treatment of entropy and its role in spontaneous processes.]

Chang, R. *Physical Chemistry with Applications to Biological Systems,* 2nd ed. New York: Macmillan, 1981. [Chapter 12 contains a detailed treatment of thermodynamics.]

Harold, F.M. *The Vital Force: A Study of Bioenergetics.* New York: W.H. Freeman, 1986. [Energetic aspects of many important life processes treated here.]

Prigogine, I., and I. Stengers. *Order Out of Chaos.* Toronto: Bantam Press, 1984. [A comparatively accessible treatment of the thermodynamics of biological systems. The first author won the 1977 Nobel Prize in chemistry for his pioneering work on the thermodynamics of complex systems.]

Tanford, C. *The Hydrophobic Effect: Formation of Micelles and Biological Membranes,* 2nd ed. New York: Wiley-Interscience, 1980. [A comprehensive treatment of the topic.]

Most textbooks of general chemistry have several chapters on thermodynamics. Of these, S. Zumdahl, *Chemistry* (Lexington, MA: D.C. Heath, 1986) has a particularly good treatment of the thermodynamics of solution formation in Sections 11.2 and 16.3.

# 9

# The Three-Dimensional Structure of Proteins

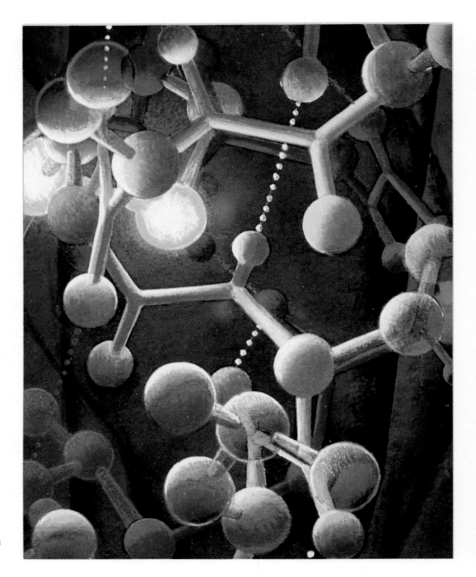

*The structure of a part of the protein crambin.*

*Amino acids joined together form a protein (polypeptide) chain. The repeating units are amide planes containing the peptide bond. These amide planes can twist about their connecting carbon atoms to create the three-dimensional conformation of proteins. Forty years ago, Linus Pauling predicted that linked amino acids would form an α helix. Years later, Pauling's prediction was confirmed when myoglobin, an oxygen-binding protein, was found to be made from Pauling's α helices. This type of local folding of the protein chain is called* secondary *structure; the linear sequence being the* primary *structure. Conformation of a complete protein chain is its* tertiary *structure. Myoglobin, a molecule that binds oxygen tightly, has a single protein chain. Hemoglobin, a protein with four myoglobin-like subunits fitted together, has a* quaternary *structure. This allows it to change its shape from the* oxy *conformation when it binds oxygen in the lungs, to the* deoxy *form when it releases oxygen to working tissues. The discovery of structure/function relationships in hemoglobin led to understanding how complex multisubunit enzymes are able to regulate metabolic pathways.*

## 9.1
## INTERPLAY OF THE FORCES THAT DETERMINE PROTEIN STRUCTURE

### Levels of Structure in Proteins

There are many different possible conformations (three-dimensional structures) for a molecule as large as a protein. Of the many possible structures, one, or at most a few, have biological activity; these are called the **native conformations.** Most proteins appear to have little regular arrangement of atoms. As a consequence, proteins are frequently referred to as having large segments of random structure. The term "random" is really a misnomer, since the same complex structure is found in all molecules of a given protein having the same native conformation. Because of the complexity of the three-dimensional structure of proteins, it has become usual to define several levels of structure to attack the problem of structure determination more efficiently.

Four levels of structural organization can be distinguished in proteins. **Primary structure** refers to the order in which the amino acids are covalently linked together. The peptide $H_2N$—Leu—Gly—Thr—Val—Arg—Asp—His—COOH has a different primary structure from that of the peptide $H_2N$—Val—His—Asp—Leu—Gly—Arg—Thr—COOH, even though both have the same number and kind of amino acids. Note that the order of amino acids can be written on one line. The primary structure is the one-dimensional first step in specifying the three-dimensional structure of a protein.

Two aspects of the three-dimensional structure of a single polypeptide chain, called the secondary and tertiary structure, can be considered separately. **Secondary structure** refers to the arrangement in space of the atoms in the backbone of the polypeptide chain. The $\alpha$-helix and $\beta$-pleated sheet hydrogen-bonded arrangements (Section 3.2) are two different types of secondary structure. **Tertiary structure** includes the three-dimensional arrangement of all the atoms in the protein, including the atoms in the side chains and any **prosthetic groups** (ones other than amino acids). In very large proteins, the folding of parts of the chain can occur independently of the folding of other parts. Such independently folded portions of proteins are referred to as **domains.**

Proteins can consist of more than one polypeptide chain; they have more than one **subunit.** Determining the arrangement of subunits with respect to one another specifies the **quaternary structure.** Interaction between subunits is mediated by noncovalent interactions, such as hydrogen bonds and hydrophobic interactions.

## The Importance of Primary Structure

The amino acid sequence (the primary structure) of a protein determines its three-dimensional structure, which in turn determines its properties. In enzymes, the complex three-dimensional structure serves to place the crucial amino acids that are directly involved in catalyzing the reaction close to each other. In all proteins the correct three-dimensional structure is needed for correct functioning.

One of the most striking examples of the importance of primary structure is found in the hemoglobin associated with **sickle-cell anemia.** In this disease, which is of genetic origin, red blood cells are unable to bind oxygen efficiently. The red cells also assume a characteristic sickled shape, giving the disease its name. The sickled cells tend to become trapped in small blood vessels, cutting off circulation and causing organ damage. A change in one amino acid residue in the sequence of the primary structure causes these drastic consequences.

The difference between sickle-cell hemoglobin (hemoglobin S) and normal hemoglobin (hemoglobin A) becomes apparent when the two proteins are cleaved into smaller peptides by the enzyme trypsin and the peptides are separated by the technique of fingerprinting (see Interchapter A, Section 2). Trypsin cleaves hemoglobin at the lysine and arginine residues. When the tryptic digest of hemoglobin S is subjected to the fingerprinting technique, the results differ from those for hemoglobin A by a change in only one peptide. The four amino acid chains of hemoglobin consist of two pairs of identical chains, called $\alpha$ and $\beta$, respectively. In order to determine whether the abnormal peptide is part of the $\alpha$ or $\beta$ chain, it is necessary to separate the two kinds of chains. This can be done by ion-exchange chromatography (see Interchapter A, Section 1). Fingerprinting can again be done, this time on the separated chains, and the abnormal peptide is found in the $\beta$ chain.

The abnormal peptide, which is eight residues long, is found at the N-terminal end of the $\beta$ chain. Sequence determination shows that valine occurs in hemoglobin S at position 6 of the $\beta$ chain, replacing the glutamic acid that is found at this position in hemoglobin A.

| Hemoglobin A | Val—His—Leu—Thr—Pro—**Glu**—Glu—Lys— |
| Hemoglobin S | Val—His—Leu—Thr—Pro—**Val**—Glu—Lys— |
| | $\beta$1   2   3   4   5   **6**   7   8 |

The highly polar side chain of glutamic acid, containing an ionizable carboxyl group, has been replaced by a nonpolar one, the isopropyl group of valine. In the three-dimensional structure of hemoglobin, this residue is on the outside of the molecule. One molecule of hemoglobin S can become

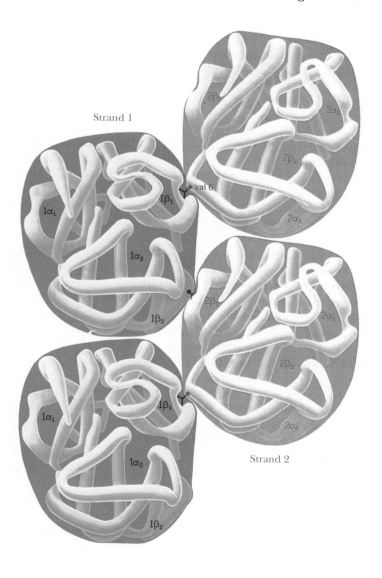

Strand 1

Strand 2

The aggregation of hemoglobin molecules in sickle-cell anemia. The yellow dots indicate the position of $\beta$ 6 valine, which fits into a hydrophobic pocket ($\beta$ 85 phenylalanine and $\beta$ 88 leucine) on an adjacent HbS molecule.

involved in hydrophobic interactions with other hemoglobin molecules because of the presence of a nonpolar residue. This interaction does not take place in hemoglobin A with a polar residue in the same position. As a result, groups of molecules of hemoglobin S aggregate with each other, and the aggregates distort the shape of the red blood cell, causing the symptoms of the disease in the form of sickling episodes.

## Forces Involved in Protein Folding

The primary structure of a protein is determined by covalent bonds, because the order of amino acids in the polypeptide chain depends on the formation of peptide bonds. Higher order levels of structure, such as the conformation of the backbone (secondary structure) and the position of all the atoms in the protein (tertiary structure), depend on noncovalent interactions; if the protein consists of several subunits, the interaction of the subunits (quaternary structure) also depends on noncovalent interactions. Noncovalent stabilizing forces contribute to the most stable structure for a given protein, the one with the lowest energy.

Several types of hydrogen bonding occur in proteins. *Backbone hydrogen bonding* is a major determinant of secondary structure; *hydrogen bonds between the side chains of amino acids* are also possible in proteins. Nonpolar residues tend to cluster together in the interior of protein molecules as a result of *hydrophobic* interactions. *Electrostatic* attraction between oppositely charged groups, which frequently occur on the surface of the molecule, results in such groups being close to one another. Several side chains can be *complexed* to a single metal ion. (Metal ions also occur as part of prosthetic groups.)

In addition to these noncovalent interactions, *disulfide bonds* form covalent links between the side chains of amino acids. When such bonds are formed, they impose restrictions on the folding patterns available to polypeptide chains. There are specialized laboratory methods for determining the number and positions of disulfide links in a given protein. Information about the location of disulfide links can then be combined with knowledge of the primary structure to give the complete covalent structure of the protein. Note the subtle difference here: the primary structure is the order of amino acids, while the complete covalent structure also specifies the positions of the disulfide bonds (Figure 9.1).

As a result of these various stabilizing forces, residues that are far apart in the primary sequence can be close to each other in the three-dimensional structure produced by the folding of the protein. When a polypeptide chain folds back on itself, it can assume a compact globular shape. A different polypeptide chain (or the same chain under different conditions) can assume a rodlike fibrous form.

Every protein does not necessarily exhibit all possible structural features. This point is especially true of disulfide bridges and metal ion complexation. There are no disulfide bridges in myoglobin and hemoglobin, which are oxygen storage and transport proteins and classic examples of protein structure, but they both contain Fe(II) ions as part of a prosthetic

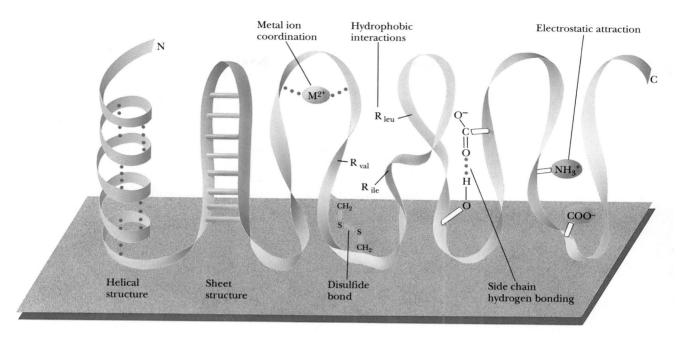

Metal ion coordination · Hydrophobic interactions · Electrostatic attraction · N · C · M²⁺ · R leu · R val · R ile · O⁻ · C · O · H · O · CH₂ · S · S · CH₂ · NH₃⁺ · COO⁻ · Helical structure · Sheet structure · Disulfide bond · Side chain hydrogen bonding

FIGURE 9.1   Forces that stabilize the tertiary structure of proteins. Note that helical structure and sheet structure are two kinds of backbone hydrogen bonding.

group. [The notation Fe(II) is preferred to $Fe^{2+}$ when metal ions occur in complexes.] The enzymes trypsin and chymotrypsin (which we will discuss in Section 9.2 in connection with sequencing of proteins) do not contain complexed metal ions, but they do have disulfide bridges. Hydrogen bonds, electrostatic interactions, and hydrophobic attractions are found in most proteins.

The three-dimensional conformation of a protein is the result of the interplay of all the stabilizing forces. It is known, for example, that proline does not fit into an $\alpha$-helix and that the presence of this amino acid can cause a polypeptide chain to turn a corner, ending an $\alpha$-helical segment. The presence of proline is not, however, a requirement for a turn in a polypeptide chain. Other residues are routinely encountered at bends in polypeptide chains. The segments of proteins at bends in the polypeptide chain and in other portions of the protein that are not involved in helical or pleated sheet structures are frequently referred to as "random," but the term is something of a misnomer. The forces that stabilize proteins are responsible for a definite conformation for each protein.

## Denaturation and Refolding

The noncovalent interactions responsible for maintaining the three-dimensional structure of a protein are weak, and it is not surprising that they can be disrupted easily. The process of unfolding a protein is called **denaturation.** Reduction of disulfide bonds leads to even more extensive unraveling of the tertiary structure. Denaturation and reduction of disulfide bonds are frequently combined when complete disruption of the tertiary structure of proteins is desired. The three-dimensional structure of proteins can be completely disrupted and, under proper experimental conditions, completely recovered. This process of denaturation and refolding is a dramatic example of the relationship between the primary structure of the protein and the forces that determine the tertiary structure.

Proteins are denatured by *heat* because vibrations within the molecule are favored by an increase in temperature. The energy of these vibrations can become large enough to disrupt the tertiary structure. *Extremes of pH,* either very high or very low, lead to changes in the charge on the protein. The electrostatic interactions that would normally stabilize the native, active form of the protein are drastically reduced at extremes of pH

**FIGURE 9.2**  Denaturation of proteins. (a) The process of denaturation. (b) The mode of action of detergents. $R^+$ is a positively charged side chain, and $R_{NP}$ is a nonpolar side chain. (c) The mode of action of urea.

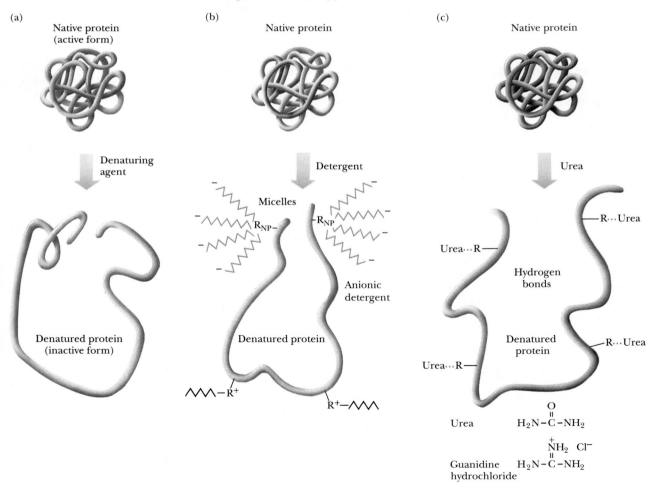

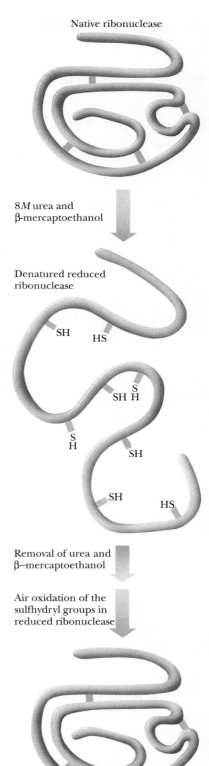

Native ribonuclease

8 M urea and
β-mercaptoethanol

Denatured reduced
ribonuclease

SH    HS

SH   S
     H

S
H        SH

SH        HS

Removal of urea and
β—mercaptoethanol

Air oxidation of the
sulfhydryl groups in
reduced ribonuclease

Native ribonuclease

**FIGURE 9.3** Denaturation and refolding in ribonuclease. The protein ribonuclease can be completely denatured by the action of urea and mercaptoethanol. When denaturing conditions are removed, activity is recovered.

because at least some of the charges are missing. The binding of *detergents* such as sodium dodecyl sulfate (SDS) also denatures proteins. Detergents tend to disrupt hydrophobic interactions. If a detergent is charged, it can also disrupt electrostatic interactions within the protein. Other reagents such as *urea* or *guanidine hydrochloride* form hydrogen bonds to the protein that are stronger than those within the protein. These two reagents can also disrupt hydrophobic interactions in much the same way as detergents (Figure 9.2).

**Mercaptoethanol** ($HS—CH_2—CH_2—OH$) is frequently used for reducing disulfide bridges to two sulfhydryl groups. Urea is usually added to the reaction mixture to facilitate unfolding of the protein and to increase the accessibility of the disulfides to the reducing agent. If experimental conditions are properly chosen, the native conformation of the protein can be recovered when both mercaptoethanol and urea are removed (Figure 9.3). Experiments of this type provide some of the strongest evidence that the amino acid sequence of the protein provides all the information required for the complete three-dimensional structure. It is a point of some interest in protein research to find conditions under which a protein can be denatured, including reduction of disulfides, and the native conformation recovered later.

## 9.2
## THE DETERMINATION OF PRIMARY STRUCTURE

### The Strategy for Determining Primary Structure

Determining the sequence of amino acids in a protein is a routine, but not trivial, operation. It consists of several parts, which must be carried out carefully to obtain accurate results. The first question to answer is, "Which of the commonly occurring amino acids are found in this protein and how many of each kind?" (We do not need to consider the occurrence of amino acids other than the usual ones for our current discussion.) The next question is, "Which amino acids occur at the N-terminal and C-terminal ends of the molecule?" The final step consists of determining the exact sequence of amino acids by cleaving the protein enzymatically or chemically at specific sites. Such a reaction produces peptides short enough to be sequenced by removing and identifying one amino acid at a time. This procedure gives the order of amino acids within the peptides, but does not specify the order in which the peptides are arranged within the protein. This difficulty is overcome by using another enzyme or chemical reagent to cleave another sample of the same protein at other, equally specific sites. The peptides obtained from the second cleavage reaction have sequences that overlap those of the peptides produced by the first reaction. The

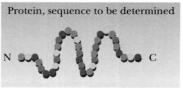
Protein, sequence to be determined

FIGURE 9.4    The strategy for determining the primary structure of a protein.

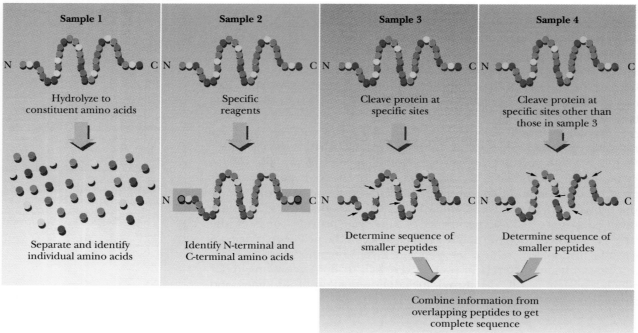

complete amino acid sequence of the protein can be deduced by comparing the overlapping sequences (Figure 9.4).

## Determining the Amino Acid Sequence

The first step in determining the primary structure of a protein is to establish which amino acids are present and in what proportions. It is relatively easy to break a protein down to its component amino acids by heating a solution of the protein in acid, usually 6 M HCl, at 100 to 110°C for 12 to 36 hours. Separation and identification of the products are somewhat more difficult and are best done by an instrument known as an **amino acid analyzer.** A detailed description of the operation of an amino acid analyzer appears in Interchapter A on experimental methods. These automated instruments give both qualitative information about the identities of the amino acids present and quantitative information about the relative amounts of these amino acids.

The identity of the N-terminal and C-terminal amino acids in a protein sequence can be determined in several ways. There are both chemical and enzymatic methods for identifying the amino acids at the ends of the molecule. These techniques are also described in Interchapter A.

When the end-group analysis has been done, it is possible to proceed with the determination of the amino acid sequence. There are automated instruments that can do a stepwise modification, cleavage, and identification of all the amino acids in a polypeptide chain, but the process becomes more difficult as the number of amino acids increases. In most proteins the chain is more than 100 residues long. It is usually necessary to break a long polypeptide chain into fragments of reasonable size for sequencing.

Proteins can be cleaved at specific sites by enzymes or by chemical reagents. The enzyme **trypsin** cleaves peptide bonds at amino acids that have positively charged R groups, such as lysine and arginine. The cleavage takes place in such a way that the amino acid with the charged side chain ends up at the C-terminal end of one of the peptides produced by the reaction. The C-terminal amino acid of the original protein can be any one of the twenty amino acids and is not necessarily one at which cleavage takes place. A peptide can be automatically identified as the C-terminal end of the original chain if its C-terminal amino acid is not one of the ones at which cleavage takes place.

Another enzyme, **chymotrypsin,** cleaves peptide bonds preferentially at aromatic amino acids such as tyrosine, tryptophan, and phenylalanine. The aromatic amino acid is found at the C-terminal end of the peptides produced by the reaction (Figure 9.5). In the case of the chemical reagent **cyanogen bromide** (CN—Br), the site of cleavage is at internal methionine residues. The sulfur of the methionine reacts with the carbon of the cyanogen bromide to produce a homoserine lactone at the C-terminal end of the fragment (Figure 9.6). The use of several such reagents on different samples of a protein to be sequenced will produce different mixtures of fragments. The sequences of a set of peptides produced by one reagent will overlap the sequences of peptides produced by another reagent (Figure

**FIGURE 9.5** Cleavage of proteins by chymotrypsin. Chymotrypsin is an endopeptidase, an enzyme that cleaves internal peptide bonds.

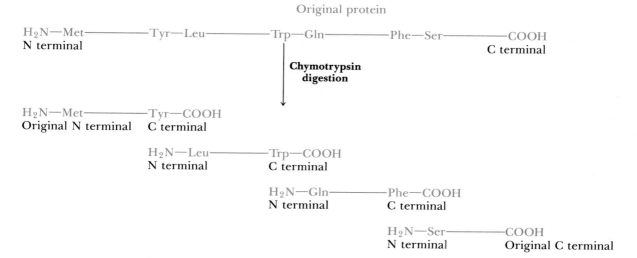

**FIGURE 9.6**  Cleavage of proteins at internal methionine residues by cyanogen bromide.

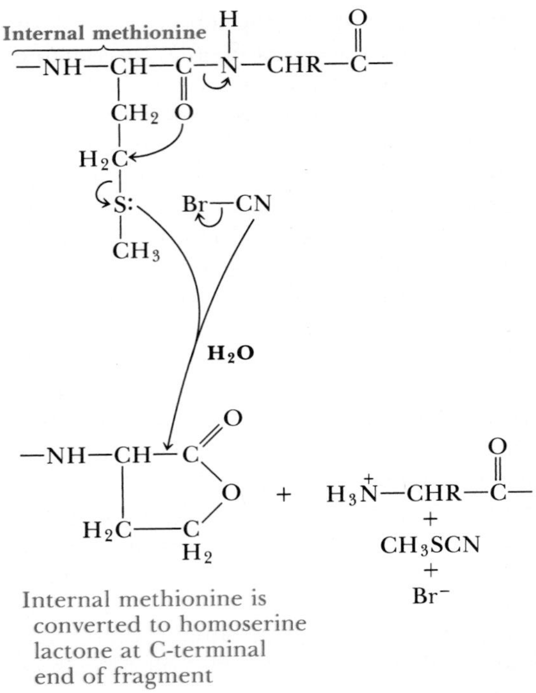

Internal methionine is converted to homoserine lactone at C-terminal end of fragment

9.7). This makes it possible to arrange the peptides in the proper order once their own sequences have been determined.

The cleavage of a protein by any of these reagents produces a mixture of peptides. The individual peptides must be separated from one another before they can be sequenced. Several methods are combined to obtain a good separation; frequently a method that uses separation on the basis of charge is combined with one that uses separation on the basis of polarity. This technique, which is called **fingerprinting,** is described in detail in Interchapter A, Section 2.

The actual sequencing of each peptide produced by specific cleavage of a protein is done by repeated application of a procedure called **Edman degradation.** In this method, the Edman reagent, **phenyl isothiocyanate,** reacts with the N-terminal residue of the peptide. The modified amino acid can be cleaved off, *leaving the rest of the peptide intact,* and can be detected as the phenylthiohydantoin derivative of the amino acid. The second amino

**FIGURE 9.7**  The use of overlapping sequences to determine sequence. Partial digestion was done using chymotrypsin and cyanogen bromide.

| Chymotrypsin | $^+H_3N-Leu-Asn-Asp-Phe$ |
| Cyanogen bromide | $^+H_3N-Leu-Asn-Asp-Phe-His-Met$ |
| Chymotrypsin | His−Met−Thr−Met−Ala−Trp |
| Cyanogen bromide | Thr−Met |
| Cyanogen bromide | Ala−Trp−Val−Lys−COO$^-$ |
| Chymotrypsin | Val−Lys−COO$^-$ |

Overall sequence     $^+H_3N-Leu-Asn-Asp-Phe-His-Met-Thr-Met-Ala-Trp-Val-Lys-COO^-$

acid of the original peptide can then be treated in the same way, then the third. The process can be repeated until the whole peptide is sequenced; this is done on an automated instrument called a **sequencer** (Figure 9.8). The sequences of peptides containing 10 to 20 residues can easily be determined by this method.

After the Edman procedure has been used to determine the sequences of all the various peptides produced as a result of partial digestion of the protein, it becomes possible to determine the sequence of the entire protein. The overlapping sequences of peptides produced by different reagents provide the key to solving the puzzle. The alignment of like

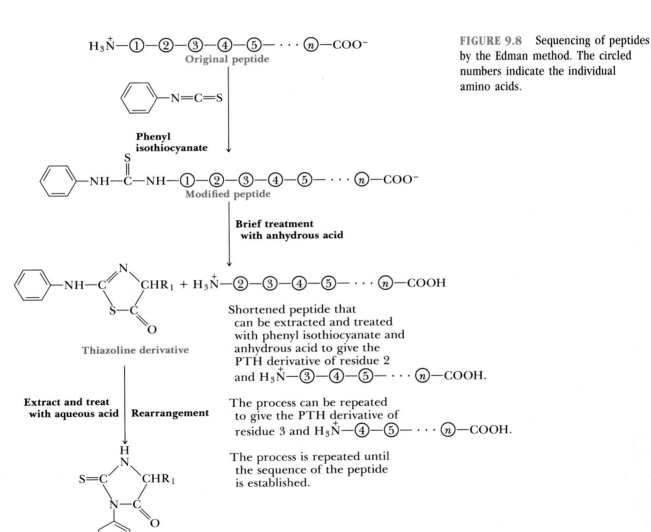

**FIGURE 9.8** Sequencing of peptides by the Edman method. The circled numbers indicate the individual amino acids.

sequences on different peptides makes it possible to deduce the overall sequence. The amino acid sequences of the individual peptides in Figure 9.7 were determined by the Edman method after the peptides were separated from one another and before the overall sequence of the larger peptide could be determined.

Another sequencing method makes use of the fact that the amino acid sequence of a protein reflects the base sequence of the DNA in the gene that coded for that protein. It is easier, using currently available methods, to obtain the sequence of the DNA than that of the protein. (See Section 20.5 and Box 20.1 for a discussion of sequencing methods for nucleic acids.) Convenient though this method may be, it does not determine the positions of disulfide bonds or detect amino acids, such as hydroxyproline, that are modified after translation.

## 9.3
## SECONDARY STRUCTURE OF PROTEINS

The secondary structure of proteins is the arrangement in space of the atoms of the backbone, the polypeptide chain itself. The spatial arrangement of the atoms of the side chains is not considered here but is part of the tertiary structure. The nature of the bonds in the peptide backbone plays an important role here. There are two bonds with reasonably free rotation within each amino acid residue. They are (1) the bond between the α-carbon of a given residue and the amino nitrogen of that residue and (2) the bond between the α-carbon and the carboxyl carbon of that residue. We have already seen that the peptide group (CO—NH) is planar (Section 5.4). The combination of the planar peptide group and the two freely rotating bonds has important implications for the three-dimensional conformations of peptides and proteins. A peptide chain backbone can be drawn as a series of playing cards, representing the planar peptide groups. The peptide groups are linked at opposite ends by swivels, representing the bonds about which there is considerable freedom of rotation (Figure 9.9). The side chains also play a vital role in determining the three-dimensional shape of proteins, but only the backbone is considered in secondary structure.

The angles $\phi$ (phi) and $\psi$ (psi) are used to designate rotations around the C—N and C—C bonds, respectively. The conformation of a protein backbone can be described by specifying the values of $\phi$ and $\psi$ for each residue of a protein. Two kinds of secondary structures that occur frequently in proteins are the α-helix and β-pleated sheet hydrogen-bonded structures. These repeating structures deserve a closer look; they are not the only possible secondary structures, but they are by far the most important.

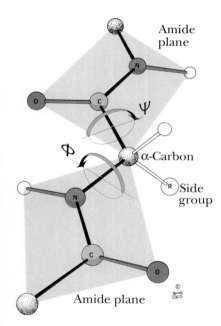

FIGURE 9.9  Definition of the angles that determine the conformation of a polypeptide chain. The planar peptide groups are shaded. The angle of rotation around the $C^\alpha$—N bond is designated as $\phi$ (phi), while the angle of rotation around the $C^\alpha$—C bond is designated as $\psi$ (psi).

## Periodic Structures in Protein Backbones

The α-helix, β-pleated sheet, and collagen triple helix are all periodic structures; their features repeat at regular intervals. The α-helix and collagen triple helix are both rodlike in nature; the α-helix involves only one polypeptide chain, while the collagen structure consists of three chains twisted together. The $\beta$-pleated sheet structure can give a two-dimensional array and can involve one or more polypeptide chains.

The $\alpha$-helix is stabilized by hydrogen bonds within the backbone of the polypeptide chain. Counting from the N-terminal end, the CO group of each amino acid residue is hydrogen bonded to the NH group of the amino acid four residues away from it in the covalently bonded sequence. The helical conformation allows a linear arrangement of the atoms involved in the hydrogen bonds. This situation is the one most favorable for strength in the hydrogen bond, affording greater stabilization for the helical conformation (Section 3.2). There are 3.6 residues for each turn of the helix, and the *pitch* of the helix (the linear distance between corresponding points on successive turns) is 5.4 Å (Figure 9.10). [The angstrom

**(a)**                **(b)**

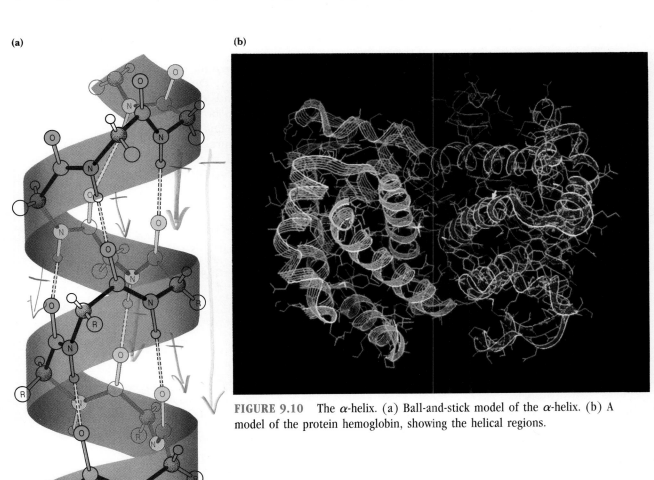

FIGURE 9.10    The $\alpha$-helix. (a) Ball-and-stick model of the $\alpha$-helix. (b) A model of the protein hemoglobin, showing the helical regions.

unit ($1\ \text{Å} = 10^{-8}\ \text{cm} = 10^{-10}\ \text{m}$) is convenient for interatomic distances in molecules, but it is not one of the Système International (SI) units. Nanometers ($1\ \text{nm} = 10^{-9}\ \text{m}$) and picometers ($1\ \text{pm} = 10^{-12}\ \text{m}$) are the SI units used for interatomic distances. In SI units the pitch of the $\alpha$-helix is 0.54 nm or 540 pm.]

Several factors tend to disrupt the $\alpha$-helix. The amino acid proline creates a bend in the backbone because of its cyclic structure; it cannot fit into the $\alpha$-helix. Other localized factors involving the side chains include strong electrostatic repulsion due to the proximity of several charged groups of the same sign, such as a group of positively charged lysine and arginine residues or a group of negatively charged glutamate and aspartate residues. Another possibility is crowding (steric repulsion) due to the proximity of several bulky side chains. In the $\alpha$-helical conformation all the side chains lie outside the helix; there is not enough room for them in the interior. The $\beta$ carbon is just outside the helix, and crowding can occur if it is bonded to two atoms other than hydrogen, as is the case with valine and threonine.

The arrangement of atoms in the $\beta$-pleated sheet conformation differs markedly from that in the $\alpha$-helix. The peptide backbone is almost completely extended. Hydrogen bonds can be formed between different parts of the same chain (**intrachain** bonds) or between different chains (**interchain** bonds). If the peptide chains run in the same direction (that is, if they are all aligned in the same direction from the N-terminal to the C-terminal end), a **parallel pleated sheet** is formed. When alternating chains run in opposite directions, an **antiparallel pleated sheet** is formed (Figure 9.11).

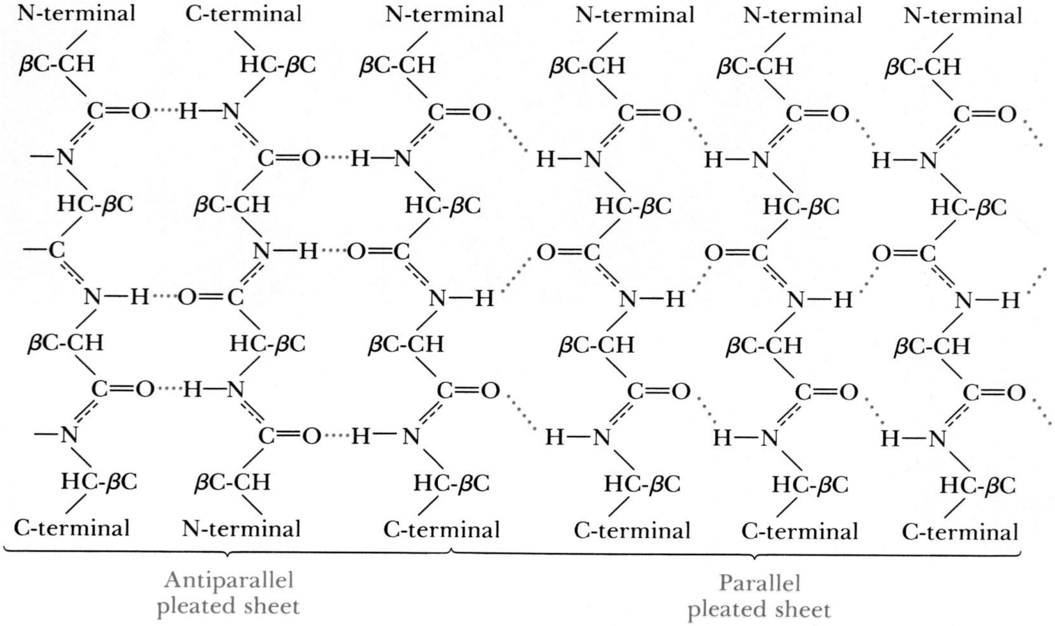

**FIGURE 9.11**  $\beta$-Pleated sheet structures. In an antiparallel pleated sheet, the peptide chains run in opposite directions from the N terminal to the C terminal end. In a parallel pleated sheet, the peptide chains run in the same direction.

The α-helix and β-pleated sheet are combined in many ways as the polypeptide chain folds back on itself in a protein. Glycine is frequently encountered in **reverse turns,** at which the polypeptide chain changes direction for steric reasons; the single hydrogen of the side chain prevents crowding (Figure 9.12). The cyclic structure of proline has the correct geometry for a reverse turn, and this amino acid is also encountered frequently in such turns (Figure 9.12(c)). The combination of α and β strands produces various kinds of **supersecondary structures** in proteins. The commonest feature of this sort is the **βαβ unit,** in which two parallel strands of β-sheet are connected by a stretch of α-helix (Figure 9.13(a)). An **αα unit** consists of two antiparallel α-helices (Figure 9.13(b)). In such an arrangement there are energetically favorable contacts between the side chains in the two stretches of helix. In a **β meander,** an antiparallel sheet is formed by a series of tight reverse turns connecting stretches of the polypeptide chain (Figure 9.13(c)). Another kind of antiparallel sheet is

**FIGURE 9.12** Structures of reverse turns. Arrows indicate the direction of the polypeptide chain. (a) Type I reverse turn. In residue 3 the side chain lies outside the loop, and any amino acid can occupy this position. (b) Type II reverse turn. The side chain of residue 3 has been rotated 180° from the position in the type I turn and is now on the inside of the loop. Only the hydrogen side chain of glycine can fit into the space available, so glycine must be the third residue in a type II reverse turn. (c) The five-membered ring of proline has the correct geometry for a reverse turn; this residue normally occurs as the second residue of a reverse turn. The turn shown here is type II, with glycine as the third residue.

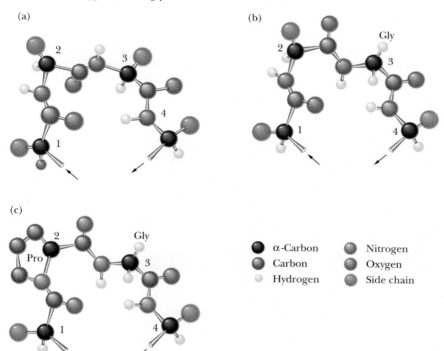

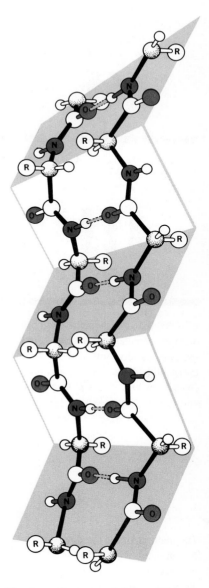

The three-dimensional form of the β-pleated sheet arrangement. The chains do not fold back on each other but are in a fully extended conformation.

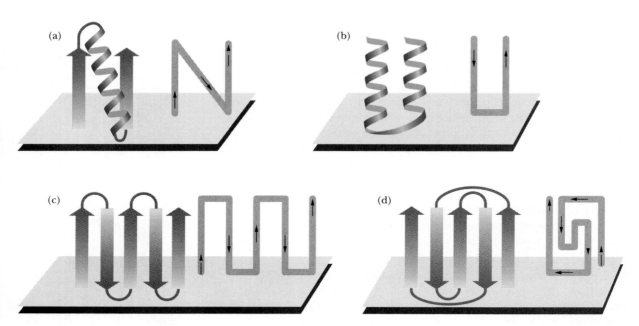

**FIGURE 9.13**    Schematic diagrams of supersecondary structures. Arrows indicate the direction of the polypeptide chain. (a) A $\beta\alpha\beta$ unit, (b) an $\alpha\alpha$ unit, (c) a $\beta$ meander, and (d) the Greek key.

formed when the polypeptide chain doubles back on itself in a pattern known as the **Greek key,** found on pottery from the classical period (Figure 9.13(d)). When $\beta$-sheets are extensive enough they can fold back on themselves, forming a **$\beta$-barrel,** a structural feature that occurs in many proteins (Figure 9.14).

The collagen triple helix has a unique fibrous structure. Each of the three chains has, within limits, a repeating amino acid sequence. The repeating sequence consists of three amino acid residues, X-Pro-Gly or X-Hyp-Gly, where Hyp stands for hydroxyproline; any amino acid can occupy the first position, designated as X. Quite frequently X is also

The main chain of the digestive enzyme carboxypeptidase A shows clusters of $\alpha$-helical and $\beta$-pleated sheet conformations. The red dot shows the position of a $Zn^{2+}$ ion complexed to three specific side chains; this metal ion is necessary for the activity of the enzyme.

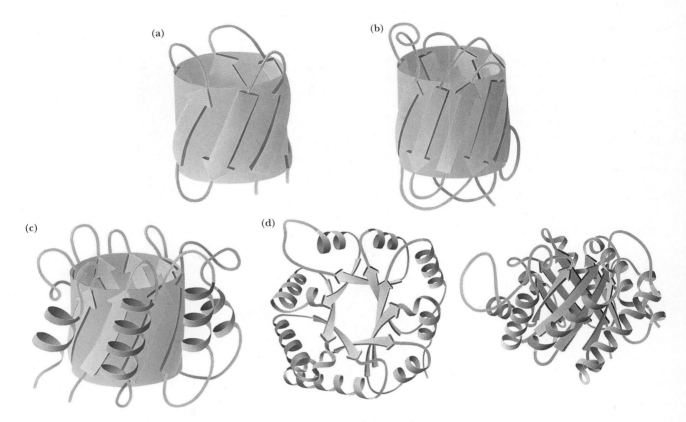

**FIGURE 9.14** Some β barrel arrangements (a) A linked series of β meanders. This arrangement occurs in the protein rubredoxin from *Clostridium pasteurianum*. (b) The Greek key pattern occurs in human prealbumin. (c) A β barrel involving alternating βαβ units. This arrangement occurs in triose phosphate isomerase from chicken muscle. (d) Top and side views of the polypeptide backbone arrangement in triose phosphate isomerase. Note that the α-helical sections lie outside the actual β barrel.

proline or hydroxyproline. Hydroxyproline is formed from proline by a specific hydroxylating enzyme after the amino acids are linked together. In the amino acid sequence of collagen, every third position must be occupied by glycine. The arrangement of the triple helix is such that every third residue on each chain is inside the helix. Only glycine is small enough to fit into the space available (Figure 9.15).

Each individual collagen chain has a helical conformation. The three individual helices are twisted around each other in a superhelical arrangement to form a stiff rod. This triple helical molecule is called **tropocollagen;** it is 300 nm (3000 Å) long and 1.5 nm (15 Å) in diameter. The three strands are held together by hydrogen bonds. The molecular weight of the triple-stranded array is about 300,000; each strand contains about 800 amino-acid residues.

Collagen in which the proline is not hydroxylated to hydroxyproline to the usual extent is less stable than normal collagen. The symptoms of

**FIGURE 9.15**   The collagen triple helix. Each of the three chains of the collagen triple helix has a helical conformation. Each chain also has an amino acid sequence that can be considered a repeat of the same tripeptide, glycine-hydroxyproline-proline. Some variations are possible (see text) but glycine must occur at every fourth position along the chain, and most of the other residues are proline or hydroxyproline. The three chains are wrapped around each other to form a rope-like structure.

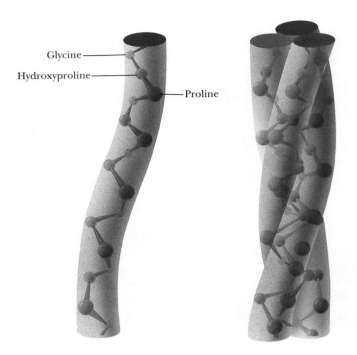

scurvy, such as bleeding gums and skin discoloration, are the result of fragile collagen. The enzyme that hydroxylates proline and thus maintains the normal state of collagen requires ascorbic acid (vitamin C) to maintain its active state. Thus, scurvy is ultimately caused by dietary vitamin C deficiency.

## 9.4
## TERTIARY STRUCTURE OF PROTEINS

Specifying the tertiary structure of a protein requires determining the three-dimensional arrangement of all the atoms in the molecule. The conformations of the side chains and positions of any prosthetic groups are a part of the tertiary structure, as is the arrangement of helical and pleated sheet sections with respect to one another. In a fibrous protein such as collagen, in which the overall shape of the protein is that of a long rod, the secondary structure provides a good deal of the information about the tertiary structure as well. The helical backbone of the protein does not fold back on itself, and the only important point of the tertiary structure that is left unspecified is the arrangement of the atoms of the side chains. In a globular protein considerably more information is needed. It is necessary to determine how the helical and pleated sheet sections fold back on each other, in addition to determining the positions of the side chain atoms and any prosthetic groups.

The experimental technique used to determine the tertiary structure of a protein is **x-ray crystallography.** Crystals of proteins can be obtained under suitable conditions. The conditions of crystal growth must be carefully controlled so that a perfect crystal will be formed. In such a crystal, all the individual protein molecules have the same three-dimensional conformation, and each molecule has the same orientation with respect to every other molecule. Crystals of this quality can be formed only from proteins of very high purity.

When a suitably mounted protein crystal is exposed to a beam of x-rays, a *diffraction pattern* is produced. This pattern is produced as a result of scattering of the x-rays by the electrons in each atom in the molecule. The number of electrons in each atom determines the intensity of scattering of x-rays from that atom; heavier atoms scatter x-rays more effectively than light atoms. The scattered x-rays from the individual atoms can reinforce each other or cancel each other (constructive and destructive interference), giving rise to the characteristic pattern for each type of molecule. A series of diffraction patterns, taken from several angles, contains the information needed to determine the tertiary structure. The information is extracted from the diffraction patterns by a mathematical analysis known as a Fourier series. Many thousands of such calculations are required to determine the structure of a protein, and even though they are done by computer the process is a fairly long one. Improving the calculation process is a subject of active research. The articles by Hauptmann and by Karle cited in the bibliography at the end of this chapter outline some of the accomplishments in the field.

## Myoglobin: An Example of Protein Structure

In many ways myoglobin, the function of which is oxygen storage in mammalian muscle, is the classic example of a globular protein. Myoglobin was the first protein for which the complete tertiary structure was determined by x-ray crystallography (Figure 9.16). The complete myoglobin molecule includes a prosthetic group, the **heme** group, which also occurs in hemoglobin. The myoglobin molecule including the heme group has a compact structure, with the interior atoms very close to each other. Its structure provides examples of many of the forces responsible for the three-dimensional shape of proteins.

There are eight $\alpha$-helical regions in myoglobin and no pleated sheet regions. Approximately 75% of the residues in myoglobin are found in these helical regions, which are designated by the letters A through H. Hydrogen bonding in the polypeptide backbone stabilizes the $\alpha$-helical regions; amino acid side chains are also involved in hydrogen bonds. The polar residues are located on the exterior of the molecule. The interior of the protein contains nonpolar amino acid residues almost exclusively. Two polar histidine residues are located in the interior; these two are involved in interactions with the heme group and bound oxygen, and thus play an important role in the function of the molecule. The planar heme group fits into a hydrophobic pocket in the protein portion of the molecule, and is

**(a)**

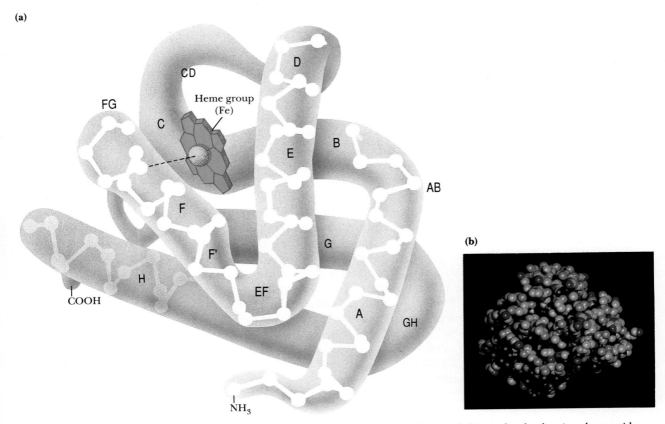

**FIGURE 9.16** (a) The structure of the myoglobin molecule showing the peptide backbone and the heme group. The helical segments are designated by the letters A through H. (b) Space-filling model of myoglobin.

held in position by hydrophobic attractions between heme's porphyrin ring and the nonpolar side chains of the protein. The presence of the heme group affects the conformation of the polypeptide drastically: the apoprotein (the polypeptide chain alone without the prosthetic group) is not as tightly folded as the complete molecule.

The heme group consists of a metal ion, Fe(II), and an organic part, protoporphyrin IX (Figure 9.17). Porphyrin in turn consists of four five-membered rings based on the pyrrole structure; these four rings are linked by bridging methine (—CH=) groups to form a square planar structure. The Fe(II) ion has six coordination sites; it forms six metal-ion complexation bonds. Four of the six sites are occupied by the nitrogen atoms of the four pyrrole-type rings of the porphyrin to give the complete heme group. The presence of the heme group is required for oxygen binding by myoglobin.

The fifth coordination site of the Fe(II) ion is occupied by one of the nitrogen atoms of the imidazole side chain of histidine residue F8 (the

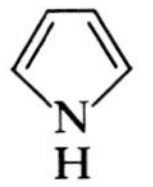

Pyrrole

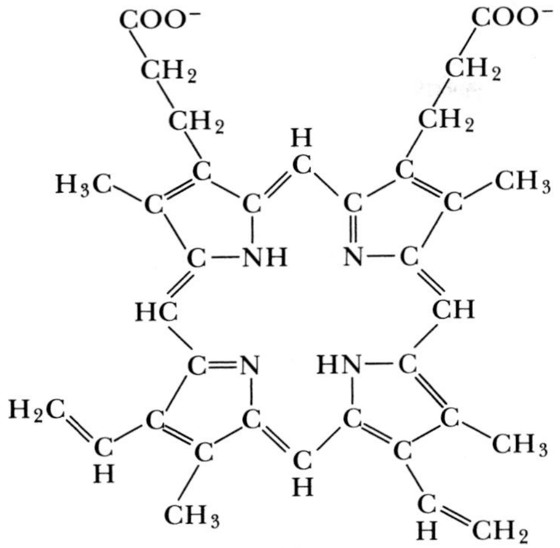

Protoporphyrin IX

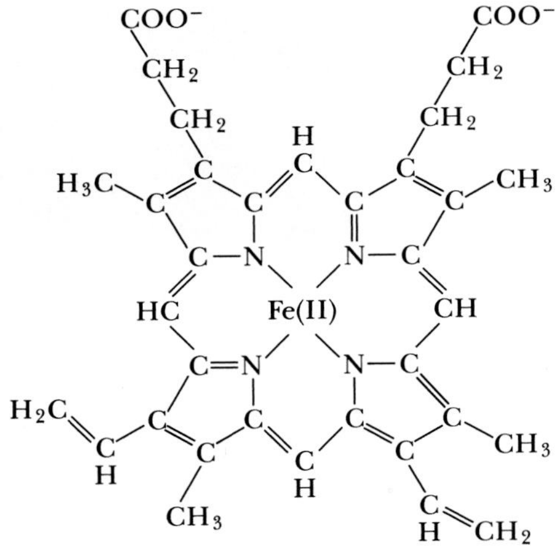

Heme
(Fe-protoporphyrin IX)

FIGURE 9.17 The structure of the heme group. Four pyrrole rings are linked by bridging groups to form a planar porphyrin ring. Several isomeric porphyrin rings are possible, depending on the nature and arrangement of the side chains. The porphyrin isomer found in heme is protoporphyrin IX. Addition of iron to protoporphyrin IX produces the heme group.

eighth residue in helical segment F). This histidine residue is one of the two located in the interior of the molecule. The oxygen is bound at the sixth coordination site of the iron, which remains in the same oxidation state (Fe(II)) whether the oxygen is bound or not. The fifth and sixth coordination sites lie on opposite sides of the plane of the porphyrin ring. The other histidine residue located in the interior of the molecule, residue E7 (the seventh residue in helical segment E), lies on the same side of the heme group as the bound oxygen (Figure 9.18). This second histidine is not bound to the iron or to any part of the heme group, but serves to stabilize the binding site for the oxygen. Oxygen does not bind to the isolated heme group; the environment provided by the protein is also necessary for binding.

(a)

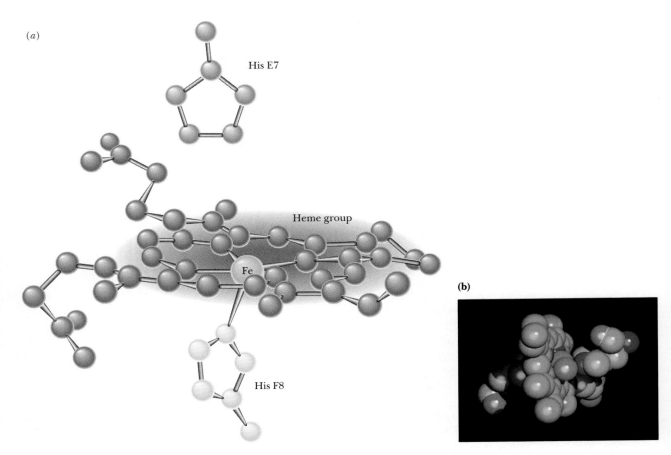

His E7

Heme group

Fe

(b)

His F8

FIGURE 9.18 The oxygen-binding site of myoglobin. (a) The porphyrin ring occupies four of the six coordination sites of the Fe(II). Histidine F8 (His F8) occupies the fifth coordination site of the iron (see text). Oxygen is bound at the sixth coordination site of the iron, while histidine E7 lies close to the oxygen. (b) Space-filling model of the oxygen-binding site of myoglobin. The oxygen is shown in red at the center, the heme group (green) is vertical, and two histidines lie to the left of the heme and the right of the oxygen.

## 9.5
## QUATERNARY STRUCTURE OF PROTEINS

Quaternary structure is a property of proteins that consist of more than one polypeptide chain. They may be **dimers, trimers,** or **tetramers,** depending on whether they consist of two, three, or four polypeptide chains. (The generic term for a molecule that consists of a small number of subunits is **oligomer.**) The chains interact with each other by noncovalent interactions. As a result of such interactions, subtle changes in structure at one site on the molecule can cause drastic changes in properties at a distant site. Proteins that exhibit this property are called **allosteric.** One of the best illustrations of quaternary structure of proteins and its effect on properties is a comparison of hemoglobin, an allosteric protein, with myoglobin, which consists of a single polypeptide chain.

### Hemoglobin

Hemoglobin is a tetramer, consisting of four polypeptide chains, two $\alpha$ chains and two $\beta$ chains (Figure 9.19). (In oligomeric proteins, different kinds of polypeptide chains are designated with Greek letters.) The two $\alpha$ chains of hemoglobin are identical, as are the two $\beta$ chains. The overall structure of hemoglobin is $\alpha_2\beta_2$ in the Greek letter notation. Both the $\alpha$

**(a)**

**(b)**

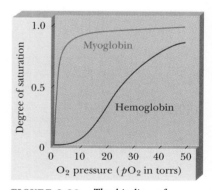

**FIGURE 9.19** The structure of hemoglobin. (a) Hemoglobin ($\alpha_2\beta_2$) is a tetramer, consisting of four polypeptide chains, two $\alpha$ chains, and two $\beta$ chains. (b) Space-filling model of hemoglobin.

and $\beta$ chains of hemoglobin are very similar to the myoglobin chain. The $\alpha$ chain is 141 residues long and the $\beta$ chain is 146 residues long; for comparison, the myoglobin chain is 153 residues long. The $\alpha$ chain, the $\beta$ chain, and myoglobin have long **homologous sequences,** that is, ones in which most or all the amino acid residues are in the same positions. The heme group is the same in myoglobin and hemoglobin.

Four molecules of oxygen can bind to one hemoglobin molecule; we have already seen that one molecule of myoglobin binds one oxygen molecule. Both hemoglobin and myoglobin bind oxygen reversibly, but the binding of oxygen to hemoglobin is cooperative while oxygen binding to myoglobin is not. **Cooperative binding** of oxygen to hemoglobin means that when one oxygen molecule is bound, it becomes easier for the next oxygen molecule to bind. A graph of the oxygen binding properties of the two proteins is one of the best ways to illustrate the point (Figure 9.20).

When the degree of saturation of myoglobin with oxygen is plotted against the oxygen pressure, a steady rise is observed until complete saturation is approached and the curve levels off. The oxygen binding curve of myoglobin is thus said to be **hyperbolic.** The shape of the oxygen binding curve is different for hemoglobin. The **sigmoidal** shape of the

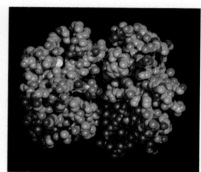

**FIGURE 9.20** The binding of oxygen by myoglobin and hemoglobin. A comparison of the oxygen-binding behavior of myoglobin and hemoglobin. The oxygen-binding curve of myoglobin is hyperbolic, while the oxygen-binding curve of hemoglobin is sigmoidal.

curve indicates that the binding of the first oxygen molecule facilitates the binding of the second oxygen. The binding of the second oxygen facilitates the binding of the third, which in turn facilitates the binding of the fourth. This is precisely what is meant by the term "cooperative binding."

The two different types of behavior are also related to the different functions of these proteins. The function of myoglobin is oxygen *storage* in muscle. It must bind strongly to oxygen at very low pressures, and it is 50% saturated at 1 torr partial pressure of oxygen. [A **torr** is a widely used unit of pressure, but it is not an SI unit. One torr is the pressure exerted by a column of mercury 1 mm high at 0°C. One atmosphere is equal to 760 torr.] The function of hemoglobin is oxygen *transport*, and it must be able both to bind strongly to oxygen and to release oxygen easily, depending upon conditions. In the alveoli of lungs (the point at which hemoglobin must bind oxygen for transport to the tissues) the oxygen pressure is 100 torr. At this pressure hemoglobin is 100% saturated with oxygen. In the capillaries of active muscles the pressure of oxygen is 20 torr, corresponding to less than 50% saturation of hemoglobin (which occurs at 26 torr). In

**FIGURE 9.21**   The motion of the subunits of hemoglobin with respect to one another in going from the oxygenated to the deoxygenated state. The $\alpha_1\beta_1$ dimer moves with respect to the $\alpha_2\beta_2$ dimer.

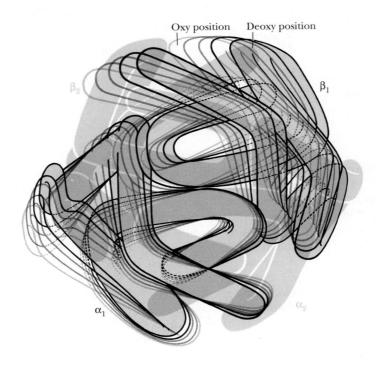

other words, hemoglobin gives up oxygen easily in capillaries, where the need for oxygen is great.

Structural changes during binding of small molecules are characteristic of allosteric proteins such as hemoglobin. The binding of a small molecule such as oxygen accompanies a change in the quaternary structure of the protein. The protein has a different quaternary structure in the bound and unbound forms, as is the case with oxygenated and deoxygenated hemoglobin. In oxygenated hemoglobin the two $\beta$ chains are much closer to each other than in deoxygenated hemoglobin. The change is so marked that the two forms of hemoglobin have different crystal structures (Figures 9.21 and 9.22).

## Conformational Changes that Accompany Hemoglobin Function

Other ligands are involved in cooperative effects when oxygen binds to hemoglobin. Both $CO_2$ and $H^+$, which themselves bind to hemoglobin, affect the affinity of hemoglobin for oxygen. This behavior is called the **Bohr effect** after its discoverer, Christian Bohr (the father of the physicist Niels Bohr). The acid-base properties of hemoglobin affect and are affected by its oxygen-binding properties. (The oxygen-binding ability of myoglobin is not affected by the presence of $H^+$ or of $CO_2$.) An increase in the concentration of $H^+$ (*i.e.,* a lowering of the pH) reduces the oxygen affinity of hemoglobin. An increase in the concentration of $CO_2$ also decreases the oxygen-binding capability of hemoglobin (Figures 9.23 and 9.24).

The normal pH of blood is 7.4; the $pK'_a$ of $H_2CO_3$ is 6.35. As a result, about 90% of dissolved $CO_2$ will be present as $HCO_3^-$. (The Henderson-Hasselbalch equation can be used to confirm this point. The *in vivo* buffer system involving $H_2CO_3$ and $HCO_3^-$ in blood was discussed in Section 3.6.) Most of the bicarbonate ion binds reversibly to hemoglobin. The presence of this charged group favors the quaternary structure characteristic of deoxygenated hemoglobin. Therefore, increased amounts of $CO_2$ lower the affinity of hemoglobin for oxygen. The acid-base properties of the hemoglobin molecule itself also have an effect. The oxygenated form of hemoglobin is a stronger acid (has a lower $pK'_a$) than the deoxygenated form. In other words, deoxygenated hemoglobin has a higher affinity for $H^+$ than does the oxygenated form. Thus, changes in the quaternary structure of hemoglobin can modulate the buffering of blood through the hemoglobin molecule itself.

In the presence of large amounts of $H^+$ and $CO_2$, hemoglobin releases oxygen to actively metabolizing tissues and binds carbon dioxide. The deoxygenated hemoglobin then transports the $CO_2$ to the lungs to be expired. The presence of large amounts of oxygen in the lungs reverses the process, causing hemoglobin to release $CO_2$ and bind $O_2$. The oxygenated hemoglobin can then transport oxygen to the tissues. The process is complex, but it allows for fine tuning of pH as well as levels of $CO_2$ and $O_2$.

(a)

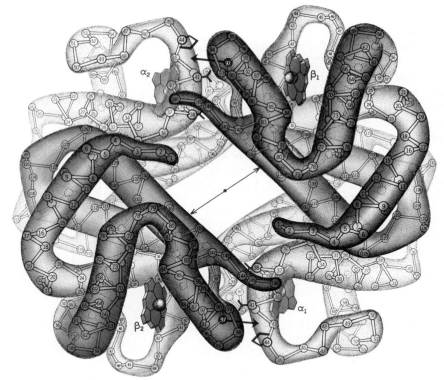

**FIGURE 9.22** The structures of (a) deoxyhemoglobin and (b) oxyhemoglobin. Note the motion of subunits with respect to one another. There is much less room at the center of oxyhemoglobin.

(b)

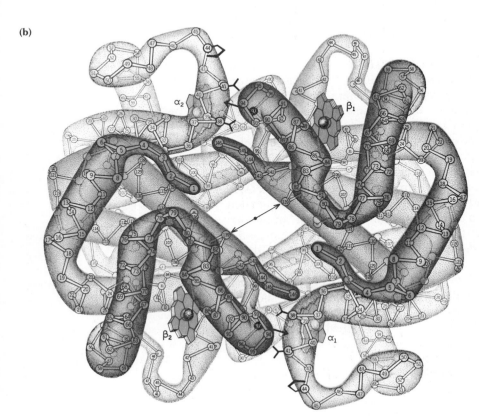

$$\text{HbO}_2 + \text{H}^+ + \text{CO}_2 \underset{\substack{\text{Actively metabolizing}\\\text{tissue (such as muscle)}}}{\overset{\text{Alveoli of lungs}}{\rightleftharpoons}} \text{O}_2 + \text{Hb} \overset{\text{CO}_2}{\underset{\text{H}^+}{<}}$$

**FIGURE 9.23**  The general features of the Bohr effect. In actively metabolizing tissue, hemoglobin releases oxygen and binds both $CO_2$ and $H^+$. In the lungs, hemoglobin releases both $CO_2$ and $H^+$ and binds oxygen.

Hemoglobin in blood is bound to another ligand, 2,3-bisphospho-glycerate (BPG), with drastic effects on its oxygen-binding capacity.

Structure of 2,3-bisphosphoglycerate (BPG)

The binding is electrostatic, with specific interactions between the negative charges on BPG and positive charges on the protein (Figure 9.25). In the presence of BPG, the partial pressure at which 50% of hemoglobin is bound to oxygen is 26 torr. If BPG were not present in blood, the oxygen-binding capacity of hemoglobin would be much higher, and little oxygen would be released in the capillaries. "Stripped" hemoglobin, which is isolated from blood and from which the endogenous BPG has been removed, displays this behavior (Figure 9.26).

BPG also plays a role in supplying a growing fetus with oxygen. The fetus obtains oxygen from the mother's bloodstream via the placenta. Fetal hemoglobin (Hb F) has a higher affinity for oxygen than does maternal hemoglobin. Two features of fetal hemoglobin contribute to its higher oxygen-binding capacity. One is the presence of two different polypeptide chains; the subunit structure of Hb F is $\alpha_2\gamma_2$, where the $\beta$ chains of Hb A, the usual hemoglobin, have been replaced by the $\gamma$ chains, which are similar but not identical in structure. The second feature is that Hb F binds less strongly to BPG than does Hb A, because there are fewer positively charged groups on Hb F to stabilize the electrostatic interaction. As a result, Hb F has a stronger affinity for oxygen than does adult hemoglobin, allowing for efficient transfer of oxygen from the mother to the fetus (Figure 9.27).

Box 9.1 discusses several other variants of hemoglobin, whose behavior illustrates the connection between protein structure and chemical properties.

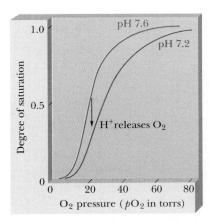

**FIGURE 9.24**  A comparison of the oxygen-binding capacity of hemoglobin at pH 7.6 and pH 7.2. An increase in the hydrogen ion concentration in blood leads to a release of oxygen from hemoglobin.

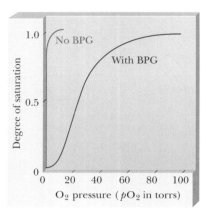

**FIGURE 9.26** A comparison of the oxygen-binding properties of hemoglobin in the presence and absence of 2,3-bisphosphoglycerate (BPG). Note that the presence of the BPG markedly decreases the oxygen affinity of hemoglobin.

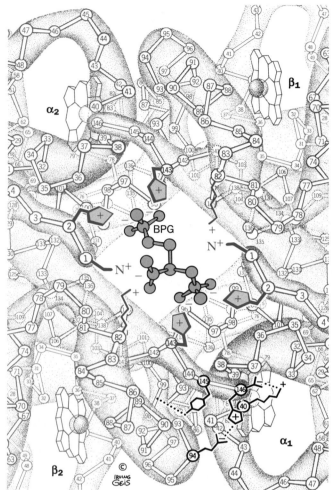

**FIGURE 9.25** The binding of BPG to deoxyhemoglobin viewed from the same perspective as Figure 9.21. Note the electrostatic interactions between the BPG and the protein.

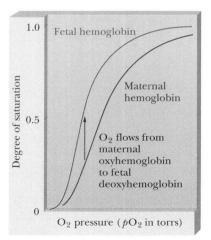

**FIGURE 9.27** A comparison of the oxygen-binding capacity of fetal and maternal hemoglobins. Fetal hemoglobin binds less strongly to BPG than does maternal hemoglobin. Consequently, fetal hemoglobin has a greater affinity for oxygen than maternal hemoglobin.

## BOX 9.1
## ABNORMAL HEMOGLOBINS

Hemoglobin S (Hb S), the mutant form characteristic of sickle-cell anemia, is by no means the only abnormal hemoglobin known, although it is the most common. More than 400 mutant hemoglobins are known, most of them resulting from substitutions of a single amino acid; these altered proteins have provided a great opportunity for studying structure-function relationships in a system consisting of a protein of known structure with a large number of variants, which are also well characterized. Not all mutant hemoglobins have clinical manifestations. For example, Hb E occurs in as many as 10% of the population of some parts of Southeast Asia. In this variant form, the glutamate at position 26 of the $\beta$ chain has been replaced by a lysine. The notation for such a change is Glu B8(26)$\beta \longrightarrow$Lys, where B8 refers to the location of this residue at position 8 of the B helix. In Hb E, the change is on the protein's surface, where there is usually little effect on the stability of hemoglobin. Sickle-cell anemia hemoglobin is a glaring exception to this statement, but there the drastic effect is due to the role of the altered residue in intermolecular interactions.

More frequently, when hemoglobin is altered at an internal residue, there is a marked decrease in the stability of the molecule. Degradation products of such hemoglobins accumulate at the cell membrane of erythrocytes (red blood cells), reducing the stability of the membrane. **Hemolytic anemia** arises from the premature cell lysis (disintegration) associated with unstable hemoglobins. An example of such unstable hemoglobin is Hb Savannah (named for the city in which it was

discovered), in which Gly B6(24)$\beta$ is replaced by Val. There is not enough room for the side chain of the valine between the B helix and the adjacent E helix, disrupting the entire structure. A similar situation is observed with Hb Bibba [Leu H19(136)$\alpha \longrightarrow$Pro], in which the proline disrupts the H helix.

Mutations that affect the binding of the heme group have very noticeable consequences. This is particularly true when such changes stabilize the Fe(III) oxidation state of the heme and thus eliminate binding of oxygen by the defective subunits. **Methemoglobin,** abbreviated as Hb M, is the name for hemoglobin in the Fe(III) state. Methemoglobin is brown and is responsible for the color of dried blood and old meat; normal hemoglobin, in which the heme iron is in the Fe(II) oxidation state, is red. Individuals whose blood contains methemoglobin are said to have **methemoglobinemia,** and their blood is chocolate brown. The presence of large concentrations of deoxygenated Hb M in their arterial blood leads to **cyanosis,** characterized by bluish skin. The underlying structural change in Hb M is the substitution of an anionic oxygen ligand for histidine at one of the binding sites of the Fe. In Hb M Iwate [His F8(87)$\alpha \longrightarrow$Tyr], the tyrosine simply replaces the histidine. In Hb Milwaukee [Val E11(67)$\beta \longrightarrow$Glu], the glutamate side chain forms an ion pair with the heme iron, stabilizing the Fe(III) oxidation state and preventing the binding of oxygen. Heterozygotes for Hb M (people having one gene for Hb M and one for normal Hb) do not suffer physical disabilities, but there are no recorded cases of persons homozygous for Hb M (both genes for Hb M). That condition appears to be lethal.

## SUMMARY

The structure of proteins is complex, with little orderly arrangement of atoms. Many three-dimensional conformations are possible for proteins, but only one, or at most a few, conformations have biological activity; these are called the native con-

formations. To facilitate structure determination, it is customary to define four levels of structural organization. Primary structure refers to the order in which the amino acids are covalently linked together. Secondary structure refers to the arrangement in space of the atoms in the backbone of the polypeptide chain. Tertiary structure includes the three-dimensional arrangement of *all* the atoms in the protein. Quaternary structure is the arrangement of subunits in multisubunit proteins.

The amino-acid sequence (the primary structure) of a protein determines its three-dimensional structure, which in turn determines its properties. A striking example of the importance of primary structure is sickle-cell anemia, a disease caused by a change in one amino acid in each of two of the four chains of hemoglobin. The higher order (secondary and tertiary) levels of structure depend on noncovalent interactions, including hydrogen bonds, hydrophobic interactions, electrostatic interactions, and complexation of metal ions. The three-dimensional structure of proteins can be completely disrupted and, under proper experimental conditions, completely recovered. This process of denaturation and refolding is a dramatic example of the relationship between the primary structure of the protein and the forces that determine the tertiary structure.

The primary structure of a protein can be determined by chemical methods. Three questions must be answered: first, which amino acids are found in the protein and how many of each kind; second, which amino acids occur at the N-terminal and C-terminal ends of the molecule; third, what is the sequence of amino acids. The secondary and tertiary structures of a protein can be determined simultaneously by x-ray crystallography. The oxygen-storage protein myoglobin was the first protein for which the complete tertiary structure was determined by crystallography.

The individual polypeptide chains of multisubunit proteins interact with each other by noncovalent interactions. Subtle changes in structure at one site on the molecule can cause drastic changes in properties at a distant site as a result of such interactions. Proteins that exhibit this property are called allosteric. A comparison of the properties of hemoglobin, an allosteric protein, with those of myoglobin, which is not allosteric, shows how they differ. In hemoglobin, which is an oxygen-transport protein, the binding of oxygen is cooperative; as each oxygen is bound, it becomes easier for the next one to bind. The binding of oxygen to myoglobin is not cooperative. The binding of oxygen to hemoglobin is modulated by such ligands as $H^+$, $CO_2$, and BPG (2,3-bisphosphoglycerate).

## EXERCISES

1. Match the following statements about protein structure with the proper level of organization.

   (a) primary structure

   (b) secondary structure

   (c) tertiary structure

   (d) quaternary stucture

   (1) the three-dimensional arrangement of all atoms

   (2) the order of amino acid residues in the polypeptide chain

   (3) the interaction between subunits in proteins that consist of more than one polypeptide chain

   (4) the arrangement in space of the polypeptide backbone

2. Show by a series of equations (with structures) the first stage of the Edman method applied to a peptide that has leucine as its N-terminal residue.

3. A sample of an unknown peptide was divided into two aliquots. One aliquot was treated with trypsin and the other with cyanogen bromide. Given the following sequences (N-terminal to C-terminal) of the resulting fragments, deduce the sequence of the original peptide.

   **Trypsin treatment**

   Asn—Thr—Trp—Met—Ile—Lys

   Gly—Tyr—Met—Gln—Phe

   Val—Leu—Gly—Met—Ser—Arg

   **Cyanogen bromide treatment**

   Gln—Phe

   Val—Leu—Gly—Met

   Ile—Lys—Gly—Tyr—Met

   Ser—Arg—Asn—Thr—Trp—Met

4. A sample of a peptide of unknown sequence was treated with trypsin; another sample of the same

peptide was treated with chymotrypsin. The sequences (N-terminal to C-terminal) of the smaller peptides produced by trypsin digestion were

Met—Val—Ser—Thr—Lys

Val—Ile—Trp—Thr—Leu—Met—Ile

Leu—Phe—Asn—Glu—Ser—Arg

The sequences of the smaller peptides produced by chymotrypsin digestion were

Asn—Glu—Ser—Arg—Val—Ile—Trp

Thr—Leu—Met—Ile

Met—Val—Ser—Thr—Lys—Leu—Phe

Deduce the sequence of the original peptide.

5. List five forces responsible for maintaining the correct three-dimensional shapes of proteins. Specify which groups on the protein are involved in each type of interaction.
6. Define denaturation in terms of the effects of secondary, tertiary, and quaternary structure.
7. List two similarities and two differences between hemoglobin and myoglobin.
8. Suggest a way in which the difference between the functions of hemoglobin and myoglobin is reflected in the shapes of their respective oxygen binding curves.
9. In oxygenated hemoglobin, $pK'_a = 6.6$ for the histidines at position 146 on the $\beta$ chain. In deoxygenated hemoglobin the $pK'_a$ of these residues is 8.2. How can this piece of information be correlated with the Bohr effect?
10. Suggest an explanation for the observation that covalently modified proteins cannot be denatured reversibly.

11. List some of the differences between the $\alpha$ and $\beta$ forms of secondary structure.
12. List some of the possible combinations of $\alpha$ and $\beta$ arrangements in supersecondary structures.
13. Rationalize the following observations.
    (a) Serine is the amino acid residue that can be replaced with the least effect on protein structure and function.
    (b) Replacement of tryptophan causes the greatest effect on protein structure and function.
    (c) Replacements such as Lys⟶Arg and Leu⟶Ile have very little effect on protein structure and function.
14. Suggest a reason for the observation that persons with sickle-cell trait sometimes encounter breathing problems on high altitude flights.
15. Describe the Bohr effect.
16. Why is proline frequently encountered at the places where the polypeptide chain turns a corner in both myoglobin and hemoglobin?
17. Describe the effect of 2,3-bisphosphoglycerate on the binding of oxygen by hemoglobin.
18. How does the oxygen-binding curve of fetal hemoglobin differ from that of adult hemoglobin?
19. List some abnormal hemoglobins and describe how their structures differ from that of normal hemoglobin.
20. Does a fetus homozygous for Hb S have normal Hb F?
21. What is the molecular basis for the observation that blood changes color from red to brown as it dries? What connection does this observation have with abnormal hemoglobins?

## ANNOTATED BIBLIOGRAPHY

Cantor, C.R, and P.R. Schimmel. *Biophysical Chemistry.* San Francisco: W.H. Freeman, 1980. [A multivolume work dealing with properties that are useful in characterizing and separating proteins. Paperback edition available.]

Changeux, J-P., A. Devillers-Thiery, and P. Chemoulli. Acetylcholine Receptor: An Allosteric Protein. *Science* **225**, 1335–1345 (1984). [A look at the importance of allosteric properties in regulating the action of one of the most important proteins in the nervous system.]

Dayhoff, M.O., ed. *Atlas of Protein Sequence and Structure.* Washington, DC: National Biomedical Research Foundation, 1978. [A listing of all known amino acid sequences. Updated periodically.]

Dickerson, R.E., and I. Geis. *The Structure and Action of Proteins,* 2nd ed. Menlo Park, CA: Benjamin/Cummings, 1981. [A well written and particularly well illustrated general introduction to protein chemistry.]

Doolittle, R.F. Proteins. *Sci. Amer.* **253** (4), 88–89 (1985). [A well illustrated discussion of protein structure with emphasis on evolutionary considerations.]

Fermi, G., and M.F. Perutz. *Atlas of Molecular Structures in Biology.* Vol. 2, *Haemoglobin and Myoglobin.* Oxford: Clarendon Press, 1981. [A detailed description of the structures of these proteins, with particular emphasis on the results of x-ray crystallography.]

Freifelder, D. *Physical Biochemistry,* 2nd ed. San Francisco: W.H. Freeman, 1982. [An introduction to methods for characterizing and isolating proteins. Paperback edition available.]

Gierasch, L.M., and J. King, eds. *Protein Folding: Deciphering the Second Half of the Genetic Code.* Waldorf, Md: AAAS Books, 1990. [A collection of articles on recent discoveries on the processes involved in protein folding. Experimental methods for studying protein folding are emphasized.]

Hauptmann, H. The Direct Methods of X-ray Crystallography. *Science* **233,** 178–183 (1986). [An article about improvements in ways of doing the calculations involved in determining protein structure. Based on a Nobel Prize address. This article should be read in connection with the one by Karle, and provides an interesting contrast to the articles by Perutz and by Kendrew.]

Jaenicke, R. Protein Folding and Protein Association. *Agnew. Chem. Int. Ed. Engl.* **23,** 395–413 (1984). [A discussion of a possible "folding code" for protein tertiary structure. The genetic code for the amino acid sequence determines the folding code.]

Karle, J. Phase Information from Intensity Data. *Science* **232,** 837–843 (1986). [Another Nobel Prize address on the subject of x-ray crystallography. See annotations for the article by Hauptmann.]

Kendrew, J.C. Myoglobin and the Structure of Proteins. *Science* **139,** 1259–1266 (1963).

_____. The Three-dimensional Structure of a Protein Molecule. *Sci. Amer.* **205** (6), 96–111 (1961). [An introduction to determination of protein structure by x-ray crystallography. The first article is based on a Nobel Prize address. These two articles and the ones by Perutz show the earliest accomplishments in protein crystallography, and are an interesting contrast to the articles by Hauptmann and by Karle.]

Leszczynski, J.F., and G.D. Rose. Loops in Globular Proteins: A Novel Category of Secondary Structure. *Science* **234,** 849–855 (1986). [Makes the point that the "random" portions of proteins are not really random.]

Monod, J., J-P. Changeux, and F. Jacob. Allosteric Proteins and Cellular Control Systems. *J. Mol. Biol.* **6,** 306–329 (1963). [The original model for the mode of action of allosteric proteins, and still one of the standard references on the subject.]

Perutz, M. The Hemoglobin Molecule. *Sci. Amer.* **211,** (5), 64–76 (1964).

_____. The Hemoglobin Molecule and Respiratory Transport. *Sci. Amer.* **239** (6), 92–125 (1978). [The relationship between molecular structure and cooperative binding of oxygen. Also see annotations to the articles by Kendrew.]

Richardson, J. The Anatomy and Taxonomy of Protein Structure. *Adv. Prot. Chem.* **34,** 168–339 (1981). [An extensive review of secondary and tertiary structure, with excellent illustrations.]

# Experimental Methods for Determining Protein Structure

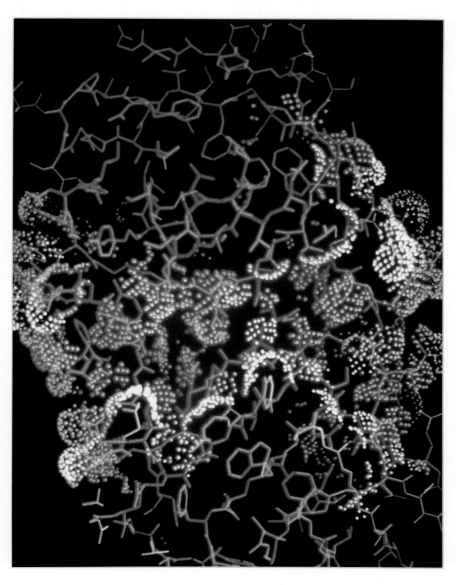

*A computer-generated model of the protein pepsin.*

197

*Since there are thousands of **different** protein molecules in a cell, the task of separating them and determining the structure of a single protein is exceedingly difficult. There are many techniques for characterizing a protein—ranging from strategies for discovering the number and type of its constituent amino acids to elucidating its complete three-dimensional structure. After a protein has been degraded to its amino acids, they can be identified according to their charge and polarity by chromatography and electrophoresis. When the first amino acid of a protein has been chemically labeled, the chain can be degraded one amino acid at a time to discover its sequence. Often this is done by identifying the sequences in related peptide fragments. In a final step of structure determination, a complete protein can be subjected to x-ray diffraction analysis to determine its three-dimensional conformation. Before this can be accomplished, however, the protein must be purified by a number of techniques, such as molecular sieve chromatography, after which it may be crystallized. The crystal is bombarded with x-rays and a diffraction pattern registered on photographic film. Analysis of the x-ray diffraction data gives the atomic positions that define the protein's three-dimensional structure.*

## A.1
## SEPARATION OF AMINO ACIDS: THE MODE OF OPERATION OF AN AMINO ACID ANALYZER

Separation of amino acids on an amino acid analyzer is done by one of the many forms of a technique called **chromatography.** The word comes from the Greek *chroma,* "color," and *graphein,* "to write," as the technique was first used around the beginning of the 20th century to separate plant pigments with easily visible colors. It has long since become possible to separate colorless compounds as long as there are methods for detecting them.

Chromatography is based on the fact that different compounds can distribute themselves to varying extents between different phases. There is always a **stationary phase** and a **mobile phase.** The mobile phase flows over the stationary material and carries the sample to be separated along with it. The various components in the sample interact with the stationary phase to different extents. The components of the sample that interact more strongly with the stationary phase are carried along more slowly by the mobile phase than are those that interact less strongly. The different mobilities of the components, which in turn are based on the different extent of interaction with the stationary phase, form the basis of the separation.

As an example let us assume there is a mixture of cations (positively charged ions) in the sample and that the stationary phase consists of an anionic substance (negatively charged ions) with counterions (cations) bound to the anionic sites by electrostatic forces. A cation with a charge of +2 in the sample will interact more strongly with an anion in the stationary phase than will a cation with a charge of +1 in the sample. The less highly charged cation is carried along more quickly by the mobile phase than the more highly charged one, and the difference in rates of migration eventually brings about the desired separation.

In one type of chromatographic experiment the stationary phase is packed in a column, and the sample is applied to the top of the column as a small volume of concentrated solution; the mobile phase, called the **eluent,** is passed through the column. The sample is diluted by the eluent, and the separation process also increases the volume occupied by the sample. All the sample eventually comes off the column in a successful experiment. An example of this procedure is shown in Figure A.1.

An amino acid analyzer is an example of **column chromatography.** There are two different types of amino acid analyzer; the difference between the two lies in the nature of the interaction that takes place between the stationary phase and components of the sample and that provides the basis for separation. In the case of **ion exchange chromatography,** the interaction in question is electrostatic attraction. In the other case, **HPLC** (high-performance liquid chromatography), the separation depends on differences in polarity. We shall use ion exchange chromatography as our example here.

**FIGURE A.1**  An example of column chromatography.

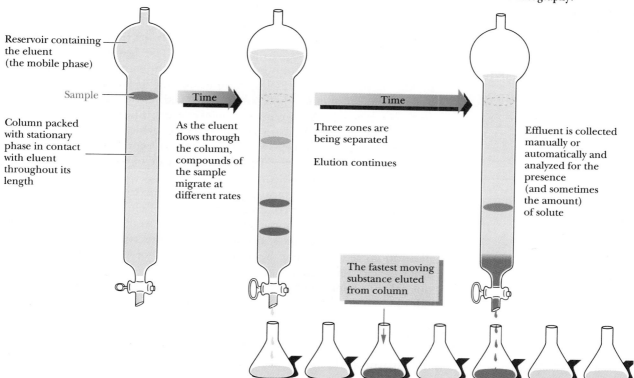

Reservoir containing the eluent (the mobile phase)

Sample

Column packed with stationary phase in contact with eluent throughout its length

Time

As the eluent flows through the column, compounds of the sample migrate at different rates

Time

Three zones are being separated

Elution continues

The fastest moving substance eluted from column

Effluent is collected manually or automatically and analyzed for the presence (and sometimes the amount) of solute

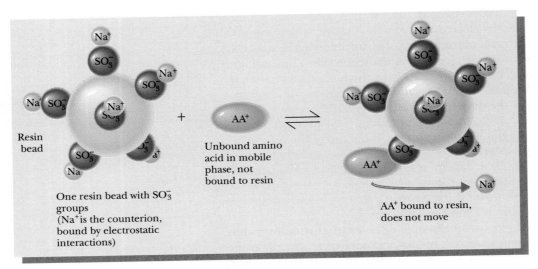

One resin bead with SO$_3^-$
groups
(Na$^+$ is the counterion,
bound by electrostatic
interactions)

Unbound amino
acid in mobile
phase, not
bound to resin

AA$^+$ bound to resin,
does not move

**FIGURE A.2**   The mode of action of cation exchange resins. In the forward reaction, the amino acid is bound to the resin, and in the reverse reaction, it is released to the mobile phase. The process occurs many times on each bead and along the entire column. The components of the mixture encounter new beads as they move.

Synthetic resins are used as ion exchangers in many chromatographic operations. These ion exchange resins are cross-linked long-chain polymers, available commercially in bead form. The beads, which are the material used to pack the column, have diameters on the order of micrometers ($\mu = 10^{-6}$ m). Each bead can contain as many as several thousand charged groups of the same sign, all positive or all negative (Figure A.2). The resins used in the columns of amino acid analyzers typically contain negatively charged sulfonate groups ($-SO_3^-$, the ionized form of the sulfonic acid group, $-SO_3H$). The beads are soaked in a solution containing the counterion desired for the start of the experiment. If Na$^+$ is the desired starting counterion, the beads are treated with sodium chloride before the column is packed. Other positively charged ions can replace the Na$^+$ by an exchange process. The separation of amino acids takes place on the basis of the extent to which each amino acid is positively charged and the extent to which it can replace the sodium ion as the counterion that balances the negative charge of each sulfonate group of the **cation exchange resin.**

At the start of a typical separation, the mixture of amino acids is applied to the top of the column, and the flow of eluting buffer is started. As the elution proceeds, some of the Na$^+$ is replaced by positively charged amino acids, AA$^+$, which become the counterions. For purposes of illustration, let us assume a sample consisting of three amino acids—aspartic acid, serine, and histidine—dissolved in a buffer of pH 3.25. At this pH the amino groups of all three amino acids are protonated, as is the side-chain imidazole of the histidine. Essentially all the $\alpha$ carboxyl groups are ionized, and about 25% of the side-chain carboxyl groups of the aspartic acid are ionized (Figure A.3). (Use the Henderson-Hasselbalch equation to confirm the degree of ionization of the side-chain carboxyl of the aspartic acid.)

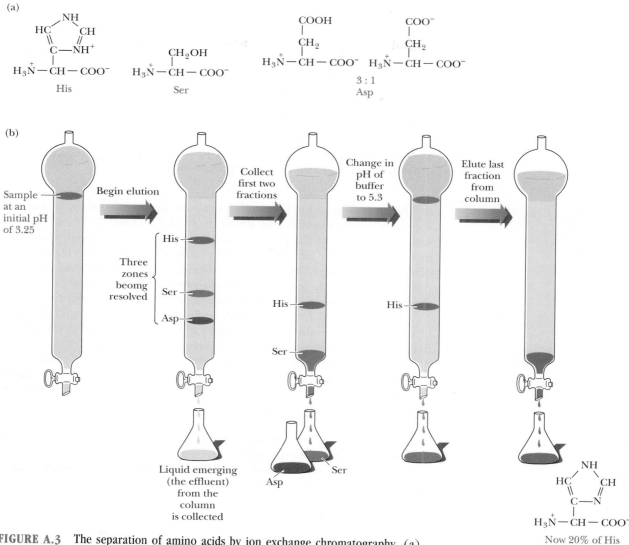

**FIGURE A.3** The separation of amino acids by ion exchange chromatography. (a) Sample components. (b) An example of the separation process by ion exchange.

    In our example the histidine has a net positive charge of 1, the serine is electrically neutral, and the aspartic acid is a 3:1 mixture of electrically neutral molecules and of molecules with a net negative charge of 1. The histidine is bound most strongly to the cation exchange resin because of its charge, the serine next, and the aspartic acid least strongly. More eluting buffer is added, and a second stage of exchange takes place in which the positively charged amino acids are replaced by cations from the buffer. The aspartic acid is eluted first, followed by the serine; the histidine remains bound to the resin. The pH of the eluting buffer is raised in stages

R
|
H—C—NH$_2$  + 2 Ninhydrin ⟶ Blue-purple substance + RCHO + CO$_2$
|
COOH

Amino acid          Ninhydrin          Blue-purple substance
                                        absorbs at 540 nm

**FIGURE A.4**   The reaction of amino acids with ninhydrin. Note that in this reaction, the original amino acid has been converted to an aldehyde with one carbon fewer than the parent acid. The carboxyl group of the original amino acid has been lost as $CO_2$, and the amino nitrogen appears in the product.

to facilitate release from the column of amino acids with positively charged R groups. In our example, the pH of the buffer is raised to 5.3. At this pH value, about 20% of the histidine side chains are deprotonated, and these molecules have no net charge. (Why? Use the Henderson-Hasselbalch equation again to confirm this point.) The histidine is then eluted from the column. Good resolution can be obtained with an amino acid analyzer, and the individual amino acids are well separated from one another.

The individual amino acids can be detected by titrating each fraction of the eluate with ninhydrin to produce a purple compound (a yellow one in the case of the amino acid proline) (Figure A.4). The absorption of light of any desired wavelength by each fraction collected from the column can then be measured. The intensity of the absorption gives information about the relative amount of each amino acid as well as serving as a means of detection. Comparison with standard samples enables accurate determination of the identity and amount of each amino acid. Amino acid analyzers are automated so that eluting buffers can be changed when needed; ninhydrin is automatically added to the eluate and the absorption recorded. Figure A.5 shows the results of an amino acid analysis. This procedure does not, however, give information about the order of amino acids in proteins, and determining the primary structure of proteins requires establishing the order of amino acids.

**FIGURE A.5**   A schematic representation of the results of an amino acid analysis. The sample shown here is a mixture of aspartic acid, serine, and histidine, shown in Figure A.3. In a protein sample there is a peak for each amino acid present. Proper choice of elution conditions ensures good separation of the peaks.

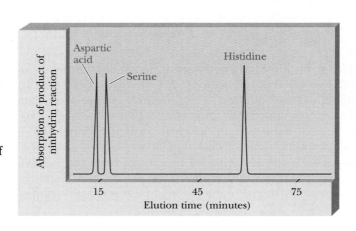

## A.2
## SEPARATION OF PEPTIDES

Ion exchange column chromatography depends on differences in electrical charge as the basis for separation of the components of a mixture. Another method, **paper chromatography,** can be used to separate peptides on the basis of polarity.

A polar organic molecule will tend to dissolve more easily in water than in a nonpolar organic solvent. The mobile phase, which is less polar than water, flows over the stationary phase, which is polar; the mobile phase carries the sample to be separated along with it. In paper chromatography the stationary phase is the water adsorbed on the cellulose fibers of the paper, which serves as an inert support. The various components in the sample interact with the stationary phase to different extents. The components of the sample that interact more strongly with the stationary phase, the more polar ones, are carried along more slowly by the mobile phase than are those that interact less strongly, the less polar ones. The mobile phase is frequently a mixture of solvents, such as n-butyl alcohol and water or n-butyl alcohol, butyric acid, and water. This mixture of solvents is less polar than water. As a result, the more polar peptides tend to partition themselves in the polar stationary phase and are carried along more slowly by the mobile phase than are the less polar peptides, which have a greater affinity for the mobile phase.

The paper acts as a wick along which the solvent travels slowly by capillary action. The mobile phase works its way through channels between the fibers of the paper. The extent to which the mobile phase migrates can be seen by the position of the **solvent front,** showing the degree to which the paper has been moistened by the solvent. The various components of a mixture can be characterized by the distance that they travel from the position at which the sample is applied, called the **origin,** compared with the distance traveled by the solvent front. The ratio of these two distances is called the $R_f$, and its numerical value can be of great use in identifying substances by comparison with standards (Figure A.6).

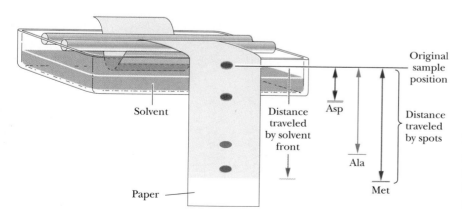

Solvent

Paper

Distance traveled by solvent front

Asp

Ala

Met

Original sample position

Distance traveled by spots

**FIGURE A.6** Paper chromatography. $R_f$ values are determined and substances are identified by comparison with standards.

$$R_f = \frac{\text{distance traveled by substance}}{\text{distance traveled by solvent front}}$$

In a typical experiment, a mixture of peptides is applied as a spot at the origin, near one end of the paper. The mixture is separated into its components by developing the chromatogram, which can be done by either **ascending** or **descending paper chromatography.** In ascending paper chromatography the paper, usually rolled into a cylinder, is placed in a chamber with a shallow layer of the mobile phase liquid at the bottom. In descending paper chromatography the upper end of the paper is placed in a trough of the solvent, which is then allowed to migrate down the paper. In both ascending and descending paper chromatography, the chamber is saturated with the vapor of the developing solvent (Figure A.7). Different peptides produced by partial digestion of proteins will have different $R_f$ values, but the differences may be small.

**FIGURE A.7**    (a) Ascending and (b) descending paper chromatography.

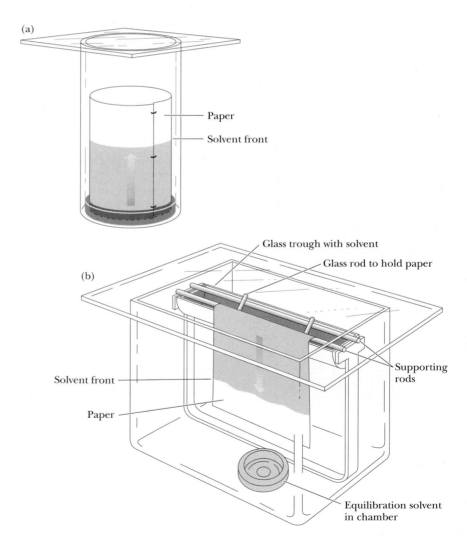

(a)

Paper

Solvent front

(b)

Glass trough with solvent

Glass rod to hold paper

Solvent front

Paper

Supporting rods

Equilibration solvent in chamber

It is usually possible to improve a separation by using two separation methods on the same mixture. **Electrophoresis** can separate peptides on the basis of charge. We have already discussed one method (ion exchange column chromatography) for separation of amino acids and peptides on the basis of charge, so we need only see how the experimental details of the two methods differ. The underlying physical principle, electrostatic attraction, remains the same in both cases. In the case of electrophoresis, though, the size of the molecules to be separated can make a difference. Size is not a major factor here, however, because the peptides are more or less the same size.

Electrophoresis depends on different rates of migration of particles (peptides in this case) of different charge in an electric field. The charged particles move through a liquid that conducts an electric current. In modern applications of electrophoresis, an inert substance such as paper or a gel is used as a supporting medium for the conducting liquid.

When peptides are to be separated, the sample is applied to a strip of paper moistened with the conducting solution, usually a buffer. The ends of the strip of paper are placed in reservoirs of buffer solution. A positive electrode is placed in one reservoir and a negative electrode in the other, and a high voltage (on the order of thousands of volts) is applied. Peptides with a net positive charge will migrate toward the negative electrode; those with a higher positive charge will move faster than those with a lower positive charge. Peptides with a net negative charge will migrate toward the positive electrode; those with a higher negative charge will move faster than those with a lower charge (Figure A.8). A peptide with no net charge will not migrate in the electric field; this statement implies that the pH of the conducting buffer is equal to the isoelectric pH of the peptide (recall the term "isoelectric pH" from Section 5.3). The net charge of each peptide depends on pH, since the charge of each of the individual titratable groups—namely, the N-terminal amino group, the C-terminal carboxyl group, and the titratable groups on the side chains—depends on pH.

A two-dimensional separation by chromatography and by electrophoresis can be done on the same piece of paper. A single spot containing the mixture is applied to the paper, and a chromatographic separation is performed. The paper is then turned 90°, and an electric field is applied so that an electrophoretic separation can be done. After the separation is complete, the spots corresponding to the various peptides can be made

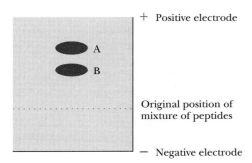

+ Positive electrode

A

B

Original position of mixture of peptides

− Negative electrode

**FIGURE A.8**  An example of paper electrophoresis. Both peptide A and peptide B are negatively charged, but peptide A has a higher negative charge than peptide B and migrates toward the positive electrode more quickly.

**FIGURE A.9** Fingerprinting. Chromatography and electrophoresis are combined to separate peptides.

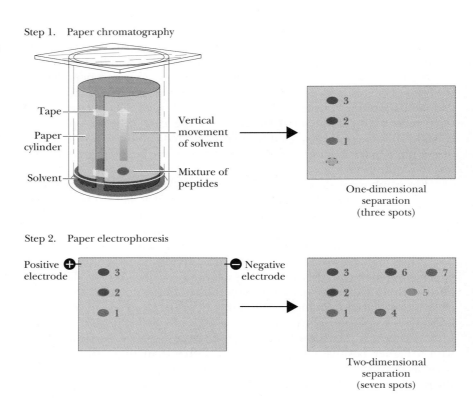

Step 1.   Paper chromatography

Tape

Paper cylinder

Solvent

Vertical movement of solvent

Mixture of peptides

One-dimensional separation (three spots)

Step 2.   Paper electrophoresis

Positive electrode

Negative electrode

Two-dimensional separation (seven spots)

visible by treatment with ninhydrin. The resulting array of separated peptides on the paper is called a **fingerprint** (Figure A.9).

In addition to two-dimensional separations of peptides that combine chromatography and electrophoresis, it is possible to do fingerprinting by **two-dimensional paper chromatography.** After a chromatogram is run in one direction with one solvent system, the paper is allowed to dry and then another chromatogram is run at right angles using another solvent system.

The actual pattern of the fingerprint for each protein will depend on the conditions under which the separation is carried out. The enzyme used for partial digestion of the protein into peptides, the solvent system used for chromatography, the pH of the buffer, and the voltage of the electric field used in electrophoresis all affect the results. However, the same protein will produce the same pattern of separated peptides under the same conditions. The fingerprinting method is sensitive enough to detect the difference between two proteins that differ by only one amino acid if the change produces a difference in the polarity or charge of the side chain, or both. The peptide containing the altered amino acid will behave differently in the separations and will produce a different spot in the fingerprint. A change of one amino acid can produce drastic changes in the properties of the protein as well. The differences between sickle-cell anemia hemoglobin and normal hemoglobin provide a particularly striking example of this point.

# A.3
# ISOLATION OF PROTEINS

Many different proteins exist in a single cell, but a detailed study of the properties of any protein requires a homogeneous sample consisting of only one kind of molecule. The separation and isolation of proteins constitute an essential first step for further experimentation. Most methods for purification of proteins use some sort of chromatographic or electrophoretic technique, including techniques already discussed, such as ion exchange chromatography. Some methods are particularly useful for large molecules, and they warrant some discussion now.

**Molecular sieve chromatography** can be used to separate molecules on the basis of size. It is a form of column chromatography in which the stationary phase, the material used to pack the column, consists of cross-linked gel particles. The gel particles are usually supplied in bead form and consist of one of two kinds of polymers. One of the two kinds includes carbohydrate polymers such as dextran or agarose, and the other kind is polyacrylamide. The cross-linked structure of these polymers produces pores in the material. The pore size can be selected for a desired

**FIGURE A.10**   Schematic representation of molecular sieve chromatography.

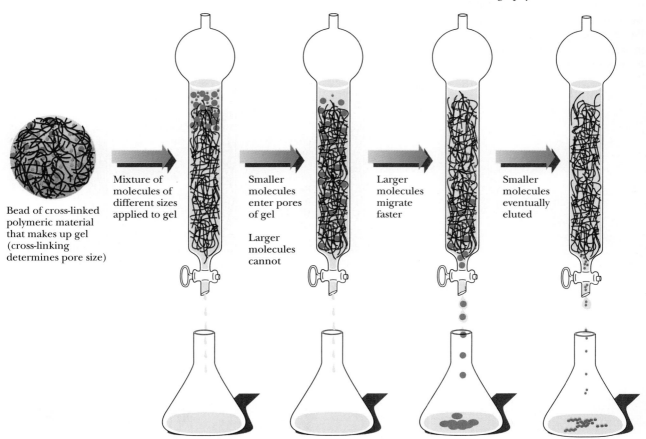

Bead of cross-linked polymeric material that makes up gel (cross-linking determines pore size)

Mixture of molecules of different sizes applied to gel

Smaller molecules enter pores of gel

Larger molecules cannot

Larger molecules migrate faster

Smaller molecules eventually eluted

value, which depends on the extent of cross-linking. When a sample is applied to the column, smaller molecules can enter the pores and tend to be delayed in their progress down the column, but the larger molecules cannot. As the sample is eluted by the mobile phase, the larger molecules are eluted first, followed later by the smaller ones, which were retained in the pores. Molecular sieve chromatography is represented schematically in Figure A.10. The particular advantages of molecular sieve chromatography are that it provides a convenient way to separate molecules on the basis of size and that it can be used to estimate molecular weight by comparison with standard samples.

**Affinity chromatography** makes use of the specific binding properties of many proteins. It is also a form of column chromatography, with some sort of polymeric material as the stationary phase. The distinguishing feature of affinity chromatography is that the polymer is covalently linked to some compound, called a **substrate,** which binds specifically to the desired protein (Figure A.11). The other proteins in the sample do not bind to the column and can easily be eluted with buffer, while the bound

**FIGURE A.11** The principle of affinity chromatography. In a mixture of proteins, only one (designated $P_1$) will bind to a substance (S) called the substrate. Ⓢ is the substance S in solution. The binding is reversible: $P_1 + S \Longleftrightarrow P_1—S$.

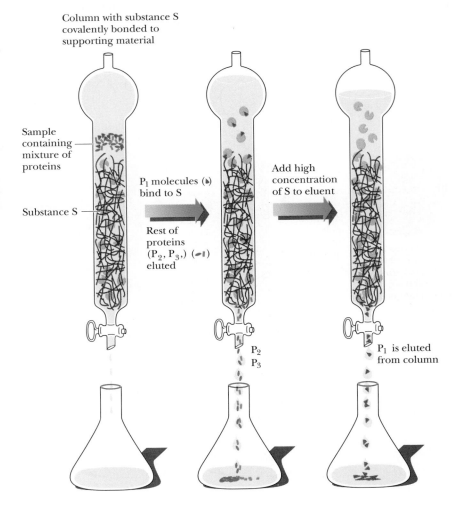

Column with substance S covalently bonded to supporting material

Sample containing mixture of proteins

Substance S

$P_1$ molecules (▸) bind to S

Rest of proteins ($P_2$, $P_3$,) (◂I) eluted

Add high concentration of S to eluent

$P_2$
$P_3$

$P_1$ is eluted from column

protein remains on the column. The bound protein can then be eluted from the column by adding high concentrations of the substrate in soluble form. The protein binds to the substrate in the mobile phase and is recovered from the column. Affinity chromatography is a convenient separation method and has the advantage of producing very pure proteins.

Polymeric gels can also be used as a supporting medium for electrophoresis. The gel, usually polyacrylamide, is prepared and cast as a continuous cross-linked matrix rather than the bead form employed in column chromatography. **Polyacrylamide gel electrophoresis** is a useful technique, and some variations can increase its usefulness (Figure A.12).

One variation of polyacrylamide gel electrophoresis involves treating the protein sample with the detergent sodium dodecyl sulfate (SDS) before applying the sample to the gel. The structure of SDS is $CH_3(CH_2)_{10}CH_2OSO_3^- Na^+$. The anion binds strongly to proteins by nonspecific adsorption. The larger the protein, the more of the anion it will adsorb. The presence of SDS completely denatures proteins, breaking all the noncovalent interactions that determine tertiary and quaternary structure. This last statement implies that multisubunit proteins can be analyzed as the component polypeptide chains. All the proteins in a sample will have a negative charge as a result of adsorption of anion. In **SDS–polyacrylamide gel electrophoresis (SDS-PAGE),** the acrylamide offers more resistance to large molecules than to small molecules. The size of the protein now becomes the determining factor in the separation because the ratio of charge to mass is approximately the same for all the proteins in the sample. Small proteins move faster than large ones, providing the basis for the separation. As is the case with molecular sieve chromatography, SDS–polyacrylamide gel electrophoresis can be used to estimate the molecular weight of proteins by comparison with standard samples.

**Isoelectric focusing** is another variation of gel electrophoresis. Since different proteins will have different titratable groups, they will have different isoelectric points. Recall that the isoelectric pH (pI) is the pH at which a protein (or amino acid or peptide) will have no net charge. At the pI the number of positive charges exactly balances the number of negative charges. In an isoelectric focusing experiment, the gel is prepared with a pH gradient that parallels the electric field gradient. As proteins migrate through the gel under the influence of the electric field, they encounter regions of different pH, and the charge on the protein changes. Eventually each protein reaches the point at which it has no net charge, its isoelectric point, and it no longer migrates. Each protein remains at the position on the gel corresponding to its pI, allowing for an effective method of separation.

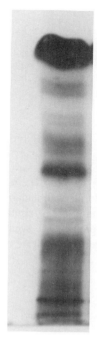

**FIGURE A.12** Separation of proteins by gel electrophoresis. Each band seen in the gel represents a different protein. In the SDS-PAGE technique, the sample is treated with detergent before being applied to the gel. In isoelectric focusing a pH gradient runs the length of the gel. The resulting bands have a similar appearance in all types of electrophoretic techniques.

## A.4
## N-TERMINAL AND C-TERMINAL AMINO ACIDS

The identity of the N-terminal and C-terminal amino acids in a protein sequence can be determined in several ways. There are both chemical and enzymatic methods for identifying the amino acids at the ends of the molecule.

A chemical reagent for labeling the N-terminal amino acid is **dansyl chloride** (Figure A.13). The protein is hydrolyzed to the constituent amino acids by acid hydrolysis. The amino acids are separated chromatographically, and the labeled amino acid is identified by comparison with standards. An advantage of this method is that the dansyl group is fluorescent, allowing detection of quantities of material at the nanogram ($10^{-9}$ g) level.

**Aminopeptidases** are enzymes that remove the N-terminal amino acid of a protein or polypeptide. These enzymes are called **exopeptidases,**

**FIGURE A.13** (a) The reaction of dansyl chloride with amino acids. (b) Dansylation of the N-terminal amino acid of a protein. The symbols ①, ②, ③, and ⓝ indicate the individual amino acids.

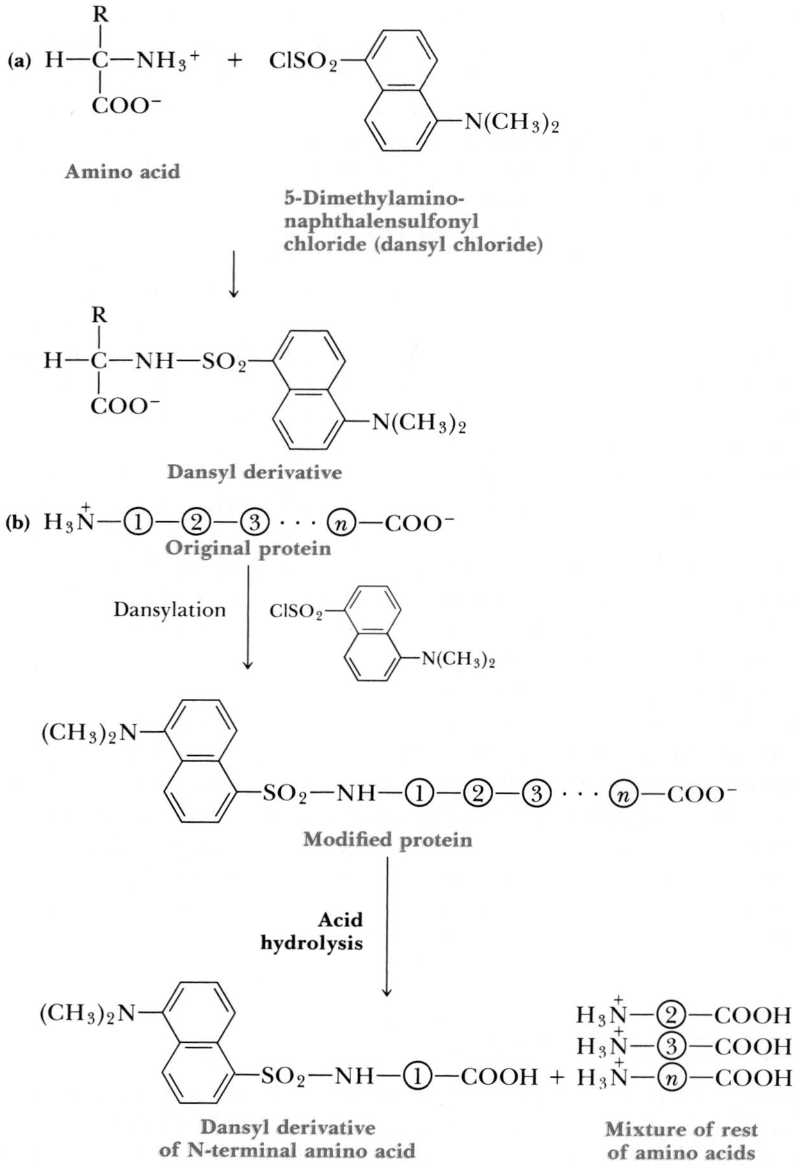

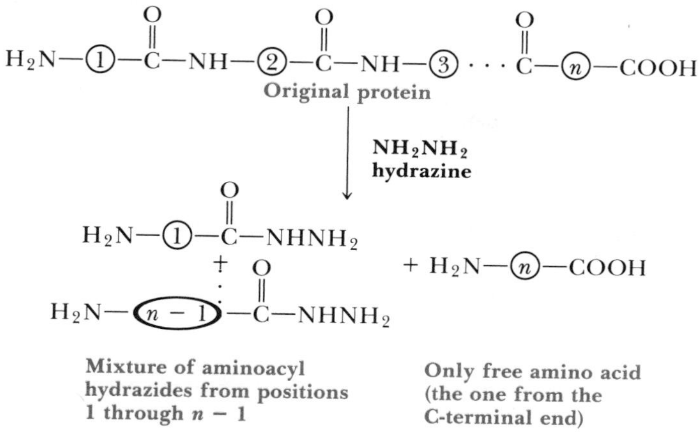

$$H_2N-①-\overset{\overset{\displaystyle O}{\|}}{C}-NH-②-\overset{\overset{\displaystyle O}{\|}}{C}-NH-③\cdots \overset{\overset{\displaystyle O}{\|}}{C}-ⓝ-COOH$$

Original protein

$$\downarrow \quad \begin{array}{l} NH_2NH_2 \\ \text{hydrazine} \end{array}$$

$$H_2N-①-\overset{\overset{\displaystyle O}{\|}}{C}-NHNH_2$$
$$+ \qquad\qquad + H_2N-ⓝ-COOH$$
$$H_2N-\overset{.}{ⓝ-1}-\overset{\overset{\displaystyle O}{\|}}{C}-NHNH_2$$

Mixture of aminoacyl          Only free amino acid
hydrazides from positions     (the one from the
1 through $n - 1$             C-terminal end)

**FIGURE A.14**  The use of hydrazine to determine C-terminal amino acids. The products of the reaction are separated chromatographically and identified by comparison against standards.

since they attack only peptide bonds at the end of a polypeptide. After they cleave the first peptide bond, however, they continue to cleave off the other amino acids in sequence. If conditions are not carefully controlled, confusing and ambiguous information can result.

The C-terminal amino acid can be determined by methods similar to those used for the N-terminal acid. A chemical method is treatment with **hydrazine,** which reacts with the carbonyl group of each peptide bond. The bond is cleaved, and each amino acid derivative is released as the hydrazide derivative, $NH_2-CHR-CO-NH-NH_2$, a compound similar to an amide. Since the C-terminal amino acid is not involved in a peptide bond, it remains in the mixture as the only unmodified amino acid. After chromatographic separation and comparison with the standards, the C-terminal amino acid can be identified (Figure A.14).

**Carboxypeptidases** are used for enzymatic determination of the C-terminal amino acid. Like aminopeptidases, they are exopeptidases, in this case cleaving polypeptides from the C-terminal end. Also like aminopeptidases, they present the difficulty of continuing to digest the polypeptide after the C-terminal residue has been removed.

Determining the N-terminal and C-terminal residues can indicate whether a given protein consists of one amino acid chain or whether two or even more chains are covalently bonded together. In the case of insulin, for example, there are two N-terminal residues, glycine and phenylalanine, and two C-terminal residues, asparagine and alanine. Insulin consists of two polypeptide chains, designated A and B, held together by two sets of disulfide linkages (Figure A.15). In such a situation it is necessary to be sure that the separation techniques we discussed earlier, such as chromatography and electrophoresis, have given correct results in identifying the products of the reaction. Careful experimentation is needed to avoid ambiguous results.

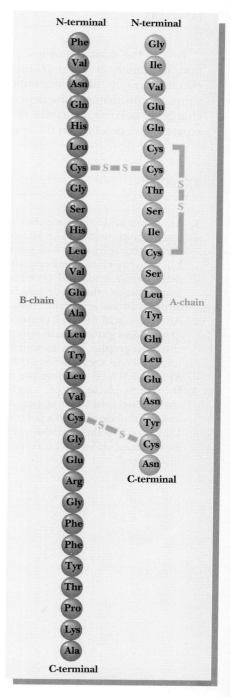

**FIGURE A.15**  Schematic representation of the two chains of insulin as determined by end-group analysis.

## SUMMARY

Two of the most important methods for separation of amino acids, peptides, and proteins are chromatography and electrophoresis. The various forms of chromatography depend on differences in charge, in polarity, or in size of the molecules to be separated, depending on the application. In electrophoresis, differences in charge and in size are the criteria for separation. Determining the N-terminal and C-terminal amino acids of proteins depends on using these separation methods after the ends of the molecule have been chemically labeled.

## EXERCISES

1. An amino acid mixture consisting of lysine, leucine, and glutamic acid is to be separated by ion exchange chromatography using a cation exchange resin at pH 3.5, with the eluting buffer at the same pH. Which of these amino acids will be eluted from the column first? Will any other treatment be needed to elute one of these amino acids from the column?

2. An amino acid mixture consisting of phenylalanine, glycine, and glutamic acid is to be separated by paper chromatography. The solvent is less polar than water. Which of these amino acids will have the largest $R_f$ value? Which will have the smallest?

3. In reverse-phase HPLC (high-performance liquid chromatography), the stationary phase is nonpolar and the mobile phase is a polar solvent at neutral pH. Which of the three amino acids in Question 2 will move fastest on an HPLC column? Which one will be the slowest?

4. Molecular sieve chromatography is a useful method for removing salts such as ammonium sulfate from protein solutions. Describe how such a separation is accomplished.

5. A newly isolated protein was treated with dansyl chloride and then subjected to complete acid hydrolysis. Two dansyl-labeled amino acids were detected, dansylalanine and dansylmethionine. What can you conclude about the structure of the protein from these results?

6. What experimental methods were used to detect the difference between normal hemoglobin and sickle-cell anemia hemoglobin?

7. How can molecular sieve column chromatography be used to arrive at an estimate of the molecular weight of a protein?

# The Behavior of Proteins: Enzymes

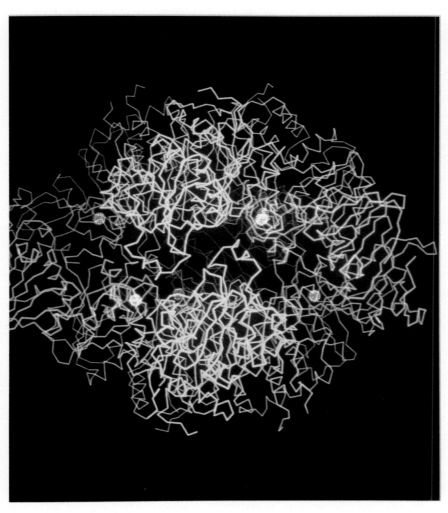

*The structure of the enzyme aspartate transcarbamolyase (ATCase) in the T form. Catalytic subunits are shown in yellow and regulatory subunits in green.*

*Your automobile is powered by oxidizing the hydrocarbon gasoline to carbon dioxide and water in a controlled explosion within an engine where hot gases can reach 4,000°F. By contrast, the living cell gets its energy by oxidizing the carbohydrate glucose to carbon dioxide and water at a temperature (in humans) of 98.6°F. The secret ingredient in living organisms is* **catalysis** *performed by protein enzymes. Their three-dimensional architecture gives them exquisite specificity to select which substrate molecules they will bind to and operate on. Each enzyme has, in fact, a miniature "operating table" where the substrate is momentarily held in a predetermined position so that it can be cut or altered with surgical precision. The scene of the operation is called the* **active site**, *usually a groove, cleft, or cavity on the surface of the protein. Enzyme surgery, cleaving molecules or stitching them together, may take as long as a second. In some cases, many thousands of operations can be performed per second. The miracle of life is that myriads of chemical reactions in the cell are occurring simultaneously with great accuracy at astonishing speed. Without the proper enzymes to process the food you eat, it might take you fifty years to digest breakfast.*

## 10.1
## ENZYMES ARE BIOLOGICAL CATALYSTS

Of all the functions of proteins, the one that is probably most important is that of **catalysis.** In the absence of catalysis, most reactions in biological systems would take place far too slowly to provide products at an adequate pace for a metabolizing organism. The catalysts that serve this function in organisms are called **enzymes;** all enzymes are globular proteins (with the exception of some recently discovered RNAs that catalyze their own splicing). Enzymes are the most efficient catalysts known and are able to increase the rate of a reaction by a factor of up to $10^{20}$ over uncatalyzed reactions; nonenzymatic catalysts, on the other hand, typically enhance the rate of reaction by factors of $10^2$ to $10^4$. Enzymes are highly specific, even to the point of being able to distinguish stereoisomers of a given compound. In many cases the action of enzymes is fine-tuned by regulatory processes.

## 10.2
## CATALYSIS: KINETIC VS. THERMODYNAMIC
## ASPECTS OF REACTIONS

Enzymes, like all catalysts, speed up a reaction. They cannot alter the equilibrium constant or the free energy change for the reaction. Reaction rates and the spontaneity of a reaction in the thermodynamic sense are two different topics, although they are closely related. This is true of all reactions, whether or not a catalyst is involved. The spontaneity of a reaction (Section 8.2) depends on the difference in free energy ($\Delta G°$) between the reactants and products, the initial and final states for the reaction. The reaction rate does not depend on the free energy change between reactants and products. Instead, it depends on the **activation energy,** which is the energy input required to initiate the reaction. The activation energy for an uncatalyzed reaction is higher than that for a catalyzed reaction; in other words, an uncatalyzed reaction requires more energy to get started, and this is why its rate is slower than that of a catalyzed reaction. An example of a reaction that requires a number of enzymatic catalysts is the reaction of glucose and oxygen gas to produce carbon dioxide and water,

$$\text{Glucose} + 6\ O_2 \longrightarrow 6\ CO_2 + 6\ H_2O$$

This reaction is spontaneous in the thermodynamic sense, because its free energy change is negative ($\Delta G° = -2880$ kJ mole$^{-1}$ = $-689$ kcal mole$^{-1}$). Energy must be supplied to start the reaction, which then proceeds with a release of energy. This energy that must be supplied to start the reaction, the activation energy (written as $\Delta G°\ddagger$), is conceptually similar to the act of pushing an object to the top of a hill so that it can then slide down the other side.

Activation energy and its relationship to the free energy change of a reaction can best be shown in graphic form. In Figure 10.1(a), the $x$ coordinate shows the extent to which the reaction has taken place, and the $y$ coordinate indicates free energy. The activation energy profile shows the intermediate stages of a reaction, those that lie between the initial and final states. The difference between the energies of the reactants (the initial state) and products (the final state) gives the standard free energy change, the $\Delta G°$, for the reaction; this figure shows the energies for an exergonic, spontaneous reaction such as the complete oxidation of glucose. At the maximum of the curve connecting the reactants and the products lies the **transition state** with the amount of energy and correct arrangement of atoms needed to produce products. The activation energy is the amount of free energy ($\Delta G°\ddagger$) required to bring the reactants to the transition state.

The anology of traveling over a hill (or, more usually, over a mountain pass between two valleys) is frequently applied to the discussion of activation energy profiles. The change in energy is analogous to the change in elevation, and the progress of the reaction is analogous to the distance traveled. The analogue of the transition state is the top of the pass. Considerable effort has gone into elucidating the intermediate stages in reactions of interest to chemists and biochemists, and into determining the pathway or reaction mechanism that lies between the initial and final states. The study of the intermediate stages of reaction mechanisms, called reaction dynamics, is currently a very active field of research. Activation energy profiles are essential in the discussion of catalysts. The activation energy directly affects the rate of reaction, and the presence of a catalyst speeds up a reaction by changing the mechanism and thus lowering the activation energy.

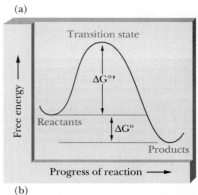

(a)

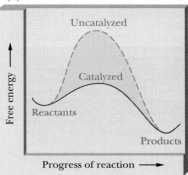

(b)

**FIGURE 10.1**   Activation energy profiles. (a) The activation energy profile for a typical reaction. The reaction shown here is exergonic (energy-releasing). Note the difference between the activation energy ($\Delta G°\ddagger$) and the standard free energy of the reaction ($\Delta G°$). (b) Comparison of activation energy profiles for catalyzed and uncatalyzed reactions. The activation energy of the catalyzed reaction is much less than that of the uncatalyzed reaction.

The most important effect of a catalyst on a chemical reaction is apparent from a comparison of the activation energy profiles of the same reaction, catalyzed and uncatalyzed, as shown in Figure 10.1(b). The standard free energy change for the reaction, $\Delta G°$, remains unchanged on addition of a catalyst, but the activation energy, $\Delta G°\ddagger$, for the reaction is lowered. In the "hill and valley" analogy, the catalyst serves as a guide who finds an easier path through the mountain pass between two valleys. A situation analogous to the difference between catalyzed and uncatalyzed pathways is the difference between two routes from San Francisco to Los Angeles. The highest point on Route 5 is Tejon Pass (elevation 4400 feet), analogous to the uncatalyzed path. The highest point on Route 101 is not much over 1000 feet, an easier route analogous to the catalyzed pathway. The initial and final points of the trip are the same, but the path between them (the mechanism) is different. The rate of the catalyzed reaction is much greater than the rate of the uncatalyzed reaction. In the case of enzymatic catalysis, the enhancement of reaction rate can be by many powers of ten.

## 10.3
## ENZYME KINETICS

The **rate of a chemical reaction** is usually expressed in terms of a change in the concentration of a reactant or of a product in a given time. Any convenient experimental method can be used to monitor changes in concentration. In a reaction of the form $A + B \longrightarrow P$, the rate of reaction can be expressed as the rate of disappearance of one of the reactants or in terms of the rate of appearance of product. The rate of disappearance of A is $-\Delta[A]/\Delta t$, where the symbol $\Delta$ refers to change, [A] refers to the concentration of A in moles liter$^{-1}$, and $t$ is time. Likewise, the rate of disappearance of B is $-\Delta[B]/\Delta t$, and the rate of appearance of P is $\Delta[P]/\Delta t$. The rate of the reaction can be expressed in terms of any of these changes, because the rates of appearance of product and disappearance of reactant are related by the stoichiometric equation for the reaction.

$$\text{Rate} = -\frac{\Delta[A]}{\Delta t} = -\frac{\Delta[B]}{\Delta t} = \frac{\Delta[P]}{\Delta t}$$

The negative signs for the changes in concentration of A and B refer to the fact that A and B are being used up in the reaction, while P is being produced.

It has been established that the rate of a reaction at a given time is proportional to the product of the concentrations of the reactants raised to the appropriate powers,

$$\text{Rate} \propto [A]^f[B]^g$$

or as an equation

$$\text{Rate} = k\,[A]^f[B]^g$$

where $k$ is a proportionality constant called the **rate constant.** The symbols $f$ and $g$ are exponents. The exponents $f$ and $g$ are *not necessarily* equal to the coefficients of the balanced equation and *must be determined experimentally.* The square brackets, as usual, denote molar concentration. When the exponents in the rate equation have been determined experimentally, a mechanism can be proposed for the reaction. The mechanism is a description of the detailed steps along the path between reactants and products. Outlining these detailed steps requires knowing how many molecules are involved in the process; these numbers are precisely those given by the experimentally determined exponents.

The exponents in the rate equation are usually small whole numbers such as 1 or 2. (There are also some cases in which the exponent 0 occurs.) The values of the exponents depend on the number of molecules involved in the detailed steps that constitute the mechanism. The **overall order** of a reaction is the sum of all the exponents. If, for example, the rate of a reaction $A \longrightarrow B$ is given by the rate equation

$$\text{Rate} = k[A]^1 \tag{10.1}$$

where $k$ is the rate constant and the exponent for the concentration of A is 1, then the reaction is **first order** with respect to A, and first order overall. The rate of radioactive decay of the widely used tracer isotope phosphorus 32 ($^{32}P$, atomic weight = 32) depends only on the concentration of $^{32}P$ present; here we have an example of a first order reaction. Only the $^{32}P$ atoms are involved in the mechanism of the radioactive decay. As an equation

$$^{32}P \longrightarrow \text{decay products}$$

$$\text{Rate} = k[^{32}P]^1 = k[^{32}P]$$

If the rate of a reaction $A + B \longrightarrow C + D$ is given by

$$\text{Rate} = k[A]^1[B]^1 \tag{10.2}$$

where $k$ is the rate constant, the exponent for the concentration of A is 1 and the exponent for the concentration of B is 1, then the reaction is said to be first order with respect to A, first order with respect to B, and **second order** overall. In the reaction of methyl bromide ($CH_3Br$) with hydroxide ion ($OH^-$) to give methyl alcohol ($CH_3OH$) and bromide ion ($Br^-$), the rate of reaction depends on the concentrations of both reactants.

$$CH_3Br + OH^- \longrightarrow CH_3OH + Br^-$$

$$\text{Rate} = k[CH_3BR]^1[OH^-]^1 = k[CH_3Br][OH^-]$$

where $k$ is the rate constant. Both the methyl bromide and hydroxide ion take part in the reaction mechanism. The reaction of methyl bromide with hydroxide ion is first order with respect to $CH_3Br$, first order with respect to $OH^-$, and second order overall.

Many common reactions are first or second order. Conclusions about the mechanism of a reaction can be drawn once the order of the reaction is determined experimentally.

The possibility exists that exponents in a rate equation may be equal to zero, with the rate for a reaction $A \longrightarrow B$ given by the equation

$$\text{Rate} = k[A]^0 = k \tag{10.3}$$

Such reactions are called **zero order,** and their rate, which is constant, depends not on concentrations of reactants, but on other factors such as the presence of catalysts. Enzyme-catalyzed reactions can exhibit zero order kinetics when the concentrations of reactants are so high that the enzyme is completely saturated with reactant molecules. This point will be discussed in more detail later in this chapter, but for the moment we can consider the situation analogous to a bottleneck in traffic with six lanes of cars trying to get across a two-lane bridge. The rate at which the cars cross is not affected by the number of waiting cars, but only by the number of lanes available on the bridge.

## 10.4
## THE TRANSITION STATE IN ENZYMATIC REACTIONS

In an enzyme-catalyzed reaction, the enzyme binds to the **substrate** (one of the reactants) to form a complex, leading in turn to formation of the transition state species, which then forms the product. The nature of transition states in enzymatic reactions is a large field of research in itself, but some general statements can be made on the subject. A substrate binds to a small portion of the enzyme called the **active site,** frequently located

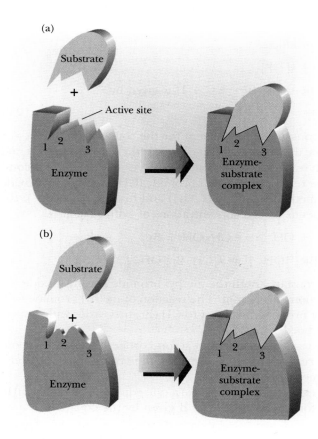

**FIGURE 10.2**   Two models for the binding of a substrate to an enzyme. (a) In the lock-and-key model for binding of a substrate to an enzyme, the shape of the substrate and the conformation of the active site are complementary to one another. (b) In the induced-fit model for the binding of a substrate to an enzyme, there is a conformational change in the enzyme on binding to substrate. The shape of the active site becomes complementary to the shape of the substrate only after the substrate binds to the enzyme.

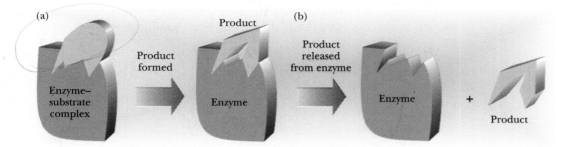

**FIGURE 10.3**    Formation of product from substrate (bound to the enzyme), followed by release of the product.

in a cleft or crevice in the protein (Figure 10.2). The catalyzed reaction takes place at the active site, usually in several steps. The first step is the binding of substrate to the enzyme, which occurs because of highly specific interactions between the substrate and the side chains of certain amino acids in the enzyme, which are essential for enzymatic activity. These essential amino acids form the active site. There are two important models that describe the binding process. In the first, the **lock-and-key** model, there is assumed to be a high degree of similarity between the shape of the substrate and the geometry of the binding site on the enzyme (Figure 10.2(a)). The substrate binds to a site into which it fits exactly, like a key in a lock or the right piece in a three-dimensional jigsaw puzzle. The second model takes into account the fact that proteins have some three-dimensional flexibility. According to this model, the **induced-fit** theory, the binding of the substrate induces a conformational change in the enzyme such that there is an exact fit once the substrate is bound (Figure 10.2(b)). The binding site has a different three-dimensional shape before the substrate is bound.

When the substrate is bound and the transition state is formed, there is a rearrangement of bonds. In the transition state, the substrate is bound in close proximity to atoms with which it is to react. Further, the substrate is placed in the correct orientation with respect to the atoms with which it is to react. Both effects, proximity and orientation, serve to speed up the reaction. As bonds are broken and new bonds are formed, the substrate is transformed into product. The product is released from the enzyme, which can then catalyze the reaction of more substrate to form more product (Figure 10.3). Each enzyme has its own unique mode of catalysis, which is not surprising in view of enzymes' great specificity. However, there are some general modes of catalysis in enzymatic reactions. Two enzymes, chymotrypsin and aspartate transcarbamoylase, are good examples of the general principles of enzymatic reactions.

## 10.5
## TWO EXAMPLES OF ENZYME-CATALYZED REACTIONS

**Chymotrypsin** is an enzyme that catalyzes the hydrolysis of peptide bonds, with some specificity for residues containing aromatic side chains. Chymotrypsin also cleaves peptide bonds at other sites, such as leucine, histidine, and glutamine, but with a lower frequency than at aromatic amino-acid residues. It also catalyzes the hydrolysis of ester bonds.

$$R_1-\overset{\overset{\displaystyle O}{\|}}{C}-\underset{\underset{\displaystyle H}{|}}{N}-R_2 + H_2O \rightleftharpoons R_1-\overset{\overset{\displaystyle O}{\|}}{C}-O^- + {}^+H_3N-R_2$$

Peptide                Acid            Amine

$$R_1-\overset{\overset{\displaystyle O}{\|}}{C}-O-R_2 + H_2O \rightleftharpoons R_1-\overset{\overset{\displaystyle O}{\|}}{C}-O^- + HO-R_2$$

Ester                Acid            Alcohol

$$\underset{\text{$p$-Nitrophenylacetate}}{\text{[structure]}} \xrightarrow[\textbf{Basic conditions}]{\textbf{H}_2\textbf{O}} H_3C-\overset{\overset{\displaystyle O}{\|}}{C}-O^- + H^+ \qquad \underset{\text{$p$-Nitrophenolate (yellow)}}{\text{[structure]}} + H^+$$

(10.4)

Reactions catalyzed by chymotrypsin

While ester hydrolysis is not important to the physiological role of chymotrypsin in the digestion of proteins, this type of reaction is a convenient model system for investigating the enzyme's catalysis of hydrolysis reactions. A usual laboratory procedure is to use $p$-nitrophenyl esters as the substrate and to monitor the progress of the reaction by the appearance of a yellow color in the reaction mixture due to the production of $p$-nitrophenolate ion.

In a typical reaction when a $p$-nitrophenyl ester is hydrolyzed by chymotrypsin, the experimental rate of the reaction depends on the concentration of the substrate, in this case the $p$-nitrophenyl ester. At low substrate concentrations the rate of reaction increases as more substrate is added. At higher substrate concentrations the rate of the reaction changes very little with addition of more substrate; a maximum rate is reached. When these results are presented in a graph, the shape of the curve is **hyperbolic** (Figure 10.4).

Another example of an enzyme-catalyzed reaction is the one catalyzed by the enzyme **aspartate transcarbamoylase** (ATCase). This reaction is the first step in a pathway that leads to the formation of cytidine triphosphate (CTP) and uridine triphosphate (UTP), which are ultimately needed for the biosynthesis of RNA and DNA. In this reaction carbamoyl phosphate reacts with aspartate to produce carbamoyl aspartate; phosphate ion is released in the course of the reaction.

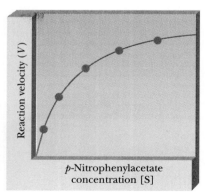

**FIGURE 10.4** Dependence of reaction velocity, $V$, on $p$-nitrophenylacetate concentration, [S], in a reaction catalyzed by chymotrypsin. The shape of the curve is hyperbolic.

$$\text{Carbamoyl phosphate} + \text{Aspartate} \longrightarrow \text{Carbamoyl aspartate} + HPO_4{}^{2-} \qquad (10.5)$$

The reaction catalyzed by aspartate transcarbamoylase

The rate of this reaction also depends on substrate concentration, in this case the concentration of aspartate, which can vary more widely in living organisms than that of carbamoyl phosphate. Experimental results show that once again the rate of the reaction depends on substrate concentration at low and moderate concentrations, and once again a maximum rate is reached at high substrate concentrations. There is, however, one very important difference. A graph that shows how the rate of the reaction catalyzed by aspartate transcarbamoylase depends on substrate concentration is **sigmoidal,** rather than hyperbolic (Figure 10.5).

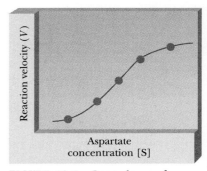

**FIGURE 10.5** Dependence of reaction velocity, *V,* on aspartate concentration, [S], in a reaction catalyzed by aspartate transcarbamoylase. The shape of the curve is sigmoidal.

The results of experiments on the reaction kinetics of chymotrypsin and aspartate transcarbamoylase are representative of experimental results obtained with many enzymes. The overall kinetic behavior of many enzymes is similar to that of chymotrypsin, while others resemble aspartate transcarbamoylase. We can use this information to draw some general conclusions about the behavior of enzymes.

The comparison between the kinetic behavior of chymotrypsin and ATCase is reminiscent of the relationship between the oxygen-binding behaviors of myoglobin and hemoglobin, discussed in Chapter 9. ATCase and hemoglobin are allosteric proteins, while chymotrypsin and myoglobin are not. (Recall from Section 9.5 that allosteric proteins are the ones whose behavior is affected by interactions between subunits. Cooperative effects, such as the fact that the binding of the first oxygen molecule to hemoglobin makes it easier for other oxygen molecules to bind, are a hallmark of allosteric proteins.) The differences in behavior between allosteric and nonallosteric proteins can be understood in terms of models based on structural differences between the two kinds of proteins. For our purposes, we need a model that will explain the hyperbolic plot of kinetic data for nonallosteric enzymes, and another model that will explain the sigmoidal plot for allosteric enzymes. We shall need both models when we discuss the mechanisms of the many enzyme-catalyzed reactions as we encounter them in subsequent chapters. The Michaelis-Menten model is widely used for nonallosteric enzymes, and there are several models for the behavior of allosteric enzymes. We must discuss both types of behavior in some detail.

## 10.6
## THE MICHAELIS-MENTEN APPROACH TO ENZYME KINETICS

A particularly useful model for the kinetics of enzyme-catalyzed reactions was devised in 1913 by Leonor Michaelis and Maud Menten. This is still the basic model for nonallosteric enzymes, one that is widely used in spite of many modifications.

A typical reaction might be the conversion of some substrate, S, to a product, P. The stoichiometric equation for the reaction is

$$S \longrightarrow P$$

The mechanism for an enzyme-catalyzed reaction can be summarized as

$$E + S \underset{k_{-1}}{\overset{k_1}{\rightleftarrows}} ES \overset{k_2}{\longrightarrow} E + P \tag{10.6}$$

In this equation $k_1$ is the rate constant for the formation of the enzyme-substrate complex, ES, from the enzyme, E, and the substrate, S; $k_{-1}$ is the rate constant for the reverse reaction, dissociation of the ES complex to free enzyme and substrate; and $k_2$ is the rate constant for the conversion of the ES complex to product P and the subsequent release of product from the enzyme. The enzyme appears explicitly in the mechanism, and the concentrations of both free enzyme, E, and enzyme-substrate complex, ES, therefore appear in the rate equations. It is a characteristic of catalysts that they are regenerated at the end of the reaction, and this is true of enzymes.

The general mechanism of the enzyme-catalyzed reaction involves binding of the enzyme E to the substrate to form a complex ES, which then forms the product. In the initial stages of the reaction there is so little product present that no reverse reaction of product to complex need be considered. It is the **initial rate** that is usually determined in enzymatic reactions, and this rate depends on the rate of breakdown of the enzyme-substrate complex into product and enzyme.

The rate of formation of the enzyme-substrate complex, ES, is

$$\text{Rate of formation} = \frac{\Delta[ES]}{\Delta t} = k_1[E][S] \tag{10.7}$$

where the notation $\Delta[ES]/\Delta t$ means the change in the concentration of the complex, $\Delta[ES]$, during a given length of time $\Delta t$, and $k_1$ is the rate constant for the formation of the complex. The complex, ES, breaks down in two reactions, by returning to enzyme and substrate, or by giving rise to product and releasing enzyme. The rate of disappearance of complex is the sum of the rates of the two reactions.

$$\text{Rate of breakdown} = \frac{-\Delta[ES]}{\Delta t} = k_{-1}[ES] + k_2[ES] \tag{10.8}$$

The negative sign in the term $\Delta[ES]/\Delta t$ means that the concentration of the complex decreases when the complex breaks down. The term $k_{-1}$ is the rate constant for the dissociation of complex to regenerate enzyme and substrate, while $k_2$ is the rate constant for the reaction of the complex to give product and enzyme. Enzymes are capable of processing the substrate very efficiently, and a **steady state** is soon reached in which the rate of formation of the enzyme-substrate complex is equal to the rate of its breakdown. There is very little complex present, and it turns over rapidly, but its concentration stays the same with time.

According to the steady state theory, then, the rate of appearance of the enzyme-substrate complex is equal to the rate of its disappearance,

$$\frac{\Delta[ES]}{\Delta t} = \frac{-\Delta[ES]}{\Delta t} \tag{10.9}$$

and

$$k_1[E][S] = k_{-1}[ES] + k_2[ES] \tag{10.10}$$

In order to solve for the concentration of the complex ES, it is necessary to know the concentration of the other species involved in the reaction. The initial concentration of substrate is a known experimental condition and, during the initial stages of the reaction, does not change significantly. The substrate concentration is much larger than the enzyme concentration. The initial concentration of the enzyme, $[E]_0$, is also known, but a comparatively large amount may be involved in the complex. The concentration of free enzyme, $[E]$, is the difference between $[E]_0$, the initial concentration, and $[ES]$, which can be written as an equation:

$$[E] = [E]_0 - [ES] \tag{10.11}$$

Substituting for the concentration of free enzyme, $[E]$, in Equation 10.10,

$$k_1 ([E]_0 - [ES]) [S] = k_{-1}[ES] + k_2[ES] \tag{10.12}$$

Collecting all the rate constants for the individual reactions,

$$\frac{([E]_0 - [ES]) [S]}{[ES]} = \frac{k_{-1} + k_2}{k_1} = K_M \tag{10.13}$$

where $K_M$ is called the **Michaelis constant.** It is now possible to solve Equation 10.13 for the concentration of enzyme-substrate complex $[ES]$,

$$[E]_0[S] = [ES](K_M + [S])$$

or

$$[ES] = \frac{[E]_0[S]}{K_M + [S]} \tag{10.14}$$

In the Michaelis-Menten model the initial rate ($V_{init}$) (in some texts the notation for initial rate is $V_0$) of formation of product depends only on the rate of the breakdown of the ES complex,

$$V_{init} = k_2[ES] \tag{10.15}$$

and substituting the expression for $[ES]$ from Equation 10.14.

$$V_{init} = \frac{k_2[E]_0[S]}{K_M + [S]} \tag{10.16}$$

If the substrate concentration is so high that the enzyme is completely saturated with substrate ($[ES] = [E]_0$), the reaction proceeds at its maximum possible rate ($V_{max}$), and substituting $[E]_0$ for $[ES]$ in Equation 10.15,

$$V_{max} = k_2[E]_0 \tag{10.17}$$

The original concentration of enzyme is a constant, which means that

$$V_{max} = \text{constant}$$

This expression for $V_{max}$ resembles that for a zero-order reaction given in Equation 10.3:

$$\text{Rate} = k[A]_0 = k$$

Note that the concentration of substrate, $[A]$, appears in Equation 10.3 rather than the concentration of enzyme, $[E]$, as is the case in Equation

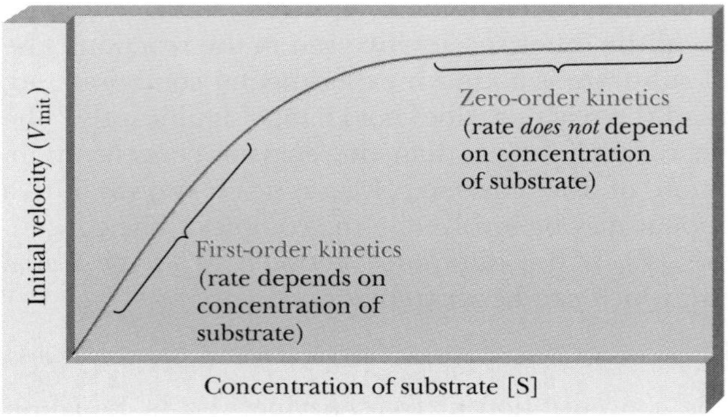


**FIGURE 10.6** The rate and the observed kinetics of an enzymatic reaction depend on substrate concentration. The concentration of enzyme, [E], is constant.

10.17. Zero-order kinetics with respect to substrate are observed when the enzyme is saturated with substrate, underlying the necessity of using comparatively low concentrations of substrate to determine initial rates of enzymatic reactions.

Substituting the expression for $V_{max}$ into Equation 10.16 enables us to relate the initial and maximum rates of an enzymatic reaction:

$$V_{init} = \frac{V_{max}\,[S]}{K_M + [S]} \tag{10.18}$$

Figure 10.6 shows the effect of increasing substrate concentration on the observed rate. In such an experiment the reaction is run at several substrate concentrations, and the rate is determined by following the disappearance of reactant or the appearance of product by any convenient method. In the early stages of the reaction and at low substrate concentrations, first-order kinetics are observed. At higher substrate concentrations, when the enzyme is saturated, the constant reaction rate characteristic of zero-order kinetics is observed. From now on we shall need to discuss the rate of the reaction in general ($V$) rather than just the initial rate ($V_{init}$). The constant rate at saturation is the $V_{max}$ for the enzyme, and the value of $V_{max}$ can be estimated from the graph. The value of $K_M$ can also be estimated from the graph. From Equation 10.18,

$$V = \frac{V_{max}[S]}{K_M + [S]}$$

When experimental conditions are adjusted so that $[S] = K_M$,

$$V = \frac{V_{max}[S]}{[S] + [S]}$$

and

$$V = \frac{V_{max}}{2}$$

In other words, when the rate of the reaction is one half its maximum value, the substrate concentration is equal to the Michaelis constant (Figure 10.7), and this point is the basis of the graphical determination of $K_M$.

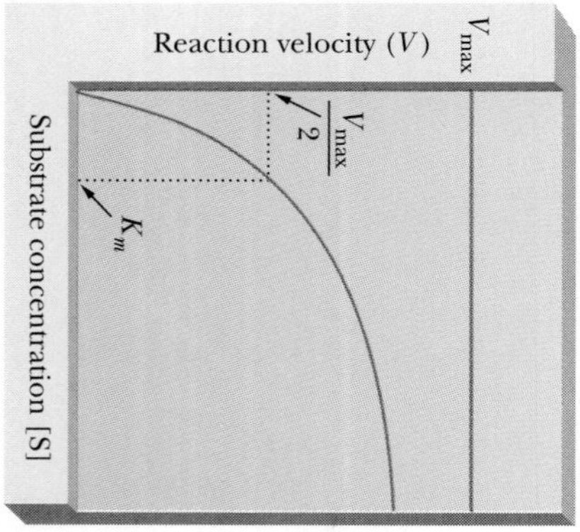


**FIGURE 10.7** Graphic determination of $V_{max}$ and $K_M$ from a plot of reaction velocity, $V$, against substrate concentration, [S]. $V_{max}$ is the constant rate reached when the enzyme is completely saturated with substrate, a value that frequently must be estimated from such a graph.

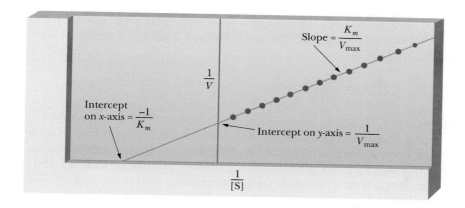

Slope $= \dfrac{K_m}{V_{max}}$

$\dfrac{1}{V}$

Intercept on x-axis $= \dfrac{-1}{K_m}$

Intercept on y-axis $= \dfrac{1}{V_{max}}$

$\dfrac{1}{[S]}$

**FIGURE 10.8**   A Lineweaver–Burk double reciprocal plot of enzyme kinetics. The reciprocal of reaction velocity, $1/V$, is plotted against the reciprocal of the substrate concentration, $1/[S]$. The slope of the line is $K_M/V_{max}$, and the y-intercept is $1/V_{max}$. The x-intercept is $-1/K_M$.

The curve that describes the rate of a nonallosteric enzymatic reaction is hyperbolic. It is quite difficult to determine a single point at which the rate levels off. This problem in turn makes it difficult to determine the $V_{max}$ and thus the $K_M$ of the enzyme. It is considerably easier to work with a straight line than a curve. It is possible to transform the equation for a hyperbola (Equation 10.18)

$$V = \frac{V_{max}[S]}{K_M + [S]}$$

into an equation for a straight line by taking the reciprocal of both sides:

$$\frac{1}{V} = \frac{K_M + [S]}{V_{max}[S]}$$

$$\frac{1}{V} = \frac{K_M}{V_{max}[S]} + \frac{[S]}{V_{max}[S]}$$

$$\frac{1}{V} = \frac{K_M}{V_{max}} \cdot \frac{1}{[S]} + \frac{1}{V_{max}} \tag{10.19}$$

The equation now has the form of a straight line, $y = mx + b$, where $1/V$ takes the place of the $y$ coordinate and $1/[S]$ takes the place of the $x$ coordinate. The slope of the line, $m$, is $K_M/V_{max}$, and the intercept, $b$, is $1/V_{max}$. Figure 10.8 presents this information graphically as a **Lineweaver-Burk double-reciprocal plot**. It is usually easier to draw the best straight line through a set of points than to estimate the best fit of points to a curve. There are convenient computer methods for drawing the best straight line through a series of experimental points.

## Significance of $K_M$ and $V_{max}$

We have already seen that when the rate of a reaction $V$ is equal to one half the maximum rate possible, $V = V_{max}/2$, then $K_M = [S]$. One interpretation of the Michaelis constant $K_M$ is that it is equal to the concentration of substrate at which 50% of the enzyme active sites are occupied by substrate. The Michaelis constant has the units of concentration.

There is another interpretation as well, one that relies on the assumptions of the original Michaelis-Menten model of enzyme kinetics. From Equation 10.6,

$$E + S \underset{k_{-1}}{\overset{k_1}{\rightleftharpoons}} ES \xrightarrow{k_2} E + P$$

As before, $k_1$ is the rate constant for the formation of the enzyme-substrate complex ES from the enzyme E and the substrate S; $k_{-1}$ is the rate constant for the reverse reaction, dissociation of the ES complex to free enzyme and substrate; and $k_2$ is the rate constant for the formation of product P and the subsequent release of product from the enzyme. Also recall that

$$K_M = \frac{k_{-1} + k_2}{k_1}$$

If $k_{-1}$ is much larger than $k_2$ ($k_{-1} \gg k_2$), as is assumed by the steady-state model, then approximately

$$K_M = \frac{k_{-1}}{k_1}$$

It is informative to compare the expression for the Michaelis constant with the equilibrium constant expression for the dissociation of the ES complex,

$$ES \underset{k_1}{\overset{k_{-1}}{\rightleftharpoons}} E + S$$

The $k$'s are the rate constants as before. The equilibrium constant expression is

$$K_{eq} = \frac{[E][S]}{[ES]} = \frac{k_{-1}}{k_1}$$

This expression is the same as that for $K_M$, and makes the point that when the assumption that $k_{-1} \gg k_2$ is valid, $K_M$ is the dissociation constant for

**TABLE 10.1    Turnover Numbers for Some Typical Enzymes**

| ENZYME | FUNCTION | TURNOVER NUMBER* |
|---|---|---|
| Catalase | Scavenges harmful free radicals | 40,000,000 |
| Carbonic anhydrase | Catalyzes the hydration of $CO_2$ to $HCO_3^-$ | 800,000 |
| Acetylcholinesterase | Regenerates an important substance in the transmission of nerve impulses | 20,000 |
| Lactate dehydrogenase | An oxidation–reduction enzyme | 1,000 |
| Chymotrypsin | A proteolytic enzyme | 100 |
| DNA polymerase I | Produces new copies of DNA | 15 |
| Tryptophan synthetase | Catalyzes the final step in the biosynthesis of an amino acid | 2 |
| Lysozyme | Hydrolyzes the material of bacterial cell walls | 0.5 |

*The units of turnover numbers are (moles substrate)(mole enzyme)$^{-1}$second$^{-1}$.

the ES complex. The $K_M$ is a measure of how tightly the substrate is bound to the enzyme.

The $V_{max}$ is a measure of the **turnover number** of an enzyme, a quantity equal to the catalytic constant, $k_{cat}$.

$$k_{cat} = \frac{V_{max}}{[E]_0} = \text{turnover number}$$

The turnover number is the number of moles of substrate that react to form product per mole of enzyme per unit time. This statement assumes that the enzyme is fully saturated with substrate and thus that the reaction is proceeding at the maximum rate. Turnover numbers for typical enzymes are given in Table 10.1. In all cases the number refers to number of moles of substrate that react per mole of enzyme *per second*. Turnover numbers are a particularly dramatic illustration of the efficiency of enzymatic catalysis.

## 10.7
## INHIBITION OF ENZYMATIC REACTIONS

An **inhibitor,** as the name implies, is a substance that interferes with the action of an enzyme and slows the rate of a reaction. A good deal of information about enzymatic reactions can be obtained by observing the changes in the reaction caused by the presence of inhibitors. There are two ways in which inhibitors can affect an enzymatic reaction. A **reversible** inhibitor can bind to the enzyme and subsequently be released, leaving the enzyme in its original condition. An **irreversible** inhibitor reacts with the enzyme to produce a protein that is not enzymatically active, and from which the original enzyme cannot be regenerated.

Two classes of reversible inhibitors can be distinguished on the basis of the site to which they bind on the enzyme. One possibility is that an inhibitor is a compound very similar in structure to the substrate. In this case the inhibitor can bind to the active site and block access of the substrate to the active site. This mode of action is called **competitive inhibition** because the inhibitor competes with the substrate for the active site on the enzyme. The other possibility is that the inhibitor binds to the enzyme at a site other than the active site, and as a result of binding causes a change in the structure of the enzyme, especially around the active site. The substrate may still be able to bind to the active site, but the enzyme cannot catalyze the reaction as efficiently as it could in the absence of the inhibitor. This second mode of action is called **noncompetitive inhibition** (Figure 10.9).

The two forms of inhibition can be distinguished from one another in the laboratory. The technique is to carry out the reaction in the presence of inhibitor at several substrate concentrations and to compare the rates obtained with those of the uninhibited reaction. The differences in the Lineweaver-Burk plots for the inhibited and uninhibited reactions provide the basis for the comparison.

In the presence of a competitive inhibitor, the slope of the Lineweaver-Burk plot changes but the intercept does not. The $V_{max}$ is unchanged, but

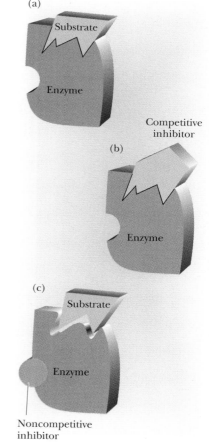

**FIGURE 10.9** Modes of action of inhibitors. The distinction between a competitive and a noncompetitive inhibitor is that a competitive inhibitor prevents binding of the substrate to the enzyme, while a noncompetitive inhibitor does not. (a) An enzyme–substrate complex in the absence of inhibitor. (b) A competitive inhibitor binds to the active site; the substrate cannot bind. (c) A noncompetitive inhibitor binds at a site other than the active site. The substrate still binds, but the enzyme cannot catalyze the reaction because of the presence of the bound inhibitor.

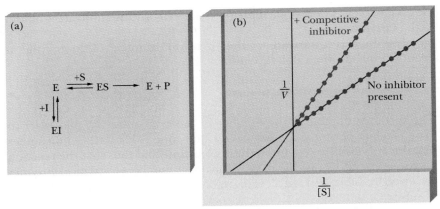

**FIGURE 10.10**    A Lineweaver–Burk double reciprocal plot of enzyme kinetics for competitive inhibition. (a) These are the binding equilibria to be considered. (b) In the plot of $1/V$ versus $1/[S]$, the black circles represent the presence of a competitive inhibitor, while the red circles represent the control reaction with no inhibitor present. The value of $V_{max}$ remains the same; the measured value of $K_M$ increases.

the $K_M$ increases. More substrate is needed to get to $V_{max}/2$ (recall that at $V_{max}/2$, the substrate concentration, [S], equals $K_M$) or to $V_{max}$ itself (Figure 10.10). Competitive inhibition can be overcome by a sufficiently high substrate concentration.

In the presence of a competitive inhibitor the equation for an enzymatic reaction becomes

$$EI \overset{I}{\rightleftharpoons} E \overset{S}{\rightleftharpoons} ES \longrightarrow E + P$$

where EI is the enzyme-inhibitor complex. The dissociation constant for the enzyme-inhibitor complex can be written:

$$EI \rightleftharpoons E + I$$

$$K_I = \frac{[E][I]}{[EI]}$$

It can be shown algebraically, although we shall not do it here, that in the presence of inhibitor the value of $K_M$ increases by the factor

$$1 + \frac{[I]}{K_I}$$

If we substitute $K_M(1 + [I]/K_I)$ for $K_M$ in Equation 10.19,

$$\frac{1}{V} = \frac{K_M}{V_{max}} \cdot \frac{1}{[S]} + \frac{1}{V_{max}}$$

we obtain

$$\frac{1}{V} = \frac{K_M}{V_{max}} \left[1 + \frac{[I]}{K_I}\right] \left[\frac{1}{[S]}\right] + \frac{1}{V_{max}}$$
$$y = \qquad\qquad m \qquad\quad x \quad + \quad b$$

(10.20)

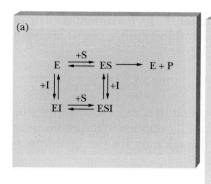

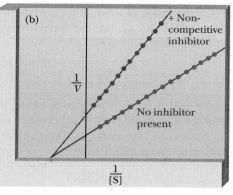

**FIGURE 10.11**   A Lineweaver–Burk double reciprocal plot of enzyme kinetics for noncompetitive inhibition. (a) These are the binding equilibria to be considered. (b) In the plot of $1/V$ versus $1/[S]$, the black triangles represent the presence of a noncompetitive inhibitor, while the red circles represent the control reaction with no inhibitor present. The value of $V_{max}$ decreases; the value of $K_M$ remains the same.

In Equation 10.20 the term $1/V$ takes the place of the $y$ coordinate, and the term $1/[S]$ takes the place of the $x$ coordinate, as was the case in Equation 10.19. The intercept $1/V_{max}$, the $b$ term in the equation for a straight line, has not changed from the earlier equation, but the slope $K_M/V_{max}$ in Equation 10.19 has increased by the factor $(1 + [I]/K_I)$. The slope, the $m$ term in the equation for a straight line, is now

$$\frac{K_M}{V_{max}} \left[ 1 + \frac{[I]}{K_I} \right]$$

accounting for the changes in the slope of the Lineweaver-Burk plot. This algebraic treatment of competitive inhibition agrees with experimental results, confirming the validity of the model for competitive inhibition, just as experimental results support the validity of the underlying Michaelis-Menten model for enzyme action.

Kinetic results are different in the case of noncompetitive inhibition. The Lineweaver-Burk plots for the reaction in the presence and absence of the noncompetitive inhibitor show that both the slope and the $y$-axis intercept have changed for the inhibited reaction (Figure 10.11). The value of $V_{max}$ has decreased but that of $K_M$ has not. Increasing the substrate concentration cannot overcome noncompetitive inhibition.

The reaction pathway has become considerably more complicated, and several equilibria have to be considered.

In the presence of a noncompetitive inhibitor, I, the maximum velocity of the reaction, $V^I_{max}$, has the form (derived by biochemists, but we shall not do the derivation here)

$$V^I_{max} = \frac{V_{max}}{1 + [I]/K_I}$$

where $K_I$ is again the dissociation constant for the enzyme-inhibitor complex EI. Recall that the maximum rate $V_{max}$ appears in the expressions for both the slope and the intercept in the equation for the Lineweaver-Burk plot (Equation 10.19):

$$\frac{1}{V} = \frac{K_M}{V_{max}} \cdot \frac{1}{[S]} + \frac{1}{V_{max}}$$
$$y = \quad m \quad\quad x \quad + \quad b$$

In noncompetitive inhibition we replace the term $V_{max}$ with the expression for $V^I_{max}$, to obtain

$$\frac{1}{V} = \frac{K_M}{V_{max}}\left[1 + \frac{[I]}{K_I}\right] \cdot \frac{1}{[S]} + \frac{1}{V_{max}}\left[1 + \frac{[I]}{K_I}\right] \qquad (10.21)$$
$$y = \qquad\qquad m \qquad\qquad x \quad + \qquad\qquad b$$

The expressions for both the slope and the intercept in the equation for a Lineweaver-Burk plot for an uninhibited reaction have been replaced by more complicated expressions in the equation that describes noncompetitive inhibition. This interpretation is borne out by the observed results.

## 10.8
## THE MICHAELIS-MENTEN MODEL DOES NOT DESCRIBE THE BEHAVIOR OF ALLOSTERIC ENZYMES

The behavior of many well-known enzymes can be described quite adequately in terms of the Michaelis-Menten model, but allosteric enzymes behave in a very different manner. Earlier in this chapter we saw that there are similarities between the reaction kinetics of an enzyme such as chymotrypsin, which does not display allosteric behavior, and the binding of oxygen by myoglobin, also an example of nonallosteric behavior. We need to carry the analogy further to show the similarity in the kinetic behavior of an allosteric enzyme such as aspartate transcarbamoylase (ATCase) and the binding of oxygen by hemoglobin. Both ATCase and hemoglobin are allosteric proteins; both show cooperative effects in their behavior caused by subtle changes in quaternary structure. (Recall that quaternary structure is the arrangement in space that results from the interaction of subunits through various noncovalent forces. Recall also that cooperativity refers to the fact that binding of substrate at low levels facilitates the action of the protein at higher levels of substrate, whether the behavior involved is catalytic action or oxygen binding.) In addition to displaying cooperative kinetics, allosteric enzymes show a response to the presence of inhibitors that is different from that of nonallosteric enzymes as described by the Michaelis-Menten model.

## Control Mechanisms That Affect Allosteric Enzymes

ATCase catalyzes the first step in a series of reactions that eventually lead to the production of cytidine triphosphate (CTP).

Cytidine triphosphate (CTP)
Allosteric inhibitor of ATCase

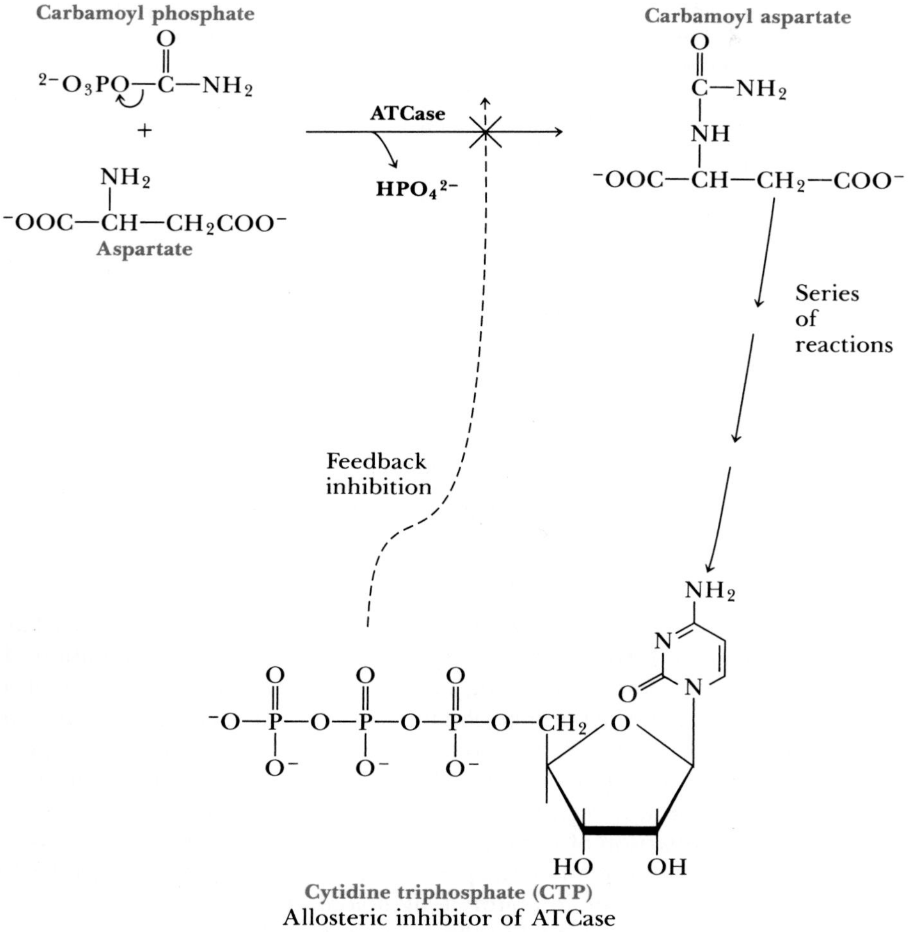

The reaction catalyzed by ATCase
leads eventually to the production of CTP

CTP, the end product of this series of reactions, is an inhibitor of ATCase, the enzyme that catalyzes the first reaction in the pathway. This behavior is an example of **feedback inhibition** (also called end-product inhibition), in which the end product of the sequence of reactions inhibits the first reaction in the series. Feedback inhibition is an efficient control mechanism, because the entire series of reactions can be shut down when there is an excess of the final product; intermediates in the pathway are not formed. Feedback inhibition is a general feature of metabolism and is not confined to allosteric enzymes. However, the observed kinetics of the ATCase reaction, including the mode of inhibition, are the typical kinetics of allosteric enzymes.

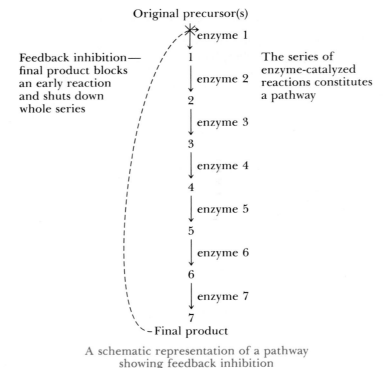

A schematic representation of a pathway
showing feedback inhibition

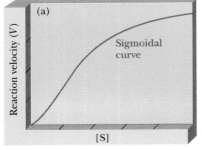

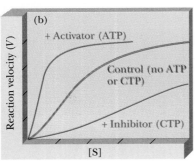

**FIGURE 10.12** The kinetics of allosteric enzymes. (a) Reaction velocity, $V$, as a function of substrate concentration, [S], in the reaction of ATCase. The substrate concentration that varies is that of aspartate. The sigmoidal curve observed here is evidence of cooperative kinetics. (b) The effect of an activator (ATP) and an inhibitor (CTP) on the kinetics of the ATCase reaction.

When ATCase catalyzes the condensation of aspartate and carbamoyl phosphate to form carbamoyl aspartate, the graphic representation of the rate as a function of increasing substrate concentration is a sigmoidal curve rather than the hyperbola obtained in the Michaelis-Menten treatment (Figure 10.12(a)). The sigmoidal curve is indicative of the cooperative behavior of allosteric enzymes. (In this reaction in organisms, aspartate is the substrate with significant variation in concentration levels.)

A comparison of the rate of the uninhibited reaction of ATCase with the rate of the reaction in the presence of CTP is shown in Figure 10.12(b). In the presence of inhibitor a sigmoidal curve still describes the rate behavior of the enzyme, but the curve is shifted to higher substrate levels. A higher concentration of aspartate is needed for the enzyme to achieve the same rate of reaction in the presence of the inhibitor than without CTP. The same maximal rate ($V_{max}$) is observed in the presence and absence of inhibitor at high substrate concentrations. In the Michaelis-Menten scheme the $V_{max}$ changes when a reaction takes place in the presence of a noncompetitive inhibitor, so noncompetitive inhibition cannot be the case here. In the same Michaelis-Menten model this sort of behavior is characteristic of competitive inhibition, but that part of the model does not give a reasonable picture. Competitive inhibitors bind to the same site as the substrate because they are very similar in structure. The CTP molecule is very different in structure from the substrate, namely aspartate. It is much more likely that CTP is bound to a different site on the ATCase molecule. There is, in fact, experimental evidence for two different binding sites. The enzyme can be modified so that CTP cannot bind, but the binding of aspartate is not affected by this modification, indicating two different binding sites.

The situation becomes "curiouser and curiouser" when the ATCase reaction takes place not in the presence of CTP, a pyrimidine nucleoside triphosphate, but in the presence of ATP, a purine nucleoside triphosphate.

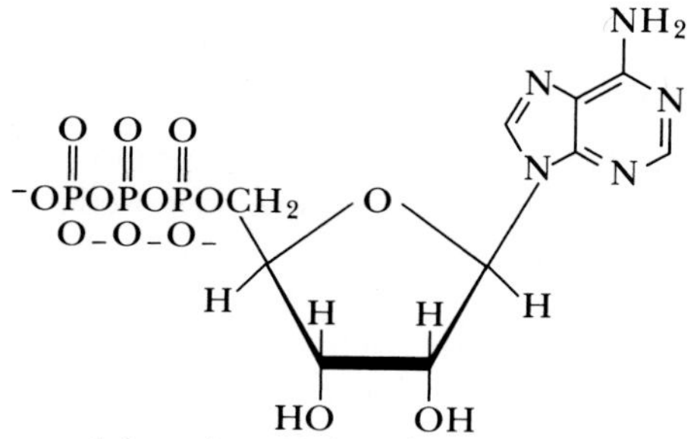

Adenosine triphosphate (ATP)
a purine nucleotide;
activator of
ATCase

The structural similarities between CTP and ATP are apparent, but ATP is not a product of the pathway that includes the reaction of ATCase and that produces CTP. Both ATP and CTP are needed for the synthesis of RNA and DNA. The relative proportions of ATP and CTP are specified by the needs of a particular organism. If there is not enough CTP compared to ATP, the enzyme needs a signal to produce more. In the presence of ATP the rate of the enzymatic reaction is increased at lower levels of aspartate and the shape of the rate curve becomes less sigmoidal and more hyperbolic. In other words, there is less cooperativity in the reaction. The binding site for ATP on the enzyme molecule is the same as that for CTP, which is not surprising in view of their structural similarity, but ATP is an activator rather than an inhibitor like CTP. When CTP is in short supply in an organism, the ATCase reaction will not be inhibited, and the binding of ATP increases the activity of the enzyme still more.

The key to allosteric behavior, including cooperativity and modifications of cooperativity, is the existence of different forms for the quaternary structure of allosteric proteins. The word "allosteric" is derived from *allo*, "other," and *steric*, "shape," referring to the fact that the various possible conformations affect the behavior of the protein. The binding of substrates, inhibitors, and activators causes changes in the quaternary structure of allosteric proteins, and the changes in structure are reflected in changes in behavior. A substance that modifies the quaternary structure, and thus the behavior, of an allosteric protein by binding to the protein is called an **allosteric effector.** The term "effector" can apply to substrates, inhibitors, and activators. Several models for the behavior of allosteric enzymes have been proposed, and it will be of interest for us to compare them. Before we do so, it is useful to define two terms. **Homotropic** effects are allosteric interactions that occur when several identical molecules are bound to the protein. The binding of substrate molecules to different sites on an enzyme, such as the binding of aspartate to ATCase, is an example of a homotropic effect. **Heterotropic** effects are allosteric interactions that occur when different substances (such as inhibitor and substrate) are bound to the protein. In the ATCase reaction, inhibition by CTP and activation by ATP are both heterotropic effects.

## 10.9
## MODELS FOR THE BEHAVIOR OF ALLOSTERIC ENZYMES

The two principal models for the behavior of allosteric enzymes are the **concerted model** and the **sequential model.** They were proposed in 1965 and 1966, respectively, and both are in current use as a basis for interpreting experimental results. The concerted model has the advantage of comparative simplicity, and it describes the behavior of some enzyme systems very well. The sequential model sacrifices a certain amount of simplicity for a more realistic picture of the structure and behavior of proteins; it also deals very well with the behavior of some enzyme systems.

**FIGURE 10.13**    The concerted model for allosteric behavior. (a) The T (inactive) and R (active) forms of the enzyme are in equilibrium. The equilibrium lies to the left, in favor of the T form. (b) The cooperatiave binding of substrate shifts the equilibrium to the right, in favor of the R form.

## The Concerted Model for Allosteric Behavior

In 1965 Jeffries Wyman, Jacques Monod, and Jean-Pierre Changeux proposed the concerted model for the behavior of allosteric proteins in a paper that has become a classic in the biochemical literature. The citation for this paper is given in the bibliography at the end of this chapter. In this

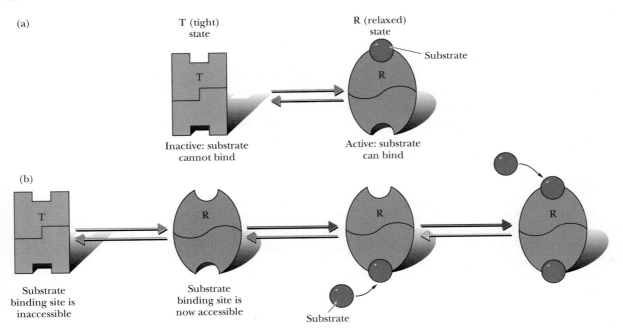

picture the protein has two conformations, the R (relaxed) conformation, which binds substrate tightly, and the T (tight, also called taut) conformation, which binds substrate less tightly. The distinguishing feature of this model is that the conformations of all subunits change simultaneously when substrate S binds to any one site on any subunit. In Figure 10.13 a hypothetical protein with two subunits is shown. Both subunits change conformation from the inactive T conformation to the active R conformation at the same time; in other words, there is a **concerted** change of conformation. When the first molecule of substrate binds to one subunit, the binding of the second substrate molecule to the other subunit is facilitated, which is exactly what is meant by cooperative binding. In the absence of substrate the enzyme exists mainly in the T form in equilibrium with small amounts of the R form. The presence of substrate shifts the equilibrium to produce more of the R form.

In this model the effects of inhibitors and activators can also be considered in terms of shifting the equilibrium between the T and R forms of the enzyme. The binding of inhibitors to allosteric enzymes is cooperative; allosteric inhibitors bind to the T form of the enzyme. The binding of activators to allosteric enzymes is also cooperative; allosteric activators bind to the R form of the enzyme.

When an activator A is present, the cooperative binding of A shifts the equilibrium between the T and R forms; the R form is favored. There is less need for substrate to shift the equilibrium in favor of the R form. Because the activator A has already shifted the T-to-R conversion to favor the R form, there is less need for cooperativity in the binding of substrate S (Figure 10.14(a)).

**FIGURE 10.14** (a) Concerted model for the cooperative binding of activator A to allosteric enzyme. The T ⇌ R equilibrium shifts to right. (b) Concerted model for the cooperative binding of inhibitor I to allosteric enzyme. The T ⇌ R equilibrium shifts to left.

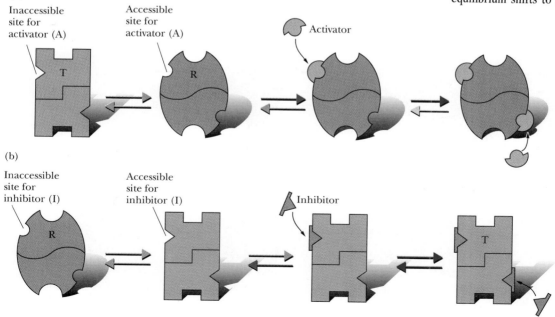

(a)

Inaccessible site for activator (A)

Accessible site for activator (A)

Activator

(b)

Inaccessible site for inhibitor (I)

Accessible site for inhibitor (I)

Inhibitor

When an inhibitor I is present, the cooperative binding of I also shifts the equilibrium between the T and R forms, but this time the T form is favored. More substrate S is needed to shift the T-to-R equilibrium in favor of the R form. A greater degree of cooperativity is needed in the binding of substrate, because the presence of inhibitor has shifted the T-to-R equilibrium to favor the T form (Figure 10.14(b)).

## The Sequential Model for Allosteric Behavior

The name of Daniel Koshland is associated with the direct **sequential** model of allosteric behavior. The distinguishing feature of the sequential model is that the binding of substrate induces the conformational change from the T to the R form. This type of behavior is that postulated by the induced-fit theory of substrate binding. The conformational change from the T to the R form in one subunit favors the same change in the other subunits; this process of the conformational changes being "passed along" from one subunit to another is the form in which cooperative binding is expressed in this model (Figure 10.15(a)).

The binding of activators and inhibitors also takes place by the induced-fit mechanism in the sequential model. The conformational change that begins with binding of inhibitor or activator to one subunit affects the conformation of other subunits. The net result is to favor the R state when activator is present and the T form in the presence of inhibitor (Figure 10.15(b)). Binding of inhibitor I to one subunit causes a conformational change such that the T form is even less likely to bind substrate than before. This conformational change is passed along to other subunits, also making them more likely to bind inhibitor and less likely to bind substrate. Here we see another example of cooperative behavior, in this case

**FIGURE 10.15** (a) Sequential model of cooperative binding of substrate S to allosteric enzyme. (b) Sequential model of cooperative binding of inhibitor I to allosteric enzyme.

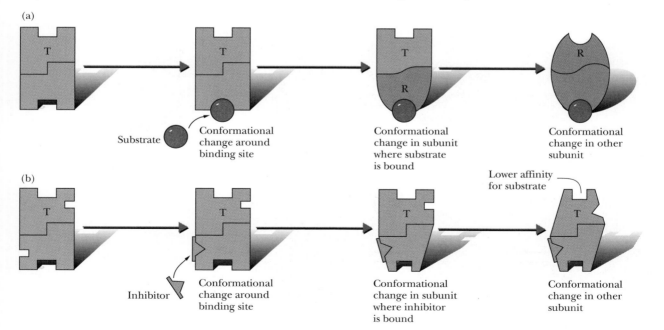

(a)

Substrate    Conformational change around binding site    Conformational change in subunit where substrate is bound    Conformational change in other subunit

(b)

Lower affinity for substrate

Inhibitor    Conformational change around binding site    Conformational change in subunit where inhibitor is bound    Conformational change in other subunit

cooperative behavior that leads to more inhibition of the enzyme. Likewise, binding of an activator will cause a conformational change that favors substrate binding, and this effect is passed from one subunit to another.

The sequential model for the binding of effectors of all types, including substrates, to allosteric enzymes contains a feature that is unique to this model. There is another possibility for the type of change in the behavior of the enzyme brought about by the binding of the first molecule. The conformational changes thus induced can make the enzyme less likely to bind more molecules of the same type. This phenomenon is called **negative cooperativity,** and it has been observed in a few enzymes. An example is the enzyme tyrosyl tRNA synthetase, which plays a role in protein synthesis. In the reaction catalyzed by this enzyme, the amino acid tyrosine forms a covalent bond to a molecule of transfer RNA (tRNA). In subsequent steps the tyrosine is passed along to its place in the sequence of the growing protein. The tyrosyl tRNA synthetase consists of two subunits; binding of the first molecule of substrate to one of the subunits inhibits binding of a second molecule to the other subunit. The sequential model has been successful in accounting for the negative cooperativity observed in the behavior of this enzyme. The concerted model makes no provision for negative cooperativity.

## A Final Comment on the Two Models

The behavior of allosteric proteins is a field of active research, and a more inclusive model that includes aspects of both the concerted and sequential models may well emerge in the future (Figure 10.16). In one inclusive

**FIGURE 10.16** A comparison of the concerted and sequential models for the binding of substrate by an allosteric enzyme with four subunits. (a) The concerted model. (b) The sequential model. (c) The general model for the conformational changes of an allosteric protein. The extreme left and right columns represent the simplification of the concerted model; the diagonal represents the simplification of the sequential model. (From G. G. Hammes and C. Wu, 1971, *Science,* **172**, 1205–1211.)

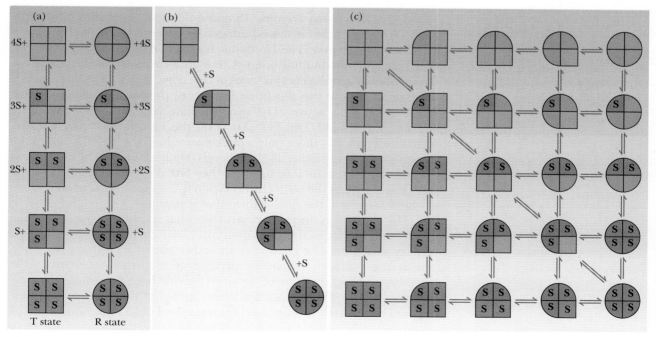

model both the concerted and sequential models are partial representations of a general binding scheme shown in Figure 10.16. The columns on the left and right of the figure represent the concerted model, while the diagonal represents the sequential model.

## 10.10
## ZYMOGENS: THE BASIS OF ANOTHER TYPE OF CONTROL MECHANISM IN ENZYME ACTION

Allosteric interactions control the behavior of proteins through changes in quaternary structure, but this mechanism, effective though it may be, is not the only one available. A **zymogen,** which is an inactive precursor of an enzyme, can be transformed into an active enzyme by cleavage of covalent bonds.

The proteolytic enzymes **trypsin** and **chymotrypsin** provide a classic example of zymogens and their activation. The inactive precursor molecules, trypsinogen and chymotrypsinogen, respectively, are formed in the pancreas, where they would cause damage if they were in an active form. In the small intestine, where their digestive properties are needed, they are activated by cleavage of specific peptide bonds. The process of blood clotting also requires a series of proteolytic activations involving several proteins, particularly the conversions of prothrombin to thrombin and of fibrinogen to fibrin.

The conversion of inactive chymotrypsinogen to active chymotrypsin occurs as a result of the cleavage of a single, quite specific peptide bond. Chymotrypsinogen consists of a single polypeptide chain 245 residues long with five disulfide (—S—S—) bonds. When chymotrypsinogen is secreted into the small intestine, trypsin present in the digestive system cleaves the peptide bond between arginine 15 and isoleucine 16, counting from the N-terminal end of the chymotrypsinogen sequence. The cleavage produces active $\pi$-chymotrypsin. The 15-residue fragment remains bound to the rest of the protein by a disulfide bond. Although $\pi$-chymotrypsin is fully active, it is not the end product of this series of reactions. The $\pi$-chymotrypsin acts on itself to remove two dipeptide fragments, producing $\alpha$-chymotrypsin, which is also fully active. The two dipeptide fragments cleaved off are Ser 14–Arg 15 and Thr 147–Asn 148; the final form of the enzyme, $\alpha$-chymotrypsin, has three polypeptide chains held together by two of the five original and still intact disulfide bonds (Figure 10.17). (The other three disulfide bonds remain intact as well; they link portions of single polypeptide chains.) When the term chymotrypsin is used without specifying the $\alpha$ or the $\pi$ form, it is the final $\alpha$ form that is meant.

The changes in the primary structure that accompany the conversion of chymotrypsinogen to $\alpha$-chymotrypsin bring about changes in the tertiary structure. The enzyme is active because of its tertiary structure, just as the zymogen is inactive because of its tertiary structure. The three-dimensional structure of chymotrypsin has been determined by x-ray crystallography. The protonated amino group of the isoleucine residue exposed by the first cleavage reaction is involved in an ionic bond with the

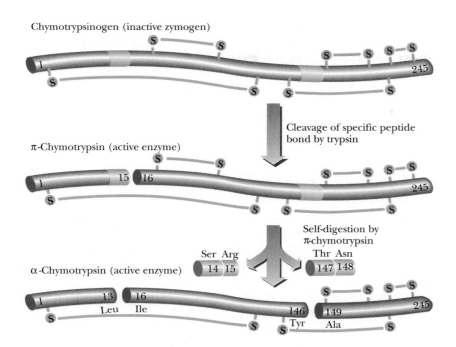

Chymotrypsinogen (inactive zymogen)

Cleavage of specific peptide bond by trypsin

π-Chymotrypsin (active enzyme)

Self-digestion by π-chymotrypsin

α-Chymotrypsin (active enzyme)

Ser Arg
14 15

Thr Asn
147 148

13   16
Leu   Ile

146   149
Tyr   Ala

**FIGURE 10.17**   The activation of chymotrypsinogen.

carboxylate side chain of aspartate residue 194. This ionic bond is necessary for the active conformation of the enzyme, since it is near the active site. Chymotrypsinogen, which lacks this bond, does not have the active conformation and cannot bind substrate.

The process of blood clotting is a complex one, but for the purposes of this discussion the important point is that activation of zymogens plays a crucial role in the process. In the final, best characterized step of clot formation, the soluble protein **fibrinogen** is converted to the insoluble protein **fibrin** as a result of the cleavage of four peptide bonds. The conversion of fibrinogen to fibrin occurs as the rsult of action of the proteolytic enzyme **thrombin,** which in turn is produced from a zymogen called **prothrombin.** The conversion of prothrombin to thrombin requires a number of proteins called **clotting factors,** as well as $Ca^{2+}$.

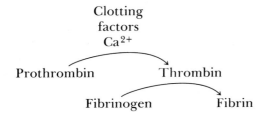

Clotting
factors
$Ca^{2+}$

Prothrombin                    Thrombin

Fibrinogen                    Fibrin

Some of the processes involved in blood clotting

The early stages of blood clotting consist of an elaborate multistep mechanism, which allows for fine-tuning of the process but which can also cause great problems if something goes wrong with one of the steps. The molecular disease **hemophilia,** for example, is typically caused by a lack of one of the clotting factors; a hemophiliac can bleed to death from a very small cut that would not trouble a normal person.

## 10.11
## ACTIVE SITE EVENTS IN ENZYMES: A LOOK AT REACTION MECHANISMS

We can ask several questions about the mode of action of enzymes. Here are some of the most important:

1. Which amino acid residues on the enzyme are located in the active site and catalyze the reaction? In other words, which are the essential amino acid residues?
2. What is the spatial relationship of the essential amino acid residues in the active site?
3. What is the mechanism by which the essential amino acid residues catalyze the reaction?

Answers to these questions are available for chymotrypsin, and we shall use its mechanism as an example of enzyme action.

Information on well-known systems such as chymotrypsin can lead to general principles applicable to all enzymes. There are many ways in which enzymes catalyze chemical reactions, but all reactions have in common the requirement that some reactive group on the enzyme must interact with the substrate. In proteins the $\alpha$-carboxyl and $\alpha$-amino groups of the amino acids are no longer free, since they have formed peptide bonds. Thus, the side-chain reactive groups are the ones involved in the action of the enzyme. Hydrocarbon side chains do not contain reactive groups and are not involved in the process. Functional groups that can play a catalytic role include the imidazole group of histidine, the hydroxyl group of serine, the carboxyl side chains of aspartate and glutamate, the sulfhydryl group of cysteine, the amino side chain of lysine, and the phenol group of tyrosine.

Chymotrypsin catalyzes the hydrolysis of peptide bonds adjacent to aromatic amino acid residues in the protein being hydrolyzed; other residues are attacked at lower frequency. In addition, chymotrypsin catalyzes the hydrolysis of esters in model studies in the laboratory. The use of model systems is a common technique in biochemistry, since the essential features of a reaction are present in a simple system that is easier to work with than the one found in nature. The amide (peptide) bond and the ester bond are similar enough that the enzyme can accept both types of compounds as substrates. Model systems based on the hydrolysis of esters are frequently used to study the peptide hydrolysis reaction. A typical model compound is *p*-nitrophenyl acetate, which is hydrolyzed in two stages. The acetyl group is covalently attached to the enzyme at the end of the first stage (Step 1) of the reaction, but the *p*-nitrophenolate ion is released. In the second stage (Step 2) the acyl-enzyme intermediate is hydrolyzed, releasing acetate and regenerating the free enzyme.

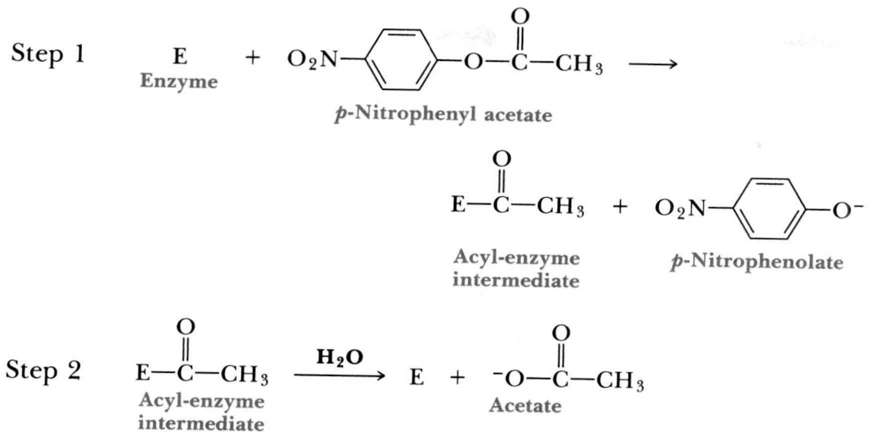

Step 1    E    +    $O_2N$—〈 〉—$O$—$\overset{\displaystyle O}{\overset{\|}{C}}$—$CH_3$    $\longrightarrow$

Enzyme

*p*-Nitrophenyl acetate

$E$—$\overset{\displaystyle O}{\overset{\|}{C}}$—$CH_3$    +    $O_2N$—〈 〉—$O^-$

Acyl-enzyme          *p*-Nitrophenolate
intermediate

Step 2    $E$—$\overset{\displaystyle O}{\overset{\|}{C}}$—$CH_3$    $\xrightarrow{\ H_2O\ }$    $E$    +    $^-O$—$\overset{\displaystyle O}{\overset{\|}{C}}$—$CH_3$

Acyl-enzyme                                 Acetate
intermediate

The hydrolysis of *p*-nitrophenyl acetate
catalyzed by chymotrypsin

## Determining the Essential Amino Acid Residues

The serine residue at position 195 is required for the activity of chymotrypsin. In this respect chymotrypsin is typical of a class of enzymes known as **serine proteases.** The enzyme is completely inactivated when this serine reacts with diisopropylphosphofluoridate (DIPF), forming a covalent bond that links the serine side chain with the DIPF.

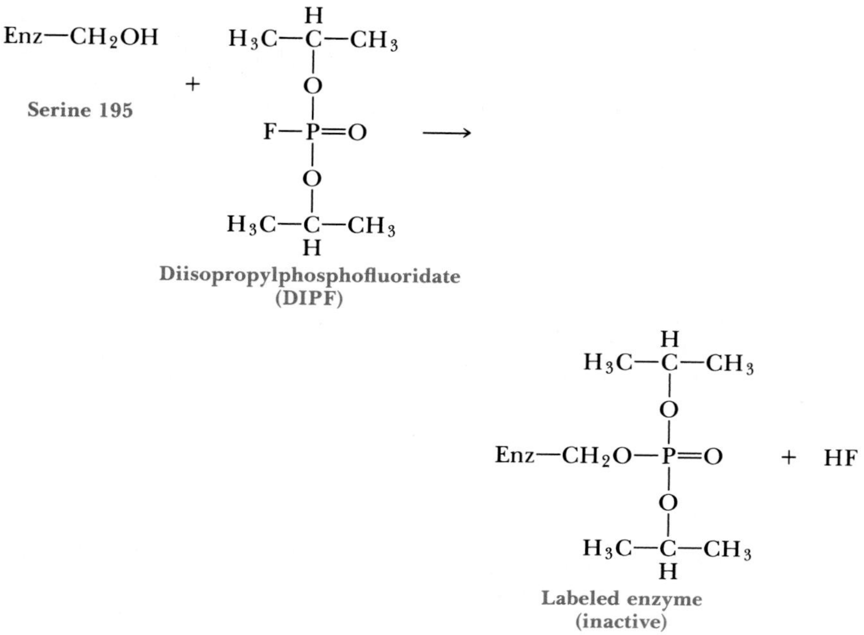

Enz—$CH_2OH$        $H_3C$—$\overset{\displaystyle H}{\underset{\displaystyle}{C}}$—$CH_3$

Serine 195

+

$\overset{\displaystyle}{\underset{\displaystyle O}{}}$

$F$—$P$=$O$        $\longrightarrow$

$\overset{\displaystyle}{\underset{\displaystyle O}{}}$

$H_3C$—$\overset{\displaystyle}{\underset{\displaystyle H}{C}}$—$CH_3$

**Diisopropylphosphofluoridate**
**(DIPF)**

$H_3C$—$\overset{\displaystyle H}{\underset{\displaystyle}{C}}$—$CH_3$

$\overset{\displaystyle}{\underset{\displaystyle O}{}}$

Enz—$CH_2O$—$P$=$O$        +    HF

$\overset{\displaystyle}{\underset{\displaystyle O}{}}$

$H_3C$—$\overset{\displaystyle}{\underset{\displaystyle H}{C}}$—$CH_3$

**Labeled enzyme**
**(inactive)**

The labeling of the active serine of chymotrypsin by
diisopropylphosphofluoridate (DIPF)

The technique of forming covalently modified versions of specific side chains on proteins is called **labeling.** The other serine residues of chymotrypsin are far less reactive and are not labeled by DIPF.

Histidine 57 is another essential amino acid residue in chymotrypsin. Chemical labeling again provides the evidence for involvement of this residue in the activity of chymotrypsin. In this case the reagent used to label the essential amino acid residue is *N*-tosylamido-L-phenylethyl chloromethyl ketone (TPCK), also called tosyl-L-phenylalanine chloromethyl ketone.

**(a)**   Phenylalanyl moiety chosen because of specificity of chymotrypsin for aromatic amino acid residues

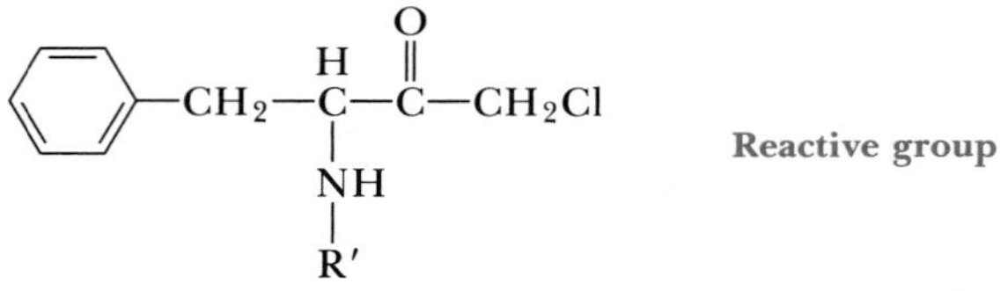

Structure of N-tosylamido-L-phenylethyl chloromethyl ketone (TPCK), a labeling reagent for chymotrypsin (R′ represents a tosyl (toluenesulfonyl) group)

**(b)**

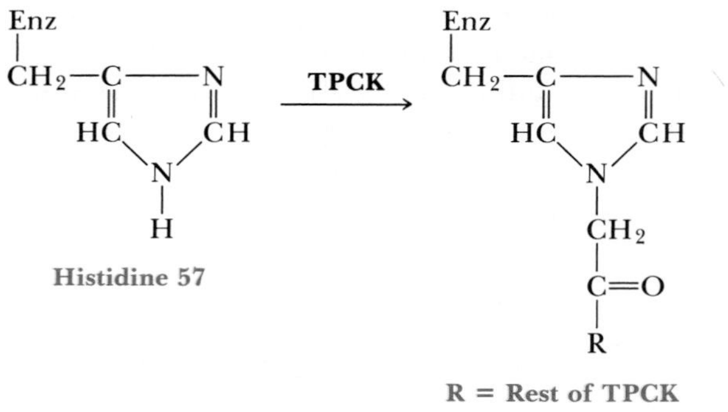

The labeling of the active site histidine of chymotrypsin by TPCK

The phenylalanine moiety is bound to the enzyme because of the specificity for aromatic amino-acid residues at the active site, and the active site histidine residue reacts because the labeling reagent is similar to the usual substrate.

## The Architecture of the Active Site

Both serine 195 and histidine 57 are required for the activity of chymotrypsin; therefore, they must be close to each other in the active site. The determination of the three-dimensional structure of the enzyme by x-ray crystallography provides evidence that the active site residues do indeed have a close spatial relationship. The folding of the chymotrypsin backbone, mostly in an antiparallel pleated sheet array, positions the essential

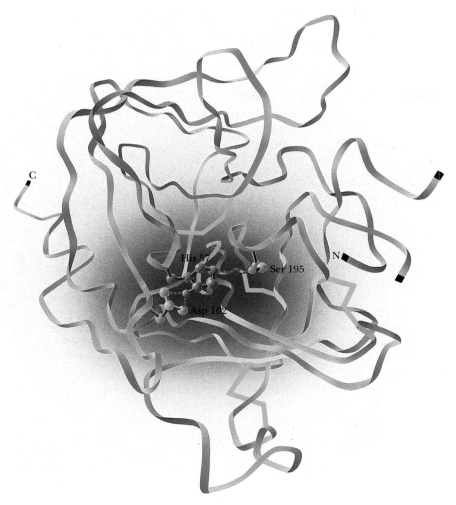

**FIGURE 10.18** The tertiary structure of chymotrypsin places the essential residues close to one another. The essential amino acid residues are shown in color. (From B. S. Hartley and D. M. Shotten, 1971, *in* P. D. Boyer, Ed., "The Enzymes," 3rd ed., Vol. 3, Academic Press, New York.)

residues around an active site pocket (Figure 10.18). Only a few residues are directly involved in the active site, but the whole molecule is necessary to provide the correct three-dimensional arrangement for those critical residues.

Other important pieces of information about the three-dimensional structure of the active site emerge when a complex is formed between chymotrypsin and a substrate analog. When one such substrate analog, formyl-L-tryptophan,

The structure of formyl-L-tryptophan

is bound to the enzyme, the tryptophan side chain fits into a hydrophobic pocket near serine 195. This type of binding is not surprising in view of the specificity of the enzyme for aromatic amino-acid residues at the cleavage site.

In addition to the binding site for aromatic amino acid side chains of substrate molecules, the results of x-ray crystallography show that there is a definite arrangement of the amino-acid side chains that are responsible for the catalytic activity of the enzyme. The residues involved in this arrangement are serine 195 and histidine 57.

## The Mechanism of Chymotrypsin Action

Any reaction mechanism must be considered a model to be modified or discarded if it is not consistent with experimental results. There is consensus, but not total agreement, on the main features of the mechanism we shall discuss in this section. At one point it had been thought that aspartate 102, which is essential for the activity of chymotrypsin, is involved in a "charge-relay" system with serine 195 and histidine 57. It is now known that the charge-relay mechanism does not apply to chymotrypsin and that the aspartate must play some other role.

The essential amino-acid residues, serine 195 and histidine 57, are involved in the mechanism of catalytic action. In the terminology of organic chemistry, the oxygen of the serine side chain is a nucleophile. A **nucleophile** (nucleus-seeking substance) tends to bond to sites of positive charge or polarization, in other words electron-poor sites, as opposed to an **electrophile** (electron-seeking substance), which tends to bond to sites of negative charge or polarization (electron-rich sites). The nucleophilic oxygen of the serine attacks the carbonyl carbon of the peptide group. The carbon now has four single bonds, and a tetrahedral intermediate is formed; the original —C=O bond becomes a single bond, and the carbonyl oxygen becomes an oxyanion. The acyl-enzyme intermediate is formed from the tetrahedral species (Figure 10.19). The histidine and the amino portion of the original peptide group are involved in this part of the reaction, as the amino group hydrogen bonds to the imidazole portion of the histidine. Note that the imidazole is already protonated and that the proton came from the hydroxyl group of the serine. The histidine behaves as a base; in the terminology of the physical organic chemist, the histidine is acting as a general base catalyst. (There is some recent evidence that the histidine does not act as a general base catalyst, but the main outline of the mechanism is the same even with the difference on this point. Modified mechanisms in which the histidine imidazole does not act as a general base catalyst still postulate a hydrogen bond between the imidazole and the serine.) The carbon–nitrogen bond of the original peptide group breaks, leaving the acyl-enzyme intermediate.

In the deacylation phase of the reaction, the last two steps are reversed, with water acting as the attacking nucleophile. In this second phase the water is hydrogen bonded to the histidine. The oxygen of water now performs the nucleophilic attack on the acyl carbon that came from the original peptide group. Once again a tetrahedral intermediate is formed. In the final step of the reaction the bond between the serine

**1st stage reaction**

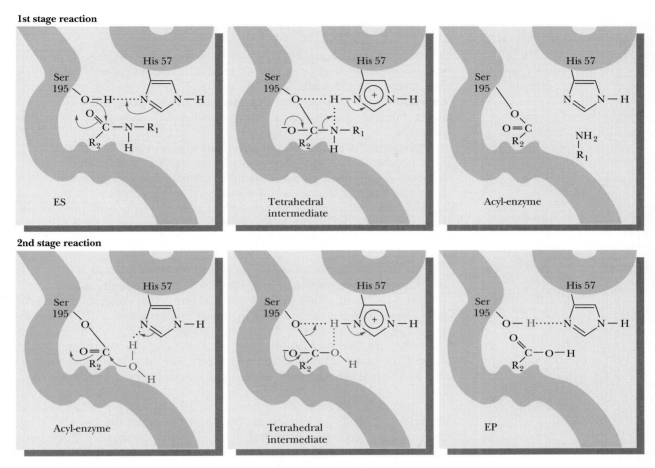

**2nd stage reaction**

**FIGURE 10.19** The mechanism of action of chymotrypsin. In the first stage of the reaction, there is a nucleophilic attack of serine 195 on the substrate. In the second stage of the reaction, water is the nucleophile that attacks the acyl-enzyme intermediate. Note the involvement of histidine 57 in both stages of the reaction. (From G. Hammes, 1982, *Enzyme Catalysis and Regulation,* Academic Press, New York.)

oxygen and the carbonyl carbon breaks, releasing the product with a carboxyl group where the original peptide group had been and regenerating the original enzyme. Note that the serine is hydrogen bonded to the histidine. This hydrogen bond increases the nucleophilicity of the serine; in the second part of the reaction the hydrogen bond between the water and the histidine increases the nucleophilicity of the water.

The mechanism of chymotrypsin action is a particularly well studied one, and in many respects is a typical one. There are numerous types of reaction mechanisms known for enzyme action, and we shall discuss them in the context of the reactions catalyzed by the enzymes in question. For future reference, it is useful at this point to discuss some general types of catalytic mechanisms and how they affect the specificity of enzymatic reactions.

## 10.12
## TYPES OF CATALYTIC MECHANISMS

The overall mechanism for a reaction may be fairly complex, as we have seen in the case of chymotrypsin, but the individual parts of a complex mechanism can themselves be fairly simple. Concepts such as nucleophilic attack and acid catalysis are very common in discussing enzymatic reactions. We can draw quite a few general conclusions from these two concepts.

**Nucleophilic substitution reactions** play a large role in the study of organic chemistry, and they provide an excellent illustration of the importance of kinetic measurements in determining the mechanism of a reaction. The distinction between the $S_N1$ and $S_N2$ reactions (first and second order nucleophilic substitution reactions, respectively) is a classic example, particularly with optically active starting materials. There are many examples of nucleophilic substitutions in enzymatic reaction mechanisms.

Before we discuss acid-base catalysis, we should recall the various definitions of acids and bases. In the Brønsted-Lowry definition, an acid is a proton donor and a base is a proton acceptor. The concept of **general acid–base catalysis** depends on donation and acceptance of protons by groups such as imidazole, hydroxyl, carboxyl, sulfhydryl, amino, and phenolic side chains of amino acids; all these functional groups can act as acids or bases. The donation and acceptance of protons gives rise to the bond breaking and re-formation that constitutes the enzymatic reaction.

There is a form of acid–base catalysis based on another, more general definition of acids and bases. In the Lewis formulation, an acid is an electron-pair acceptor, and a base is an electron-pair donor. Metal ions, including such biologically important ones as $Mn^{2+}$, $Mg^{2+}$, and $Zn^{2+}$, are Lewis acids. Thus they can play a role in **Lewis acid–base catalysis.** The role of $Zn^{2+}$ in the enzymatic activity of carboxypeptidase A is an example of this type of behavior. This enzyme catalyzes the hydrolysis of C-terminal peptide bonds of proteins. The Zn(II), which is required for the activity of the enzyme, is complexed to the imidazole side chains of histidines 69 and 196 and to the carboxylate side chain of glutamate 72. (The notation Zn(II) is used because the zinc ion is involved in a complex.) The zinc ion is also complexed to the substrate.

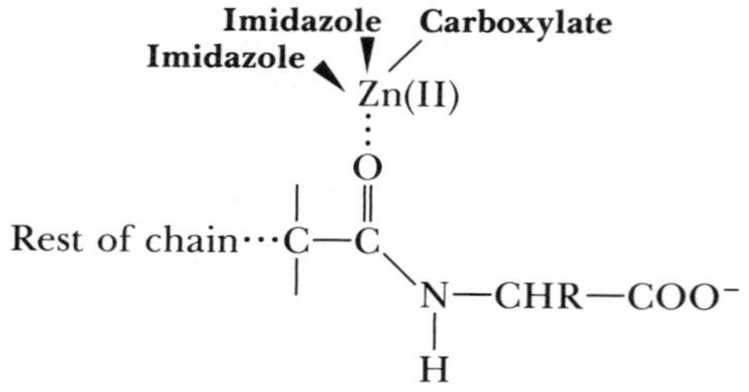

A zinc ion is complexed to three side chains of carboxypeptidase and to a carbonyl group on the substrate

The type of binding involved in the complex is similar to that which links iron to the large ring involved in the heme group. Binding of the substrate to the zinc ion polarizes the carbonyl group, making it susceptible to attack by water and allowing the hydrolysis to proceed more rapidly than in the uncatalyzed reaction.

**(b)**

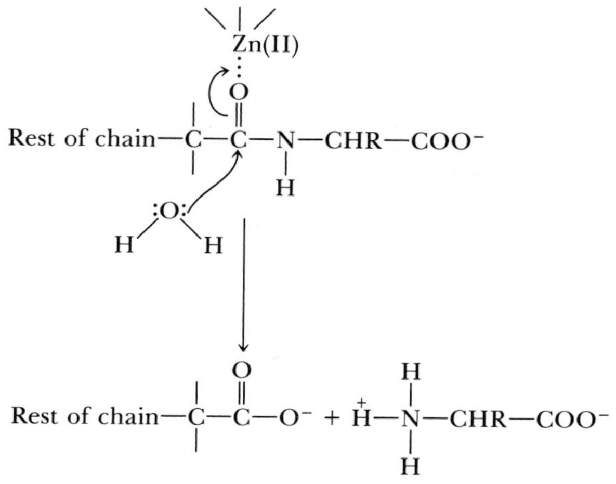

The reaction catalyzed by carboxypeptidase A

There is a definite connection between acid–base concepts and the idea of nucleophiles and their complementary substances, electrophiles. An electrophile is an electron-seeking group; a Lewis acid is an electrophile. A nucleophile is a nucleus-seeking group; a Lewis base is a nucleophile. Catalysis by enzymes, including their remarkable specificity, is based on these well-known chemical principles operating in a complex environment.

The nature of the active site plays a particularly important role in the specificity of enzymes. Enzymes that display **absolute specificity,** catalyzing the reaction of one, and only one, substrate to a particular product are likely to have a fairly rigid active site that is best described by the lock-and-key model of substrate binding. The many enzymes that display **relative specificity,** catalyzing the reaction of structurally related substrates to related products, apparently have more flexibility in their active sites and are better characterized by the induced-fit model of enzyme–substrate binding. Chymotrypsin is a good example here. Finally, there are enzymes for which optical activity plays a role in their specificity; in other words, such enzymes are **stereospecific.** The binding site itself must be asymmetric in this situation. If the enzyme is to bind specifically to an optically active substrate, the binding site must have the shape of the substrate and not its mirror image. There are even enzymes that introduce a center of optical acitivity into the product. The substrate itself is not optically active, but the product is only one of two possible isomers; there is

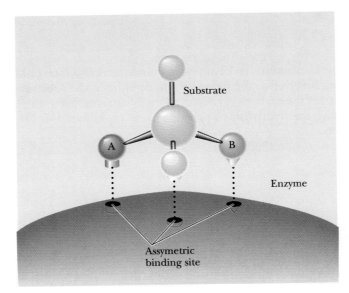

FIGURE 10.20 An asymmetric binding site on an enzyme can distinguish between identical groups such as A and B. Note that the binding site consists of three parts, giving rise to asymmetric binding because one part is different from the other two.

only one product, not a mixture of optical isomers. The binding site on such an enzyme is asymmetric (Figure 10.20).

## 10.13
## COENZYMES

Coenzymes are nonprotein substances that take part in enzymatic reactions and are regenerated for further reaction. There are two important classes of coenzymes. Metal ions frequently play such a role. The other important class consists of a mixed bag of organic compounds; many of them are vitamins or are metabolically related to vitamins.

Because metal ions are Lewis acids (electron-pair acceptors), they can act as Lewis acid–base catalysts. They can also form coordination compounds by acting as Lewis acids, while the groups to which they bind act as Lewis bases. Coordination compounds are an important part of the chemistry of metal ions in biological systems, as shown by Zn(II) in carboxypeptidase and by Fe(II) in hemoglobin. The coordination com-

**TABLE 10.2 Coenzymes, Their Reactions and Vitamin Precursors**

| COENZYME | REACTION TYPE | VITAMIN PRECURSOR | SEE SECTION |
|---|---|---|---|
| Biotin | $CO_2$ fixation | Biotin | 13.7, 16.6 |
| Coenzyme A | Acyl transfer | Pantothenic acid | 12.5, 14.3, 16.6 |
| Flavin coenzymes | Oxidation–reduction | Riboflavin ($B_2$) | 12.3, 14.3 |
| Lipoic acid | Acyl transfer | — | 14.3 |
| Nicotinamide adenine coenzymes | Oxidation–reduction | Niacin | 12.3, 13.3, 14.3 |
| Pyridoxal phosphate | Transamination | Pyridoxine ($B_6$) | 18.4 |
| Tetrahydrofolic acid | Transfer of one-carbon units | Folic acid | 18.4 |
| Thiamine pyrophosphate | Aldehyde transfer | Thiamine ($B_1$) | 13.5, 13.6 |

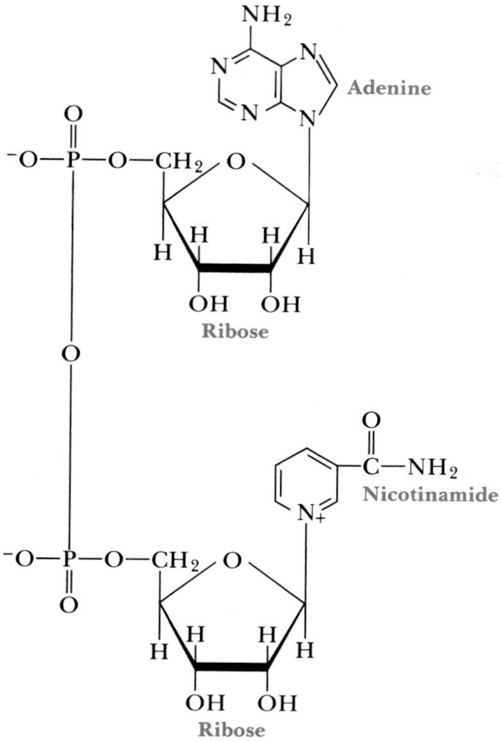

**FIGURE 10.21**   The structure of nicotinamide adenine dinucleotide (NAD+).

pounds formed by metal ions tend to have quite specific geometries, which aid in positioning groups involved in a reaction so that optimum catalysis can be achieved.

Some of the most important organic coenzymes are vitamins or their derivatives, especially B vitamins. Quite a few of them are involved in oxidation–reduction reactions, which provide energy for the organism. Others serve as group-transfer agents in metabolic processes (Table 10.2). We shall see these coenzymes again when we discuss the reactions in which they are involved. For the present we shall look at one example of a particularly important oxidation–reduction coenzyme and one example of a group-transfer coenzyme.

Nicotinamide adenine dinucleotide (NAD+) is a coenzyme in many oxidation–reduction reactions. Its structure (Figure 10.21) consists of three parts: a nicotinamide ring, an adenine ring, and two sugar–phosphate groups linked together. The nicotinamide ring contains the site at which oxidation and reduction reactions occur (Figure 10.22). Nicotinic acid is another name for the vitamin niacin. The adenine-sugar-phosphate portion of the molecule is structurally related to nucleotides.

The $B_6$ vitamins (pyridoxal, pyridoxamine, and pyridoxine and their phosphorylated forms, which are the coenzymes) are involved in the transfer of amino groups from one molecule to another, an important step in the biosynthesis of amino acids (Figure 10.23). In the reaction the amino group is transferred from the donor to the coenzyme and then from the coenzyme to the ultimate acceptor (Figure 10.24).

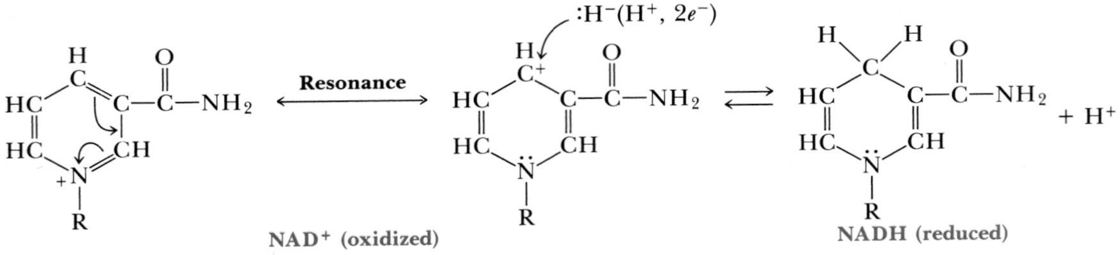

**FIGURE 10.22** The role of the nicotinamide ring in oxidation–reduction reactions. R is the rest of the molecule. In reactions of this sort, an $H^+$ is transferred along with the two electrons.

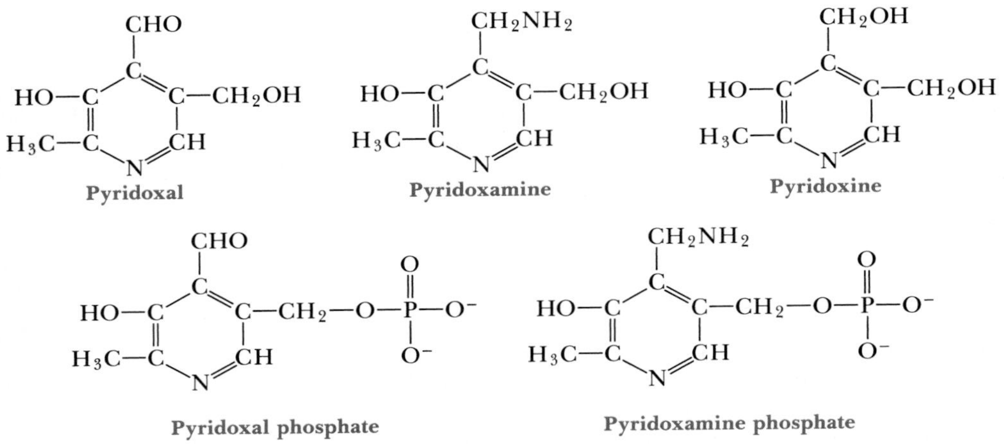

**FIGURE 10.23** Forms of vitamin $B_6$. The first three structures are vitamin $B_6$ itself, while the last two structures show the modifications that give rise to the metabolically active coenzyme.

**FIGURE 10.24** The role of pyridoxal phosphate as a coenzyme in a transamination reaction. PyrP is pyridoxal phosphate, P is the apoenzyme (the polypeptide chain alone), and E is the active holoenzyme (polypeptide plus coenzyme).

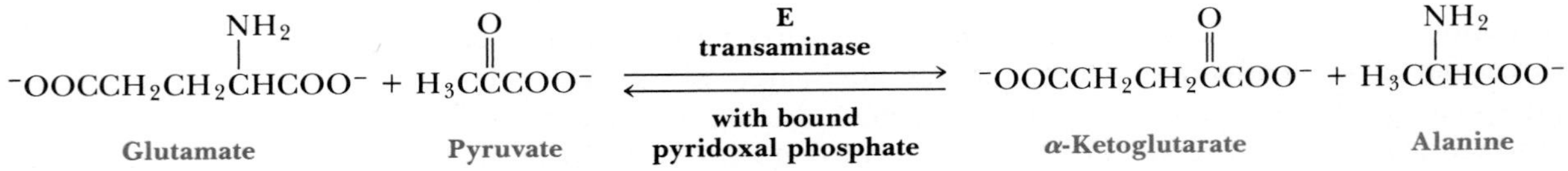

This amino ($NH_2$) group transfer reaction occurs in two stages:

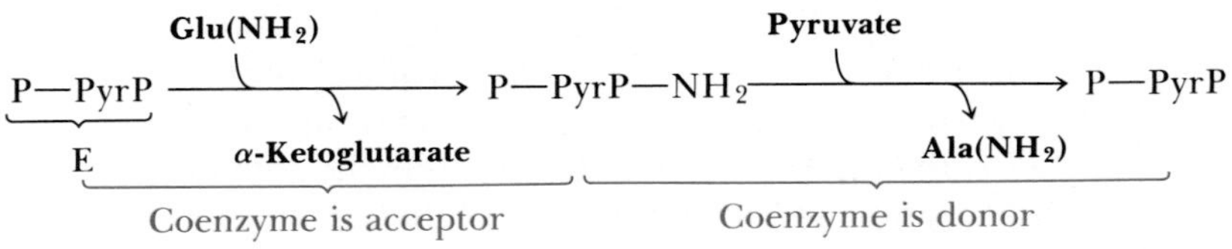

250

## SUMMARY

Of all the functions of proteins, the one that is probably most important is that of catalysis. Biological catalysts are called enzymes; all enzymes are globular proteins, with the exception of some recently discovered RNAs that catalyze their own self-splicing. Enzymes are the most efficient catalysts known. Enzymes speed up a reaction by lowering the activation energy, a kinetic parameter. Catalysts do not affect the thermodynamics of the reaction.

The first step in an enzyme-catalyzed reaction is the binding of the enzyme to the substrate to form a complex, leading in turn to formation of the transition-state species, which then forms the product. A substrate binds to a small portion of the enzyme called the active site. Two models have been proposed to describe enzyme-substrate binding: the lock-and-key model, in which there is an exact fit between the enzyme and substrate, and the induced-fit model, in which the enzyme is considered to have conformational flexibility and in which there is an exact fit only when the substrate is bound.

The kinetics of many enzyme-catalyzed reactions can be described by the Michaelis-Menten model. In this model the concept of the steady state, with a constant concentration of the enzyme-substrate complex, plays a vital role.

Inhibitors can give a considerable amount of information about enzymatic reactions. A reversible inhibitor can bind to the enzyme and subsequently be released. An irreversible inhibitor reacts with the enzyme to produce a protein that is not enzymatically active. There are two kinds of reversible inhibitors. Competitive inhibitors bind to the active site and block access of the substrate to the active site. Noncompetitive inhibitors bind to the enzyme at a site other than the active site and cause a change in the structure of the enzyme, especially around the active site, as a result of binding. In the Michaelis-Menten model, competitive inhibitors increase the $K_M$ but leave the $V_{max}$ unchanged. Noncompetitive inhibitors change the $V_{max}$ but leaves the $K_M$ unchanged.

The Michaelis-Menten model does not describe the behavior of allosteric enzymes. Changes in quaternary structure on binding of substrates, inhibitors, and activators all affect the observed kinetics of such enzymes. In the concerted model for allosteric behavior, the binding of substrate, inhibitor, or activator to one subunit shifts the equilibrium between an active form of the enzyme (which binds substrate) and an inactive form (which does not bind substrate). The conformational change takes place in all subunits at the same time. In the sequential model, the binding of substrate induces the conformational change in one subunit, and this conformational change is subsequently passed along to other subunits. Both models are useful; they may eventually be incorporated in a more inclusive model.

Another type of control mechanism in enzyme action is zymogen activation, in which an inactive precursor of an enzyme is transformed to an active enzyme by cleavage of covalent bonds. The proteolytic enzymes trypsin and chymotrypsin arise from the zymogens trypsinogen and chymotrypsinogen, respectively. Similar protein activations take place in the blood clotting process.

Several questions arise about the events that take place at the active site of an enzyme in the course of a reaction. The most important include the nature of the essential amino acid residues, their spatial arrangement, and the mechanism of the reaction. Chymotrypsin is a good example of an enzyme for which most of these questions have been answered. The essential amino acid residues have been determined to be serine 195, histidine 57, and aspartate 102. The complete three-dimensional structure of chymotrypsin, including the architecture of the active site, has been determined by x-ray crystallography. Nucleophilic attack by serine is the main feature of the mechanism, with histidine hydrogen-bonded to serine in the course of the reaction. Common organic reaction mechanisms such as nucleophilic substitution and general acid–base catalysis are known to play a role in enzymatic catalysis.

Coenzymes are nonprotein substances that take part in enzymatic reactions and are regenerated for further reaction. Metal ions can serve as coenzymes, frequently by acting as Lewis acids. There are also many organic coenzymes, most of which are vitamins or structurally related to vitamins.

## EXERCISES

1. For the reaction of glucose with oxygen to produce carbon dioxide and water,

$$\text{Glucose} + 6\,O_2 \longrightarrow 6\,CO_2 + 6\,H_2O$$

the $\Delta G^\circ$ is $-2880$ kJ mol$^{-1}$, a strongly exergonic reaction. However, a sample of glucose can be maintained indefinitely in an oxygen-containing atmosphere. Reconcile these two statements.

2. For a hypothetical reaction

$$3\,A + 2\,B \longrightarrow 2\,C + 3\,D$$

the rate was determined experimentally to be

$$\text{Rate} = k[A]^1[B]^1$$

What is the order of the reaction with respect to A? With respect to B? What is the overall order of the reaction? Suggest how many molecules each of A and B are likely to be involved in the detailed mechanism of the reaction.

3. Distinguish between the lock-and-key and induced-fit models for binding of a substrate to an enzyme.

4. Show graphically the dependence of reaction velocity on substrate concentration for an enzyme that follows Michaelis-Menten kinetics and for an allosteric enzyme.

5. Define "steady state" and comment on the relevance of this concept to theories of enzyme reactivity.

6. For an enzyme that displays Michaelis-Menten kinetics, what is the reaction velocity $V$ (as a percentage of $V_{max}$) observed at (a) $[S] = K_M$; (b) $[S] = 0.5\,K_M$; (c) $[S] = 0.1\,K_M$; (d) $[S] = 2\,K_M$; (e) $[S] = 10\,K_M$?

7. How is the turnover number of an enzyme related to $V_{max}$?

8. How can competitive and noncompetitive inhibition be distinguished in terms of $K_M$?

9. Draw Lineweaver-Burk plots for the behavior of an enzyme for which the following experimental data are available.

| [S] (mM) | V—no inhibitor (mmol min$^{-1}$) | V—inhibitor present (mmol min$^{-1}$) |
|---|---|---|
| 3.0 | 4.58 | 3.66 |
| 5.0 | 6.40 | 5.12 |
| 7.0 | 7.72 | 6.18 |
| 9.0 | 8.72 | 6.98 |
| 11.0 | 9.50 | 7.60 |

What are the $K_M$ and $V_{max}$ values for the inhibited and uninhibited reactions? Is the inhibitor competitive or noncompetitive?

10. Distinguish between the molecular mechanisms of competitive and noncompetitive inhibition.

11. What features distinguish enzymes that undergo allosteric control from those that obey the Michaelis-Menten equation?

12. Distinguish between the concerted and sequential models for the behavior of allosteric enzymes.

13. Name three proteins subject to the control mechanism of zymogen activation.

14. List three coenzymes and their function.

15. Is this statement true or false, and why? The mechanisms of enzymic catalysis have nothing in common with those encountered in organic chemistry.

16. An experiment is performed to test a suggested mechanism for an enzyme-catalyzed reaction. The results fit the model exactly (to within experimental error). Do the results prove that the mechanism is correct? Give a reason for your answer.

17. What properties of metal ions make them useful coenzymes?

18. Briefly describe the role of nucleophilic catalysis in the mechanism of the chymotrypsin reaction.

19. Explain why cleavage of the bond between arginine 15 and isoleucine 16 of chymotrypsinogen activates the zymogen.

20. An inhibitor that specifically labels chymotrypsin at histidine 57 is N-tosyl-L-phenylalanyl chloromethyl ketone (TPCK). How would you modify the structure of this inhibitor to label the active site of trypsin?

21. The enzyme lactate dehydrogenase catalyzes the reaction

$$\text{Pyruvate} + \text{NADH} + H^+ \longrightarrow \text{lactate} + \text{NAD}^+$$

NADH absorbs light at 340 nm in the near ultraviolet region of the electromagnetic spectrum, but NAD$^+$ does not. Suggest an experimental method for following the rate of this reaction, assuming that you have available a spectrophotometer capable of measuring light at this wavelength.

22. Methanol (wood alcohol) is highly toxic because it is converted to formaldehyde in a reaction catalyzed by the enzyme alcohol dehydrogenase:

$$\text{NAD}^+ + \text{methanol} \longrightarrow \text{NADH} + H^+ + \text{formaldehyde}$$

Part of the medical treatment for methanol poisoning is to administer ethanol (ethyl alcohol) in amounts large enough to cause intoxication under normal circumstances. Give a reason for the effectiveness of this treatment.

# ANNOTATED BIBLIOGRAPHY

Bachmair, A., D. Finley, and A. Varshavsky. In Vivo Half-Life of a Protein Is a Function of Its Amino-Terminal Residue. *Science* **234,** 179–186 (1986). [A particularly striking example of the relationship between structure and stability in proteins.]

Bender, M.L., R.L. Bergeron, and M. Komiyama. *The Bioorganic Chemistry of Enzymatic Catalysis.* New York: J. Wiley, 1984. [A discussion of mechanisms in enzymatic reactions.]

Dugas, H., and C. Penney. *Bioorganic Chemistry: A Chemical Approach to Enzyme Action.* New York: Springer-Verlag, 1981. [Discusses model systems as well as enzymes.]

Fersht, A. *Enzyme Structure and Mechanism,* 2nd ed. New York: W.H. Freeman, 1985. [A thorough coverage of enzyme action.]

Hammes, G. *Enzyme Catalysis and Regulation.* New York: Academic Press, 1982. [A good basic text on enzyme mechanisms.]

Koshland, D.E. Correlation of Structure and Function of Enzyme Action. *Science* **142,** 1533–1541 (1963). [The definitive statement of the induced-fit theory. Mainly of historic interest, but a classic.]

Kraut, J. How Do Enzymes Work? *Science* **242,** 533–540 (1988). [An advanced discussion of the role of transition states in enzymatic catalysis.]

Monod, J., J.P. Changeux, and F. Jacob. Allosteric Proteins and Cellular Control Systems. *J. Mol. Biol.* **6,** 306–329 (1963). [The original article on allosterism, describing the concerted model.]

Moore, J.W., and R.G. Pearson. *Kinetics and Mechanism,* 3rd ed. New York: John Wiley Interscience, 1980. [Quite advanced, but a classic treatment of the use of kinetic data to determine mechanisms.]

Neurath, H. Evolution of Proteolytic Enzymes. *Science* **224,** 350–357 (1984). [A classification of enzymes based on essential residues found in active sites.]

Page, M.I., ed. *The Chemistry of Enzyme Action.* New York: Elsevier, 1984. [Treatment includes energetics; Chapter 5 concentrates on the relationship between free energy and reaction mechanisms.]

Steitz, T.A., and R.G. Shulman. Crystallographic and NMR Studies of the Serine Proteases. *Ann. Rev. Biophys. Bioeng.* **11,** 419–444 (1982). [A review article on the experimental evidence on which mechanisms are based.]

# The Dynamics of Membrane Structure

## 11

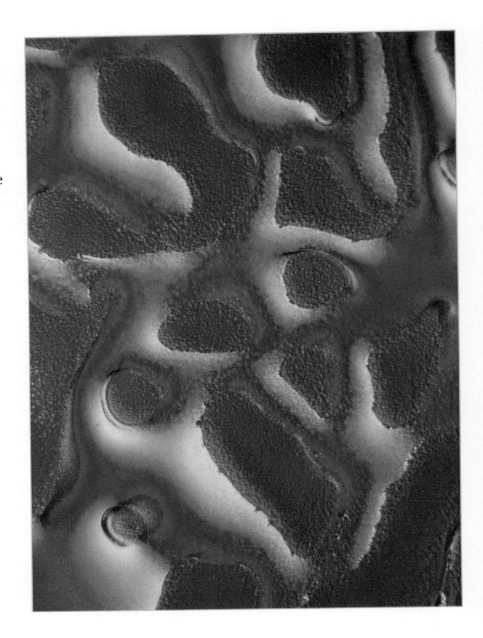

*Crystals of lecithin, an important component of membranes, viewed under polarized light.*

*Think of the cell as a country with a well defined border. Within the country are states, each with its own border. In the eukaryotic cell, the "border" of the country is the plasma membrane. State "borders" are the membranes of specialized organelles such as the nucleus, mitochondrion, and the Golgi apparatus. In the actual cell, membranes are made of phospholipids with charged polar heads and long hydrocarbon tails. Some small molecules can migrate through the membrane from a high concentration on one side to a low concentration on the other side by simple diffusion. The membrane is not a homogeneous barrier. At least half of its structure consists of protein molecules associated with the lipid bilayer. Some proteins form pores allowing specified ions and small molecules to pass through the membrane. Heart muscle cells, which act in close synchrony, are connected by gap junctions—gated tubes that join the cells through their outer membranes. Cells have on their surfaces glycoproteins and lipoproteins that recognize other molecules, as well as receptors that act as gates which allow the passage of ions and molecules into the cell. Altogether there is a lively commerce going on through the boundaries of "states" as well as between "countries."*

# 11.1
# THE NATURE OF BIOLOGICAL MEMBRANES

All cells have a cell membrane (also called a plasma membrane); eukaryotic cells also have membrane-bounded organelles such as nuclei and mitochondria. We are now in a position to discuss the molecular basis of membrane structure and to see how the lipid and protein components determine membrane function.

**Phosphoglycerides** are prime examples of *amphiphilic* molecules (ones with both hydrophilic and hydrophobic parts) and they are the principal lipid components of membranes. The existence of **lipid bilayers** depends on hydrophobic interactions (Section 8.6). These bilayers are frequently used as models for membranes because they have many features in common with biological membranes, such as their hydrophobic interior and their ability to control the transport of small molecules and ions. Bilayer systems are useful models because they are simpler and easier to work with in the laboratory than cell membranes. The most important difference between lipid bilayers and cell membranes is that biological membranes contain proteins as well as lipids. The protein component of a membrane can make up 20 to 80 percent of its total weight. An understanding of membrane structure requires knowledge of how the protein and lipid components contribute to the properties of the membrane.

(a)

(b)

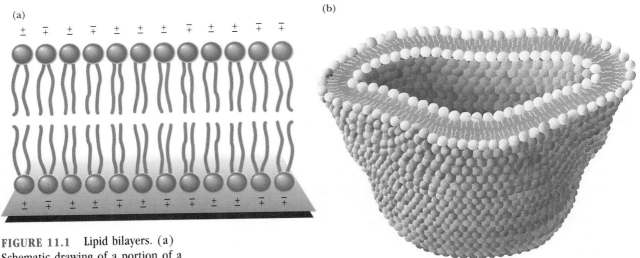

**FIGURE 11.1**    Lipid bilayers. (a) Schematic drawing of a portion of a bilayer consisting of phospholipids. The polar surface of the bilayer contains charged groups. The hydrocarbon "tails" lie in the interior of the bilayer. (b) Cutaway view of a lipid bilayer vesicle. Note that there is an aqueous inner compartment and that the inner layer is more tightly packed than the outer layer. (From M. S. Bretscher, "The Molecules of the Cell Membrane." *Scientific American,* October 1985, p. 103.)

## Lipid Bilayers

In addition to phosphoglycerides, biological membranes contain glycolipids and cholesterol as part of the lipid component. In the lipid-bilayer part of the membrane (Figure 11.1), the polar head groups are in contact with water, and the nonpolar tails lie in the interior. The whole bilayer arrangement is held together by noncovalent interactions such as van der Waals and hydrophobic interactions (Section 3.1). The surface of the bilayer is polar and contains charged groups. The nonpolar hydrocarbon interior of the bilayer consists of the saturated and unsaturated chains of fatty acids and the fused ring system of cholesterol. There is a mixture of lipids in both the inner and outer layers of the bilayer, but the compositions of the two layers are different; the inner and outer layers can be distinguished from each other on the basis of composition (Figure 11.2). Bulkier molecules tend to occur in the outer layer, while smaller molecules tend to occur in the inner layer.

The arrangement of the hydrocarbon interior of the bilayer can be ordered and rigid or disordered and fluid. The fluidity of the bilayer depends on its composition. In saturated fatty acids, a linear arrangement of the hydrocarbon chains leads to close packing of the molecules in the bilayer, and thus to rigidity. In unsaturated fatty acids, there is a kink in the hydrocarbon chain that does not exist in saturated fatty acids (Figure 11.3). These kinks cause disorder in the packing of the chains, with a more open structure than would be possible for straight saturated chains (Figure 11.4). The disordered structure due to the presence of unsaturated fatty acids in turn causes greater fluidity in the bilayer.

Greater order and rigidity may also result from the presence of cholesterol. The fused ring structure of cholesterol is itself quite rigid, and the presence of cholesterol stabilizes the extended straight chain arrangement of saturated fatty acids by van der Waals interactions (Figure 11.5).

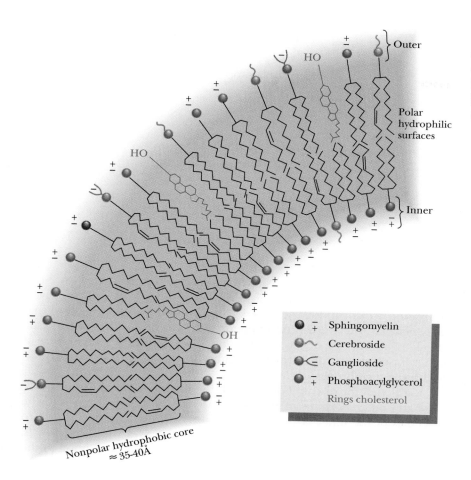

**FIGURE 11.2** Lipid bilayer asymmetry. The composition of the outer layer differs from that of the inner layer; the concentration of bulky molecules is higher in the outer layer, which has more room.

The lipid portion of plant membranes has a higher percentage of unsaturated fatty acids, and especially polyunsaturated (containing two or more double bonds) fatty acids, than the lipid portion of animal membranes. Further, the presence of cholesterol is characteristic of animal, rather than plant, membranes. As a result, animal membranes are less fluid (more rigid) than plant membranes.

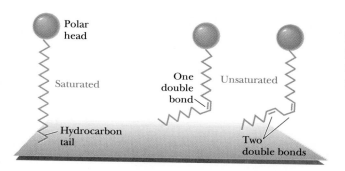

**FIGURE 11.3** The effect of double bonds on the conformation of the hydrocarbon tails of fatty acids. Unsaturated fatty acids have kinks in the tail region.

**FIGURE 11.4** Schematic drawing of a portion of a highly fluid phospholipid bilayer. The kinks in the unsaturated side chains prevent close packing of the hydrocarbon portions of the phospholipids.

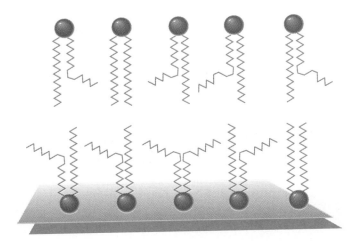

The degree of order in lipid bilayers can be decreased by heat. Ordered bilayers become less ordered; bilayers that are comparatively disordered become more disordered. This is a **cooperative transition** that takes place at a characteristic temperature, like the melting of a crystal (which is also a cooperative transition). The transition temperature is higher for more rigid and ordered membranes than it is for more fluid and disordered membranes.

Recall that the distribution of the different lipids is not the same in the inner and outer portions of the bilayer. Since the bilayer is curved, the molecules of the inner layer are more tightly packed together. (Refer to Figure 11.1.) Bulkier molecules such as cerebrosides (see Section 7.2; cerebrosides are glycolipids that contain a sugar moiety in addition to the amino alcohol and a long chain fatty acid) tend to be located in the outer layer. There is very little tendency for "flip-flop" migration of lipid molecules from one layer of the bilayer to another. Lateral motion of lipid molecules within one of the two layers frequently takes place, however, especially in more fluid bilayers. Several methods exist for monitoring the motion of molecules within a lipid bilayer. These methods have in common

**FIGURE 11.5** "Stiffening" of the lipid bilayer by cholesterol. The presence of cholesterol in a membrane reduces fluidity by stabilizing extended chain conformations of hydrocarbon tails of fatty acids as a result of van der Waals interactions.

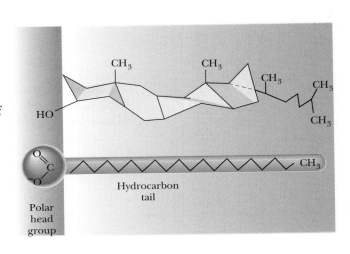

the use of "reporter groups" on analogues of the phospholipids normally found in membranes. In one frequently used method, called **spin labeling,** the reporter group carries an unpaired electron. The motion of the reporter group in the bilayer can be determined from the magnetic properties that arise from the spin of the unpaired electron. In another method, the reporter group is fluorescent; the motion of such a group can be followed easily, since fluorescence is a highly sensitive technique.

## 11.2
## MEMBRANE PROTEINS

Proteins in a biological membrane can be associated with the lipid bilayer in either of two ways: as **peripheral proteins** on the surface of the membrane, or as **integral proteins** within the lipid bilayer (Figure 11.6). Peripheral proteins are usually bound to the charged head groups of the lipid bilayer by polar or electrostatic interactions, or both. Peripheral proteins can be removed by such mild treatment as raising the ionic strength of the medium. The larger number of charged particles present in a medium of higher ionic strength undergo more electrostatic interactions with the lipid and with the protein, "swamping out" the comparatively fewer electrostatic interactions between the protein and the lipid. It is much more difficult to remove integral proteins from membranes. Harsh conditions, such as treatment with detergents or extensive sonication (exposure to ultrasonic vibrations), are usually required. Such treatment frequently denatures the protein, which often remains bound to lipids in spite of all efforts to obtain the protein in pure form. The denatured protein is of course inactive, whether or not it remains bound to lipids. Fortunately, it is becoming possible to study proteins of this sort in living tissue by using nuclear magnetic resonance techniques. The structural integrity of the whole membrane system appears to be necessary for the activity of the protein.

Membrane proteins have a variety of functions. Most, but not all, of the important functions of the whole membrane are those of the protein

**FIGURE 11.6**  Some types of associations of proteins with membranes. The proteins marked 1, 2, and 4 are integral proteins, while protein 3 is a peripheral protein. Note that the integral proteins can be associated with the lipid bilayer in several ways. Protein 1 transverses the membrane, protein 2 lies entirely within the membrane, and protein 4 projects into the membrane.

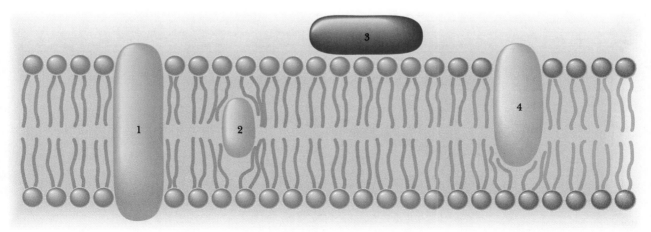

component. **Transport proteins** help move substances in and out of the cell, and **receptor proteins** are important in the specific uptake of biologically active substances such as insulin. In addition, some enzymes are tightly bound to membranes; examples include many of the enzymes responsible for aerobic oxidation reactions, which are located in specific parts of mitochondrial membranes. Some of these enzymes are located on the inner surface of the membrane and some on the outer surface. There is an uneven distribution of proteins of all types on the inner and outer layers of all cell membranes, just as there is an asymmetric distribution of lipids.

## 11.3
## THE FLUID MOSAIC MODEL OF MEMBRANE STRUCTURE

We have seen that biological membranes have both lipid and protein components. The next question is how these two parts combine to produce a biological membrane. The **fluid mosaic model** is the description of biological membranes that is most widely accepted at the moment. The term "mosaic" implies that the two components exist side by side without forming some other substance of intermediate nature. There is no extensive formation of lipid–protein complexes, for example. Instead, the basic structure of biological membranes is that of the lipid bilayer, with the proteins embedded in the bilayer structure (Figure 11.7). The term "fluid mosaic" implies that there is the same sort of lateral motion in membranes that we have already seen in lipid bilayers. The proteins "float" in the lipid bilayer and can move along the plane of the membrane.

One aspect of this model that may have to be modified as time goes on is the statement that lipid–protein complexes do not exist to any important degree. There have been reports of external proteins covalently bonded to

**FIGURE 11.7** Fluid mosaic model of membrane structure. Membrane proteins can be seen embedded in the lipid bilayer. (From S. J. Singer, *in* G. Weismann and R. Claiborne, eds., *Cell Membranes: Biochemistry, Cell Biology, and Pathology.* New York: HP Pub., 1975, p. 37.)

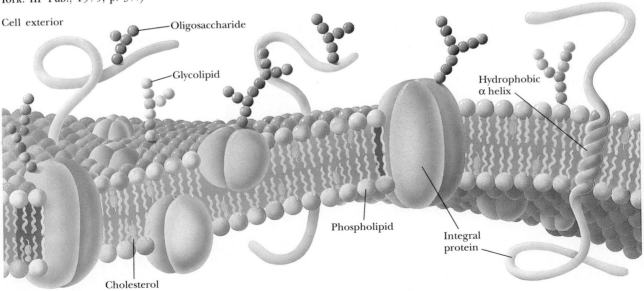

Cell exterior

Oligosaccharide

Glycolipid

Hydrophobic α helix

Phospholipid

Integral protein

Cholesterol

Cytosol

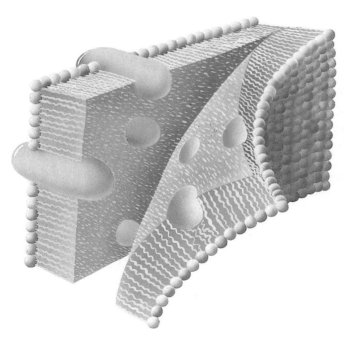

FIGURE 11.8 Replica of a freeze-fractured membrane. In the freeze-fracture technique, the lipid bilayer is split parallel to the surface of the membrane. The hydrocarbon tails of the two layers are separated from each other, and the proteins can be seen as "hills" in the replica shown. In the other layer, seen edge on, there are "valleys" where the proteins had been. (From S. J. Singer, *in* G. Weismann and R. Claiborne, eds., *Cell Membranes: Biochemistry, Cell Biology, and Pathology.* New York: HP Pub., 1975, p. 38.)

a glycolipid that is part of the lipid bilayer. The article by Kolata listed in the bibliography at the end of this chapter has more information on this point.

It is possible to obtain electron micrographs of membranes that have been frozen and then fractured along the interface between the two layers. The outer layer is removed, exposing the interior of the membrane. The interior has a granular appearance because of the presence of the integral membrane proteins (Figures 11.8 and 11.9).

FIGURE 11.9 Electron micrograph of freeze-fractured thylakoid membrane of pea (magnified 110,000 ×). The grains protruding from the surface are integral membrane proteins.

## 11.4
## MEMBRANE FUNCTION

As we have already mentioned, three important functions take place in or on membranes, in addition to the structural role of membranes as the boundary and container of all cells and of the organelles within eukaryotic cells. The first of these functions is **transport.** Membranes are semipermeable barriers to the flow of substances into and out of cells and organelles. Transport through the membrane can involve the lipid bilayer as well as the membrane proteins; the other two important functions primarily involve the membrane proteins. The second of these functions is **catalysis.** As we have seen, enzymes can be bound to membranes, in some cases very tightly bound, and the enzymatic reaction takes place on the membrane. The third is the **receptor** property, in which proteins bind specific biologically important substances that trigger biochemical responses in the cell. We shall discuss enzymes bound to membranes in subsequent chapters, especially in our treatment of aerobic oxidation reactions. The other two functions we shall now consider in turn.

### Membrane Transport

The most important question about transport of substances across biological membranes is whether or not the process requires expenditure of energy by the cell. In **passive transport** a substance moves from a region of higher concentration to one of lower concentration. In other words, the movement of the substance is in the same direction as a *concentration gradient,* and there is no expenditure of energy by the cell. In **active transport** a substance moves from a region of lower concentration to one of higher concentration (against a concentration gradient), and the process does require expenditure of energy by the cell.

We can write an expression for the free energy change that occurs when a substance is transported from compartment 1 across a membrane to compartment 2. This expression follows directly from the laws of thermodynamics, but we shall not give the derivation here. The equation is

$$\Delta G = 2.303 \ RT \log \frac{C_2}{C_1}$$

where $\Delta G$ is the free energy change, $R$ is the gas constant (8.31 J mol$^{-1}$ K$^{-1}$), $T$ is the absolute temperature, and $C_1$ and $C_2$ are the concentrations of the substance in the two compartments. Let us assume that $C_1$ is outside the cell and $C_2$ is inside the cell. When $C_1 > C_2$ (higher concentration of the substance outside the cell than inside), the log term is negative, since it is the log of a fraction; as a result the $\Delta G$ term is negative. The process of transporting the substance from a region of higher concentration outside the cell to a region of lower concentration inside the cell (in the same direction as the concentration gradient) is spontaneous ($\Delta G < 0$). When $C_1 < C_2$, the log term is positive, since it is the log of a number greater than 1; the $\Delta G$ term is also positive. The process of transporting a substance from a region of lower concentration outside the cell to a region of higher concentration inside the cell (against a concentration gradient) is not

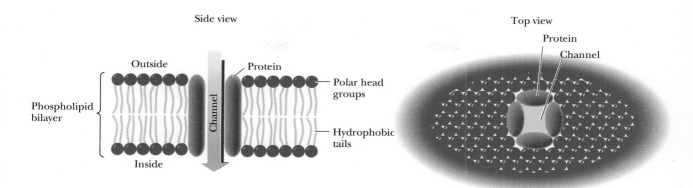

Side view                          Top view

FIGURE 11.10 Schematic representation of a channel protein. This channel protein is shown as having four subunits. Both channel proteins, which play a role in simple diffusion, and carrier proteins, which mediate facilitated diffusion, have a pore through which substances can pass. The difference between the two is that substances entering the cell bind to the carrier protein in facilitated diffusion, while in simple diffusion the ions or small molecules simply pass through the pore in the channel protein.

spontaneous ($\Delta G > 0$) and requires energy if it is to take place. A cell must expend energy to transport substances against a concentration gradient. To give an indication of the quantity of energy involved for most organisms, the process of active transport against concentration gradients accounts for about 30% of the energy consumption of a resting animal.

The process of passive transport can be subdivided into two categories, simple diffusion and facilitated diffusion. In **simple diffusion** a molecule moves directly through an opening or pore in the membrane without interacting with another molecule. The molecule being transported either goes directly through the lipid bilayer or through an opening in a **channel protein** (Figure 11.10). Small molecules such as water, oxygen, and carbon dioxide can pass directly through the lipid bilayer, but larger molecules, especially polar ones, require a channel protein. Ions cannot pass through the lipid bilayer because of their charge, and they also require channel proteins. The sizes and shapes of the openings in channel proteins make them specific for given ions or molecules.

**Facilitated diffusion** involves a **carrier protein.** The molecule to be transported across the membrane binds to the carrier protein, rather than simply passing through it, as is the case with channel proteins in simple diffusion. In both channel proteins and carrier proteins a pore is created by the helical folding of the backbone and side chains. In this picture the helical portion of the protein spans the membrane. The exterior of the helix, which is in contact with the lipid bilayer, is hydrophobic, while the interior of the helix, through which ions pass, is hydrophilic.

**Active transport** requires moving substances against a concentration gradient, a situation similar to pumping water uphill. The similarity to a pump is so marked that one of the most extensively studied examples of active transport, moving potassium ions into a cell and simultaneously moving sodium ions out of the cell, is referred to as the **sodium-potassium ion pump.**

Under normal circumstances the concentration of $K^+$ is higher inside the cell than in extracellular fluids ($[K^+]_{inside} > [K^+]_{outside}$), but the concentration of $Na^+$ is lower inside the cell than out ($[Na^+]_{inside} < [Na^+]_{outside}$). The energy required to move these ions against their gradients comes from an exergonic (energy-releasing) reaction, the hydrolysis of ATP to ADP and $P_i$ (phosphate ion). There can be no transport of ions without hydrolysis of ATP. The same protein appears to be both the

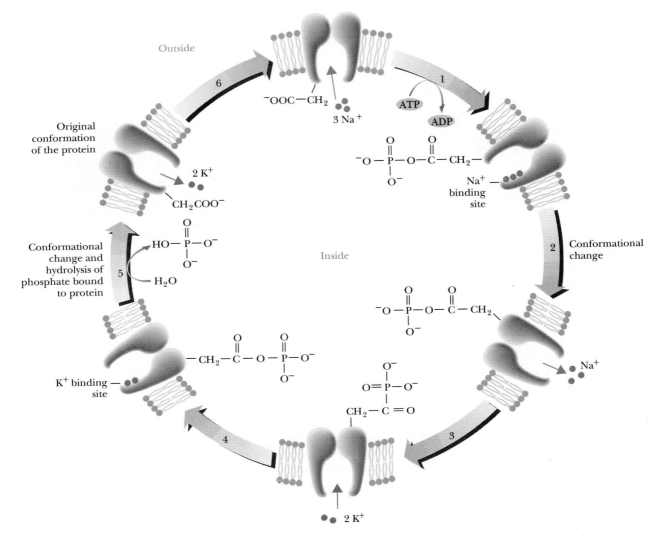

**FIGURE 11.11** The sodium–potassium pump.

enzyme that hydrolyzes the ATP (the ATPase) and the transport protein; it consists of several subunits. The reactants and products of this hydrolysis reaction—ATP, ADP, and phosphate ion—remain within the cell, and the phosphate becomes covalently bonded to the transport protein for part of the process.

The $Na^+$–$K^+$ pump operates in several steps (Figure 11.11). One subunit of the protein hydrolyzes the ATP and transfers the phosphate group to an aspartate side chain on another subunit. There is simultaneous binding of 3 $Na^+$ from the interior of the cell. The phosphorylation of one subunit causes a conformational change in the protein, which opens a channel or pore through which the 3 $Na^+$ can be released to the extracellular fluid. Outside the cell, 2 $K^+$ bind to the pump enzyme, which is still phosphorylated. Another conformational change takes place when the bond between the enzyme and the phosphate group is hydrolyzed. This second conformational change regenerates the original form of the enzyme and allows the 2 $K^+$ to enter the cell (Figure 11.12). The process transports three $Na^+$ ions out of the cell for every two $K^+$ ions transported into the cell.

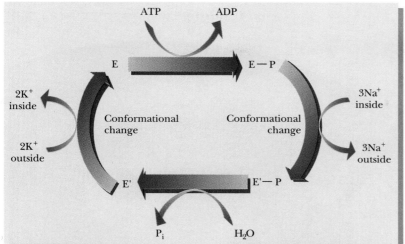

**FIGURE 11.12**   Another way of looking at the sodium–potassium pump. E is the enzyme (original conformation). E—P is the enzyme with a covalently bound phosphate. E′ is the enzyme after hydrolysis of phosphate (second conformation). E′—P is the enzyme in the second conformation with covalently bound phosphate.

It is possible to reverse the operation of the pump when there is no $K^+$ and a high concentration of $Na^+$ in the extracellular medium; in this case ATP is produced by the phosphorylation of ADP. The actual operation of the $Na^+$–$K^+$ pump may well be even more complicated and is not completely understood; unanswered questions about the detailed mechanism of active transport provide opportunities for future research.

## Membrane Receptors

The first step in the effects of some biologically active substances is binding of the substance to a protein receptor site on the exterior of the cell. The interaction between receptor proteins and the active substances to which they bind is very similar to enzyme–substrate recognition. There is a requirement for essential functional groups that have the right three-dimensional conformation with respect to one another. The binding site, whether on a receptor or an enzyme, must provide a good fit for the substrate. In receptor binding, as in enzyme behavior, there is a possibility for inhibition of the action of the protein by some sort of "poison" or inhibitor. The study of receptor proteins is less advanced than the study of enzymes, since many receptors are tightly bound integral proteins, and their activity depends on the membrane environment. Receptors are often large oligomeric proteins (ones with several subunits), with molecular weights of the order of hundreds of thousands. Also, quite frequently there are very few molecules of the receptor in each cell, adding to the difficulties of isolating and studying this type of protein.

An important type of receptor is that for low-density lipoprotein (LDL). The principal carrier of cholesterol in the bloodstream is LDL, a particle that consists of cholesterol, phosphoglycerides, and a protein. The protein portion of the LDL particle binds to the LDL receptor of a cell. The complex formed between the LDL and the receptor is pinched off into the cell by a process called **endocytosis.** (Endocytosis is described in detail in the articles by Brown and Goldstein and by Dautry-Varsat and Lodish

**FIGURE 11.13**    The mode of action of the LDL receptor.

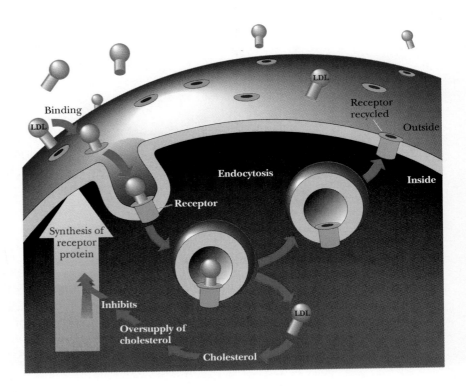

listed in the bibliography at the end of this chapter. The process of endocytosis is an important aspect of receptor action.) The receptor protein is then recycled back to the surface of the cell (Figure 11.13). The cholesterol portion of the LDL is used in the cell, but an oversupply of cholesterol causes problems. The synthesis of LDL receptor is inhibited by an excess of cholesterol. If there are fewer receptors for LDL, the level of cholesterol in the bloodstream increases. Eventually the excess cholesterol is deposited in the arteries, blocking them severely. This blocking of the arteries, called **atherosclerosis,** can eventually lead to heart attacks and strokes. In many industrialized countries, the usual blood cholesterol level is high, and the incidence of heart attacks and strokes is correspondingly high.

## 11.5
## A CASE STUDY IN MEMBRANE BEHAVIOR: THE NEUROMUSCULAR JUNCTION

The action of the neuromuscular junction between a nerve cell (a neuron) and a muscle cell is complex, but it gives examples of important aspects of receptor behavior in conjunction with membrane transport. One of the most important features is the operation of **gated channels.** The channels in proteins, which serve as an avenue for transport into the cell, are not necessarily open all the time. Some channels are open continuously, but others are opened only transiently by gating mechanisms. **Ligand-gated**

(a)

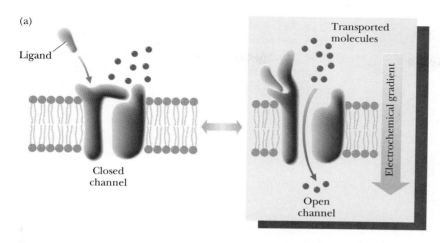

(b)

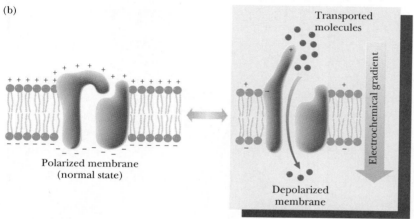

FIGURE 11.14    Two kinds of gated channels for the passage of molecules. (a) Ligand-gated channel. (b) Voltage-gated channel.

channels are controlled by the binding to the receptor of some substance in the extracellular fluid; **voltage-gated** channels are controlled by the voltage across the membrane (membrane potential) generated by the concentration gradients of various ions (Figure 11.14). (The external side of the plasma membrane has a positive charge with respect to the internal side in the normal situation of **membrane polarization.**) In the neuromuscular junction there is a specific sequence of opening and closing of four such gated channels.

Figure 11.15 shows a resting neuromuscular junction and an activated junction with muscular contraction. Of the four gated channels that play a role in the process, three are voltage-gated and one is ligand-gated. The three voltage-gated channels are:

1. A $Ca^{2+}$ channel in the plasma membrane of the nerve cell ($Ca^{2+}$ enters the nerve cell through this channel).
2. A $Na^+$ channel in the plasma membrane of the muscle cell ($Na^+$ enters the muscle cell).
3. A $Ca^{2+}$ channel in the internal membrane system (sarcoplasmic reticulum) of the muscle cell ($Ca^{2+}$ flows from the sarcoplasmic reticulum into the cytosol of the muscle cell).

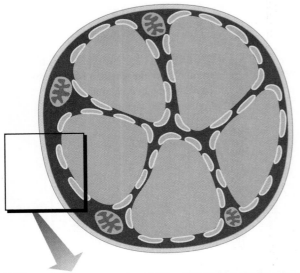

FIGURE 11.15 Comparison of a resting and an activated neuromuscular junction. (a) Resting neuromuscular junction. (b) Activated neuromuscular junction causing muscular contraction. The various gated channels are numbered in the sequence in which they open.

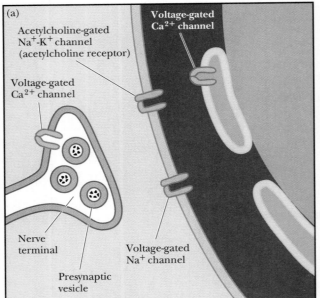

(a)

Acetylcholine-gated Na⁺-K⁺ channel (acetylcholine receptor)

Voltage-gated Ca²⁺ channel

Voltage-gated Ca²⁺ channel

Nerve terminal

Voltage-gated Na⁺ channel

Presynaptic vesicle

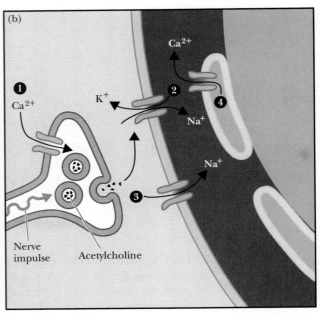

(b)

Nerve impulse

Acetylcholine

The ligand-gated channel in the plasma membrane of the muscle cell is gated by binding of the neurotransmitter acetylcholine.

$$CH_3 \overset{\overset{\displaystyle O}{\|}}{-C} -O-CH_2CH_2-\overset{+}{N}(CH_3)_3$$

**Acetylcholine**

When open, the acetylcholine-gated cation channel allows the inflow of $Na^+$ into the muscle cell and the outflow of $K^+$, both in the directions of their respective concentration gradients.

The activation of the neuromuscular junction is initiated by the nerve impulse. A characteristic of the impulse is a decrease in the membrane potential (depolarization of the membrane). The depolarization leads to the transient opening of the $Ca^{2+}$ channel in the nerve cell. (This is the first of the four gated channels involved in the process.) The $Ca^{2+}$ concentration is about 1000 times higher outside the cell than inside, so $Ca^{2+}$ flows freely into the cell.

The presence of increased levels of $Ca^{2+}$ inside the nerve cell triggers the release of acetylcholine from membrane-bounded vesicles within the nerve cell. The vesicles fuse with the plasma membrane of the nerve cell, releasing acetylcholine into the space between the nerve cell and the muscle cell. Acetylcholine is taken up by receptors on the muscle cell plasma membrane. These receptors are the second of the four gated channels involved in the operation of the neuromuscular junction; they are acetylcholine-gated channels for $Na^+$ and $K^+$. Note that this protein is the only receptor with a binding site for a molecule among the four gated channels that control the action of the neuromuscular junction, but the receptor action depends on all four proteins acting in concert. The flow of $Na^+$ into and the flow of $K^+$ out of the muscle cell causes local depolarization of the muscle membrane.

The depolarization of the muscle membrane leads in turn to the transient opening of voltage-gated $Na^+$ channels, leading to a greater inflow of $Na^+$ and still further depolarization of the muscle membrane. (The $Na^+$ channel is the third of the four channels.) The depolarization (also called the action potential) eventually spreads to the whole muscle membrane.

The action potential leads to the transient opening of the $Ca^{2+}$ channels of the sarcoplasmic reticulum within the muscle cell, the fourth of the gated channels in the process. (In cells, the internal membrane system is connected to the plasma membrane.) The outflow of $Ca^{2+}$ from the sarcoplasmic reticulum into the cytosol increases the cytosolic concentration of $Ca^{2+}$ drastically, triggering the $Ca^{2+}$-dependent process of contraction of the myofibrils within the muscle cell (muscle contraction).

## SUMMARY

Biological membranes consist of a lipid part and a protein part. The lipid part is a bilayer, with the polar head groups in contact with the aqueous interior and exterior of the cell and the nonpolar portions of the lipid in the interior of the membrane. Because membranes are curved, the inner and outer layers of the membrane differ in composition; bulkier molecules are found in the outer layer where there is more room to accommodate them. Lateral motion of lipid molecules within one layer of a membrane occurs frequently, but there is very little tendency for "flip-flop" migration of lipids from one layer to another. The proteins that occur in membranes can be peripheral proteins, found on the surface of the membrane, or integral proteins, which lie within the lipid bilayer. The fluid mosaic model describes the interaction of lipids and proteins in biological membranes. There is no extensive formation of lipid-protein complexes in this model. The proteins "float" in the lipid bilayer.

Three important functions take place in or on membranes. The first, transport across the membrane, can involve the lipid bilayer as well as the membrane proteins. The second, catalysis, is carried out by enzymes bound to the membrane. Finally, there are receptor proteins that bind biologically important substances which trigger a biochemical response in the cell.

The most important question about transport of substances across biological membranes is whether the process requires expenditure of energy by the cell. In passive transport a substance moves from a region of higher concentration to one of lower concentration, requiring no expenditure of energy by the cell. The process of passive transport can be divided into two categories, simple diffusion and facilitated diffusion. In simple diffusion a molecule moves directly through an opening or pore in the membrane without interacting with another molecule. The molecule being transported either goes directly through the lipid bilayer or through an opening in a channel protein. Facilitated diffusion involves a carrier protein; the molecule to be transported across the membrane binds to the carrier protein, rather than simply passing through it. Active transport requires moving substances against a concentration gradient, a situation similar to pumping water up a hill. Energy, as well as a carrier protein, is required for active transport. The sodium–potassium pump is an example of active transport.

The first step in the effect of some biologically active substances is binding to a protein receptor site on the exterior of the cell. The interaction between receptor proteins and the active substances to which they bind is very similar to enzyme–substrate recognition. The action of receptors frequently depends on a conformational change in the receptor protein. Receptors and the substances bound to them can be taken into the interior of the cell by the process of endocytosis, in which the membrane folds in on itself to engulf the receptor–ligand complex. Receptors can also be ligand-gated channel proteins, in which the binding of ligand transiently opens a channel protein through which substances such as ions can flow in the direction of a concentration gradient. The activation of the neuromuscular junction involves the action of such a protein, in which the neurotransmitter acetylcholine is the ligand. The binding of the ligand controls the operation of a $Na^+$–$K^+$ channel. Ligand-gated proteins also offer an interesting contrast with voltage-gated channel proteins, in which the voltage across the cell membrane, rather than receptor action, controls the operation of the channel.

# EXERCISES

1. There is an order-disorder transition in lipid bilayers similar to the melting of a crystal. Would you expect this transition to occur at a higher temperature, a lower temperature, or the same temperature in a lipid bilayer in which most of the fatty acids are unsaturated, compared to the transition temperature of a lipid bilayer in which most of the fatty acids are saturated? Give the reasons for your answer.

2. A membrane consists of 50% protein by weight and 50% phosphoglycerides by weight. The average molecular weight of the lipids is 800 daltons, while the average molecular weight of the proteins is 50,000 daltons. Calculate the molar ratio of lipid to protein.

3. Which statements are consistent with what is known about membranes?
   (a) Membranes consist of a layer of proteins sandwiched between two layers of lipids.
   (b) The composition of the inner and outer lipid layers is the same in all membranes.
   (c) Membranes contain glycolipids and glycoproteins.
   (d) Lipid bilayers are an important component of membranes.

   (e) There is covalent bonding between lipids and proteins in most membranes.
4. Inorganic ions such as $K^+$, $Na^+$, $Ca^{2+}$, and $Mg^{2+}$ do not cross biological membranes by simple diffusion. Suggest a reason for this behavior.
5. Which statements are consistent with the fluid mosaic model of membranes?
   (a) All membrane proteins form an integral part of the membrane.
   (b) Both proteins and lipids undergo transverse ("flip-flop") diffusion from the inside to the outside of the membrane.
   (c) Both proteins and lipids undergo lateral diffusion along the inner or outer surface of the membrane.
   (d) Carbohydrates are covalently bonded to the outside of the membrane.
   (e) The term "mosaic" refers to the arrangement of the lipids alone.
6. Which statements are consistent with the known facts about membrane transport?
   (a) Active transport moves a substance from a region where its concentration is lower to one where its concentration is higher.

(b) Transport does not involve any pores or channels in membranes.

(c) Transport proteins may be involved in bringing substances into cells.

7. The cell membranes of bacteria grown at 20°C tend to have a higher proportion of unsaturated fatty acids than the membranes of bacteria of the same species grown at 37°C. (In other words, the bacteria grown at 37°C have a higher proportion of saturated fatty acids in their cell membranes.) Suggest a reason for this observation.

8. What types of amino acid residues are likely to be found in the binding site of the LDL receptor? Give a reason for your answer.

9. What is the thermodynamic driving force for the formation of phospholipid bilayers?

10. Animals that live in cold climates tend to have a higher proportion of polyunsaturated fatty acid residues in their lipids than do animals that live in warm climates. Suggest a reason for this observation.

11. Succulent plants from arid regions generally have a waxy surface coating. Suggest why this coating is valuable for the survival of the plant.

12. In the preparation of sauces that involve mixing water and melted butter, egg yolks are added to prevent separation. How do the egg yolks prevent separation? Hint: egg yolks are rich in phosphatidylcholine (lecithin).

## ANNOTATED BIBLIOGRAPHY

Alberts, B., D. Bray, J. Lewis, M. Raff, K. Roberts, and J.D. Watson. *Molecular Biology of the Cell. 2nd ed.* New York: Garland Publishing, Inc., 1988. [Chapter 6 contains a lucid and well illustrated description of membrane structure and function from the standpoint of cell biology.]

Bittar, E.E., ed. *Membrane Structure and Function,* 3 vols. New York: Wiley-Interscience, 1980. [Thorough coverage of all aspects of membrane structure and function.]

Bretscher, M.S. The Molecules of the Cell Membrane. *Sci. Amer.* **253** (4), 100–108 (1985). [A particularly well illustrated description of the role of lipids and proteins in cell membranes.]

Brown, M.S., and J.L. Goldstein. How LDL Receptors Influence Cholesterol and Atherosclerosis. *Sci. Amer.* **251** (5), 58–66 (1984). [A description of lipid metabolism and the role of membrane receptors.]

Brown, M.S., and J.L. Goldstein. A Receptor-Mediated Pathway for Cholesterol Homeostasis. *Science* **232**, 34–47 (1986). [A more recent description of the role of cholesterol in heart disease.]

Dautry-Varsat, A., and H.F. Lodish. How Receptors Bring Proteins and Particles into Cells. *Sci. Amer.* **250** (5), 52–58 (1984). [A detailed description of endocytosis.]

Dunant, Y., and M. Israel. The Release of Acetylcholine. *Sci. Amer.* **252** (4), 58–66 (1985). [A description of the role of neurotransmitters and receptors for them in the propagation of nerve impulses.]

Kolata, G. Novel Protein/Membrane Attachment. *Science* **229**, 850 (1985). [A Research News article on the discovery of surface proteins covalently bonded to glycolipids in membranes.]

Marx, J.L. A New View of Receptor Action. *Science* **224**, 271–274 (1984). [A possible connection between calcium ion-linked receptors and cancer.]

Ostro, M.J. Liposomes. *Sci. Amer.* **256** (1), 102–111 (1987). [A description of possible uses of lipid bilayers as a vehicle for delivery of drugs to tissues.]

Rasmussen, H. The Cycling of Calcium as an Intracellular Messenger. *Sci. Amer.* **261** (4), 66–73 (1989). [An article on how receptors for calcium ion mediate its intracellular effects.]

Singer, S.J., and G.L. Nicholson. The Fluid Mosaic Model of the Structure of Membranes. *Science* **175**, 720–731 (1972). [The article in which the fluid mosaic model was first introduced.]

Tanford, C. The Mechanism of Free Energy Coupling in Active Transport. *Ann. Rev. Biochem.* **52**, 379–409 (1983). [A review on thermodynamic aspects of active transport.]

Tanford, C. *The Hydrophobic Effect: Formation of Micelles and Biological Membranes,* 2nd ed. New York: Wiley-Interscience, 1980. [A standard text on the physical chemistry of membranes.]

Unwin, N., and R. Henderson. The Structure of Proteins in Biological Membranes. *Sci. Amer.* **250** (2), 78–94 (1984). [The results of electron microscopic studies on integral proteins tightly bound in membranes.]

# Metabolism

## OUTLINE

# INTERVIEW

# Ponzy Lu

Ponzy Lu was born in Shanghai, China, and did his primary and secondary schooling in southern California and London, England. He is currently Professor of Chemistry and Chair of the Biochemistry Major Program at The University of Pennsylvania, where he often teaches undergraduate biochemistry courses. He received his bachelor of science degree (1964) in chemistry from the California Institute of Technology and his Ph.D. (1970) in biophysics from the Massachusetts Institute of Technology, followed by three postdoctoral years in Germany and Switzerland. The choice of Germany was motivated by a southern Californian fantasy of owning a Porsche and using it without speed limits. This dream was realized in Göttingen, where quantum physics and organic chemistry originated. In Switzerland he joined Jeffrey Miller, now at UCLA, to learn bacterial genetics first hand, which also began a longstanding collaboration on work with the lactose operon repressor. He has been at The University of Pennsylvania since 1973, with one year at l'Un-

iversité d'Aix-Marseille, France, in 1980. He has published over 60 articles and abstracts in the field of molecular biology. His research interests involve a combination of nuclear magnetic resonance (NMR) methods and genetics directed at repressor proteins. Professor Lu is often invited to lecture at universities and biochemical conferences, such as the international Conference on Magnetic Resonance in Biological Systems. Professor Lu's primary interests are biochemical: food, wine, and gene regulation.

**How did you get interested in science and choose Caltech for your undergraduate training?**

In high school my idea of fun was mechanical things: toys, model airplanes, cars, and explosions—firecrackers during Chinese New Year. I really didn't choose Caltech. Being a Californian and good at high school math, it was an obvious place to go. It wasn't until I arrived at Caltech's "Frosh Camp," a

weekend with faculty in the San Bernardino Mountains, before classes began, that I realized what science was all about. Science was everything (Caltech was not yet coed): individual challenge, self-aggrandizement, even human progress; but most important, the laboratory and computational equipment offered a vast range of toys.

Much of what we now call molecular biology was initiated at Caltech. Pauling's lectures in freshman chemistry on DNA, RNA, protein structure, and his theories on anesthesia with inert gases made me excited about studying biological science. Even today, learning about biochemical topics in freshman chemistry is unconventional.

**How did graduate school at MIT affect your research interests?**

MIT was never a serious consideration for graduate school;

one institute of technology was enough. I visited several other graduate programs, including those of liberal arts schools in Massachusetts. At one, several disgruntled graduate students, especially a violin-playing geneticist, suggested that I visit MIT because of the breadth of research interests in the biology department.

During the time I was a graduate student, all of the central dogma—information flow from DNA to protein, the genetic code, and its implementation—was being discovered. As I was finishing my Ph.D. at MIT, it was clear to me that there were two research directions I could take: macromolecular structure and interactions or cells and organisms. Given my simplistic, mechanical approach to biochemical research, cells and organisms seemed too complicated, so I chose structure and interactions. This was at a time before restriction enzymes and the whole array of methods using recombinant DNA that now allow dissection of any organism with Petri dishes and gels. I suspect that I chose to work on molecular structure because I could play with more research toys.

### How did your postdoctoral experience shape your career in biochemistry?

Most of my postdoctoral work was done at the Max-Planck Institute for Biophysical Chemistry in Göttingen. The most important thing I learned was that you need pure, active proteins or nucleic acids in order to do structure analysis. No matter how fancy your x-ray or NMR instrumentation, if you put "garbage in" you get "garbage out." First, you use Petri dishes and gels to verify genetics, obtain expression, and purify macromolecules; then you can do biophysical studies.

### What are your research interests in your labs at The University of Pennsylvania?

We use a combination of modern genetic techniques and high-field NMR spectroscopy to study the mechanism of gene expression. My research group is interested in the regulatory proteins that control gene expression. In particular, we are probing the structure and dynamics of repressors, activators, and RNA polymerase, both alone and bound to their specific DNA sites. Since we are isolating and manipulating both the DNA sites and the proteins at the gene level, all the techniques that we use are applicable to any cloned structural gene. The Penn Cancer Center's DNA synthesis service is housed in our laboratory.

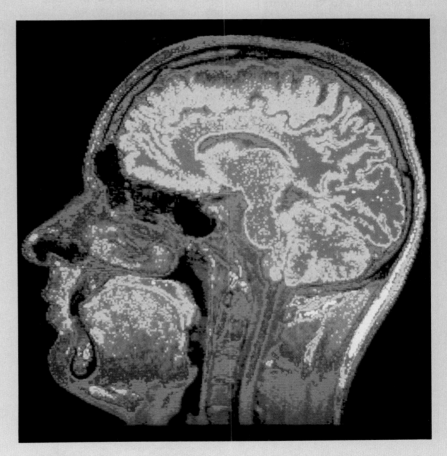

*Magnetic resonance image of a human head, showing the structures of a normal brain, airways, and facial tissues.*

### Can you explain how NMR works?

NMR detects differences in the energy levels of atomic nuclei when they are in magnetic fields. The individual atomic nucleus in a molecule, placed in a magnet, sees the field through its own and the neighboring atoms' electrons. Each nucleus in a molecule sees a different magnetic field—and thus has its own set of energy levels. NMR spectrometers not only can measure these energy level differences but also can identify individual atoms in the molecule. These are the "peaks" and groups of "peaks" seen in NMR spectra presented in organic chemistry texts. In addition, the spectrometer can slectively send energy into one nucleus and affect the energy levels of a neighboring nucleus. These effects follow defined rules of nuclear physics, without changing any of the electronic structure that holds the molecules together. Some effects travel though chemical bonds (spin-spin coupling) and some through space (nuclear Overhauser effect, NOE). NMR determines the geometric relationship of the atomic nuclei, and we learn how the molecule looks.

### How does NMR differ from MRI (magnetic resonance imaging) used in medical diagnosis?

NMR and MRI observe the same physical phenomenon. The molecular compositions of the different tissues, including $H_2O$, are quite different. If you place a whole person in a magnet, the same molecules in different tissues have different nuclear energy levels because they are in different molecular environments. The MRI instrument records not only the energy level differences but also the spatial position in the magnet and therefore in the organism. The detection electronics and computations of MRI are identical to those of molecular NMR. The presentation of results shows up as a color picture with areas outlined by anatomical shape. NMR works in the same electromagnetic frequency range as broadcast radio and television ($10^8$ Hz), so the energies involved are $10^{-10}$ that of x-rays ($10^{18}$ Hz) used for looking at bones or in computed tomographic (CT) scans. The name "MRI" was chosen to remove the word "nuclear," which many associate with radioactivity and thus danger.

### Generally, what can physical methods like NMR tell us about the nature of biological macromolecules?

Biochemists, in large part, work on an assumption that the structure of a molecule is related to function. To see the three-dimensional model of a molecule is the first step in this approach. NMR and x-ray crystallography are the only methods that yield comprehensive three-dimensional information. Macromolecules are not static and homogeneous. They are more like a partially wet sponge; parts are soft and parts are firm, and they are constantly undergoing thermal motion. NMR in solution allows one to see this heterogeneity and how it changes as the pH, ionic strength, and temperature are varied.

### Specifically, what can NMR tell us about how molecules interact with DNA?

Again, we assume we need to see the structure first. NMR will tell us the structure of the protein-DNA or DNA-ligand complex. DNA has

considerable sequence-dependent variation from the average helical structure. NMR analysis can tell us how this variation is exploited by specific interactions of the DNA with proteins.

## Will your research have any medical applications, such as in diagnosis or the development of new drugs?

We are back to the assumption that structure is important. Most drugs have been found by accident. The term "rational drug design" is now used a lot. To design a drug, we need to see where it has to go. First, we need to know the structure of the target biological macromolecule. This structure can be elucidated by using NMR. This target site can be an enzyme-active site, or it can be the surface between two proteins or a protein and nucleic acid.

## Is NMR the best technique for determining protein structure?

NMR has limitations. Complete structure determination of proteins by NMR measurements beyond 200 amino acids, which are multimeric, is at the present time not possible. We have been interested in *lac* repressor, a tetramer of 360 amino acids, for some time. Our NMR experiments are being used to probe large proteins, RNA polymerases, which in *Escherichia coli* have five subunits, each ranging from 300 to 1400 amino acids. We will not get detailed atomic structures, but useful overviews.

## What would you most want to tell students who are interested in a career in biochemical research?

Don't let introductory science courses discourage you. They are usually boring; unfortunately, the course content has not changed very much in the past 50 years.

However, learning science and doing it are very different. Take lab courses and do independent research with one of your professors. Chemistry and biochemistry can be fun; learn how to play with the toys in the lab. To test if you enjoy learning about science or biochemical science, read *Scientific American* or the Tuesday *New York Times.* Read business publications such as the *Wall Street Journal, Fortune,* or *Business Week,* since new products use new technology. Once you start reading regularly, you will not stop. New discoveries are being made constantly. There are many areas of science to participate in. For example, biochemical science is very broad, ranging through biomass conversion for fuel, agriculture, food processing, cosmetics, drug abuse, DNA forensics, medicine, and pharmaceuticals.

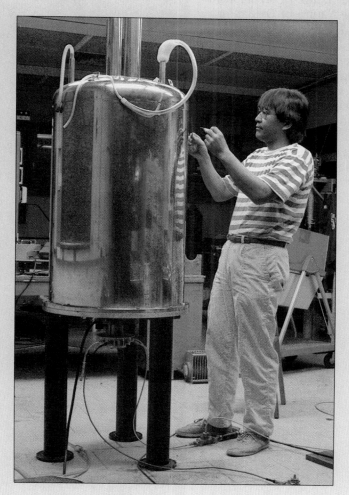

*Professor Ponzy Lu making measurements on an NMR superconducting magnet.*

# CHAPTER

## 12

# Metabolism and Electron Transfer

OUTLINE

12.1 The Nature of Metabolism
12.2 The Nature of Oxidation and Reduction
12.3 Biologically Important Oxidation–Reduction Reactions are Mediated by Coenzymes
12.4 Coupling of Production and Use of Energy
12.5 Metabolism Proceeds in Stages: The Role of Coenzyme A in Activation Processes

*Crystals of glucose, a key molecule in metabolism, viewed under polarized light.*

*The source of energy for life on earth is the sun. In photosynthesis, light splits $H_2O$ molecules in green plants, giving off oxygen into the atmosphere. The energy of sunlight is stored in starch granules in the plant cell. Animals (including humans) capture the nutritional energy of starch by converting it to glucose, which is oxidized to carbon dioxide and water. The energy extraction process takes place in a series of many small steps in which electron donors transfer energy to electron acceptors. This process, called oxidation–reduction reaction, is fundamental to deriving energy from the glucose molecule. The principal electron carriers are NADH (the reduced form of Nicotinamide adenine dinucleotide) and NAD$^+$, its oxidized form. NADH is oxidized when it transfers an electron to NAD$^+$; and NAD$^+$ is reduced to NADH when it accepts an electron. Oxidation of a glucose molecule is complete when two electrons and two protons join an oxygen atom to form $H_2O$. Energy generated in this reaction is conserved by transforming low energy ADP to higher energy ATP. The ADP–ATP system is like a very active checking account where deposits and withdrawals are in dynamic equilibrium. The energy from ATP is never used up, only transferred in the myriad chemical reactions in the cell that require energy.*

## 12.1
## THE NATURE OF METABOLISM

Up to now we have discussed some basic chemical principles and have also looked at the nature of molecules of which living cells are composed. We have yet to discuss the chemical reactions of biomolecules themselves. The reactions of the biomolecules in the cell constitute **metabolism.** The molecules of carbohydrates, fats, and proteins taken into an organism are processed in a variety of ways (Figure 12.1). The breakdown of larger molecules to smaller ones is called **catabolism.** Small molecules are used as the starting point of a variety of reactions to produce larger and more complex molecules, including proteins and nucleic acids; this process is called **anabolism.** Catabolism and anabolism are separate pathways; catabolic reactions are not simply the reverse of anabolic reactions. Metabolism is the biochemical basis of all life processes.

Catabolism is an oxidative process that releases energy; anabolism is a reductive process that requires energy. We shall need several chapters to explore some of the implications of this statement. In this chapter we shall discuss the nature of oxidation and reduction (electron-transfer reactions) and their relation to the use of energy by living cells.

## 12.2
## THE NATURE OF OXIDATION AND REDUCTION

Oxidation–reduction reactions are those in which electrons are transferred from a donor to an acceptor. **Oxidation** is the loss of electrons, and **reduction** is the gain of electrons. The substance that loses electrons (the

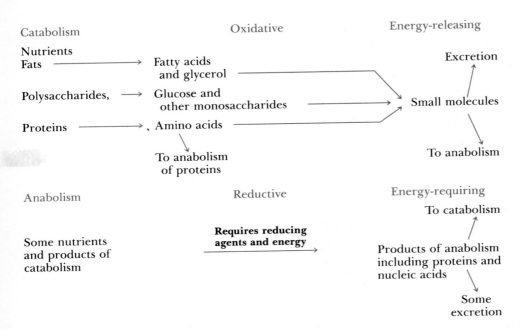

FIGURE 12.1    A comparison of catabolism and anabolism.

electron donor), the one that is oxidized, is called the **reducing agent** or reductant. The substance that gains electrons (the electron acceptor), the one that is reduced, is called the **oxidizing agent** or oxidant. The presence of both an oxidizing agent and a reducing agent is necessary for the transfer of electrons (an oxidation–reduction reaction) to take place. Such reactions are also referred to as *redox* reactions.

An example of an oxidation–reduction reaction is the one that occurs when a strip of metallic zinc is placed in an aqueous solution containing copper ions. This reaction does not occur in living organisms, although both zinc and copper ions play a role in life processes. In this reaction, though, it is fairly easy to follow where the electrons are going. It is not always quite as easy to keep track of the details in some biological redox reactions, so this comparatively simple reaction is a good place to start our discussion of electron transfer. The experimental observation is that the zinc metal disappears and zinc ions go into solution, while copper ions are removed from the solution and copper metal is deposited. The equation for this reaction is

$$Zn(s) + Cu^{2+}(aq) \longrightarrow Zn^{2+}(aq) + Cu(s)$$

The notation (s) refers to a solid and (aq) to a solute in aqueous solution.

In the reaction between zinc metal and copper ion, the Zn lost two electrons to become $Zn^{2+}$ ion and was oxidized. A separate equation can be written for this part of the overall reaction, and it is called the **half reaction** of oxidation:

$$Zn \longrightarrow Zn^{2+} + 2e^-$$

Zn is the reducing agent (loses electrons, electron donor, is oxidized).

Likewise, the $Cu^{2+}$ ion gained two electrons to form Cu and was reduced. An equation can also be written for this part of the overall reaction; this part is called the half reaction of reduction.

$$Cu^{2+} + 2e^- \longrightarrow Cu$$

$Cu^{2+}$ is the oxidizing agent (gains electrons, electron acceptor, is reduced).

If the two equations for the half reactions are combined, the result is an equation for the overall reaction,

$$Zn \longrightarrow Zn^{2+} + 2e^-$$
$$\underline{Cu^{2+} + 2e^- \longrightarrow Cu}$$
$$Zn + Cu^{2+} \longrightarrow Zn^{2+} + Cu$$

This reaction is a particularly clear example of electron transfer; it will be useful to keep these basic principles in mind when we look at the flow of electrons in the more complex redox reactions of aerobic metabolism.

## 12.3
## BIOLOGICALLY IMPORTANT OXIDATION–REDUCTION REACTIONS ARE MEDIATED BY COENZYMES

Oxidation–reduction reactions are discussed at length in textbooks of general and inorganic chemistry, but the oxidation of nutrients by living organisms to provide energy requires its own special treatment. The description of redox reactions in terms of oxidation numbers, widely used with inorganic compounds, can be used to deal with the oxidation of carbon-containing molecules. However, the discussion will be more pictorial and easier to follow if we write equations for the half reactions, and then concentrate on the functional groups of the reactants and products and on the number of electrons transferred. We shall use the latter approach in our discussion. An example is the oxidation half reaction for the conversion of ethanol to acetaldehyde.

The half reaction of oxidation of
ethanol to acetaldehyde

$$\overset{\displaystyle H}{\underset{\displaystyle H}{CH_3 - \overset{\displaystyle ..}{\underset{\displaystyle ..}{C}} \!:\! \overset{..}{\underset{..}{O}} \!:\! H}} \rightleftharpoons \overset{\displaystyle H}{CH_3 - \overset{\displaystyle ..}{C} \!::\! \overset{..}{\underset{..}{O}} \!:\! + 2H^+ + 2e^-}$$

Ethanol (12 electrons in          Acetaldehyde (10 electrons
groups involved in reaction)     in groups involved in reaction)    (12.1)

Writing the Lewis electron-dot structures for the functional groups involved in the reaction helps in keeping track of the electrons being transferred. In the oxidation of ethanol, there are 12 electrons in the part of the ethanol molecule involved in the reaction, and 10 electrons in the corresponding part of the acetaldehyde molecule; two electrons are transferred to an electron acceptor (an oxidizing agent). This type of bookkeeping is useful in dealing with biochemical reactions. Many biological oxidation reactions, like this example, are accompanied by the transfer of a proton ($H^+$). The oxidation half reaction has been written as a reversible reaction because the occurrence of oxidation or reduction depends on what other reagents are present.

Another example of an oxidation half reaction is that for the conversion of NADH, the reduced form of *nicotinamide adenine dinucleo-tide*, to the oxidized form, $NAD^+$. This substance is an important coenzyme in many reactions. The structure of NADH is given in Figure 12.2; the nicotinamide portion, the functional group involved in the reaction, is indicated in the structure. Nicotinamide is a derivative of nicotinic acid (also called niacin), one of the B-complex vitamins. A similar compound is NADPH (for which the oxidized form is $NADP^+$). It differs from NADH by having an additional phosphate group; the site of attachment of this phosphate group is also indicated in Figure 12.2. To simplify writing the equation, only the nicotinamide ring will be shown explicitly, with the rest of the molecule designated as R. The two electrons that are lost can be considered to come from the bond between carbon and the lost hydrogen, with the nitrogen lone pair electrons becoming involved in a bond.

**FIGURE 12.2**   (a) The structure of NADH. (b) Space-filling model of $NAD^+$, the oxidized form of NADH.

The half reaction of oxidation of NADH to NAD$^+$

$$H-C \overset{\overset{\displaystyle H \quad H}{\diagup}}{\underset{\underset{\displaystyle N}{\diagdown}}{C}} \quad -CONH_2 \quad \rightleftharpoons \quad H-C \overset{\overset{\displaystyle H}{\diagup}}{\underset{\underset{\displaystyle N^+}{\diagdown}}{C}} \quad -CONH_2 + H^+ + 2e^- \quad (12.2)$$

The equations for both examples given here have been written as oxidation half reactions. If ethanol and NADH were mixed in a test tube, no reaction could take place because there is no electron acceptor. If, however, NADH were mixed with acetaldehyde, which is an oxidized species, a transfer of electrons could take place, producing ethanol and NAD$^+$.

$$NADH + H^+ + \underset{\textbf{Acetaldehyde}}{CH_3CHO} \longrightarrow NAD^+ + \underset{\textbf{Ethanol}}{CH_3CH_2OH}$$

Such a reaction does take place in some organisms as the last step of alcoholic fermentation.

Another important electron acceptor is FAD (*f*lavin *a*denine *d*inucleotide), which is the oxidized form (Figure 12.3). The reduced form is

**FIGURE 12.3**   (a) The structure of FAD, the oxidized form of *f*lavin *a*denine *d*inucleotide. (b) Space-filling model of FAD.

(a)

(b)

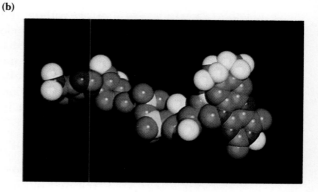

FADH$_2$, a form of notation that explicitly recognizes that protons (hydrogen ions) as well as electrons are accepted by FAD.

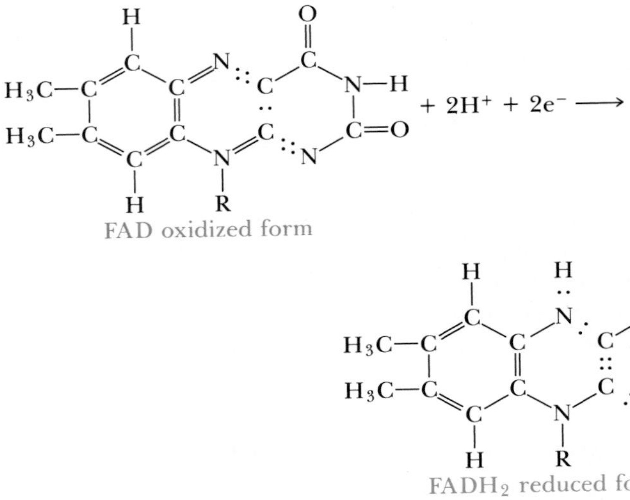

FAD oxidized form

$+ 2H^+ + 2e^- \longrightarrow$

FADH$_2$ reduced form                  (12.3)

The half reaction of reduction of FAD to FADH$_2$

The structures shown in this equation again point out the electrons that are transferred in the reaction. There are several other coenzymes that contain the flavin group; they are derived from the vitamin riboflavin (vitamin B$_2$).

Oxidation of nutrients to provide energy for an organism cannot take place without reduction of some electron acceptor (*i.e.*, an oxidizing agent). The ultimate electron acceptor in aerobic oxidation is oxygen; we shall meet intermediate electron acceptors as we discuss metabolic processes. Reduction of metabolites plays an important role in living organisms in anabolic processes. Important biomolecules are synthesized in organisms by many reactions in which a metabolite is reduced while the reduced form of a coenzyme is oxidized.

## 12.4
## COUPLING OF PRODUCTION AND USE OF ENERGY

Another important question about metabolism is, "How is the energy released by the oxidation of nutrients trapped and used?" This energy cannot be used directly; it must be shunted into an easily accessible form of chemical energy. In Section 8.2 we saw that several phosphorus-containing compounds such as ATP can be hydrolyzed easily, and that the reaction releases energy. Formation of ATP is intimately linked with the release of energy from oxidation of nutrients. The **coupling** of reactions that produce energy and reactions that require energy is a central feature in the metabolism of all organisms.

The phosphorylation of ADP (*a*denosine *di*phosphate) to produce ATP (*a*denosine *tri*phosphate) requires energy, which can be supplied by the oxidation of nutrients.

$$ADP + P_i + H^+ \longrightarrow ATP + H_2O \qquad \Delta G^{\circ\prime} = 30.5 \text{ kJ mol}^{-1} = 7.3 \text{ kcal mol}^{-1}$$

or in structural form,

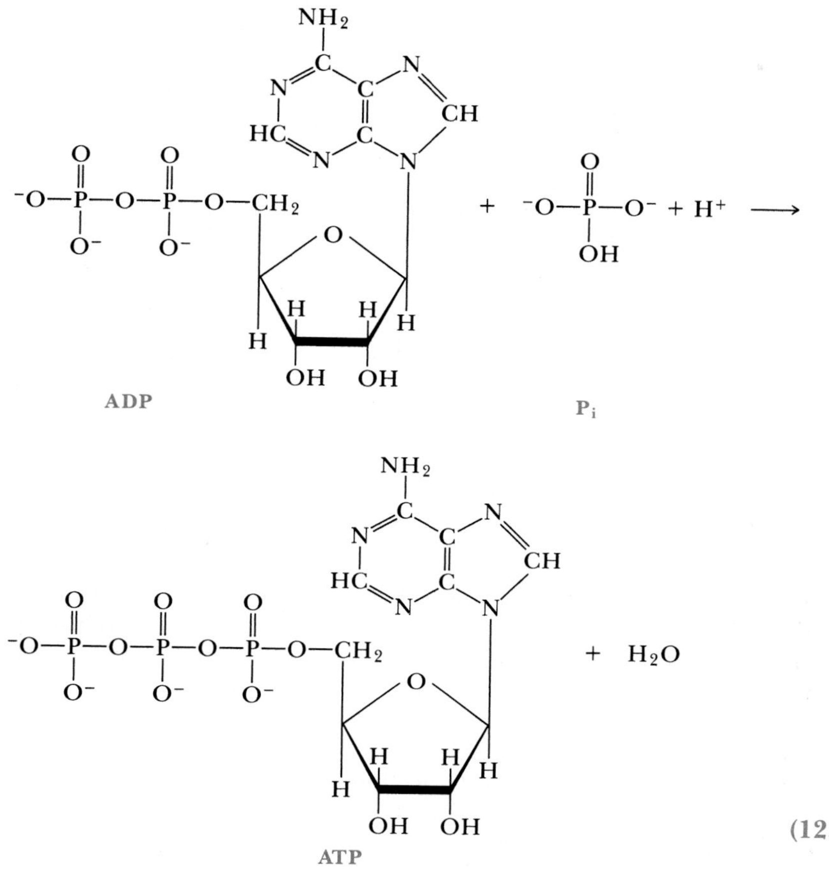

ADP

$P_i$

ATP

$$+ \ H_2O$$

(12.4)

The forms of ADP and ATP shown in the structural equation are in their ionization states for pH 7. The symbol $P_i$ for phosphate ion comes from its name in biochemical jargon, "inorganic phosphate." Note that there are four negative charges on ATP and three on ADP; electrostatic repulsion makes ATP less stable than ADP. Energy must be expended to put an additional negatively charged phosphate group on ADP by forming a covalent bond to the phosphate group being added. The $\Delta G^{\circ\prime}$ for the reaction refers to the usual biochemical convention of pH 7 as the standard state for hydrogen ion (Section 8.4).

The reverse reaction, the hydrolysis of ATP to ADP and phosphate ion, releases 30.5 kJ mol$^{-1}$ (7.3 kcal mol$^{-1}$) when energy is needed:

$$ATP + H_2O \longrightarrow ADP + P_i + H^+ \qquad \Delta G^{\circ\prime} = -30.5 \text{ kJ mol}^{-1} = -7.3 \text{ kcal mol}^{-1}$$

The bond that is hydrolyzed when this reaction takes place is sometimes called a "high-energy bond," a shorthand terminology for a reaction in

**TABLE 12.1    Free Energies of Hydrolysis of Selected Organophosphates**

| | $\Delta G^{\circ\prime}$ (mol$^{-1}$) | |
| COMPOUND | kJ | kcal |
| --- | --- | --- |
| Phosphoenolpyruvate | −61.9 | −14.8 |
| Carbamoyl phosphate | −51.4 | −12.3 |
| Creatine phosphate | −43.1 | −10.3 |
| Acetyl phosphate | −42.2 | −10.1 |
| ATP (to ADP) | −30.5 | −7.3 |
| Glucose 1-phosphate | −20.9 | −5.0 |
| Glucose 6-phosphate | −12.5 | −3.0 |
| Glycerol 3-phosphate | −9.7 | −2.3 |

which hydrolysis of a specific bond releases a useful amount of energy. There are numerous organophosphate compounds with "high-energy bonds" that play a role in metabolism, but ATP is by far the most important (Table 12.1). In some cases the free energy of hydrolysis of organophosphates is high enough to drive the phosphorylation of ADP.

The energy of hydrolysis of ATP is not stored energy, just as an electric current does not represent stored energy. Both ATP and electric current must be produced when they are needed, by organisms or by a power plant as the case may be. The cycling of ATP and ADP in metabolic processes is a way of shunting energy from its production (by oxidation of nutrients) to use when needed (in processes such as biosynthesis of essential compounds or muscle contraction). The oxidation processes take place when the organism needs the energy that can be generated by the hydrolysis of ATP. The usual way of storing chemical energy is in the form of fats and carbohydrates, which are metabolized as needed. (Small biomolecules such as creatine phosphate can also serve as vehicles for storing chemical energy.) The energy that must be supplied for many endergonic reactions that take place in life processes comes directly from the hydrolysis of ATP and indirectly from oxidation of nutrients, which produces the energy needed to phosphorylate ADP to ATP (Figure 12.4).

At this point we shall look at some biological reactions that release energy, and see how some of that energy is used to phosphorylate ADP to ATP. The multistep conversion of glucose to lactate ions is an exergonic and anaerobic process. Two molecules of ADP are phosphorylated to ATP for each molecule of glucose metabolized. The basic reactions are the production of lactate, which is exergonic,

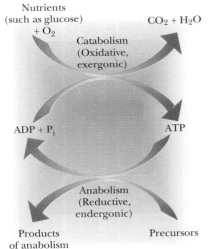

$$\text{Glucose} \longrightarrow 2 \text{ lactate ions} \qquad \Delta G^{\circ\prime} = -184.5 \text{ kJ mol}^{-1} = -44.1 \text{ kcal mol}^{-1}$$

and phosphorylation of two moles of ADP for each mole of glucose, which is endergonic.

$$2 \text{ ADP} + 2 \text{ P}_i \longrightarrow 2 \text{ ATP} \qquad \Delta G^{\circ\prime} = 61.0 \text{ kJ} = 14.6 \text{ kcal}$$

(In the interest of simplicity we shall write the equation for phosphorylation of ADP in terms of ADP, $P_i$, and ATP only.) The overall reaction is

**FIGURE 12.4**  The role of ATP as energy currency in processes that require energy and processes that use energy.

$$\text{Glucose} + 2\ \text{ADP} + 2\ \text{P}_i \longrightarrow 2\ \text{lactate ions} + 2\ \text{ATP}$$

$$\Delta G^{\circ\prime}\ \text{overall} = -184.5 + 61.0 = -123.5\ \text{kJ mol}^{-1} = -29.5\ \text{kcal mol}^{-1}$$

Not only can we add the two chemical reactions to obtain an equation for the overall reaction, we can also add the free energy changes for the two reactions to find the overall free energy change. The exergonic reaction provides energy, which drives the endergonic reaction. This phenomenon is called **coupling.** The percentage of the energy released that is used to phosphorylate ADP is the efficiency of energy use in anaerobic metabolism; it is $(61.0/184.5) \times 100$, or about 33 percent.

The breakdown of glucose under aerobic conditions goes further than under anaerobic conditions. The end products of aerobic oxidation are six molecules of carbon dioxide and six molecules of water for each molecule of glucose. Up to 38 molecules of ADP can be phosphorylated to ATP when one molecule of glucose is broken down completely to carbon dioxide and water. The exergonic reaction for the complete oxidation of glucose is

$$\text{Glucose} + 6\ \text{O}_2 \longrightarrow 6\ \text{CO}_2 + 6\ \text{H}_2\text{O} \quad \Delta G^{\circ\prime} = -2867\ \text{kJ mol}^{-1} = -685.9\ \text{kcal mol}^{-1}$$

The endergonic reaction for phosphorylation is

$$38\ \text{ADP} + 38\ \text{P}_i \longrightarrow 38\ \text{ATP} \qquad \Delta G^{\circ\prime} = 1159\ \text{kJ} = 277.3\ \text{kcal}$$

The net reaction is

$$\text{Glucose} + 6\ \text{O}_2 + 38\ \text{ADP} + 38\ \text{P}_i \longrightarrow 6\ \text{CO}_2 + 6\ \text{H}_2\text{O} + 38\ \text{ATP}$$

$$\Delta G^{\circ\prime} = -2867 + 1159 = -1708\ \text{kJ mol}^{-1} = -408.6\ \text{kcal mol}^{-1}$$

(Note that once again we add the two reactions and their respective free energy changes to obtain the overall reaction and its free energy change.) The efficiency of aerobic oxidation of glucose is $(1159/2867) \times 100$, about 40 percent. More ATP is produced by the coupling process in aerobic oxidation of glucose than by the coupling process in anaerobic oxidation. The hydrolysis of ATP produced by breakdown (aerobic or anaerobic) of glucose can be coupled to endergonic processes, such as muscle contraction in exercise. As any jogger or long distance swimmer knows, aerobic metabolism involves large quantities of energy, processed in highly efficient fashion. We have now seen two examples of coupling of exergonic and endergonic processes, aerobic oxidation of glucose and anaerobic fermentation of glucose, involving different amounts of energy.

## 12.5
## METABOLISM PROCEEDS IN STAGES: THE ROLE OF COENZYME A IN ACTIVATION PROCESSES

The metabolic oxidation of glucose that we saw in the last section does not take place in one step. The anaerobic breakdown of glucose requires many steps; the complete aerobic oxidation of glucose to carbon dioxide and

**(a)**

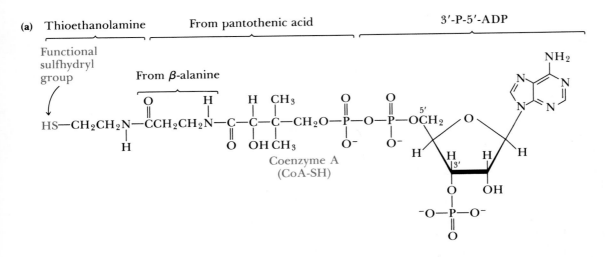

**(b)**

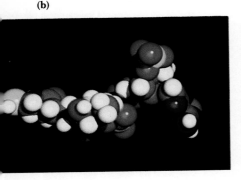

FIGURE 12.5    (a) The structure of coenzyme A. (b) Space-filling model of coenzyme A.

water has still more steps. One of the most important points about the multistep nature of all metabolic processes, including the oxidation of glucose, is that the many stages allow for efficient production and use of energy. The electrons produced by the oxidation of glucose are passed along to oxygen, the ultimate electron acceptor, by intermediate electron acceptors. Many of the intermediate stages of the oxidation of glucose are coupled to ATP production by phosphorylation of ADP.

One kind of step frequently encountered in metabolism is the process of **activation**. In a reaction of this sort, a metabolite (a component of a metabolic pathway) is bonded to some other molecule such as a coenzyme, and the free energy change for breaking this new bond is negative. In other words, the next reaction in the metabolic pathway is exergonic. For example, if A is the metabolite and it reacts with substance B to give AB, the following series of reactions might take place.

$$A + Coenzyme \longrightarrow A\text{—}Coenzyme \qquad \text{Activation step}$$

$$A\text{—}Coenzyme + B \longrightarrow$$
$$AB + Coenzyme \qquad \Delta G^{\circ\prime} < 0 \text{ (exergonic reaction)}$$

The formation of a more reactive substance in this fashion is called activation. There are many examples of activation in metabolic processes. We can look at one of the most useful of them right now. It involves forming a covalent bond to a compound known as coenzyme A (CoA).

The structure of CoA is complex. It consists of several smaller components linked together covalently (Figure 12.5). One part is 3′-P-5′-ADP, a derivative of adenosine with phosphate groups esterified to the sugar as shown in the structure. Another part is derived from the vitamin pantothenic acid, and the part of the molecule involved in activation reactions contains a thiol group. In fact, coenzyme A is frequently written as CoA-SH to emphasize that the thiol group is the reactive portion of the molecule. For example, carboxylic acids form thioester linkages with CoA-SH. The metabolically active form of a carboxylic acid is the

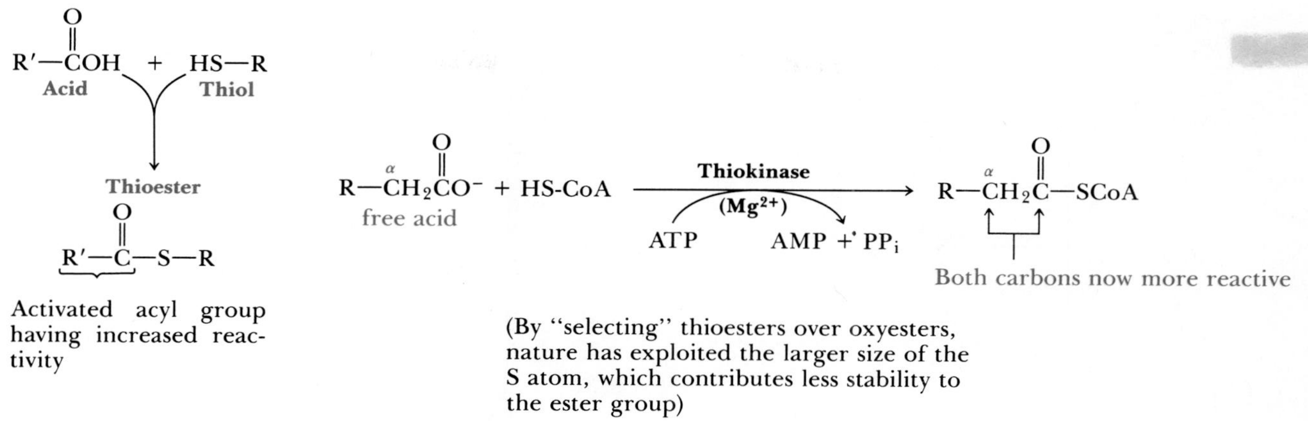

$$R'-\overset{\overset{\displaystyle O}{\|}}{C}OH \ + \ HS-R$$

Acid               Thiol

Thioester

$$R'-\overset{\overset{\displaystyle O}{\|}}{C}-S-R$$

Activated acyl group having increased reactivity

$$R-\overset{\alpha}{CH_2}\overset{\overset{\displaystyle O}{\|}}{C}O^- \ + \ HS\text{-}CoA \ \xrightarrow[\substack{(Mg^{2+}) \\ ATP \qquad AMP +' PP_i}]{\text{Thiokinase}} \ R-\overset{\alpha}{CH_2}\overset{\overset{\displaystyle O}{\|}}{C}-SCoA$$

free acid

Both carbons now more reactive

(By "selecting" thioesters over oxyesters, nature has exploited the larger size of the S atom, which contributes less stability to the ester group)

**FIGURE 12.6** The formation of thioester bonds between CoA and carboxylic acids.

corresponding acyl-CoA thioester, where the thioester linkage is a "high-energy" bond (Figure 12.6). Acetyl-CoA is a particularly important metabolic intermediate; other acyl-CoA species figure prominently in lipid metabolism.

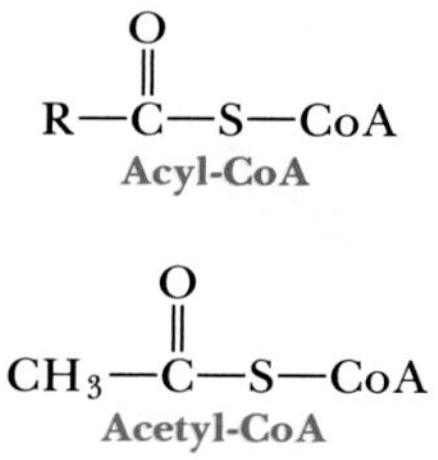

$$R-\overset{\overset{\displaystyle O}{\|}}{C}-S-CoA$$

Acyl-CoA

$$CH_3-\overset{\overset{\displaystyle O}{\|}}{C}-S-CoA$$

Acetyl-CoA

(12.5)

Parenthetically, the important coenzymes we have met in this chapter—NAD$^+$, NADP$^+$, FAD, and coenzyme A—share an important structural feature: all these substances contain ADP. In NADP$^+$, there is an additional phosphate group at the 2' position of the ribose group of ADP. In CoA, the additional phosphate group is at the 3' position.

Like catabolism, anabolism proceeds in stages. Unlike catabolism, which releases energy, anabolism requires energy. The ATP produced as a result of catabolism is hydrolyzed to release the needed energy. Reactions in which metabolites are reduced are a part of anabolism; they require reducing agents such as NADH, NADPH, and FADH$_2$, all of which are the reduced forms of coenzymes we met in this chapter. In their oxidized form these coenzymes serve as the intermediate oxidizing agents needed in catabolism. In their reduced form the same coenzymes provide the "reducing power" needed for the anabolic processes of biosynthesis; in this case the coenzymes act as reducing agents.

We are now in a position to expand on our earlier statements on the nature of anabolism and catabolism. We can show an outline of metabolic pathways that explicitly points out two important features of metabolism: first, the role of electron transfer, and second, the role of ATP in the

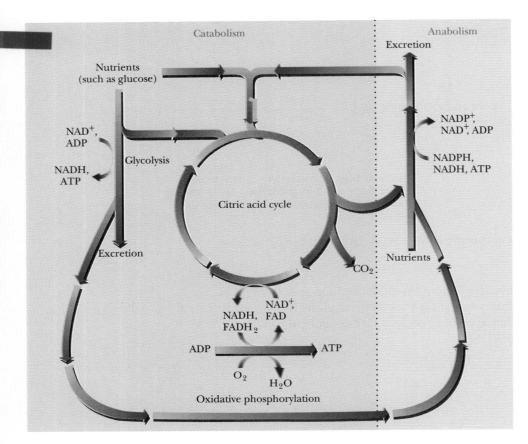

**FIGURE 12.7** The role of electron transfer and ATP production in metabolism. $NAD^+$, FAD, and ATP are constantly recycled.

release and utilization of energy (Figure 12.7). Even this more extended outline is a very general one; the more important specific pathways have been studied in detail, and some are still the subject of active research. We shall discuss some of the most important metabolic pathways in the remainder of this text.

The fermentation of glucose, a process called **glycolysis,** is a series of reactions in which glucose undergoes oxidation anaerobically, producing compounds that contain three carbon atoms. The **citric acid cycle** is a series of reactions that are a part of aerobic metabolism and in which the oxidation of glucose gives rise to carbon dioxide. The citric acid cycle is **amphibolic:** it plays a role both in catabolism and in anabolism. **Electron transport to oxygen** is the final (and aerobic) stage of glucose metabolism. Also, at this stage the process of **oxidative phosphorylation** (coupling of aerobic oxidation of nutrients and phosphorylation of ADP to ATP) is particularly important, because it provides an efficient mechanism for the organism to use the energy released by the oxidation of foodstuffs. Most of the ATP production that accompanies the complete oxidation of glucose occurs as a result of oxidative phosphorylation. The metabolism of lipids is an important source of energy in catabolism and provides a form of energy storage in anabolism. The anabolic process of **photosynthesis** is essential to the well-being of all organisms on this planet (except bacteria that are obligate anaerobes). Many details of the biosynthesis of nucleic acids and proteins are well understood, but other points, particularly about the details of expression of the genetic code, are the subject of active research.

## SUMMARY

The reactions of the biomolecules in the cell constitute metabolism. The breakdown of larger molecules to smaller ones is called catabolism. The reaction of small molecules to produce larger and more complex molecules is called anabolism. Catabolism and anabolism are separate pathways; catabolic reactions are not simply the reverse of anabolic reactions. Metabolism is the biochemical basis of all life processes.

Catabolism is an oxidative process that releases energy; anabolism is a reductive process that requires energy. Oxidation–reduction (redox) reactions are those in which electrons are transferred from a donor to an acceptor. Oxidation is the loss of electrons, and reduction is the gain of electrons. Many biologically important redox reactions involve coenzymes such as NADH and $FADH_2$.

The coupling of reactions that produce energy and reactions that require energy is a central feature in the metabolism of all organisms. In catabolism, oxidative reactions are coupled to the endergonic production of ATP by phosphorylation of ADP. Aerobic metabolism is a more efficient means of making use of the chemical energy of nutrients than is anaerobic metabolism. In anabolism, the exergonic hydrolysis of the "high-energy" bond of ATP releases the energy needed to drive endergonic reductive reactions.

Metabolism proceeds in stages, and the many stages allow for the efficient production and use of energy. The process of activation, producing "high-energy" intermediates, occurs in many metabolic pathways. The formation of thioester linkages by reaction of carboxylic acids with coenzyme A is an example of the activation process.

## EXERCISES

1. The following half reactions play an important role in metbolism.

$$\frac{1}{2} O_2 + 2\ H^+ + 2e^- \longrightarrow H_2O$$

$$NADH + H^+ \longrightarrow NAD^+ + 2\ H^+ + 2e^-$$

Which of these two is a half reaction of oxidation? Which one is a half reaction of reduction? Write the equation for the overall reaction. Which reagent is the oxidizing agent (electron acceptor)? Which reagent is the reducing agent (electron donor)?

2. All the organophosphate compounds listed in Table 12.1 undergo hydrolysis reactions in the same way as ATP. The following equation illustrates the situation for glucose 1-phosphate.

$$\text{Glucose 1-phosphate} + H_2O \longrightarrow \text{glucose} + P_i$$

$$\Delta G^{\circ\prime} = -20.9\ \text{kJ mol}^{-1}$$

Using the free energy values in Table 12.1, predict whether the following reactions will proceed in the direction written and calculate the $\Delta G^{\circ\prime}$ for the reaction, assuming that the reactants are initially present in a 1:1 molar ratio.
(a) ATP + creatine $\longrightarrow$ creatine phosphate + ADP
(b) ATP + glycerol $\longrightarrow$ glycerol 3-phosphate + ADP
(c) ATP + pyruvate $\longrightarrow$ phosphoenolpyruvate + ADP
(d) ATP + glucose $\longrightarrow$ glucose 6-phosphate + ADP

3. Short periods of exercise such as sprints are characterized by lactic acid production and the condition known as oxygen debt. Comment on this point in light of the material we have discussed in this chapter.

4. Using the data in Table 12.1, calculate the value of $\Delta G^{\circ\prime}$ for the reaction

Creatine phosphate + glycerol $\longrightarrow$
creatine + glycerol 3-phosphate

Hint: this reaction proceeds in stages. ATP is formed in the first step, and the phosphate group is transferred from ATP to glycerol in the second step.

5. Calculate the value of $\Delta G^{\circ\prime}$ for the following reaction using information from Table 12.1.

Glucose 1-phosphate $\longrightarrow$ glucose 6-phosphate

6. Show that the hydrolysis of ATP to AMP and 2 $P_i$ releases the same amount of energy by either of the two following pathways.

*Pathway 1*

$$ATP + H_2O \longrightarrow ADP + P_i$$

$$ADP + H_2O \longrightarrow AMP + P_i$$

*Pathway 2*

$$ATP + H_2O \longrightarrow AMP + PP_i$$
$$\text{(pyrophosphate)}$$

$$PP_i + H_2O \longrightarrow 2\ P_i$$

7. The standard free energy change for the reaction

Arginine + ATP $\longrightarrow$ phosphoarginine + ADP

is $+1.7$ kJ mol$^{-1}$. From this information and that in Table 12.1, calculate the $\Delta G^{\circ\prime}$ for the reaction

Phosphoarginine + $H_2O$ $\longrightarrow$ arginine + $P_i$

8. There is a reaction in carbohydrate metabolism in which glucose 6-phosphate reacts with NADP$^+$ to give 6-phosphoglucono-δ-lactone and NADPH. (Recall the conversion of sugars to lactones from Section 6.2.)

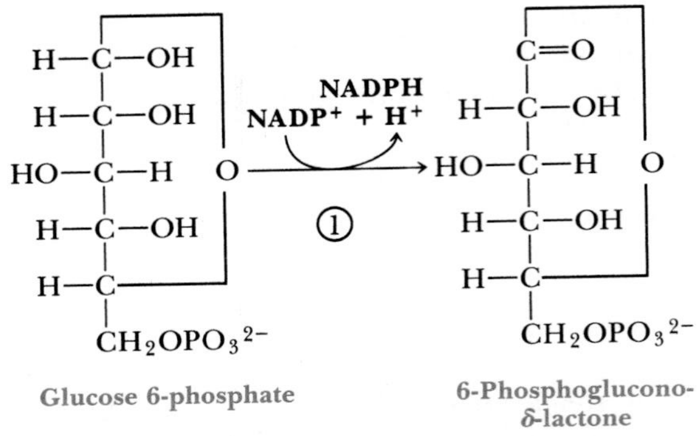

Glucose 6-phosphate        6-Phosphoglucono-
                           δ-lactone

In this reaction, which substance is oxidized, and which one is reduced? Which substance is the oxidizing agent, and which one is the reducing agent?

9. There is a reaction in which succinate reacts with FAD$^+$ to give fumarate and FADH$_2$.

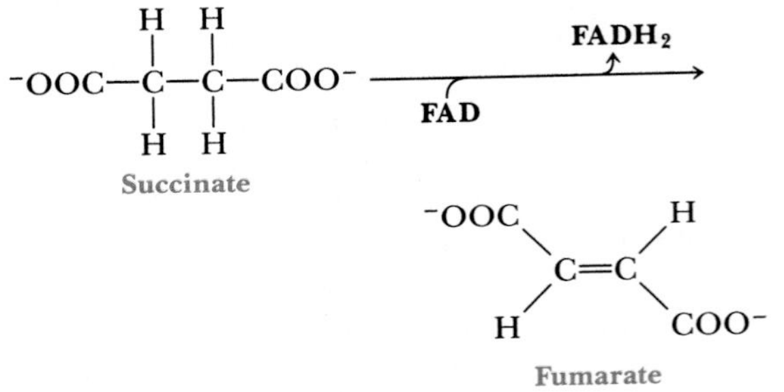

Succinate

Fumarate

In this reaction, which substance is oxidized, and which one is reduced? Which substance is the oxidizing agent, and which one is the reducing agent?

10. What structural feature do NAD$^+$, NADP$^+$, and FAD have in common?

## ANNOTATED BIBLIOGRAPHY

Edsall, J.T., and H. Gutfreund. *Biothermodynamics: The Study of Biochemical Processes at Equilibrium.* New York: Wiley, 1983. [A short book with useful examples.]

Fasman, G.D., ed. *Handbook of Biochemistry and Molecular Biology,* 3rd. ed. Sec. D, *Physical and Chemical Data.* Cleveland: CRC Press, 1976. [Volume 1 contains data on the free energies of hydrolysis of many important compounds, especially organophosphates.]

Hinkle, P.C., and R.E. McCarty. How Cells Make ATP. *Sci. Amer.* **238** (3), 104–125 (1978). [Getting old, but a particularly good article on energy coupling.]

Two standard multivolume references cover specific aspects of metabolism in detail. One of these, *The Enzymes,* 3rd ed. (P. D. Boyer, ed. New York: Academic Press) is a series that has been in production since 1970. The other, *Comprehensive Biochemistry* (M. Florkin and E.H. Stotz, eds. New York: Elsevier) has been in production since 1962.

# Glycolysis: Anaerobic Oxidation of Glucose

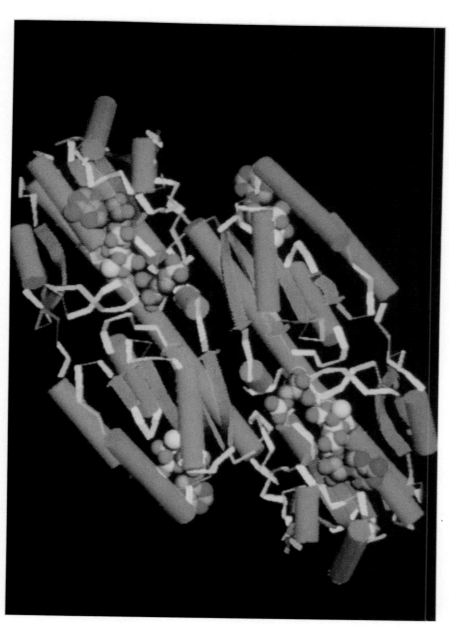

*The structure of phosphofructokinase, a regulatory enzyme in glycolysis, showing bound substrate.*

*The complete oxidation of glucose to carbon dioxide and water (involving glycolysis, the citric acid cycle, and oxidative phosphorylation) yields the energy equivalent of 38 molecules of ATP. The first stage of glucose metabolism, glycolysis, an anaerobic process, yields only two molecules of ATP. Nevertheless, in sudden bursts of energy, such as the hundred yard dash, the body temporarily prefers glycolysis. In sustained activity, glucose, stored as glycogen in muscle cells, may be quickly depleted. In such cases, pyruvate, the end product of glycolysis, can be converted to lactate and exported from muscle to the liver. There, with the help of its energy reserves of ATP, the liver can convert lactate to "new glucose" and return the glucose to muscle for another round of glycolysis. This process, called gluconeogenesis, is essentially (with some exceptions) glycolysis in reverse. New glucose can also be made from pyruvate combined with molecules from the citric acid cycle. This "home made" glucose produced by gluconeogenesis is available to supply glucose to the brain, which has little reserve of its own. The brain requires a steady stream of energy in the form of glucose (usually its only fuel), no matter what level of mental activity, awake or asleep.*

## 13.1
## AN OVERVIEW OF THE GLYCOLYTIC PATHWAY

In the **glycolytic pathway,** one molecule of glucose (a six-carbon compound) gives rise to two molecules of pyruvate ion (a three-carbon compound). The glycolytic pathway (also called the Embden-Meyerhoff pathway) involves many steps, including the reactions in which metabolites of glucose are oxidized; there are other steps as well. Each reaction in the pathway is catalyzed by an enzyme specific for that reaction. There are two reactions in the pathway in which one molecule of ATP is hydrolyzed for each molecule of glucose metabolized; the energy released makes coupled endergonic reactions possible. There are also two reactions in which two molecules of ATP are produced by phosphorylation of ADP for each molecule of glucose, giving a total of four ATP molecules produced. Comparing the number of ATP molecules used by hydrolysis and the number produced, there is a net gain of two ATP molecules for each molecule of glucose processed in glycolysis (Section 12.4).

When pyruvate is formed, it can have one of several fates (Figure 13.1). In aerobic metabolism, pyruvate loses carbon dioxide; the remaining two carbon atoms become linked to coenzyme A (Section 12.5) as an acetyl group to form acetyl-CoA, which then enters the citric acid cycle (Chapter 14). There are two fates for pyruvate in anaerobic metabolism. In organisms capable of alcoholic fermentation, pyruvate loses carbon dioxide, this time producing acetaldehyde, which in turn is reduced to produce

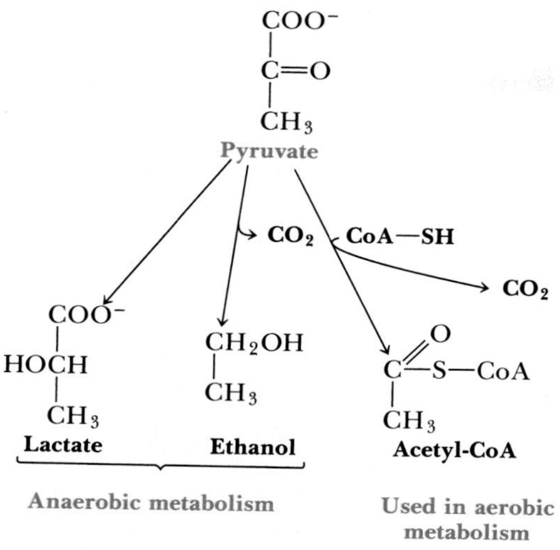

$$COO^-$$
$$|$$
$$C=O$$
$$|$$
$$CH_3$$

**Pyruvate**

**FIGURE 13.1**   The alternate metabolic fates of pyruvate.

$CO_2$    CoA—SH

$CO_2$

$$COO^-$$
$$|$$
$$HOCH$$
$$|$$
$$CH_3$$

**Lactate**

$$CH_2OH$$
$$|$$
$$CH_3$$

**Ethanol**

$$C\!\!\!\diagup^{O}_{S—CoA}$$
$$|$$
$$CH_3$$

**Acetyl-CoA**

**Anaerobic metabolism**          **Used in aerobic metabolism**

ethanol (Section 13.5). The more common fate of pyruvate in anaerobic metabolism is reduction to lactate. In this chapter we shall concentrate on the conversion of glucose to lactate, called **anaerobic glycolysis** to distinguish it from conversion of glucose to pyruvate, which is simply called glycolysis. The anaerobic breakdown of glucose to lactate can be summarized as follows:

Glucose (six carbon atoms) $\longrightarrow$ 2 lactate (three carbon atoms)

$$\frac{2\ ATP + 4\ ADP + 2\ P_i \longrightarrow 2\ ADP + 4\ ATP\ \text{(phosphorylation)}}{\text{Glucose} + 2\ ADP + 2\ P_i \longrightarrow 2\ \text{lactate} + 2\ ATP\ \text{(net reaction)}}$$

Figure 13.2 shows the reaction sequence with the names of the compounds. All sugars in the pathway have the D configuration; we shall assume this point throughout this chapter.

## A Summary of the Reactions of Glycolysis

**Step 1.** **Phosphorylation** of glucose to give glucose 6-phosphate (ATP is the source of the phosphate group). (See Equation 13.1, p. 298.)

Glucose + ATP $\longrightarrow$ glucose 6-phosphate + ADP

**Step 2.** **Isomerization** of glucose 6-phosphate to give fructose 6-phosphate. (See Equation 13.2, p. 299.)

Glucose 6-phosphate $\rightleftharpoons$ fructose 6-phosphate

**Step 3.** **Phosphorylation** of fructose 6-phosphate to give fructose 1,6-*bis*phosphate (ATP is the source of the phosphate group). (See Equation 13.3, p. 300.)

Fructose 6-phosphate + ATP $\longrightarrow$ fructose 1,6-*bis*phosphate + ADP

**FIGURE 13.2** The pathway of anaerobic glycolysis.

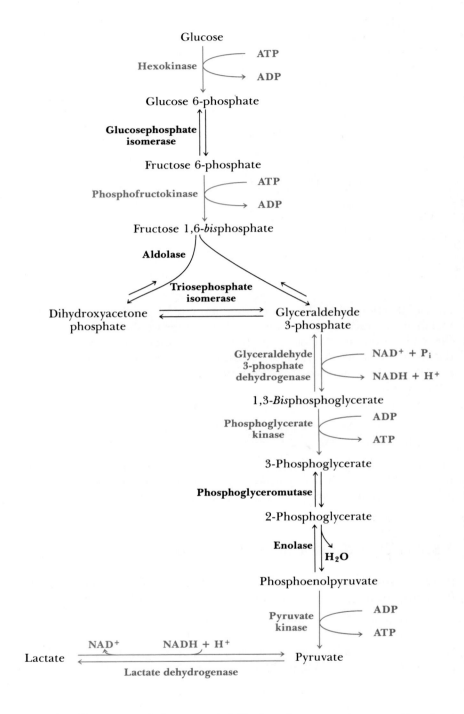

Step 4. **Cleavage** of fructose 1,6-*bis*phosphate to give two 3-carbon fragments, glyceraldehyde 3-phosphate and dihydroxyacetone phosphate. (See Equation 13.4, p. 301.)

Fructose 1,6-*bis*phosphate ⇌ glyceraldehyde 3-phosphate + dihydroxyacetone phosphate

Step 5.   **Isomerization** of dihydroxyacetone phosphate to give glyceraldehyde 3-phosphate. (See Equation 13.5, p. 302.)

Dihydroxyacetone phosphate $\rightleftharpoons$ glyceraldehyde 3-phosphate

Step 6.   **Oxidation** of glyceraldehyde 3-phosphate to give 1,3-*bis*phosphoglycerate. (See Equation 13.6, p. 303.)

Glyceraldehyde 3-phosphate + NAD$^+$ + P$_i$ $\longrightarrow$ NADH + 1,3-*bis*phosphoglycerate + H$^+$

Step 7.   **Transfer of a phosphate group** from 1,3-*bis*phosphoglycerate to ADP (phosphorylation of ADP to ATP) to give 3-phosphoglycerate. (See Equation 13.7, p. 304.)

1,3-*Bis*phosphoglycerate + ADP $\rightleftharpoons$ 3-phosphoglycerate + ATP

Step 8.   **Isomerization** of 3-phosphoglycerate to give 2-phosphoglycerate. (See Equation 13.8, p. 306.)

3-Phosphoglycerate $\rightleftharpoons$ 2-phosphoglycerate

Step 9.   **Dehydration** of 2-phosphoglycerate to give phosphoenolpyruvate. (See Equation 13.9, p. 309.)

2-Phosphoglycerate $\rightleftharpoons$ phosphoenolpyruvate + H$_2$O

Step 10.   **Transfer a phosphate group** from phosphoenolpyruvate to ADP (phosphorylation of ADP to ATP) to give pyruvate. (See Equation 13.10, p. 310.)

Phosphoenolpyruvate + ADP $\longrightarrow$ pyruvate + ATP

Step 11.   **Reduction** of pyruvate to lactate. (See Equation 13.11, p. 310.)

Pyruvate + NADH + H$^+$ $\rightleftharpoons$ lactate + NAD$^+$

Note that only two of the eleven steps in this pathway involve electron-transfer reactions. We shall now look at each of these reactions in detail.

## 13.2
## REACTIONS OF GLYCOLYSIS I: CONVERSION OF GLUCOSE TO GLYCERALDEHYDE 3-PHOSPHATE

The first steps of the glycolytic pathway prepare for the electron transfer and the eventual phosphorylation of ADP; these reactions make use of the free energy of hydrolysis of ATP.

Step 1.   Glucose is phosphorylated to give glucose 6-phosphate. The phosphorylation of glucose is an endergonic reaction.

Glucose + P$_i$ $\longrightarrow$ glucose 6-phosphate + H$_2$O

$\Delta G^{\circ\prime} = 13.8$ kJ mol$^{-1}$ = 3.3 kcal mol$^{-1}$

The hydrolysis of ATP is exergonic,

$$ATP + H_2O \longrightarrow ADP + P_i$$

$$\Delta G^{\circ\prime} = -30.5 \text{ kJ mol}^{-1} = -7.3 \text{ kcal mol}^{-1}$$

These two reactions are coupled, so the overall reaction is the sum of the two and is exergonic.

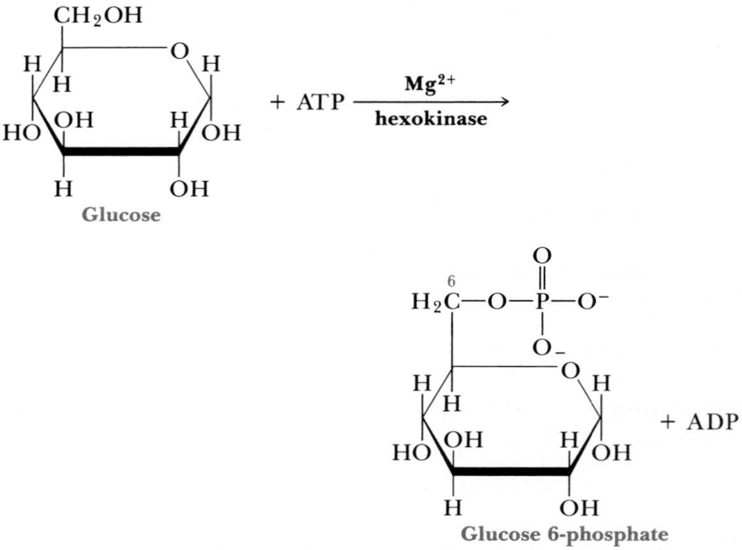

Glucose + ATP $\longrightarrow$ glucose 6-phosphate + ADP

$$\Delta G^{\circ\prime} = 13.8 + (-30.5) = -16.7 \text{ k J mol}^{-1} = -4.0 \text{ kcal mol}^{-1}$$

This reaction illustrates the use of chemical energy originally produced by the oxidation of nutrients and trapped by phosphorylation of ADP to ATP.

The enzyme that catalyzes this reaction is **hexokinase.** The term "kinase" is applied to the class of ATP-dependent enzymes that transfer a phosphate group from the ATP to a substrate. In the case of hexokinase the substrate need not be glucose, but can be any one of a number of hexoses such as glucose, fructose, and mannose. In some organisms there is a **glucokinase,** an enzyme that specifically phosphorylates glucose. The glucokinase in the human liver lowers blood glucose levels after one has eaten a meal.

The rate of phosphorylation of glucose by hexokinase controls the rate of glycolysis. In the terminology of chemical kinetics, the hexokinase reaction is the **rate-limiting step** in glycolysis. This type of control refers to the kinetic rate of the reaction, not to its thermodynamic energetics. Kinetic control means that the production of glucose 6-phosphate controls glycolysis by determining the availability of the starting material for the pathway.

A large conformational change takes place in hexokinase when substrate is bound. It has been shown by x-ray crystallography that, in the absence of substrate, two lobes of the enzyme that surround the binding

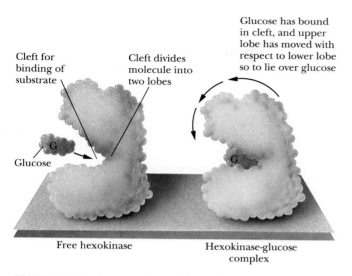

Cleft for binding of substrate

Cleft divides molecule into two lobes

Glucose has bound in cleft, and upper lobe has moved with respect to lower lobe so to lie over glucose

Glucose

Free hexokinase

Hexokinase-glucose complex

**FIGURE 13.3**   A comparison of the conformations of hexokinase and the hexokinase–glucose complex.

site are quite far apart. When glucose is bound, the two lobes move closer together and the glucose becomes almost completely surrounded by protein (Figure 13.3). This type of behavior is consistent with the induced-fit theory of enzyme action (Section 10.5). In all kinases for which the structure is known, there is a cleft that closes when substrate is bound.

Step 2.   Glucose 6-phosphate isomerizes to give fructose 6-phosphate. **Glucosephosphate isomerase** is the enzyme that catalyzes this reaction. The C-1 aldehyde group of glucose 6-phosphate is reduced to a hydroxyl group, and the C-2 hydroxyl group is oxidized to give the ketone group of fructose 6-phosphate, with no net oxidation or reduction. (Recall from Section 6.1 that glucose is an aldose, a sugar whose open-chain, noncyclic structure contains an aldehyde group, while fructose is a ketose, a sugar whose corresponding structure contains a ketone group.) The phosphory- lated forms, glucose 6-phosphate and fructose 6-phosphate, are an aldose and a ketose, respectively.

Glucose 6-phosphate    $\xrightleftharpoons{\text{glucosephosphate isomerase}}$    Fructose 6-phosphate    (13.2)

Step 3.    Fructose 6-phosphate is further phosphorylated, producing fructose 1,6-*bis*phosphate.

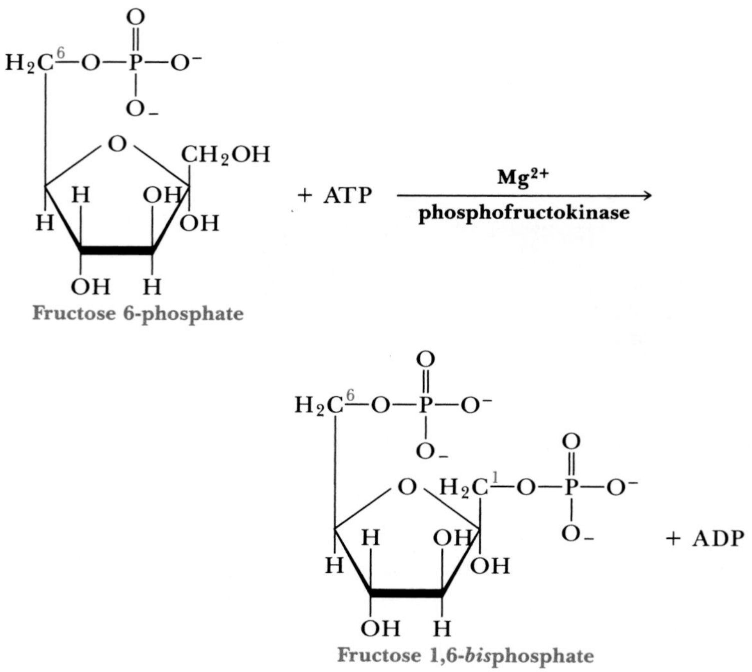

Fructose 6-phosphate    + ATP    $\xrightarrow[\text{phosphofructokinase}]{\text{Mg}^{2+}}$

Fructose 1,6-*bis*phosphate    + ADP                                   (13.3)

The endergonic reaction of phosphorylation of fructose 6-phosphate is coupled to the exergonic reaction of hydrolysis of ATP, and the overall reaction is exergonic, as was the case in step 1. See Table 13.1 (p. 313).

Fructose 6-phosphate + ATP $\longrightarrow$ fructose 1,6-*bis*phosphate + ADP

The reaction in which fructose 6-phosphate is phosphorylated to give fructose 1,6-*bis*phosphate is the one in which the sugar is committed to glycolysis. Glucose 6-phosphate and fructose 6-phosphate can play roles in other pathways, but fructose 1,6-*bis*phosphate does not. Once fructose 1,6-*bis*phosphate is formed from the original sugar, no other pathways are available and the molecule must undergo the rest of the reactions of glycolysis. The phosphorylation of fructose 6-phosphate is highly exergonic and irreversible, and **phosphofructokinase,** the enzyme that catalyzes it, is the key regulatory enzyme in glycolysis.

Phosphofructokinase is a tetramer (in other words, it consists of four subunits) that is subject to allosteric regulation of the type we discussed in Chapter 10. This enzyme exists in slightly different forms in muscle and in liver; the two forms, designated M and L respectively, are referred to as **isozymes.** The subunits differ slightly in amino acid composition, so the two isozymes can be separated from each other by electrophoresis (Interchapter A, Section 2). The tetrameric form that occurs in muscle is designated $M_4$, while that in liver is designated $L_4$. In red blood cells, all possible combinations of monomers combine to form the tetramer: $M_4$, $M_3L$, $M_2L_2$, $ML_3$, and $L_4$ (Figure 13.4). Individuals who lack the gene that directs the synthesis of the M form of the enzyme can carry on glycolysis in their livers but suffer muscle weakness because they lack the enzyme in muscle.

When the rate of the phosphofructokinase reaction is observed at varying concentrations of substrate (fructose 6-phosphate), the sigmoidal curve typical of allosteric enzymes is obtained. ATP is an allosteric effector in the reaction. High levels of ATP depress the rate of the reaction, and low levels of ATP stimulate the reaction. When there is a high level of ATP in the cell, there is a good deal of chemical energy immediately available from hydrolysis of ATP. The cell does not need to metabolize glucose for energy, so the presence of ATP inhibits the glycolytic pathway at this point.

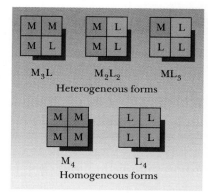

**FIGURE 13.4** The possible isozymes of phosphofructokinase. The symbol M refers to the monomeric form that predominates in muscle, while the symbol L refers to the form that predominates in liver.

**Step 4.** Fructose 1,6-*bis*phosphate is then split into two 3-carbon fragments.

Fructose 1,6-*bis*phosphate

$\xrightleftharpoons{\text{aldolase}}$

Dihydroxyacetone phosphate    +    D-Glyceraldehyde 3-phosphate    (13.4)

Fructose 1,6-*bis*phosphate $\rightleftharpoons$ glyceraldehyde 3-phosphate + dihydroxyacetone phosphate

The cleavage reaction here is the reverse of an aldol condensation; the enzyme that catalyzes it is called **aldolase.** An aldol condensation reaction takes place between two molecules that contain carbonyl groups; in other words, aldehydes and ketones are the starting materials. The product of an aldol condensation contains a carbonyl group and a hydroxyl group on the $\beta$ carbon counting from the carbonyl carbon:

## ALDOL CONDENSATIONS

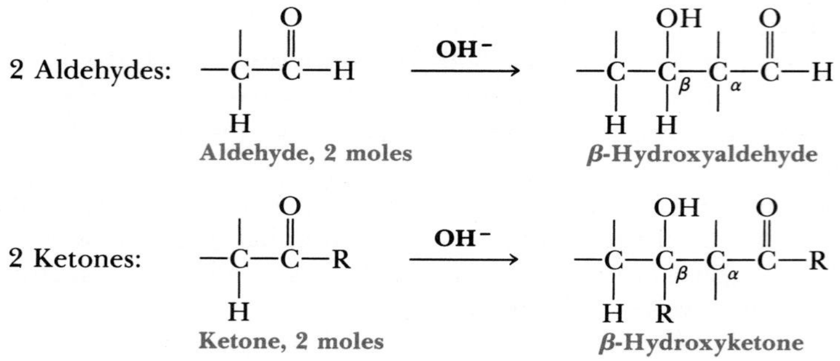

2 Aldehydes:

$$\underset{\substack{| \\ H}}{\overset{\overset{\displaystyle O}{\parallel}}{-C}-C}-H \quad \xrightarrow{\ \textbf{OH}^-\ } \quad \underset{\substack{| \\ H}}{-C}-\underset{\substack{\beta \\ |}}{\overset{\overset{\displaystyle OH}{|}}{C}}-\underset{\substack{\alpha \\ |}}{\overset{\overset{\displaystyle O}{\parallel}}{C}}-C-H$$

Aldehyde, 2 moles                β-Hydroxyaldehyde

2 Ketones:

$$\underset{\substack{| \\ H}}{\overset{\overset{\displaystyle O}{\parallel}}{-C}-C}-R \quad \xrightarrow{\ \textbf{OH}^-\ } \quad -C-\underset{\beta}{C}-\underset{\alpha}{C}-C-R$$

Ketone, 2 moles                β-Hydroxyketone

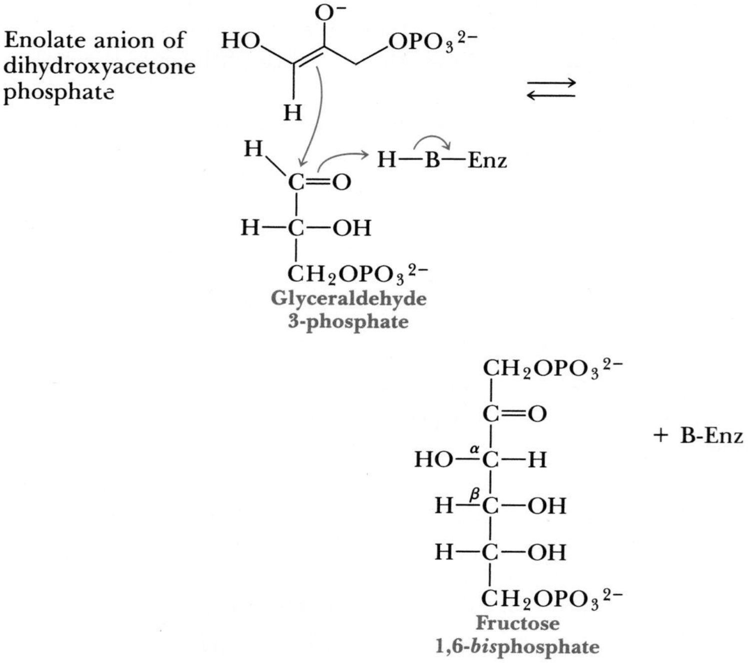

Enolate anion of dihydroxyacetone phosphate

$$\text{HO}\diagdown\overset{\overset{\displaystyle O^-}{|}}{C}\diagup\text{OPO}_3{}^{2-}$$

$$\text{H}-\text{B}-\text{Enz}$$

$$\underset{\substack{| \\ CH_2OPO_3{}^{2-}}}{\overset{\overset{\displaystyle H}{\diagdown}}{C}}=O$$
$$\text{H}-\text{C}-\text{OH}$$

Glyceraldehyde
3-phosphate

$$\rightleftharpoons$$

$$\begin{array}{c}
CH_2OPO_3{}^{2-} \\
| \\
C=O \\
| \\
HO\overset{\alpha}{-}C-H \\
| \\
H\overset{\beta}{-}C-OH \\
| \\
H-C-OH \\
| \\
CH_2OPO_3{}^{2-}
\end{array} \quad + \text{B-Enz}$$

Fructose
1,6-*bis*phosphate                                                    (13.5)

Base-catalyzed aldol condensation of glyceraldehyde 3-phosphate and dihydroxyacetone phosphate (basic group is part of the enzyme, B-Enz) (see Figure 13.5)

(We saw this terminology, which designates the positions of carbon atoms by letters of the Greek alphabet, in Section 5.1. The carbon atom next to the site of interest is the $\alpha$ carbon, the second one is the $\beta$ carbon, and so on.) Even though the reaction takes its name from the two functional groups, *ald*ehyde and alcoh*ol*, the carbonyl group can be an aldehyde or ketone. An aldol condensation is frequently base-catalyzed. Equation 13.5 shows the base-catalyzed aldol condensation of glyceraldehyde 3-phosphate and dihydroxyacetone phosphate to form fructose 1,6-*bis*phosphate. An enolate anion is an important intermediate.

An aldol cleavage is the reverse of an aldol condensation. Except for the direction, everything about the reaction is the same, including the

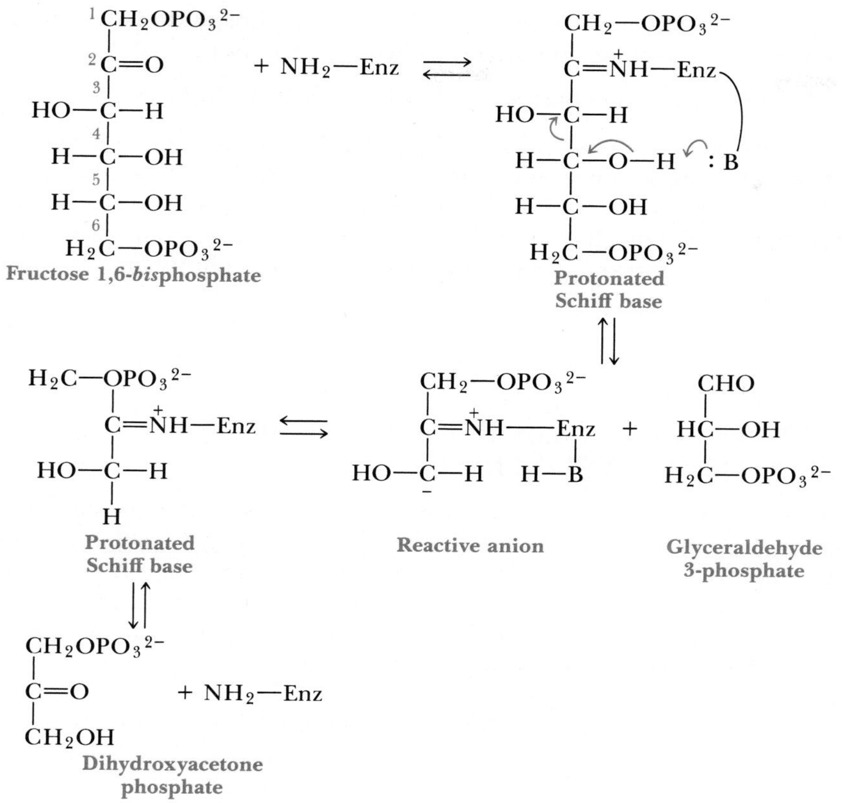

$$
\begin{array}{l}
^1CH_2OPO_3{}^{2-} \\
^2C=O \\
HO-^3C-H \\
H-^4C-OH \\
H-^5C-OH \\
^6H_2C-OPO_3{}^{2-}
\end{array}
\quad + NH_2-Enz \rightleftharpoons
\begin{array}{l}
CH_2-OPO_3{}^{2-} \\
C=\overset{+}{N}H-Enz \\
HO-C-H \\
H-C-O-H \quad :B \\
H-C-OH \\
H_2C-OPO_3{}^{2-}
\end{array}
$$

Fructose 1,6-*bis*phosphate                     Protonated Schiff base

$$
\begin{array}{l}
H_2C-OPO_3{}^{2-} \\
C=\overset{+}{N}H-Enz \\
HO-C-H \\
H
\end{array}
\rightleftharpoons
\begin{array}{l}
CH_2-OPO_3{}^{2-} \\
C=\overset{+}{N}H-Enz \\
HO-C-H \quad H-B \\
-
\end{array}
+
\begin{array}{l}
CHO \\
HC-OH \\
H_2C-OPO_3{}^{2-}
\end{array}
$$

Protonated Schiff base              Reactive anion              Glyceraldehyde 3-phosphate

$$
\begin{array}{l}
CH_2OPO_3{}^{2-} \\
C=O \\
CH_2OH
\end{array}
\quad + NH_2-Enz
$$

Dihydroxyacetone phosphate

FIGURE 13.5   The aldol cleavage of fructose 1,6-*bis*phosphate. The ε-amino group of a lysine residue reacts with the substrate to form a Schiff base. (Enz stands for enzyme, and B for the basic group on the enzyme.)

mechanism. The only effect a catalyst has on a reaction is to increase the rate at which both the forward and reverse reactions occur. Because of this fact, the mechanism is necessarily the same for an aldol cleavage as for an aldol condensation. In the enzyme isolated from most animal sources (the one from muscle is the most extensively studied), the basic side chain of an essential lysine residue plays the key role in catalyzing this reaction. A compound known as a **Schiff base** is the key intermediate; it involves covalent bonding of the substrate to the ε-amino group of the essential lysine (Figure 13.5).

Step 5.   The dihydroxyacetone phosphate is converted to glyceraldehyde 3-phosphate.

$$
\begin{array}{l}
H_2C-OH \\
C=O \qquad\quad O \\
\qquad\qquad \| \\
H_2C-O-P-O^- \\
\qquad\quad | \\
\qquad\quad O_-
\end{array}
\xrightarrow[]{\text{triosephosphate isomerase}}
\begin{array}{l}
HC=O \\
H-C-OH \quad O \\
\qquad\qquad \| \\
H_2C-O-P-O^- \\
\qquad\quad | \\
\qquad\quad O_-
\end{array}
$$

Dihydroxyacetone phosphate              Glyceraldehyde 3-phosphate

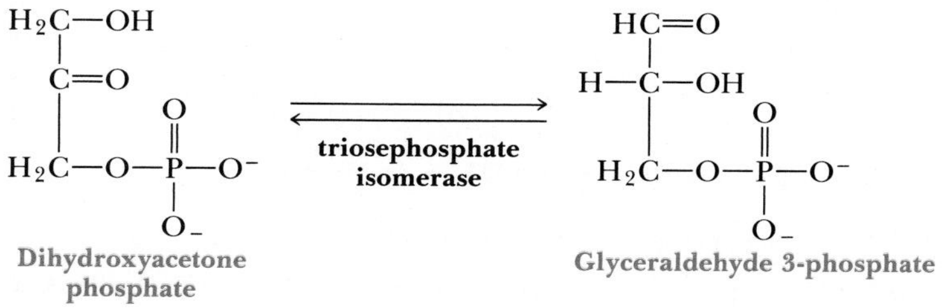

(13.6)

The enzyme that catalyzes this reaction is **triosephosphate isomerase.** (Both dihydroxyacetone and glyceraldehyde are trioses.)

Dihydroxyacetone phosphate $\rightleftharpoons$ glyceraldehyde 3-phosphate

One molecule of glyceraldehyde 3-phosphate has already been produced by the aldolase reaction; we now have a second molecule of glyceraldehyde 3-phosphate, produced by the triosephosphate isomerase reaction. The original molecule of glucose, which contains six carbon atoms, has now been converted to two molecules of glyceraldehyde 3-phosphate, each of which contains three carbon atoms.

## 13.3
## REACTIONS OF GLYCOLYSIS II: CONVERSION OF GLYCERALDEHYDE 3-PHOSPHATE TO LACTATE

At this point a molecule of glucose (a six-carbon compound) that enters the pathway has been converted to two molecules of glyceraldehyde 3-phosphate. We have not seen any oxidation reactions yet, but now we shall encounter them. Keep in mind that in the rest of the pathway two molecules of each of the three-carbon compounds take part in every reaction for each original glucose molecule.

Step 6.   The next step is the oxidation of glyceraldehyde 3-phosphate to 1,3-*bis*phosphoglycerate.

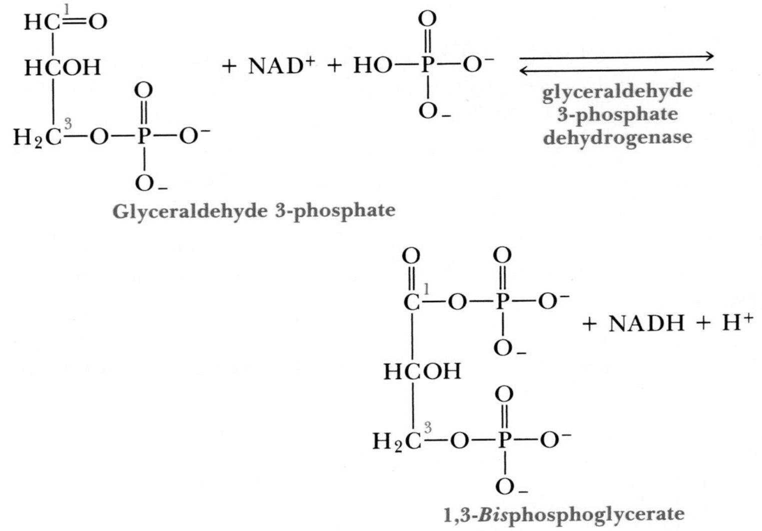

Glyceraldehyde 3-phosphate

1,3-*Bis*phosphoglycerate                    (13.7)

Glyceraldehyde 3-phosphate + NAD$^+$ + P$_i$ $\longrightarrow$

NADH + 1,3-*bis*phosphoglycerate + H$^+$

This reaction, *the* characteristic reaction of glycolysis, should be looked at more closely. It involves the addition of a phosphate group to glyceraldehyde-3-phosphate as well as an electron-transfer reaction, from glyceraldehyde 3-phosphate to $NAD^+$. It will simplify discussion to consider the two parts separately.

The half reaction of oxidation is that of an aldehyde to a carboxylic acid group, where water can be considered to take part in the reaction.

$$RCHO + H_2O \longrightarrow RCOOH + 2\ H^+ + 2\ e^-$$

The half reaction of reduction is that of $NAD^+$ to NADH (Section 12.3).

$$NAD^+ + 2\ H^+ + 2\ e^- \longrightarrow NADH + H^+$$

The overall redox reaction is thus

$$RCHO + H_2O + NAD^+ \longrightarrow RCOOH + H^+ + NADH$$

**glyceraldehyde**                **3-phosphoglycerate**
**3-phosphate**

where R indicates the portions of the molecule other than the aldehyde and carboxylic acid groups, respectively. The oxidation reaction is exergonic under standard conditions ($\Delta G^{\circ\prime} = -43.1$ kJ mol$^{-1}$ = $-10.3$ kcal mol$^{-1}$), but oxidation is only part of the overall reaction.

The phosphate group that is linked to the carboxyl group does not form an ester, since an ester linkage requires an alcohol and an acid. Instead, the carboxylic acid group and phosphoric acid form a mixed anhydride of two acids by loss of water (Section 4.1),

$$\text{3-Phosphoglycerate} + P_i \longrightarrow \text{1,3-\textit{bis}phosphoglycerate} + H_2O$$

where the substances involved in the reaction are in the ionized form appropriate at pH 7. Note that ATP and ADP do not appear in the equation. The source of the phosphate group is phosphate ion itself, rather than ATP. A reaction of this type in which $P_i$ rather than ATP is the source of the phosphate group is called **substrate-level phosphorylation.** The phosphorylation reaction is endergonic under standard conditions ($\Delta G^{\circ\prime} = 49.3$ kJ mol$^{-1}$ = 11.8 kcal mol$^{-1}$).

The overall reaction, including electron transfer and phosphorylation, is

$$RCHO + HPO_3{}^{2-} + NAD^+ \rightleftharpoons RC{-}OPO_3{}^{2-} +$$
$$NADH + H^+$$

Glyceraldehyde 3-phosphate + $P_i$ + $NAD^+ \longrightarrow$
1,3-*bis*phosphoglycerate + NADH + $H^+$

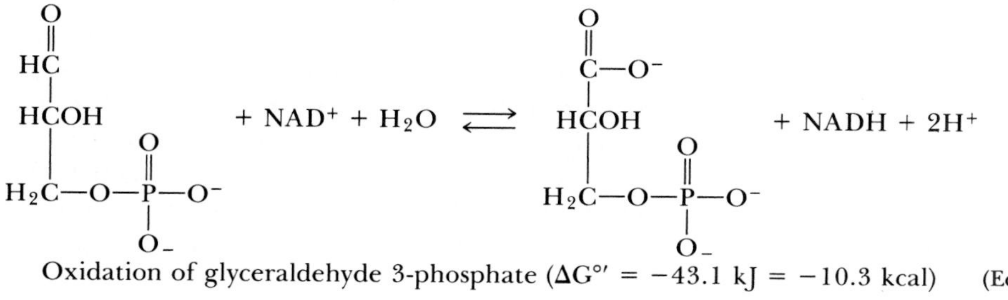

Oxidation of glyceraldehyde 3-phosphate ($\Delta G^{\circ\prime} = -43.1$ kJ = $-10.3$ kcal)     (Equation continues on next page.)

$$
\begin{array}{c}
\text{O} \\
\parallel \\
\text{C}-\text{O}^- \\
| \\
\text{HCOH} \qquad\qquad\qquad\qquad \text{O} \\
| \qquad\qquad\qquad\qquad\qquad \parallel \\
\text{H}_2\text{C}-\text{O}-\overset{\displaystyle\text{O}}{\underset{\displaystyle|}{\overset{\parallel}{\text{P}}}}-\text{O}^- \qquad + \text{HO}-\text{P}-\text{O}^- + \text{H}^+ \;\rightleftharpoons \\
| \qquad\qquad\qquad\qquad\qquad | \\
\text{O}_- \qquad\qquad\qquad\qquad\qquad \text{O}_-
\end{array}
$$

$$
\begin{array}{c}
\text{O} \qquad\quad \text{O} \\
\parallel \qquad\quad \parallel \\
\text{C}-\text{O}-\text{P}-\text{O}^- \\
| \qquad\quad | \\
\text{HCOH} \qquad\;\; \text{O}_- \qquad\qquad + \text{H}_2\text{O} \\
| \qquad\qquad\quad \text{O} \\
\text{H}_2\text{C}-\text{O}-\overset{\parallel}{\text{P}}-\text{O}^- \\
| \\
\text{O}_-
\end{array}
$$

Phosphorylation of 3-phosphoglycerate ($\Delta G^{\circ\prime}$ = 49.3 kJ = 11.8 kcal)

Sum = 6.2 kJ = 1.5 kcal    **(13.8)**

The standard free energy change for the overall reaction is the sum of the values for the oxidation and phosphorylation reactions. The overall reaction is not far from equilibrium, being only slightly endergonic.

$$\Delta G^{\circ\prime}{}_{\text{overall}} = \Delta G^{\circ\prime}{}_{\text{oxidation}} + \Delta G^{\circ\prime}{}_{\text{phosphorylation}}$$

$$= (-43.1) + (49.3)\ \text{kJ mol}^{-1}$$

$$= 6.2\ \text{kJ mol}^{-1} = 1.5\ \text{kcal mol}^{-1}$$

This value of the standard free energy change is for the reaction of one mole of glyceraldehyde 3-phosphate; the value must be multiplied by two to get the value for each mole of glucose ($\Delta G^{\circ\prime}$ = 12.4 kJ = 3.0 kcal).

The enzyme that catalyzes the conversion of glyceraldehyde 3-phosphate to 1,3-*bis*phosphoglycerate is **glyceraldehyde 3-phosphate dehydrogenase.** This enzyme is one of a class of similar enzymes, the NADH-linked dehydrogenases. The structures of a number of dehydrogenases of this type have been studied by x-ray crystallography. The overall structures are not strikingly similar, but the structure of the binding site for NADH is quite similar in all these enzymes (Figure 13.6). (The oxidizing agent is NAD$^+$; both oxidized and reduced forms of the coenzyme bind to the enzyme.) One portion of the binding site is specific for the nicotinamide ring, and one portion is specific for the adenine ring.

The molecule of glyceraldehyde 3-phosphate dehydrogenase is a tetramer, consisting of four identical subunits. Each subunit binds one molecule of NAD$^+$, and each subunit contains an essential cysteine residue. A thioester involving the cysteine residue is the key intermediate in this reaction. There is a **hydride ion (H$^-$) transfer** from the aldehyde to the nicotinamide ring of the NAD$^+$, forming NADH and the thioester. (An ester is a derivative of an acid; the first part of the reaction, namely the oxidation of an aldehyde to an acid, has already taken place.) NADH is released by the enzyme, and NAD$^+$ binds in its place (Figure 13.7).

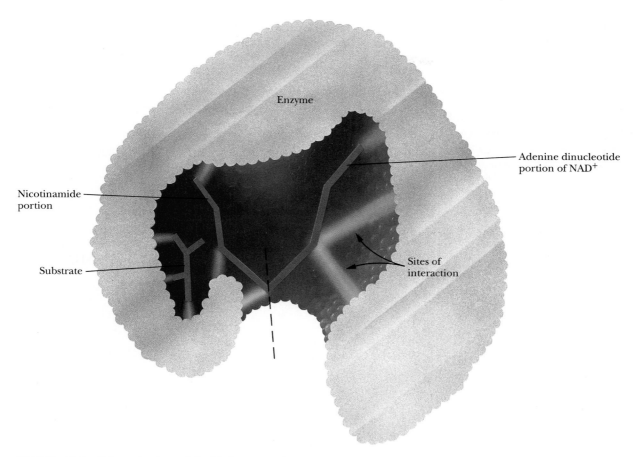

Enzyme

Adenine dinucleotide
portion of NAD$^+$

Nicotinamide
portion

Substrate

Sites of
interaction

**FIGURE 13.6**  Schematic view of the binding site of an NADH-linked
dehydrogenase. There are specific binding sites for the adenine nucleotide portion
(shown in black) and nicotinamide portion (shown in red) of the coenzyme, in
addition to the binding site for the substrate. Specific interactions with the enzyme
hold the substrate and coenzyme in the proper position. Sites of interaction are
shown as a series of red lines.

In the phosphorylation step, the thioester acts as a "high-energy"
intermediate. Phosphate ion attacks the thioester, forming a mixed
anhydride of the carboxylic and phosphoric acids, which is also a "high-
energy" compound (Figure 13.8). This compound is 1,3-*bis*phosphoglycerate, the product of the reaction. Production of ATP requires a "high-
energy" compound as starting material. The 1,3-*bis*phosphoglycerate
fulfills this requirement and transfers a phosphate group to ADP in a
highly exergonic reaction (*i.e.,* it has a high phosphate-group transfer
potential).

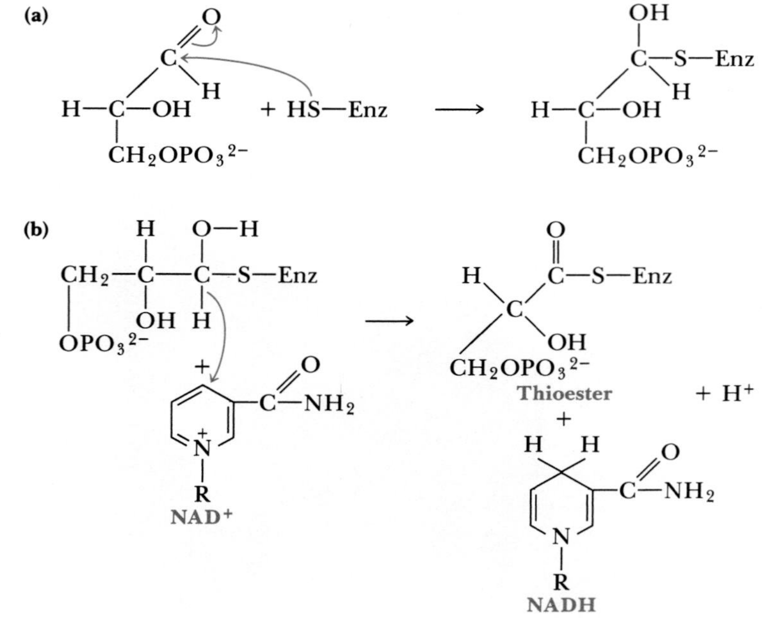

**FIGURE 13.7**    Hydride ion transfer from an enzyme-bound derivative of glyceraldehyde 3-phosphate to NAD$^+$ produces a thioester and NADH. (a) Formation of enzyme-bound derivative of glyceraldehyde 3-phosphate. Enz is glyceraldehyde 3-phosphate dehydrogenase. (b) Hydride ion transfer. Note that H$^-$ and H: are alternate ways of writing the formula for the hydride ion.

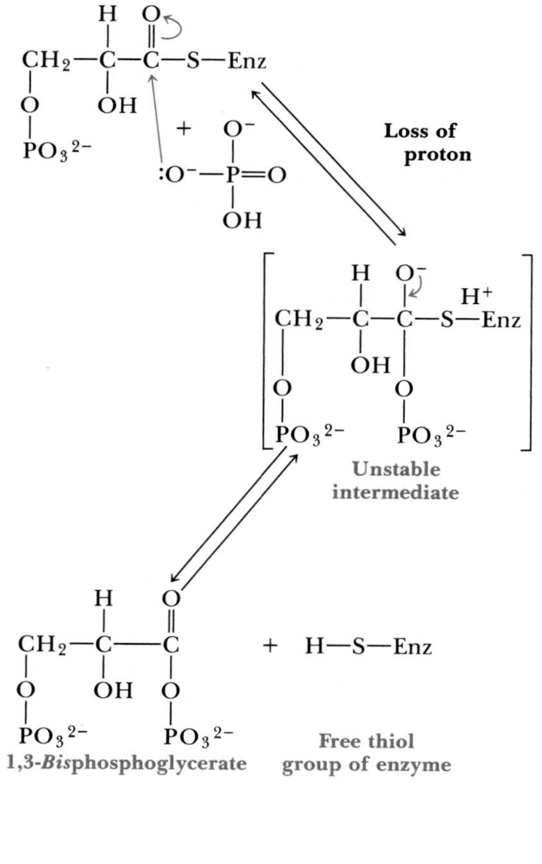

**FIGURE 13.8**    Phosphate ion attacks the thioester derivative of glyceraldehyde 3-phosphate dehydrogenase (Enz) to produce 1,3-*bis*phosphoglycerate and regenerate the thiol group of cysteine.

Step 7. The next step is one of the two reactions in which ATP is produced by phosphorylation of ADP:

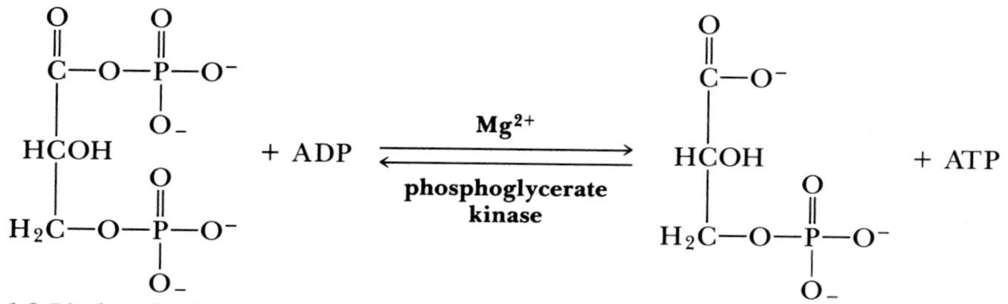


**1,3-*Bis*phosphoglycerate** **3-Phosphoglycerate** **(13.9)**

$$\text{1,3-}Bis\text{phosphoglycerate} + \text{ADP} \rightleftharpoons \text{3-phosphoglycerate} + \text{ATP}$$

The enzyme that catalyzes this reaction is **phosphoglycerate kinase.** By now the term "kinase" should be familiar as the generic name for a class of ATP-dependent phosphate-group transfer enzymes. The most striking feature of the reaction has to do with energetics of phosphate-group transfer. In this step in glycolysis, a phosphate group is transferred from 1,3-*bis*phosphoglycerate to a molecule of ADP, producing ATP, the first of two such reactions in the glycolytic pathway. We already mentioned that 1,3-*bis*phosphoglycerate can easily transfer a phosphate group to other substances. The only requirement is that the standard free energy of the hydrolysis reaction is more negative than that for hydrolysis of the new phosphate compound being formed. Recall that the standard free energy of hydrolysis of 1,3-*bis*phosphoglycerate is $-49.3$ kJ mol$^{-1}$. We have already seen that the standard free energy of hydrolysis of ATP is $-30.5$ kJ mol$^{-1}$, and we must change the sign of the free energy change when the reverse reaction occurs:

$$\text{ADP} + \text{P}_i + \text{H}^+ \longrightarrow \text{ATP} + \text{H}_2\text{O}$$

$$\Delta G^{\circ\prime} = 30.5 \text{ kJ mol}^{-1} = 7.3 \text{ kcal mol}^{-1}$$

The net reaction is

$$\text{1,3-}Bis\text{phosphoglycerate} + \text{ADP} \rightleftharpoons \text{3-phosphoglycerate} + \text{ATP}$$

$$\Delta G^{\circ\prime} = -18.8 \text{ kJ mol}^{-1} = -4.5 \text{ kcal mol}^{-1}$$

Two molecules of ATP are produced by this reaction for each molecule of glucose that enters the glycolytic pathway. In the earlier stages of the pathway two molecules of ATP were invested to produce fructose 1,6-*bis*phosphate, and now they have been recovered. At this point the balance of ATP use and production is exactly even. The next few reactions will bring about the production of two more molecules of ATP for each original molecule of glucose, leading eventually to the net gain of two ATP molecules in glycolysis.

Step 8. In the next reaction the phosphate group is transferred from carbon 3 to carbon 2 of the glyceric acid backbone, setting the stage for the reactions that follow.

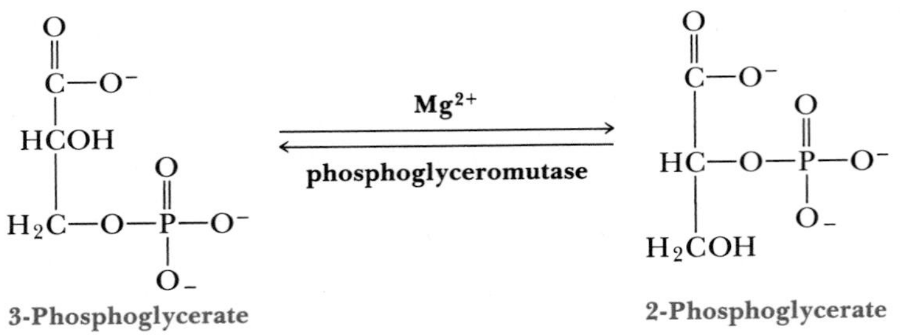

$$
\text{3-Phosphoglycerate} \underset{\text{phosphoglyceromutase}}{\overset{Mg^{2+}}{\rightleftharpoons}} \text{2-Phosphoglycerate} \tag{13.10}
$$

3-Phosphoglycerate $\rightleftharpoons$ 2-phosphoglycerate

The enzyme that catalyzes this reaction is **phosphoglyceromutase.**

Step 9. The 2-phosphoglycerate molecule loses one molecule of water, producing phosphoenolpyruvate. This reaction does not involve electron transfer; it is a dehydration reaction. **Enolase,** the enzyme that catalyzes this reaction, requires $Mg^{2+}$ as a cofactor. The water molecule that is eliminated binds to $Mg^{2+}$ in the course of the reaction.

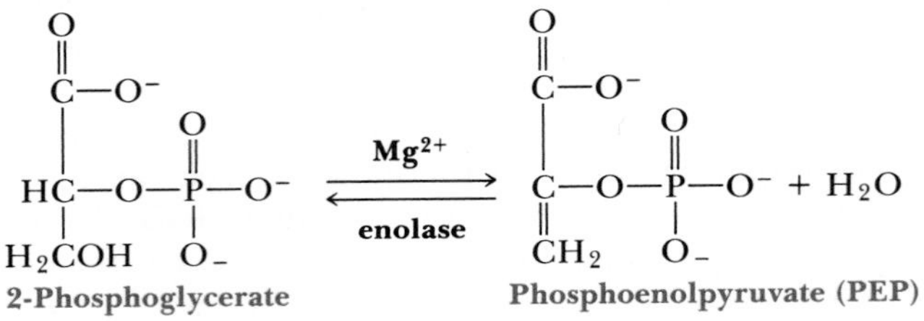

$$
\text{2-Phosphoglycerate} \underset{\text{enolase}}{\overset{Mg^{2+}}{\rightleftharpoons}} \text{Phosphoenolpyruvate (PEP)} + H_2O \tag{13.11}
$$

2-Phosphoglycerate $\rightleftharpoons$ phosphoenolpyruvate + $H_2O$

Step 10. Phosphoenolpyruvate then transfers its phosphate group to ADP, producing ATP and pyruvate,

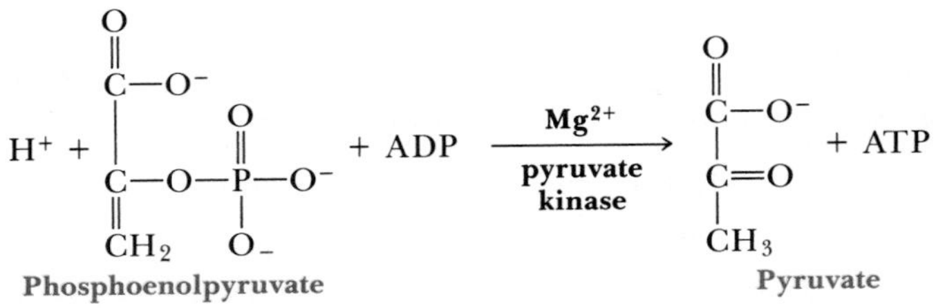

$$
H^+ + \text{Phosphoenolpyruvate} + \text{ADP} \underset{\text{kinase}}{\overset{Mg^{2+}}{\xrightarrow{\quad\text{pyruvate}\quad}}} \text{Pyruvate} + \text{ATP} \tag{13.12}
$$

Phosphoenolpyruvate + ADP $\longrightarrow$ pyruvate + ATP

The double bond shifts to the oxygen on carbon 2 and a hydrogen shifts to carbon 3. Phosphoenolpyruvate is a "high-energy compound" with a high "phosphate-group transfer potential." The free energy of hydrolysis of this compound is more negative than that of ATP ($-61.9$ kJ mol$^{-1}$ vs. $-30.5$ kJ mol$^{-1}$, or $-14.8$ kcal mol$^{-1}$ vs. $-7.3$ kcal mol$^{-1}$). The reaction that occurs in this step can be considered to be the sum of the hydrolysis of phosphoenolpyruvate and the phosphorylation of ADP.

$$\text{Phosphoenolpyruvate} \longrightarrow \text{pyruvate} + P_i \qquad \Delta G^{\circ\prime} = -61.9 \text{ kJ mol}^{-1} = -14.8 \text{ kcal mol}^{-1}$$

$$\text{ADP} + P_i \longrightarrow \text{ATP} \qquad \Delta G^{\circ\prime} = 30.5 \text{ kJ mol}^{-1} = 7.3 \text{ kcal mol}^{-1}$$

$$\text{Phosphoenolpyruvate} + \text{ADP} \longrightarrow \text{pyruvate} + \text{ATP}$$

$$\Delta G^{\circ\prime} = -31.4 \text{ kJ mol}^{-1} = -7.5 \text{ kcal mol}^{-1}$$

Since two moles of pyruvate are produced for each mole of glucose, twice as much energy is released for each mole of starting material.

**Pyruvate kinase** is the enzyme that catalyzes this reaction. Like phosphofructokinase, it is an allosteric enzyme consisting of four subunits. Pyruvate kinase is inhibited by ATP. The conversion of phosphoenolpyruvate to pyruvate slows down when the cell has a high concentration of ATP, that is to say, when the cell does not have a great need for energy in the form of ATP.

Step 11.    The final reaction of anaerobic glycolysis is the reduction of pyruvate to lactate.

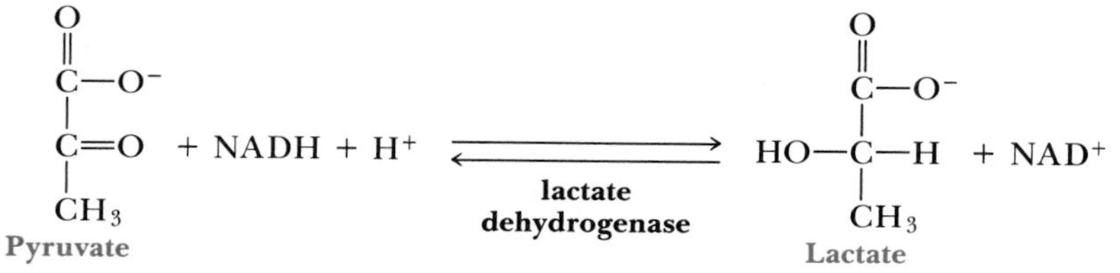

$$\text{Pyruvate} + \text{NADH} + \text{H}^+ \rightleftarrows \text{lactate} + \text{NAD}^+$$

This reaction is also exergonic ($\Delta G^{\circ\prime} = -25.1$ kJ mol$^{-1}$ = $-6.0$ kcal mol$^{-1}$); as before, we need to multiply this value by two to find the energy yield for each molecule of glucose that enters the pathway. Lactate is a dead end in metabolism, but it can be recycled to form pyruvate and even glucose by a pathway called gluconeogenesis ("new synthesis of glucose"), which we will discuss in Section 13.7.

**Lactate dehydrogenase** (LDH) is the enzyme that catalyzes this reaction. Like glyceraldehyde 3-phosphate dehydrogenase, LDH is an NADH-linked dehydrogenase. LDH is an allosteric enzyme with four subunits. There are two kinds of subunits, designated M and H, which vary slightly in amino acid composition. The quaternary structure of the tetramer can vary according to the relative amounts of the two kinds of subunits. These variant forms of the enzyme are called isozymes (see Section 13.2 for a discussion of the isozymic forms of phosphofructokinase). In human skeletal muscle the homogeneous tetramer of the $M_4$ type predominates, and in heart the other homogeneous possibility, the $H_4$ tetramer, is the predominant form. The heterogeneous forms, $M_3H$, $M_2H_2$, and $MH_3$, occur in blood serum. A very sensitive clinical test for

**FIGURE 13.9** The recycling of NAD$^+$ and NADH in anaerobic glycolysis.

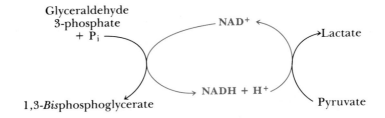

heart disease is based on the existence of the various isozymic forms of this enzyme. The relative amounts of the H$_4$ and MH$_3$ isozymes in blood serum increase drastically after myocardial infarction (heart attack) compared with normal serum.

At this point one might ask why the reduction of pyruvate to lactate (a waste product in aerobic organisms) is the last step in anaerobic glycolysis, a pathway that provides energy for the organism by oxidation of nutrients. There is another point to consider about the reaction, one that involves the relative amounts of NAD$^+$ and NADH in a cell. The half reaction of reduction can be written

$$\text{Pyruvate} + 2\ \text{H}^+ + 2\ \text{e}^- \longrightarrow \text{lactate}$$

and the half reaction of oxidation is

$$\text{NADH} + \text{H}^+ \longrightarrow \text{NAD}^+ + 2\ \text{e}^- + 2\ \text{H}^+$$

The overall reaction is, as we saw earlier,

$$\text{Pyruvate} + \text{NADH} + \text{H}^+ \longrightarrow \text{lactate} + \text{NAD}^+$$

The NADH produced from NAD$^+$ by the earlier oxidation of glyceraldehyde 3-phosphate is used up with no net change in the relative amounts of NADH and NAD$^+$ in the cell (Figure 13.9). This regeneration is needed under anaerobic conditions in the cell so that NAD$^+$ will be present for further glycolysis to take place. Without this regeneration, the oxidation reactions in anaerobic organisms would soon come to a halt because of the lack of NAD$^+$ to serve as an oxidizing agent in fermentative processes. On the other hand, NADH is a frequently encountered reducing agent in many reactions, and it is lost to the organism in lactate production. Aerobic metabolism makes more efficient use of reducing agents ("reducing power") such as NADH, because the conversion of pyruvate to lactate does not occur in aerobic metabolism. The NADH produced in the stages of glycolysis leading to the production of pyruvate is available for use in reactions in which a reducing agent is needed.

## 13.4
## ENERGY CONSIDERATIONS IN GLYCOLYSIS

Now that we have seen the reactions of the glycolytic pathway, we can do some bookkeeping and determine the standard free energy change for the entire pathway. This calculation is shown in Table 13.1. The standard free

**TABLE 13.1   The Reactions of Glycolysis and Their Standard Free Energy Changes**

| REACTION | $\Delta G^{\circ\prime}$ mol$^{-1}$ | |
|---|---|---|
| | *kJ* | *kcal* |
| Glucose + ATP $\longrightarrow$ glucose 6-phosphate + ADP | −16.7 | −4.0 |
| Glucose 6-phosphate $\longrightarrow$ fructose 6-phosphate | +1.7 | +0.4 |
| Fructose 6-phosphate + ATP $\longrightarrow$ fructose 1,6-*bis*phosphate + ADP | −13.8 | −3.3 |
| Fructose 1,6-*bis*phosphate $\longrightarrow$ dihydroxyacetone phosphate + glyceraldehyde 3-phosphate | +23.8 | +5.7 |
| Dihydroxyacetone phosphate $\longrightarrow$ glyceraldehyde 3-phosphate | +7.5 | +1.8 |
| 2(Glyceraldehyde 3-phosphate + NAD$^+$ + P$_i$ $\longrightarrow$ 1,3-*bis*phosphoglycerate + NADH + H$^+$) | +12.4 | +3.0 |
| 2(1,3-*Bis*phosphoglycerate + ADP $\longrightarrow$ 3-phosphoglycerate + ATP) | −37.6 | −9.0 |
| 2(3-Phosphoglycerate $\longrightarrow$ 2-phosphoglycerate) | +8.8 | +2.1 |
| 2(2-Phosphoglycerate $\longrightarrow$ phosphoenolpyruvate + H$_2$O) | +3.4 | +0.8 |
| 2(Phosphoenolpyruvate + ADP $\longrightarrow$ pyruvate + ATP) | −62.8 | −15.0 |
| Glucose + 2 ADP + 2 P$_i$ $\longrightarrow$ 2 pyruvate + 2 ATP | −73.3 | −17.5 |
| 2(Pyruvate + NADH + H$^+$ $\longrightarrow$ lactate + NAD$^+$) | −50.2 | −12.0 |
| Glucose + 2 ADP + 2 P$_i$ $\longrightarrow$ 2 lactate + 2 ATP | −123.5 | −29.5 |

energy change for the conversion of pyruvate to lactate is listed separately because pyruvate can have several alternate fates in metabolism.

The overall process of glycolysis is exergonic. The energy released in the exergonic phases of the process drives the endergonic reactions. The net reaction of glycolysis explicitly includes an important endergonic process, that of phosphorylation of two molecules of ADP.

$$2\ \text{ADP} + 2\ \text{P}_i \longrightarrow 2\ \text{ATP} \qquad \Delta G^{\circ\prime}_{\text{reaction}} = 61\ \text{kJ} = 14.6\ \text{kcal}$$

Without the production of ATP the reaction of one molecule of glucose to produce two molecules of lactate in anaerobic glycolysis would be even more exergonic.

$$
\begin{array}{ll}
\text{Glucose} + 2\ \text{ADP} + 2\ \text{P}_i \longrightarrow 2\ \text{lactate} + 2\ \text{ATP} & \Delta G^{\circ\prime} = \\
& -123.5\ \text{kJ} = \\
& -29.5\ \text{kcal} \\[4pt]
-(2\ \text{ADP} + 2\ \text{P}_i \longrightarrow 2\ \text{ATP}) & -(\Delta G^{\circ\prime}_{\text{reaction}} = \\
& 61\ \text{kJ} = 14.6 \\
& \text{kcal}) \\ \hline
\text{Glucose} \longrightarrow 2\ \text{lactate} & \Delta G^{\circ\prime} = -184.5 \\
& \text{kJ} = -44.1 \\
& \text{kcal mol}^{-1}
\end{array}
$$

Without production of ATP, the energy released by the conversion of glucose to lactate would be lost to the organism and dissipated as heat. The energy required to produce the two molecules of ATP for each molecule of glucose can be recovered by the organism when the ATP is hydrolyzed in some metabolic process. We discussed this point briefly in Chapter 12 when we compared the thermodynamic efficiency of anaerobic and aerobic metabolism. The percentage of the energy released by the breakdown of glucose that is "captured" by the organism when ADP is phosphorylated to ATP is the efficiency of energy use in glycolysis; it is $(61.0/184.5) \times 100$, or about 33 percent. The net release of energy in glycolysis, 123.5 kJ (29.5 kcal) for each mole of glucose, is not retained for use by the organism. Without the production of ATP to serve as a source of energy for other metabolic processes, the energy released by glycolysis would serve no purpose for the organism, except to help maintain body temperature in warm-blooded animals. A soft drink with ice can help keep you warm even on the coldest day of winter (if it is not a diet drink) because of its high sugar content.

The free energy changes we have listed in this chapter are the standard values, assuming the standard conditions such as 1 M concentration of all solutes except hydrogen ion. Concentrations under physiological conditions can differ markedly from standard values. Fortunately, there are well known methods (Section 8.3) for calculating the difference in the free energy change. Also, large changes in concentrations frequently lead to relatively small differences in the free energy change, about a few kJ $mol^{-1}$. Some of the free energy changes may be different under physiological conditions from the values listed here for standard conditions, but the underlying principles and the conclusions drawn from them remain the same.

## 13.5
## SOME REACTIONS RELATED TO THE MAIN GLYCOLYTIC PATHWAY

In addition to the reactions of the main glycolytic pathway we have just discussed, there are several other reactions that have to do with glycolysis. Even though these reactions are not part of the main pathway, they are of some interest to biochemists.

One of these reactions uses an alternate starting material for glycolysis. **Glycogen** is a polymer of glucose (Section 6.4) that plays a role in energy storage. Glycogen is degraded in an enzymatic reaction in which one glucose residue is cleaved from glycogen; the glucose residue reacts with phosphate ion to produce glucose 1-phosphate, which isomerizes to give glucose 6-phosphate, bypassing the first step of the pathway.

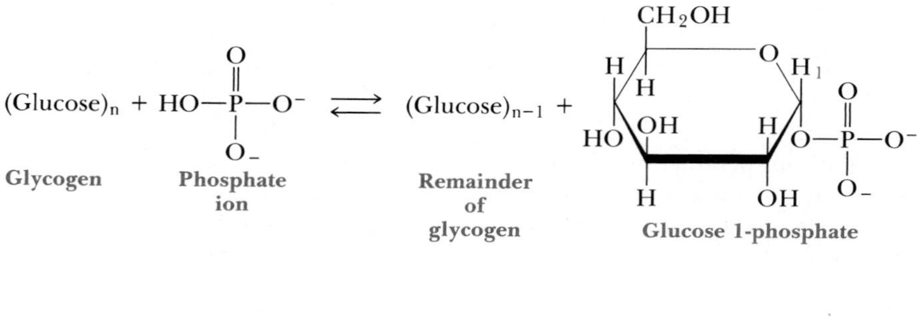

(Glucose)$_n$ + HO—P—O⁻  ⇌  (Glucose)$_{n-1}$ +  Glucose 1-phosphate

Glycogen    Phosphate ion    Remainder of glycogen    Glucose 1-phosphate

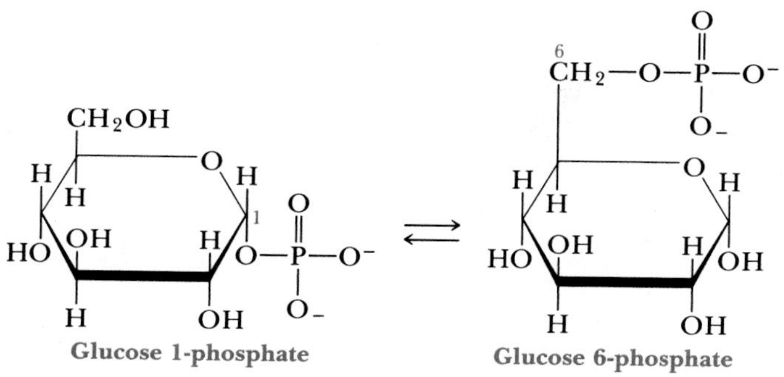

Glucose 1-phosphate          Glucose 6-phosphate          (13.14)

$$\text{Glycogen} + P_i \rightleftarrows \text{glucose 1-phosphate} + \text{remainder of glycogen}$$

$$\text{Glucose 1-phosphate} \rightleftarrows \text{glucose 6-phosphate}$$

The enzyme that catalyzes the first of these reactions is **glycogen phosphorylase;** the second reaction is catalyzed by **phosphoglucomutase.** Note that no ATP is hydrolyzed in the first reaction. In the main glycolytic pathway we saw another example of phosphorylation of a substrate directly by phosphate ion without involvement of ATP: the phosphorylation of glyceraldehyde 3-phosphate to 1,3-*bis*phosphoglycerate. As we saw earlier, a phosphorylation reaction that does not involve the "high-energy phosphate" of ATP is called substrate-level phosphorylation. This alternate mode of entry to the glycolytic pathway also "saves" one molecule of ATP for each molecule of glucose because it bypasses the first step in glycolysis. When glycogen rather than glucose is the starting material for glycolysis, there is a net gain of 3 ATP molecules for each glucose monomer, rather than 2 ATP when glucose itself is the starting point.

Two other reactions related to the glycolytic pathway lead to the production of ethanol by **alcoholic fermentation.** This process is one of the alternate fates of pyruvate (Section 13.1). In the first of the two reactions that lead to the production of ethanol, pyruvate is decarboxylated (loses carbon dioxide) to produce acetaldehyde. The enzyme that catalyzes this reaction is **pyruvate decarboxylase.**

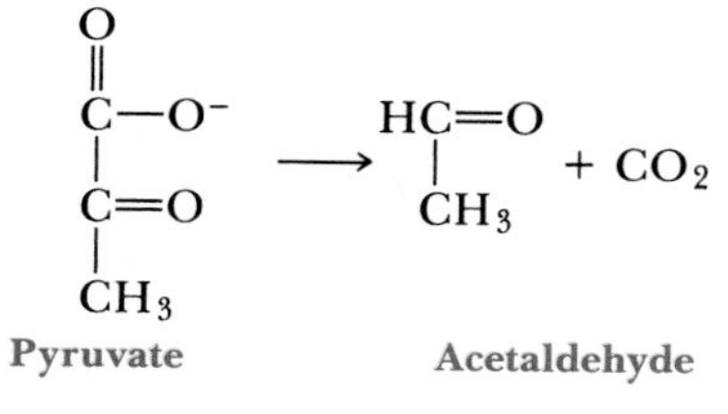

Pyruvate          Acetaldehyde          (13.15[a])

This enzyme requires $Mg^{2+}$ and a cofactor we have not met before, **thiamine pyrophosphate** (TPP). (Thiamine itself is vitamin $B_1$.) In TPP the carbon atom between the nitrogen and the sulfur in the thiazole ring (Figure 13.10) is highly reactive. It forms a carbanion (an ion with a negative charge on a carbon atom) quite easily, and the carbanion in turn attacks the carbonyl group of pyruvate to form an adduct. Carbon dioxide

**FIGURE 13.10**   The structures of thiamine (vitamin $B_1$) and thiamine pyrophosphate (TPP), the active form of the coenzyme.

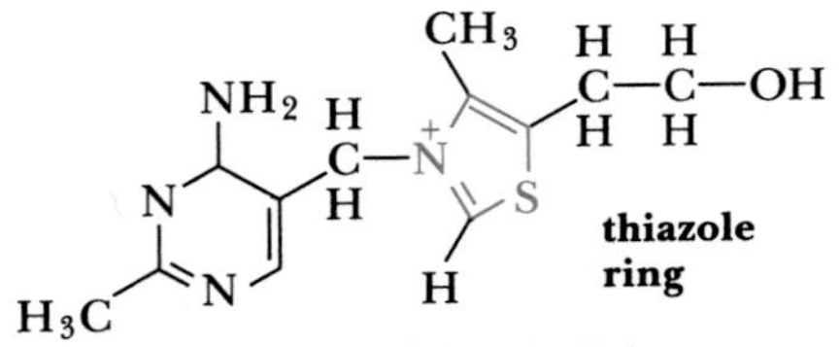

Thiamine (vitamin $B_1$)

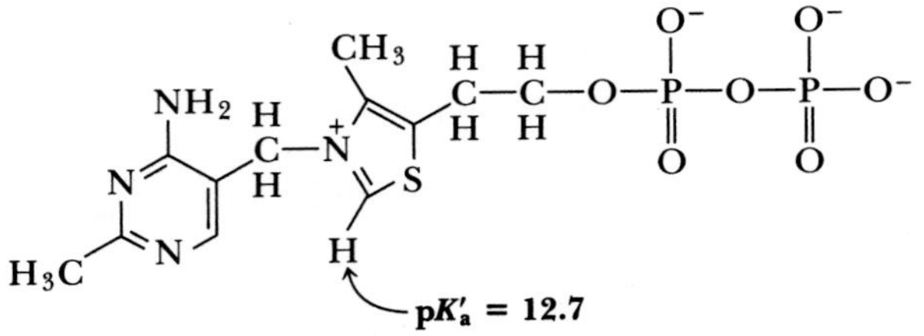

Thiamine pyrophosphate (TPP)

splits off, leaving a two-carbon fragment covalently bonded to TPP. There is a shift of electrons, and the two-carbon fragment splits off, producing acetaldehyde (Figure 13.11). (The two-carbon fragment bonded to TPP is sometimes called activated acetaldehyde.)

$$Pyruvate \longrightarrow acetaldehyde + CO_2$$

The carbon dioxide produced is responsible for the bubbles in beer and sparkling wines. Acetaldehyde is then reduced to produce ethanol, and at the same time one molecule of NADH is oxidized to $NAD^+$ for each molecule of ethanol produced.

**FIGURE 13.11**   The mechanism of the pyruvate decarboxylase reaction. The carbanion form of the thiazole ring of TPP is strongly nucleophilic. The carbanion attacks the carbonyl carbon of pyruvate to form an adduct. Carbon dioxide splits out, leaving a two-carbon fragment (activated acetaldehyde) covalently bonded to the coenzyme. A shift of electrons releases acetaldehyde, regenerating the carbanion.

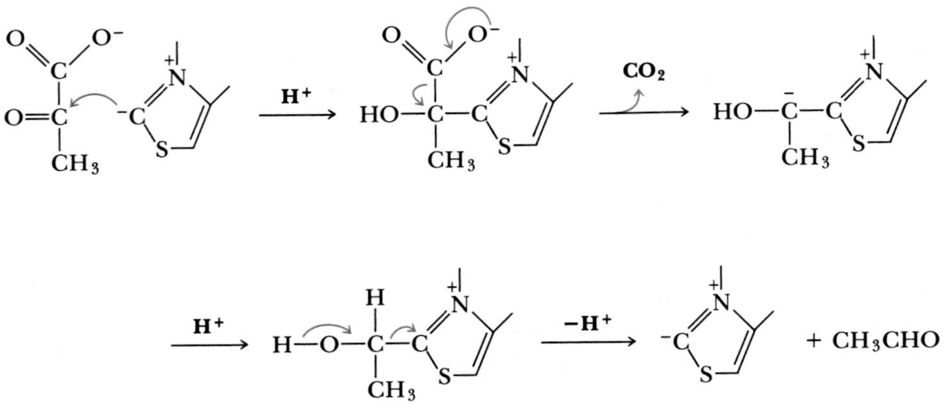

$$\underset{\text{Acetaldehyde}}{\overset{\displaystyle HC=O}{\underset{\displaystyle CH_3}{|}}} + NADH + H^+ \longrightarrow \underset{\text{Ethanol}}{\overset{\displaystyle H_2C-OH}{\underset{\displaystyle CH_3}{|}}} + NAD^+ \qquad (13.15[b])$$

Box 13.1 discusses one connection between acetaldehyde and ethanol.

## BOX 13.1
## FETAL ALCOHOL SYNDROME

The complex of injuries to a fetus caused by maternal consumption of ethanol is called fetal alcohol syndrome. In catabolism of ethanol by the body, the first step is conversion to acetaldehyde—the reverse of the last reaction of alcoholic fermentation. The level of acetaldehyde in the blood of a pregnant woman is the key to detecting fetal alcohol syndrome. It has recently been shown that the acetaldehyde is transferred across the placenta and accumulates in the liver of the fetus. The acetaldehyde in turn is easily oxidized to breakdown products and is responsible for the harmful effects.

The reduction reaction of alcoholic fermentation is similar to the reduction of pyruvate to lactate, in the sense that it provides for recycling of $NAD^+$ and thus allows further anaerobic oxidation (fermentation) reactions. The net reaction for alcoholic fermentation is

$$\text{Glucose} + 2\ \text{ADP} + 2\ P_i + 2\ H^+ \longrightarrow 2\ \text{ethanol} + 2\ \text{ATP} + 2\ CO_2 + 2\ H_2O$$

$NAD^+$ and NADH do not appear explicitly in the net equation. It is essential that the recycling of NADH to $NAD^+$ takes place here, just as it does when lactate is produced, so that there can be further anaerobic oxidation. **Alcohol dehydrogenase,** the enzyme that catalyzes the conversion of acetaldehyde to ethanol, is similar to lactate dehydrogenase in many ways. The most striking similarity is that both are NADH-linked dehydrogenases. Both enzymes are allosteric enzymes, and both are tetramers.

## 13.6
## THE PENTOSE PHOSPHATE PATHWAY—AN ALTERNATE PATHWAY FOR GLUCOSE METABOLISM

In glycolysis, one of our most important concerns was the production of ATP. In the process NADH is produced and $NAD^+$ is regenerated. In the **pentose phosphate** pathway, the production of ATP is not the crux of the matter. As the name of the pathway indicates, five-carbon sugars including ribose are produced from glucose. Ribose plays an important role in the structure of nucleic acids. Another important facet of the pentose phosphate pathway is the production of NADPH, a compound that differs from NADH by having one extra phosphate group esterified to the ribose ring of the adenine nucleotide portion of the molecule (Figure 13.12). A more important difference is the way these two coenzymes function. NADH is produced in the oxidative reactions that give rise to ATP. NADPH is a reducing agent in biosynthesis, which by its very nature is a reductive process.

The pentose phosphate pathway begins with a series of oxidation reactions that produce NADPH and five-carbon sugars. The remainder of the pathway involves nonoxidative reshuffling of the carbon skeletons of

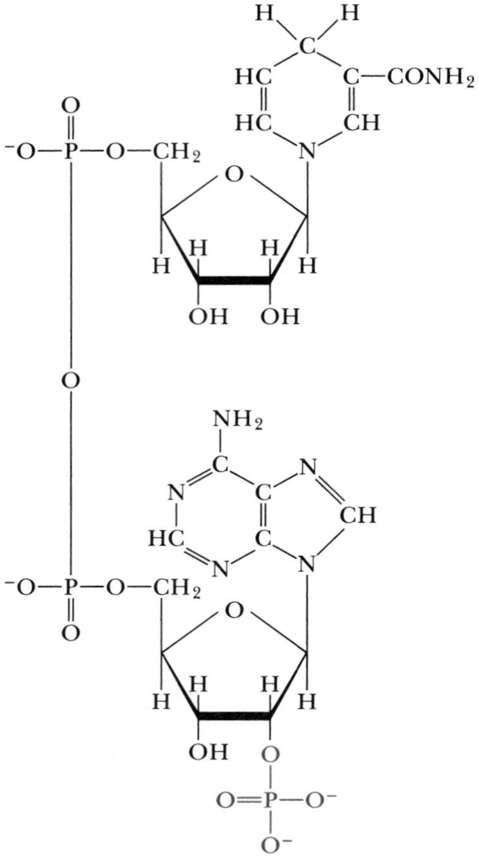

**FIGURE 13.12**   The structure of
reduced nicotinamide adenine
dinucleotide phosphate (NADPH).

the sugars involved. The products of these nonoxidative reactions include
substances such as fructose 6-phosphate and glyceraldehyde 3-phosphate,
which play a role in glycolysis.

## Oxidative Reactions of the Pentose Phosphate Pathway

In the first reaction of the pathway, glucose 6-phosphate is oxidized to
6-phosphoglucono-δ-lactone (Figure 13.13). The enzyme that catalyzes
this reaction is **glucose 6-phosphate dehydrogenase.** Note that NADPH
is produced by the reaction. The same sort of reaction takes place in the
oxidation of sugars by metal ions (Section 6.2). The hemiacetal form of the
sugar is a cyclic structure resulting from the addition of one of the hydroxyl
groups to the aldehyde group. It is oxidized to a lactone, which is a cyclic
ester formed between the resulting carboxylic acid group and an alcohol
group elsewhere in the molecule.

The cyclic ester bond of 6-phosphoglucono-δ-lactone is hydrolyzed in
the next reaction. The open-chain compound 6-phosphogluconate (the
glucuronic acid is present in its ionized form) is the product of this reaction,
which is catalyzed by the enzyme **lactonase** (Figure 13.13).

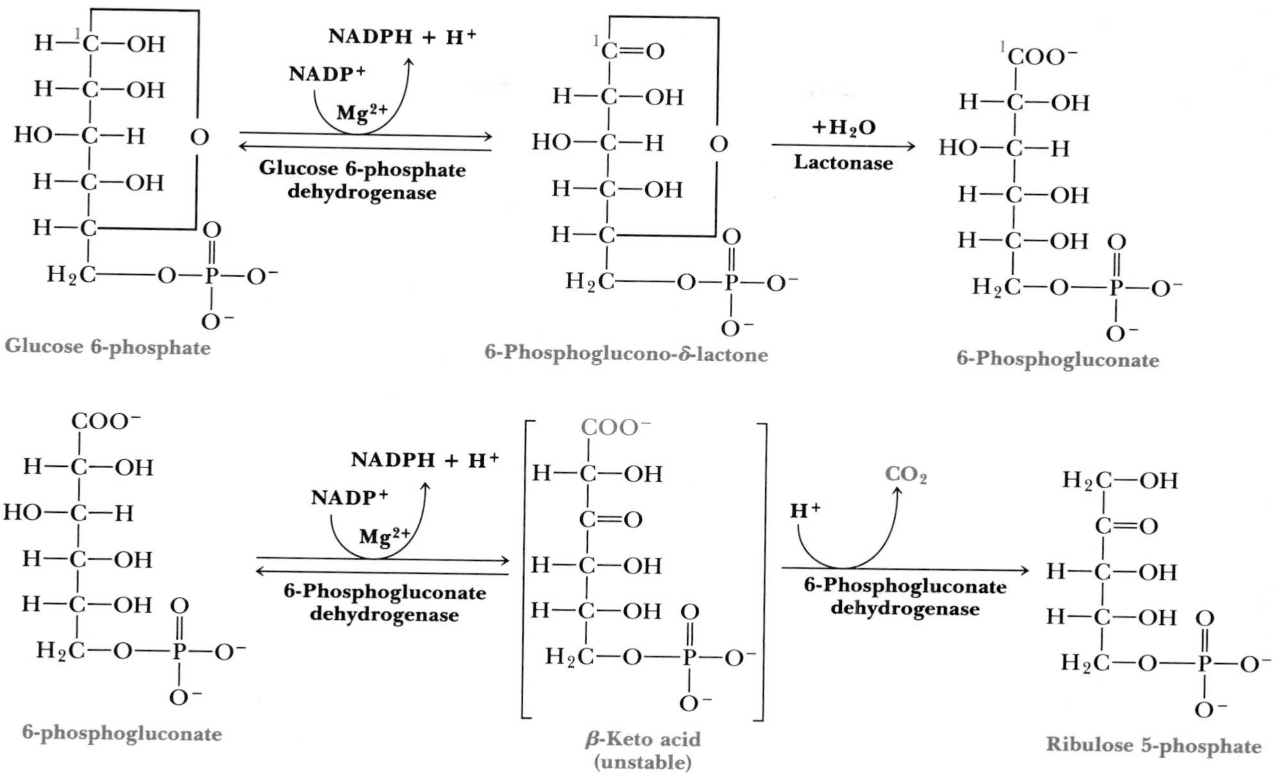

NADPH + H$^+$

NADP$^+$

Mg$^{2+}$

**Glucose 6-phosphate dehydrogenase**

+H$_2$O

**Lactonase**

Glucose 6-phosphate

6-Phosphoglucono-δ-lactone

6-Phosphogluconate

NADPH + H$^+$

NADP$^+$

Mg$^{2+}$

**6-Phosphogluconate dehydrogenase**

CO$_2$

H$^+$

**6-Phosphogluconate dehydrogenase**

6-phosphogluconate

β-Keto acid (unstable)

Ribulose 5-phosphate

**FIGURE 13.13** The oxidative reactions of the pentose phosphate pathway.

The next reaction is an oxidative decarboxylation, and NADPH is produced once again. The 6-phosphogluconate molecule loses its carboxyl group, which is released as carbon dioxide, and the five-carbon keto-sugar (ketose) ribulose 5-phosphate is the other product. The enzyme that catalyzes this reaction is **6-phosphogluconate dehydrogenase.** Note that in the process the C-3 hydroxyl group of the 6-phosphogluconate is oxidized to form a β-keto acid, which is unstable and readily decarboxylates to form ribulose 5-phosphate.

## Nonoxidative Reactions of the Pentose Phosphate Pathway

In the remaining steps of the pentose phosphate pathway, several reactions involve transfer of two- and three-carbon units. In order to keep track of the carbon backbone of the sugars and their aldehyde and ketone functional groups, we shall write the formulas in the open-chain form.

There are two different reactions in which ribulose 5-phosphate isomerizes. In one of these reactions, catalyzed by **phosphopentose 3-epimerase,** there is an inversion of configuration around carbon atom 3, producing xylulose 5-phosphate, which is also a ketose (Figure 13.14). (See Section 6.1; the term epimer refers to two sugars that differ from each other only by configuration around one carbon atom. In this reaction the enzyme catalyzes the conversion of one epimer to the other.) The other

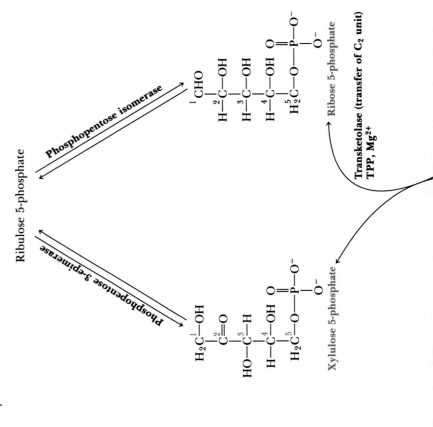

**FIGURE 13.14** The nonoxidative reactions of the pentose phosphate pathway. Carbons from xylulose are numbered in color.

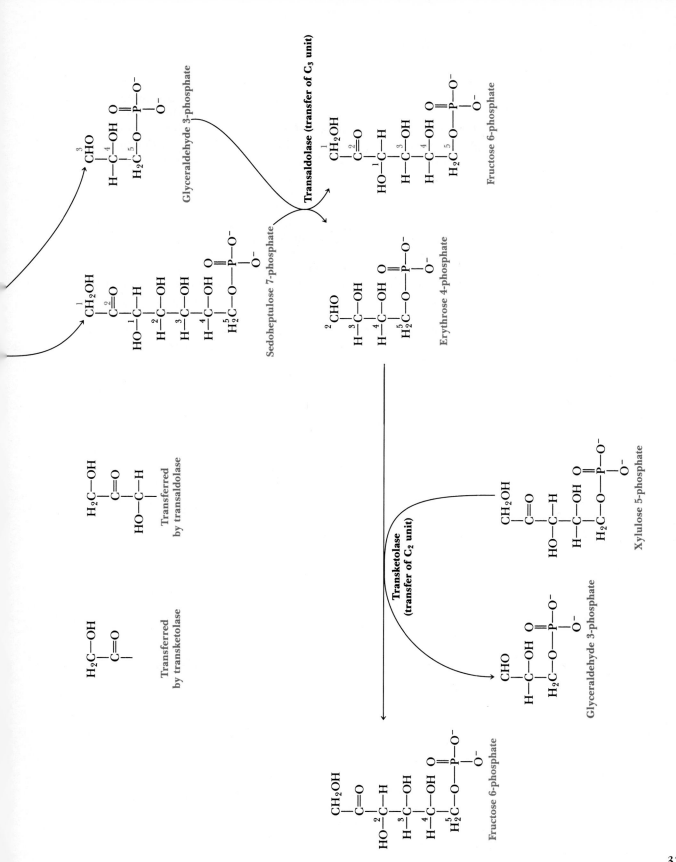

Glyceraldehyde 3-phosphate

Transaldolase (transfer of C₃ unit)

Fructose 6-phosphate

Sedoheptulose 7-phosphate

Erythrose 4-phosphate

Transferred by transaldolase

Transferred by transketolase

Transketolase (transfer of C₂ unit)

Xylulose 5-phosphate

Glyceraldehyde 3-phosphate

Fructose 6-phosphate

321

isomerization reaction, catalyzed by **phosphopentose isomerase,** produces a sugar with an aldehyde group (an aldose) rather than a ketone. In this second reaction, ribulose 5-phosphate isomerizes to ribose 5-phosphate (Figure 13.14). Ribose 5-phosphate is a necessary building block for the synthesis of nucleic acids and coenzymes such as NADH.

The group-transfer reactions that link the pentose phosphate pathway with glycolysis require the two five-carbon sugars produced by the isomerization of ribulose 5-phosphate. Two molecules of xylulose 5-phosphate and one molecule of ribose 5-phosphate rearrange to give two molecules of fructose 6-phosphate and one molecule of glyceraldehyde 3-phosphate. In other words, three molecules of pentose (with five carbon atoms each) give two molecules of hexose (with six carbon atoms each) and one molecule of a triose (with three carbon atoms). The total number of carbon atoms (15) does not change, but there is considerable rearrangement as a result of group transfer.

Two enzymes, transketolase and transaldolase, are responsible for the reshuffling of the carbon atoms of sugars such as ribose 5-phosphate and xylulose 5-phosphate in the remainder of the pathway, which consists of three reactions. Transketolase transfers a two-carbon unit, and transaldolase transfers a three-carbon unit. Transketolase catalyzes the first and third reactions in the rearrangement process, and transaldolase catalyzes the second reaction. The results of these transfers can be summarized in Table 13.2.

In the first of these reactions a two-carbon unit from xylulose 5-phosphate (a five-carbon ketose) is transferred to ribose 5-phosphate (a five carbon-aldose) to give sedoheptulose 7-phosphate (a seven-carbon ketose) and glyceraldehyde 3-phosphate (a three-carbon aldose), as shown in Figure 13.14. This reaction is catalyzed by transketolase. Carbon atoms 1 and 2 of the xylulose unit are transferred to carbon atom 1 of the ribose in this reaction. It is a general rule in these three reactions that a ketose is always the donor in the group-transfer reaction, and an aldose is always the acceptor.

In the reaction catalyzed by transaldolase, a three-carbon unit is transferred from the seven-carbon ketose sedoheptulose 7-phosphate to the three-carbon aldose glyceraldehyde 3-phosphate (Figure 13.14). Car-

**TABLE 13.2   Group Transfer Reactions in the Pentose Phosphate Pathway**

| | REACTANTS | ENZYME | PRODUCTS |
|---|---|---|---|
| | $C_5 + C_5$ | Transketolase ⇌ Two-carbon shift | $C_7 + C_3$ |
| | $C_7 + C_3$ | Transaldolase ⇌ Three-carbon shift | $C_6 + C_4$ |
| | $C_5 + C_4$ | Transketolase ⇌ Two-carbon shift | $C_6 + C_3$ |
| Net Reaction | $3\,C_5$ | ⟶ | $2\,C_6 + C_3$ |

bon atoms 1 through 3 of the sedoheptulose 7-phosphate are transferred to carbon atom 1 of the glyceraldehyde 3-phosphate. The products of the reaction are fructose 6-phosphate (a six-carbon ketose) and erythrose 4-phosphate (a four-carbon aldose).

In the final reaction of this type in the pathway, the donor ketose is xylulose 5-phosphate and the acceptor aldose is erythrose 4-phosphate. This reaction is catalyzed by transketolase, and once again a two-carbon unit is transferred. The products of the reaction are fructose 6-phosphate and glyceraldehyde 3-phosphate (Figure 13.14).

In the pentose phosphate pathway, glucose 6-phosphate can be converted to fructose 6-phosphate and glyceraldehyde 3-phosphate by a means other than the glycolytic pathway. For this reason the pentose phosphate pathway is also called the **hexose monophosphate shunt,** and this name is used in some texts. A major feature of the pentose phosphate pathway is the production of ribose 5-phosphate and NADPH. The control mechanisms of the pentose phosphate pathway can respond to the varying needs of organisms for either or both of these compounds.

## Control of the Pentose Phosphate Pathway

As we have seen, the reactions catalyzed by transketolase and transaldolase are reversible. We shall discuss some of the distinctive properties of these enzymes as we see how the pentose phosphate pathway responds to the needs of an organism. The starting material, glucose 6-phosphate, will undergo different reactions depending on whether there is a greater need for ribose 5-phosphate or for NADPH. The operation of the oxidative portion of the pathway depends strongly on the organism's requirement for NADPH. The need for ribose 5-phosphate can be met in other ways, since ribose 5-phosphate can be obtained from glycolytic intermediates without the oxidative reactions of the pentose phosphate pathway (Figure 13.15).

**FIGURE 13.15**    Relationships between the pentose phosphate pathway and glycolysis. If the organism needs NADPH more than ribose 5-phosphate, the entire pentose phosphate pathway is operative. If the organism needs ribose 5-phosphate more than NADPH, the nonoxidative reactions of the pentose phosphate pathway, operating in reverse, produce ribose 5-phosphate (see text).

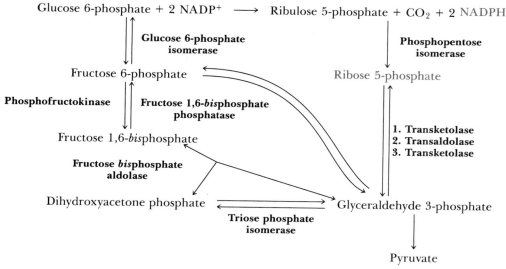

If the organism needs more NADPH than ribose 5-phosphate, the reaction series goes through the complete pathway we have just discussed. The oxidative reactions at the beginning of the pathway are needed to produce NADPH. The net reaction for the oxidative portion of the pathway is

$$6 \text{ Glucose 6-phosphate} + 12 \text{ NADP}^+ + 6 \text{ H}_2\text{O} \longrightarrow$$
$$6 \text{ ribose 5-phosphate} + 6 \text{ CO}_2 + 12 \text{ NADPH} + 12 \text{ H}^+$$

The ribose 5-phosphate then undergoes the nonoxidative reactions to give glyceraldehyde 3-phosphate and fructose 6-phosphate. In this series of reactions, there is no further production or use of NADPH, nor is there production or use of ATP. The two sugar phosphates, glyceraldehyde 3-phosphate and fructose 6-phosphate, are recycled to glucose 6-phosphate by a reversal of the reactions of glycolysis, rather than undergoing further stages of glycolysis. One molecule of $CO_2$ is released by each molecule of glucose 6-phosphate processed by the complete pathway. If six molecules of glucose 6-phosphate are processed, six molecules of $CO_2$ are produced and the net result is the same as if one molecule of glucose had been completely oxidized to $CO_2$. In fact, one molecule of $CO_2$ is contributed by each of six molecules of glucose 6-phosphate.

A likely alternate fate for the glyceraldehyde 3-phosphate produced by the pentose phosphate pathway is that it can undergo further glycolytic reactions to produce pyruvate, rather than being recycled to glucose 6-phosphate. In such a situation the organism produces ATP from glyceraldehyde 3-phosphate, just as we have seen in the later steps of the glycolytic pathway, as well as NADPH from the pentose phosphate pathway. The net reaction for all stages of conversion of glucose 6-phosphate to pyruvate—involving both the reactions of the pentose phosphate pathway and the glycolytic reactions involving glyceraldehyde 3-phosphate—is

$$3 \text{ Glucose 6-phosphate} + 6 \text{ NADP}^+ + 5 \text{ NAD}^+ + 5 \text{ P}_i +$$
$$8 \text{ ADP} \longrightarrow 5 \text{ pyruvate} + 3 \text{ CO}_2 + 6 \text{ NADPH} + 5 \text{ NADH} +$$
$$8 \text{ ATP} + 2 \text{ H}_2\text{O} + 8 \text{ H}^+$$

Note that six, rather than twelve, NADPH molecules are produced when glyceraldehyde 3-phosphate undergoes further glycolytic reactions rather than being recycled to glucose, but the organism also gains ATP.

Box 13.2 discusses a clinical manifestation of an enzyme malfunction in the pentose phosphate pathway.

If the organism has a greater need for ribose 5-phosphate than for NADPH, fructose 6-phosphate and glyceraldehyde 3-phosphate can give rise to ribose 5-phosphate by the successive operation of the transketolase and transaldolase reactions, bypassing the oxidative portion of the pentose phosphate pathway (Figure 13.15). The reactions catalyzed by transketolase and transaldolase are reversible, and this fact plays an important role in the ability of the organism to adjust its metabolism to changes in conditions. We shall now take a look at the mode of action of these two enzymes.

## BOX 13.2
## THE PENTOSE PHOSPHATE PATHWAY AND HEMOLYTIC ANEMIA

T he pentose phosphate pathway is the only source of NADPH in red blood cells, which as a result are highly dependent on the proper functioning of the enzymes involved. A glucose 6-phosphate dehydrogenase deficiency leads to an NADPH deficiency, which can in turn lead to **hemolytic anemia** because of wholesale destruction of red blood cells.

The relationship between NADPH deficiency and anemia is an indirect one. NADPH is required to reduce the peptide glutathione from the disulfide to the free thiol form. The presence of the reduced form of gluta-

thione is necessary for the maintenance of the sulfhydryl groups of hemoglobin and other proteins in their reduced form, as well as for keeping the Fe(II) of hemoglobin in its reduced form. Glutathione also maintains the integrity of red cells, by a mechanism that is not well understood. About 11% of African-Americans are affected by glucose 6-phosphate dehydrogenase deficiency. This condition, like the sickle-cell trait, leads to increased resistance to malaria, accounting for some of its persistence in the gene pool in spite of its otherwise deleterious consequences.

Glutathione and its reactions. (a) The structure of glutathione. (b) The role of NADPH in the production of glutathione. (c) The role of glutathione in maintaining the reduced form of protein sulfhydryl groups.

(a)

$$+H_3N-CH$$ 

Reduced glutathione
($\gamma$-glutamylcysteinylglycine)

(b) $\gamma$-Glu–Cys–Gly
|
S
|
S
|
$\gamma$-Glu–Cys–Gly
$+ \text{NADPH} + \text{H}^+ \xrightarrow{\text{Glutathione reductase}} 2\ \gamma\text{-Glu–Cys–Gly} + \text{NADP}^+$
|
SH

(c) $2\ \gamma$-Glu–Cys–Gly $+ \text{R-S-S-R} \longrightarrow \gamma\text{-Glu–Cys–Gly} + 2\ \text{RSH}$
|
SH
|
S
|
S
|
$\gamma$-Glu–Cys–Gly

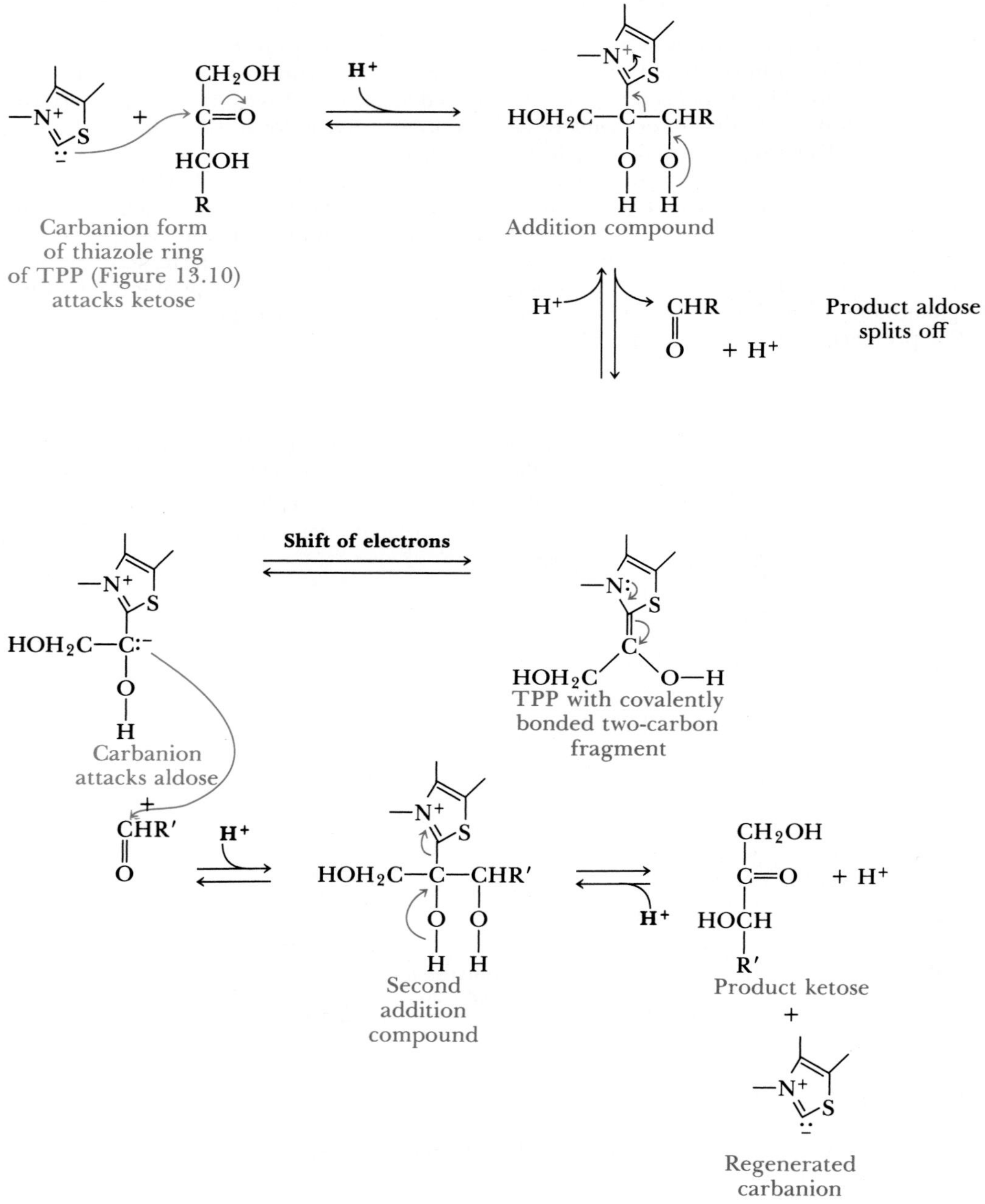

**FIGURE 13.16**    The mechanism of the transketolase reaction.

Transaldolase has many features in common with the enzyme aldolase, which we met in the glycolytic pathway. Both an aldol cleavage and an aldol condensation occur at different stages of the reaction. We have already seen the mechanisms of aldol cleavage and aldol condensation, involving the formation of a Schiff base, when we discussed the aldolase reaction in glycolysis, and we need not discuss this point further.

Transketolase resembles pyruvate decarboxylase, the enzyme that converts pyruvate to acetaldehyde (Section 13.5), in that it also requires $Mg^{2+}$ and thiamine pyrophosphate (TPP). As in the pyruvate decarboxylase reaction, a carbanion plays a crucial role in the reaction mechanism. The carbanion attacks the carbonyl group of a ketose to form an addition compound (Figure 13.16). An aldose splits off from the addition compound, leaving behind a two-carbon fragment covalently bonded to the TPP. A shift of electrons produces a second carbanion, in which the charge is localized on one of the newly bonded carbons. The carbanion attacks a new aldose to form a second addition compound. The ketose product is formed and subsequently released from the enzyme.

## 13.7
## GLUCONEOGENESIS

The conversion of pyruvate to glucose occurs by a process called **gluconeogenesis.** Gluconeogenesis is not the exact reversal of glycolysis. Some of the reactions of glycolysis are essentially irreversible; these reactions are bypassed in gluconeogenesis. An analogy is a hiker who goes directly down a steep slope but who climbs back up the hill by an alternate, easier route. We shall see that the biosynthesis and the degradation of many important biomolecules follow different pathways.

There are three irreversible steps in glycolysis, and it is in these three reactions that the differences between glycolysis and gluconeogenesis are found. The first is the production of pyruvate (and ATP) from phosphoenolpyruvate. The second is the production of fructose 1,6-*bis*phosphate from fructose 6-phosphate, and the third is the production of glucose 6-phosphate from glucose. The first of these reactions is exergonic, and the reverse reaction is endergonic. Reversing the second and third reactions requires the production of ATP from ADP, which is also an endergonic reaction. The net result of gluconeogenesis is the reversal of these three glycolytic reactions, but the pathway is different, with different reactions and different enzymes (Figure 13.17).

## Oxaloacetate is an Intermediate in the Production of Phosphoenolpyruvate in Gluconeogenesis

The conversion of pyruvate to phosphoenolpyruvate in gluconeogenesis takes place in two steps. The first step is the reaction of pyruvate and carbon dioxide to give oxaloacetate. This step requires energy, which is available from the hydrolysis of ATP.

$$
\begin{array}{c}
\overset{\displaystyle O}{\underset{\displaystyle }{\|}} \\
\text{C}-\text{O}^- \\
| \\
\text{C}=\text{O} \\
| \\
\text{CH}_3 \\
\textbf{Pyruvate}
\end{array}
+ \text{ATP} + \text{CO}_2 + \text{H}_2\text{O}
\xrightleftharpoons[\substack{\textbf{acetyl-CoA}\\ \textbf{biotin}\\ \\ \textbf{pyruvate}\\ \textbf{carboxylase}}]{\textbf{Mg}^{2+}}
\begin{array}{c}
\overset{\displaystyle O}{\underset{\displaystyle }{\|}} \\
\text{C}-\text{O}^- \\
| \\
\text{C}=\text{O} \\
| \\
\text{CH}_2 \\
| \\
\text{C}-\text{O}^- \\
| \\
\text{O} \\
\textbf{Oxaloacetate}
\end{array}
+ \text{ADP} + \text{P}_i
$$

(13.16)

$$\text{Pyruvate} + \text{ATP} + \text{CO}_2 + \text{H}_2\text{O} \rightleftharpoons$$
$$\text{oxaloacetate} + \text{ADP} + \text{P}_i + 2\text{H}^+$$

The enzyme that catalyzes this reaction is **pyruvate carboxylase,** an allosteric enzyme. Acetyl-CoA is an allosteric effector that activates pyruvate carboxylase. This reaction is a control point in the pathway because the enzyme that catalyzes it is subject to allosteric regulation. If high levels of acetyl-CoA are present (in other words, if there is more acetyl-CoA than is needed to supply the citric acid cycle), pyruvate (a precursor of acetyl-CoA) can be diverted to gluconeogenesis. Magnesium ion ($\text{Mg}^{2+}$) and biotin are also required for effective catalysis. We have seen $\text{Mg}^{2+}$ as a cofactor before, but we have not seen biotin and it requires some discussion.

**Biotin** is a carrier of carbon dioxide; it has a specific site for covalent attachment of $\text{CO}_2$ (Figure 13.18). The carboxyl group of the biotin forms an amide bond with the $\epsilon$-amino group of a specific lysine side chain of pyruvate carboxylase. The $\text{CO}_2$ is attached to the biotin, which in turn is covalently bonded to the enzyme, and then the $\text{CO}_2$ is shifted to pyruvate to form oxaloacetate (Figure 13.19).

The conversion of oxaloacetate to phosphoenolpyruvate is catalyzed by the enzyme **phosphoenolpyruvate carboxykinase.** This reaction also involves hydrolysis of a nucleoside triphosphate, GTP in this case rather than ATP.

$$
\begin{array}{c}
\overset{\displaystyle O}{\underset{\displaystyle }{\|}} \\
\text{C}-\text{O}^- \\
| \\
\text{C}=\text{O} \\
| \\
\text{CH}_2 \\
| \\
\text{C}-\text{O}^- \\
\| \\
\text{O} \\
\textbf{Oxaloacetate}
\end{array}
+ \text{GTP}
\xrightarrow[\substack{\textbf{phosphoenolpyruvate}\\ \textbf{carboxykinase}}]{\textbf{Mg}^{2+}}
\begin{array}{c}
\overset{\displaystyle O}{\underset{\displaystyle }{\|}} \\
\text{C}-\text{O}^- \\
| \\
\text{C}-\text{O}-\overset{\displaystyle O}{\underset{\displaystyle O_-}{\overset{\|}{\text{P}}}}-\text{O}^- \\
\| \\
\text{CH}_2 \\
\textbf{Phosphoenolpyruvate}
\end{array}
+ \text{CO}_2 + \text{GDP}
$$

(13.17)

$$\text{Oxaloacetate} + \text{GTP} \overset{\text{Mg}^{2+}}{\rightleftharpoons} \text{phosphoenolpyruvate} + \text{CO}_2 + \text{GDP}$$

The successive carboxylation and decarboxylation reactions are both close to equilibrium and, as a result, the conversion of pyruvate to phosphoenolpyruvate is also close to equilibrium ($\Delta G^{\circ\prime} = 2.1$ kJ mol$^{-1}$ = 0.5 kcal

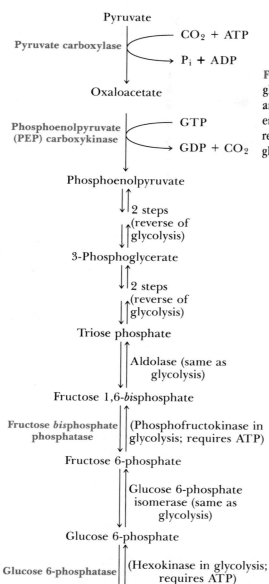

Pyruvate
Pyruvate carboxylase
$CO_2$ + ATP
$P_i$ + ADP
Oxaloacetate
Phosphoenolpyruvate (PEP) carboxykinase
GTP
GDP + $CO_2$
Phosphoenolpyruvate
2 steps (reverse of glycolysis)
3-Phosphoglycerate
2 steps (reverse of glycolysis)
Triose phosphate
Aldolase (same as glycolysis)
Fructose 1,6-*bis*phosphate
Fructose *bis*phosphate phosphatase
(Phosphofructokinase in glycolysis; requires ATP)
Fructose 6-phosphate
Glucose 6-phosphate isomerase (same as glycolysis)
Glucose 6-phosphate
Glucose 6-phosphatase
(Hexokinase in glycolysis; requires ATP)
Glucose

**FIGURE 13.17** The pathway of gluconeogenesis. The enzymes in red are unique to this pathway. The enzymes that catalyze reversible reactions are shared with the glycolytic pathway.

**FIGURE 13.18** The structure of biotin and its mode of attachment to pyruvate carboxylate.

Lysine residue of enzyme

**Biotin**

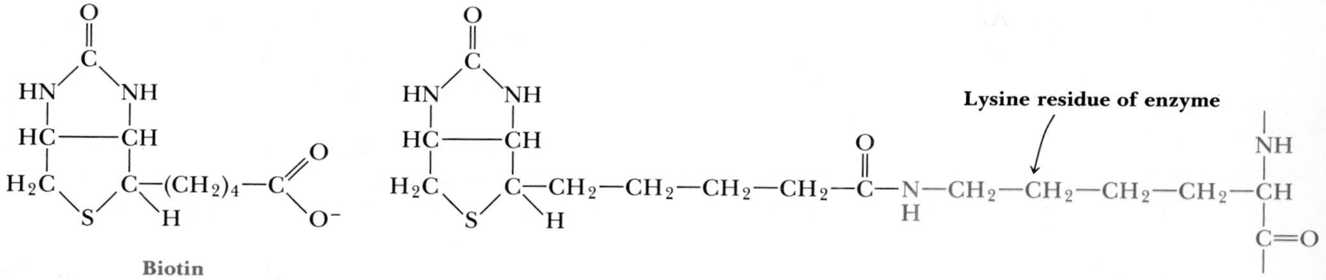

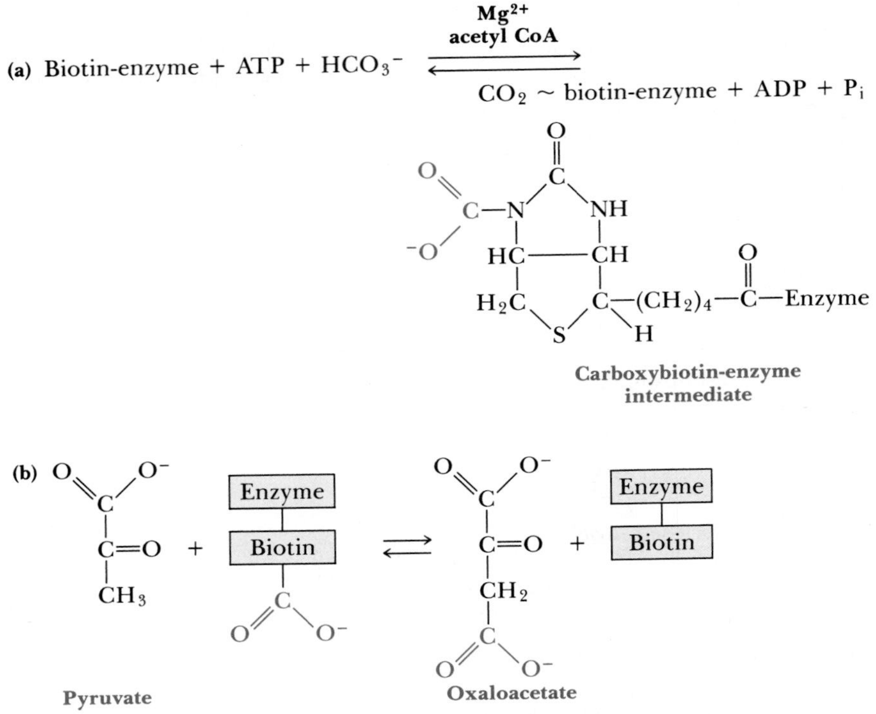

(a) Biotin-enzyme + ATP + $HCO_3^-$ $\xrightleftharpoons{\overset{Mg^{2+}\ \text{acetyl CoA}}{}}$ $CO_2 \sim$ biotin-enzyme + ADP + $P_i$

Carboxybiotin-enzyme
intermediate

(b) Pyruvate + Enzyme–Biotin $\rightleftharpoons$ Oxaloacetate + Enzyme Biotin

Pyruvate                    Oxaloacetate

**FIGURE 13.19**   The two stages of the pyruvate carboxylate reaction. (a) $CO_2$ is attached to the biotinylated enzyme. (b) $CO_2$ is transferred from the biotinylated enzyme to pyruvate, forming oxaloacetate.

$mol^{-1}$). A small increase in the level of oxaloacetate can drive the equilibrium to the right, while a small increase in the level of phosphoenol-pyruvate can drive it to the left. Such behavior is an example of a concept well known in general chemistry, the **law of mass action,** which states that a reaction will proceed to the right on addition of products and to the left on addition of reactants (Section 11.4).

Pyruvate + ATP + GTP $\rightleftharpoons$ Phosphoenolpyruvate + ADP + GDP + $P_i$

## The Role of Sugar Phosphates in Gluconeogenesis

The other two reactions in which gluconeogenesis differs from glycolysis are ones in which a phosphate-ester bond to a sugar-hydroxyl group is hydrolyzed. Both reactions are catalyzed by phosphatases, and both reactions are exergonic. The first reaction is the hydrolysis of fructose 1,6-*bis*phosphate to produce fructose 6-phosphate and phosphate ion ($\Delta G^{\circ\prime} = -16.7$ kJ $mol^{-1} = -4.0$ kcal $mol^{-1}$).

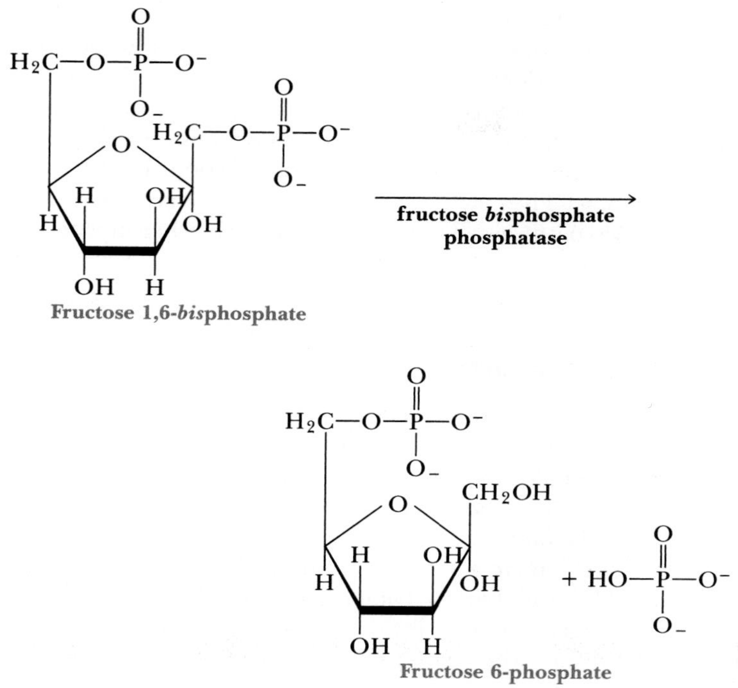

Fructose 1,6-*bis*phosphate $\longrightarrow$ fructose 6-phosphate + $P_i$

This reaction is catalyzed by the enzyme **fructose *bis*phosphate phosphatase,** an allosteric enzyme strongly inhibited by adenosine monophosphate (AMP) but stimulated by ATP. Because of allosteric regulation, this reaction is also a control point in the pathway. When the cell has an ample supply of ATP, the formation rather than the breakdown of glucose is favored.

The second reaction is the hydrolysis of glucose 6-phosphate to glucose and phosphate ion ($\Delta G^{\circ\prime} = -13.8$ kJ mol$^{-1}$ = $-3.3$ kcal mol$^{-1}$). The enzyme that catalyzes this reaction is glucose 6-phosphate phosphatase.

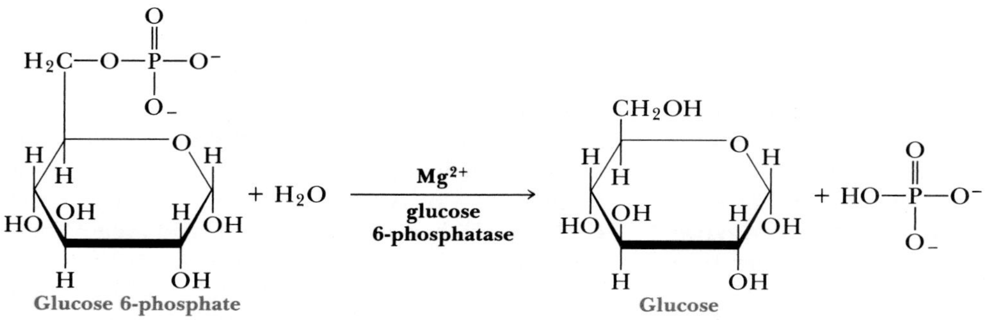

Glucose 6-phosphate $\longrightarrow$ glucose + $P_i$

When we discussed glycolysis earlier in this chapter, we saw that both of the phosphorylation reactions that are the reverse of the two phosphatase-catalyzed reactions are endergonic. In glycolysis the phosphorylation reactions must be coupled to the hydrolysis of ATP to make them exergonic and thus energetically allowed. In gluconeogenesis the organism can make direct use of the fact that the hydrolysis reactions of the sugar phosphates are exergonic. The corresponding reactions are not the reverse of each other in the two pathways. They differ from each other in whether they require ATP, and in the enzymes involved.

## A Last Look at Gluconeogenesis

Glycolysis in skeletal muscle produces lactate under conditions of oxygen debt such as a sprint. Skeletal muscle has comparatively few mitochondria, so metabolism is largely anaerobic in this tissue. The buildup of lactate is responsible for the muscular aches that follow strenuous exercise. Gluconeogenesis recycles the lactate that is produced (lactate is first oxidized to pyruvate). The process occurs to a great extent in the liver after the lactate is transported there by the blood. Glucose produced in the liver is transported back to skeletal muscle by the blood, where it is stored as glycogen, an energy store for the next burst of exercise. Note that we have here a division of labor between two different types of organs, muscle and liver. In the same cell (of whatever type) these two metabolic pathways, glycolysis and gluconeogenesis, are not highly active simultaneously. When the cell needs ATP, glycolysis is more active; when there is little need for ATP, gluconeogenesis is more active. Because of the hydrolysis of ATP and GTP in the reactions of gluconeogenesis that differ from those of glycolysis, the overall pathway from two molecules of pyruvate back to one molecule of glucose is exergonic ($\Delta G°' = -37.6$ kJ mol$^{-1}$ = $-9.0$ kcal mol$^{-1}$, for one mole of glucose). The conversion of pyruvate to lactate is exergonic, which means that the reverse reaction is endergonic. The energy released by the exergonic conversion of pyruvate to glucose by gluconeogenesis facilitates the conversion of lactate to pyruvate.

## SUMMARY

In glycolysis, one molecule of glucose gives rise after a long series of reactions to two molecules of pyruvate. There are two reactions in the pathway in which one molecule of ATP is hydrolyzed for each molecule of glucose metabolized. There are also two reactions in which two molecules of ATP are produced by phosphorylation of ADP for each molecule of glucose, giving a total of four ATP molecules produced. There is a net gain of two ATP molecules for each molecule of glucose processed in glycolysis.

Several metabolic fates are possible for pyruvate. In aerobic metabolism, pyruvate loses carbon dioxide; the remaining two carbon atoms become linked to coenzyme A as an acetyl group to form acetyl-CoA, which then enters the citric acid cycle. There are two fates for pyruvate in anaerobic metabolism. In organisms capable of alcoholic fermentation, pyruvate loses carbon dioxide, this time producing acetaldehyde, which in turn is reduced to produce ethanol. The common fate of pyruvate in anaerobic metabolism is reduction to lactate; in this chapter we concentrated on the conversion of glucose to lactate, called anaerobic glycolysis to distinguish it from conversion of glucose to pyruvate.

The anaerobic breakdown of glucose to lactate can be summarized as follows:

$$\text{Glucose} + 2\ \text{ADP} + 2\ \text{P}_i \longrightarrow 2\ \text{lactate} + 2\ \text{ATP}$$

The overall process of glycolysis is exergonic.

|  | $\Delta G^{\circ\prime}$, mol$^{-1}$ | |
| --- | --- | --- |
|  | kJ | kcal |
| Glucose + 2 ADP + 2 P$_i$ $\longrightarrow$ 2 pyruvate + 2 ATP | −73.3 | −17.5 |
| Glucose + 2 ADP + 2 P$_i$ $\longrightarrow$ 2 lactate + 2 ATP | −123.5 | −29.5 |

Without production of ATP, glycolysis would be still more exergonic but the energy released would be lost to the organism and dissipated as heat.

The pentose phosphate pathway is an alternate pathway for glucose metabolism. In this pathway five-carbon sugars, including ribose, are produced from glucose. In the oxidative reactions of the pathway, NADPH is produced as well. Control of the pathway allows the organism to adjust the relative levels of production of five-carbon sugars and of $\cdot$ NADPH according to its needs.

The conversion of pyruvate to glucose takes place by a process called gluconeogenesis. Gluconeogenesis is not the exact reversal of glycolysis. There are three irreversible steps in glycolysis, and it is in these three reactions that gluconeogenesis differs from glycolysis. The net result of gluconeogenesis is the reversal of these three glycolytic reactions, but the pathway is different, with different reactions and different enzymes. In the same cell, glycolysis and gluconeogenesis are not highly active simultaneously. When the cell needs ATP, glycolysis is more active; when there is little need for ATP, gluconeogenesis is more active.

## EXERCISES

1. What does the material of this chapter have to do with beer? With tired and aching muscles?
2. Which reaction or reactions we have met in this chapter require ATP? Which reaction or reactions produce ATP? List the enzymes that catalyze the reactions that require and that produce ATP.
3. Which reaction or reactions we have met in this chapter require NADH? Which reaction or reactions require NAD$^+$? List the enzymes that catalyze the reactions that require NADH and that require NAD$^+$.
4. List three differences in structure or function between NADH and NADPH.
5. Explain the following statement: in the absence of glycogen breakdown, the phosphorylation of glucose, catalyzed by hexokinase, is the rate-determining step of glycolysis.
6. Distinguish between the rate-determining step of glycolysis (see Exercise 5) and the committed step of glycolysis, namely the phosphorylation of fructose 6-phosphate, catalyzed by phosphofructokinase.
7. Which of the enzymes discussed in this chapter are NADH-linked dehydrogenases?
8. What requirements for acetyl-CoA and for biotin have we discussed in this chapter?

9. Explain the origin of the name of the enzyme aldolase.
10. What is the connection between material in this chapter and hemolytic anemia?
11. Show how the estimate of 33 percent efficiency of energy use in anaerobic glycolysis is derived.
12. Which steps of glycolysis are irreversible? What bearing does this observation have on the reactions in which gluconeogenesis differs from glycolysis?
13. Show that the reaction

    $$\text{Glucose} \longrightarrow 2\ \text{glyceraldehyde 3-phosphate}$$

    is slightly endergonic ($\Delta G^{\circ\prime} = 2.5$ kJ mol$^{-1}$ = 0.6 kcal mol$^{-1}$), i.e., it is not too far from equilibrium. Use the data in Table 13.1.
14. Show by a series of equations the energetics of phosphorylation of ADP by phosphoenolpyruvate.
15. Using the Lewis electron-dot notation, show explicitly the transfer of electrons in the following redox reactions.
    (a) Pyruvate + NADH + H$^+$ $\longrightarrow$ lactate + NAD$^+$
    (b) Glucose 6-phosphate + NADP$^+$ $\longrightarrow$ 6-phosphoglucono-$\delta$-lactone + NADPH + H$^+$
    (c) Acetaldehyde + NADH + H$^+$ $\longrightarrow$ ethanol + NAD$^+$

(d) Glyceraldehyde 3-phosphate + NAD$^+$ $\longrightarrow$ 3-phosphoglycerate + NADH + H$^+$ (redox reaction only)

16. What is the net gain of ATP molecules derived from the reactions of glycolysis?

17. What are the possible metabolic fates of pyruvate?

18. Is the reaction of 2-phosphoglycerate to phosphoenolpyruvate a redox reaction? Give the reason for your answer.

19. In what way is the observed mode of action of hexokinase consistent with the induced-fit theory of enzyme action?

20. How does ATP act as an allosteric effector in the mode of action of phosphofructokinase?

21. Define substrate-level phosphorylation and give an example from the reactions discussed in this chapter.

22. Define isozymes and give an example from the material discussed in this chapter.

23. Does the net gain of ATP in glycolysis differ when glycogen, rather than glucose, in the starting material? If so, what is the change?

24. What is a major difference between transketolase and transaldolase?

25. Briefly discuss the role of thiamine pyrophosphate in enzymatic reactions, using material from this chapter to illustrate your points.

## ANNOTATED BIBLIOGRAPHY

Bodner, G.M. Metabolism: Part I Glycolysis, or the Embden-Meyerhoff Pathway. *J. Chem. Ed.* **63**, 566–570 (1986). [A clear, concise summary of the pathway. Part of a series on metabolism of carbohydrates and lipids.]

Boyer, P.D., ed. *The Enzymes,* Vols 5–9. New York: Academic Press, 1972. [A standard reference with review articles on the glycolytic enzymes; lactate dehydrogenase and alcohol dehydrogenase appear in Volume 10.]

Florkin, M., and E.H. Stotz, eds. *Comprehensive Biochemistry.* New York: Elsevier, 1967. [Another standard reference. Volume 17, *Carbohydrate Metabolism,* deals with glycolysis.]

Hers, H.G., and L. Hue. Gluconeogenesis and Related Aspects of Glycolysis. *Ann. Rev. Biochem.* **52**, 617–653 (1983). [A review article that concentrates on the relationship between glycolysis and gluconeogenesis.]

Horecker, B.L. Transaldolase and Transketolase. *Comprehensive Biochemistry,* Volume 15 (1973). [A review article on these two enzymes and their mechanism of action.]

Karl, P.I., B.H.J. Gordon, C.S. Lieber, and S.E. Fisher. Acetaldehyde Production and Transfer by the Perfused Human Placental Cotyledon. *Science* **242,** 273–275 (1988). [A report describing some of the processes involved in fetal alcohol syndrome.]

Lipmann, F. A Long Life in Times of Great Upheaval. *Ann. Rev. Biochem.* **53**, 1–33 (1984). [The reminiscences of a Nobel laureate whose research contributed greatly to the understanding of carbohydrate metabolism. Very interesting reading from the standpoint of autobiography and the author's contributions to biochemistry.]

Meister, A., and M.E. Anderson. Glutathione. *Ann. Rev. Biochem.* **52**, 711–760 (1983). [A review article on several aspects of the action of this peptide.]

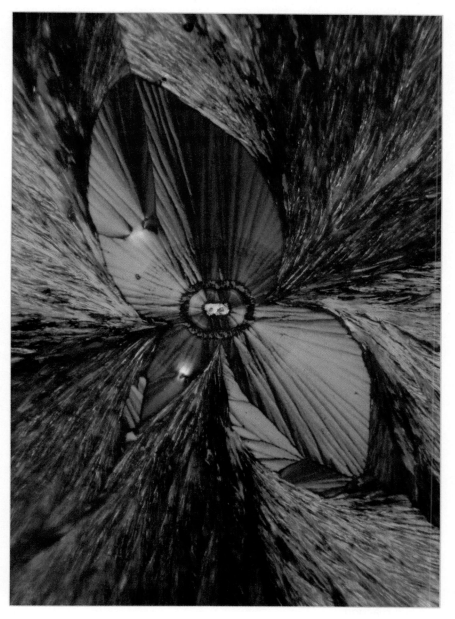

# The Citric Acid Cycle

*Crystals of citric acid viewed under polarized light.*

*If the mitochondrion is the power plant of the cell, then the citric acid cycle operating inside the mitochondrion is its engine room. Here, metabolic fuels, especially glucose derived from carbohydrates, amino acids from proteins and fatty acids from lipids, are all fed into the cycle to be oxidized to carbon dioxide and water. Their energy is transferred to electron carriers and ultimately to the terminal electron acceptor— oxygen. All metabolic fuels enter the citric acid cycle as acetyl-CoA. In the first stage of energy extraction from carbohydrates, glucose is catabolized to pyruvate in the ten anaerobic steps of glycolysis. In* **aerobic** *catabolism, pyruvate is converted to "high energy" two-carbon acetyl-CoA, which then enters the citric acid cycle. As the first step of the cycle, acetyl-CoA combines with oxaloacetate to form citric acid. Each turn of the cycle releases two molecules of $CO_2$, regenerates the starting molecule oxaloacetate, and delivers reducing agents to the electron transport chain. In the final stage of glucose metabolism, proton flow across the inner mitochondrial membrane creates ATP, the energy currency of the cell. The citric acid cycle not only furnishes energy, it also provides intermediates for the biosynthesis of proteins, lipids, and the heme group.*

## 14.1
## THE ROLE OF THE CITRIC ACID CYCLE IN METABOLISM

The evolution of aerobic metabolism, by which nutrients are oxidized to carbon dioxide and water, was an important step in the history of life on earth. Organisms can obtain far more energy from nutrients by aerobic oxidation than by anaerobic oxidation. Three processes play a role in aerobic metabolism: the **citric acid cycle,** which we shall discuss in this chapter, and **electron transport** and **oxidative phosphorylation,** both of which we shall discuss in the next chapter.

Metabolism consists of catabolism, which is oxidative breakdown of nutrients, and anabolism, which is reductive synthesis of biomolecules. The citric acid cycle is **amphibolic,** meaning that it plays a role in both catabolism and anabolism. While the citric acid cycle is a part of the pathway of aerobic oxidation of nutrients (a catabolic pathway), some of the molecules that are included in this cycle are the starting points of biosynthetic (anabolic) pathways.

There are two other common names for the citric acid cycle. One is the **Krebs cycle,** after Sir Hans Krebs, who first investigated the pathway (work for which he received a Nobel Prize in 1953). The other name is the **tricarboxylic acid cycle,** from the fact that some of the molecules involved are acids with three carboxyl groups. We shall start our discussion with a general overview of the pathway and then go on to discussion of specific reactions.

## 14.2
## OVERVIEW OF THE CITRIC ACID CYCLE

An important difference between glycolysis and the citric acid cycle is the part of the cell in which these pathways occur. In prokaryotes glycolysis takes place in the cytosol, while the citric acid cycle takes place on the plasma membrane. In eukaryotes glycolysis occurs in the cytosol, while the citric acid cycle takes place in mitochondria. Most of the enzymes of the citric acid cycle are present in the mitochondrial matrix.

Under aerobic conditions, pyruvate ion produced by glycolysis is oxidized further, with carbon dioxide and water as the final products. First the pyruvate is oxidized to one carbon dioxide molecule and to one acetyl group, which becomes linked to an intermediate, coenzyme A (CoA) (Section 12.5). The acetyl-CoA enters the citric acid cycle. In the citric acid cycle two more molecules of carbon dioxide are produced for each molecule of acetyl-CoA that enters the cycle, and electrons are lost in the process (Figure 14.1). The immediate electron acceptor in all cases but one is $NAD^+$. In the one case where there is another intermediate electron acceptor, it is FAD (*f*lavin *a*denine *d*inucleotide), which takes up electrons and hydrogen to produce $FADH_2$, the reduced form of flavin derived from riboflavin (vitamin $B_2$). The electrons are passed from NADH and $FADH_2$ through several stages of an electron transport chain with a different redox reaction at each step. The final electron acceptor is oxygen, with water as the product.

A quick review of some aspects of mitochondrial structure is in order here, since we shall want to describe the exact location of each of the components of the citric acid cycle and the electron transport chain. Recall from Chapter 2 that mitochondria have an inner and an outer membrane (Figure 14.2). The region enclosed by the inner membrane is called the **matrix,** and there is an **intermembrane space** between the inner and outer membranes. The reactions of the citric acid cycle take place in the matrix, except for the one in which the intermediate electron acceptor is FAD. The enzyme that catalyzes the FAD-linked reaction is an integral part of the inner mitochondrial membrane.

The citric acid cycle is shown in schematic fashion in Figure 14.1. In the first reaction of the cycle the two-carbon acetyl group condenses with the four-carbon oxaloacetate ion to produce the six-carbon citrate ion. In the next few steps the citrate isomerizes, and then it both loses carbon dioxide and is oxidized. This process, called **oxidative decarboxylation,** produces the five-carbon compound α-ketoglutarate, which again is oxidatively decarboxylated to produce the four-carbon compound succinate. The cycle is completed by regeneration of oxaloacetate from succinate in several steps.

There are four oxidation steps in the citric acid cycle, steps 3, 4, 6, and 8 (see Figure 14.1). The oxidizing agent is $NAD^+$ in all except step 6, in which FAD plays the same role. In step 5 a molecule of GDP (guanosine *di*phosphate) is phosphorylated to produce GTP (guanosine *tri*phosphate). This reaction is equivalent to the production of ATP, since the phosphate group is easily transferred to ADP, producing GDP and ATP. GTP differs from ATP only in the substitution of guanine for adenine.

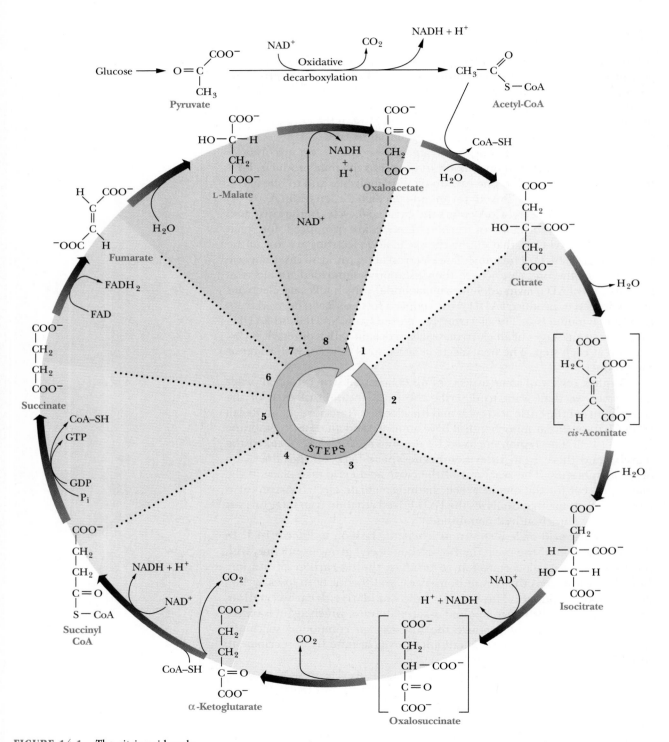

**FIGURE 14.1**    The citric acid cycle.

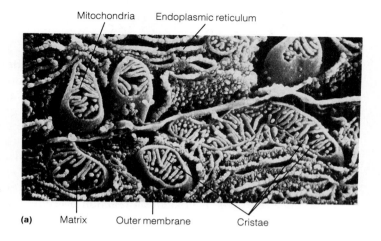

(a)

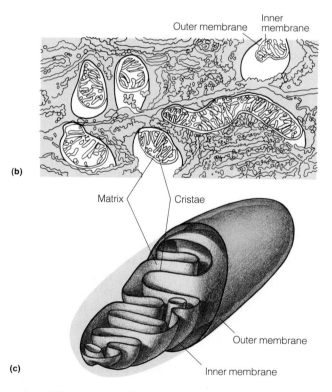

(b)

(c)

FIGURE 14.2    The structure of a mitochondrion. (a) Scanning electron micrograph showing internal structure of mitochondrion (magnified 24, 187 x). (b) Interpretative drawing of the SEM. (c) Perspective drawing of a mitochondrion. (For an electron micrograph of mitochondrial structure, see Figure 2.4.)

## 14.3
## INDIVIDUAL REACTIONS OF THE CITRIC ACID CYCLE

### Conversion of Pyruvate to Acetyl-CoA

An enzyme system called the **pyruvate dehydrogenase complex** is responsible for the conversion of pyruvate to carbon dioxide and the acetyl portion of acetyl-CoA. There is an —SH group at one end of the CoA molecule, which is the point at which the acetyl group is attached. As a result, CoA is frequently shown in equations as CoA-SH. Because CoA is a thiol [the sulfur (thio) analog of an alcohol], acetyl-CoA is a **thioester,** with a sulfur atom replacing an oxygen of the usual carboxylic ester. This difference is important, since thioesters are "high energy" compounds. In other words, the hydrolysis of thioesters releases enough energy to drive other reactions. An oxidation reaction precedes the transfer of the acetyl group to the CoA. The whole process involves several enzymes, all of which are part of the pyruvate dehydrogenase complex. The overall reaction

Pyruvate + CoA-SH + NAD$^+$ ⟶ acetyl-CoA + CO$_2$ + H$^+$ + NADH

is exergonic ($\Delta G^{\circ\prime} = -33.4$ kJ mol$^{-1}$ = $-8.0$ kcal mol$^{-1}$), and NADH is recycled for further use.

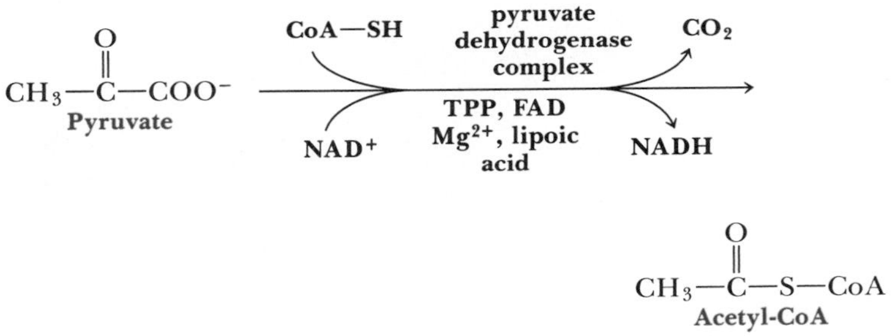

The overall reaction of the pyruvate dehydrogenase complex

(14.1)

Three enzymes make up the pyruvate dehydrogenase complex. They are **pyruvate dehydrogenase, dihydrolipoyl transacetylase,** and **dihydrolipoyl dehydrogenase.** The reaction takes place in five steps. The last two enzymes catalyze reactions of **lipoic acid,** a compound that has a disulfide group in its oxidized form and two sulfhydryl groups in its reduced form.

Lipoic acid can act as an oxidizing agent; the reaction involves hydrogen transfer, which frequently accompanies biological oxidation–reduction reactions (Section 12.3). Another reaction of lipoic acid is the formation of a thioester linkage with the acetyl group before it is transferred to the acetyl-CoA. Lipoic acid can act simply as an oxidizing agent, or it can simultaneously take part in two reactions, a redox reaction and the shift of an acetyl group by transesterification.

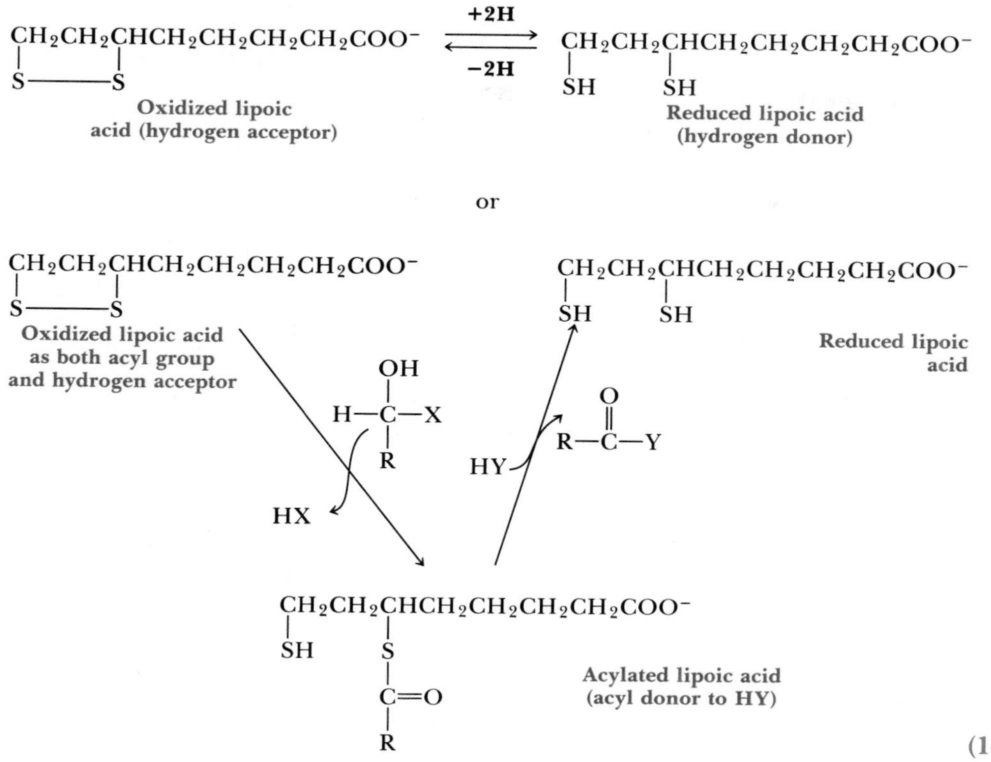

$$CH_2CH_2CHCH_2CH_2CH_2CH_2COO^- \;\; \underset{-2H}{\overset{+2H}{\rightleftharpoons}} \;\; CH_2CH_2CHCH_2CH_2CH_2CH_2COO^-$$

Oxidized lipoic
acid (hydrogen acceptor)        SH      SH
                                Reduced lipoic acid
                                (hydrogen donor)

or

$$CH_2CH_2CHCH_2CH_2CH_2CH_2COO^- \qquad\qquad CH_2CH_2CHCH_2CH_2CH_2CH_2COO^-$$

Oxidized lipoic acid
as both acyl group
and hydrogen acceptor

        OH
        |
     H—C—X
        |
        R

HX

HY

$$R—\overset{O}{\overset{\|}{C}}—Y$$

SH      SH
Reduced lipoic
acid

$$CH_2CH_2CHCH_2CH_2CH_2CH_2COO^-$$
SH      S
        |
        C=O
        |
        R

Acylated lipoic acid
(acyl donor to HY)

(14.2)

The dual role of lipoic acid as hydrogen acceptor
(oxidizing agent) and acyl group transfer agent in the
pyruvate dehydrogenase reaction

The first step in the reaction sequence that converts pyruvate to carbon dioxide and acetyl-CoA is catalyzed by pyruvate dehydrogenase. This enzyme requires thiamine pyrophosphate (TPP) as a coenzyme. The coenzyme is not covalently bonded to the enzyme; they are held together by noncovalent interactions. $Mg^{2+}$ is also required. We saw the action of TPP as a coenzyme in the conversion of pyruvate to acetaldehyde, catalyzed by pyruvate decarboxylase (Section 13.5), and in the transketolase reaction (Section 13.6). Since the mechanism is essentially the same here, we shall not discuss it in detail. In the pyruvate dehydrogenase reaction an $\alpha$-keto acid, pyruvate, loses carbon dioxide; the remaining two-carbon unit becomes covalently bonded to TPP. The form in which the two-carbon unit is attached to the thiazole ring is a hydroxyethyl group.

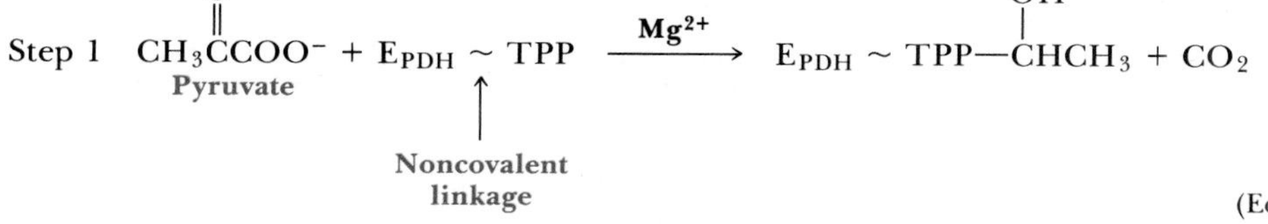

$$\text{Step 1}\quad CH_3\overset{O}{\overset{\|}{C}}COO^- + E_{PDH} \sim TPP \xrightarrow{Mg^{2+}} E_{PDH} \sim TPP—\overset{OH}{\overset{|}{C}}HCH_3 + CO_2$$

Pyruvate

Noncovalent
linkage

(Equation continues on next page.)

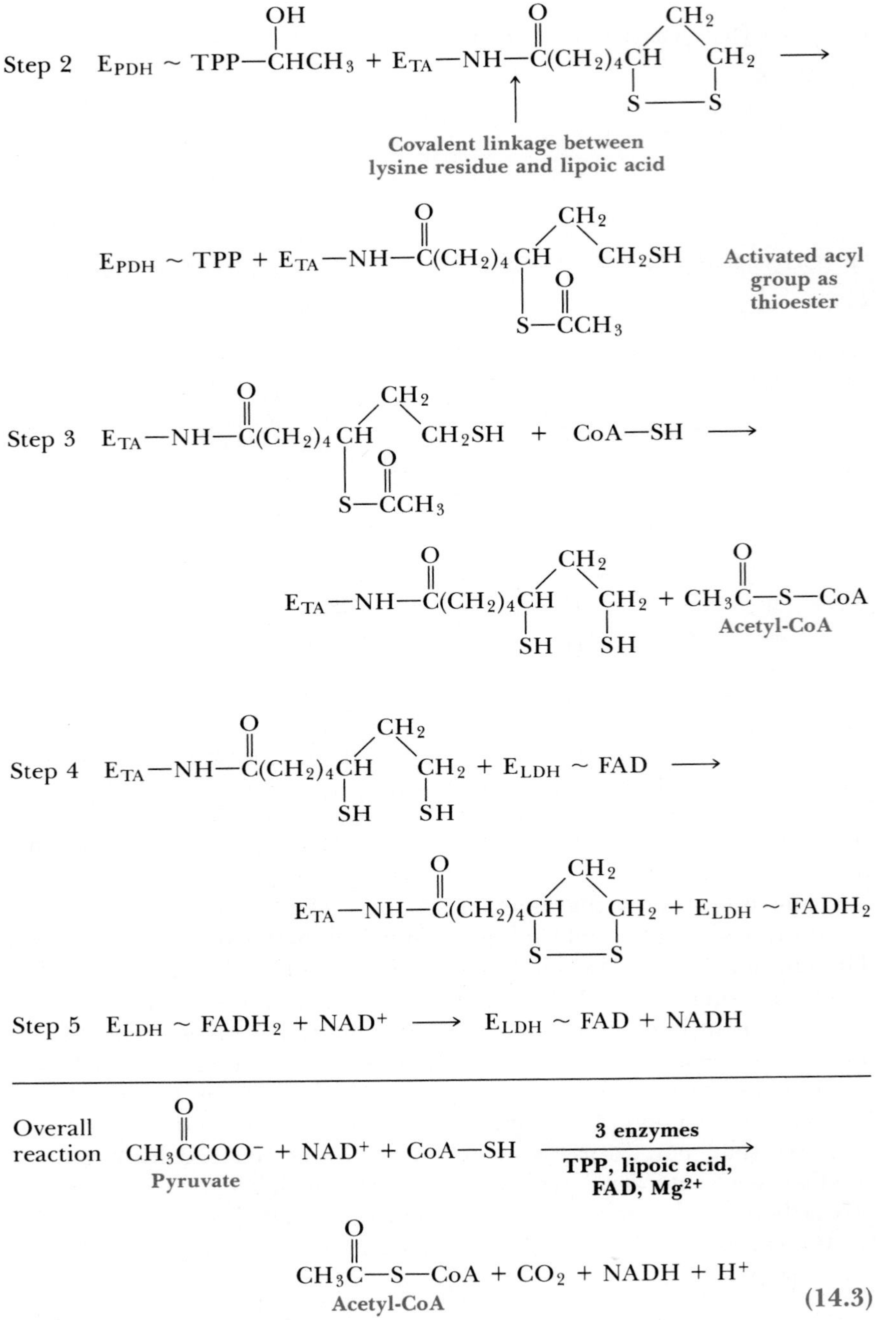

Step 2    $E_{PDH} \sim TPP-\overset{\underset{|}{OH}}{C}HCH_3 + E_{TA}-NH-\overset{\underset{\|}{O}}{C}(CH_2)_4\overset{\overset{CH_2}{\diagup \diagdown}}{CH \quad\quad CH_2} \longrightarrow$

$$\underset{\substack{\uparrow \\ \text{Covalent linkage between} \\ \text{lysine residue and lipoic acid}}}{S-\!\!\!-S}$$

$$E_{PDH} \sim TPP + E_{TA}-NH-\overset{\underset{\|}{O}}{C}(CH_2)_4\overset{\overset{CH_2}{\diagup \diagdown}}{CH \quad\quad CH_2SH} \quad\quad \textbf{Activated acyl group as thioester}$$

$$\underset{S-\overset{\underset{\|}{O}}{C}CH_3}{}$$

Step 3    $E_{TA}-NH-\overset{\underset{\|}{O}}{C}(CH_2)_4\overset{\overset{CH_2}{\diagup \diagdown}}{CH \quad\quad CH_2SH} \quad + \quad CoA-SH \longrightarrow$

$$\underset{S-\overset{\underset{\|}{O}}{C}CH_3}{}$$

$$E_{TA}-NH-\overset{\underset{\|}{O}}{C}(CH_2)_4\underset{\underset{SH}{|}}{\overset{\overset{CH_2}{\diagup \diagdown}}{CH \quad\quad CH_2}}\underset{SH}{\phantom{|}} + CH_3\overset{\underset{\|}{O}}{C}-S-CoA$$
$$\textbf{Acetyl-CoA}$$

Step 4    $E_{TA}-NH-\overset{\underset{\|}{O}}{C}(CH_2)_4\underset{\underset{SH}{|}}{\overset{\overset{CH_2}{\diagup \diagdown}}{CH \quad\quad CH_2}}\underset{SH}{\phantom{|}} + E_{LDH} \sim FAD \longrightarrow$

$$E_{TA}-NH-\overset{\underset{\|}{O}}{C}(CH_2)_4\overset{\overset{CH_2}{\diagup \diagdown}}{CH \quad\quad CH_2} + E_{LDH} \sim FADH_2$$
$$S-\!\!\!-S$$

Step 5    $E_{LDH} \sim FADH_2 + NAD^+ \longrightarrow E_{LDH} \sim FAD + NADH$

---

Overall reaction    $CH_3\overset{\underset{\|}{O}}{C}COO^- + NAD^+ + CoA-SH \xrightarrow[\substack{\textbf{TPP, lipoic acid,} \\ \textbf{FAD, Mg}^{2+}}]{\textbf{3 enzymes}}$
$$\textbf{Pyruvate}$$

$$CH_3\overset{\underset{\|}{O}}{C}-S-CoA + CO_2 + NADH + H^+$$
$$\textbf{Acetyl-CoA} \hspace{4cm} (14.3)$$

The five individual steps and the overall reaction catalyzed by the pyruvate dehydrogenase complex

$E_{PDH}$ = pyruvate dehydrogenase          $E_{LDH}$ = dihydrolipoyl dehydrogenase
$E_{TA}$ = dihydrolipoyl transacetylase          TPP = thiamine pyrophosphate

The second step of the reaction is catalyzed by dihydrolipoyl transacetylase. This enzyme requires lipoic acid as a coenzyme. The lipoic acid is covalently bonded to the enzyme by an amide bond to the $\epsilon$-amino group of a lysine side chain. The two-carbon hydroxyethyl unit that originally came from pyruvate is transferred from the thiamine pyrophosphate to the lipoic acid. In the process the two-carbon unit is oxidized to produce an acetyl group. The disulfide group of the lipoic acid is the oxidizing agent, and the product of the reaction is a thioester. In other words, the acetyl group is now covalently bonded to the lipoic acid by a thioester linkage (see Equation 14.3).

The third step of the reaction is also catalyzed by dihydrolipoyl transacetylase. A molecule of CoA-SH attacks the thioester linkage, and the acetyl group is transferred to it. The acetyl group remains bound in a thioester linkage, this time as acetyl-CoA rather than esterified to lipoic acid. The reduced form of lipoic acid remains covalently bound to dihydrolipoyl transacetylase (see Equation 14.3). The reaction of pyruvate and CoA-SH has now reached the stage of the products, carbon dioxide and acetyl-CoA, but the lipoic acid coenzyme is in a reduced form. If the transacetylase is to catalyze further reactions, the lipoic acid must be regenerated.

In the fourth step of the overall reaction, the enzyme dihydrolipoyl dehydrogenase reoxidizes the reduced lipoic acid from the sulfhydryl to the disulfide form. The lipoic acid still remains covalently bonded to the transacetylase enzyme. The dehydrogenase also has a coenzyme, FAD (Section 12.3), which is bound to the enzyme by noncovalent interactions. An electron acceptor is simply another name for an oxidizing agent, and the coenzyme serves this function here. As a result, FAD is reduced to $FADH_2$. In the fifth step of the reaction $FADH_2$ is reoxidized in turn. The oxidizing agent is $NAD^+$, and NADH is the product along with reoxidized FAD.

The reduction of $NAD^+$ to NADH accompanies the oxidation of pyruvate to the acetyl group, and the overall equation shows that there has been a transfer of two electrons from pyruvate to $NAD^+$ (Equation 14.3). The electrons gained by $NAD^+$ in generating NADH in this step are passed to the electron transport chain (the next step in aerobic metabolism).

The reaction leading from pyruvate to acetyl-CoA is a complex one that requires three enzymes, each of which has its own coenzyme. The spatial orientation of the individual enzyme molecules with respect to one another is itself complex. In the enzyme isolated from *E. coli* the arrangement is quite compact so that the various steps of the reaction can be thoroughly coordinated. There is a core of 24 dihydrolipoyl transacetylase molecules. There is some evidence that the 24 polypeptide chains are arranged in eight trimers, with each trimer occupying the corner of a cube; this point is not definitely established, however. There are 12 dimers of pyruvate dehydrogenase, and they occupy the edges of the cube. Finally, six dimers of dihydrolipoyl dehydrogenase lie on the six faces of the cube (Figure 14.3). Note that many levels of structure combine to produce a

(a)

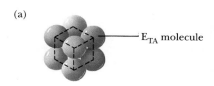

$E_{TA}$ molecule

(b)

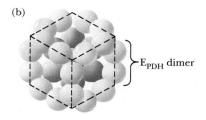

$E_{PDH}$ dimer

(c)

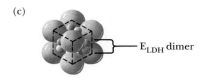

$E_{LDH}$ dimer

(d)

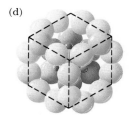

**FIGURE 14.3**  The molecular architecture of the pyruvate dehydrogenase complex from *Escherichia coli.* (a) Cubic cluster of 24 transacetylase ($E_{TA}$) molecules, possibly arranged as eight trimers. (b) Twelve pyruvate dehydrogenase ($E_{PDH}$) dimers on the edges of the transacetylase core. (c) Six dimers of dihydrolipoyl dehydrogenase ($E_{LDH}$) on each face of the transacetylase core; three dimers shown here. (d) The complete pyruvate dehydrogenase aggregate.

suitable environment for the conversion of pyruvate to acetyl-CoA. Each of the enzyme molecules in this array has its own tertiary structure, and the array itself has the cubical structure we have just seen.

A compact arrangement such as the one in the pyruvate dehydrogenase multienzyme complex has two great advantages over an arrangement in which the various components are more widely dispersed. First, the various stages of the reaction can take place more efficiently because the reactants and the enzymes are in such close proximity to each other. The role of lipoic acid is particularly important here. Recall that the lipoic acid is covalently attached to the transacetylase enzyme that occupies a central position in the complex. The lipoic acid and the lysine side chain to which it is bonded are long enough to act as a "swinging arm," which can move to the site of each of the steps of the reaction (Figure 14.4). As a result of the swinging arm action, the lipoic acid can move to the pyruvate dehydrogenase site to accept the two-carbon unit and then transfer it to the active site of the transacetylase. The acetyl group can then be transesterified to CoA-SH from the lipoic acid. Finally, the lipoic acid can swing to the active site of the dehydrogenase so that the sulfhydryl groups can be reoxidized to a disulfide.

A second advantage of a multienzyme complex is that regulatory controls can be applied in such a system more efficiently than with a single enzyme molecule. In the case of the pyruvate dehydrogenase complex, controlling factors are intimately associated with the multienzyme complex itself. The overall reaction is part of a pathway that releases energy. It is not surprising that the enzyme which catalyzes it is inhibited by ATP and NADH, since both compounds are abundant when a cell has a good deal of energy readily available. The end products of a series of reactions inhibit the first reaction of the series, and the intermediate reactions do not take place when their products are not needed. It is consistent with this picture that the pyruvate dehydrogenase complex is activated by ADP, which is abundant when a cell needs energy. In mammals the actual mechanism by which the inhibition takes place is the phosphorylation of pyruvate dehydrogenase. A phosphate group is covalently bound to the enzyme in a reaction catalyzed by the enzyme **pyruvate dehydrogenase kinase.** When the need arises for pyruvate dehydrogenase to be activated, the hydrolysis of the phosphate ester linkage (dephosphorylation) is catalyzed by another enzyme, **phosphoprotein phosphatase.** This latter enzyme is itself activated by $Ca^{2+}$. Both these enzymes are associated with the intact pyruvate dehydrogenase complex, permitting effective control of the overall reaction from pyruvate to acetyl-CoA.

## The Citric Acid Cycle Proper

The first step of the citric acid cycle proper is the reaction of acetyl-CoA and oxaloacetate to form citrate and CoA-SH. This reaction is called a condensation because a new carbon-carbon bond is formed. The condensation reaction of acetyl-CoA and oxaloacetate to form citryl-CoA takes

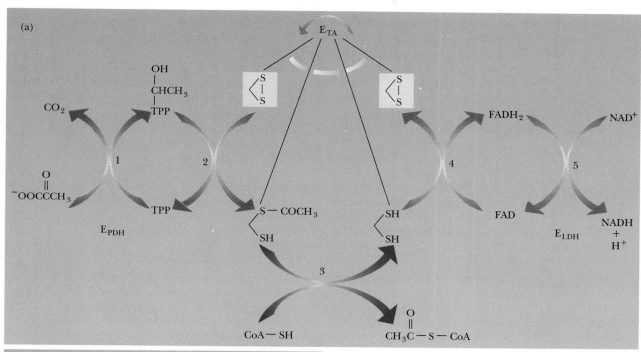

**FIGURE 14.4** The role of the "swinging arm" in the reactions of the pyruvate dehydrogenase complex. (a) The numerals refer to the steps in Equation 14.3. The abbreviations for the enzymes are also those used in Equation 14.3. Steps 1 and 2 take place on the $E_{PDH}$ subunit, step 3 on the $E_{TA}$, and steps 4 and 5 on the $E_{LDH}$. (b) The symbol —S—$_S$ refers to the lysyl–lipoamide "swinging arm," which can take different positions in the course of the reaction. Its structure is shown here.

place in the first stage of the reaction; a basic group on the enzyme (Enz—B:) plays a major role here. The condensation is followed by the hydrolysis of citryl-CoA to give citrate and CoA-SH.

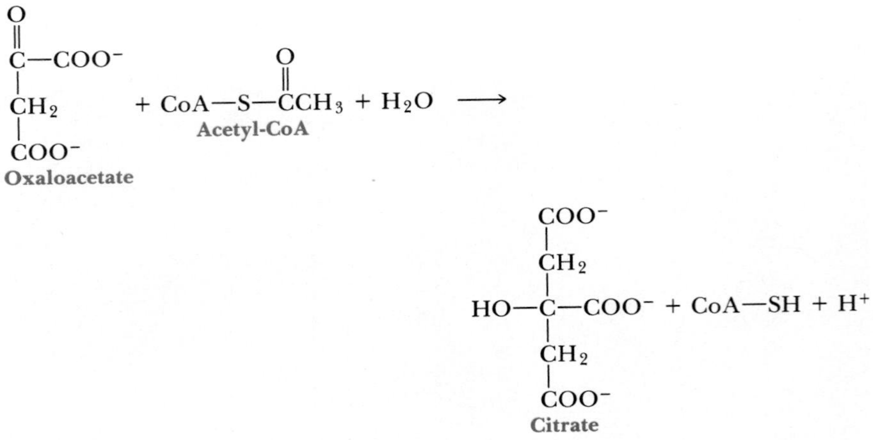

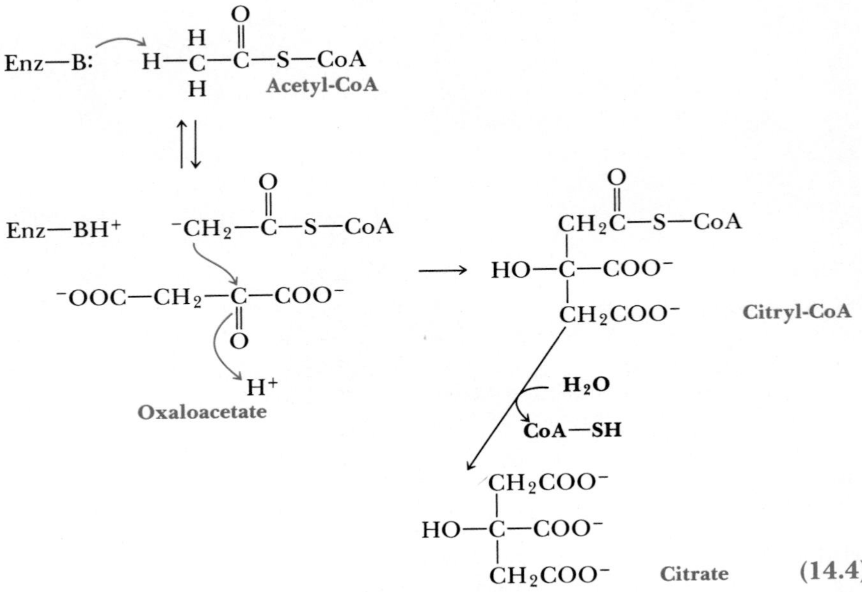

$$
\begin{array}{c}
\text{O} \\
\parallel \\
\text{C—COO}^- \\
\mid \\
\text{CH}_2 \qquad + \text{ CoA—S—CCH}_3 + \text{H}_2\text{O} \longrightarrow \\
\mid \\
\text{COO}^- \\
\text{Oxaloacetate} \qquad\qquad \text{Acetyl-CoA}
\end{array}
$$

$$
\begin{array}{c}
\text{COO}^- \\
\mid \\
\text{CH}_2 \\
\mid \\
\text{HO—C—COO}^- + \text{CoA—SH} + \text{H}^+ \\
\mid \\
\text{CH}_2 \\
\mid \\
\text{COO}^- \\
\text{Citrate}
\end{array}
$$

The condensation of acetyl-CoA and oxaloacetate to form citrate

(14.4)

The reaction is catalyzed by the enzyme **citrate synthetase,** originally called "condensing enzyme." It is an exergonic reaction ($\Delta G^{\circ\prime} = -32.8$ kJ mol$^{-1}$ = $-7.8$ kcal mol$^{-1}$) because the hydrolysis of a thioester releases energy; thioesters are considered "high energy" compounds.

Citrate synthetase is an allosteric enzyme; thus the inhibitor mechanisms we discussed in Section 10.9 apply here. Both ATP and NADH inhibit the citrate synthetase reaction, as we saw earlier with the pyruvate dehydrogenase reaction. Succinyl-CoA, a substance that appears later in the cycle, is an inhibitor as well. This is another example of the type of control mechanism in which products of a long series of reactions "turn off" the first reaction of the series.

A possible alternate substrate for citrate synthetase is **fluoroacetyl-CoA.** The source of the fluoroacetyl-CoA is fluoroacetate, which is found in the leaves of various types of poisonous plants, including locoweeds. Animals that ingest these plants form fluoroacetyl-CoA, which in turn is converted to fluorocitrate by their citrate synthetase.

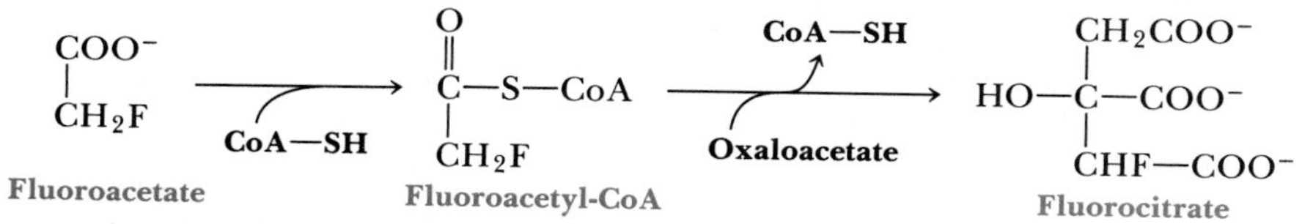

$$
\begin{array}{ccccc}
\underset{\text{Fluoroacetate}}{\overset{\displaystyle COO^-}{\underset{\displaystyle CH_2F}{|}}}
& \xrightarrow[\text{CoA—SH}]{}
& \underset{\text{Fluoroacetyl-CoA}}{\overset{\displaystyle O}{\underset{\displaystyle \underset{\displaystyle CH_2F}{|}}{\overset{\displaystyle \|}{C}\text{—S—CoA}}}}
& \xrightarrow[\text{Oxaloacetate}]{\text{CoA—SH}}
& \underset{\text{Fluorocitrate}}{\overset{\displaystyle CH_2COO^-}{\underset{\displaystyle CHF\text{—}COO^-}{HO\text{—}\overset{\displaystyle |}{\underset{\displaystyle |}{C}}\text{—}COO^-}}}
\end{array}
\qquad (14.5)
$$

The formation of fluorocitrate from fluoroacetate

Fluorocitrate in turn is a potent inhibitor of **aconitase,** the enzyme that catalyzes the next reaction of the citric acid cycle. These plants are poisonous because they produce a potent inhibitor of life processes.

The second reaction of the citric acid cycle, the one catalyzed by aconitase, is the isomerization of citrate to isocitrate. The enzyme requires $Fe^{2+}$. One of the most interesting features of the reaction is that citrate, a symmetrical (achiral) compound, is converted into isocitrate, a chiral compound, a molecule that cannot be superimposed on its mirror image.

It is often possible for a chiral compound to have several different isomers. Isocitrate has four possible isomers, but only one of the four is produced by this reaction. (We shall not discuss nomenclature of the isomers of isocitrate here. See the exercises for a question about the other isomers.) Aconitase, the enzyme that catalyzes the conversion of citrate to isocitrate, is able to select one end of the citrate molecule in preference to the other.

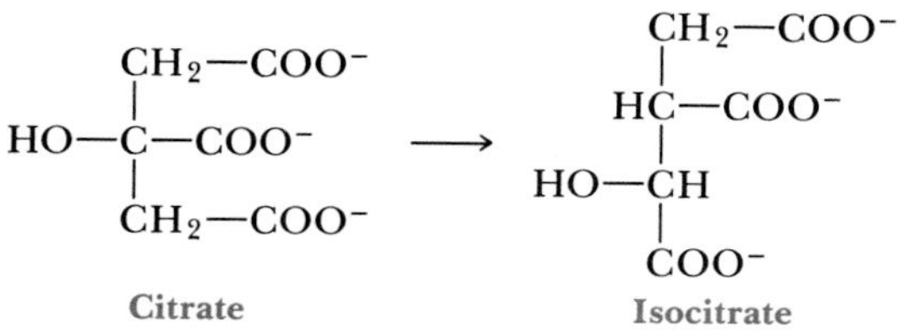

$$
\underset{\text{Citrate}}{\overset{\displaystyle CH_2\text{—}COO^-}{\underset{\displaystyle CH_2\text{—}COO^-}{HO\text{—}\overset{\displaystyle |}{\underset{\displaystyle |}{C}}\text{—}COO^-}}}
\longrightarrow
\underset{\text{Isocitrate}}{\overset{\displaystyle CH_2\text{—}COO^-}{\underset{\displaystyle \underset{\displaystyle COO^-}{HO\text{—}CH}}{\underset{\displaystyle |}{HC\text{—}COO^-}}}}
\qquad (14.6)
$$

The formation of isocitrate (a chiral compound) from citrate (an achiral compound)

This type of behavior means that the enzyme can bind a symmetrical substrate in an unsymmetrical binding site. In Section 10.13 we mentioned that this possibility exists, and here we have an example of it. The enzyme forms an unsymmetrical three-point attachment to the citrate molecule. The reaction proceeds by removal of a water molecule from the citrate to produce *cis*-aconitate, and then water is added back to the *cis*-aconitate to give isocitrate.

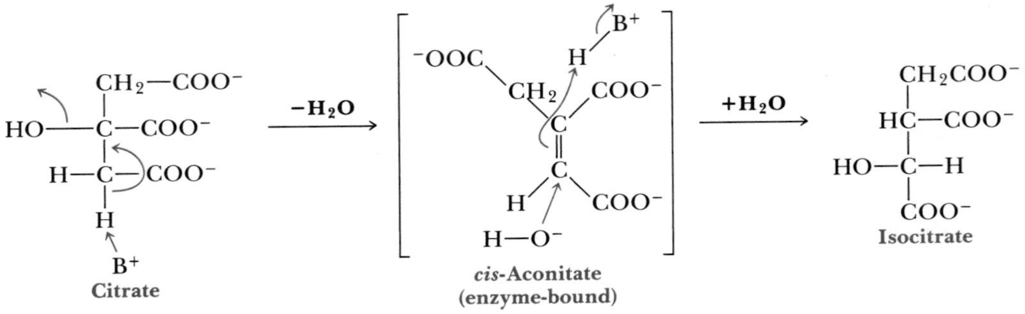

*cis*-aconitate as an intermediate in the conversion
of citrate to isocitrate
(B is a basic group on the enzyme)

The intermediate, *cis*-aconitate, remains bound to the enzyme during the course of the reaction. There is some evidence that the citrate is complexed to the Fe(II) in the active site of the enzyme in such a way that the citrate curls back on itself in a nearly circular conformation. Several authors have been unable to resist the temptation to call this situation the "ferrous wheel."

Step 3 in the citric acid cycle is the oxidative decarboxylation of isocitrate to $\alpha$-ketoglutarate and carbon dioxide. This reaction is the first of two oxidative decarboxylations of the citric acid cycle; the enzyme that catalyzes it is **isocitrate dehydrogenase.** The reaction takes place in two steps. First, isocitrate is oxidized to oxalosuccinate, which remains bound to the enzyme. Then oxalosuccinate is decarboxylated, and the carbon dioxide and $\alpha$-ketoglutarate are released. This is the first of the reactions in which NADH is produced. One molecule of NADH is produced from $NAD^+$ at this stage by the loss of two electrons in the oxidation.

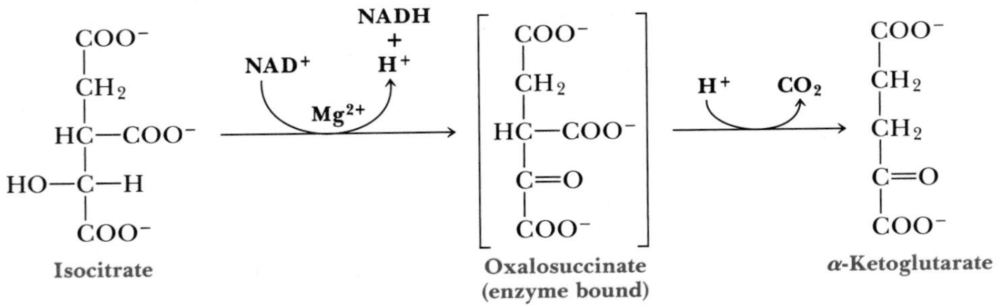

The oxidative decarboxylation of isocitrate to $\alpha$-ketoglutarate

This reaction is a control site in the cycle. The enzyme is an oligomer, with a molecular weight on the order of $10^5$. The reaction is controlled by the allosteric mechanism discussed in Chapter 10. The enzyme is inhibited by ATP and NADH and is activated by ADP and $NAD^+$. This pattern of inhibition by ATP and NADH and activation by ADP and $NAD^+$ is frequently encountered in catabolic reactions.

The second oxidative decarboxylation takes place in step 4 of the citric acid cycle, in which carbon dioxide and succinyl-CoA are formed from

$\alpha$-ketoglutarate and CoA. This reaction is similar to the one in which acetyl-CoA is formed from pyruvate, and once again NADH is produced from $NAD^+$.

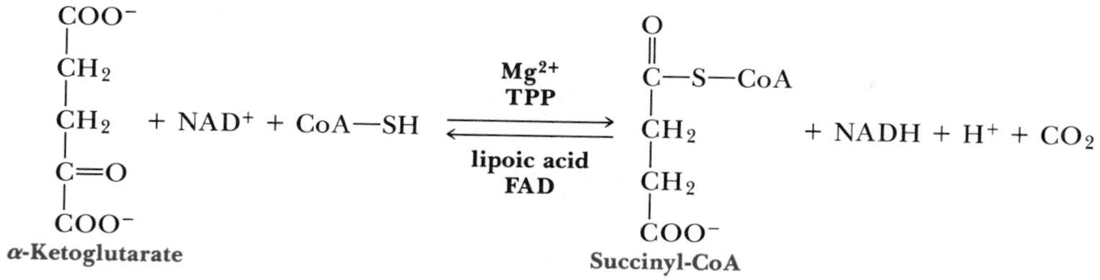

$$
\begin{array}{l}
\text{COO}^- \\
| \\
\text{CH}_2 \\
| \\
\text{CH}_2 \quad + \text{NAD}^+ + \text{CoA}-\text{SH} \\
| \\
\text{C}=\text{O} \\
| \\
\text{COO}^- \\
\alpha\text{-Ketoglutarate}
\end{array}
\quad
\underset{\substack{\text{lipoic acid} \\ \text{FAD}}}{\overset{\substack{\text{Mg}^{2+} \\ \text{TPP}}}{\rightleftharpoons}}
\quad
\begin{array}{l}
\text{O} \\
\| \\
\text{C}-\text{S}-\text{CoA} \\
| \\
\text{CH}_2 \quad + \text{NADH} + \text{H}^+ + \text{CO}_2 \\
| \\
\text{CH}_2 \\
| \\
\text{COO}^- \\
\text{Succinyl-CoA}
\end{array}
\tag{14.9}
$$

The conversion of $\alpha$-ketoglutarate to succinyl-CoA

The reaction occurs in several stages and is catalyzed by an enzyme system called the **$\alpha$-ketoglutarate dehydrogenase complex,** which is very similar to the pyruvate dehydrogenase complex. Both these multienzyme systems consist of three enzymes. The reaction takes place in several steps, and there is again a requirement for thiamine pyrophosphate (TPP), FAD, lipoic acid, and $Mg^{2+}$. This reaction is highly exergonic ($\Delta G^{\circ\prime} = -33.4$ kJ $mol^{-1} = -8.0$ kcal $mol^{-1}$), as is the one catalyzed by pyruvate dehydrogenase.

At this point, two molecules of $CO_2$ have been produced by the citric acid cycle. We should also mention that the $\alpha$-ketoglutarate dehydrogenase complex reaction is the fourth one in which we have encountered an enzyme that requires TPP. In all four cases—the pyruvate dehydrogenase complex, the $\alpha$-ketoglutarate dehydrogenase complex, pyruvate decarboxylase (Section 13.5), and transketolase (Section 13.6)—the reaction involves transfer of an activated two-carbon group. (Refer to the exercises for a treatment of the nature of the activated group in each reaction.)

In the next step of the cycle, the thioester bond of succinyl-CoA is hydrolyzed to produce succinate and CoA-SH; an accompanying reaction is the phosphorylation of GDP to GTP. The whole reaction is catalyzed by the enzyme **succinate thiokinase.** (This is a kinase linked to GTP, rather than to ATP.) In the reaction mechanism, a phosphate group covalently bonded to the enzyme is directly transferred to the GDP. The phosphorylation of GDP to GTP is endergonic, as is the corresponding ADP–ATP reaction ($\Delta G^{\circ\prime} = 30.5$ kJ $mol^{-1} = 7.3$ kcal $mol^{-1}$).

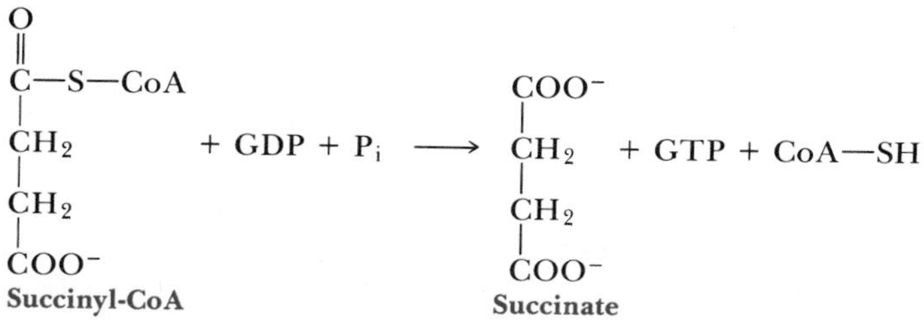

$$
\begin{array}{l}
\text{O} \\
\| \\
\text{C}-\text{S}-\text{CoA} \\
| \\
\text{CH}_2 \quad + \text{GDP} + \text{P}_i \\
| \\
\text{CH}_2 \\
| \\
\text{COO}^- \\
\text{Succinyl-CoA}
\end{array}
\quad \longrightarrow \quad
\begin{array}{l}
\text{COO}^- \\
| \\
\text{CH}_2 \quad + \text{GTP} + \text{CoA}-\text{SH} \\
| \\
\text{CH}_2 \\
| \\
\text{COO}^- \\
\text{Succinate}
\end{array}
\tag{14.10}
$$

The conversion of succinyl-CoA to succinate

The energy required for the phosphorylation of GDP to GTP is provided by the hydrolysis of succinyl-CoA to produce succinate and CoA. The free energy of hydrolysis ($\Delta G^{\circ\prime}$) of succinyl-CoA is $-33.4$ kJ mol$^{-1}$ ($-8.0$ kcal mol$^{-1}$). The overall reaction is slightly exergonic ($\Delta G^{\circ\prime} = -3.3$ kJ mol$^{-1}$ = $-0.8$ kcal mol$^{-1}$), and, as a result, does not contribute greatly to the overall production of energy by the mitochondrion.

The enzyme **nucleoside diphosphokinase** catalyzes the transfer of a phosphate group from GTP to ADP to give GDP and ATP.

$$GTP + ADP \longrightarrow GDP + ATP$$

This reaction step is called substrate-level phosphorylation to distinguish it from the type of reaction for production of ATP that is coupled to the electron transport chain. The production of ATP in this reaction is the only place in the citric acid cycle in which chemical energy in the form of ATP is made available to the cell. Except for this reaction, the generation of ATP characteristic of aerobic metabolism is associated with the electron transport chain, the subject of the next chapter. As many as 38 molecules of ATP can be obtained from the oxidation of a single molecule of glucose by the combination of anaerobic and aerobic oxidation, compared to only two molecules of ATP produced by anaerobic glycolysis alone. The combined reactions that occur in mitochondria are of great importance to aerobic organisms.

In steps 6 through 8 of the citric acid cycle the four-carbon succinate ion is converted to oxaloacetate ion to complete the cycle.

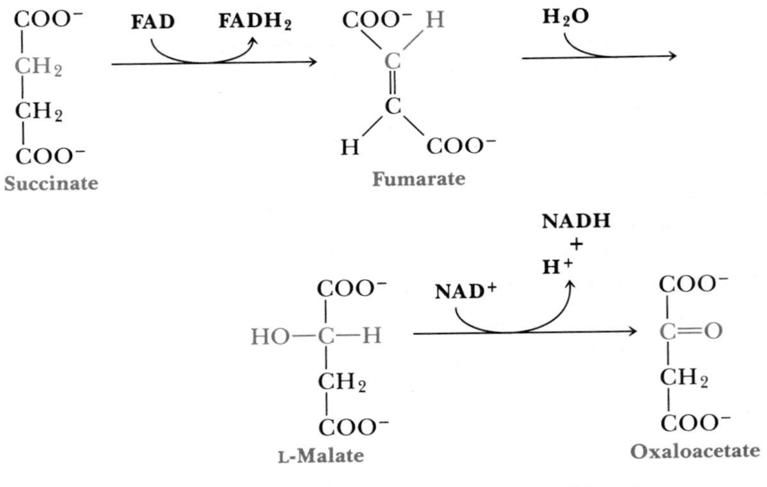

The final stages of the citric acid cycle
succinate is converted to oxaloacetate
(conversion of a methylene group to a carbonyl group)

(14.11)

In step 6, succinate is oxidized to fumarate, a reaction that is catalyzed by the enzyme **succinate dehydrogenase.** This enzyme is bound to the inner mitochondrial membrane. The other individual enzymes of the citric acid cycle are in the mitochondrial matrix. The electron acceptor, which is

FAD rather than $NAD^+$, is covalently bonded to the enzyme; succinate dehydrogenase is called a flavoprotein because of the presence of FAD with its flavin moiety. In the succinate dehydrogenase reaction, FAD is reduced to $FADH_2$ and succinate is oxidized to fumarate.

$$
\begin{array}{c}
COO^- \\
| \\
CH_2 \\
| \\
CH_2 \\
| \\
COO^- \\
\text{Succinate}
\end{array}
\quad + \text{ FAD} \longrightarrow
\begin{array}{c}
COO^- \quad H \\
\diagdown \quad \diagup \\
C \\
\| \\
C \\
\diagup \quad \diagdown \\
H \quad\quad COO^- \\
\text{Fumarate}
\end{array}
\quad + \text{ FADH}_2
\qquad (14.12)
$$

The conversion of succinate to fumarate

The overall reaction is

Succinate + E—FAD $\longrightarrow$ fumarate + E—FADH$_2$

The E—FAD and E—FADH$_2$ in the equation indicate that the electron acceptor is covalently bonded to the enzyme. The FADH$_2$ group also passes electrons on to the electron transport chain and eventually to oxygen.

Succinate dehydrogenase contains four iron atoms, as well as four sulfur atoms that are bonded to the iron atoms. The enzyme does not contain a heme group; it is also referred to as a **nonheme iron protein** or an **iron–sulfur protein.** This enzyme exhibits stereospecificity. The double bond produced by this reaction could have either the *trans* configuration (fumarate) or the *cis* configuration (maleate), but only the *trans* configuration is actually produced. This reaction is not a regulatory control point of the citric acid cycle.

In reaction 7, which is catalyzed by the enzyme **fumarase,** water is added across the double bond of fumarate in a hydration reaction to give malate. Again there is stereospecificity in the reaction. Malate has two enantiomers, L- and D-malate, but only L-malate is produced. This reaction is another in which no control of the cycle is apparent.

$$
\begin{array}{c}
COO^- \quad H \\
\diagdown \quad \diagup \\
C \\
\| \\
C \\
\diagup \quad \diagdown \\
H \quad\quad COO^- \\
\text{Fumarate}
\end{array}
\quad + \text{ H}_2O \longrightarrow
\begin{array}{c}
COO^- \\
| \\
HO—C—H \\
| \\
CH_2 \\
| \\
COO^- \\
\text{L-Malate}
\end{array}
\qquad (14.13)
$$

The conversion of fumarate to L-Malate

In step 8, malate is oxidized to oxaloacetate, and another molecule of $NAD^+$ is reduced to NADH. This reaction is catalyzed by the enzyme **malate dehydrogenase** and is not a control point for the cycle. This reaction is similar to that of other NADH-linked dehydrogenases, which we have seen previously. The oxaloacetate can then react with another molecule of acetyl-CoA to start another round of the cycle.

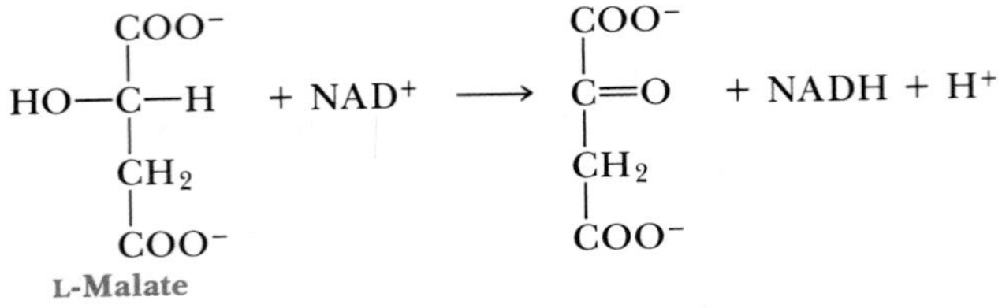

$$
\begin{array}{ccc}
\text{COO}^- & & \text{COO}^- \\
| & & | \\
\text{HO--C--H} \;\; + \text{NAD}^+ \longrightarrow & \text{C=O} \;\; + \text{NADH} + \text{H}^+ \\
| & & | \\
\text{CH}_2 & & \text{CH}_2 \\
| & & | \\
\text{COO}^- & & \text{COO}^-
\end{array}
$$
$$\text{L-Malate}$$

(14.14)

The conversion of L-Malate to oxaloacetate

The oxidation of pyruvate by the pyruvate dehydrogenase complex and the citric acid cycle results in the production of three molecules of $CO_2$. As a result of these oxidation reactions, one molecule of GDP is phosphorylated to GTP, one molecule of FAD is reduced to $FADH_2$, and four molecules of $NAD^+$ are reduced to NADH. Of the four molecules of NADH produced, three come from the citric acid cycle and one from the reaction of the pyruvate dehydrogenase complex. The overall stoichiometry of the oxidation reactions is the sum of the pyruvate dehydrogenase reaction and the citric acid cycle. Note that there is no production of ATP from ADP *directly* from the citric acid cycle.

*Pyruvate dehydrogenase complex:*

Pyruvate + CoA-SH + $NAD^+$ $\longrightarrow$ acetyl-CoA + NADH + $CO_2$ + $H^+$

*Citric acid cycle:*

Acetyl-CoA + 3 $NAD^+$ + FAD + GDP + $P_i$ + 2 $H_2O$ $\longrightarrow$ 2 $CO_2$ + CoA-SH + 3 NADH + 3 $H^+$ + $FADH_2$ + GTP

*Overall reaction:*

Pyruvate + 4 $NAD^+$ + FAD + GDP + $P_i$ + 2 $H_2O$ $\longrightarrow$ 3 $CO_2$ + 4 NADH + $FADH_2$ + GTP + 4 $H^+$

## 14.4
## ENERGETICS AND CONTROL OF THE CITRIC ACID CYCLE

The reaction of pyruvate to acetyl-CoA is exergonic, as we have seen ($\Delta G^{\circ\prime}$ = $-33.4$ kJ mol$^{-1}$ = $-8.0$ kcal mol$^{-1}$). The citric acid cycle itself is also exergonic ($\Delta G^{\circ\prime}$ = $-44.3$ kJ mol$^{-1}$ = $-10.6$ kcal mol$^{-1}$), and you are asked in Exercise 17 to confirm this point. The standard free energy changes for the individual reactions are listed in Table 14.1. Of the individual reactions of the cycle, only one is strongly endergonic, the oxidation of malate to oxaloacetate ($\Delta G^{\circ\prime}$ = $+29.2$ kJ mol$^{-1}$ = $+7.0$ kcal mol$^{-1}$). This endergonic reaction is, however, coupled to one of the strongly exergonic reactions of the cycle, the condensation of acetyl-CoA and oxaloacetate to produce citrate and acetyl-CoA ($\Delta G^{\circ\prime}$ = $-33.4$ kJ mol$^{-1}$ = $-8.0$ kcal mol$^{-1}$). In addition to the energy released by the oxidation reactions, there is more release of energy to come in the electron transport chain. When the four NADH and single $FADH_2$ produced by the pyruvate dehydrogenase

**TABLE 14.1    The Energetics of Conversion of Pyruvate to $CO_2$**

| REACTION | $(\Delta G°'$ $(mol^{-1})$ | |
| --- | --- | --- |
| | kJ | kcal |
| Pyruvate + CoA-SH + $NAD^+$ ⟶ acetyl-CoA + NADH + $CO_2$ + $H^+$ | −33.4 | −8.0 |
| Acetyl-CoA + oxaloacetate + $H_2O$ ⟶ citrate + CoA-SH + $H^+$ | −32.2 | −7.7 |
| Citrate ⟶ isocitrate | +6.3 | +1.5 |
| Isocitrate + $NAD^+$ ⟶ α-ketoglutarate + NADH + $CO_2$ | −7.1 | −1.7 |
| α-Ketoglutarate + $NAD^+$ + CoA-SH ⟶ succinyl-CoA + NADH + $CO_2$ + $H^+$ | −33.4 | −8.0 |
| Succinyl-CoA + GDP + $P_i$ ⟶ succinate + GTP + CoA-SH | −3.3 | −0.8 |
| Succinate + FAD ⟶ fumarate + $FADH_2$ | ≃ 0 | ≃ 0 |
| Fumarate + $H_2O$ ⟶ L-malate | −3.8 | −0.9 |
| L-Malate + $NAD^+$ ⟶ oxaloacetate + NADH + $H^+$ | +29.2 | +7.0 |
| **Overall reaction:** | | |
| Pyruvate + 4 $NAD^+$ + FAD + GDP + $P_i$ + 2 $H_2O$ ⟶ | | |
| 3 $CO_2$ + 4 NADH + $FADH_2$ + GTP + 4 $H^+$ | −77.7 | −18.6 |

complex and citric acid cycle are reoxidized by the electron transport chain, considerable quantities of ATP are produced.

Control of the citric acid cycle is exercised at three points; that is, there are three enzymes within the citric acid cycle that play a regulatory role (Figure 14.5). There is also a control point outside the cycle itself, namely the reaction in which pyruvate is oxidatively decarboxylated to produce the acetyl-CoA needed for the first reaction of the citric acid cycle proper. As we have already seen, the pyruvate dehydrogenase complex is inhibited by ATP and NADH. (There is also product inhibition by acetyl-CoA in this reaction.)

Within the citric acid cycle itself the three control points are the reactions catalyzed by citrate synthetase, isocitrate dehydrogenase, and the α-ketoglutarate dehydrogenase complex. We have already mentioned that the first reaction of the cycle is one in which regulatory control appears, as is to be expected in the first reaction of any pathway. Citrate synthetase is an allosteric enzyme inhibited by ATP, NADH, and succinyl-CoA.

The second regulatory site is the isocitrate dehydrogenase reaction. In this case, ADP is an allosteric activator of the enzyme. We have called attention to the recurring pattern in which ATP and NADH inhibit enzymes of the pathway, and ADP and $NAD^+$ activate these enzymes.

The α-ketoglutarate dehydrogenase complex is the third regulatory site. As before, ATP and NADH are inhibitors and ADP and $NAD^+$ are activators. This recurring theme in metabolism reflects the way in which a cell can adjust to an active state or to a resting state.

When a cell is metabolically active it uses ATP and NADH at a great rate, producing large amounts of ADP and $NAD^+$ (Table 14.2). In other words, when the ratio ATP/ADP is low, the cell is using energy and needs to release more energy from stored nutrients. A low NADH/$NAD^+$ ratio is also characteristic of an active metabolic state. On the other hand, a

**FIGURE 14.5** Control points in the conversion of pyruvate to acetyl-CoA and in the citric acid cycle.

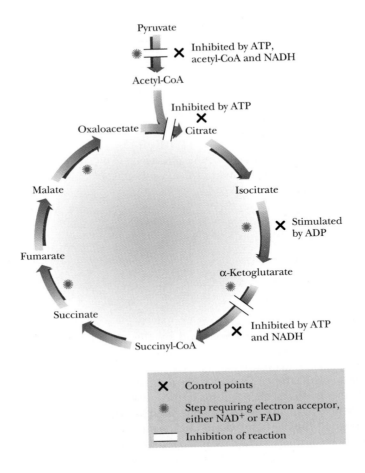

resting cell has fairly high levels of ATP and NADH. The ratios of ATP/ADP and NADH/NAD$^+$ are also high in resting cells, which do not need to maintain a high level of oxidation to produce energy.

When cells have low energy requirements (they have a high "energy charge") with high ATP/ADP and NADH/NAD$^+$ ratios, the presence of so much ATP and NADH serves as a signal to "shut down" the enzymes responsible for oxidative reactions. When cells have a low energy charge,

**Relationship Between the Metabolic State of a Cell and the (ATP/ADP) and (NADH/NAD$^+$) Ratios**

Cells in a resting metabolic state
  Need and use comparatively little energy
  High ATP, low ADP levels imply high (ATP/ADP)
  High NADH, low NAD$^+$ levels imply high (NADH/NAD$^+$)
Cells in a highly active metabolic state
  Need and use more energy than resting cells
  Low ATP, high ADP levels imply low (ATP/ADP)
  Low NADH, high NAD$^+$ levels imply low (NADH/NAD$^+$)

characterized by low ATP/ADP and NADH/NAD$^+$ ratios, the need to release more energy and to generate more ATP serves as a signal to "turn on" the oxidative enzymes. This relationship of energy requirements to enzyme activity is the basis for the overall regulatory mechanism exerted at a few key control points in metabolic pathways.

## 14.5
## THE GLYOXYLATE CYCLE: A RELATED PATHWAY

In plants and in some bacteria, but not in animals, acetyl-CoA can serve as the starting material for the biosynthesis of carbohydrates. Animals can convert carbohydrates to fats, but not fats to carbohydrates. (Acetyl-CoA is produced in the catabolism of fatty acids.) Two enzymes are responsible for the ability of plants and bacteria to produce glucose from fatty acids. **Isocitrate lyase** cleaves isocitrate, producing glyoxylate and succinate. **Malate synthetase** catalyzes the reaction of glyoxylate with acetyl-CoA to produce malate.

(a)

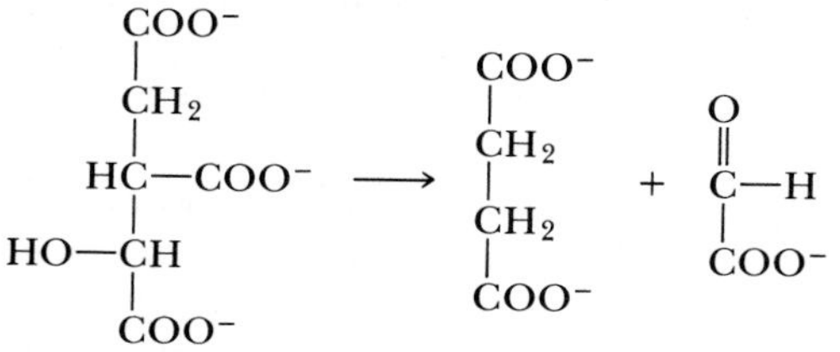

$$
\begin{array}{c}
\text{COO}^- \\
| \\
\text{CH}_2 \\
| \\
\text{HC}-\text{COO}^- \\
| \\
\text{HO}-\text{CH} \\
| \\
\text{COO}^- \\
\text{Isocitrate}
\end{array}
\longrightarrow
\begin{array}{c}
\text{COO}^- \\
| \\
\text{CH}_2 \\
| \\
\text{CH}_2 \\
| \\
\text{COO}^- \\
\text{Succinate}
\end{array}
\;+\;
\begin{array}{c}
\text{O} \\
\| \\
\text{C}-\text{H} \\
| \\
\text{COO}^- \\
\text{Glyoxylate}
\end{array}
$$

The conversion of isocitrate to glyoxylate and succinate

(b)

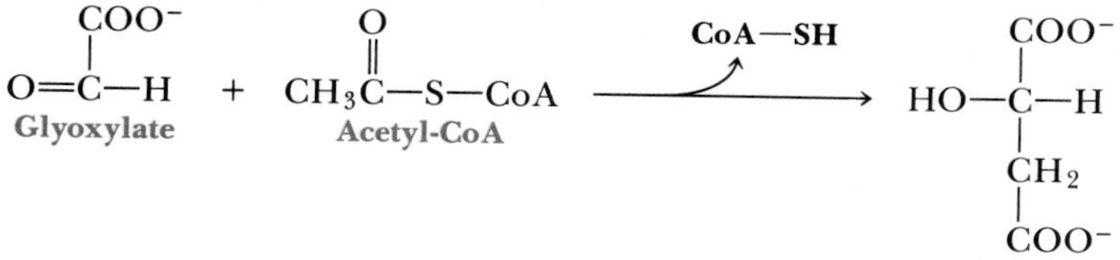

$$
\begin{array}{c}
\text{COO}^- \\
| \\
\text{O}=\text{C}-\text{H} \\
\text{Glyoxylate}
\end{array}
\;+\;
\begin{array}{c}
\text{O} \\
\| \\
\text{CH}_3\text{C}-\text{S}-\text{CoA} \\
\text{Acetyl-CoA}
\end{array}
\xrightarrow{\;\text{CoA}-\text{SH}\;}
\begin{array}{c}
\text{COO}^- \\
| \\
\text{HO}-\text{C}-\text{H} \\
| \\
\text{CH}_2 \\
| \\
\text{COO}^- \\
\text{Malate}
\end{array}
\qquad (14.15)
$$

The reaction of glyoxylate with acetyl-CoA to produce malate

The unique reactions of the glyoxylate cycle

These two reactions in succession bypass the two oxidative decarboxylation steps of the citric acid cycle. The net result is an alternate pathway, the **glyoxylate cycle** (Figure 14.6). Two molecules of acetyl-CoA enter the glyoxylate cycle; they give rise to one molecule of malate and eventually to one molecule of oxaloacetate. Two two-carbon units (the acetyl groups of acetyl-CoA) give rise to a four-carbon unit (malate), which is then converted to oxaloacetate (also a four-carbon compound). Glucose can then be produced from oxaloacetate by gluconeogenesis.

**FIGURE 14.6**   The glyoxylate cycle.

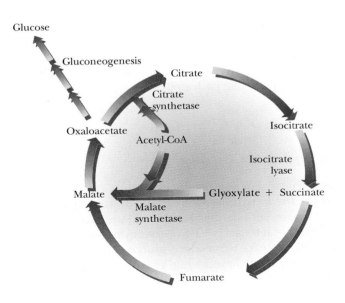

Specialized organelles in plants, called **glyoxysomes,** are the site of the glyoxylate cycle. This pathway is particularly important in germinating seeds. The fatty acids stored in the seeds are broken down for energy during germination. First the fatty acids give rise to acetyl-CoA, which can enter the citric acid cycle and go on to release energy in the ways we have already seen. The citric acid cycle and the glyoxylate cycle can operate simultaneously. Acetyl-CoA also serves as the starting point for the synthesis of glucose and any other compounds needed by the growing seedling. (Recall that carbohydrates play an important structural, as well as energy-producing, role in plants.)

The glyoxylate cycle also occurs in bacteria. This point is far from surprising, since many types of bacteria can live on very limited carbon sources. They have metabolic pathways that can produce all the biomolecules they need from quite simple molecules. The glyoxylate cycle is one example of how they manage this feat.

## 14.6
## A FINAL NOTE

The citric acid cycle is considered part of aerobic metabolism, but we have not encountered any reactions in this chapter in which oxygen takes part. The reactions of the citric acid cycle are intimately related to those of electron transport and oxidative phosphorylation, which do eventually lead to oxygen. The citric acid cycle provides a vital link between the chemical energy of nutrients and the chemical energy of ATP. Many molecules of ATP can be generated as a result of coupling to oxygen, and we shall see that the number depends on the NADH and $FADH_2$ generated in the citric acid cycle.

# SUMMARY

The citric acid cycle plays a central role in metabolism. It is the first part of aerobic metabolism; it is also amphibolic (both catabolic and anabolic). Unlike glycolysis, which takes place in the cytosol, the citric acid cycle occurs in mitochondria. Most of the enzymes of the citric acid cycle are in the mitochondrial matrix. (Succinate dehydrogenase is localized in the inner mitochondrial membrane.)

Pyruvate produced by glycolysis is transformed by oxidative decarboxylation into acetyl-CoA in the presence of Coenzyme A. Acetyl-CoA then enters the citric acid cycle by reacting with oxaloacetate to produce citrate. The reactions of the citric acid cycle include two other oxidative decarboxylations, which transform the six-carbon compound citrate into the four-carbon compound succinate. The cycle is completed by regeneration of oxaloacetate from succinate in a multistep process that includes two other oxidation reactions. The overall reaction, starting with pyruvate, is

$$\text{Pyruvate} + 4\ NAD^+ + FAD + GDP + P_i + 2\ H_2O \longrightarrow 3\ CO_2 + 4\ NADH + FADH_2 + GTP + 4\ H^+$$

$NAD^+$ and FAD are the electron acceptors in the oxidation reactions. The cycle is strongly exergonic.

Control of the citric acid cycle is exercised at three points. There is also a control point outside the cycle itself, the reaction in which pyruvate is oxidatively decarboxylated to produce acetyl-CoA. Within the citric acid cycle itself the three control points are the reactions catalyzed by citrate synthetase, isocitrate dehydrogenase, and the $\alpha$-ketoglutarate dehydrogenase complex. In general, ATP and NADH are inhibitors and ADP and $NAD^+$ are activators of the enzymes at the control points.

In plants and bacteria there is a pathway related to the citric acid cycle, the glyoxylate cycle. The two oxidative decarboxylations of the citric acid cycle are bypassed. This pathway plays a role in the ability of plants to convert acetyl-CoA to carbohydrates, a process that does not occur in animals.

# EXERCISES

1. We have seen one of the four possible isomers of isocitrate, the one produced in the aconitase reaction. Draw the configurations of the other three.
2. Draw the structures of the activated two-carbon groups bound to thiamine pyrophosphate in three enzymes that contain this coenzyme. Hint: keto-enol tautomerism may enter into the picture.
3. Why is the citric acid cycle considered part of aerobic metabolism even though molecular oxygen does not appear in any reaction?
4. ATP is a competitive inhibitor of NADH binding to malate dehydrogenase, as are ADP and AMP. Suggest a structural basis for this inhibition.
5. How does an increase in the ADP/ATP ratio affect the activity of isocitrate dehydrogenase?
6. How does an increase in the $NADH/NAD^+$ ratio affect the activity of pyruvate dehydrogenase?
7. Would you expect the citric acid cycle to be more or less active when a cell has a high ATP/ADP and a high $NADH/NAD^+$ ratio? Give the reason for your answer.
8. Show by Lewis electron-dot structures of the appropriate portions of the molecule where electrons are lost in the following conversions:
   (a) pyruvate to acetyl-CoA
   (b) isocitrate to $\alpha$-ketoglutarate
   (c) $\alpha$-ketoglutarate to succinyl-CoA
   (d) succinate to fumarate
   (e) malate to oxaloacetate
9. Is the conversion of fumarate to malate a redox (electron-transfer) reaction or not? Give the reason for your answer.
10. Prepare a sketch showing how the individual reactions of the three enzymes of the pyruvate dehydrogenase complex give rise to the overall reaction.
11. Why is the reaction catalyzed by citrate synthetase considered a condensation reaction?
12. Would you expect the $\Delta G^{\circ\prime}$ for the hydrolysis of a

thioester to be (a) large and negative, (b) large and positive, (c) small and negative, or (d) small and positive? Give the reason for your answer.

**13.** In what part of the cell does the citric acid cycle take place? Does this differ from the part of the cell where glycolysis occurs?

**14.** What electron acceptors play a role in the citric acid cycle?

**15.** Briefly describe the dual role of lipoic acid in the pyruvate dehydrogenase complex.

**16.** Discuss oxidative decarboxylation, using a reaction from this chapter to illustrate your points.

**17.** Some reactions of the citric acid cycle are endergonic; show how the overall cycle is exergonic. (See Table 14.1.)

**18.** Describe the conversion of acetyl-CoA to oxaloacetate in the glyoxylate cycle.

## ANNOTATED BIBLIOGRAPHY

Bodner, G.M. The Tricarboxylic Acid (TCA), Citric Acid, or Krebs Cycle. *J. Chem. Ed.* **63,** 673–677 (1986). [A concise and well written summary of the citric acid cycle. Part of a series on metabolism.]

Boyer, P.D., ed. *The Enzymes,* 3rd ed. New York: Academic Press, 1975. [There are reviews on aconitase in Volume 5 and on dehydrogenases in Volume 11.]

Krebs, H.A. *Reminiscences and Reflections.* New York: Oxford University Press, 1981. [A review of the citric acid cycle along with the autobiography.]

Popjak, G. Stereospecificity of Enzyme Reactions. In Boyer, P.D., ed., *The Enzymes,* 3rd ed., Vol. 2. New York: Academic Press, 1970. [A review article on stereochemical aspects of the citric acid cycle.]

See also the references for Chapter 15.

# Electron Transport and Oxidative Phosphorylation

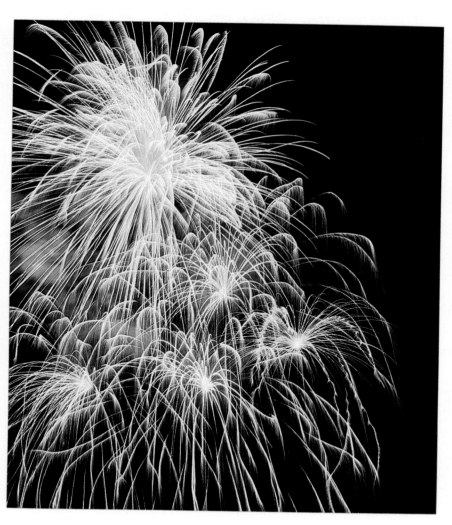

*Fireworks are a spectacular example of chemical reactions that take place in air. The reactions of aerobic metabolism use oxygen from the air in a less spectacular, but highly essential, way.*

359

*Energy derived from the oxidation of metabolic fuels is ultimately converted to ATP, the quick energy currency of the cell. In eukaryotic cells, under aerobic conditions, ATP is generated by the power of electron transport **along** the inner membrane of the mitochondrion coupled with proton transport **across** the inner membrane. The electron transport chain is actually four closely related enzyme complexes embedded in the inner mitochondrial membrane. In a series of oxidation-reduction transfers, they conduct electrons along the membrane from one complex to another until the electrons reach their final destination where they combine with molecular oxygen to reduce $O_2$ to $2H_2O$. The energy of electron transport can then be used by these same enzyme complexes to pump **protons** across the membrane out into the intermembrane space. The reverse flow of protons back through the membrane into the inner matrix can be used to generate ATP. Also embedded in the inner membrane is an ATP synthase complex that binds ADP and phosphate ion to synthesize ATP. The flow of protons through the ATP synthase from the intermembrane space to the inner matrix releases the new ATP that has been synthesized. This process is very similar to the production of ATP by photosynthesis in the thylakoid membrane of the chloroplast in green plants.*

## 15.1
## THE ROLE OF ELECTRON TRANSPORT IN METABOLISM

Aerobic metabolism is a highly efficient way for an organism to extract energy from nutrients. In eukaryotic cells the aerobic processes (including conversion of pyruvate to acetyl-CoA, the citric acid cycle, and electron transport) all occur in the mitochondria, while the anaerobic process, glycolysis, takes place outside the mitochondria in the cytosol. We have not yet seen any reactions in which oxygen plays a part, but in this chapter we shall discuss the role of oxygen in metabolism as the final acceptor of electrons in the **electron transport chain.** The reactions of the electron transport chain take place in the inner mitochondrial membrane.

The energy released by the oxidation of nutrients is used by organisms in the form of chemical energy of ATP. Production of ATP in the cell requires the process of **oxidative phosphorylation,** in which ADP is phosphorylated to give ATP. The production of ATP by oxidative phosphorylation is a process separate from electron transport, but the reactions of the electron transport chain are strongly linked to one another and are tightly coupled to synthesis of ATP by phosphorylation of ADP. The operation of the electron transport chain leads to pumping of hydrogen ions (protons) across the inner mitochondrial membrane, creating a pH gradient (also called a **proton gradient**); this proton gradient represents stored energy and provides the basis of the coupling mechanism

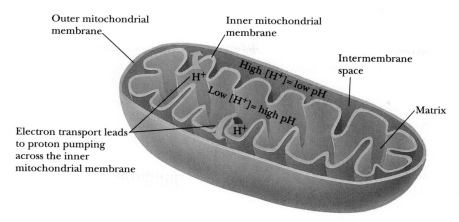

Outer mitochondrial membrane

Inner mitochondrial membrane

Intermembrane space

High [H⁺]= low pH

Low [H⁺]= high pH

H⁺

H⁺

Matrix

Electron transport leads to proton pumping across the inner mitochondrial membrane

**FIGURE 15.1**    A proton gradient is established across the inner mitochondrial membrane as a result of electron transport. Transfer of electrons through the electron transport chain leads to the pumping of protons from the matrix to the intermembrane space. The proton gradient (also called the pH gradient), together with the membrane potential, provides the basis of the coupling mechanism that drives ATP synthesis.

(Figure 15.1). Oxidative phosphorylation gives rise to most of the ATP production associated with the complete oxidation of glucose.

The NADH and $FADH_2$ molecules generated in glycolysis and the citric acid cycle transfer electrons to oxygen in the series of reactions known collectively as the electron transport chain. The NADH and $FADH_2$ are oxidized to $NAD^+$ and FAD, and can be used again in various metabolic pathways. Oxygen, the ultimate electron acceptor, is reduced to water; this completes the process by which glucose is completely oxidized to carbon dioxide and water. We have already seen how carbon dioxide is produced from pyruvate, which in turn is produced from glucose, by the pyruvate dehydrogenase complex and the citric acid cycle. In this chapter we shall see how water is produced.

The complete series of oxidation–reduction reactions of the electron transport chain is given in schematic form in Figure 15.2. A particularly noteworthy point about electron transport is that three molecules of ATP are generated for each molecule of NADH that enters the electron transport chain, and two molecules of ATP are produced for each molecule of $FADH_2$. The general outline of the process is that NADH passes

**FIGURE 15.2**    Schematic representation of the electron transport chain, showing sites of proton pumping coupled to oxidative phosphorylation. FMN is the flavin coenzyme *flavin mononucleotide*, which differs from FAD in not having an adenine nucleotide. CoQ is coenzyme Q (see Figure 15.3). Cyt **b**, cyt $c_1$, cyt **c**, and cyt $aa_3$ are the heme-containing proteins cytochrome **b**, cytochrome $c_1$, cytochrome **c**, and cytochrome $aa_3$, respectively.

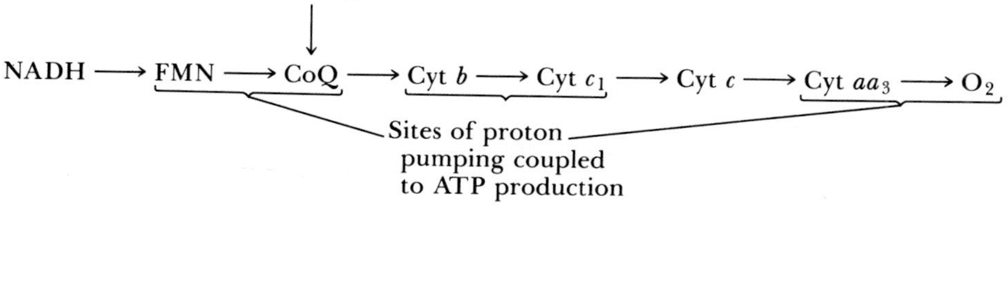

FADH₂

NADH ⟶ FMN ⟶ CoQ ⟶ Cyt *b* ⟶ Cyt $c_1$ ⟶ Cyt *c* ⟶ Cyt $aa_3$ ⟶ $O_2$

Sites of proton pumping coupled to ATP production

electrons to coenzyme Q, as does $FADH_2$, providing an alternate mode of entry into the electron transport chain. Electrons are then passed from coenzyme Q to a series of proteins called cytochromes, designated by lower case letters, and eventually to oxygen.

## 15.2
## STANDARD REDUCTION POTENTIALS

When we discussed the overall pathway of electron transport in Section 15.1, we did not treat the relationship between energy changes and electron transfer in detail. The change in energy for each reaction can, of course, be treated in terms of the $\Delta G^{\circ\prime}$, the standard free energy change, but there is another way of looking at the situation. Oxidation–reduction reactions by definition involve electron transfer. A flow of electrons under proper conditions is an electric current. It is possible to consider the energy changes in oxidation–reduction reactions in terms of easily measureable electrical quantities.

A voltage or emf (*electromotive force*) is a measure of the tendency of electrons to flow. To return to the original example of an oxidation–reduction reaction, which we first met in Section 12.2, there is a flow of electrons from zinc metal to copper ion, producing zinc ion and copper metal. We shall use this example again because of its clarity.

$$Zn(s) \longrightarrow Zn^{2+}(aq) + 2e^-$$
$$\underline{Cu^{2+}(aq) + 2e^- \longrightarrow Cu(s)}$$
$$Zn(s) + Cu^{2+}(aq) \longrightarrow Zn^{2+}(aq) + Cu(s)$$

A voltage (measured under standard conditions) can be determined for this reaction. More importantly for our purposes, the half reactions of oxidation and reduction can be considered separately, and a voltage can be assigned to each of the half reactions. The assignment of a voltage for each half reaction is done by comparison with a standard value.

If a given half reaction is chosen as a reference, and its voltage—its **standard potential—** is assigned a value of zero ($E^\circ = 0$), voltages for other half reactions can be measured against the reference. (We need not concern ourselves with the details of how the measurement is done.) The reaction

$$2H^+(aq) + 2e^- \longrightarrow H_2(g) \qquad E^\circ = 0$$

has been chosen for the zero value.

In our example for the copper–zinc reaction, the measured voltages are:

|  | $E^\circ$ (volts) |
|---|---|
| $Zn(s) \longrightarrow Zn^{2+}(aq) + 2e^-$ | 0.76 |
| $Cu^{2+}(aq) + 2e^- \longrightarrow Cu(s)$ | 0.34 |
| $Zn(s) + Cu^{2+}(aq) \longrightarrow Zn^{2+}(aq) + Cu(s)$ | 1.10 |

Note that the potential for the overall reaction is the sum of the potentials of the two half reactions.

In the interests of consistency, standard potentials are listed in tables only for reduction half reactions. The standard potential for an oxidation reaction is numerically the same as that for the reverse reduction reaction, but with the sign changed. The half reaction for zinc given in a table is, for example,

$$Zn^{2+}(aq) + 2e^- \longrightarrow Zn(s) \qquad E° = -0.76 \text{ volts}$$

If metallic copper were placed in a solution of zinc ion, there would be no reaction. Zinc is a better reducing agent than copper, since zinc is oxidized at the expense of copper ion being reduced. The better the reducing agent, the more negative is its reduction potential.

$$Cu(s) + Zn^{2+}(aq) \longrightarrow \text{NO REACTION}$$

An overall voltage can be calculated even though the reaction does not proceed as written.

$$
\begin{array}{ll}
Zn^{2+} + 2e^- \longrightarrow Zn(s) & E° = -0.76 \\
\underline{Cu(s) \longrightarrow Cu^{2+} + 2e^-} & \underline{E° = -0.34} \\
Zn^{2+} + Cu(s) \longrightarrow Zn(s) + Cu^{2+} & E°_{cell} = -1.10 \text{ volts}
\end{array}
$$

Note that the voltage is negative for this reaction, which does not actually occur. A positive voltage always indicates an exergonic reaction that will occur as written, and a negative voltage indicates an endergonic one that will not occur as written unless energy is supplied.

The term "standard potential" also includes the standard state conditions (Section 8.3). Standard potentials are measured under slightly different standard conditions in biochemical, as opposed to chemical, practice. The standard state for biochemical usage is defined to be the same as that used in chemical practice except for the hydrogen ion activity. Instead of unit activity (~1 M) of hydrogen ion, which is pH = 0, the biochemical standard state refers to pH = 7. Standard half-cell potentials measured under these conditions are given the symbol $E°'$, similar to the $\Delta G°'$ notation for standard free-energy changes with pH 7 as a standard state.

## 15.3
## STANDARD POTENTIALS AND ENERGETICS OF BIOCHEMICAL OXIDATION–REDUCTION REACTIONS

Standard reduction potentials have been determined for the half reactions in glycolysis, the citric acid cycle, and the electron transport chain. When the reactions take place, there is of course a half reaction of oxidation (for which the voltage has the opposite sign from the reduction potential) as well as the half reaction of reduction. Some standard reduction potentials $E°'$ referred to pH 7 are listed in Table 15.1. In general, the further the reaction lies along the path of electron transport to oxygen, the more

**TABLE 15.1   Some Standard Reduction Potentials**

| REACTION | $E^{\circ\prime}$ (volts) |
|---|---|
| $\alpha$-Ketoglutarate + 2e$^-$ $\longrightarrow$ succinate + $CO_2$ | $-0.67$ |
| $2H^+ + 2e^- \longrightarrow H_2$ | $-0.42$ |
| $NAD^+ + 2H^+ + 2e^- \longrightarrow NADH + H^+$ | $-0.32$ |
| Acetaldehyde + $2H^+$ + 2e$^-$ $\longrightarrow$ ethanol | $-0.20$ |
| Pyruvate + $2H^+$ + 2e$^-$ $\longrightarrow$ lactate | $-0.19$ |
| Fumarate + $2H^+$ + 2e$^-$ succinate | $0.03$ |
| 2 Cytochrome $b_{oxid}$(Fe(III)) + 2e$^-$ $\longrightarrow$ 2 cytochrome $b_{red}$(Fe(II)) | $0.07$ |
| 2 Cytochrome $c_{oxid}$(Fe(III)) + 2e$^-$ $\longrightarrow$ 2 cytochrome $c_{red}$(Fe(II)) | $0.22$ |
| 2 Cytochrome $a/a_{3\,oxid}$(Fe(III)) + 2e$^-$ $\longrightarrow$ 2 cytochrome $a/a_3$(Fe(II)) | $0.38$ |
| $\frac{1}{2}O_2 + 2H^+ + 2e^- \longrightarrow H_2O$ | $0.82$ |

positive the reduction potential. For an electron acceptor to oxidize another substance that has already been oxidized, the electron acceptor must itself be very easily reduced (i.e., be a good oxidizing agent).

The $E^{\circ\prime}$ for the electron transport chain can be calculated from the two half reactions,

$$\begin{array}{ll} & E^{\circ\prime}\text{ (volts)} \\ \frac{1}{2}O_2 + 2H^+ + 2e^- \longrightarrow H_2O & +0.82 \\ NAD^+ + 2H^+ + 2e^- \longrightarrow NADH + H^+ & -0.32 \end{array}$$

Both reactions are written here as reductions. However, NADH is oxidized when the reaction actually takes place, so that half reaction must be reversed and the sign for its standard reduction potential must be changed to give the overall equation for electron transport.

$$\begin{array}{ll} & E^{\circ\prime}\text{ (volts)} \\ \frac{1}{2}O_2 + 2H^+ + 2e^- \longrightarrow H_2O & +0.82 \\ NADH + H^+ \longrightarrow NAD^+ + 2H^+ + 2e^- & -(-0.32) \\ \hline NADH + H^+ + \frac{1}{2}O_2 \longrightarrow NAD^+ + H_2O & +1.14 \end{array}$$

As this calculation shows, electron transport is exergonic.

A standard free energy change, $\Delta G^{\circ\prime}$, can also be calculated for each mole of NADH that enters the electron transport chain. We shall not derive the equation that allows us to do so, but we shall make use of it. The relationship between the free energy change and potential is

$$\Delta G^{\circ\prime} = -nFE^{\circ\prime}$$

where $n$ is the number of electrons transferred in the reaction and $F$ is the Faraday constant, 96,400 joules/volt per mole of substance oxidized.

In the electron transport equation, two electrons are transferred for each molecule of NADH, and thus

$$\Delta G^{\circ\prime} = -2 \times 96{,}400 \text{ J/volt/mol NADH} \times 1.14 \text{ volt}$$

$$\Delta G^{\circ\prime} = -220 \text{ kJ/mol NADH} = -52.6 \text{ kcal/mol}$$

Again, you can see that electron transport is an exergonic process.

Not all reactions take place under standard conditions. It has already been shown (Section 8.3) that there is a relationship between the free energy change, $\Delta G$, for a reaction under any conditions and the standard free energy change, $\Delta G°$. There is likewise a relationship between the potential measured under nonstandard conditions, $E$, and the potential measured under standard conditions $E°'$. This relationship, which we shall not derive, is called the Nernst equation.

For the general half reaction of reduction

$$\text{Oxidizing agent} + ne^- \longrightarrow \text{reducing agent}$$

the Nernst equation takes the form

$$E = E°' + \frac{2.3\ RT}{nF} \log \frac{[\text{oxidizing agent}]}{[\text{reducing agent}]}$$

where $R$ (the universal gas constant) is 8.31 J K$^{-1}$ mol$^{-1}$, $T$ is absolute temperature in K, $n$ is the number of electrons transferred in the reaction, and $F$ is the Faraday constant. For 25°C and $n = 1$,

$$E = E°' + 0.059 \log \frac{[\text{oxidizing agent}]}{[\text{reducing agent}]}$$

and when $n = 2$,

$$E = E°' + 0.03 \log \frac{[\text{oxidizing agent}]}{[\text{reducing agent}]}$$

A change in the relative concentrations of oxidizing agent and reducing agent can obviously change the value of $E$ from that calculated for standard conditions.

## 15.4
## ELECTRON TRANSPORT FROM NADH TO $O_2$ REQUIRES THREE MEMBRANE-BOUND COMPLEXES

Intact mitochondria isolated from cells can carry out all the reactions of the electron transport chain; the electron transport apparatus can also be resolved into its component parts by a process called fractionation. Three separate **respiratory complexes** can be isolated from the inner mitochondrial membrane. These complexes are multienzyme systems; in the last chapter we encountered other examples of such multienzyme complexes such as the pyruvate dehydrogenase complex and the $\alpha$-ketoglutarate dehydrogenase complex. Each of the respiratory complexes can carry out the reactions of a portion of the electron transport chain.

The first complex, **NADH-COQ oxidoreductase,** catalyzes the first steps of electron transport, namely the transfer of electrons from NADH to coenzyme Q. The complex is an integral part of the inner mitochondrial membrane and includes an iron–sulfur protein and the flavoprotein that oxidizes NADH. The flavoprotein has a flavin coenzyme, FMN, which differs from FAD in not having an adenine nucleotide.

The structure of FMN
(*Flavin* mono*nucleotide*)

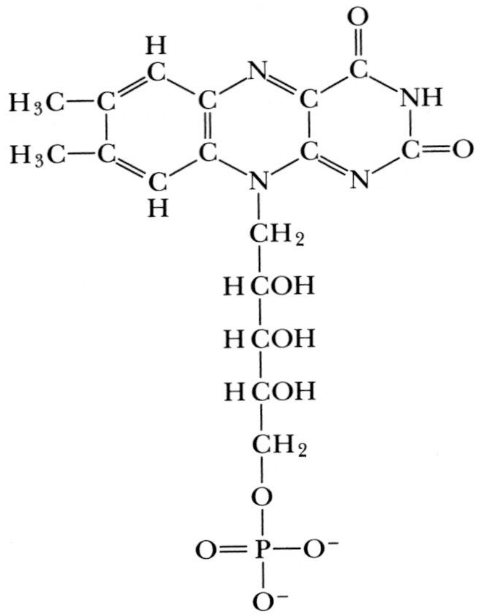

The reaction occurs in several steps with successive oxidation and reduction of the flavoprotein and the iron–sulfur protein. The first step is the transfer of electrons from NADH to the flavin portion of the flavoprotein:

$$NADH + H^+ + E\text{—}FMN \longrightarrow NAD^+ + E\text{—}FMNH_2$$

where the notation E—FMN indicates that the flavin is covalently bonded to the enzyme. In the second step the reduced flavoprotein is reoxidized and coenzyme Q (represented simply as CoQ) is reduced to $CoQH_2$ as a pair of electrons is passed on (Figure 15.3). Coenzyme Q is also called ubiquinone. The electrons are passed first to the iron–sulfur protein and then to coenzyme Q.

$$E\text{—}FMNH_2 + Fe\text{—}S_{oxidized} \longrightarrow E\text{—}FMN + Fe\text{—}S_{reduced} + 2H^+$$

$$Fe\text{—}S_{reduced} + CoQ + 2H^+ \longrightarrow Fe\text{—}S_{oxidized} + CoQH_2$$

The notation Fe—S refers to the iron–sulfur protein. The overall equation for the reaction is

$$NADH + H^+ + CoQ \longrightarrow NAD^+ + CoQH_2$$

This reaction is one of the three responsible for the proton pumping (Figure 15.4) that creates the pH (proton) gradient. The standard reduction potential ($E^{\circ\prime}$) for the overall reaction catalyzed by complex I is 0.42 volt; the corresponding standard free energy change ($\Delta G^{\circ\prime} = -81$ kJ $mol^{-1} = -19.4$ kcal $mol^{-1}$) indicates that the reaction is strongly exergonic, releasing enough energy to drive the phosphorylation of ADP to ATP (Figure 15.5).

In further steps of the electron transport chain, electrons are passed from coenzyme Q, which is then reoxidized, to the first of a series of very similar proteins called cytochromes. Each of these proteins contains a

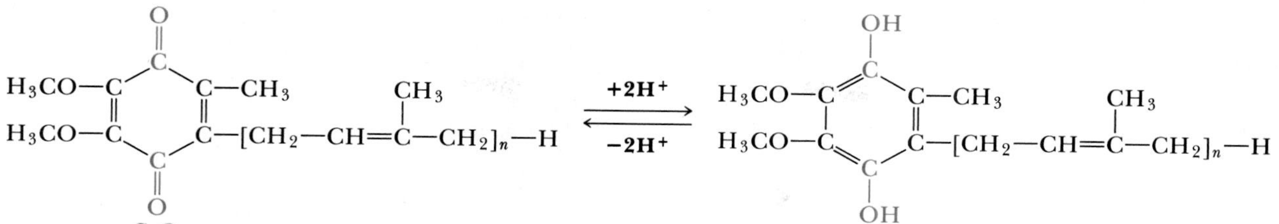

**CoQ**
(Oxidized quinone form)

**CoQH$_2$**
(Reduced hydroquinone form)

**FIGURE 15.3**   The oxidized and reduced forms of coenzyme Q. Coenzyme Q is also called ubiquinone.

heme group, and in each heme group the iron ion is successively reduced to Fe(II) and reoxidized to Fe(III). This situation differs from that of the iron in the heme group of hemoglobin, which remains in the reduced form as Fe(II) through the entire process of oxygen transport in the bloodstream. There are also some structural differences between the heme group in hemoglobin and the heme groups in the various types of cytochromes.

**FIGURE 15.4**   The electron transport chain, showing the respiratory complexes. In the reduced cytochromes the iron is in the Fe(II) oxidation state, while in the oxidized cytochromes the oxygen is in the Fe(III) oxidation state.

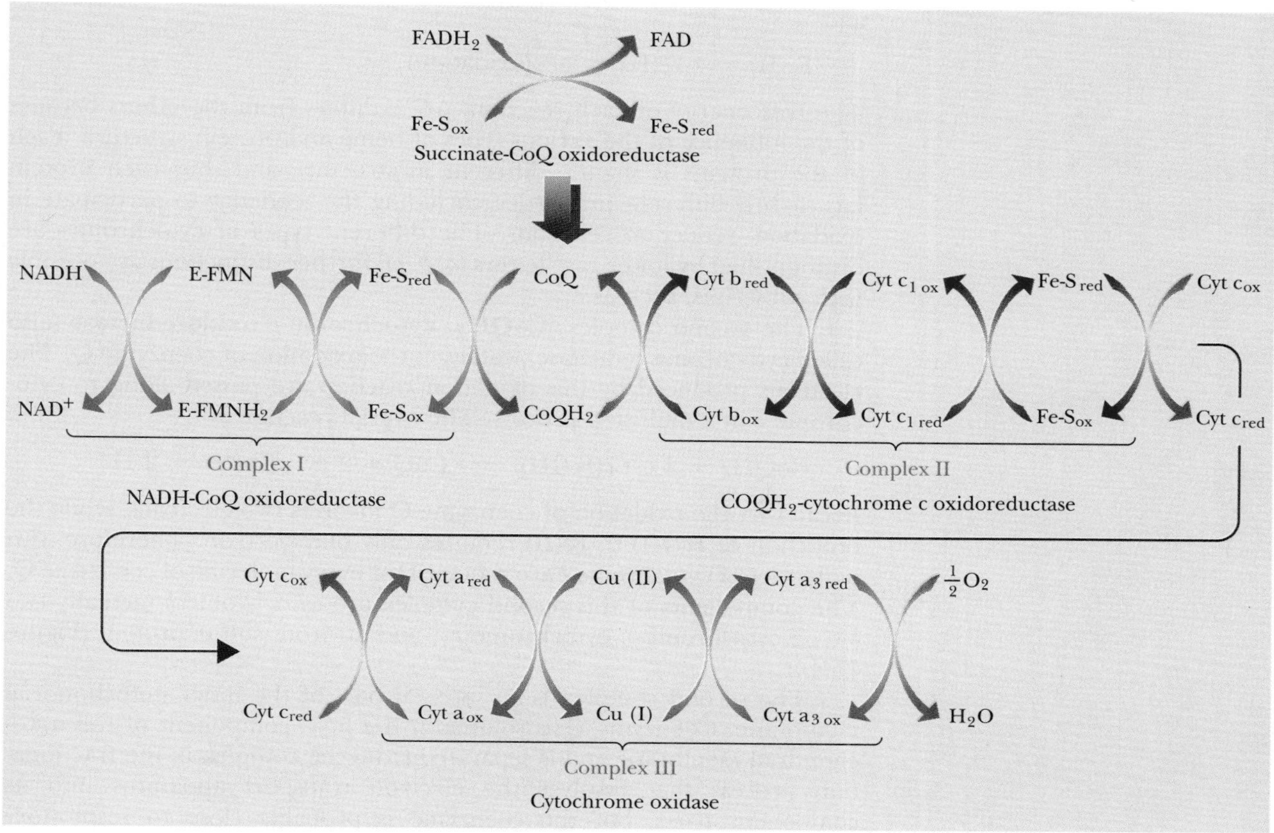

**FIGURE 15.5** The energetics of electron transport.

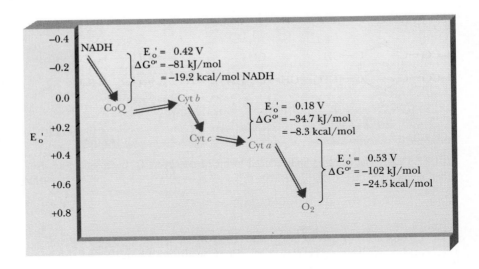

To be strictly accurate, the successive oxidation–reduction reactions of the cytochromes are not simply

$$Fe(III) + e^- \longrightarrow Fe(II) \text{ (reduction)}$$

and

$$Fe(II) \longrightarrow Fe(III) + e^- \text{ (oxidation)}$$

The free energy of each reaction, $\Delta G^{\circ\prime}$, differs from the others because of the influence of the various types of heme and protein structure. Each of the proteins is slightly different in structure and thus each protein has slightly different properties, including the tendency to participate in oxidation–reduction reactions. The different types of cytochromes are distinguished by lower case letters ($a$, $b$, $c$); further distinctions are possible with subscripts, such as $c_1$.

The second complex, **CoQH$_2$–cytochrome $c$ oxidoreductase** (also called cytochrome reductase), catalyzes the oxidation of coenzyme Q. The electrons produced by this oxidation reaction are passed along to cytochrome $c$ in a multistep process. The overall reaction is

$$CoQH_2 + 2 \text{ cyt } c(Fe(III)) \longrightarrow CoQ + 2 \text{ cyt } c(Fe(II)) + 2 \text{ H}^+$$

Recall that the oxidation of coenzyme Q involves two electrons, while the reduction of Fe(III) to Fe(II) requires only one electron. Therefore, two molecules of cytochrome $c$ are required for every molecule of coenzyme Q. The components of this second complex are cytochrome $b$ (actually two $b$-type cytochromes), cytochrome $c_1$, and an iron–sulfur protein (Figure 15.4).

The second complex is an integral part of the inner mitochondrial membrane. Coenzyme Q is soluble in the lipid component of the mitochondrial membrane and is separated from the complex in the fractionation process that resolves the electron transport apparatus into its component parts, but the coenzyme is probably close to respiratory

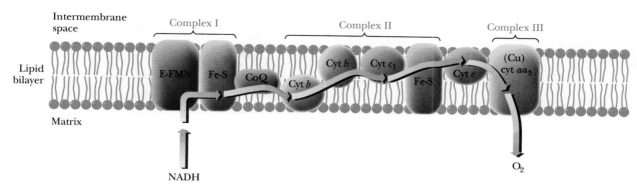

**FIGURE 15.6**  Composition and location of respiratory complexes in the inner mitochondrial membrane. NADH has accepted electrons from substrates such as pyruvate, isocitrate, $\alpha$-ketoglutarate, and malate. Note that the binding site for NADH is on the matrix side of the membrane. Coenzyme Q is soluble in the lipid bilayer and tends to be located in the middle of the membrane. Complex II contains two $b$-type cytochromes, one located on the side of the membrane toward the intermembrane space and the other on the matrix side. Cytochrome $c$ is located on the side of the membrane toward the intermembrane space. In Complex III the binding site for oxygen lies on the side toward the matrix.

complexes in the intact membrane (Figure 15.6). Cytochrome $c$ itself is not part of the complex but is located on the outer surface of the inner mitochondrial membrane, facing the intermembrane space. It is noteworthy that these two important electron carriers, coenzyme Q and cytochrome $c$, are not part of the respiratory complexes but are free to move freely in the membrane.

The flow of electrons is from reduced coenzyme Q to the $b$-type cytochromes and then to the other components of the complex:

$$CoQH_2 \longrightarrow \text{cyt } b \longrightarrow \text{cyt } c_1 \longrightarrow Fe\text{—}S \longrightarrow \text{cyt } c$$

The notation Fe—S refers to the iron–sulfur protein. The series of reactions involving coenzyme Q and the cytochromes, but omitting the iron–sulfur protein, can be written as follows:

Coenzyme Q (reduced) + cytochrome $b$ (oxidized) $\longrightarrow$
cytochrome $b$ (reduced) + coenzyme Q (oxidized)

Cytochrome $b$ (reduced, Fe(II)) + cytochrome $c_1$ (oxidized, Fe(III)) $\longrightarrow$ cytochrome $c_1$ (reduced, Fe(II)) + cytochrome $b$ (oxidized, Fe(III))

Cytochrome $c_1$ (reduced, Fe(II)) + cytochrome $c$ (oxidized, Fe(III)) $\longrightarrow$ cytochrome $c$ (reduced, Fe(II)) + cytochrome $c_1$ (oxidized, Fe(III))

Proton pumping, to which ATP production is coupled, occurs as a result of the reactions of this complex. The standard reduction potential ($E^{\circ\prime}$) for the flow of electrons from the $b$-type cytochromes to cytochrome

$c_1$ is 0.18 volt, which corresponds to a standard free energy change ($\Delta G^{\circ\prime}$) of $-34.2$ kJ $= -8.2$ kcal for each mole of NADH that enters the electron transport chain (see Figure 15.5). The phosphorylation of ADP requires 30.5 kJ mol$^{-1}$ = 7.3 kcal mol$^{-1}$, and the reaction catalyzed by the second complex supplies enough energy to drive the production of ATP.

The third complex, **cytochrome oxidase,** catalyzes the final steps of electron transport, the transfer of electrons from cytochrome $c$ to oxygen. The overall reaction is

$$2 \text{ cyt } c(\text{Fe(II)}) + 2 \text{ H}^+ + \tfrac{1}{2}O_2 \longrightarrow 2 \text{ cyt } c(\text{Fe(III)}) + H_2O$$

Proton pumping also takes place as a result of this reaction. Like the other respiratory complexes, cytochrome oxidase is an integral part of the inner mitochondrial membrane and contains cytochromes $a$ and $a_3$, as well as two $Cu^{2+}$ ions that are involved in the electron transport process. In the flow of electrons, the copper ions are intermediate electron acceptors that lie between the two $a$-type cytochromes in the sequence

$$\text{cyt } c \longrightarrow \text{cyt } a \longrightarrow Cu^{2+} \longrightarrow \text{cyt } a_3 \longrightarrow O_2$$

To show the reactions of the cytochromes more explicitly,

Cytochrome $c$ (reduced, Fe(II)) + cytochrome ($a/a_3$) (oxidized, Fe(III)) $\longrightarrow$ cytochrome ($a/a_3$) (reduced, Fe(II)) + cytochrome $c$ (oxidized, Fe(III))

Cytochromes $a$ and $a_3$ taken together form the complex known as cytochrome oxidase. The reduced cytochrome oxidase is then oxidized by oxygen, which is itself reduced to water. The half reaction for the reduction of oxygen (oxygen acts as an oxidizing agent) is

$$\tfrac{1}{2}O_2 + 2 \text{ H}^+ + 2e^- \longrightarrow H_2O$$

The overall reaction is

2 Cytochrome ($a/a_3$) (reduced, Fe(II)) + $\tfrac{1}{2}O_2$ + 2 H$^+$
$\longrightarrow$ 2 cytochrome ($a/a_3$) (oxidized, Fe(III)) + H$_2$O

Note that in this final reaction we have finally seen the link to molecular oxygen in aerobic metabolism.

The standard reduction potential ($E^{\circ\prime}$) for the overall reaction of cytochrome oxidase is 0.57 volt, which corresponds to a standard free-energy change ($\Delta G^{\circ\prime}$) of $-110$ kJ $= -26.3$ kcal for each mole of NADH that enters the electron transport chain (see Figure 15.5). We have now seen the three places in the respiratory chain where electron transport is coupled to ATP production by proton pumping. These three places are the NADH dehydrogenase reaction, the oxidation of cytochrome $b$, and the reaction of cytochrome oxidase with oxygen. Table 15.2 summarizes the energetics of electron transport reactions.

Before we leave the subject of respiratory complexes we should mention that another membrane-bound enzyme, succinate-CoQ oxidoreductase, also catalyzes the transfer of electrons to coenzyme CoQ. However, its source of electrons (in other words, the substance being oxidized) differs from the oxidizable substrate (NADH) acted on by NADH-CoQ oxidoreductase, the first of the three respiratory complexes. In this case

**TABLE 15.2  The Energetics of Electron Transport Reactions**

| REACTION | $E^{\circ\prime}$ (volts) | $\Delta G^{\circ\prime}$ (mol NADH)$^{-1}$ | |
|---|---|---|---|
| | | kJ | kcal |
| $NADH + H^+ + E—FMN \longrightarrow NAD^+ + E—FMNH_2$ | +0.20 | −38.6 | −9.2 |
| $E—FMNH_2 + CoQ + 2H^+ \longrightarrow E—FMN + CoQH_2$ | +0.22 | −42.5 | −10.2 |
| $CoQH_2 + 2$ cyt $b(Fe(III)) \longrightarrow$ <br> $CoQ + 2H^+ + 2$ cyt $b(Fe(II))$ | −0.06 | +11.6 | +2.8 |
| 2 cyt $b(Fe(II)) + 2$ cyt $c_1(Fe(III)) \longrightarrow$ <br> 2 cyt $b(Fe(III)) + 2$ cyt $c_1(Fe(II))$ | +0.18 | −34.7 | −8.3 |
| 2 cyt $c_1(Fe(II)) + 2$ cyt $c(Fe(III)) \longrightarrow$ <br> 2 cyt $c_1(Fe(III)) + 2$ cyt $c(Fe(II))$ | +0.03 | −5.8 | −1.4 |
| 2 cyt $c(Fe(II)) + 2$ cyt $a/a_3(Fe(III)) \longrightarrow$ <br> 2 cyt $c(Fe(III)) + 2$ cyt $a/a_3(Fe(II))$ | +0.04 | −7.7 | −1.8 |
| 2 cyt $a/a_3(Fe(II)) + \frac{1}{2}O_2 + 2H^+ \longrightarrow$ <br> 2 cyt $a/a_3(Fe(III)) + H_2O$ | +0.53 | −102.3 | −24.5 |
| Overall reaction: <br> $NADH + H^+ + \frac{1}{2}O_2 \longrightarrow NAD^+ + H_2O$ | +1.14 | −220 | −52.6 |

the substrate is succinate from the citric acid cycle, which is oxidized to fumarate by a flavin enzyme (see Figure 15.4).

$$\text{Succinate} + E—FAD \longrightarrow \text{fumarate} + E—FADH_2$$

The notation E—FAD indicates that the flavin portion is covalently bonded to the enzyme. The flavin group is reoxidized in the next stage of the reaction, and coenzyme Q is reduced.

$$E—FADH_2 + CoQ \longrightarrow E—FAD + CoQH_2$$

The overall reaction is

$$\text{succinate} + CoQ \longrightarrow \text{fumarate} + CoQH_2$$

We have already seen the first step of this reaction when we discussed the oxidation of succinate to fumarate as part of the citric acid cycle. Succinate dehydrogenase, the enzyme that catalyzes the oxidation of succinate to fumarate (Section 14.3), is one of the components of this enzyme complex. The other components are a *b*-type cytochrome and an iron–sulfur protein. The whole complex is an integral part of the inner mitochondrial membrane. The standard reduction potential ($E^{\circ\prime}$) for the overall reaction is 0.07 volt; the standard free energy change ($\Delta G^{\circ\prime}$) is −13.5 kJ mol$^{-1}$ = −3.2 kcal mol$^{-1}$. The overall reaction is exergonic, but there is not enough energy from this reaction to drive ATP production.

## 15.5
## THE COUPLING OF OXIDATION TO PHOSPHORYLATION

Some of the energy released by the oxidation reactions in the electron transport chain is used to drive the phosphorylation of three molecules of ADP. The phosphorylation of each mole of ADP requires 30.5 kJ = 7.3

kcal, and we have seen how each of the reactions catalyzed by the three respiratory complexes provides more than enough energy to drive this reaction. It is a common theme in metabolism that energy to be used by cells is converted to the chemical energy of ATP as needed. The energy-releasing oxidation reactions give rise to proton pumping and thus to the pH gradient across the inner mitochondrial membrane. The energy of the electrochemical potential (voltage drop) across the membrane is converted to the chemical energy of ATP by the coupling process. The process of phosphorylation is not the same as that of electron transport, even though the two are intimately associated.

A coupling factor is needed to link oxidation and phosphorylation. A complex protein oligomer, separate from the electron transport complexes, serves this function; the complete protein spans the inner mitochondrial membrane and projects into the matrix as well. The portion of the protein that spans the membrane is called $F_0$; it consists of four different kinds of polypeptide chains. The portion that projects into the matrix is called $F_1$; it consists of five different kinds of polypeptide chains in the ratio $\alpha_3\beta_3\gamma\delta\epsilon$. Electron micrographs of mitochondria show the projections into the matrix from the inner mitochondrial membrane (Figure 15.7). The $F_1$ sphere is the site of ATP synthesis. The whole protein complex is called **ATP synthetase** (or ATP synthase). It is also known as mitochondrial ATPase, because the reverse reaction of ATP hydrolysis, as well as phosphorylation, can be catalyzed by the enzyme. The hydrolytic reaction was discovered before that of the synthesis of ATP, hence the name.

Compounds known as **uncouplers** inhibit the phosphorylation of ADP without affecting electron transport. A well known example of an uncoupler is **2,4-dinitrophenol.** Various antibiotics such as **valinomycin** and **gramicidin A** are also uncouplers (Figure 15.8). When mitochondrial oxidation processes are operating normally, electron transport from NADH or $FADH_2$ to oxygen results in the production of ATP. When an uncoupler is present, oxygen is still reduced to $H_2O$, but ATP is not produced. If the uncoupler is removed, ATP synthesis linked to electron transport resumes.

A term called the **P/O ratio** is used to indicate the coupling of ATP production to electron transport. The P/O ratio gives the number of moles of $P_i$ consumed in the reaction ADP + $P_i \longrightarrow$ ATP for each mole of oxygen atoms consumed in the reaction $\frac{1}{2}O_2 + 2\,H^+ + 2\,e^- \longrightarrow H_2O$. As we have already seen, three moles of ATP are produced when one mole of NADH is oxidized to $NAD^+$. Recall that oxygen is the ultimate acceptor of the electrons from NADH and that $\frac{1}{2}$ mole of $O_2$ molecules (one mole of oxygen atoms) is reduced for each mole of NADH oxidized. The experimentally determined P/O ratio is 3 when NADH is the substrate oxidized.

**FIGURE 15.7**    The $F_0$–$F_1$ complex (mitochondrial ATPase), the site of ATP synthesis. (a) A model for the $F_0$–$F_1$ complex subunits shown schematically. (b) Electron micrograph of projections into matrix space of mitochondrion (magnified 50,000 x).

**(a)**

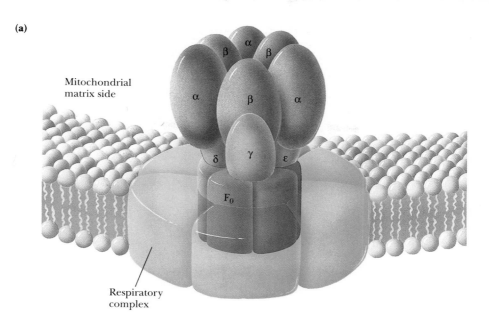

Mitochondrial
matrix side

β

α   α

β   α

β

δ   γ   ε

$F_0$

Respiratory
complex

**(b)**

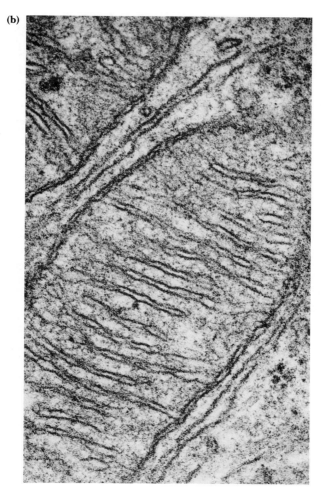

**FIGURE 15.8**   Some uncouplers of oxidative phosphorylation.

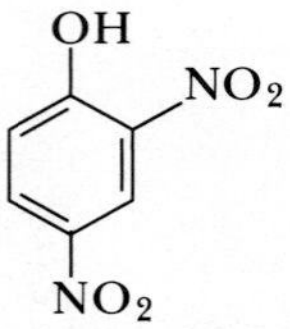

2,4-Dinitrophenol (DNP)

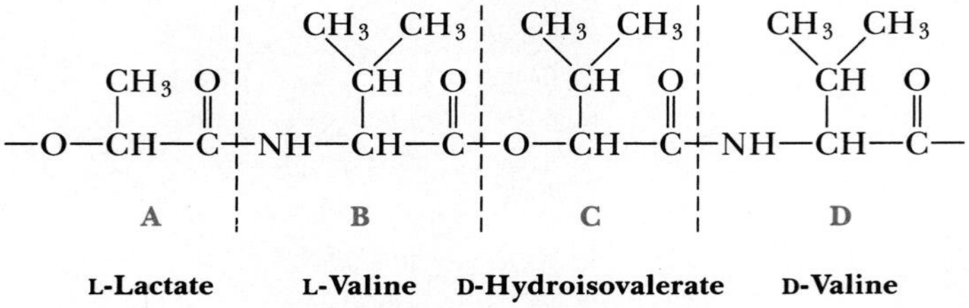

$$-O-CH-C-NH-CH-C-O-CH-C-NH-CH-C-$$

| A | B | C | D |
|---|---|---|---|
| L-Lactate | L-Valine | D-Hydroisovalerate | D-Valine |

**Repeating unit of valinomycin**

(Valinomycin is a cyclic trimer of repeating units.)

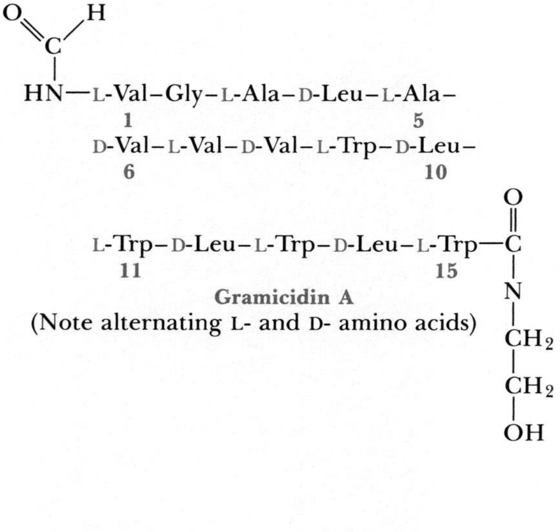

HN—L-Val–Gly–L-Ala–D-Leu–L-Ala–
　　　1　　　　　　　　　　5
　　D-Val–L-Val–D-Val–L-Trp–D-Leu–
　　　6　　　　　　　　　　10

L-Trp–D-Leu–L-Trp–D-Leu–L-Trp—C
　11　　　　　　　　　　15
**Gramicidin A**
(Note alternating L- and D- amino acids)

N
|
$CH_2$
|
$CH_2$
|
OH

The P/O ratio is 2 when $FADH_2$ is the substrate oxidized (also an experimentally determined value); we have seen that only two moles of ATP are produced for each mole of $FADH_2$ which is oxidized.

## 15.6
## THE MECHANISM OF COUPLING IN OXIDATIVE PHOSPHORYLATION

Several mechanisms have been proposed to account for the coupling of electron transport and ATP production; two mechanisms currently under consideration are **chemiosmotic coupling** and **conformational coupling.**

### Chemiosmotic Coupling

The chemiosmotic coupling mechanism is based entirely on the difference in hydrogen ion concentration between the intermembrane space and the matrix of actively respiring mitochondria. In other words, the proton (hydrogen ion, $H^+$) gradient across the inner mitochondrial membrane is the crux of the matter. The proton gradient exists because the various proteins that serve as electron carriers in the respiratory chain are not symmetrically oriented with respect to the two sides of the inner mitochondrial membrane, nor do they react in the same way with respect to the matrix and the intermembrane space (Figure 15.9). In the process of electron transport these proteins take up hydrogen ions from the matrix to transfer them in redox reactions; these electron-carrier proteins subsequently release hydrogen ions into the intermembrane space when they are reoxidized, creating the proton gradient. The reactions of NADH, coenzyme Q, and molecular oxygen ($O_2$) all require hydrogen ions, which come from the matrix side of the membrane. The iron–sulfur proteins and the $b$ cytochromes transfer the protons to the intermembrane space. As a result, there is a higher concentration of hydrogen ions in the intermembrane space than in the matrix; this condition is precisely what we mean by

**FIGURE 15.9** The creation of a proton gradient in chemiosmotic coupling. The overall effect of the electron transport reaction series is to move protons ($H^+$) out of the matrix into the intermembrane space, creating a difference in pH across the membrane.

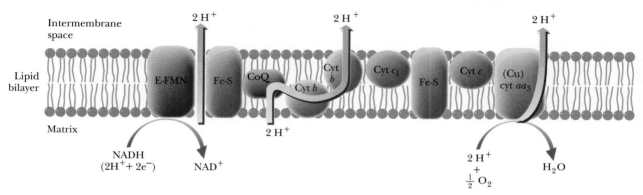

**FIGURE 15.10**  Closed vesicles prepared from mitochondria can pump protons and produce ATP.

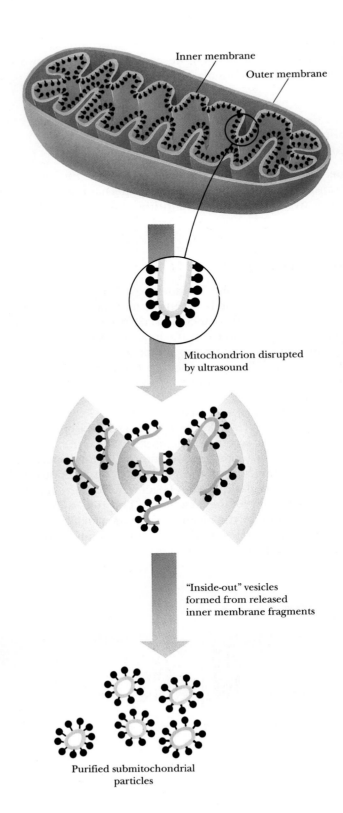

Inner membrane

Outer membrane

Mitochondrion disrupted by ultrasound

"Inside-out" vesicles formed from released inner membrane fragments

Purified submitochondrial particles

a proton gradient. It is known that the intermembrane space has a lower pH than the matrix, which is another way of saying that there is a higher concentration of hydrogen ions in the intermembrane space than in the matrix. The proton gradient in turn can drive the production of ATP, which occurs when the protons flow back into the matrix.

Since chemiosmotic coupling was first suggested in 1961, a considerable body of experimental evidence has accumulated to support it.

1.  A system with definite inside and outside compartments (closed vesicles) is essential for oxidative phosphorylation. The process does not occur in soluble preparations or in membrane fragments without compartmentalization.
2.  Submitochondrial preparations that contain closed vesicles can be prepared; such vesicles can carry out oxidative phosphorylation, and the asymmetrical orientation of the respiratory complexes with respect to the membrane can be demonstrated (Figure 15.10).
3.  A model system for oxidative phosphorylation can be constructed with proton pumping in the absence of electron transport. The model system consists of reconstituted membrane vesicles, mitochondrial ATP synthetase, and a proton pump. The pump is bacteriorhodopsin, a protein found in the purple membrane of halobacteria. The proton pumping takes place when the protein is illuminated (Figure 15.11).
4.  The existence of the pH gradient has been demonstrated and confirmed experimentally.

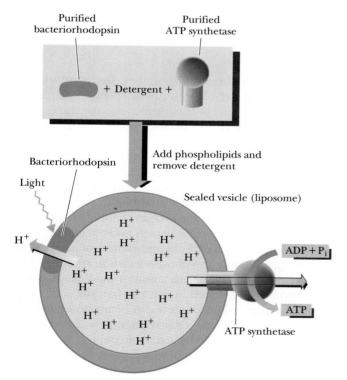

**FIGURE 15.11**   ATP can be produced by closed vesicles with bacteriorhodopsin as a proton pump.

**FIGURE 15.12**  Formation of ATP accompanies flow of protons back into the mitochondrial matrix.

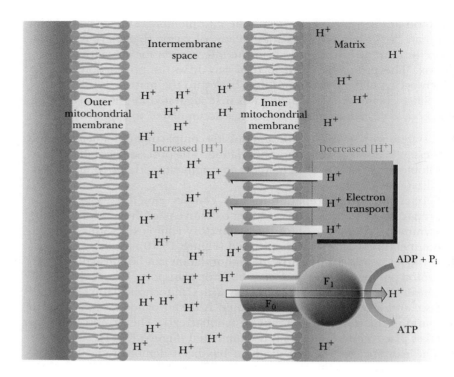

An equation has been written to describe the proton gradient and thus to characterize the driving force for phosphorylation:

$$\Delta\mu_{H^+} = \Delta\Psi - Z(\Delta pH)$$

where $\Delta\mu_{H^+}$ is the electrochemical proton potential, the quantity that describes the proton gradient; $\Delta\Psi$ is the membrane potential, which is the voltage difference between the two sides of the membrane; and the $Z(\Delta pH)$ term refers to the difference in pH between the two sides of the membrane.

The way in which the proton gradient leads to the production of ATP depends on ion channels through the inner mitochondrial membrane; these channels are a feature of the structure of mitochondrial ATPase. Protons flow back into the matrix through ion channels in the mitochondrial ATPase; the $F_0$ part of the protein is the proton channel. The flow of protons is accompanied by formation of ATP, which takes place in the $F_1$ unit (Figure 15.12). The unique feature of chemiosmotic coupling is the direct linkage of the proton gradient to the phosphorylation reaction. The details of how phosphorylation takes place as a result of the linkage to the proton gradient are not explicitly specified in this mechanism.

A reasonable mode of action for uncouplers can be proposed in light of the existence of a proton gradient. Dinitrophenol is an acid; its conjugate base, dinitrophenolate anion, is the actual uncoupler, since it can react with hydrogen ions in the intermembrane space, reducing the difference in hydrogen ion concentration between the two sides of the

## BOX 15.1
## BROWN ADIPOSE TISSUE: A CASE OF USEFUL INEFFICIENCY

There are two situations in which dissipation of energy as heat is useful to organisms: cold-induced nonshivering thermogenesis (production of heat) and diet-induced thermogenesis. Cold-induced nonshivering thermogenesis enables animals to survive in the cold once they have become adapted to such conditions, and diet-induced thermogenesis prevents the development of obesity in spite of prolonged overeating. These two processes may be the same biochemically; it is firmly established that they occur principally, if not exclusively, in brown adipose tissue (BAT), which is rich in mitochondria. (Brown fat takes its color from the large number of mitochondria present in it, unlike the usual white fat cells.) The key to this "inefficient" use of energy in brown adipose tissue appears to be a mitochondrial protein called thermogenin, also referred to as the "uncoupling protein." When this membrane-bound protein is activated in thermogenesis, it serves as a proton channel through the inner mitochondrial membrane. Like all other uncouplers it "punches a hole" in the mitochondrial membrane and decreases the effect of the proton gradient. Protons flow back into the matrix through thermogenin, bypassing the ATP synthetase complex.

Very little research on the biochemistry or physiology of brown adipose tissue has been done in humans. Most of the work on both obesity and adaptation to cold stress has been done on small mammals such as rats, mice, and hamsters. What role, if any, brown fat deposits play in the development of obesity in humans is an open question for researchers.

inner mitochondrial membrane. The antibiotic uncouplers such as gramicidin A and valinomycin are **ionophores,** creating a channel through which ions such as $H^+$, $K^+$, and $Na^+$ can pass through the membrane; the proton gradient is overcome, resulting in the uncoupling of oxidation and phosphorylation. A natural uncoupler is discussed in Box 15.1.

## Conformational Coupling

In the conformational coupling mechanism the proton gradient is indirectly related to ATP production. The proton gradient leads to conformational changes in a number of proteins, particularly in the ATPase (ATP synthetase) itself. It appears from recent evidence that the effect of the proton gradient is not the actual formation of ATP but the release of tightly bound ATP from the synthetase as a result of the conformational change (Figure 15.13). There are three sites for substrate on the synthetase

**FIGURE 15.13** The role of conformational change in releasing ATP from ATP synthetase. According to the binding-change mechanism, the effect of the proton flux is to cause a conformational change that leads to the release of already formed ATP from ATP synthetase. [From R. L. Cross, D. Cunningham, and J. K. Tamura, 1984. *Curr. Top. Cell. Regul.* 24:336.]

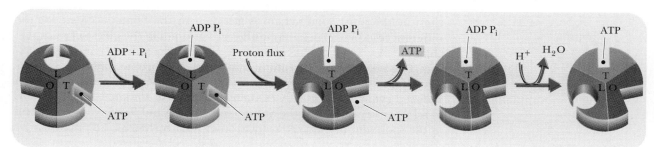

and three possible conformational states: open (O), with low affinity for substrate; loose binding (L), which is not catalytically active; and tight binding (T), which is catalytically active. At any given time, each of the sites is in one of three different conformational states. These states interconvert as a result of the proton flux through the synthetase. ATP already formed by the synthetase is bound at a site in the T conformation, while ADP and P$_i$ bind at a site in the L conformation. A proton flux converts the site in the T conformation to the O conformation, releasing the ATP. The site at which ATP and P$_i$ are bound assumes the T conformation, which can then give rise to ATP.

Electron micrographs have shown that the conformation of the inner mitochondrial membrane and of the cristae is distinctly different in the resting and active states. It is quite possible that, in time, a hybrid mechanism for oxidative phosphorylation may be developed, one that may include features of both the chemiosmotic and conformational coupling mechanisms.

## 15.7
## CYTOCHROMES AND OTHER IRON-CONTAINING PROTEINS OF ELECTRON TRANSPORT

In contrast to the electron carriers in the early stages of electron transport such as NADH, FMN, and CoQ, the cytochromes are macromolecules. These proteins are found in all types of organisms and are typically located in membranes. In eukaryotes the usual site is the inner mitochondrial membrane, but cytochromes can also occur in the endoplasmic reticulum.

The amino-acid sequences of cytochromes from many different types of organisms, particularly cytochrome *c*, have been determined. The evolutionary implications of sequence similarities and differences among organisms constitute a topic of considerable interest to biochemists. There is a discussion of the topic in the article by Dickerson listed in the bibliography at the end of this chapter.

All cytochromes contain the heme group, a part of the structure of hemoglobin and myoglobin (Section 9.4). In the cytochromes the iron of the heme group does not bind to oxygen; instead, the iron is involved in the series of redox reactions, which we have already seen. There are differences in the side chains of the heme group of the cytochromes involved in the various stages of electron transport (Figure 15.14). These structural differences, combined with the variations in the polypeptide chain and in the way the polypeptide chain is attached to the heme, account for the differences in properties among the cytochromes in the electron transport chain.

The absorption spectrum of a reduced cytochrome is not the same as that of the same cytochrome in the oxidized form. The spectrum of a reduced cytochrome usually contains three peaks, designated $\alpha$, $\beta$, and $\gamma$. (The $\gamma$ peak is also called the Soret band.) The $\alpha$ and $\beta$ peaks are not found in the spectrum of oxidized cytochromes (Figure 15.15). The presence or

(a)

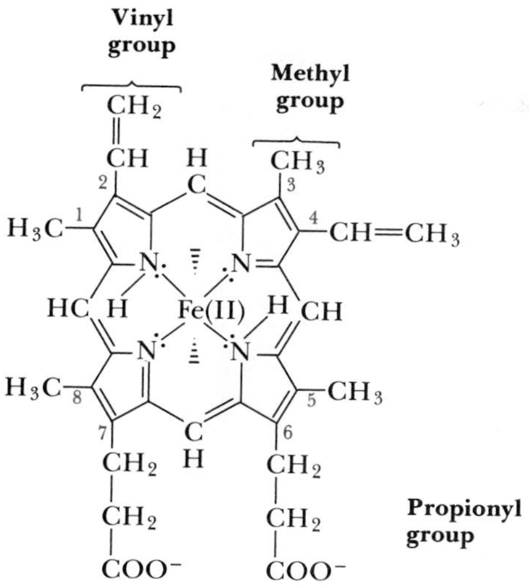

**Vinyl group**

**FIGURE 15.14** The heme group of cytochromes. (a) Structure of heme of all *b* cytochromes and of hemoglobin and myoglobin. The wedge bonds show the fifth and sixth coordination sites of the iron atom. (b) Comparison of the side chains of *a* and *c* cytochromes to those of *b* cytochromes.

(b)

| POSITION | *a* CYTOCHROMES | *c* CYTOCHROMES |
|---|---|---|
| 1 | Same | Same |
| 2 (in *a*) | $-CHCH_2CH(CH_2)_3CH(CH_2)_3CH(CH_3)_2$<br>    $\vert$       $\vert$       $\vert$<br>   OH    $CH_3$   $CH_3$ | |
| 2 (in *c*) | | $-CHCH_3$<br>    $\vert$<br>    S—protein<br>(Covalent attachment) |
| 3 | Same | Same |
| 4 | Same | $-CHCH_3$<br>    $\vert$<br>    S—protein |
| 5 | Hydrogen (—H) | Same |
| 6 | Same | Same |
| 7 | Same | Same |
| 8 | $-C=O$ (Formyl group)<br>   $\vert$<br>   H | Same |

absence of the $\alpha$ and $\beta$ peaks enables us to distinguish the oxidized and reduced forms, and the relative intensities of all the peaks make it possible to determine the relative amounts of the two forms.

**Non-heme iron proteins** do not contain a heme group, as their name indicates. Many of the most important proteins in this category contain sulfur, as is the case with the non-heme iron proteins involved in electron transport; these are the iron-sulfur proteins that are components of the respiratory complexes. The iron is usually bound to cysteine or to $S^{2-}$ (Figure 15.16). There are still many questions about the location and mode of action of iron-sulfur proteins in mitochondria.

FIGURE 15.15   Characteristic absorption peaks of cytochromes. (a) Typical absorption spectrum. The wavelength ranges of absorption of light in the $\alpha$, $\beta$, and $\gamma$ peaks of reduced cytochromes are indicated. (b) Table of wavelengths of maximum absorption of light by selected cytochromes.

**(a)**

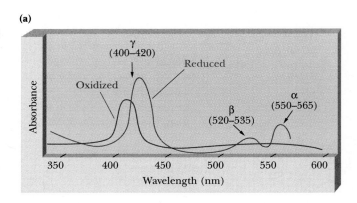

**(b)**

| CYTOCHROME (SOURCE) | Absorption Bands (nm) | | |
|---|---|---|---|
| | $\alpha$ | $\beta$ | $\gamma$ |
| $b$ | 563 | 532 | 429 |
| $c$ | 550 | 521 | 415 |
| $c_1$   Mitochondria of animals, plants, | 554 | 524 | 418 |
| $a$   yeasts, and fungi | 600 | Absent | 439 |
| $a_3$ | 604 | Absent | 443 |
| $b_1$ (*E. coli*) | 558 | 528 | 425 |
| $b_2$ (yeast) | 557 | 528 | 424 |

## 15.8
## RESPIRATORY INHIBITORS BLOCK THE FLOW OF ELECTRONS IN ELECTRON TRANSPORT

If a pipeline is blocked, there will be a backup. Liquid will accumulate upstream of the blockage point, but there will be less liquid downstream. In electron transport the flow of electrons is from one compound to another rather than along a pipe, but the analogy of a blocked pipeline can be useful in understanding the workings of the pathway. When a flow of electrons is blocked in a series of redox reactions, there will be an accumulation of reduced compounds before the blockage point in the pathway. Recall that reduction is a gain of electrons, and oxidation represents a loss of electrons. The compounds that come after the blockage point will be lacking electrons and will tend to be found in the oxidized form (Figure 15.17). As we have already seen, the successive components of the electron transport chain have increasingly positive reduction potentials ($E^{\circ\prime}$, Table 15.1), but here we can gather additional evidence to establish the order of the electron transport pathway.

The use of respiratory inhibitors to determine the order of the electron transport chain depends on determining the relative amounts of oxidized and reduced forms of the various electron carriers in intact

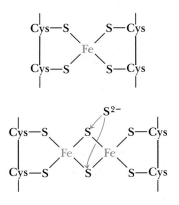

FIGURE 15.16   Iron–sulfur bonding in non-heme iron proteins.

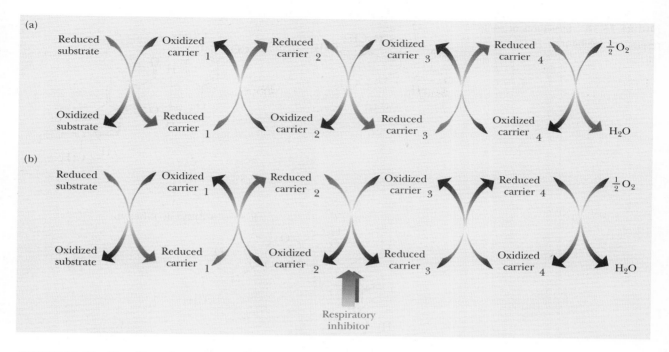

**FIGURE 15.17**    The effect of respiratory inhibitors. (a) No inhibitor present. Schematic view of electron transport. The red arrows indicate the flow of electrons. (b) Inhibitor present. The flow of electrons from carrier 2 to carrier 3 is blocked by the respiratory inhibitor. Reduced carrier 2 accumulates, as does oxidized carrier 3, since they cannot react with one another.

mitochondria. The logic of the experiment can be seen from the analogy to the blocked pipe. In this case the reduced form of the carrier upstream (e.g., cytochrome $b$) will accumulate because it cannot pass electrons further in the chain. Likewise, the oxidized form of the carrier downstream (cytochrome $c_1$ in our example) will also accumulate because the supply of electrons that it could accept has been cut off (Figure 15.17). By use of careful techniques, intact mitochondria can be isolated from cells and can carry out electron transport if an oxidizable substrate is available. If electron transport in mitochondria occurs in the presence and absence of a respiratory inhibitor, there will be different relative amounts of oxidized and reduced forms of the electron carriers.

The type of experiment done to determine the relative amounts of oxidized and reduced forms of electron carriers depends on the spectroscopic properties of these substances. The oxidized and reduced forms of cytochromes can be distinguished from one another by the presence of the $\alpha$ and $\beta$ peaks characteristic of the reduced form. Specialized spectroscopic techniques exist to detect the presence of electron carriers in intact mitochondria. The individual types of cytochromes can be identified by

**FIGURE 15.18**    Structures of some respiratory inhibitors.

**Amytal**

**Rotenone**

R ⟵ *n*-Butyl or *n*-hexyl

**Antimycin A**

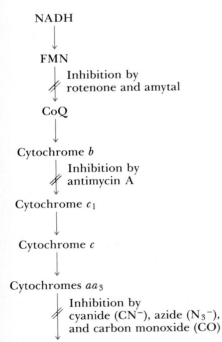

NADH

FMN

 Inhibition by
 rotenone and amytal

CoQ

Cytochrome *b*

 Inhibition by
 antimycin A

Cytochrome $c_1$

Cytochrome *c*

Cytochromes $aa_3$

 Inhibition by
 cyanide ($CN^-$), azide ($N_3^-$),
 and carbon monoxide (CO)

$O_2$

**FIGURE 15.19**    Sites of action of some respiratory inhibitors.

the wavelength at which the peak appears, and the relative amounts can be determined from the intensities of the peaks.

There are three sites in the electron transport chain at which inhibitors have an effect. At the first site, barbiturates (of which amytal is an example) block the transfer of electrons from the flavoprotein NADH reductase to coenzyme Q. Rotenone is another inhibitor that is active at this site. The second site at which blockage can occur is that of electron transfer from cytochrome *b* to cytochrome $c_1$; the inhibitor responsible for this blockage is the antibiotic antimycin A (Figure 15.18). The third site subject to blockage is the transfer of electrons from the cytochrome $a/a_3$ complex to oxygen. Several potent inhibitors operate at this site (Figure 15.19), such as cyanide ($CN^-$), azide ($N_3^-$), and carbon monoxide (CO). Note that each of the three sites of action of respiratory inhibitors corresponds to one of the respiratory complexes.

## 15.9
## SHUTTLE MECHANISMS MEDIATE TRANSPORT OF METABOLITES BETWEEN MITOCHONDRIA AND CYTOPLASM

NADH is produced by glycolysis, which occurs in the cytosol, but NADH in the cytosol cannot cross the mitochondrial membrane to enter the electron transport chain. However, the electrons can be transferred to a carrier that

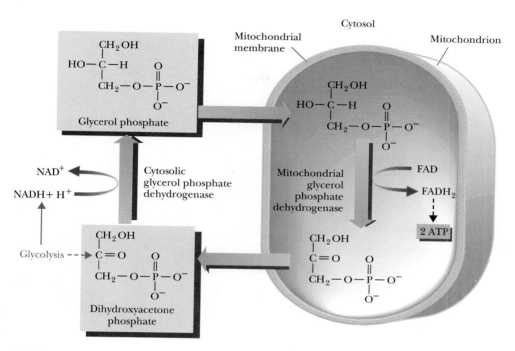

**FIGURE 15.20**   The glycerol phosphate shuttle.

can cross the membrane. The number of ATP molecules generated depends on the nature of the carrier, which varies according to the type of cell in which it occurs.

One carrier system that has been extensively studied in insect flight muscle is the **glycerol phosphate shuttle.** This mechanism makes use of the fact that glycerol phosphate can cross the mitochondrial membrane. The glycerol phosphate is produced by the reduction of dihydroxyacetone phosphate; in the course of the reaction NADH is oxidized to $NAD^+$. Inside the mitochondrion the glycerol phosphate is reoxidized to dihydroxy-acetone phosphate, which can pass back out into the cytosol. In this reaction the oxidizing agent (which is itself reduced) is FAD, and the product is $FADH_2$ (Figure 15.20). The $FADH_2$ then passes electrons through the electron transport chain, leading to the production of two molecules of ATP for each molecule of cytosolic NADH. This mechanism has also been observed in mammalian muscle and brain.

A more complex and more efficient shuttle mechanism is the **malate–aspartate shuttle,** which has been found in mammalian kidney, liver, and heart. This shuttle makes use of the fact that malate can cross the mitochondrial membrane, while oxaloacetate cannot do so. The notewor-thy point about this shuttle mechanism is that the transfer of electrons from NADH in the cytosol produces NADH in the mitochondrion. In the cytosol, oxaloacetate is reduced to malate by the cytosolic malate dehydro-

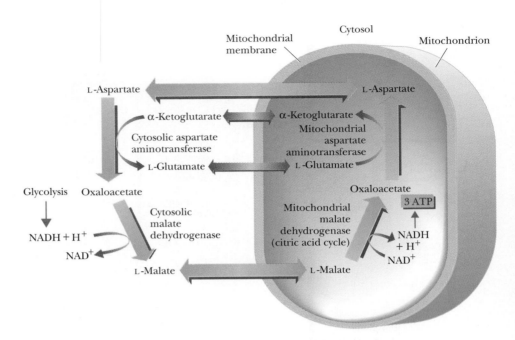

**FIGURE 15.21**   The malate–aspartate shuttle.

genase, accompanied by the oxidation of cytoplasmic NADH to NAD$^+$ (Figure 15.21). The malate then crosses the mitochondrial membrane. In the mitochondrion, the conversion of malate back to oxaloacetate is catalyzed by the mitochondrial malate dehydrogenase (one of the enzymes of the citric acid cycle). Oxaloacetate is converted to aspartate, which can also cross the mitochondrial membrane. Aspartate is converted to oxalacetate in the cytosol, completing the cycle of reactions. The NADH that is produced in the mitochondrion thus passes electrons to the electron transport chain. With the malate–aspartate shuttle, three molecules of ATP are produced for each molecule of cytosolic NADH rather than two molecules of ATP in the glycerol phosphate shuttle, which uses FADH$_2$ as a carrier.

## 15.10
## THE ATP YIELD FROM COMPLETE OXIDATION OF GLUCOSE

In Chapters 13 through 15 we have discussed the complete oxidation of glucose to carbon dioxide and water. At this point it is useful to do some bookkeeping to see how many molecules of ATP are produced for each molecule of glucose oxidized. Recall that some ATP is produced in glycolysis, but that far more ATP is produced by aerobic metabolism. Table 15.3 summarizes ATP production, and also follows the recycling of NADH and FADH$_2$.

**TABLE 15.3  The Balance Sheet for Oxidation of One Molecule of Glucose**

| | NADH MOLECULES | FADH$_2$ MOLECULES | ATP MOLECULES |
|---|---|---|---|
| **CYTOPLASMIC REACTIONS** | | | |
| 1. Glucose $\longrightarrow$ glucose 6-phosphate | | | −1 |
| 2. Fructose 6-phosphate $\longrightarrow$ fructose 1,6-*bis*phosphate | | | −1 |
| 3. 1,3-*bis*phospho-glycerate $\longrightarrow$ 3-phospho-glycerate (2 molecules) | | | +2 |
| 4. Phosphoenol-pyruvate $\longrightarrow$ pyruvate (2 molecules) | | | +2 / +2 |
| Oxidation of glyceraldehyde 3-phosphate (2 molecules) | +2 | | |
| **MITOCHONDRIAL REACTIONS** | | | |
| Pyruvate $\longrightarrow$ acetyl-CoA (2 molecules) | +2 | | |
| **CITRIC ACID CYCLE** | | | |
| 1. Succinyl-CoA formed GDP $\longrightarrow$ GTP (2 molecules) | | | +2 |
| 2. Oxidation of succinate (2 molecules) | | +2 | |
| 3. Oxidation of 2 molecules each of isocitrate, α-ketoglutarate, and malate | +6 | | |
| **OXIDATIVE PHOSPHORYLATION-ELECTRON TRANSPORT** | | | |
| 1. Reoxidation of NADH produced in glycolysis* | −2 | | +6* |
| 2. Reoxidation of NADH from the pyruvate $\longrightarrow$ acetyl-CoA step | −2 | | +6 |
| 3. Reoxidation of FADH$_2$ produced in citric acid cycle | | −2 | +4 |
| 4. Reoxidation of NADH from citric acid cycle | −6 / 0 | 0 | +18 / 38 |

*There are 4, not 6, ATP molecules produced in this step in muscle and brain, where the glycerol phosphate, rather than the malate–aspartate shuttle, predominates. In that case the total ATP produced is reduced to 36 molecules.

Note that there is no net change in the number of molecules of NADH or FADH$_2$.

## SUMMARY

In the final stages of aerobic metabolism, electrons are transferred from NADH to oxygen (the ultimate electron acceptor) in a series of oxidation–reduction reactions known as the electron transport chain. This series of reactions creates a pH gradient across the inner mitochondrial membrane. The stored energy of the pH gradient drives the process of oxidative phosphorylation, which is separate from electron transport. The two processes, electron transport and oxidative phosphorylation, are coupled by the mechanism of chemiosmotic coupling, which ultimately owes its existence to the pH gradient.

Three molecules of ATP are generated for each molecule of NADH that enters the electron transport chain and two molecules of ATP for each molecule of $FADH_2$. The general outline of the process is that NADH passes electrons to coenzyme Q, as does $FADH_2$, providing an alternate mode of entry into the electron transport chain. Electrons are then passed from coenzyme Q to the cytochromes and eventually to oxygen. The energy involved in each reaction can be characterized by a standard reduction potential ($E°'$) as well as by a standard free energy change ($\Delta G°'$).

Three separate respiratory complexes can be isolated from the inner mitochondrial membrane. Each of the respiratory complexes can carry out the reactions of a portion of the electron transport chain. In addition to the respiratory complexes, two electron carriers, coenzyme Q and cytochrome *c*, are not bound to the complexes but are free to move within the membrane. Many of the workings of the electron transport chain have been elucidated by experiments using respiratory inhibitors. Each of the respiratory complexes carries out proton pumping across the inner mitochondrial membrane, creating the pH gradient.

A complex protein oligomer is the coupling factor that links oxidation and phosphorylation; the complete protein spans the inner mitochondrial membrane and projects into the matrix as well. The portion of the protein that spans the membrane is called $F_0$; it consists of four different kinds of polypeptide chains. The portion that projects into the matrix is called $F_1$; it consists of five different kinds of polypeptide chains in the ratio $\alpha_3\beta_3\gamma\delta\epsilon$. The $F_1$ sphere is the site of ATP synthesis. The whole protein complex is called ATP synthetase (or ATP synthase). It is also known as mitochondrial ATPase.

Chemiosmotic coupling is the mechanism most widely used to explain the manner in which electron transport and oxidative phosphorylation are coupled to one another. In this mechanism the proton gradient is directly linked to the phosphorylation process. The way in which the proton gradient leads to the production of ATP depends on ion channels through the inner mitochondrial membrane; these channels are a feature of the structure of mitochondrial ATPase. Protons flow back into the matrix through ion channels in the mitochondrial ATPase; the $F_0$ part of the protein is the proton channel. The flow of protons is accompanied by formation of ATP, which occurs in the $F_1$ unit. In the conformational coupling mechanism the proton gradient is indirectly related to ATP production. It appears from recent evidence that the effect of the proton gradient is not the formation of ATP but the release of tightly bound ATP from the synthetase as a result of the conformational change.

Two shuttle mechanisms—the glycerol phosphate shuttle and the malate–aspartate shuttle—transfer the electrons, but not the NADH, produced in cytosolic reactions into the mitochondrion. In the first of the two shuttles, which is found in muscle and brain, the electrons are transferred to FAD; in the second, which is found in kidney, liver, and heart, the electrons are transferred to $NAD^+$. With the malate–aspartate shuttle, three molecules of ATP are produced for each molecule of cytosolic NADH rather than two ATP in the glycerol phosphate shuttle, a point that affects the overall yield of ATP in these tissues.

# EXERCISES

1. Briefly summarize the steps in the electron transport chain from NADH to oxygen.
2. What yield of ATP can be expected from complete oxidation of each of the following substrates by the reactions of glycolysis, the citric acid cycle, and oxidative phosphorylation?
   (a) Fructose 1,6-*bis*phosphate
   (b) Glucose
   (c) Phosphoenolpyruvate
   (d) Glyceraldehyde 3-phosphate
   (e) NADH
   (f) Pyruvate
3. Comment on the fact that the reduction of pyruvate to lactate, catalyzed by lactate dehydrogenase, is strongly exergonic (recall this from Chapter 13) even though the standard reduction potential for the half reaction

$$\text{Pyruvate} + 2\,H^+ + 2\,e^- \longrightarrow \text{lactate}$$

   is negative ($E^{\circ\prime} = -0.19$ V), indicating an endergonic reaction.
4. The free-energy change ($\Delta G^{\circ\prime}$) for the oxidation of the cytochrome $a/a_3$ complex by molecular oxygen is $-102.3$ kJ $= -24.5$ kcal for each mole of electron pairs transferred. What is the maximum number of moles of ATP that could be produced in the process? How many moles of ATP are actually produced? What is the efficiency of the process, expressed as a percentage?
5. What is the effect of each of the following substances on electron transport and production of ATP? Be specific about which reaction is affected.
   (a) Azide
   (b) Antimycin A
   (c) Amytal
   (d) Rotenone
   (e) Dinitrophenol
   (f) Gramicidin A
   (g) Carbon monoxide
6. What is the approximate P/O ratio that can be expected if intact mitochondria are incubated in the presence of oxygen, along with added succinate?

7. Cytochrome oxidase and succinate-CoQ oxidoreductase are isolated from mitochondria and are incubated in the presence of oxygen, along with cytochrome $c$, coenzyme Q, and succinate. What is the overall oxidation–reduction reaction that can be expected to take place?
8. Two biochemistry students are about to use mitochondria isolated from rat liver for an experiment on oxidative phosphorylation. The directions for the experiment specify addition of purified cytochrome $c$ from any source to the reaction mixture. Why is the added cytochrome $c$ needed and why does the source not have to be the same as that of the mitochondria?
9. Briefly summarize the main arguments of the chemiosmotic coupling hypothesis.
10. Describe the role of the $F_1$ portion of mitochondrial ATPase in oxidative phosphorylation.
11. How does the yield of ATP from complete oxidation of one molecule of glucose in muscle and brain differ from that in liver, heart, and kidney? What is the underlying reason for this difference?
12. Using the information in Table 15.1, calculate $E^{\circ\prime}$ for each of the following reactions. Also calculate $\Delta G^{\circ\prime}$ for each reaction.
   (a) $NADH + H^+ + \text{pyruvate} \longrightarrow NAD^+ + \text{lactate}$
   (b) 2 Cytochrome $a/a_{3\,\text{oxid}}(\text{Fe(III)}) + 2$ cytochrome $b_{\text{red}}(\text{Fe(II)}) \longrightarrow 2$ cytochrome $a/a_{3\,\text{red}}(\text{Fe(II)}) + 2$ cytochrome $b_{\text{oxid}}(\text{Fe(III)})$
   (c) $NADH + H^+ + \text{acetaldehyde} \longrightarrow NAD^+ + \text{ethanol}$
13. List the reactions of electron transport from NADH to oxygen. Show how a P/O ratio of 3 is obtained when NADH is the starting point of the electron transport chain.
14. Show how the reactions of the electron transport chain differ from those in Exercise 13 when $FADH_2$ is the starting point for electron transport. Show how the P/O ratio is 2 for electron transport starting from $FADH_2$.

# ANNOTATED BIBLIOGRAPHY

Cannon, B., and J. Nedergaard. The Biochemistry of an Inefficient Tissue: Brown Adipose Tissue. *Essays in Biochemistry* **20**, 110–164 (1985). [A review describing the usefulness to mammals of the "inefficient" production of heat in brown fat.]

Dickerson, R.E. Cytochrome $c$ and the Evolution of

Energy Metabolism. *Sci. Amer.* **242** (3), 136–152 (1980). [An account of the evolutionary implications of cytochrome *c* structure.]

Fillingame, R. The Proton-Translocating Pumps of Oxidative Phosphorylation. *Ann. Rev. Biochem.* **49**, 1079–1114 (1980). [A review article on chemiosmotic coupling.]

Hatefi, Y. The Mitochondrial Electron Transport and Oxidative Phosphorylation System. *Ann. Rev. Biochem.* **54**, 1015–1069 (1985). [A review article that emphasizes the coupling between oxidation and phosphorylation.]

Hinkle, P.C., and R.E. McCarty. How Cells Make ATP. *Sci. Amer.* **238** (3), 104–123 (1978). [An article on chemiosmotic coupling and the mode of action of uncouplers. Getting old, but very good.]

Lane, M.D., P.L. Pedersen, and A.S. Mildvan. The Mitochondrion Updated. *Science* **234**, 526–527 (1986). [A report on an international conference on bioenergetics and energy coupling.]

Mitchell, P. Keilin's Respiratory Chain Concept and its Chemiosmotic Consequences. *Science* **206**, 1148–1159 (1979). [A Nobel Prize lecture by the scientist who first proposed the chemiosmotic coupling hypothesis.]

Vignais, P.V., and J. Lunardi. Chemical Probes of the Mitochondrial ATP Synthesis and Translocation. *Ann. Rev. Biochem.* **54**, 977–1014 (1985). [A review article about the synthesis and use of ATP.]

# Lipid Metabolism

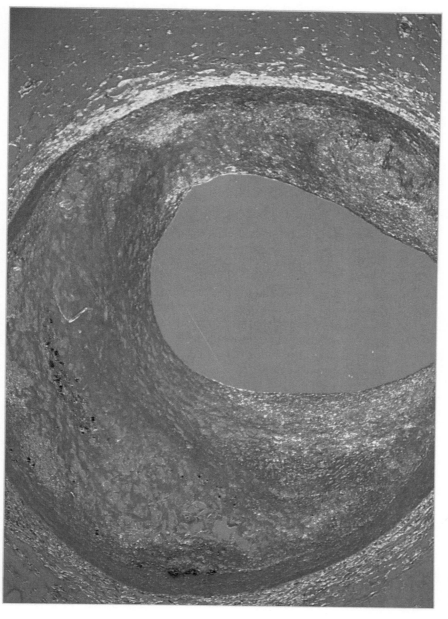

*A human heart artery blocked by cholesterol (atherosclerosis) viewed under polarized light.*

391

*In the energy economy of the cell, glucose reserves are like ready cash, while lipid reserves are like a fat savings account. The potential energy of lipids resides in the fatty acid chains of triacylglycerols. When there are excess calories, fatty acids are **synthesized** and stored in fat cells. When energy demands are great, fatty acids are **catabolized** to liberate energy. The synthesis of fatty acids begins with acetyl Co-A, after which carbon atoms are added to the growing hydrocarbon chain, usually two at a time. Catabolism proceeds in the opposite direction—beginning with the terminal $CH_3$ group and ending with acetyl-CoA. The fragmented hydrocarbon chain and acetyl-CoA are both oxidized in the citric acid cycle to provide energy that is temporarily stored as ATP. Pathways of synthesis and catabolism of lipids occur simultaneously, but in different parts of the cell. If these pathways were to operate at the same time in the same place, the lipids would be taken apart as soon as they were synthesized. To avoid such a futile cycle, the pathways of lipid synthesis and catabolism differ in important ways: (1) The reactions proceed in different directions; (2) synthesis takes place in the cytosol while catabolism takes place in the mitochondrion; and (3) NADPH is the donor of high energy electrons in lipid synthesis, while FAD and $NAD^+$ are electron acceptors in lipid catabolism.*

## 16.1
## THE METABOLISM OF LIPIDS PROVIDES PATHWAYS FOR THE GENERATION AND STORAGE OF ENERGY

In the past few chapters we have seen how energy can be released by the catabolic breakdown of carbohydrates in aerobic and anaerobic processes. Earlier, in Chapter 6, we saw that there are carbohydrate polymers, such as starch in plants and glycogen in animals, which represent stored energy in the sense that these carbohydrates can be hydrolyzed to monomers and then oxidized to provide energy in response to the needs of an organism. In this chapter we shall see how the metabolic oxidation of lipids releases large quantities of energy and how lipids represent an even more efficient way of storing chemical energy.

## 16.2
## CATABOLISM OF LIPIDS

The oxidation of fatty acids is the chief source of energy in the catabolism of lipids. Both triacylglycerols, which are the main storage form of the chemical energy of lipids, and phosphoacylglycerols, which are important components of biological membranes, have fatty acids as part of their covalently bonded structure. In both types of compounds, the bond between the fatty acid and the rest of the molecule can be hydrolyzed (Figure 16.1), with the reaction catalyzed by suitable groups of enzymes,

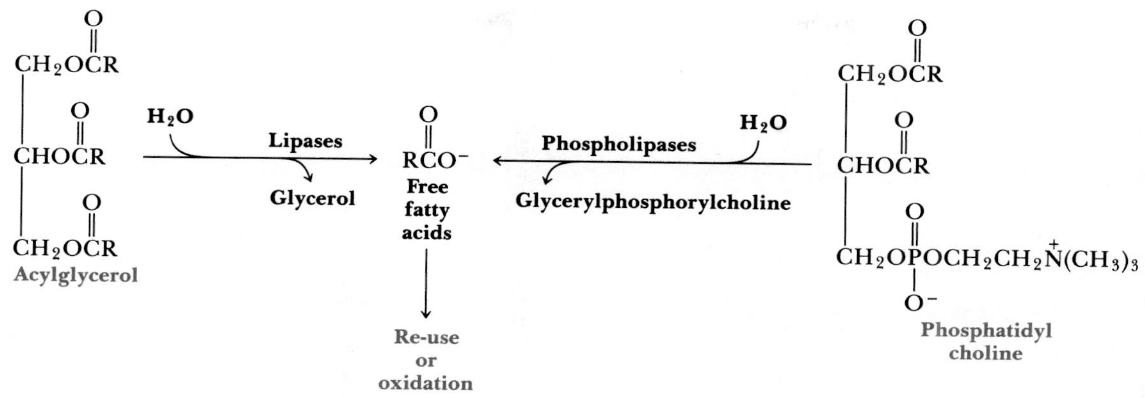

**FIGURE 16.1**    The release of fatty acids for future use.

**lipases** in the case of triacylglycerols and **phospholipases** in the case of phosphoacylglycerols. These hydrolytic reactions take place in the cytosol, as does the activation of the fatty acid, the step that prepares for the series of oxidation reactions.

Fatty acid oxidation begins with **activation** of the molecule. In this reaction a thioester bond is formed between the carboxyl group of the fatty acid and the thiol group of coenzyme A (CoA-SH). The activated form of the fatty acid is an acyl-CoA, the exact nature of which depends on the nature of the fatty acid itself. Keep in mind throughout this discussion that all acyl-CoA molecules are thioesters, since the fatty acid is esterified to the thiol group of CoA. The enzyme that catalyzes formation of the ester bond, an **acyl-CoA synthetase,** requires ATP for its action. In the course of the reaction, an acyl adenylate intermediate is formed. The acyl group is then transferred to CoA-SH. ATP is converted to AMP and $PP_i$, rather than to ADP and $P_i$. The $PP_i$ is hydrolyzed to two $P_i$; the hydrolysis of two "high-energy" phosphate bonds provides energy for the activation of the fatty acid. Note also that the hydrolysis of ATP to AMP and two $P_i$ represents an increase in entropy (Figure 16.2). There are several enzymes of this type, some specific for longer and some for shorter chain fatty acids.

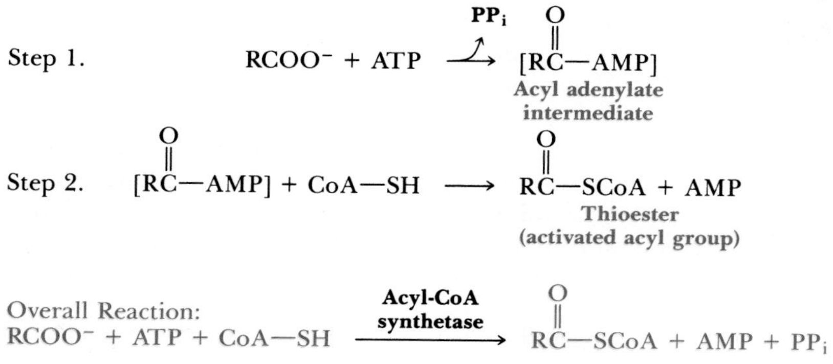

**FIGURE 16.2**    The formation of an acyl-CoA.

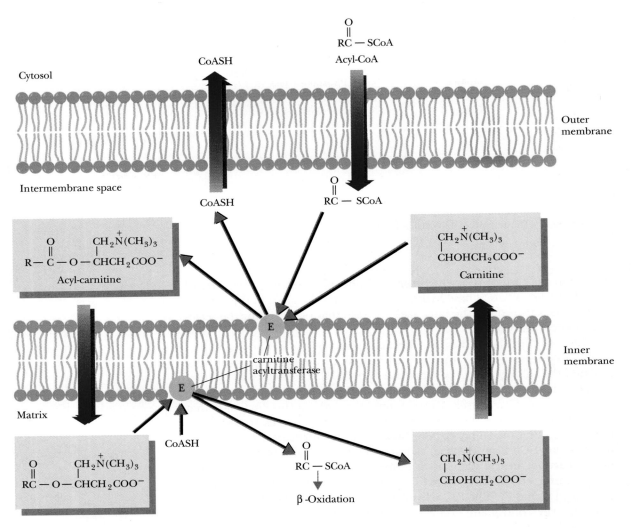

**FIGURE 16.3** The role of carnitine in the transfer of acyl groups to the mitochondrial matrix.

Both saturated and unsaturated fatty acids can serve as substrates for these enzymes. The esterification takes place in the cytosol, but the rest of the reactions of fatty acid oxidation occur in the mitochondrial matrix. The activated fatty acid must be transported into the mitochondrion so that the rest of the oxidation process can proceed.

The acyl-CoA can cross the outer mitochondrial membrane but not the inner membrane (Figure 16.3). In the intermembrane space the acyl group is transferred to carnitine by transesterification; this reaction is catalyzed by the enzyme carnitine acyltransferase, which is located in the inner membrane. Acyl carnitine, a compound that can cross the inner mitochondrial membrane, is formed. In the matrix the acyl group is transferred from carnitine to mitochondrial CoA-SH by another transesterification reaction.

In the matrix a repeated sequence of reactions successively cleaves two-carbon units from the fatty acid, starting from the carboxyl end. This

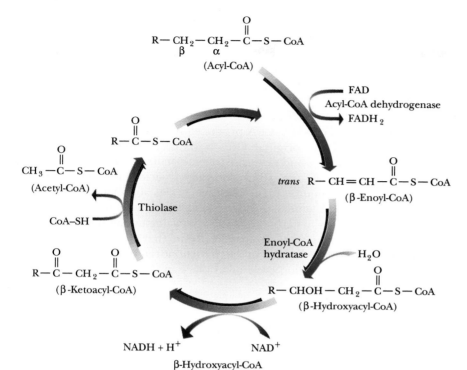

**FIGURE 16.4**  The β-oxidation cycle for fatty acids.

process is called **β-oxidation,** since the oxidative cleavage takes place at the β-carbon of the acyl group esterified to CoA. The β-carbon of the original fatty acid becomes the carboxyl carbon in the next stage of degradation. The whole cycle requires four reactions (Figure 16.4).

1. The acyl-CoA is *oxidized* to an α,β unsaturated acyl-CoA (also called a β-enoyl-CoA). The product has the *trans* arrangement at the double bond. This reaction is catalyzed by an FAD-dependent acyl-CoA dehydrogenase.
2. The unsaturated acyl-CoA is *hydrated* to produce a β-hydroxyacyl-CoA. This reaction is catalyzed by the enzyme enoyl-CoA hydratase.
3. A second *oxidation* reaction is catalyzed by β-hydroxyacyl-CoA dehydrogenase, an NADH-dependent enzyme. The product is β-ketoacyl-CoA.
4. The enzyme thiolase catalyzes the *cleavage* of the β-ketoacyl-CoA; a molecule of CoA is required for the reaction. The products are acetyl-CoA and an acyl-CoA that is two carbons shorter than the original molecule that entered the β-oxidation cycle. The CoA is needed in this reaction to form the new thioester bond in the smaller acyl-CoA molecule. This smaller molecule then undergoes another round of the β-oxidation cycle.

When a fatty acid with an even number of carbon atoms undergoes successive rounds of the β-oxidation cycle, the product is acetyl-CoA. (Fatty acids with even numbers of carbon atoms are the ones normally found in nature, so acetyl-CoA is the usual product of fatty acid catabolism.) The

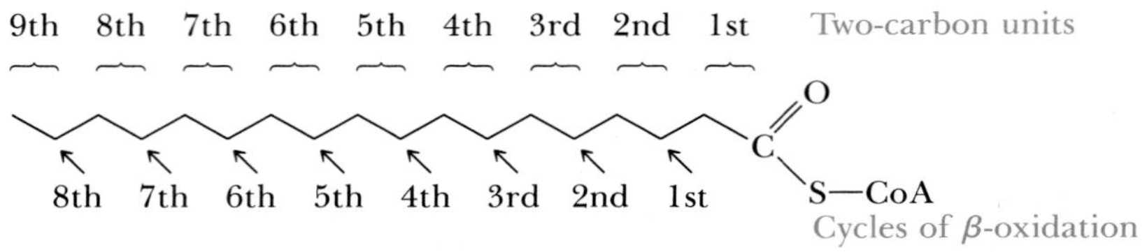

**FIGURE 16.5**  Stearic acid (18 carbons) gives rise to nine two-carbon units after eight cycles of $\beta$-oxidation. The ninth two-carbon unit remains esterified to CoA after eight cycles of $\beta$-oxidation have removed eight successive two-carbon units, starting at the carboxyl end on the right.

number of molecules of acetyl-CoA produced is equal to half the number of carbon atoms in the original fatty acid. For example, stearic acid contains 18 carbon atoms and gives rise to 9 molecules of acetyl-CoA. Note that the conversion of one 18-carbon stearic acid molecule to nine 2-carbon acetyl units requires eight, not nine, cycles of $\beta$-oxidation (Figure 16.5). The acetyl-CoA enters the citric acid cycle, with the rest of the oxidation of fatty acids to carbon dioxide and water taking place through the citric acid cycle and electron transport. Recall that most of the enzymes of the citric acid cycle are located in the mitochondrial matrix, and we have just seen that the $\beta$-oxidation cycle takes place in the matrix as well.

## 16.3
## THE ENERGY YIELD FROM THE OXIDATION OF FATTY ACIDS

In carbohydrate metabolism the energy released by oxidation reactions is used to drive the production of ATP, with most of the ATP produced in aerobic processes. In the same aerobic processes—namely, the citric acid cycle and oxidative phosphorylation—the energy released by the oxidation of fatty acids can also be used to produce ATP. There are two sources of ATP to keep in mind when calculating the overall yield of ATP. The first source is the reoxidation of the NADH and FADH$_2$ produced by the $\beta$-oxidation of the fatty acid to acetyl-CoA. The second source is ATP production from the processing of the acetyl-CoA through the citric acid cycle and oxidative phosphorylation. We shall use the oxidation of stearic acid, which contains 18 carbon atoms, as our example.

Eight cycles of $\beta$-oxidation are required to convert 1 mole of stearic acid to 9 moles of acetyl-CoA; in the process 8 moles of FAD are reduced to FADH$_2$, and 8 moles of NAD$^+$ are reduced to NADH.

$$CH_3(CH_2)_{16}CO-S-CoA + 8\ FAD + 8\ NAD^+ + 8\ H_2O +$$
$$8\ CoA\text{-}SH \longrightarrow$$

$$9\ CH_3CO-S-CoA + 8\ FADH_2 + 8\ NADH + 8\ H^+$$

The 9 moles of acetyl-CoA produced from each mole of stearic acid enter the citric acid cycle. One mole of FADH$_2$ and 3 moles of NADH are produced for each mole of acetyl-CoA that enters the citric acid cycle. At the same time, 1 mole of GDP is phosphorylated to produce GTP for each turn of the citric acid cycle.

$$9\ CH_3CO-S-CoA + 9\ FAD + 27\ NAD^+ + 9\ GDP + 9\ P_i +$$
$$27\ H_2O \longrightarrow$$

$$18\ CO_2 + 9\ CoA\text{-}SH + 9\ FADH_2 + 27\ NADH + 9\ GTP +$$
$$27\ H^+$$

The $FADH_2$ and NADH produced by $\beta$-oxidation and by the citric acid cycle enter the electron transport chain, and ATP is produced by oxidative phosphorylation. In our example there are 17 moles of $FADH_2$, 8 from $\beta$-oxidation and 9 from the citric acid cycle; there are also 35 moles of NADH, 8 from $\beta$-oxidation and 27 from the citric acid cycle. Three moles of ATP are produced for each mole of NADH that enters the electron transport chain, and 2 moles of ATP result from each mole of $FADH_2$.

$$17\ FADH_2 + 8\tfrac{1}{2}O_2 + 34\ ADP + 34\ P_i \longrightarrow 17\ FAD + 34\ ATP +$$
$$17\ H_2O$$

$$35\ NADH + 35\ H^+ + 17\tfrac{1}{2}O_2 + 105\ ADP + 105\ P_i \longrightarrow$$
$$35\ NAD^+ + 105\ ATP + 35\ H_2O$$

The overall yield of ATP from the oxidation of stearic acid can be obtained by adding the equations for $\beta$-oxidation, for the citric acid cycle, and for oxidative phosphorylation. In this calculation we take GDP as equivalent to ADP and GTP as equivalent to ATP, which means that the equivalent of nine ATP must be added to those produced in the reoxidation of $FADH_2$ and NADH. There are 9 ATP equivalent to the 9 GTP from the citric acid cycle, 34 ATP from the reoxidation of $FADH_2$, and 105 ATP from the reoxidation of NADH, for a grand total of 148 ATP.

$$CH_3(CH_2)_{16}CO-S-CoA + 26\ O_2 + 148\ ADP + 148\ P_i \longrightarrow$$
$$18\ CO_2 + 17\ H_2O + 148\ ATP + CoA\text{-}SH$$

The activation step in which stearyl-CoA was formed is not included in this calculation, and we must subtract the ATP that was required for that step. Even though only one ATP was required, two "high-energy" phosphate bonds are lost because of the production of AMP and $PP_i$. The pyrophosphate must be hydrolyzed to phosphate ($P_i$) before it can be recycled in metabolic intermediates. As a result we must subtract the equivalent of two ATP for the activation step. The net yield of ATP becomes 146 moles of ATP for each mole of stearic acid that is completely oxidized.

As a comparison, note that 38 moles of ATP can be obtained from the complete oxidation of 1 mole of glucose; but glucose contains 6, rather than 18, carbon atoms. Three glucose molecules contain 18 carbon atoms, and a more interesting comparison is the ATP yield from the oxidation of 3 glucose molecules, which is $3 \times 38 = 114$ ATP for the same number of carbon atoms. The yield of ATP from the oxidation of the lipid is still higher than that from the carbohydrate, even for the same number of carbon atoms. The reason for this is that a fatty acid is all hydrocarbon except for the carboxyl group, that is, it exists in a highly reduced state. A sugar is already partly oxidized because of the presence of its oxygen-containing groups.

Another point of interest is that water is produced in the oxidation of fatty acids. We have already seen that water is also produced in the complete oxidation of carbohydrates. The production of **metabolic water**

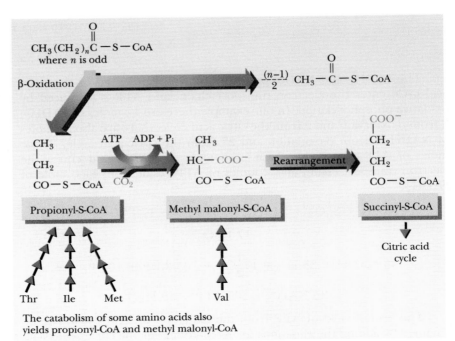

**FIGURE 16.6** The oxidation of fatty acids containing an odd number of carbon atoms.

The catabolism of some amino acids also yields propionyl-CoA and methyl malonyl-CoA

is a common feature of aerobic metabolism. This process can be a source of water for organisms that live in desert environments. Camels are a well-known example; the stored lipids in their humps are a source of both energy and water during long trips through the desert. The kangaroo rat is an even more striking example of adaptation to an arid environment. This animal has been observed to live indefinitely without having to drink water. It lives on a diet of seeds, which are rich in lipids but which contain little water. The metabolic water that the kangaroo rat produces is adequate for all its water needs. This metabolic response to arid conditions is usually accompanied by a reduced output of urine.

## 16.4
## SOME ADDITIONAL REACTIONS IN THE OXIDATION OF FATTY ACIDS

Fatty acids with odd numbers of carbon atoms are not as frequently encountered in nature as are the ones with even numbers of carbon atoms. Odd-numbered fatty acids also undergo the β-oxidation process (Figure 16.6). The last cycle of β-oxidation produces one molecule of propionyl-CoA. An enzymatic pathway exists to convert propionyl-CoA to succinyl-CoA, which then enters the citric acid cycle. In this pathway, propionyl-CoA is first carboxylated to methyl malonyl-CoA, which then undergoes rearrangement to form succinyl-CoA; since propionyl-CoA is also a product of the catabolism of several amino acids, the conversion of propionyl-CoA to succinyl-CoA also figures in amino acid metabolism (Section 18.7).

The conversion of unsaturated fatty acids to acetyl-CoA requires two reactions that are not encountered in the oxidation of saturated acids, a *cis-trans* isomerization and an epimerization (Figure 16.7). Successive

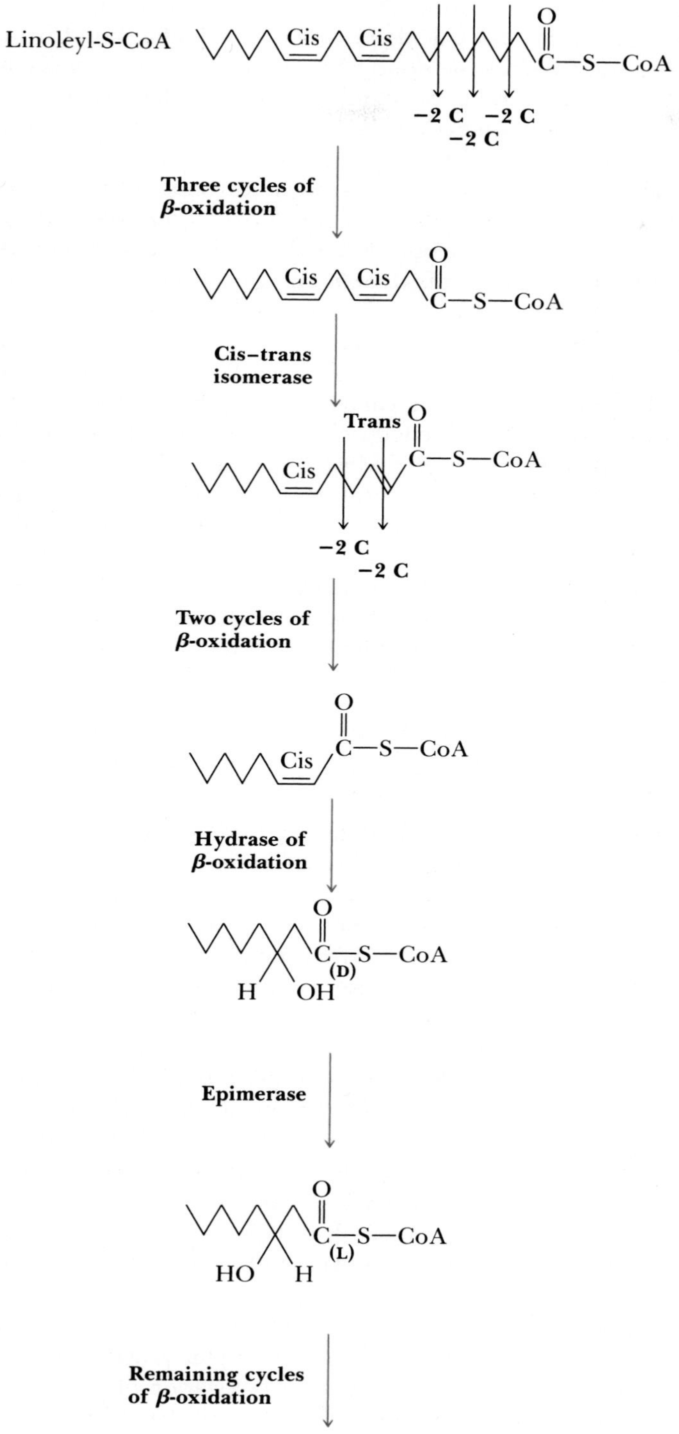

FIGURE 16.7   The oxidation of unsaturated fatty acids.

rounds of β-oxidation of linoleic acid, which has two double bonds, can provide an example of these reactions. The process of β-oxidation gives rise to unsaturated fatty acids in which the double bond is in the *trans* arrangement, whereas the double bonds in most naturally occurring fatty acids are in the *cis* arrangement; in the case of linoleic acid there are two *cis* double bonds, between carbons 9 and 10 and between carbons 12 and 13. Three rounds of β-oxidation produce a 12-carbon unsaturated fatty acid with the two *cis* double bonds between carbons 3 and 4 and between carbons 6 and 7. The hydrase of the β-oxidation cycle requires a *trans* double bond between carbon atoms 2 and 3 as a substrate. A **cis-trans isomerase** produces a *trans* double bond between carbons 2 and 3 from the *cis* double bond between carbons 3 and 4.

Two more cycles of β-oxidation give rise to an eight-carbon fatty acid with a *cis* double bond between carbons 2 and 3. The position of the double bond makes the eight-carbon intermediate a possible substrate for the hydrase, but the stereochemistry of the product presents a problem in the next step of the cycle. The β-hydroxy product has the D configuration, and the dehydrogenase that catalyzes the next stage of β-oxidation requires the L configuration. An **epimerase** catalyzes the required change of configuration, and the oxidation of the fatty acid can proceed. Unsaturated fatty acids make up a large enough portion of the fatty acids in storage fat (40% for oleic acid alone) to make the reactions of the *cis-trans* isomerase and the epimerase of some importance.

## 16.5
## THE FORMATION OF "KETONE BODIES"

Substances related to acetone ("ketone bodies") are produced when an excess of acetyl-CoA arises from β-oxidation. This condition occurs when there is not enough oxaloacetate available to react with the large amounts of acetyl-CoA that could enter the citric acid cycle. Oxaloacetate in turn arises from glycolysis, since it is formed from pyruvate in a reaction catalyzed by pyruvate carboxylase. A situation like this can come about when an organism has a high intake of lipids and a low intake of carbohydrates, but there are other possible causes as well, such as starvation and diabetes. Starvation conditions will cause an organism to break down fats for energy, leading to the production of large amounts of acetyl-CoA by β-oxidation. The amount of acetyl-CoA is excessive by comparison with the amount of oxaloacetate available to react with it. In the case of diabetics, the problem is not inadequate intake of carbohydrates but rather the inability to metabolize them that causes the imbalance.

The reactions that result in "ketone bodies" start with the condensation of two molecules of acetyl-CoA to produce acetoacetyl-CoA. **Acetoacetate** arises from acetoacetyl-CoA and can have two possible fates. A reduction reaction can produce **β-hydroxybutyrate** from acetoacetate. The other possible reaction is the decarboxylation of acetoacetate to give **acetone** (Figure 16.8). The odor of acetone can frequently be detected on the breath of diabetics.

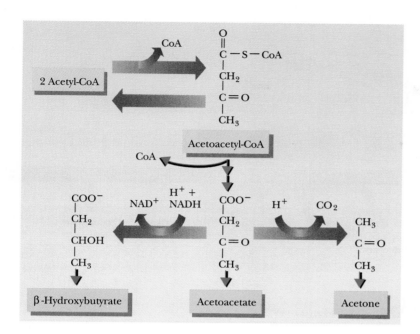

**FIGURE 16.8**  The formation of acetoacetate and related compounds from acetyl-CoA.

Even though glucose is the usual fuel in most tissues and organs, acetoacetate can be and is used as a fuel. In heart muscle and the renal cortex, acetoacetate is the preferred source of energy. Even in organs such as the brain, in which glucose is the preferred fuel, starvation conditions can lead to the use of acetoacetate for energy.

## 16.6
## THE ANABOLISM OF FATTY ACIDS

The anabolism of fatty acids is not simply a reversal of the reactions of β-oxidation. Anabolism and catabolism are not, in general, the exact reverse of each other; for instance, gluconeogenesis (Section 13.7) is not simply a reversal of the reactions of glycolysis. A first example of the differences between the degradation and the biosynthesis of fatty acids is that the anabolic reactions take place in the cytosol. We have just seen that the degradative reactions of β-oxidation take place in the mitochondrial matrix. The first step in fatty acid biosynthesis is transport of acetyl-CoA to the cytosol.

Acetyl-CoA can be formed by β-oxidation of fatty acids or by decarboxylation of pyruvate. (Degradation of certain amino acids also produces acetyl-CoA; see Section 18.7.) Most of these reactions take place in the mitochondria, requiring a transport mechanism to export acetyl-CoA to the cytosol for fatty acid biosynthesis. The transport mechanism in question is based on the fact that citrate can cross the mitochondrial membrane. Acetyl-CoA condenses with oxaloacetate (which cannot cross the mitochondrial membrane) to form citrate (recall that this is the first

**FIGURE 16.9** The transport of acetyl groups from the mitochondrion to the cytosol.

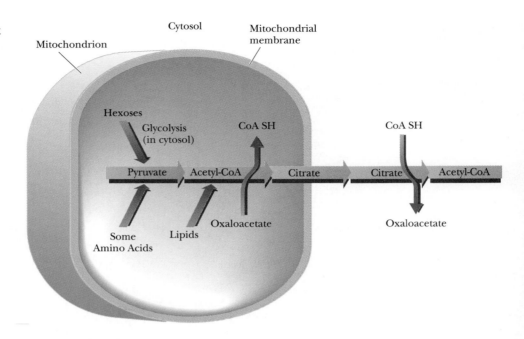

reaction of the citric acid cycle). The citrate that is exported to the cytosol can undergo the reverse reaction, producing oxaloacetate and acetyl-CoA (Figure 16.9). Acetyl-CoA enters the pathway for fatty acid biosynthesis, while oxaloacetate undergoes a series of reactions in which there is a substitution of NADPH for NADH (see Lipid Anabolism in Section 19.3). This substitution exercises control over the pathway, since NADPH is required for fatty acid anabolism.

In the cytosol acetyl-CoA is carboxylated, producing **malonyl-CoA,** a key intermediate in fatty acid biosynthesis (Figure 16.10). This reaction is catalyzed by the **acetyl-CoA carboxylase** complex, which consists of three enzymes and which requires $Mn^{2+}$ and biotin for activity, in addition to ATP. We have already seen that enzymes catalyzing reactions that take place in several steps frequently consist of several separate protein molecules, and this enzyme follows that pattern. In this case acetyl-CoA carboxylase consists of the three proteins **biotin carboxylase,** the **biotin carrier protein,** and **carboxyl transferase.** Biotin carboxylase catalyzes the transfer of the carboxyl group to biotin. The "activated $CO_2$" (the carboxyl group derived from the bicarbonate ion $HCO_3^-$) is covalently bound to biotin. Biotin (whether carboxylated or not) is bound to the

**FIGURE 16.10** The formation of malonyl-CoA.

$$\underset{\text{Acetyl-CoA}}{H_3C-\overset{\overset{\displaystyle O}{\|}}{C}-S-CoA} + ATP + HCO_3^- \xrightarrow[\text{Mn}^{2+}]{\text{Biotin}}$$

$$\underset{\text{Malonyl-CoA}}{{}^-OOC-CH_2-\overset{\overset{\displaystyle O}{\|}}{C}-S-CoA} + ADP + P_i + H^+$$

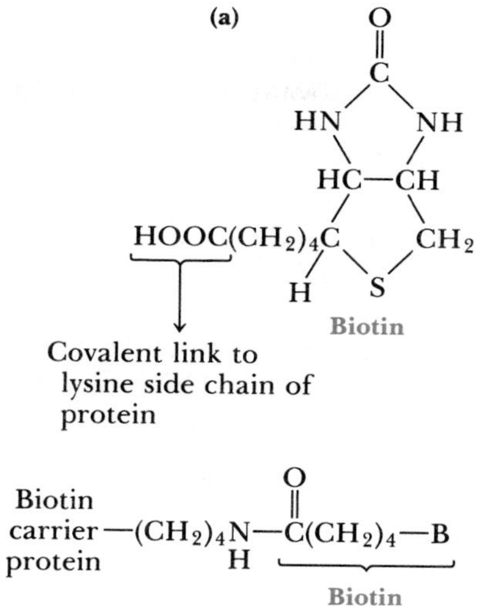

**FIGURE 16.11**   The mode of action of the three proteins that make up acetyl-CoA carboxylase. (a) The structure of biotin showing the site of its covalent link to the carrier protein. (b) In the actual reaction, biotin moves from one side to another and remains covalently linked to the biotin carrier protein.

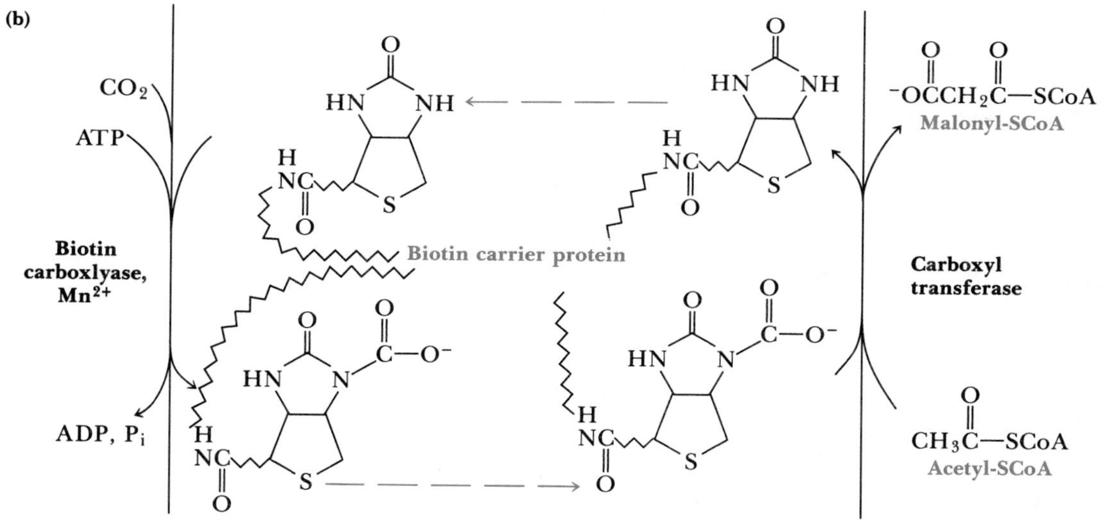

biotin carrier protein by an amide linkage to the $\epsilon$-amino group of a lysine side chain. The amide linkage to the side chain that bonds biotin to the carrier protein is long enough and flexible enough to move the carboxylated biotin into position to transfer the carboxyl group to acetyl-CoA in the reaction catalyzed by carboxyl transferase, producing malonyl-CoA (Figure 16.11).

The biosynthesis of fatty acids involves the successive addition of two-carbon units to the growing chain. Two of the three carbon atoms of the malonyl group of malonyl-CoA are added to the growing fatty acid chain at each stage of the biosynthetic reaction. This reaction, like the formation of the malonyl-CoA itself, requires a multienzyme complex located in the cytosol and not attached to any membrane. The complex, made up of the individual enzymes, is called fatty acid synthetase.

The usual product of fatty acid anabolism is **palmitate,** the 16-carbon saturated fatty acid. All 16 carbons come from the acetyl group of acetyl-CoA; we have already seen how malonyl-CoA, the immediate precursor, arises from acetyl-CoA. But first there is a priming step in which one molecule of acetyl-CoA itself is required for each molecule of palmitate produced. In this priming step the acetyl group from acetyl-CoA is transferred to an acyl carrier protein (ACP), which is considered a part of the fatty acid synthetase complex (Figure 16.12, Step 1a). The acetyl group is bound to the protein as a thioester. The group on the protein to which the acetyl group is bonded is the 4'-phosphopantetheine group, which in

**FIGURE 16.12**   The first cycle of palmitate synthesis. ACP is the acyl carrier protein.

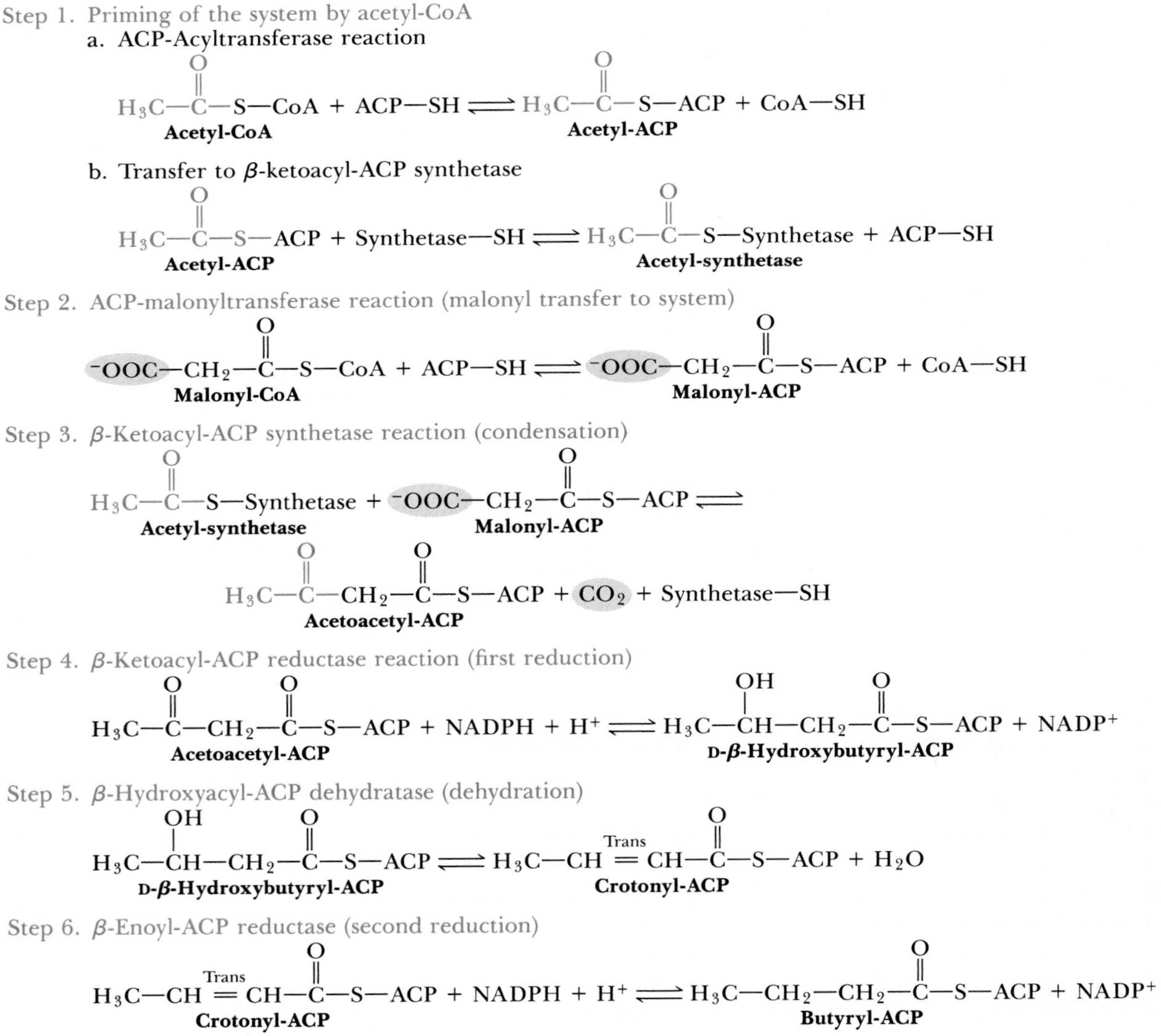

Step 1.  Priming of the system by acetyl-CoA
  a.  ACP-Acyltransferase reaction

$$H_3C-\overset{\overset{\displaystyle O}{\|}}{C}-S-CoA + ACP-SH \rightleftharpoons H_3C-\overset{\overset{\displaystyle O}{\|}}{C}-S-ACP + CoA-SH$$
**Acetyl-CoA**                    **Acetyl-ACP**

  b.  Transfer to $\beta$-ketoacyl-ACP synthetase

$$H_3C-\overset{\overset{\displaystyle O}{\|}}{C}-S-ACP + Synthetase-SH \rightleftharpoons H_3C-\overset{\overset{\displaystyle O}{\|}}{C}-S-Synthetase + ACP-SH$$
**Acetyl-ACP**                         **Acetyl-synthetase**

Step 2.  ACP-malonyltransferase reaction (malonyl transfer to system)

$$^-OOC-CH_2-\overset{\overset{\displaystyle O}{\|}}{C}-S-CoA + ACP-SH \rightleftharpoons {}^-OOC-CH_2-\overset{\overset{\displaystyle O}{\|}}{C}-S-ACP + CoA-SH$$
**Malonyl-CoA**                         **Malonyl-ACP**

Step 3.  $\beta$-Ketoacyl-ACP synthetase reaction (condensation)

$$H_3C-\overset{\overset{\displaystyle O}{\|}}{C}-S-Synthetase + {}^-OOC-CH_2-\overset{\overset{\displaystyle O}{\|}}{C}-S-ACP \rightleftharpoons$$
**Acetyl-synthetase**              **Malonyl-ACP**

$$H_3C-\overset{\overset{\displaystyle O}{\|}}{C}-CH_2-\overset{\overset{\displaystyle O}{\|}}{C}-S-ACP + CO_2 + Synthetase-SH$$
**Acetoacetyl-ACP**

Step 4.  $\beta$-Ketoacyl-ACP reductase reaction (first reduction)

$$H_3C-\overset{\overset{\displaystyle O}{\|}}{C}-CH_2-\overset{\overset{\displaystyle O}{\|}}{C}-S-ACP + NADPH + H^+ \rightleftharpoons H_3C-\overset{\overset{\displaystyle OH}{|}}{C}H-CH_2-\overset{\overset{\displaystyle O}{\|}}{C}-S-ACP + NADP^+$$
**Acetoacetyl-ACP**                              **D-$\beta$-Hydroxybutyryl-ACP**

Step 5.  $\beta$-Hydroxyacyl-ACP dehydratase (dehydration)

$$H_3C-\overset{\overset{\displaystyle OH}{|}}{C}H-CH_2-\overset{\overset{\displaystyle O}{\|}}{C}-S-ACP \rightleftharpoons H_3C-\overset{Trans}{C}H = CH-\overset{\overset{\displaystyle O}{\|}}{C}-S-ACP + H_2O$$
**D-$\beta$-Hydroxybutyryl-ACP**                **Crotonyl-ACP**

Step 6.  $\beta$-Enoyl-ACP reductase (second reduction)

$$H_3C-\overset{Trans}{C}H = CH-\overset{\overset{\displaystyle O}{\|}}{C}-S-ACP + NADPH + H^+ \rightleftharpoons H_3C-CH_2-CH_2-\overset{\overset{\displaystyle O}{\|}}{C}-S-ACP + NADP^+$$
**Crotonyl-ACP**                                **Butyryl-ACP**

turn is bonded to a serine side chain; note in Figure 16.13 that this group is structurally similar to CoA-SH itself. The acetyl group is transferred from CoA-SH, to which it is bound by a thioester linkage, to the ACP; the acetyl group is bound to the ACP by a thioester linkage.

The acetyl group is transferred in turn from the ACP to another protein, to which it is bound by a thioester linkage to a cysteine-SH; the other protein is β-ketoacyl-S-ACP-synthetase (Figure 16.12, Step 1b). The first of the successive additions of two of the three malonyl carbons to the fatty acid starts at this point. The malonyl group itself is transferred from a thioester linkage with CoA-SH to another thioester bond to the ACP (Figure 16.12, Step 2).

The next step is a condensation reaction that produces acetoacetyl-S-ACP (Figure 16.12, Step 3). In other words, the principal product of this reaction is an acetoacetyl group bound to the ACP by a thioester linkage. Two of the four carbons of acetoacetate come from the priming acetyl group, and the other two come from the malonyl group. The carbon atoms that arise from the malonyl group are the one directly bonded to the sulfur and the one in the —$CH_2$— group next to it. The $CH_3CO$— group comes from the priming acetyl group. The other carbon of the malonyl group is released as $CO_2$; the $CO_2$ that is lost is the original $CO_2$ that was used to carboxylate the acetyl-CoA to produce malonyl-CoA. The synthetase is no longer involved in a thioester linkage.

Acetoacetyl-ACP is converted to butyryl-ACP by a series of reactions involving two reductions and a dehydration (Figure 16.12, Steps 4–6). In the first reduction the β-keto group is reduced to an alcohol, giving rise to D-β-hydroxybutyryl-ACP. In the process NADPH is oxidized to $NADP^+$; the enzyme that catalyzes this reaction is β-ketoacyl-ACP reductase (Figure 16.12, Step 4). The dehydration step, catalyzed by β-hydroxyacyl-ACP dehydratase, produces crotonyl-ACP (Figure 16.12, Step 5). Note that the double bond is in the *trans* configuration. A second reduction reaction, catalyzed by β-enoyl-ACP reductase, produces butyryl-ACP (Figure 16.12, Step 6). In this reaction NADPH is the coenzyme, as was the case with the first reduction reaction in this series.

**FIGURE 16.13** Structural similarities between coenzyme A and the phosphopantetheine group of ACP.

Phosphopantetheine group of ACP

Phosphopantetheine group of coenzyme A

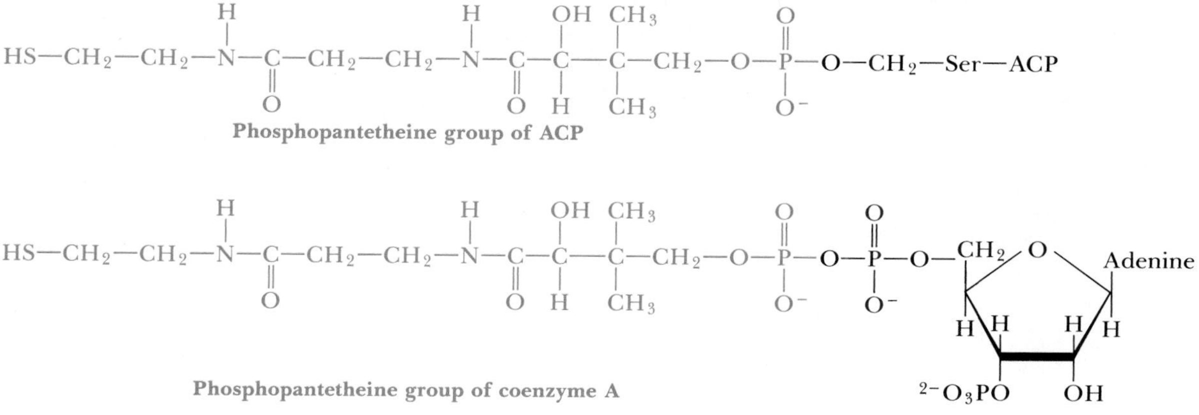

In the second round of fatty acid biosynthesis, butyryl-ACP plays the same role as acetyl-ACP in the first round. The butyryl group is transferred to the synthetase, and a malonyl group is transferred to the ACP. Once again there is a condensation reaction with malonyl-ACP (Figure 16.14). In this second round the condensation produces a six-carbon $\beta$-ketoacyl-ACP. The two added carbon atoms come from the malonyl group, as they did in the first round. The reduction and dehydration reactions take place as before, giving rise to hexanoyl-ACP. The same series of reactions is repeated until palmitoyl-ACP is produced. In mammalian systems the process stops at $C_{16}$, since the fatty acid synthetase does not produce longer chains. Mammals produce longer chain fatty acids by modification of the fatty acids formed by the synthetase reaction.

Fatty acid synthetases from different types of organisms have markedly different characteristics. In *Escherichia coli* the multienzyme system consists of an aggregate of separate enzymes, including a separate ACP (Figure

**FIGURE 16.14** The second cycle of palmitate synthesis in *Escherichia coli,* showing the role of the multienzyme system. Enzymes:
1. ACP-acyltransferase;
2. $\beta$-ketoacyl-ACP synthase;
3. ACP-malonyltransferase;
4. $\beta$-ketoacyl-ACP reductase;
5. $\beta$-hydroxyacyl-ACP dehydratase;
6. $\beta$-enoyl-ACP reductase. (Steps coincide with those in Figure 16.12.)

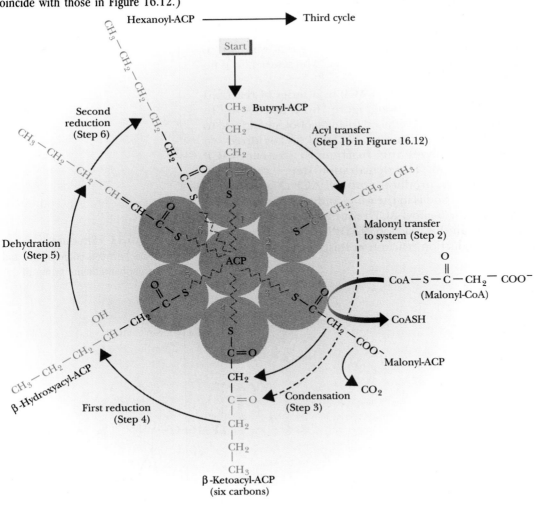

16.14). The ACP is of primary importance to the complex and is considered to occupy a central position in it. The phosphopantetheine group plays the role of a "swinging arm," much like that of biotin, which was discussed earlier in this chapter. This bacterial system has been extensively studied and has been considered a typical example of a fatty acid synthetase. In mammals and in yeast, however, the fatty acid synthetase consists of only two subunits (two different kinds of polypeptide chain), in contrast to the larger number in the bacterial system. Each of the subunits is a **multifunctional enzyme** that catalyzes reactions requiring several different proteins in the *E. coli* system. The ACP and the enzyme for the condensation reaction appear to be the two subunits in the mammalian and yeast systems. The growing fatty acid chain swings back and forth between the two on the "swinging arm," and different parts of these two polypeptide chains catalyze the other stages of the addition of each two-carbon unit. Like the bacterial system, the eukaryotic system keeps all the components of the reaction in proximity to one another.

Several additional reactions are required for the elongation of fatty acid chains and the introduction of double bonds. When mammals produce fatty acids with longer chains than that of palmitate, the reaction does not involve cytosolic fatty acid synthetase. There are two sites for the chain-lengthening reactions: the endoplasmic reticulum (ER) and the mitochondrion. In the chain-lengthening reactions in the mitochondrion, the intermediates are of the acyl-CoA type rather than the acyl-S-ACP type. In other words, the chain-lengthening reactions in the mitochondrion are the reverse of the catabolic reactions of fatty acids, with acetyl-CoA as the source of added carbon atoms; this is a difference between the main pathway of fatty acid biosynthesis and these modification reactions. In the ER the source of additional carbon atoms is malonyl-CoA. The modification reactions in the ER also differ from the biosynthesis of palmitate in that, like the mitochondrial reaction, there are no intermediates bound to ACP.

Reactions in which a double bond is introduced in fatty acids mainly take place on the ER. The insertion of the double bond is catalyzed by an oxidase that requires molecular oxygen ($O_2$) and NADH. Reactions linked to molecular oxygen are comparatively rare (Section 14.5). Mammals cannot introduce a double bond beyond carbon atom 9 (counting from the carboxyl end) of the fatty acid chain. As a result, linoleate ($CH_3—(CH_2)_4—CH=CH—CH_2—CH=CH—(CH_2)_7—COO^-$) and linolenate ($CH_3—(CH_2)_4—CH=CH—CH_2—CH=CH—CH_2—CH=CH—(CH_2)_4—COO^-$) must be included in the diet of mammals. They are **essential fatty acids,** since they are precursors of other lipids, including prostaglandins.

Even though both the anabolism and the catabolism of fatty acids require successive reactions of two-carbon units, the two pathways are not the exact reversal of each other. The differences between the two pathways can be summarized in the following list of points.

1. The product of fatty acid degradation is acetyl-CoA; the precursor in fatty acid biosynthesis is acetyl-CoA.

2. In fatty acid biosynthesis malonyl-CoA, derived from acetyl-CoA in a reaction that requires biotin, is the immediate source of two-carbon units; malonyl-CoA is not involved in fatty acid degradation, and there is no requirement for biotin.

3. Fatty acid degradation is an oxidative process that requires $NAD^+$ and FAD and produces ATP; fatty acid biosynthesis is a reductive process that requires NADPH and ATP.

4. Fatty acids form thioesters with CoA-SH during catabolism; fatty acids form thioesters with acyl carrier proteins (ACP-SH) during anabolism.

5. Fatty acids are degraded from the carboxyl end ($CH_3CO_2$—), but they are built up from the methyl end ($CH_3CH_2$—).

6. Fatty acid biosynthesis occurs in the cytosol, catalyzed by an ordered multienzyme complex; fatty acid catabolism occurs in the mitochondrial matrix, with no ordered aggregate of enzymes.

7. In fatty acid catabolism the $\beta$-hydroxyacyl intermediates have the L configuration; in anabolism, $\beta$-hydroxyacyl intermediates with the D configuration are involved.

## 16.7
## THE ANABOLISM OF ACYLGLYCEROLS AND COMPOUND LIPIDS

Other lipids, including triacylglycerols, phosphoacylglycerols, and steroids, are derived from fatty acids and metabolites of fatty acids such as acetoacetyl-CoA. Free fatty acids do not occur in the cell to any great extent; they are normally found incorporated in triacylglycerols and phosphoacylglycerols. The biosynthesis of these two types of compounds takes place principally on the ER of liver cells or fat cells (adipocytes).

### Triacylglycerols

The glycerol portion of lipids is derived from glycerol 3-phosphate, a compound available from glycolysis. Another source is glycerol released by degradation of acylglycerols. An acyl group of a fatty acid is transferred from an acyl-CoA in a reaction catalyzed by the enzyme **glycerol phosphate acyltransferase.** The products of this reaction are CoA-SH and a **lysophosphatidate** (a monoacylglycerol phosphate) (Figure 16.15). The acyl group is shown as esterified at carbon atom 2 (C-2) in this series of equations, but it is equally likely that it is esterified at C-1. A second acylation reaction takes place, catalyzed by the same enzyme, producing a **phosphatidate** (a diacylglyceryl phosphate). Phosphatidates occur in membranes and are precursors of other phospholipids. The phosphate group of the phosphatidate is removed by hydrolysis in a reaction catalyzed by **phosphatidate phosphatase,** producing a **diacylglycerol.** A third acyl group is added in a reaction catalyzed by **diacylglycerol acyltransferase.** As before, the source of the acyl group is an acyl-CoA rather than the free fatty acid.

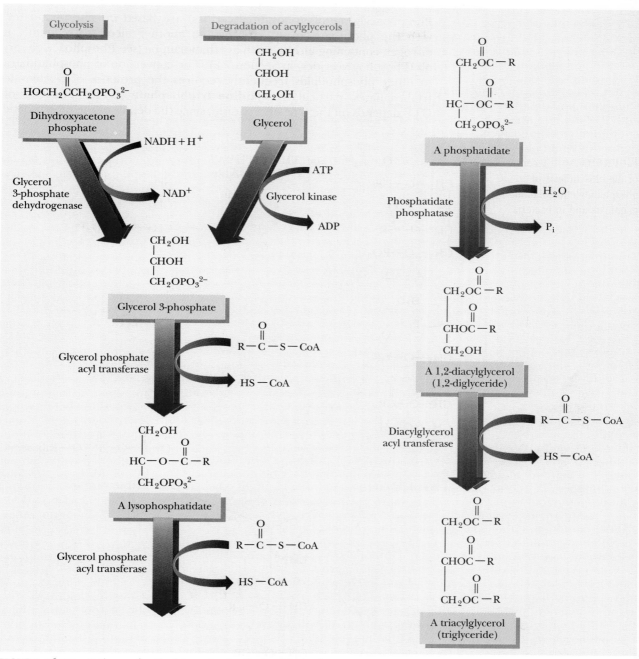

**FIGURE 16.15**  Pathways for the biosynthesis of triacylglycerols.

## Phosphoacylglycerols

Phosphoacylglycerols (phosphoglycerides) are based on phosphatidates, with the phosphate group esterified to another alcohol, frequently a nitrogen-containing alcohol such as ethanolamine (see Phosphoacylglycerols (Phosphoglycerides) in Section 7.2). The conversion of phosphatidates to other phospholipids frequently requires the presence of nucleoside triphosphates, particularly **cytidine triphosphate** (CTP). The role of CTP depends on the type of organism, since the details of the biosynthetic

FIGURE 16.16   A comparison of the biosynthesis of phosphatidylethanolamine in mammals and in bacteria.

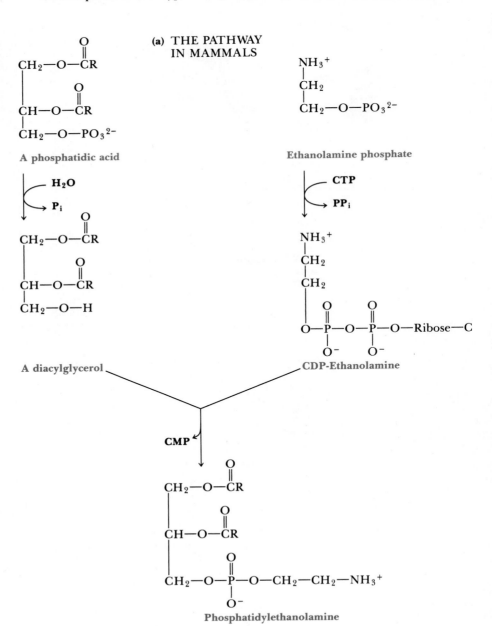

**(a) THE PATHWAY IN MAMMALS**

Phosphatidylethanolamine

pathway are not the same in mammals and bacteria. We shall use a comparison of the synthesis of phosphatidylethanolamine in mammals and in bacteria (Figure 16.16) as a case study of the kinds of reactions commonly encountered in phosphoglyceride biosynthesis.

In mammals the synthesis of **phosphatidylethanolamine** requires two preceding steps in which the component parts are processed. The first of these two steps is the removal by hydrolysis of the phosphate group of

**(b) THE PATHWAY IN BACTERIA**          *(Figure 16.16 cont'd)*

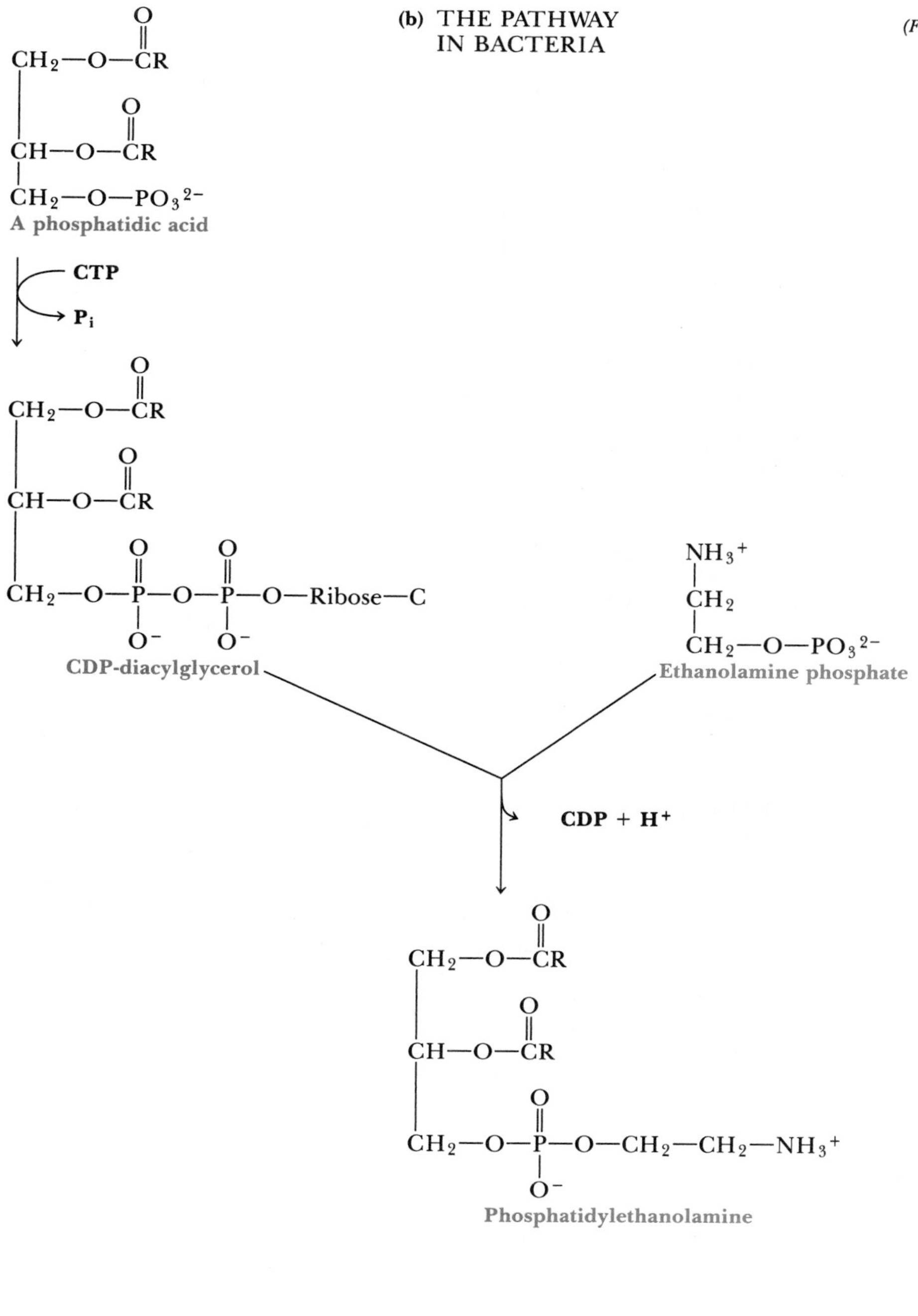

$$
\begin{array}{l}
CH_2-O-\overset{\overset{\displaystyle O}{\|}}{C}R \\
| \\
CH-O-\overset{\overset{\displaystyle O}{\|}}{C}R \\
| \\
CH_2-O-PO_3{}^{2-}
\end{array}
$$

A phosphatidic acid

CTP → P_i

$$
\begin{array}{l}
CH_2-O-\overset{\overset{\displaystyle O}{\|}}{C}R \\
| \\
CH-O-\overset{\overset{\displaystyle O}{\|}}{C}R \\
| \\
CH_2-O-\overset{\overset{\displaystyle O}{\|}}{P}-O-\overset{\overset{\displaystyle O}{\|}}{P}-O-\text{Ribose}-C \\
\quad\quad\quad O^- \quad\quad O^-
\end{array}
$$

CDP-diacylglycerol

$$
\begin{array}{l}
NH_3{}^+ \\
| \\
CH_2 \\
| \\
CH_2-O-PO_3{}^{2-}
\end{array}
$$

Ethanolamine phosphate

CDP + H⁺

$$
\begin{array}{l}
CH_2-O-\overset{\overset{\displaystyle O}{\|}}{C}R \\
| \\
CH-O-\overset{\overset{\displaystyle O}{\|}}{C}R \\
| \\
CH_2-O-\overset{\overset{\displaystyle O}{\|}}{P}-O-CH_2-CH_2-NH_3{}^+ \\
\quad\quad\quad O^-
\end{array}
$$

Phosphatidylethanolamine

the phosphatidate, producing a diacylglycerol; the second step is the reaction of ethanolamine phosphate with CTP to produce pyrophosphate ($PP_i$) and cytidine diphosphate ethanolamine (CDP-ethanolamine). The CDP-ethanolamine and diacylglycerol react to form phosphatidylethanolamine. In bacteria CTP reacts instead with phosphatidate itself to produce cytidine diphosphodiacylglycerol (a CDP diglyceride). The CDP diglyceride reacts with ethanolamine phosphate to form phosphatidylethanolamine.

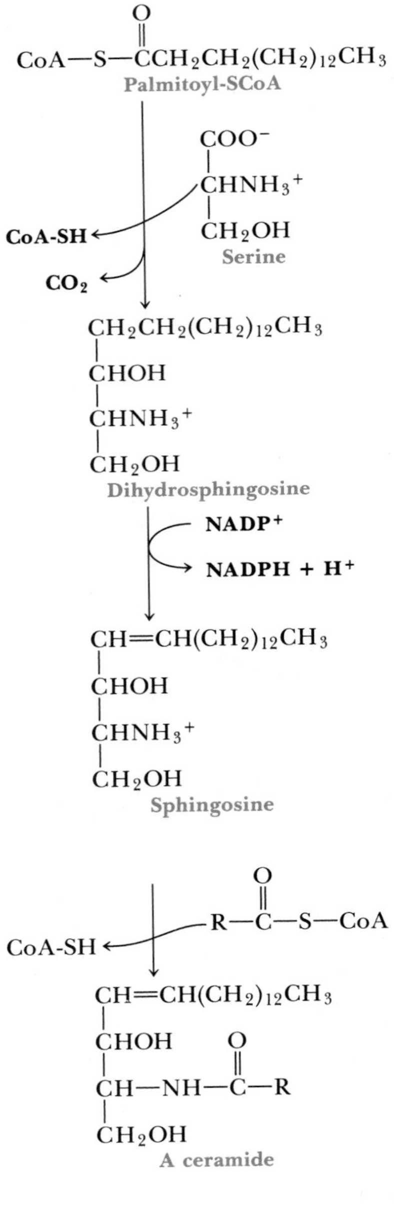

**FIGURE 16.17** The biosynthesis of sphingolipids. When ceramides are formed, they can react (a) with choline to yield sphingomyelins, (b) with sugars to yield cerebrosides, or (c) with sugars and sialic acid to yield gangliosides.

$$\underset{\text{Palmitoyl-SCoA}}{\text{CoA}-\text{S}-\overset{\overset{\text{O}}{\|}}{\text{C}}\text{CH}_2\text{CH}_2(\text{CH}_2)_{12}\text{CH}_3}$$

Serine
$$\begin{array}{c} \text{COO}^- \\ | \\ \text{CHNH}_3{}^+ \\ | \\ \text{CH}_2\text{OH} \end{array}$$

CoA-SH ←
CO$_2$ ←

$$\begin{array}{c} \text{CH}_2\text{CH}_2(\text{CH}_2)_{12}\text{CH}_3 \\ | \\ \text{CHOH} \\ | \\ \text{CHNH}_3{}^+ \\ | \\ \text{CH}_2\text{OH} \end{array}$$
**Dihydrosphingosine**

NADP$^+$
NADPH + H$^+$

$$\begin{array}{c} \text{CH}=\text{CH}(\text{CH}_2)_{12}\text{CH}_3 \\ | \\ \text{CHOH} \\ | \\ \text{CHNH}_3{}^+ \\ | \\ \text{CH}_2\text{OH} \end{array}$$
**Sphingosine**

$$\text{R}-\overset{\overset{\text{O}}{\|}}{\text{C}}-\text{S}-\text{CoA}$$

CoA-SH ←

$$\begin{array}{c} \text{CH}=\text{CH}(\text{CH}_2)_{12}\text{CH}_3 \\ | \\ \text{CHOH} \\ | \\ \text{CH}-\text{NH}-\overset{\overset{\text{O}}{\|}}{\text{C}}-\text{R} \\ | \\ \text{CH}_2\text{OH} \end{array}$$
**A ceramide**

## Sphingolipids

The structural basis of sphingolipids is not glycerol but **sphingosine,** a long-chain amine (see Sphingolipids in Section 7.2). The precursors of sphingosine are palmitoyl-CoA and the amino acid serine, which react to produce dihydrosphingosine. The carboxyl group of the serine is lost as $CO_2$ in the course of this reaction (Figure 16.17). An oxidation reaction introduces a double bond, with sphingosine as the resulting compound. Reaction of the amino group of sphingosine with another acyl-CoA to form an amide bond results in an **N-acylsphingosine,** also called a **ceramide.** Ceramides in turn are the parent compounds of **sphingomyelins, cere-**

## BOX 16.1
## TAY–SACHS DISEASE

Tay–Sachs disease is an inborn error of lipid metabolism with particularly tragic consequences. In this disease, there is a blockage in the catabolism of gangliosides (see Sphingolipids in Section 7.2). The enzyme hexosaminidase A, responsible for the hydrolysis of N-acetylgalactosamine from ganglioside $GM_2$, is missing.

Inclusions of accumulated $GM_2$ appear in the neurons of affected individuals. Newborns affected with Tay–Sachs disease appear normal at first, but by the age of 1 year, they exhibit the characteristic symptoms of weakness, retardation, and blindness. This disease is fatal by age 3 or 4. It is possible to detect the disease during fetal development by amniocentesis, a technique based on assay of amniotic fluid or amniotic cells for various enzyme activities.

The enzyme that catalyzes the hydrolysis of the bond indicated is missing in Tay-Sachs disease.

Ganglioside $GM_2$

brosides, and **gangliosides.** Attachment of choline to the primary alcohol group of a ceramide produces a sphingomyelin, while attachment of sugars such as glucose at the same site produces cerebrosides. Gangliosides are formed from ceramides by attachment of oligosaccharides that contain a sialic acid residue, also at the primary alcohol group. See Sphingolipids in Section 7.2 for the structures of these compounds.

Gangliosides play a part in **Tay–Sachs disease** (Box 16.1). Because the enzyme hexosaminidase A is missing, the catabolism of gangliosides is blocked, creating conditions that prove uniformly fatal to affected individuals.

## 16.8
## THE ANABOLISM OF CHOLESTEROL

The ultimate precursor of all the carbon atoms in cholesterol and in the other steroids that are derived from cholesterol is the acetyl group of acetyl-CoA. There are many steps in the biosynthesis of steroids. The condensation of three acetyl groups produces mevalonate, which contains six carbons. Decarboxylation of mevalonate produces the five-carbon isoprene unit frequently encountered in the structure of lipids (Section 7.3). Six isoprene units condense to form squalene, which contains 30 carbon atoms. Finally squalene is converted to cholesterol, which contains 27 carbon atoms (Figure 16.18).

$$\text{Acetate} \longrightarrow \text{mevalonate} \longrightarrow \text{[isoprene]} \longrightarrow \text{squalene} \longrightarrow \text{cholesterol}$$
$$C_2 \qquad\qquad C_6 \qquad\qquad C_5 \qquad\qquad C_{30} \qquad\qquad C_{27}$$

It is well established that 12 of the carbon atoms of cholesterol arise from the carboxyl carbon of the acetyl group; these are the carbon atoms labeled c in Figure 16.19. The other 15 carbon atoms arise from the methyl carbon of the acetyl group; these are the carbon atoms labeled m. We shall now look at the individual steps of the process in more detail.

The conversion of three acetyl groups of acetyl-CoA to **mevalonate** takes place in several steps (Figure 16.20). We have already seen the first of these steps, the production of acetoacetyl-CoA from two molecules of acetyl-CoA, when we discussed the formation of ketone bodies and the anabolism of fatty acids. A third molecule of acetyl-CoA condenses with acetoacetyl-CoA to produce **β-hydroxy-β-methylglutaryl-CoA** (also called HMG-CoA and 3-hydroxy-3-methylglutaryl-CoA). This reaction is catalyzed by the enzyme hydroxymethylglutaryl-CoA synthetase; one molecule of CoA-SH is released in the process. In the next reaction the production of mevalonate from hydroxymethylglutaryl-CoA is catalyzed by the enzyme hydroxymethyglutaryl-CoA reductase. A carboxyl group, the one esterified to CoA-SH, is reduced to a hydroxyl group, and the CoA-SH is released.

Mevalonate is then converted to an isoprenoid unit by a combination of phosphorylation, decarboxylation, and dephosphorylation reactions (Figure 16.21). Three successive phosphorylation reactions, each of which

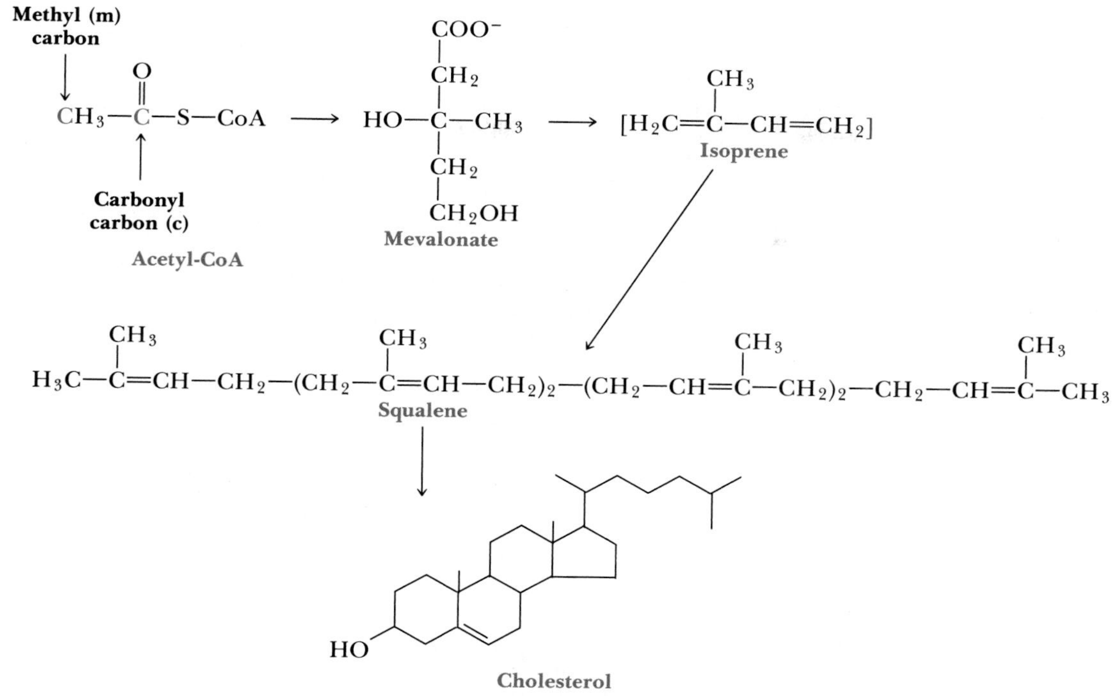

FIGURE 16.18   Outside of the biosynthesis of cholesterol.

is catalyzed by an enzyme that requires ATP, give rise to **3-phospho-5-pyrophosphomevalonate.** This last compound is unstable, and it undergoes a decarboxylation and a dephosphorylation reaction to produce **isopentenyl pyrophosphate,** a five-carbon isoprenoid derivative. Isopentenyl pyrophosphate and **dimethylallyl pyrophosphate,** another isoprenoid derivative, can be interconverted in a rearrangement reaction catalyzed by the enzyme isopentenyl pyrophosphate isomerase (Figure 16.22).

Condensation of isoprenoid units then leads to the production of squalene (Figure 16.23) and ultimately of cholesterol. Both of the isoprenoid derivatives we have met so far are required; isopentenyl pyrophos-

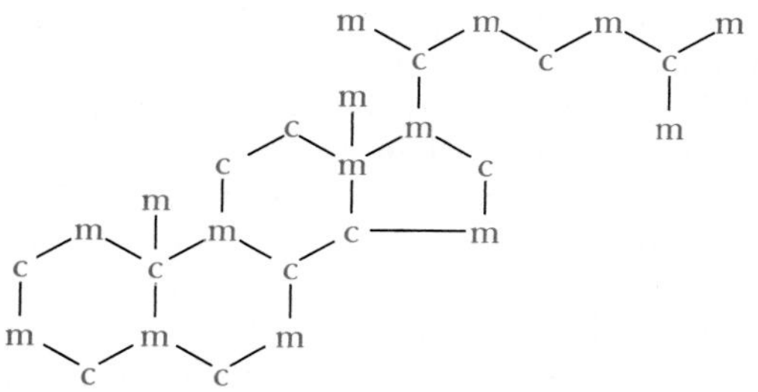

FIGURE 16.19   The labeling pattern of cholesterol. The m is methyl carbon and the c is carbonyl carbon.

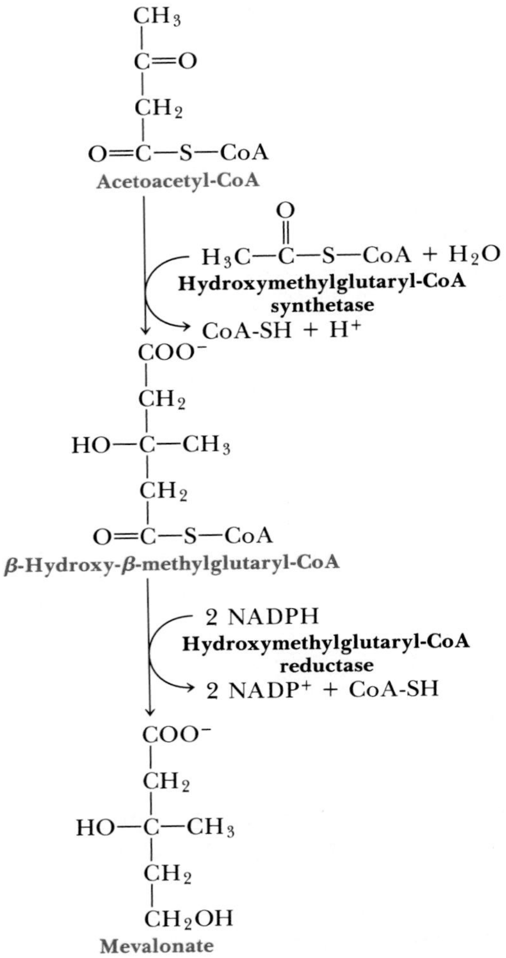

**FIGURE 16.20**    The biosynthesis of mevalonate.

$$
\begin{array}{c}
CH_3 \\
| \\
C=O \\
| \\
CH_2 \\
| \\
O=C-S-CoA
\end{array}
$$

**Acetoacetyl-CoA**

$$
H_3C-\overset{\overset{\displaystyle O}{\|}}{C}-S-CoA + H_2O
$$

**Hydroxymethylglutaryl-CoA synthetase**

$$
CoA\text{-}SH + H^+
$$

$$
\begin{array}{c}
COO^- \\
| \\
CH_2 \\
| \\
HO-C-CH_3 \\
| \\
CH_2 \\
| \\
O=C-S-CoA
\end{array}
$$

**β-Hydroxy-β-methylglutaryl-CoA**

2 NADPH

**Hydroxymethylglutaryl-CoA reductase**

$$
2\ NADP^+ + CoA\text{-}SH
$$

$$
\begin{array}{c}
COO^- \\
| \\
CH_2 \\
| \\
HO-C-CH_3 \\
| \\
CH_2 \\
| \\
CH_2OH
\end{array}
$$

**Mevalonate**

phate and dimethylallyl pyrophosphate condense with each other in a reaction catalyzed by dimethylallyl transferase (also called prenyl transferase) to produce **geranyl pyrophosphate,** a ten-carbon compound. Another condensation reaction takes place, this time between geranyl pyrophosphate and isopentenyl pyrophosphate. This reaction is again catalyzed by dimethylallyl transferase; **farnesyl pyrophosphate,** a 15-carbon compound, is produced. Two molecules of farnesyl pyrophosphate condense to form **squalene,** a 30-carbon compound. The reaction is catalyzed by squalene synthetase, and NADPH is required for the reaction.

Figure 16.24 shows the conversion of squalene to cholesterol. The details of this conversion, however, are far from completely understood. It is known that squalene is converted to **squalene epoxide** in a reaction that requires both NADPH and molecular oxygen ($O_2$). This reaction is catalyzed by squalene monooxygenase. Squalene epoxide then undergoes a complex cyclization reaction to form **lanosterol.** This remarkable reaction is catalyzed by squalene epoxide cyclase. The mechanism of the reaction is

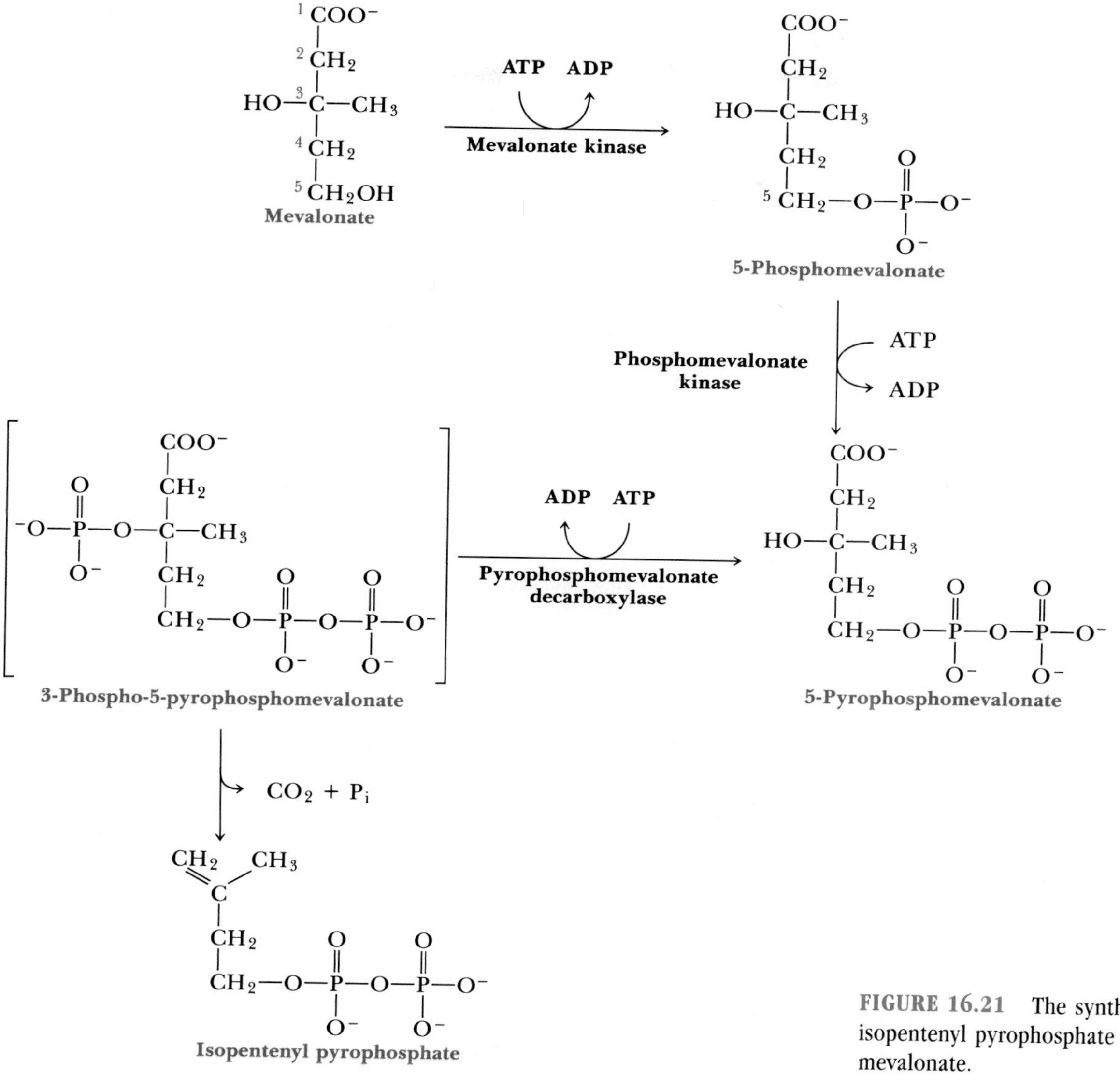

FIGURE 16.21    The synthesis of isopentenyl pyrophosphate from mevalonate.

FIGURE 16.22    Isomerization of isopentenyl pyrophosphate.

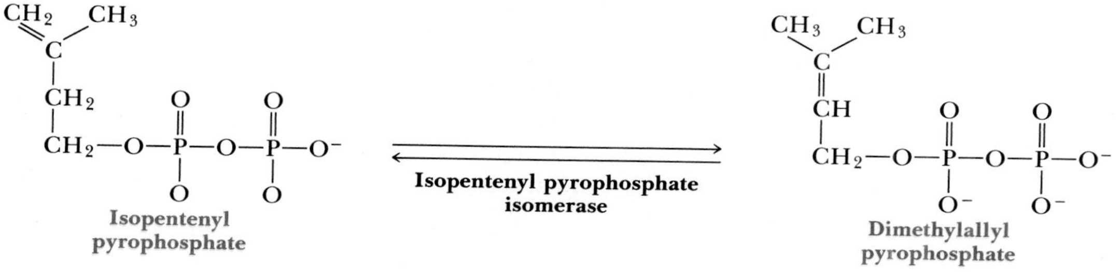

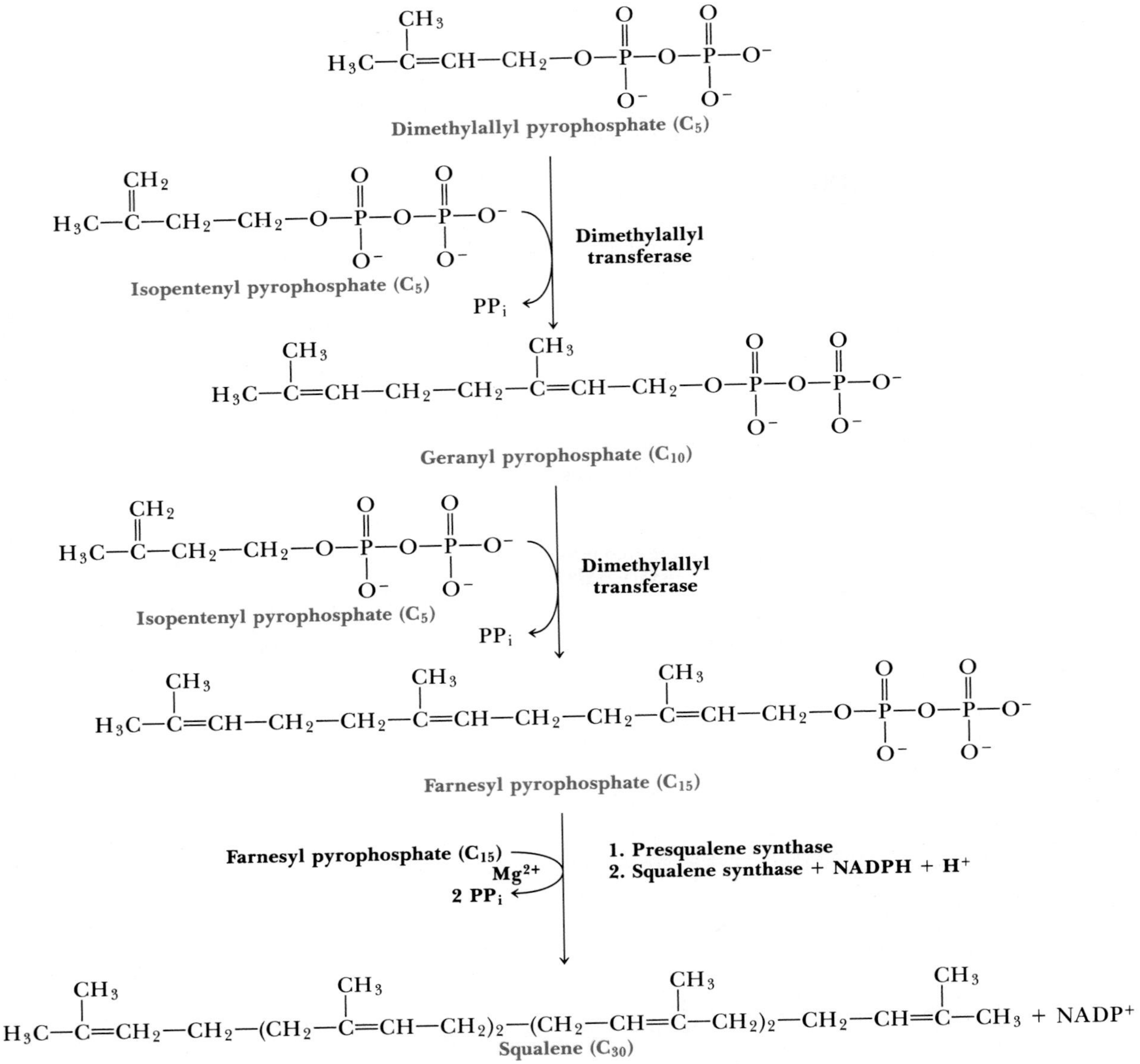

**FIGURE 16.23**   The synthesis of squalene.

a concerted reaction, that is, one in which each part is essential for any other part to take place. No portion of a concerted reaction can be left out or changed, because it all takes place in one simultaneous reaction rather than a sequence of steps. The conversion of lanosterol to cholesterol is a complex process. It is known that more than a dozen steps are required to remove three methyl groups and to move a double bond, and these are the reactions necessary for this conversion.

**FIGURE 16.24** The conversion of squalene to cholesterol.

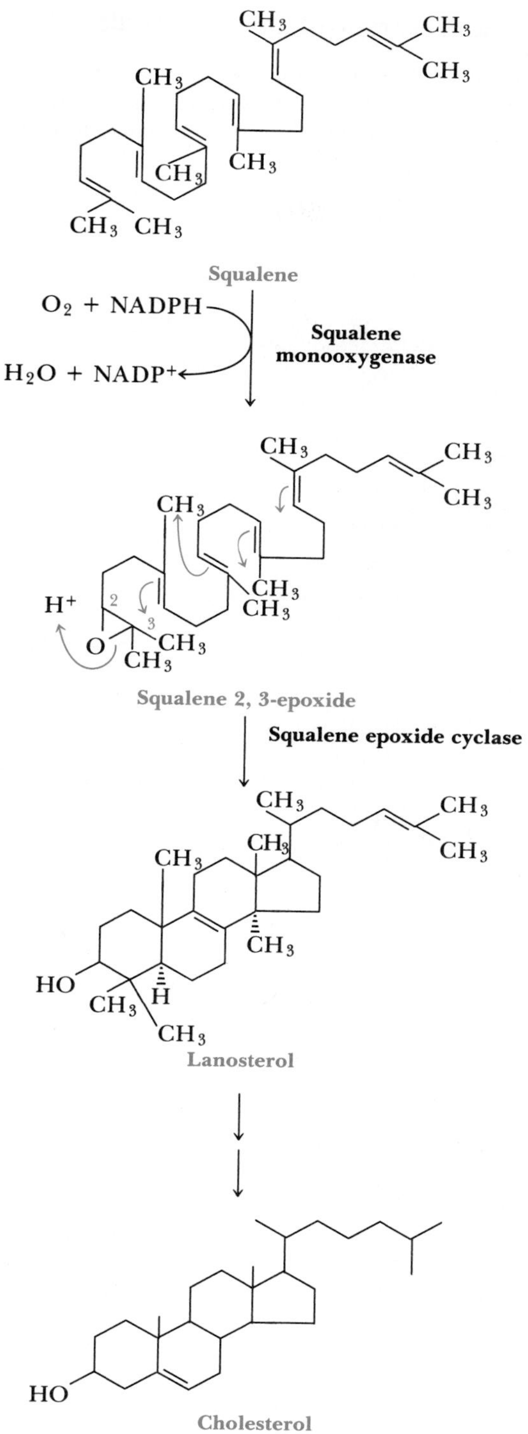

Squalene

$O_2$ + NADPH

Squalene
monooxygenase

$H_2O$ + NADP$^+$

$H^+$

Squalene 2, 3-epoxide

Squalene epoxide cyclase

Lanosterol

Cholesterol

## Cholesterol Is a Precursor of Other Steroids

Once cholesterol is formed, it can be converted to other steroids of widely varying physiological function. The smooth ER is an important site for both the synthesis of cholesterol and its converison to other steroids. Most

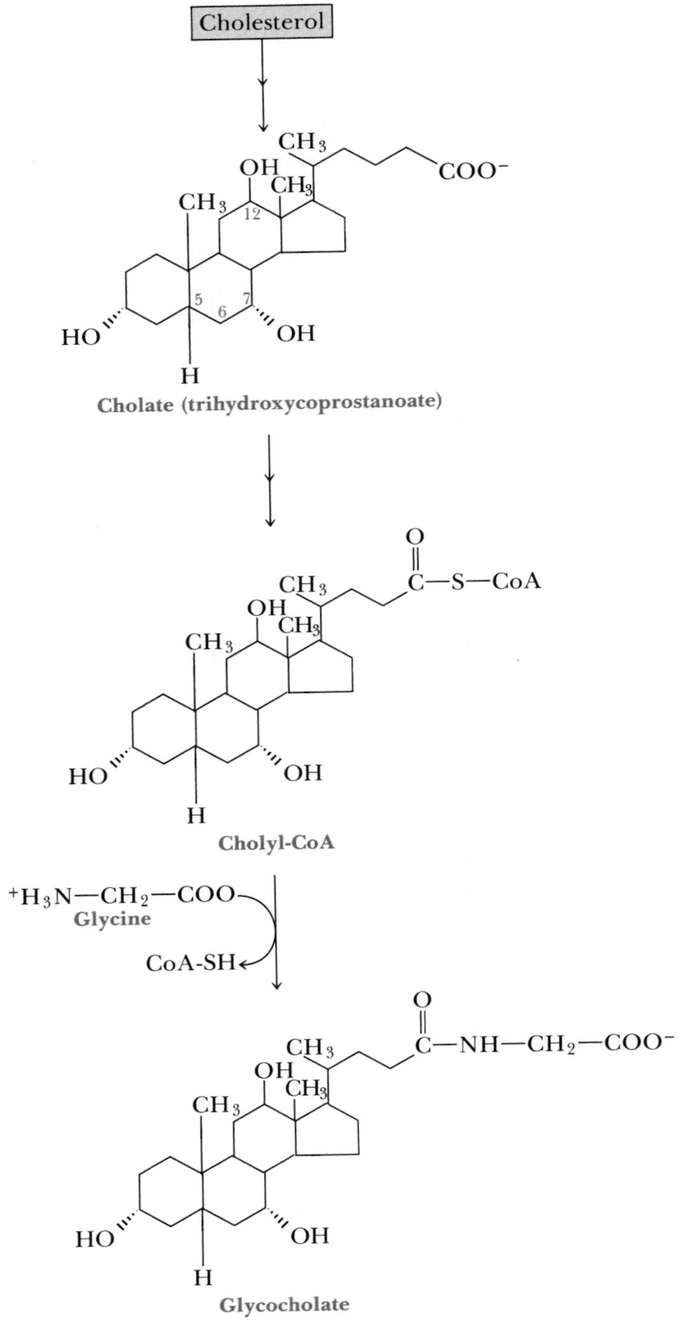

FIGURE 16.25   The synthesis of bile acids from cholesterol.

of the cholesterol formed in the liver, which is the principal site of cholesterol synthesis in mammals, is converted to **bile acids** such as cholate and glycocholate (Figure 16.25). These compounds aid in the digestion of lipid droplets by emulsifying them and rendering them more accessible to enzyme attack.

Cholesterol is also the precursor of important steroid hormones (Figure 16.26), in addition to the bile acids. **Pregnenolone** is formed from

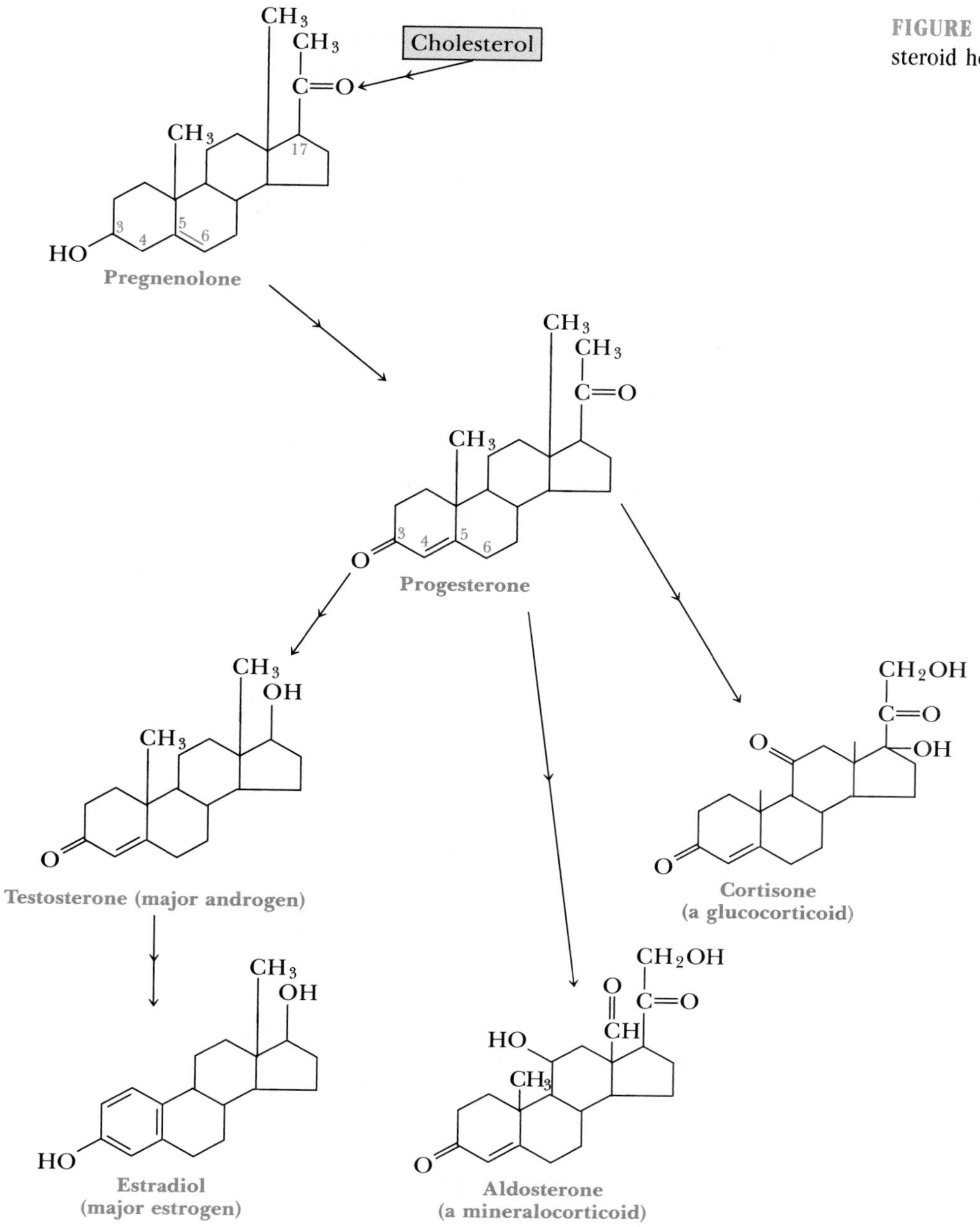

**FIGURE 16.26** The synthesis of steroid hormones from cholesterol.

cholesterol, and **progesterone** is formed from pregnenolone. Progesterone is a sex hormone and is a precursor for other sex hormones such as **testosterone** and **estradiol** (an estrogen). In addition, other types of steroid hormones also arise from progesterone. **Cortisone** is an example of **glucocorticoids,** a group of hormones that play a role in carbohydrate metabolism, as the name implies, as well as in the metabolism of proteins and fatty acids. **Mineralocorticoids** constitute another class of hormones that are involved in the metabolism of electrolytes, including metal ions ("minerals"), and water. **Aldosterone** is an example of a mineralocorticoid. In cells in which cholesterol is converted to steroid hormones, an enlarged smooth ER is frequently observed, providing a site for the process to take place.

## The Role of Cholesterol in Heart Disease

Atherosclerosis is a condition in which arteries are blocked to a greater or lesser extent by the deposition of cholesterol plaques, leading to heart attacks. The process by which the clogging of arteries occurs is a complex one. Both diet and genetics are instrumental in the development of atherosclerosis. A diet high in cholesterol will lead to a high level of cholesterol in the bloodstream. Cholesterol must be packaged for transport in the bloodstream. It is packaged in the form of low-density lipoprotein (LDL) particles, which consist of many molecules of cholesterol and cholesterol esters (the hydroxyl group of the cholesterol is esterified to an unsaturated fatty acid such as linoleate) and a group of surface proteins (Figure 16.27). The protein portions of LDL particles bind to receptor sites on the surface of a typical cell. Refer to Membrane Receptors in Section 11.4 for a discussion of the process by which LDL particles are taken into the cell as one aspect of receptor action.

LDL is degraded in the cell. The protein portion is hydrolyzed to the component amino acids, while the cholesterol esters are hydrolyzed to cholesterol and fatty acids. Free cholesterol can then be used directly as a component of membranes; the fatty acids can have any of the catabolic or anabolic fates that we reviewed earlier in this chapter (Figure 16.28). Cholesterol not needed for membrane synthesis can be stored as oleate or palmitoleate esters. The production of these esters is catalyzed by acyl-CoA cholesterol acyltransferase (ACAT), and the presence of free cholesterol increases the enzymatic activity of ACAT. In addition, cholesterol inhibits both the synthesis and the activity of the enzyme hydroxymethylglutaryl-CoA reductase (HMG-CoA reductase). This enzyme catalyzes the production of mevalonate, the reaction that is the committed step in cholesterol biosynthesis. This point has important implications. Dietary cholesterol suppresses the synthesis of cholesterol by the body, especially in tissues other than the liver. A third effect of the presence of free cholesterol in the cell is inhibition of synthesis of LDL receptors. As a result of reduction in the number of receptors, the level of LDL in the blood increases, leading to the deposition of atherosclerotic plaques.

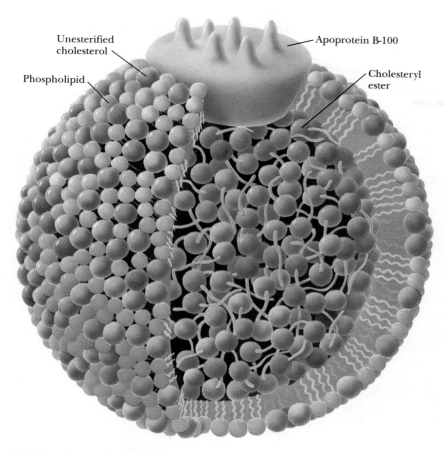

Unesterified
cholesterol

Phospholipid

Apoprotein B-100

Cholesteryl
ester

**FIGURE 16.27**    Schematic diagram of an LDL particle. (From M. S. Brown and J. L.
Goldstein, 1984. How LDL Receptors Influence Cholesterol and Atherosclerosis. *Sci.
Amer.* **251** (5), 58–66.)

The crucial role of LDL receptors in maintaining the level of
cholesterol in the bloodstream is especially clear in the case of **familial
hypercholesterolemia,** which results from a defect in the gene that codes
for active receptors. An individual who has one gene that codes for active
receptor and one defective gene is heterozygous for this trait (a heterozy-
gote). Heterozygotes have blood cholesterol levels that are above average
and therefore are at higher risk for heart disease than the general
population. An individual with two defective genes and thus no active
LDL receptor is homozygous (a homozygote). Homozygotes have very
high blood cholesterol levels from birth, and there are recorded cases of
heart attacks in 2-year-olds with this condition. Patients who are homozygous for familial hypercholesterolemia usually die before they reach
age 20.

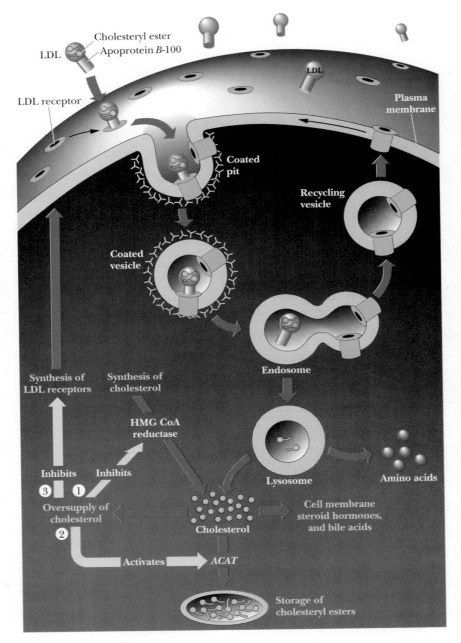

**FIGURE 16.28**    The fate of cholesterol in the cell. ACAT is the enzyme that esterifies cholesterol for storage. (From M. S. Brown and J. L. Goldstein, 1984. How LDL Receptors Influence Cholesterol and Atherosclerosis. *Sci. Amer.* **251** (5), 58–66.)

## SUMMARY

The catabolic oxidation of lipids releases large quantities of energy, while the anabolic formation of lipids represents an efficient way of storing chemical energy.

The oxidation of fatty acids is the chief source of energy in the catabolism of lipids. After an initial activation step in the cytosol, the breakdown of fatty acids takes place in the mitochondrial matrix by the process of $\beta$-oxidation. In this process, two-carbon units are successively removed from the carboxyl end of the fatty acid to produce acetyl-CoA, which subsequently enters the citric acid cycle. There is a net yield of 146 ATP molecules for each molecule of stearic acid (an 18-carbon compound) that is completely oxidized to carbon dioxide and water. The pathway of catabolism of fatty acids includes reactions in which unsaturated, as well as saturated, fatty acids can be metabolized. Odd-numbered fatty acids can also be metabolized by converting their unique breakdown product, propionyl-CoA, to succinyl-CoA, an intermediate of the citric acid cycle. "Ketone bodies" are substances related to acetone that are produced when an excess of acetyl-CoA results from $\beta$-oxidation. This situation can arise from a large intake of lipids and a low intake of carbohydrates or can occur in diabetes, in which the inability to metabolize carbohydrates causes an imbalance in the breakdown products of carbohydrates and lipids.

The anabolism of fatty acids proceeds by a different pathway from $\beta$-oxidation. Some of the most important differences between the two processes are the requirement for biotin in anabolism and not in catabolism; and the requirement for NADPH in anabolism, rather than $NAD^+$, required in catabolism. Fatty acid biosynthesis occurs in the cytosol, catalyzed by an ordered multienzyme complex; fatty acid catabolism occurs in the mitochondrial matrix, with no ordered aggregate of enzymes.

Most compound lipids, such as triacylglycerols, phosphoacylglycerols, and sphingolipids, have fatty acids as precursors. In the case of steroids, the starting material is acetyl-CoA. Isoprene units are formed from acetyl-CoA in the early stages of a lengthy process that leads ultimately to cholesterol. Cholesterol in turn is the precursor of the other steroids. Both dietary cholesterol and genetic factors influence the role of cholesterol in heart disease.

## EXERCISES

1. Compare and contrast the pathways of fatty acid breakdown and biosynthesis. What features do these two pathways have in common? How do they differ?
2. Calculate the ATP yield for the complete oxidation of one molecule of palmitic acid (16 carbons).
3. Why does the degradation of palmitic acid (see Exercise 2) to eight molecules of acetyl-CoA require seven, rather than eight, rounds of the $\beta$-oxidation process?
4. Outline the role of carnitine in the transport of acyl-CoA molecules into the mitochondrion.
5. You hear a fellow student say that the oxidation of unsaturated fatty acids requires exactly the same group of enzymes as the oxidation of saturated fatty acids. Is the statement true or false, and why?
6. It is frequently said that camels store water in their humps for long desert journeys. How would you modify this statement on the basis of information in this chapter?
7. You meet someone who has a pronounced odor of acetone on his breath. Why is it not surprising to you to discover that he is diabetic?
8. Outline the steps involved in the production of malonyl-CoA from acetyl-CoA.
9. Why are linoleate and linolenate considered essential fatty acids?
10. Is it possible to convert fatty acids to other lipids without acyl-CoA intermediates?
11. Discuss the role of isoprenoid units in the biosynthesis of cholesterol.
12. A cholesterol sample is prepared using acetyl-CoA

labeled with $^{14}$C at the carboxyl group as precursor. Which carbon atoms of cholesterol are labeled?

**13.** What is the role of citrate in the transport of acetyl groups from the mitochondrion to the cytosol?

**14.** What structural feature do all steroids have in

common? What are the biosynthetic implications of this common feature?

**15.** Why must cholesterol be packaged for transport rather than occurring freely in the bloodstream?

## ANNOTATED BIBLIOGRAPHY

Bodner, G.M. Lipids. *J. Chem. Ed.* **63**, 772–775 (1986). [Part of a series of concise and clearly written articles on metabolism.]

Brown, M.S., and J.L. Goldstein. How LDL Receptors Influence Cholesterol and Atherosclerosis. *Sci. Amer.* **251** (5), 58–66 (1984). [A description of the role of cholesterol in heart disease by the winners of the 1985 Nobel Prize in Medicine.]

Krutch, J.W. *The Voice of the Desert.* New York: Morrow, 1975. [Chapter 7, "The Mouse That Never Drinks," is a description, primarily from the naturalist's point of view, of the kangaroo rat, but it does make the point that metabolic water is this animal's only source of water.]

McGarry, J.D., and D.W. Foster. Regulation of Hepatic Fatty Acid Oxidation and Ketone Body Production. *Ann. Rev. Biochem.* **49**, 395–420 (1980). [A review article.]

Wakil, S.J., and E.M. Barnes. Fatty Acid Metabolism. *Compr. Biochem.* **18**, 57–104 (1971). [Extensive coverage of the topic.]

# Photosynthesis

CHAPTER

## 17

## OUTLINE

17.1  The Reactions of Photosynthesis
17.2  Chloroplasts and Chlorophylls
17.3  The Reactions of Photosystems I and II
17.4  A Proton Gradient Drives the Production of ATP in Photosynthesis
17.5  A Comparison of Photosynthesis With and Without Oxygen: Evolutionary Implications
17.6  The Dark Reaction of Photosynthesis: Path of Carbon
17.7  An Alternate Pathway for Carbon Dioxide Fixation

*Photosynthesis linked to oxygen plays an essential role in all life, plant and animal.*

427

*The drama of photosynthesis, converting sunlight to energy-rich carbohydrates, is played out in the chloroplast "theater" of the green plant. In each chloroplast there are stacks of **thylakoid** discs. The thylakoid membrane, inside each disc, is the lighted stage where the drama of Act I is performed. Here the energy of light is captured by the electrons of chlorophyll molecules. The excited electrons are passed along a series of acceptors in an electron transport chain. In the process, a molecule of water is split and oxygen is released into the atmosphere. At the same time, protons pumped out of the thylakoid membrane drive the production of ATP. Excited electrons reduce NADP$^+$ to NADPH and the stored energy is used in Act II for the biosynthesis of glucose, which takes place in the dark of the stroma outside the thylakoid membrane. Carbon dioxide from the atmosphere is combined with a five-carbon sugar to produce, through an intermediate, two three-carbon sugars, and eventually the energetic six-carbon molecule of **glucose**. The energy to drive this biosynthesis comes from ATP and the reducing power of NADPH, **N**icotinamide **A**denine **D**inucleotide **P**hosphate. Plants at the bottom of the food chain toil in the sun to store energy for the benefit of all us animals.*

## 17.1
## THE REACTIONS OF PHOTOSYNTHESIS

It is well known that the photosynthetic organisms, such as green plants, convert carbon dioxide ($CO_2$) and water to carbohydrates such as glucose (written here as $C_6H_{12}O_6$) and molecular oxygen ($O_2$).

$$6\ CO_2 + 6\ H_2O \longrightarrow C_6H_{12}O_6 + 6\ O_2$$

The equation actually represents two processes. One, the splitting of water to produce oxygen, is a reaction that requires light energy from the sun, and the other, the fixation of $CO_2$ to give sugars, is one that can and does take place in the dark but only because it uses solar energy indirectly. The actual storage form of the carbohydrates produced is not glucose, but oligosaccharides (*e.g.*, sucrose in sugar cane) and polysaccharides (starch and cellulose). However, it is customary and convenient to write the carbohydrate product as glucose, and we shall follow this time-honored practice.

In the light reaction, water is converted to oxygen by oxidation, and NADP$^+$ is reduced to NADPH. The light reaction is coupled to the phosphorylation of ADP to ATP in a process called **photophosphorylation.**

$$H_2O + NADP^+ \longrightarrow NADPH + H^+ + \frac{1}{2}O_2$$

$$ADP + P_i \longrightarrow ATP$$

The light reaction of photosynthesis in turn consists of two parts, accomplished by two distinct but related photosystems. One part of the reaction

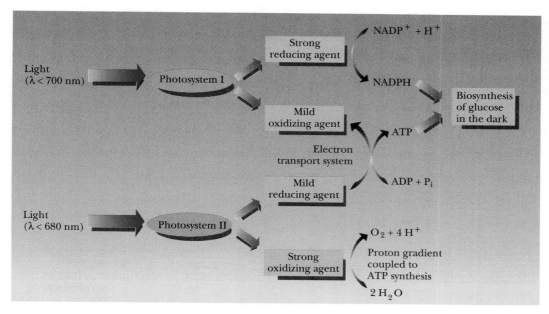

**FIGURE 17.1** Photosystems I and II. NADPH is generated by Photosystem I for subsequent use in the dark reaction of photosynthesis. ATP is generated by photophosphorylation linked to electron transport between Photosystems I and II. Photosystem II splits water, giving rise to a proton gradient coupled to ATP synthesis.

is the reduction of $NADP^+$ to NADPH, carried out by **Photosystem I.** The second part of the reaction is the splitting of water to produce oxygen, carried out by **Photosystem II.** Both photosystems carry out redox (electron transfer) reactions. Photosystem I generates a mild oxidizing agent, and Photosystem II generates a mild reducing agent (Figure 17.1). The mild oxidizing agent and mild reducing agent interact with each other indirectly through an electron transport chain that links the two photosystems. The production of ATP is linked to electron transport in a process similar to that seen in the production of ATP by mitochondrial electron transport.

In the dark reaction, the ATP and NADPH produced in the light reaction provide the energy and reducing power for the fixation of $CO_2$. The dark reaction is also a redox process, since the carbon in carbohydrates is in a more reduced state than the highly oxidized carbon in $CO_2$.

## 17.2
## CHLOROPLASTS AND CHLOROPHYLLS

In prokaryotes such as cyanobacteria, photosynthesis takes place in granules bound to the plasma membrane. The site of photosynthesis in eukaryotes such as green plants and green algae is the **chloroplast** (Figure 17.2), a membrane-bounded organelle that we discussed in Section 2.3.

**FIGURE 17.2**    Membrane structures in chloroplasts.

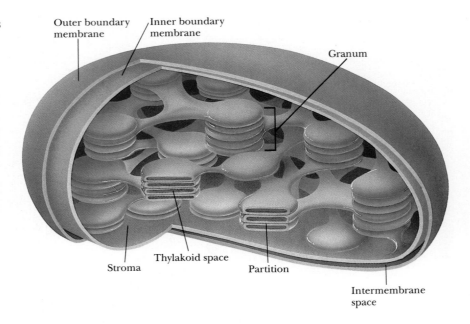

Like the mitochondrion, the chloroplast has an inner and outer membrane and an intermembrane space. Within the chloroplast there are bodies called **grana,** which consist of stacks of flattened membranes called **thylakoid disks.** The thylakoid disks are formed by the folding of the inner membrane of the chloroplast; in this respect the thylakoid disks resemble the cristae of mitochondria. The grana are connected by membranes called intergranal lamellae. The trapping of light and production of oxygen take place in the thylakoid disks. The dark reaction, in which $CO_2$ is fixed to carbohydrates, takes place in the soluble portion of the chloroplast, called the **stroma.** The stroma plays the same role in the structure of the chloroplast as does the matrix in the mitochondrion. In addition to the stroma, there is a **thylakoid space** within the thylakoid disks themselves.

It is well established that the primary event in photosynthesis is the absorption of light by **chlorophyll.** The higher energy states (excited states) of chlorophyll are useful in photosynthesis because the light energy can be passed along and converted to chemical energy in the light reaction. There are two principal types of chlorophyll, **chlorophyll *a*** and **chlorophyll *b*.** Eukaryotes such as green plants and green algae contain both chlorophyll *a* and chlorophyll *b*. Prokaryotes such as cyanobacteria (formerly called blue-green algae) contain only chlorophyll *a*. Photosynthetic bacteria other than cyanobacteria have bacteriochlorophylls, with **bacteriochlorophyll *a*** the most common. Organisms such as green and purple sulfur bacteria, which contain bacteriochlorophylls, do not use water as the ultimate source of electrons for the redox reactions of photosynthesis, nor

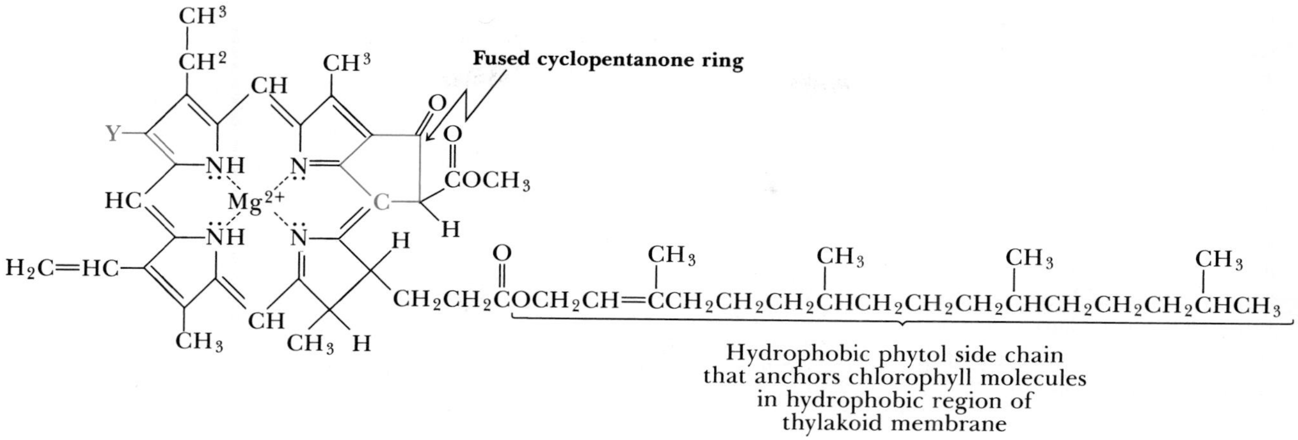

**Fused cyclopentanone ring**

Hydrophobic phytol side chain
that anchors chlorophyll molecules
in hydrophobic region of
thylakoid membrane

**Y is —CH₃ in chlorophyll *a***
**Y is —CHO in chlorophyll *b***
**Y is —CH₃ in bacteriochlorophyll *a***
**(and highlighted bond is saturated)**

**FIGURE 17.3** Molecular structures of chlorophyll *a,* chlorophyll *b,* and bacteriochlorophyll *a.*

do they produce oxygen. Instead, they use other electron sources, such as $H_2S$, which produces elemental sulfur instead of oxygen. Organisms that contain bacteriochlorophyll are anaerobic and have only one photosystem.

The structure of chlorophyll is based on the tetrapyrrole ring of porphyrins, which occurs in the heme group of myoglobin, hemoglobin, and the cytochromes (Figure 17.3). (See Section 9.4.) The metal ion bound to the tetrapyrrole ring is magnesium, Mg(II), rather than iron, which occurs in heme. Another difference between chlorophyll and heme is the presence of a cyclopentanone ring fused to the tetrapyrrole ring. There is a long hydrophobic side chain, the phytol group, which contains four isoprenoid units (five-carbon units that are basic building blocks in many lipids) (Section 16.8) and which binds to the thylakoid membrane by hydrophobic interactions. The phytol group is bound to the rest of the chlorophyll molecule by an ester linkage between the alcohol group of the phytol and a propionic acid side chain on the porphyrin ring. The difference between chlorophyll *a* and chlorophyll *b* lies in the substitution of an aldehyde group for a methyl group on the porphyrin ring. The difference between bacteriochlorophyll *a* and chlorophyll *a* is that a double bond in the porphyrin ring of chlorophyll *a* is saturated in bacteriochlorophyll *a*. The lack of a conjugated system (alternating double and single bonds) in the porphyrin ring of bacteriochlorophylls causes a significant difference in the absorption of light by bacteriochlorophyll *a* compared with chlorophyll *a* and *b*.

The absorption spectra of chlorophyll *a* and chlorophyll *b* differ slightly (Figure 17.4). Both absorb light in the red and blue portions of the visible spectrum (600–700 nm and 400–500 nm, respectively), and the presence of both types of chlorophyll guarantees that more wavelengths of the visible spectrum are absorbed than would be the case with either one individually. Recall that chlorophyll *a* is found in all photosynthetic

**FIGURE 17.4** The absorption of visible light by chlorophylls *a* and *b*. The areas marked I, II, and III are regions of the spectrum that give rise to chloroplast activity. There is greater activity in regions I and III, which are close to major absorption peaks. There are high levels of $O_2$ production when light from regions I and III is absorbed by chloroplasts. Lower (but measurable) activity is seen in region II, where some of the accessory pigments absorb.

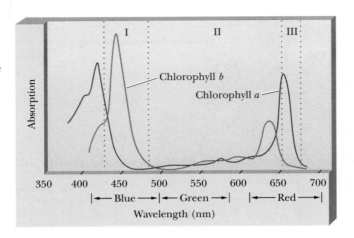

organisms that produce oxygen. Chlorophyll *b* is found in eukaryotes such as green plants and green algae, but it occurs in smaller amounts than chlorophyll *a*. The presence of chlorophyll *b*, however, increases the portion of the visible spectrum that is absorbed and thus enhances the efficiency of photosynthesis in green plants compared with cyanobacteria. Bacteriochlorophylls, the molecular form characteristic of photosynthetic organisms that do not produce oxygen, absorb light at longer wavelengths. The wavelength of maximum absorption of bacteriochlorophyll *a* is 780 nm, while others have absorption maxima at still longer wavelengths, such as 870 or 1050 nm. Light of wavelength longer than 800 nm is part of the infrared, rather than the visible, region of the spectrum. The wavelength of light absorbed plays a critical role in the light reaction of photosynthesis because the energy of light is inversely related to wavelength (see Box 17.1).

Chlorophyll molecules are arranged in **photosynthetic units.** Most of the chlorophyll molecules in the unit simply gather light (antennae chlorophylls). The light-harvesting molecules then pass along their excitation energy to a specialized pair of chlorophyll molecules at a **reaction center.** When the light energy reaches the reaction center, the chemical reactions of photosynthesis begin. The different environments of the antennae chlorophylls and the reaction-center chlorophylls give different properties to the two different kinds of molecules. In a typical photosynthetic unit, there are several hundred light-harvesting antennae chlorophylls for each unique chlorophyll at a reaction center.

The precise nature of reaction centers is the subject of active research. The most extensively studied system is that from bacteria of the genus *Rhodopseudomonas.* These bacteria do not produce molecular oxygen as a result of their photosynthetic activities, but enough similarities exist between the photosynthetic reactions of *Rhodopseudomonas* and photosynthesis linked to oxygen to draw conclusions about the nature of reaction centers in all organisms. The detailed process that goes on at the reaction center of *Rhodopseudomonas* is important enough to warrant further discussion.

## BOX 17.1
## THE RELATIONSHIP BETWEEN WAVELENGTH AND ENERGY OF LIGHT

The wavelength of light is related to its energy, a point of crucial importance for our purposes. It was established by Max Planck in the early 20th century that the energy of light is directly proportional to its frequency.

$$E = h\nu$$

where $E$ is energy, $h$ is a constant (Planck's constant), and $\nu$ is the frequency of the light. The wavelength of light is related to the frequency.

$$\nu = \frac{c}{\lambda}$$

where $\lambda$ is wavelength, $\nu$ is frequency, and $c$ is the velocity of light. We can rewrite the expression for the energy of light in terms of wavelength rather than frequency.

$$E = h\nu = h\frac{c}{\lambda}$$

Light of shorter wavelength (higher frequency) is higher in energy than light of longer wavelength (lower frequency).

It is well established that there is a pair of bacteriochlorophyll molecules in the reaction center of *Rhodopseudomonas viridis;* the critical pair of chlorophylls is embedded in a protein complex that is in turn an integral part of the photosynthetic membrane. (We shall refer to the bacteriochlorophylls simply as chlorophylls in the interest of simplifying the discussion.) **Accessory pigments,** which also play a role in the light-trapping process, have specific positions close to the special pair of chlorophylls. The absorption of light by the special pair of chlorophylls raises them to a higher energy level (Figure 17.5 (a)). An electron is passed from the chlorophylls to a series of accessory pigments (Figure 17.5 (b)). The first of these accessory pigments is pheophytin, which is structurally similar to chlorophyll, differing only by having two hydrogens in place of the magnesium. The electron is passed along to the pheophytin, raising it in turn to an excited energy level. The final electron acceptor, which is also raised to an excited state, is plastoquinone, a quinone similar in structure to coenzyme Q, which plays a role in the mitochondrial electron transport chain. The electron that the chlorophyll pair had lost is replaced by an electron donated by a cytochrome, which acquires a positive charge in the process (Figure 17.5 (c)). The cytochrome is not bound to the membrane and diffuses away, carrying its positive charge with it. The whole process takes place in less than $10^{-3}$ second. The positive and negative charges have traveled in opposite directions from the chlorophyll pair and are separated from each other. This situation is similar to the proton gradient in mitochondria, where the existence of the proton gradient is ultimately responsible for oxidative phosphorylation. The separation of charge is equivalent to a battery, a form of stored energy. The reaction center has acted as a transducer, converting light energy to a form usable by the cell to carry out the energy-requiring reactions of photosynthesis. The processes that take place in *Rhodopseudomonas* serve as a model for reaction centers in photosynthesis linked to oxygen.

**FIGURE 17.5**  Molecular events that take place at the photosynthetic reaction center of *Rhodopseudomonas*. (a) The special pair of chlorophylls is raised to a higher energy level by absorption of light. The chlorophylls in the reaction center are bacteriochlorophyll, rather than chlorophyll *a* or *b*. (b) An excited electron is passed to the accessory pigments, leaving the special pair of chlorophylls with a positive charge. A cytochrome not bound to the membrane diffuses into the region of the special pair of chlorophylls. (c) The cytochrome donates an electron to the special pair of chlorophylls, leaving them without charge. The cytochrome, which is now positively charged, diffuses away. The excited electron is passed to another accessory pigment. The resulting separation of charge represents stored energy.

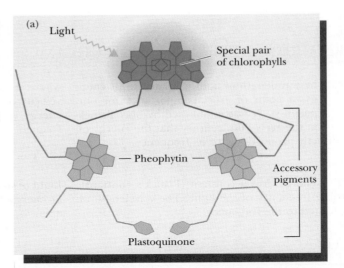
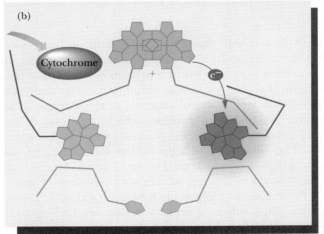
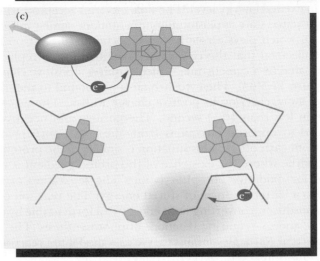

## 17.3
## THE REACTIONS OF PHOTOSYSTEMS I AND II

The two different photosystems carry out different reactions and depend on light of different wavelengths as the energy source for each reaction. Photosystem I can be excited by light of wavelengths shorter than 700 nm, but Photosystem II requires light of wavelengths shorter than 680 nm for excitation. Both photosystems must operate for the chloroplast to produce NADPH, ATP, and $O_2$, because the two photosystems are connected by an electron transport chain. The two systems are, however, structurally distinct in the chloroplast; Photosystem I can be released preferentially from the thylakoid membrane by treatment with detergents. The reaction centers of the two photosystems provide different environments for the unique chlorophylls involved. The unique chlorophyll of Photosystem I is referred to as $P_{700}$, P for pigment and 700 for the longest wavelength of absorbed light (700 nm) that initiates the reaction. Similarly, the reaction-center chlorophyll of Photosystem II is designated $P_{680}$ because the longest wavelength of absorbed light that initiates the reaction is 680 nm.

The absorption of light at the reaction center of Photosystems I and II produces a high-energy excited state in their chlorophylls. This event in turn triggers a series of redox reactions. Electron transfer reactions can be characterized in terms of their tendency to occur and their reaction energies by a standard reduction potential ($E°'$) (Figure 17.6). (See Section

**FIGURE 17.6** Electron flow in Photosystems I and II. The vertical axis shows standard reduction potentials. The energy needed to transfer electrons from $H_2O$ to $NADP^+$ is provided by the absorption of light by Photosystems I and II (vertical [up] arrows). After each absorption of light, the electrons can then flow "downhill" (diagonal [down] arrows). Photophosphorylation of ADP to yield ATP is coupled to the electron transport chain that links Photosystem I to Photosystem II. (Chl is chlorophyll; Phe is pheophytin; PQ is plastoquinone; PC is plastocyanin.) The electron carriers that mediate the transfer of electrons from $H_2O$ to Photosystem II include a manganese-containing protein and a protein with an essential tyrosine residue, referred to as component Z.

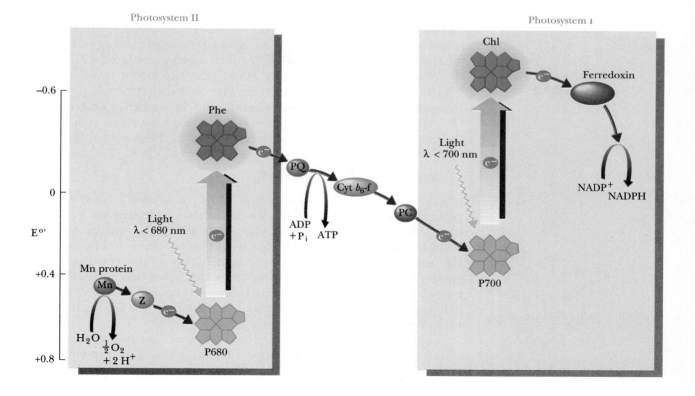

15.3.) Each substance can be assigned a standard potential referring to its tendency to gain electrons (to be reduced) compared with the tendency of other substances to gain electrons. We discussed this point in detail when we treated mitochondrial electron transport, which is an exergonic process. The net electron transport reaction of the two photosystems taken together is, except for the substitution of NADPH for NADH, the reverse of mitochondrial electron transport. The half reaction of reduction is that of $NADP^+$ to NADPH, while the half reaction of oxidation is that of water to oxygen.

$$NADP^+ + 2\,H^+ + 2\,e^- \longrightarrow NADPH + H^+ \qquad E^{\circ\prime} = -0.32\ \text{V}$$

$$H_2O \longrightarrow \frac{1}{2}\,O_2 + 2\,H^+ + 2\,e^- \qquad E^{\circ\prime} = -0.82\ \text{V}$$

$$NADP^+ + H_2O \longrightarrow NADPH + H^+ + \frac{1}{2}\,O_2 \qquad E^{\circ\prime} = -1.14\ \text{V}$$

A negative $E^{\circ\prime}$ indicates an endergonic reaction with a corresponding positive $\Delta G^{\circ\prime} = +220\ \text{kJ} = +52.6\ \text{kcal mol}^{-1}$. The light energy absorbed by the chlorophylls in both photosystems provides the energy that allows this endergonic reaction to take place.

The absorption of light by the chlorophyll of Photosystem I supplies the energy that ultimately allows the photoreduction of $NADP^+$ to take place. The absorption process can be written

$$Chl_I + h\nu\ (700\ \text{nm}) \longrightarrow Chl_I^*$$

where $Chl_I$ and $Chl_I^*$ are the $P_{700}$ chlorophyll in the ground (unexcited) and excited states, respectively, and $h\nu$ is the light absorbed. $Chl_I^*$ becomes a strong reducing agent; it can easily give electrons to another substance. After a series of electron transfers, the final substance reduced is $NADP^+$.

Since electrons have been transferred to $NADP^+$, producing NADPH, $Chl_I^+$ now lacks an electron. The needed electron is supplied by Photosystem II. The absorption of light by the chlorophyll of Photosystem II provides the energy that ultimately allows the photooxidation of water, the final electron source. The absorption process in Photosystem II can be written

$$Chl_{II} + h\nu\ (680\ \text{nm}) \longrightarrow Chl_{II}^*$$

where $Chl_{II}$ is the $P_{680}$ chlorophyll. $Chl_{II}^*$ becomes a strong enough reducing agent to pass electrons to $Chl_I^+$ through an electron transport chain coupled to the phosphorylation of ADP. The energy for the reactions that allow electrons to flow from Photosystem II to Photosystem I is supplied by the absorption of light by $Chl_{II}$, the reaction-center chlorophyll of Photosystem II. $Chl_{II}^+$, now lacking an electron, is a strong enough oxidizing agent to oxidize water to oxygen. The water, as it is oxidized to oxygen, is the ultimate source of the electrons flowing through the system.

There are two places in the reaction scheme of the two photosystems where the absorption of light supplies energy to make endergonic reactions take place (Figure 17.6). Neither reaction-center chlorophyll is a strong enough reducing agent to pass electrons to the next substance in the reaction sequence, but the absorption of light by the chlorophylls of both photosystems provides enough energy for such reactions to take place. When $Chl_I$ absorbs light, enough energy is provided to allow the ultimate reduction of $NADP^+$ to take place. (Note that the reduction potentials of the substances involved are shown on the energy axis of Figure 17.6. This type of diagram is also called a "Z scheme.") Similarly, the absorption of light by $Chl_{II}$ allows electrons to be passed to the electron transport chain that links Photosystem II and Photosystem I and generates a strong enough oxidizing agent to split water, producing oxygen. In both photosystems, the result of supplying energy is analogous to pumping water uphill.

## Photosystem I: Reduction of $NADP^+$

As we just saw, the absorption of light by $Chl_I$ leads to the series of electron transfer reactions of Photosystem I. The substance to which the excited-state chlorophyll, $Chl_I^*$, gives an electron is apparently a molecule of chlorophyll a; this transfer of electrons is mediated by processes that take place in the reaction center. The next electron acceptor in the series is bound ferredoxin, an iron-sulfur protein occurring in the membrane in Photosystem I. The bound ferredoxin passes its electron to a molecule of soluble ferredoxin. Soluble ferredoxin in turn reduces an FAD enzyme called ferredoxin-NADP reductase. The FAD portion of the enzyme reduces $NADP^+$ to NADPH (Figure 17.6). We can summarize the main features of the process in two equations, in which the notation ferredoxin refers to the soluble form of the protein.

$$Chl_I^* + ferredoxin_{oxidized} \longrightarrow Chl_I^+ + ferredoxin_{reduced}$$

$$\text{Ferredoxin-NADP reductase}$$

$$2 \, Ferredoxin_{red} + H^+ + NADP^+ \longrightarrow 2 \, ferredoxin_{ox} + NADPH$$

Two points can be noted about this series of equations. The first is a matter of bookkeeping. $Chl_I^*$ donates one electron to ferredoxin, but the electron transfer reactions of FAD and $NADP^+$ involve two electrons. Thus an electron from each of two ferredoxins is required for the production of NADPH. The second point is the fate of the oxidized chlorophyll ($Chl_I^+$), which has lost an electron. This pigment is colorless in its oxidized form, and it must be reduced back to the light-absorbing form to continue photosynthesis. The electrons needed to reduce the oxidized chlorophyll are supplied by Photosystem II.

## Photosystem II: Water Is Split to Produce Oxygen

Photosystem II is more complex than Photosystem I, and the components of Photosystem II are more difficult to isolate from the thylakoid membranes than are those of Photosystem I. The oxidation of water by Photosystem II to produce oxygen is the ultimate source of electrons in photosynthesis. These electrons are passed from Photosystem II to Photosystem I by the electron transport chain. The oxidized $Chl_I^+$ produced by Photosystem I is reduced by electrons donated by Photosystem II.

In Photosystem II, as in Photosystem I, the absorption of light by chlorophyll in the reaction center produces an excited state of chlorophyll. In this case the wavelength of light is 680 nm, rather than 700 nm; the reaction-center chlorophyll of Photosystem II is also referred to as $P_{680}$.

$$Chl_{II} + h\nu \ (680 \ nm) \longrightarrow Chl_{II}^*$$

As in Photosystem I, the excited chlorophyll is a reducing agent and passes an electron to a primary acceptor. In Photosystem II the primary electron acceptor is a molecule of **pheophytin** (Phe), one of the accessory pigments of the photosynthetic apparatus. The structure of pheophytin differs from that of chlorophyll *a* by the substitution of two hydrogens for the magnesium in the ring system. As was the case in Photosystem I, the transfer of electrons is mediated by events that take place at the reaction center. The next electron acceptor is **plastoquinone** (PQ). The structure of plastoquinone (Figure 17.7) is similar to that of coenzyme Q (ubiquinone), a part of the respiratory electron transport chain (Section 15.4).

The electron transport chain that links the two photosystems consists of pheophytin, plastoquinone, a complex of plant cytochromes, a copper-containing protein called **plastocyanin** (PC), and the oxidized form of $P_{700}$ ($Chl_I^+$) (see Figure 17.6). The complex of plant cytochromes is called the $b_6$-f complex; it consists of a b-type cytochrome (cytochrome $b_6$) and a c-type cytochrome (cytochrome f). This complex is similar in structure to the b-$c_1$ complex in mitochondria and occupies a similar central position in an electron transport chain. In plastocyanin the copper ion is the actual electron carrier; the copper ion exists as Cu(II) and Cu(I) in the oxidized and reduced forms, respectively. ATP is generated in a process coupled to this electron transport chain, as is the case in respiration.

When the oxidized chlorophyll of $P_{700}$ accepts electrons from the electron transport chain, it is reduced. The reaction-center chlorophyll of Photosystem II ($P_{680}$) is now in an oxidized state, but it gains electrons and is reduced as a result of the oxidation of water.

$$2 \ H_2O + 4 \ Chl_{II}^+ \longrightarrow O_2 + 4 \ H^+ + 4 \ Chl_{II}$$

There are intermediate steps in this reaction, since four electrons are required for the oxidation of water and $Chl_{II}^+$ ($P_{680}^+$) can accept only one electron at a time. A manganese-containing protein and several other protein components are required. The immediate electron donor, designated Z, to the $P_{680}$ chlorophyll is a tyrosine residue of one of the protein components that does not contain manganese. Several quinones serve as intermediate electron transfer agents to accommodate four electrons donated by one water molecule. Redox reactions of manganese also play a role here. (See the article by Govindjee and Coleman cited in the bibliography at the end of this chapter for a discussion of the workings of this complex.)

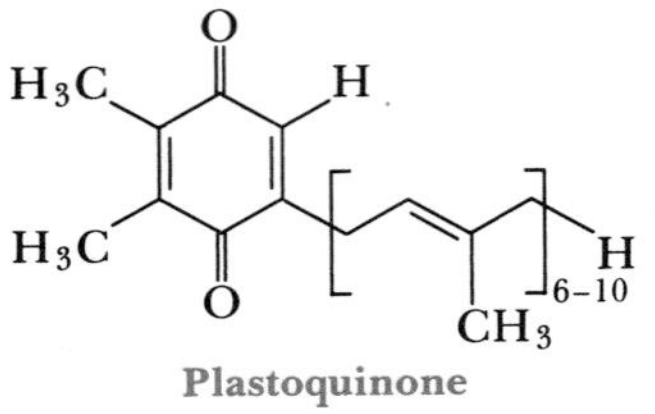

**Plastoquinone**

**FIGURE 17.7** The structure of plastoquinone.

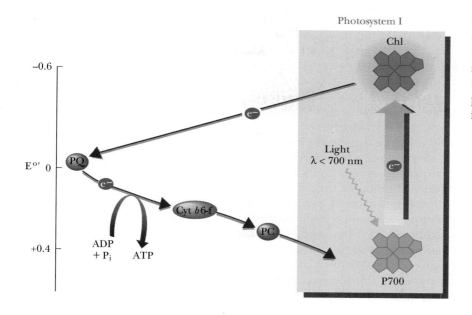

**FIGURE 17.8**   Cyclic electron flow coupled to photophosphorylation in Photosystem I. Note that water is not split and that no NADPH is produced. (Chl is chlorophyll; Phe is pheophytin; PQ is plastoquinone; PC is plastocyanin.)

The net reaction for the two photosystems together is the flow of electrons from $H_2O$ to $NADP^+$ (see Figure 17.6).

$$2\ H_2O + 2\ NADP^+ \longrightarrow O_2 + 2\ NADPH + 2\ H^+$$

## Cyclic Electron Transport in Photosystem I

In addition to the electron transfer reactions we have just described, it is possible for cyclic electron transport in Photosystem I to be coupled to the production of ATP (Figure 17.8). No NADPH is produced in this process. Photosystem II is not involved, and no $O_2$ is generated. Cyclic phosphorylation takes place when there is a high $NADPH/NADP^+$ ratio in the cell: there is not enough $NADP^+$ present in the cell to accept all the electrons generated by the excitation of $Chl_I$.

## 17.4
## A PROTON GRADIENT DRIVES THE PRODUCTION OF ATP IN PHOTOSYNTHESIS

In Chapter 15 we saw that a proton gradient across the inner mitochondrial membrane drives the phosphorylation of ADP in respiration. The mechanism of photophosphorylation is essentially the same as that of the production of ATP in the respiratory electron transport chain. In fact, some of the strongest evidence for the chemiosmotic coupling of phosphorylation to electron transport has been obtained from experiments on

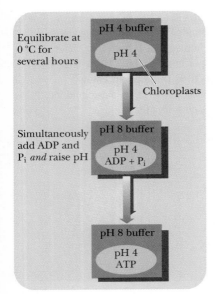

**FIGURE 17.9** ATP is synthesized by chloroplasts in the dark in the presence of a proton gradient, ADP, and $P_i$.

chloroplasts rather than mitochondria. Chloroplasts can synthesize ATP from ADP and $P_i$ *in the dark* if they are provided with a pH gradient.

If isolated chloroplasts are allowed to equilibrate in a pH 4 buffer for several hours, their internal pH will be equal to 4. If the pH of the buffer is raised rapidly to 8 and if ADP and $P_i$ are added simultaneously, ATP will be produced (Figure 17.9). The production of ATP does not require the presence of light; the proton gradient produced by the pH difference supplies the driving force for phosphorylation. This experiment provides solid evidence for the chemiosmotic coupling mechanism.

Several reactions contribute to the generation of a proton gradient in chloroplasts in an actively photosynthesizing cell. The splitting of water releases $H^+$ into the thylakoid space. Electron transport from Photosystem II and Photosystem I also helps create the proton gradient by involving plastoquinone and cytochromes in the process. Then Photosystem I reduces $NADP^+$ by using $H^+$ in the stroma to produce NADPH. As a result, the pH of the thylakoid space is lower than that of the stroma (Figure 17.10). We saw a similar situation in Chapter 15 when we discussed the pumping of protons from the mitochondrial matrix into the intermembrane space. The coupling factor in chloroplasts is similar to the mitochondrial coupling factor $F_1$. The chloroplast coupling factor is designated $CF_1$, where the C serves to distinguish it from its mitochondrial counterpart. Some evidence exists that the components of the electron chain in chloroplasts are arranged asymmetrically in the thylakoid membrane, as is the case in mitochondria. An important consequence of this asymmetric arrangement is the release of the ATP and NADPH produced by the light reaction into the stroma, where they provide energy and reducing power for the dark reaction of photosynthesis.

**FIGURE 17.10** The relationship between photophosphorylation and the proton gradient in chloroplasts. Photosynthetic electron transport pumps $H^+$ out of the stroma to the inthrathylakoid space to form the proton gradient (high pH in the stroma, low pH in the intrathylakoid space). The flow of $H^+$ back to the stroma through the ATP synthetase provides the energy for synthesis of ATP from ADP and $P_i$.

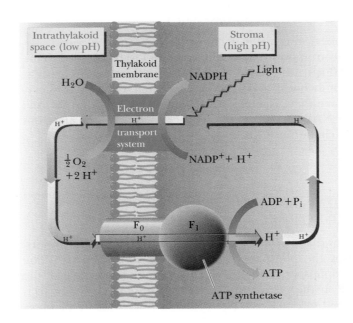

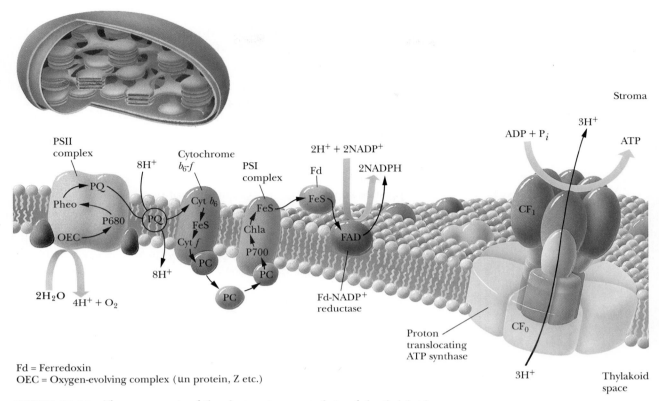

Stroma

PSII complex

Cytochrome $b_6$-$f$

PSI complex

$2H^+ + 2NADP^+$

2NADPH

$ADP + P_i$

$3H^+$

ATP

$8H^+$

Fd

PQ

Pheo

$CF_1$

P680

Cyt $b_6$

FeS

FeS

OEC

PQ

FeS

Chla

FAD

$8H^+$

Cyt $f$

PC

P700

PC

$2H_2O$

$4H^+ + O_2$

PC

Fd-NADP$^+$ reductase

Proton translocating ATP synthase

$CF_0$

$3H^+$

Thylakoid space

Fd = Ferredoxin
OEC = Oxygen-evolving complex (un protein, Z etc.)

**FIGURE 17.11**   The components of the electron transport chain of the thylakoid membrane. This schematic representation shows Photosystem II (PSII), the cytochrome $b_6$-$f$ complex, and Photosystem I (PSI), along with the soluble electron carriers plastoquinone (PQ) and plastocyanin (PC). The action of the electron transport chain sets up a proton gradient across the thylakoid membrane, coupled to synthesis of ATP by the $CF_0$-$CF_1$ ATP synthase. (After D. R. Ort and N. E. Good, 1988. *Trends Biochem. Sci.* **13,** 469.)

In mitochondrial electron transport there are three respiratory complexes connected by soluble electron carriers. The electron transport apparatus of the thylakoid membrane is similar in that it consists of several large membrane-bound complexes. They are PSII (the Photosystem II complex), the cytochrome $b_6$-f complex, and PSI (the Photosystem I complex). As in mitochondrial electron transport, several soluble electron carriers form the connection between the protein complexes. In the thylakoid membrane the soluble carriers are plastoquinone and plastocyanin, which have a role similar to that of coenzyme Q and cytochrome c in mitochondria (Figure 17.11). The proton gradient created by electron transport drives the synthesis of ATP in chloroplasts, as in mitochondria.

## 17.5
## A COMPARISON OF PHOTOSYNTHESIS WITH AND WITHOUT OXYGEN: EVOLUTIONARY IMPLICATIONS

Photosynthetic prokaryotes other than cyanobacteria have only one photo-system and do not produce oxygen. The chlorophyll in these organisms is different from that found in photosystems linked to oxygen (Figure 17.12). Anaerobic photosynthesis is not as efficient as photosynthesis linked to oxygen, but the anaerobic version of the process appears to be an evolutionary way-station. Anaerobic photosynthesis is a means for organisms to use solar energy to satisfy their needs for food and energy, even though its efficiency is less than that of aerobic photosynthesis.

A possible scenario for the development of photosynthesis starts with heterotrophic (*heterotrophs:* organisms that depend on their environment for organic nutrients and for energy) bacteria that contain some form of chlorophyll, probably bacteriochlorophyll. In such organisms the light energy absorbed by chlorophyll can be trapped in the form of ATP and NADPH. The important point about such a series of reactions is that photophosphorylation takes place, assuring an independent supply of ATP for the organism. In addition, the supply of NADPH facilitates synthesis of biomolecules from simple sources such as $CO_2$. Under conditions of limited food supply, organisms that can synthesize their own nutrients have a selective advantage. Organisms of this sort are *autotrophs* (not dependent on an external source of biomolecules) but are also anaerobes. The ultimate electron source that they use is not water, but some more easily oxidized substance, such as $H_2S$, as is the case with present-day green sulfur bacteria (and purple sulfur bacteria), or various organic compounds, as is

**FIGURE 17.12** The two possible electron transfer pathways in a photosynthetic anaerobe. Both cyclic and noncyclic forms of photophosphorylation are shown. HX is any compound (such as $H_2S$) that can be a hydrogen donor. (From L. Margulis, 1985. *Early Life*. Boston: Science Books International, p. 45.)

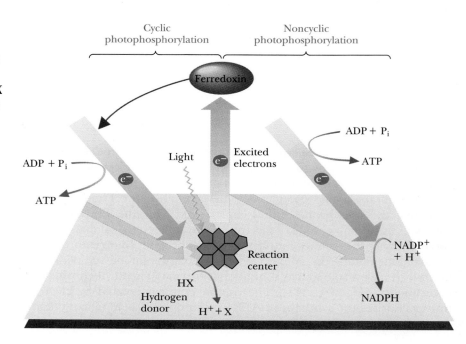

the case with present-day purple nonsulfur bacteria. These organisms do not possess an oxidizing agent powerful enough to split water, which is a far more abundant electron source than $H_2S$ or organic compounds. The ability to use water as an electron source confers a further evolutionary advantage.

Cyanobacteria were apparently the first organisms that developed the ability to use water as the ultimate reducing agent in photosynthesis. As we have seen, this feat required the development of a second photosystem as well as a new variety of chlorophyll, chlorophyll *a* rather than bacteriochlorophyll in this case. Chlorophyll *b* had not yet appeared on the scene, since it occurs only in eukaryotes, but with cyanobacteria the basic system of aerobic photosynthesis is in place. As a result of aerobic photosynthesis by cyanobacteria, the earth acquired its present atmosphere with its high levels of oxygen. The existence of all other aerobic organisms depends ultimately on the activities of cyanobacteria.

## 17.6
## THE DARK REACTION OF PHOTOSYNTHESIS: PATH OF CARBON

Carbon dioxide fixation takes place in the stroma. The equation for the overall reaction, like all equations for photosynthetic processes, is deceptively simple.

$$6\ CO_2 + 12\ NADPH \xrightarrow[\text{Enzymes}]{\text{ATP}} C_6H_{12}O_6 + 12\ NADP^+$$

The actual reaction pathway has some features in common with glycolysis and some in common with the pentose phosphate pathway.

The **net reaction** of 6 molecules of carbon dioxide to produce 1 molecule of glucose requires the carboxylation of 6 molecules of a five-carbon key intermediate, **ribulose 1,5-*bis*phosphate,** to form 6 molecules of an unstable six-carbon intermediate, which then splits to give 12 molecules of **3-phosphoglycerate.** Of these, two molecules of 3-phosphoglycerate react in turn, ultimately producing glucose. The remaining ten molecules of 3-phosphoglycerate are used to regenerate the six molecules of ribulose 1,5-*bis*phosphate. The overall reaction pathway is cyclic and is called the **Calvin cycle** (Figure 17.13) after the scientist who first investigated it, Melvin Calvin, winner of the 1961 Nobel Prize in chemistry.

The first reaction of the Calvin cycle is the condensation of ribulose 1,5-*bis*phosphate with carbon dioxide to form a six-carbon intermediate, 2-carboxy-3-ketoribitol 1,5-*bis*phosphate, which quickly hydrolyzes to give two molecules of 3-phosphoglycerate (Figure 17.14). The reaction is catalyzed by the enzyme **ribulose 1,5-*bis*phosphate carboxylase** (also called ribulose 1,5-*bis*phosphate carboxylase: oxygenase). This enzyme is located on the stromal side of the thylakoid membrane and is probably one of the most abundant proteins in nature, since it accounts for about 15% of the total protein in chloroplasts. The molecular weight of ribulose

**FIGURE 17.13**  The main features of the Calvin cycle. Glucose is produced, and ribulose 1,5-*bis*phosphate is regenerated.

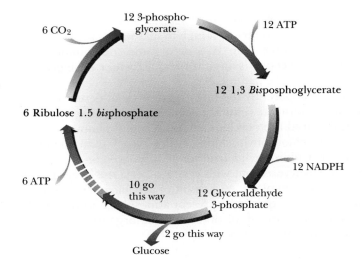

**FIGURE 17.14**  The reaction of ribulose 1,5-*bis*phosphate with $CO_2$ ultimately produces two molecules of 3-phosphoglycerate.

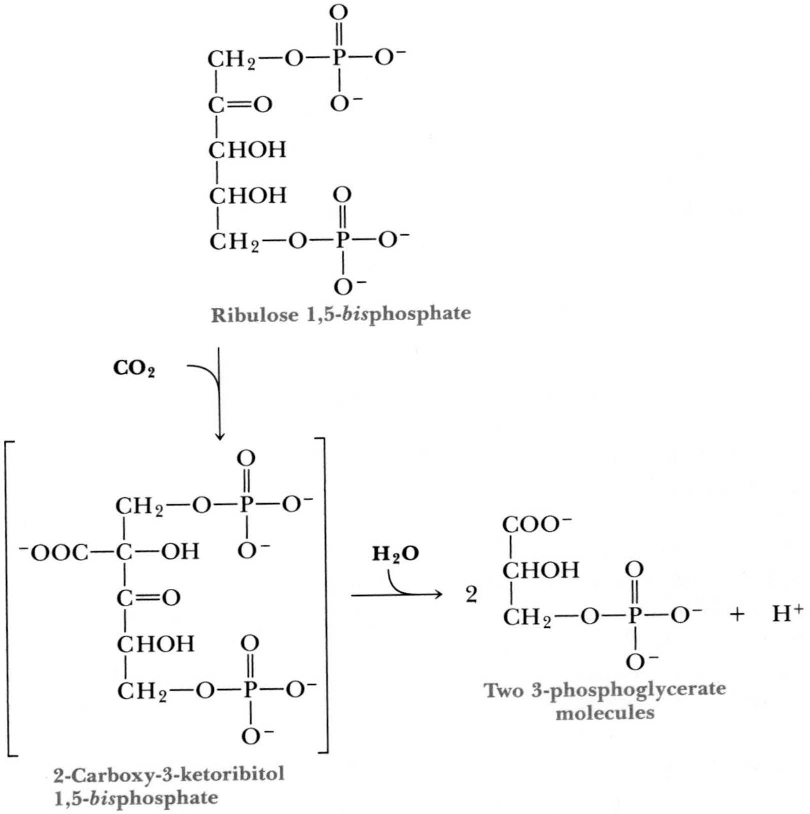

Ribulose 1,5-*bis*phosphate

$CO_2$

$H_2O$

2-Carboxy-3-ketoribitol 1,5-*bis*phosphate

Two 3-phosphoglycerate molecules

1,5-*bis*phosphate carboxylase is about 560,000, and it consists of eight large subunits (molecular weight, 55,000) and eight small subunits (molecular weight, 15,000) (Figure 17.15). The sequence of the large subunit is encoded by a chloroplast gene, while that of the small subunit is encoded by a nuclear gene. The endosymbiotic theory for the development of eukaryotes (Section 2.6) favors the idea of independent genetic material in organelles. The large subunit (chloroplast gene) is catalytic, while the small subunit (nuclear gene) plays a regulatory role, an observation that is consistent with an endosymbiotic origin for organelles such as chloroplasts.

The incorporation of $CO_2$ into 3-phosphoglycerate represents the actual fixation process; the remaining reactions are those of carbohydrates. The next two reactions lead to the reduction of 3-phosphoglycerate to form glyceraldehyde 3-phosphate. The reduction takes place in the same fashion as in gluconeogenesis, except for one unique feature (Figure 17.16 (a)): the reactions in chloroplasts require NADPH rather than NADH for the reduction of 1,3-*bis*phosphoglycerate to glyceraldehyde 3-phosphate. When glyceraldehyde 3-phosphate is formed, it can have two alternative fates: one is the production of six-carbon sugars, and the other is the regeneration of ribulose 1,5-*bis*phosphate.

**FIGURE 17.15** The subunit structure of ribulose 1,5-*bis*phosphate carboxylase.

## Production of Six-Carbon Sugars

The formation of glucose from glyceraldehyde 3-phosphate takes place in the same manner as in gluconeogenesis (Figure 17.16 (b)). The conversion of glyceraldehyde 3-phosphate to dihydroxyacetone phosphate takes place easily (Section 13.2). Dihydroxyacetone phosphate in turn reacts with glyceraldehyde 3-phosphate, in a series of reactions we have already seen, to give rise to fructose 6-phosphate and ultimately to glucose. Since we have already seen these reactions, we shall not discuss them again.

## Regeneration of Ribulose 1,5-*Bis*phosphate

This process is readily divided into four steps: *preparation, reshuffling, isomerization,* and *phosphorylation.* The preparation begins with conversion of some of the glyceraldehyde 3-phosphate to dihydroxyacetone phosphate (catalyzed by triosephosphate isomerase). (This reaction also functions in the production of six-carbon sugars.) A portion of both the glyceraldehyde 3-phosphate and the dihydroxyacetone phosphate is then condensed to form fructose 1,6-*bis*phosphate (catalyzed by aldolase). Fructose 1,6-*bis*phosphate is hydrolyzed to fructose 6-phosphate (catalyzed by diphosphatase). With a supply of glyceraldehyde 3-phosphate, dihydroxyacetone phosphate, and fructose 6-phosphate now available, the reshuffling can begin.

It is most enlightening to keep track of the number of carbon atoms involved in the reshuffling process. The actual reshuffling begins with the reaction of glyceraldehyde 3-phosphate (three carbons, or $C_3$) and fructose 6-phosphate ($C_6$) to give erythrose 4-phosphate ($C_4$) and xylulose 5-

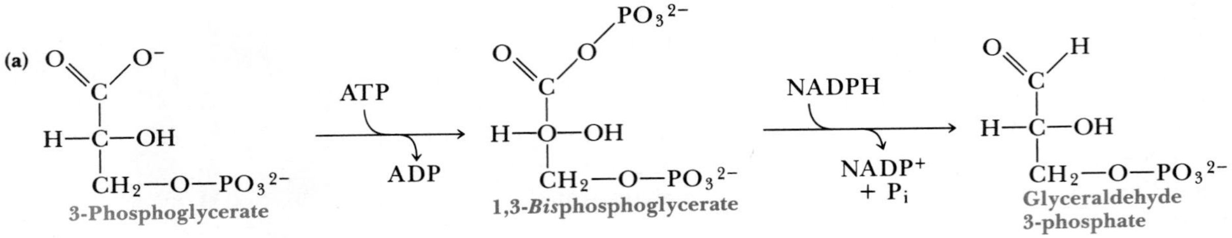

**(a)**

$$3\text{-Phosphoglycerate} \xrightarrow[\text{ADP}]{\text{ATP}} 1,3\text{-}Bis\text{phosphoglycerate} \xrightarrow[\substack{\text{NADP}^+ \\ + \text{ P}_i}]{\text{NADPH}} \text{Glyceraldehyde 3-phosphate}$$

**FIGURE 17.16** (a) Reduction of 3-phosphoglycerate to glyceraldehyde 3-phosphate. (b) The production of glucose from 3-phosphoglycerate in the Calvin cycle. Note the use of NADPH and ATP generated in the light reaction to provide energy for the dark reaction.

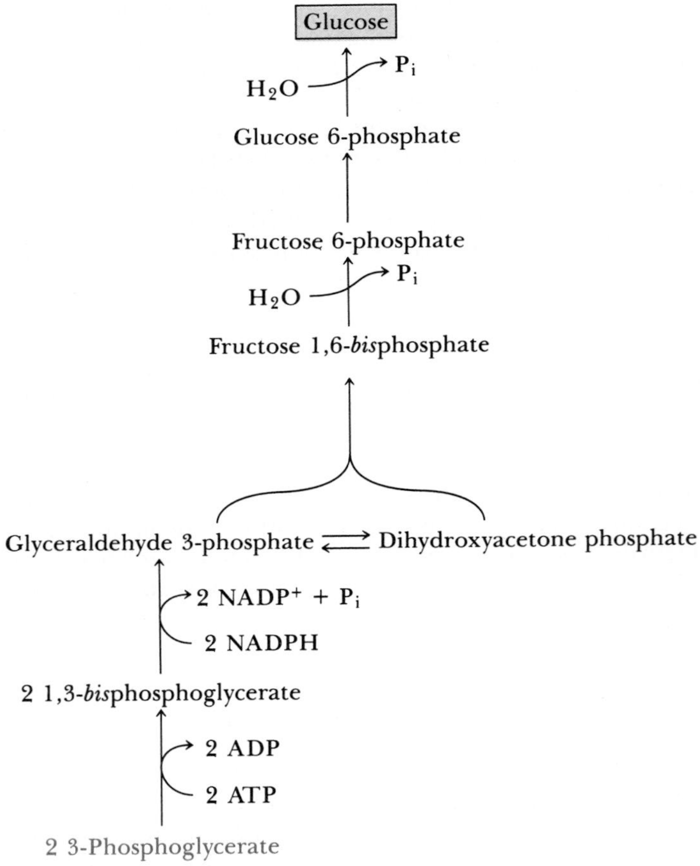

Glucose

↑ $H_2O \longrightarrow P_i$

Glucose 6-phosphate

↑

Fructose 6-phosphate

↑ $H_2O \longrightarrow P_i$

Fructose 1,6-*bis*phosphate

↑

Glyceraldehyde 3-phosphate ⇌ Dihydroxyacetone phosphate

↑ $2\ NADP^+ + P_i$ / $2\ NADPH$

2 1,3-*bis*phosphoglycerate

↑ $2\ ADP$ / $2\ ATP$

2 3-Phosphoglycerate

phosphate ($C_5$) (Figure 17.17). Note that there are nine carbons both in the reactants and in the products, but they are organized differently.

$$C_3 + C_6 \longrightarrow C_4 + C_5$$

The reaction is catalyzed by **transketolase,** an enzyme that also catalyzes some of the reactions of the pentose phosphate pathway. Erythrose 4-phosphate ($C_4$) reacts in turn with dihydroxyacetone phosphate ($C_3$) to give sedoheptulose 1,7-*bis*phosphate ($C_7$) (Figure 17.18).

$$C_4 + C_3 \longrightarrow C_7$$

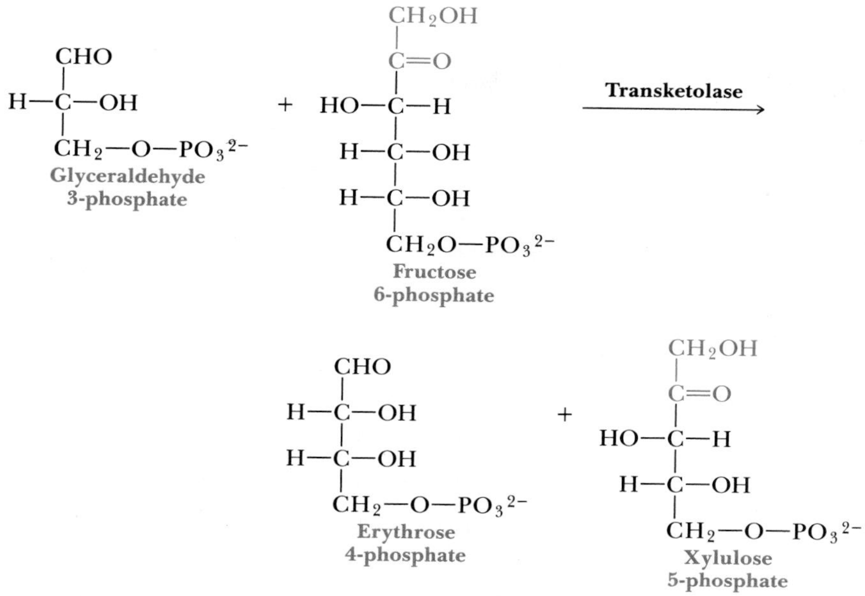

$$
\begin{array}{c}
\text{CHO} \\
| \\
\text{H}-\text{C}-\text{OH} \\
| \\
\text{CH}_2-\text{O}-\text{PO}_3{}^{2-} \\
\textbf{Glyceraldehyde} \\
\textbf{3-phosphate}
\end{array}
\quad + \quad
\begin{array}{c}
\text{CH}_2\text{OH} \\
| \\
\text{C}=\text{O} \\
| \\
\text{HO}-\text{C}-\text{H} \\
| \\
\text{H}-\text{C}-\text{OH} \\
| \\
\text{H}-\text{C}-\text{OH} \\
| \\
\text{CH}_2\text{O}-\text{PO}_3{}^{2-} \\
\textbf{Fructose} \\
\textbf{6-phosphate}
\end{array}
\quad \xrightarrow{\textbf{Transketolase}}
$$

$$
\begin{array}{c}
\text{CHO} \\
| \\
\text{H}-\text{C}-\text{OH} \\
| \\
\text{H}-\text{C}-\text{OH} \\
| \\
\text{CH}_2-\text{O}-\text{PO}_3{}^{2-} \\
\textbf{Erythrose} \\
\textbf{4-phosphate}
\end{array}
\quad + \quad
\begin{array}{c}
\text{CH}_2\text{OH} \\
| \\
\text{C}=\text{O} \\
| \\
\text{HO}-\text{C}-\text{H} \\
| \\
\text{H}-\text{C}-\text{OH} \\
| \\
\text{CH}_2-\text{O}-\text{PO}_3{}^{2-} \\
\textbf{Xylulose} \\
\textbf{5-phosphate}
\end{array}
$$

**FIGURE 17.17**    The reaction of glyceraldehyde 3-phosphate with fructose 6-phosphate to give erythrose 4-phosphate and xylulose 5-phosphate. Note that a two-carbon unit (highlighted) is transferred in this reaction.

**FIGURE 17.18**    The reactions that produce sedoheptulose 7-phosphate. A three-carbon unit is transferred in the aldolase reaction.

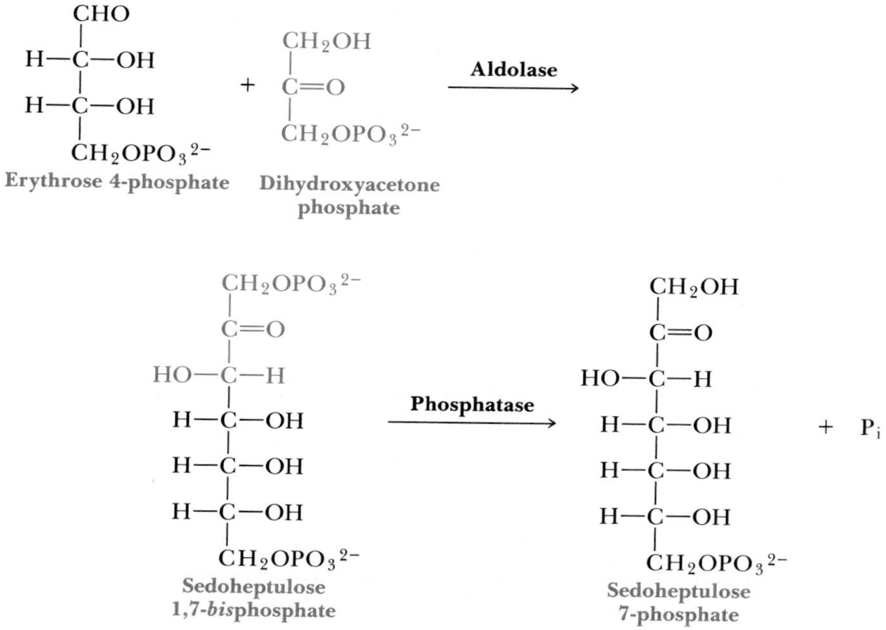

$$
\begin{array}{c}
\text{CHO} \\
| \\
\text{H}-\text{C}-\text{OH} \\
| \\
\text{H}-\text{C}-\text{OH} \\
| \\
\text{CH}_2\text{OPO}_3{}^{2-} \\
\textbf{Erythrose 4-phosphate}
\end{array}
\quad + \quad
\begin{array}{c}
\text{CH}_2\text{OH} \\
| \\
\text{C}=\text{O} \\
| \\
\text{CH}_2\text{OPO}_3{}^{2-} \\
\textbf{Dihydroxyacetone} \\
\textbf{phosphate}
\end{array}
\quad \xrightarrow{\textbf{Aldolase}}
$$

$$
\begin{array}{c}
\text{CH}_2\text{OPO}_3{}^{2-} \\
| \\
\text{C}=\text{O} \\
| \\
\text{HO}-\text{C}-\text{H} \\
| \\
\text{H}-\text{C}-\text{OH} \\
| \\
\text{H}-\text{C}-\text{OH} \\
| \\
\text{H}-\text{C}-\text{OH} \\
| \\
\text{CH}_2\text{OPO}_3{}^{2-} \\
\textbf{Sedoheptulose} \\
\textbf{1,7-\textit{bis}phosphate}
\end{array}
\quad \xrightarrow{\textbf{Phosphatase}}
\begin{array}{c}
\text{CH}_2\text{OH} \\
| \\
\text{C}=\text{O} \\
| \\
\text{HO}-\text{C}-\text{H} \\
| \\
\text{H}-\text{C}-\text{OH} \\
| \\
\text{H}-\text{C}-\text{OH} \\
| \\
\text{H}-\text{C}-\text{OH} \\
| \\
\text{CH}_2\text{OPO}_3{}^{2-} \\
\textbf{Sedoheptulose} \\
\textbf{7-phosphate}
\end{array}
\quad + \quad \text{P}_i
$$

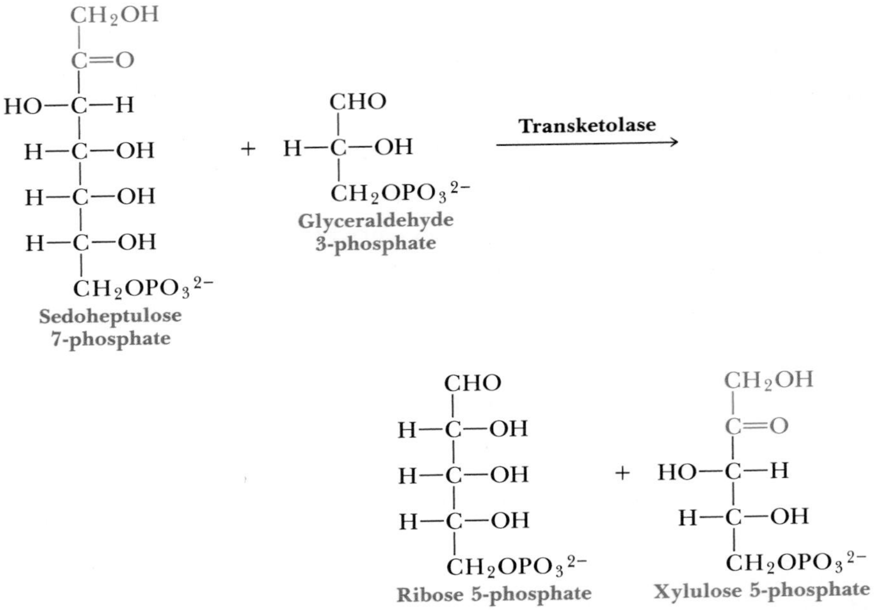

**FIGURE 17.19**   The reaction of sedoheptulose 7-phosphate and glyceraldehyde 3-phosphate to give ribose 5-phosphate and xylulose 5-phosphate. In this reaction, a two-carbon unit is transferred, as occurred in the reaction of Figure 17.16. Both reactions are catalyzed by the same enzyme.

This reaction is catalyzed by **aldolase.** In a phosphatase-catalyzed reaction, sedoheptulose 1,7-*bis*phosphate is hydrolyzed, producing sedoheptulose 7-phosphate and phosphate ion. In the final reaction involving reshuffling of the carbon skeleton of sugar phosphates, sedoheptulose 7-phosphate ($C_7$) reacts with another molecule of glyceraldehyde 3-phosphate ($C_3$) to give ribose 5-phosphate ($C_5$) and xylulose 5-phosphate ($C_5$) (Figure 17.19).

$$C_7 + C_3 \longrightarrow C_5 + C_5$$

This reaction is also catalyzed by transketolase. The reactions of rearrangement of carbon skeletons in the reshuffling phase of the Calvin cycle can be summarized as

$$C_6 + C_3 \longrightarrow C_4 + C_5$$
$$C_4 + C_3 \longrightarrow C_7$$
$$\underline{C_7 + C_3 \longrightarrow C_5 + C_5}$$
Net: $C_6 + 3\ C_3 \longrightarrow 3\ C_5$

The $C_6$ is fructose 6-phosphate, and the three $C_3$ are two molecules of glyceraldehyde 3-phosphate and one molecule of dihydroxyacetone phosphate. The three $C_5$ are two molecules of xylulose 5-phosphate and one molecule of ribose 5-phosphate.

The isomerization step involves the conversion of both ribose 5-phosphate and xylulose 5-phosphate to ribulose 5-phosphate. **Ribose 5-phosphate isomerase** catalyzes the conversion of ribose 5-phosphate to ribulose 5-phosphate, and **xylulose 5-phosphate epimerase** catalyzes the conversion of xylulose 5-phosphate to ribulose 5-phosphate (Figure 17.20

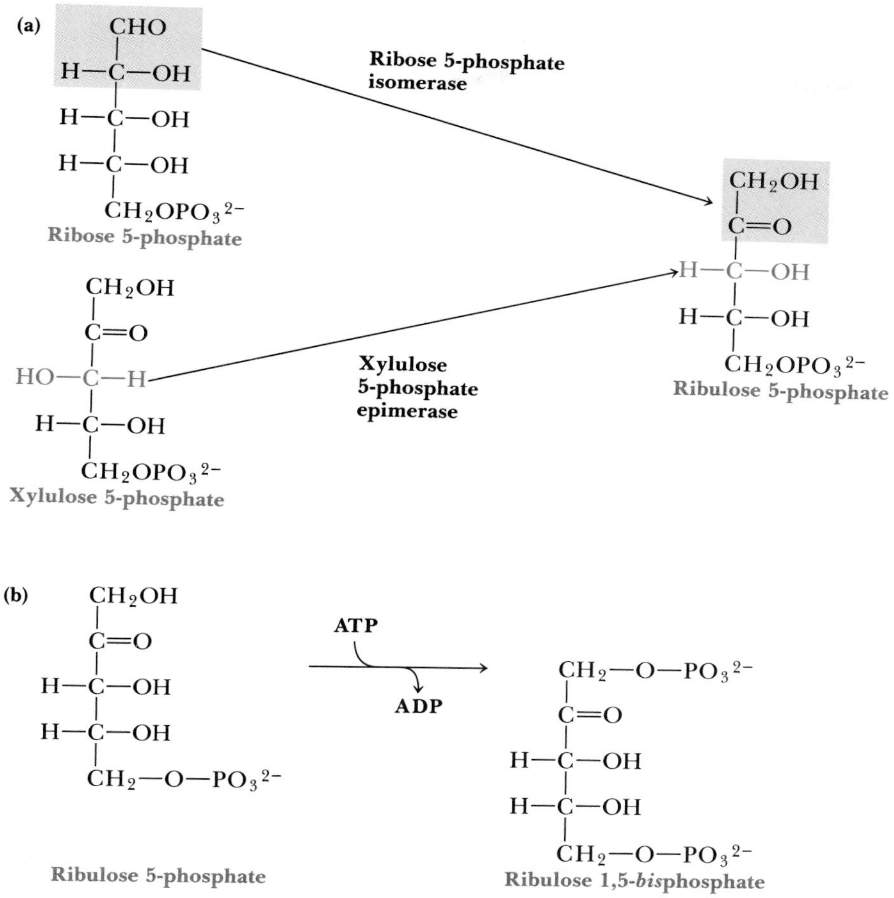

**(a)**

CHO
|
H—C—OH
|
H—C—OH
|
H—C—OH
|
$CH_2OPO_3{}^{2-}$
Ribose 5-phosphate

**Ribose 5-phosphate isomerase**

CH₂OH
|
C=O
|
HO—C—H
|
H—C—OH
|
$CH_2OPO_3{}^{2-}$
Xylulose 5-phosphate

**Xylulose 5-phosphate epimerase**

$CH_2OH$
|
C=O
|
H—C—OH
|
H—C—OH
|
$CH_2OPO_3{}^{2-}$
Ribulose 5-phosphate

**(b)**

CH₂OH
|
C=O
|
H—C—OH
|
H—C—OH
|
$CH_2—O—PO_3{}^{2-}$
Ribulose 5-phosphate

ATP → ADP

$CH_2—O—PO_3{}^{2-}$
|
C=O
|
H—C—OH
|
H—C—OH
|
$CH_2—O—PO_3{}^{2-}$
Ribulose 1,5-*bis*phosphate

FIGURE 17.20  (a) the production of ribulose 5-phosphate from ribose 5-phosphate and from xylulose 5-phosphate. (b) The reaction catalyzed by phosphoribulokinase.

(a)). The reverse of both these reactions takes place in the pentose phosphate pathway, catalyzed by the same enzymes.

In the final step, ribulose 1,5-*bis*phosphate is regenerated by the phosphorylation of ribulose 5-phosphate (Figure 17.20 (b)). This reaction requires ATP and is catalyzed by the enzyme **phosphoribulokinase.** The reactions leading to the regeneration of ribulose 1,5-*bis*phosphate are summarized in Figure 17.21, in which a net equation is obtained by adding all the reactions. Now we are in a position to examine the stoichiometry of the dark reaction of photosynthesis.

## Stoichiometry of the Calvin Cycle

It is convenient to refer to Figures 17.13 and 17.22 during our discussion. We shall follow what happens to six molecules of $CO_2$ in the course of one turn of the Calvin cycle.

For each $CO_2$ that reacts with one molecule of ribulose 1,5-*bis*phosphate, two molecules of 3-phosphoglycerate are produced. Conversion of

1. $2$ Glyceraldehyde $3\text{-}PO_3{}^{2-} \longrightarrow 2$ Dihydroxyacetone $PO_3{}^{2-}$
   $2 \times 3$ carbons                                 $2 \times 3$ carbons

2. Glyceraldehyde $3\text{-}PO_3{}^{2-}$ + Dihydroxyacetone $PO_3{}^{2-} \longrightarrow$
   3 carbons                              3 carbons
   Fructose $1,6\text{-}bis\text{-}PO_3{}^{4-}$
   6 carbons

3. Fructose $1,6\text{-}bis\text{-}PO_3{}^{4-} \longrightarrow$ Fructose $6\text{-}PO_3{}^{2-}$ + $P_i$
   6 carbons                                6 carbons

4. Fructose $6\text{-}PO_3{}^{2-}$ + Glyceraldehyde $3\text{-}PO_3{}^{2-} \longrightarrow$
   6 carbons                          3 carbons
   Erythrose $4\text{-}PO_3{}^{2-}$ + Xylulose $5\text{-}PO_3{}^{2-}$
   4 carbons                5 carbons

5. Erythrose $4\text{-}PO_3{}^{2-}$ + Dihydroxyacetone $PO_3{}^{2-} \longrightarrow$
   4 carbons                          3 carbons
   Sedoheptulose $1,7\text{-}bis\text{-}PO_3{}^{4-}$
   7 carbons

6. Sedoheptulose $1,7\text{-}bis\text{-}PO_3{}^{4-} \longrightarrow$
   7 carbons
   Sedoheptulose $7\text{-}PO_3{}^{2-}$ + $P_i$
   7 carbons

7. Sedoheptulose $7\text{-}PO_3{}^{2-}$ + Glyceraldehyde $3\text{-}PO_3{}^{2-} \longrightarrow$
   7 carbons                              3 carbons
   Ribose $5\text{-}PO_3{}^{2-}$ + Xylulose $5\text{-}PO_3{}^{2-}$
   5 carbons                5 carbons

8. Ribose $5\text{-}PO_3{}^{2-} \longrightarrow$ Ribulose $5\text{-}PO_3{}^{2-}$
   5 carbons                        5 carbons

9. $2$ Xylulose $5\text{-}PO_3{}^{2-} \longrightarrow 2$ Ribulose $5\text{-}PO_3{}^{2-}$
   $2 \times 5$ carbons                       $2 \times 5$ carbons

10. $3$ Ribulose $5\text{-}PO_3{}^{2-}$ + $3$ ATP $\longrightarrow$
    $3 \times 5$ carbons
    $3$ Ribulose $1,5\text{-}bis\text{-}PO_3{}^{4-}$ + $3$ ADP
    $3 \times 5$ carbons

---

NET: $5$ Glyceraldehyde $3\text{-}PO_3{}^{2-}$ + $3$ ATP $\longrightarrow$
     $5 \times 3$ carbons
     $3$ Ribulose $1,5\text{-}bis\text{-}PO_3{}^{4-}$ + $3$ ADP + $2$ $P_i$
     $3 \times 5$ carbons

FIGURE 17.21    The reactions that regenerate ribulose 1,5-*bis*phosphate in the Calvin cycle. In one turn of the cycle, the net process occurs twice.

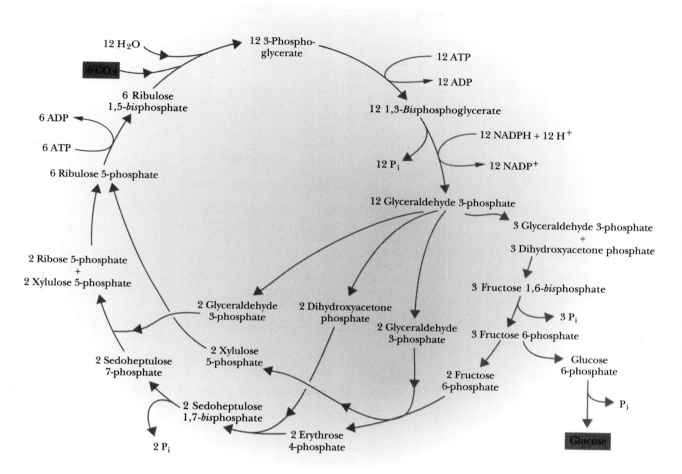

**FIGURE 17.22** The complete Calvin cycle showing the regeneration of ribulose 1,5-*bis*phosphate. Note that when glyceraldehyde 3-phosphate is formed, it (or dihydroxyacetone phosphate, to which it is easily converted) can have all four possible fates. (See Figure 17.21 for the balanced equation.)

each molecule of 3-phosphoglycerate to glyceraldehyde 3-phosphate requires one ADP and one NADPH. For six molecules of $CO_2$ we write the equation

$$6\ CO_2 + 6\ \text{ribulose 1,5-}bis\text{phosphate} + 12\ ATP\ +$$
(30 carbons)

$$12\ NADPH + 12\ H^+ + 12\ H_2O \longrightarrow$$

$$12\ \text{glyceraldehyde 3-phosphate} + 12\ ADP + 12\ P_i + 12\ NADP^+$$
(36 carbons)

Ten of the 12 glyceraldehyde 3-phosphate ($C_3$) molecules are converted to fructose 6-phosphate ($C_6$) and dihydroxyacetone phosphate ($C_3$) in the regeneration of ribulose 1,5-*bis*phosphate (Figure 17.22). The regeneration of six molecules of ribulose 1,5-*bis*phosphate accounts for 30 of the 36

**TABLE 17.1    Summary of the Path of Carbon in Photosynthesis**

**Equation 1**

$6 CO_2 + 6$ ribulose 1,5-*bis*phosphate $+ 12$ ATP $+$
(30 carbons)

$12$ NADPH $+ 12$ H$^+$ $+ 12$ H$_2$O $\longrightarrow$

$12$ glyceraldehyde 3-phosphate $+ 12$ ADP $+ 12$ P$_i$ $+ 12$ NADP$^+$
(36 carbons)

**Equation 2**

$10$ glyceraldehyde 3-phosphate $+ 6$ ATP $\longrightarrow$
(30 carbons)

$6$ ribulose 1,5-*bis*phosphate $+ 6$ ADP $+ 4$ P$_i$
(30 carbons)

**Equation 3**

$2$ glyceraldehyde 3-phosphate $\longrightarrow$ glucose $+ 2$ P$_i$
(6 carbons)                           (6 carbons)

**Overall Equation**

$6 CO_2 + 6$ ribulose 1,5-*bis*phosphate $+ 18$ ATP $+$
(30 carbons)

$12$ NADPH $+ 12$ H$^+$ $+ 12$ H$_2$O $\longrightarrow$

$6$ ribulose 1,5-*bis*phosphate $+$ glucose $+$
(30 carbons)            (6 carbons)

$12$ NADP$^+$ $+ 18$ ADP $+ 18$ P$_i$

**Net Equation**

$6 CO_2 + 18$ ATP $+ 12$ NADPH $+ 12$ H$^+$ $+ 12$ H$_2$O $\longrightarrow$
glucose $+ 12$ NADP$^+$ $+ 18$ ADP $+ 18$ P$_i$

carbon atoms in 12 molecules of glyceraldehyde 3-phosphate. The remaining six carbon atoms (two glyceraldehyde 3-phosphates) are converted to glucose.

$10$ glyceraldehyde 3-phosphate $+ 6$ ATP $\longrightarrow$
(30 carbons)

$6$ ribulose 1,5-*bis*phosphate $+ 6$ ADP $+ 4$ P$_i$
(30 carbons)

$2$ glyceraldehyde 3-phosphate $\longrightarrow$ glucose $+ 2$ P$_i$
(6 carbons)                           (6 carbons)

The P$_i$ are needed to balance the individual equations with respect to phosphorus.

The overall equation for the path of carbon in photosynthesis is the sum of these intermediate equations. Table 17.1 summarizes these equations.

$6 CO_2 + 6$ ribulose 1,5-*bis*phosphate $+ 18$ ATP $+$
(30 carbons)

$12$ NADPH $+ 12$ H$^+$ $+ 12$ H$_2$O $\longrightarrow$

$6$ ribulose 1,5-*bis*phosphate $+$ glucose $+$
(30 carbons)            (6 carbons)

$12$ NADP$^+$ $+ 18$ ADP $+ 18$ P$_i$

Finally, we can subtract the ribulose 1,5-*bis*phosphate to obtain the net equation for the path of carbon in photosynthesis.

$$6\ CO_2 + 18\ ATP + 12\ NADPH + 12\ H^+ + 12\ H_2O \longrightarrow$$

$$glucose + 12\ NADP^+ + 18\ ADP + 18\ P_i$$

The efficiency of energy use in photosynthesis can be calculated fairly easily. The $\Delta G°'$ for the reduction of $CO_2$ to glucose is +478 kJ (+114 kcal) for each mole of $CO_2$ (see Exercise 3), and the energy of light of 600-nm wavelength is 1593 kJ (381 kcal) mol$^{-1}$. We shall not explain in detail here how this figure for the energy of the light is obtained, but it comes ultimately from the equation $E = h\nu$. Light of wavelength 680 or 700 nm has lower energy than light at 600 nm. Thus the efficiency of photosynthesis is at least $(477/1593) \times 100$, or 30%.

## 17.7
## AN ALTERNATE PATHWAY FOR CARBON DIOXIDE FIXATION

In tropical plants there is a $C_4$ pathway (Figure 17.23), so named because it involves four-carbon compounds. The operation of this pathway (also called the Hatch–Slack pathway) ultimately leads to the $C_3$ (based on 3-phosphoglycerate) pathway of the Calvin cycle.

**FIGURE 17.23**   The $C_4$ pathway.

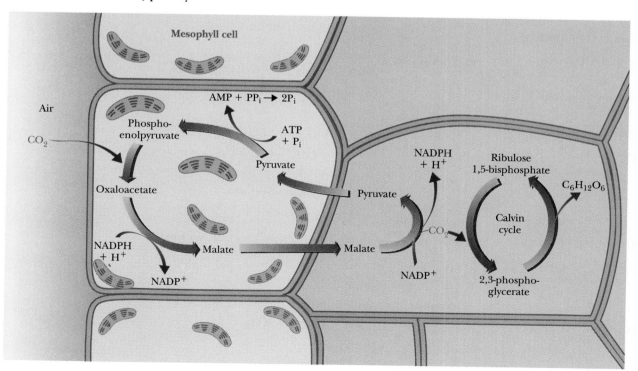

When $CO_2$ enters the outermost cells of the plant (the mesophyll cells) through the leaf pores, it reacts with phosphoenolpyruvate to produce oxaloacetate and $P_i$. Oxaloacetate is reduced to malate, with the concomitant oxidation of NADPH. Malate then enters the bundle-sheath cells (the next layer) through channels that connect two kinds of cells.

In the bundle-sheath cells malate is decarboxylated to give pyruvate and $CO_2$. In the process $NADP^+$ is reduced to NADPH (Figure 17.24). The $CO_2$ reacts with ribulose 1,5-*bis*phosphate to enter the Calvin cycle. Pyruvate is transported back to the mesophyll cells, where it is phosphorylated to phosphoenolpyruvate, which can react with $CO_2$ to start another round of the $C_4$ pathway. When pyruvate is phosphorylated, ATP is hydrolyzed to AMP and $PP_i$. This situation represents a loss of two "high-energy" phosphate bonds, equivalent to the use of two ATP. Consequently, the $C_4$ pathway requires two more ATP than the Calvin cycle alone for each $CO_2$ incorporated into glucose. The net equation for the path of carbon in the $C_4$ pathway in photosynthesis is

$$6\ CO_2 + 30\ ATP + 12\ NADPH + 12\ H^+ + 24\ H_2O \longrightarrow$$

$$glucose + 30\ ADP + 30\ P_i + 12\ NADP^+$$

Even though more ATP is required for the $C_4$ pathway than for the Calvin cycle, there is abundant light to produce the extra ATP by the light reaction of photosynthesis.

Note that the $C_4$ pathway fixes $CO_2$ in the mesophyll cells only to unfix it in the bundle-sheath cells, where $CO_2$ then enters the $C_3$ pathway. This observation raises the question of the advantage to tropical plants in

**FIGURE 17.24**   The characteristic reactions of the $C_4$ pathway.

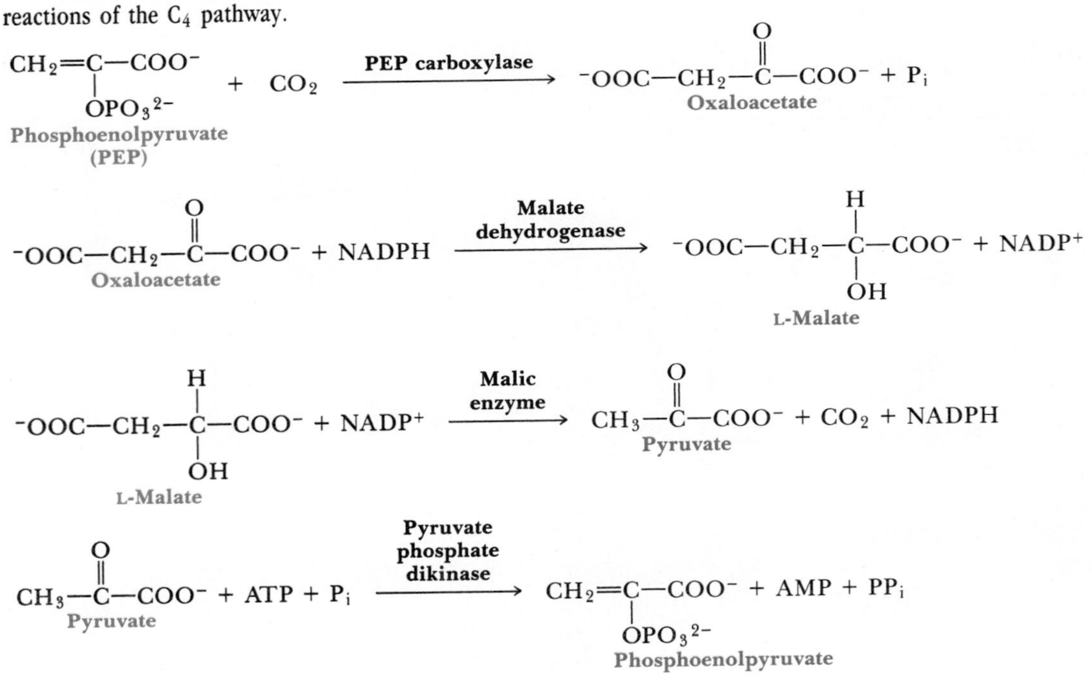

using the $C_4$ pathway. The conventional wisdom on the subject focuses on the role of $CO_2$, but there is more to the question than that. According to the conventional view, the point of the $C_4$ pathway is that it concentrates $CO_2$ and, as a result, accelerates the process of photosynthesis. Leaves of tropical plants have small pores to minimize water loss, and these small pores decrease $CO_2$ entry into the plant. In tropical areas, where there is abundant light, the amount of $CO_2$ available to plants controls the rate of photosynthesis. The $C_4$ pathway deals with the situation, allowing tropical plants to grow more quickly and to produce more biomass per unit of leaf area than plants that use the $C_3$ pathway. A more comprehensive view of the subject includes a consideration of the role of oxygen and the process of **photorespiration.**

In the dark, green leaf cells of both $C_3$ and $C_4$ plants carry out respiration and phosphorylation in their mitochondria. The energy source is the store of compounds produced by photosynthesis in the light. In the light, $C_3$ plants also carry on respiration. This respiration is not entirely a

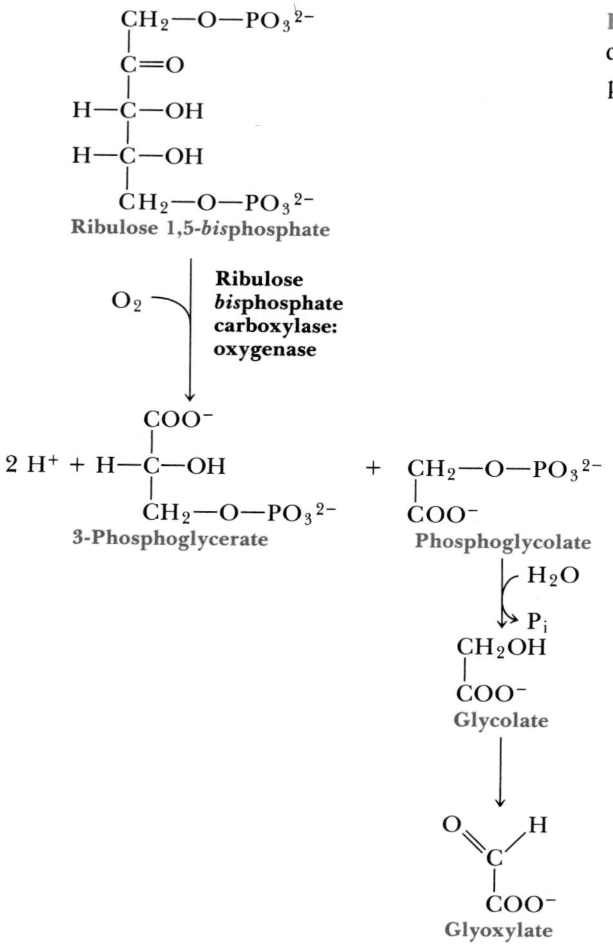

FIGURE 17.25    The characteristic reactions of photorespiration.

mitochondrial process because it continues at a lower rate in the presence of cyanide, an inhibitor of the mitochondrial electron transport chain (Section 15.6). Photorespiration is the cyanide-insensitive respiration of $C_3$ plants. The process is almost completely absent in $C_4$ plants.

Although the actual biological role of photorespiration is not known, several points are well established. The process in many ways is wasteful of reducing power and ATP. Oxidative phosphorylation does not accompany photorespiration, and reducing agents such as NADH are needed for the series of redox reactions linked to oxygen. The principal substrate oxidized in photorespiration is **glycolate** (Figure 17.25). The product of the oxidation reaction, which takes place in peroxisomes of leaf cells (Section 2.3), is **glyoxylate.** Glycolate arises ultimately from the oxidative breakdown of ribulose 1,5-*bis*phosphate. The enzyme that catalyzes this reaction is ribulose 1,5-*bis*phosphate carboxylase:oxygenase, acting as an oxygenase (linked to $O_2$) rather than as the carboxylase (linked to $CO_2$) that fixes $CO_2$ into 3-phosphoglycerate.

When $O_2$ levels are high compared with those of $CO_2$, ribulose 1,5-*bis*phosphate is oxygenated to produce phosphoglycolate (which gives rise to glycolate) and 3-phosphoglycerate by photorespiration, rather than the two molecules of 3-phosphoglycerate that arise from the carboxylation reaction. This situation occurs in $C_3$ plants. In $C_4$ plants, the small pores decrease the entry not only of $CO_2$ but also of $O_2$ into the leaves. The ratio of $CO_2$ to $O_2$ in the bundle-sheath cells is relatively high as a result of the operation of the $C_4$ pathway, favoring the carboxylation reaction. The $C_4$ pathway is an advantageous one for tropical plants because it allows such plants to dispense with photorespiration.

## SUMMARY

The equation for photosynthesis

$$6\ CO_2 + 6\ H_2O \longrightarrow C_6H_{12}O_6 + 6\ O_2$$

actually represents two processes. One, the splitting of water to produce oxygen, is a reaction that requires light energy from the sun, and the other, the fixation of $CO_2$ to give sugars, is one that can and does take place in the dark, but only because it uses solar energy indirectly. In the light reaction, water is oxidized to produce oxygen, accompanied by the reduction of $NADP^+$ to NADPH. The light reaction is coupled to the phosphorylation of ADP to ATP. In the dark reaction, the ATP and NADPH produced in the light reaction provide the energy and reducing power for the fixation of $CO_2$.

The light reaction of photosynthesis takes place in eukaryotes in the thylakoid membranes of chloroplasts. The trapping of light takes place at a reaction center within the chloroplast; the process requires a pair of chlorophylls in a unique environment.

The light reaction consists of two parts, each carried out by a separate photosystem. The reduction of $NADP^+$ to NADPH is accomplished by Photosystem I, while the splitting of water to produce oxygen is done by Photosystem II. The two photosystems are linked by an electron transport chain coupled to the production of ATP. A proton gradient drives the production of ATP in photosynthesis, as is the case in mitochondrial respiration.

The dark reaction of photosynthesis involves the net synthesis of one molecule of glucose from six molecules of $CO_2$. The net reaction of 6 molecules of $CO_2$ to produce 1 molecule of glucose requires the carboxylation of 6 molecules of a five-carbon key intermediate, ribulose 1,5-*bis*phosphate, ultimately forming 12 molecules of 3-phosphoglycerate. Of these, two molecules of 3-phosphoglycerate react to give rise to glucose. The remaining ten molecules of 3-phosphoglycerate are used to regenerate the six molecules of ribulose 1,5-*bis*phosphate. The overall reaction pathway is cyclic and is called the Calvin cycle.

In addition to the Calvin cycle, there is an alternate pathway for $CO_2$ fixation in tropical plants, called the $C_4$ pathway because it involves four-carbon compounds. In this pathway, $CO_2$ reacts in the outer (mesophyll) cells with phosphoenolpyruvate to produce oxaloacetate and $P_i$. Oxaloacetate in turn is reduced to malate. Malate is transported from mesophyll cells, where it is produced, to inner (bundle-sheath) cells, where it is ultimately passed to the Calvin cycle. Plants in which the $C_4$ pathway operates grow more quickly and produce more biomass per unit of leaf area than $C_3$ plants, in which only the Calvin cycle operates. In $C_3$ plants the process of photorespiration, which wastes energy and reducing power, is operative. In $C_4$ plants photorespiration is almost totally absent.

## EXERCISES

1. In cyclic photophosphorylation in Photosystem I, ATP is produced, even though water is not split. Explain how the process takes place.
2. Uncouplers of oxidative phosphorylation in mitochondria also uncouple photoelectron transport and ATP synthesis in chloroplasts. Give an explanation for this observation.
3. Using information from Section 12.4, show how the $\Delta G°'$ of 477 kJ (114 kcal) is obtained for each mole of $CO_2$ fixed in photosynthesis.
4. Chlorophyll is green because it absorbs green light less than it absorbs light of other wavelengths. The accessory pigments in the leaves of deciduous trees tend to be red and yellow, but their color is masked by that of the chlorophyll. Suggest a connection between these points and the display of fall foliage colors in many sections of the country.
5. How can a proton gradient be created in cyclic photophosphorylation in Photosystem I?
6. Suppose that a prokaryotic organism that contains both chlorophyll *a* and chlorophyll *b* has been discovered. Comment on the evolutionary implications of such a discovery.
7. Outline the events that take place at the photosynthetic reaction center in *Rhodopseudomonas*.
8. What are the two places where light energy is required in the light reaction of photosynthesis? Why must energy be supplied at precisely these points?
9. If photosynthesizing plants are grown in the presence of $^{14}CO_2$, is every carbon atom of the glucose that is produced labeled with the radioactive carbon? Give the reason for your answer.
10. How does the production of sugars by tropical plants differ from the same reactions in the Calvin cycle?

## ANNOTATED BIBLIOGRAPHY

Barber, J. Has the Mangano-protein of the Water Splitting Reaction of Photosynthesis Been Isolated? *Trends Biochem. Sci.* **9,** 79–80 (1984). [A report on an important topic in research on Photosystem II.]

Bering, C.L. Energy Interconversions in Photosynthesis. *J. Chem. Ed.* **62,** 659–664 (1985). [A short article on basic concepts of photosynthesis, concentrating on the light reaction and photosystems.]

Bishop, M.B., and C.B. Bishop. Photosynthesis and Carbon Dioxide Fixation. *J. Chem. Ed.* **64,** 302–305 (1987). [An article that concentrates on the Calvin cycle.]

Danks, S.M., E.H. Evans, and P.A. Whittaker. *Photosynthetic Systems: Structure, Function and Assembly.* New York: Wiley, 1983. [A short book with excellent electron micrographs of chloroplasts and related structures in Chapter 1.]

Deisenhofer, J., and H. Michel. The Photosynthetic Reaction Center from the Purple Bacterium *Rhodopseudomonas viridis. Science* **245,** 1463–1473 (1989). [The authors describe their work on the structure of the reaction center in their Nobel Prize address.]

Deisenhofer, J., H. Michel, and R. Huber. The Structural Basis of Photosynthetic Light Reactions in Bacteria. *Trends Biochem. Sci.* **10**, 243–248 (1985). [A discussion of the photosynthetic reaction center in bacteria.]

Dennis, D.T. *The Biochemistry of Energy Utilization in Plants.* New York: Chapman and Hall, 1987. [A short book that concentrates exclusively on plant biochemistry.]

Govindjee, ed. *Photosynthesis. Vol. I: Energy Conversion in Plants and Bacteria; Vol II: Development, Carbon Metabolism and Plant Productivity.* New York: Academic Press, 1982. [A fairly advanced discussion of photosynthesis.]

Govindjee and W.J. Coleman. How Plants Make Oxygen. *Sci. Amer.* **262** (2) 50–58 (1990). [An article that focuses on the water-splitting apparatus of Photosystem II.]

Halliwell, B. *Chloroplast Metabolism: The Structure and Function of Chloroplasts in Green Leaf Cells.* New York: Oxford University Press, 1981. [A detailed description of chloroplast activity.]

Hipkins, M.F., and N. R. Baker, eds. *Photosynthesis: Energy Transduction: A Practical Approach.* Oxford, England: IRL Press, 1986. [A collection of articles about research methods used to study photosynthesis.]

Margulis, L. *Early Life.* Boston: Science Books International, 1982. [Chapters 2 and 3 discuss the evolutionary development of photosynthesis.]

Youvan, D.C., and B.L. Marrs. Molecular Mechanisms of Photosynthesis. *Sci. Amer.* **256** (6) 42–48 (1987). [A detailed description of a bacterial photosynthetic reaction center and the molecular events that take place there.]

Zuber, H. Structure of Light-Harvesting Antenna Complexes of Photosynthetic Bacteria, Cyanobacteria and Red Algae. *Trends Biochem. Sci.* **11,** 414–419 (1986). [An article that concentrates on the protein portion of the photosynthetic reaction center.]

# The Metabolism of Nitrogen

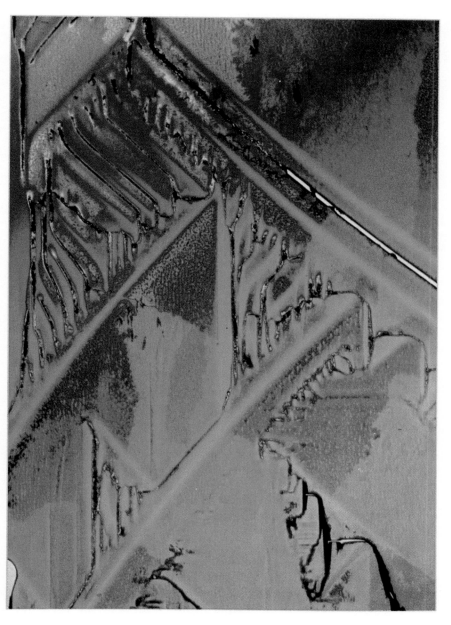

Crystals of urea viewed under polarized light.

459

*Surprisingly, nitrogen found in living organisms does not come from the atmosphere, where it comprises 80 percent of atmospheric gas. Instead, organic nitrogen enters the biological realm through bacteria living in the root nodules of leguminous plants such as peas and beans. They are able to convert the triple-bonded nitrogen molecule $N_2$ into ammonia, $NH_3$, which is incorporated into the amino acid, glutamate, which, in turn, is a precursor of proline and arginine. Thus, the 20 amino acids may be formed by the transformation from one amino acid to another, often by long and complex metabolic pathways. Although amino acids act principally as the building blocks of proteins, they also contribute to the synthesis of a variety of biologically important molecules. These molecules include the nitrogenous bases of DNA and RNA, the nucleotide coenzyme electron carriers $NAD^+$ and $NADP^+$, the oxygen-binding pyrrole ring of hemoglobin, as well as small hormones and neurotransmitters. When amino acids are deaminated (removal of the $\alpha$ amino group, $NH_3{}^+$), their carbon skeletons can be fed into the citric acid cycle to be oxidized to carbon dioxide and water. Alternately, they may be used as precursors of other biomolecules, including glucose and fatty acids. In many ways then, amino acids can be seen as connecting links between metabolic pathways.*

## 18.1
## AN OVERVIEW OF NITROGEN METABOLISM

We have seen the structures of many types of compounds that contain nitrogen, including amino acids, porphyrins, and nucleotides, but we have not discussed their metabolism. The metabolic pathways we have dealt with up to now have mainly involved compounds of carbon, hydrogen, and oxygen, such as sugars and fatty acids. Several important topics can be included in our discussion of the metabolism of nitrogen. The first of these is **nitrogen fixation,** the process by which inorganic molecular nitrogen from the atmosphere ($N_2$) is first incorporated into ammonia and then into organic compounds of use to organisms. Nitrate ion ($NO_3{}^-$), another kind of inorganic nitrogen, is the form in which nitrogen is found in the soil, and many fertilizers contain nitrates, frequently potassium nitrate. The process of **nitrification** (nitrate reduction) provides another way for organisms to obtain nitrogen. Nitrate ion and nitrite ion ($NO_2{}^-$) are also involved in **denitrification** reactions, which return nitrogen to the atmosphere (Figure 18.1).

Ammonia formed by either pathway, nitrogen fixation or nitrification, enters the biosphere. Ammonia is converted to organic nitrogen by plants, and organic nitrogen is passed to animals through food chains. Finally, waste products such as urea are excreted and degraded to ammonia by microorganisms. The word "ammonia" comes from the term "sal ammo-

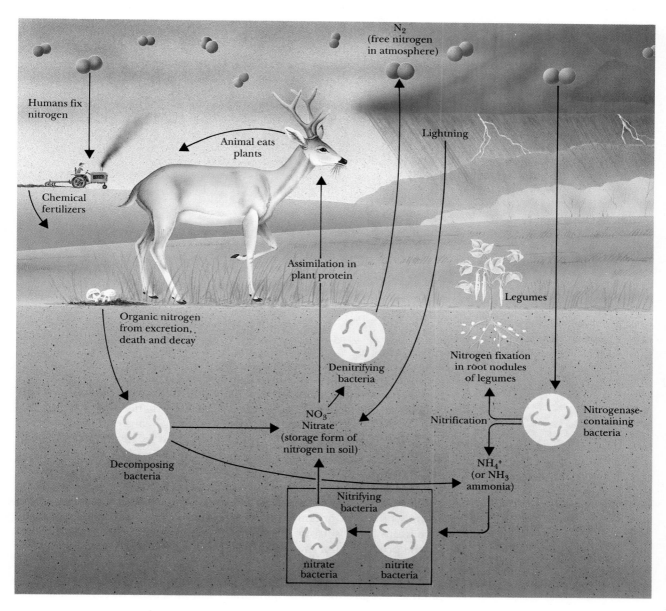

**FIGURE 18.1**    The flow of nitrogen to the biosphere.

niac'' (ammonium chloride), first prepared from the dung of camels at the temple of Jupiter Ammon in North Africa. The process of death and decay releases ammonia in both plants and animals. Denitrifying bacteria reverse the conversion of ammonia to nitrate and then recycle the $NO_3^-$ as free $N_2$ (Figure 18.1).

The topic of nitrogen metabolism includes the biosynthesis and breakdown of **amino acids, purines,** and **pyrimidines;** also, the metabolism of **porphyrins** is related to that of amino acids.

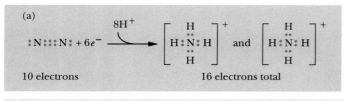

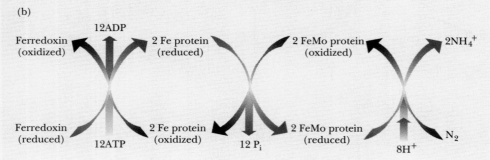

FIGURE 18.2   Some aspects of the nitrogenase reaction. (a) The reduction of $N_2$ to $2 NH_4^+$. (b) The path of electrons from ferredoxin to $N_2$.

## 18.2
## NITROGEN FIXATION

Bacteria are responsible for the reduction of $N_2$ to ammonia ($NH_3$). Typical nitrogen-fixing bacteria are symbiotic organisms that form nodules on the roots of leguminous plants such as beans and alfalfa. Many free-living microbes and some cyanobacteria also fix nitrogen. Plants and animals cannot carry out nitrogen fixation; this conversion of molecular nitrogen to ammonia is the only source of nitrogen in the biosphere except for that provided by nitrates. The conjugate acid form of $NH_3$, ammonium ion ($NH_4^+$), is the form of nitrogen that is used in the first stages of the synthesis of organic compounds. Parenthetically, $NH_3$ obtained by chemical synthesis from nitrogen and hydrogen is the starting point for the production of many synthetic fertilizers, which frequently contain nitrates.

The **nitrogenase** enzyme complex found in nitrogen-fixing bacteria catalyzes the production of ammonia from molecular nitrogen. The half reaction of reduction (Figure 18.2 (a)) is

$$N_2 + 6 e^- + 12 ATP + 12 H_2O + 8 H^+ \longrightarrow 2 NH_4^+ + 12 ADP + 12 P_i$$

Several proteins are included in the nitrogenase complex. Ferredoxin is one of them (this protein also plays an important role in electron transfer in photosynthesis; Section 17.3). There are also two proteins specific to the nitrogenase reaction. One is an iron-molybdenum (Fe-Mo) protein, and the other is a nonheme (iron-sulfur) protein, usually referred to as the Fe protein. The flow of electrons is from ferredoxin to the Fe protein to the Fe-Mo protein to nitrogen (Figure 18.2 (b)). The nature of the nitrogenase complex is a subject of active research.

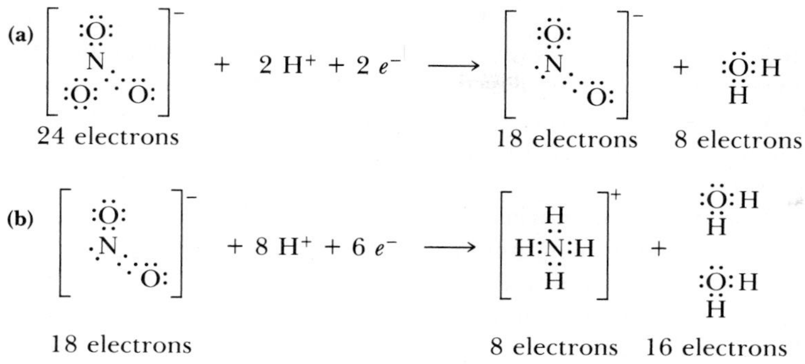

**(a)**

24 electrons        18 electrons    8 electrons

**(b)**

18 electrons            8 electrons   16 electrons

**FIGURE 18.3**  Nitrification reactions. (a) The reduction of nitrate to nitrite. (b) The reduction of nitrite to ammonium ion.

## Nitrate Reduction

In the process of **nitrification,** $NO_3^-$ is converted to $NH_3$. This process is carried out by nitrifying bacteria in the soil; the ammonia thus produced is first used by plants and then passed on to animals. The actual process of nitrate reduction takes place in several stages and involves nitrite as an intermediate. The half reaction of reduction of nitrate to nitrite (Figure 18.3 (a)) is

$$NO_3^- + 2\ H^+ + 2\ e^- \longrightarrow NO_2^- + H_2O$$

This reaction is catalyzed by **nitrate reductase,** a large multisubunit enzyme. Nitrate reductase contains several electron transport agents, including NADH and FAD. The reaction is a multistep one, but we shall not discuss the details here. The ultimate source of electrons for the reduction of nitrate is NADH, and the overall reaction is

$$NO_3^- + NADH + H^+ \longrightarrow NO_2^- + NAD^+ + H_2O$$

The reduction of nitrite to ammonia in the form of ammonium ion requires six electrons. The half reaction of reduction (Figure 18.3 (b)) is

$$NO_2^- + 6\ e^- + 8\ H^+ \longrightarrow NH_4^+ + 2\ H_2O$$

The reaction is catalyzed by **nitrite reductase,** another enzyme that contains several electron transfer agents. The reaction goes through several stages, as is the case with nitrate reduction. The ultimate source of electrons is NADPH. Three NADPH are required for each nitrite, since each NADPH transfers only two electrons. The equation for the overall reaction is

$$3\ NADPH + NO_2^- + 5\ H^+ \longrightarrow NH_4^+ + 3\ NADP^+ + 2\ H_2O$$

## 18.3
## FEEDBACK CONTROL: A UNIFYING THEME IN NITROGEN METABOLISM

The biosynthetic pathways that produce amino acids and nucleobases (purines and pyrimidines) are long and complex, requiring a large investment of energy by the organism. If there is a high level of some end product, such as an amino acid or a nucleotide, the cell saves energy by not making that compound. However, the cell needs some signal not to produce more of that particular compound. The signal is frequently a **feedback inhibition** mechanism, in which the end product of a metabolic pathway inhibits an enzyme at the beginning of the pathway. We saw an example of such a control mechanism when we discussed the allosteric enzyme aspartate transcarbamoylase in Section 10.9. This enzyme catalyzes one of the early stages of pyrimidine nucleotide biosynthesis, and it is inhibited by the end product of that pathway, namely cytidine triphosphate (CTP). Feedback inhibition is frequently encountered in the biosynthesis of amino acids and nucleotides.

## 18.4
## THE METABOLISM OF AMINO ACIDS: ANABOLISM

### General Features

Ammonia is toxic in high concentrations, and so it must be incorporated into biologically useful compounds when it is formed by the reactions of nitrogen fixation, which we discussed earlier in this chapter. The amino acids **glutamate** and **glutamine** are of central importance in the process. Glutamate arises from $\alpha$-ketoglutarate, and glutamine from glutamate (Figure 18.4). Since these reactions involve the shift of an amino group, they are frequently referred to as **transamination** reactions. (Strictly speaking, the production of glutamate is a reductive amination, and the production of glutamine is amidation.) The $\alpha$-amino group of glutamate and the side-chain amino group of glutamine are shifted to other compounds in other transamination reactions.

The biosynthesis of amino acids involves a common set of reactions. In addition to transamination reactions, **transfer of one-carbon units,** such as formyl or methyl groups, occurs frequently. We are not going to discuss all the details of the reactions that give rise to amino acids. We can, however, organize this material by grouping amino acids into families based on common precursors (Figure 18.5). The reactions of some of the individual families of amino acids provide good examples of those reactions that are of general importance, transamination and one-carbon transfer. The biosynthesis of amino acids in other families can be found in Interchapter B, which follows this chapter.

FIGURE 18.4   (a) The production of glutamate from $\alpha$-ketoglutarate. (b) The production of glutamine from glutamate.

(a) $NH_4^+ + {}^-OOC-CH_2-CH_2-\overset{\overset{\displaystyle O}{\|}}{C}-COO^- \underset{NADP^+}{\overset{NADPH + H^+}{\rightleftharpoons}} H_2O + {}^-OOC-CH_2-CH_2-\overset{\overset{\displaystyle NH_3^+}{|}}{CH}-COO^-$

$\alpha$-Ketoglutarate          Glutamate

(b) $NH_4^+ + {}^-OOC-CH_2-CH_2-\overset{\overset{\displaystyle NH_3^+}{|}}{CH}-COO^- \xrightarrow{\quad ATP \quad ADP + P_i \quad} H_2O + H_2N-\overset{\overset{\displaystyle O}{\|}}{C}-CH_2-CH_2-\overset{\overset{\displaystyle NH_3^+}{|}}{CH}-COO^-$

Glutamate                                      Glutamine

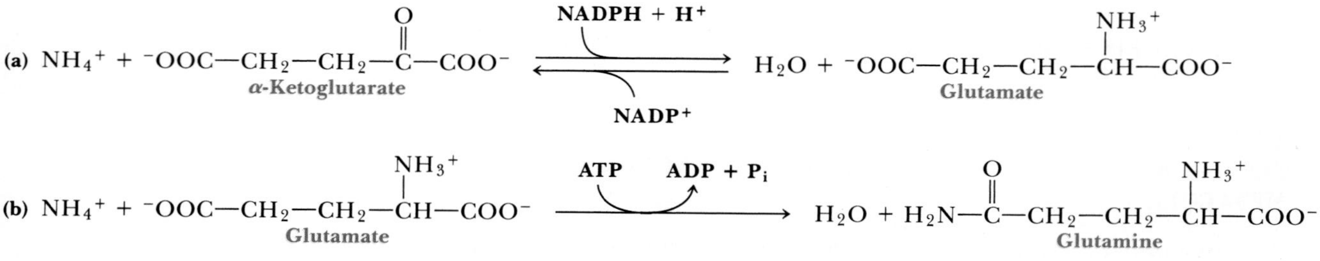

We can also make some generalizations about amino acid metabolism in terms of the relationship of the carbon skeleton to the citric acid cycle and the related reactions of pyruvate and acetyl-CoA (Figure 18.6). The citric acid cycle is amphibolic; it has a part in both catabolism and anabolism. The anabolic aspect of the citric acid cycle is of interest in amino acid biosynthesis. The catabolic aspect is apparent in the breakdown

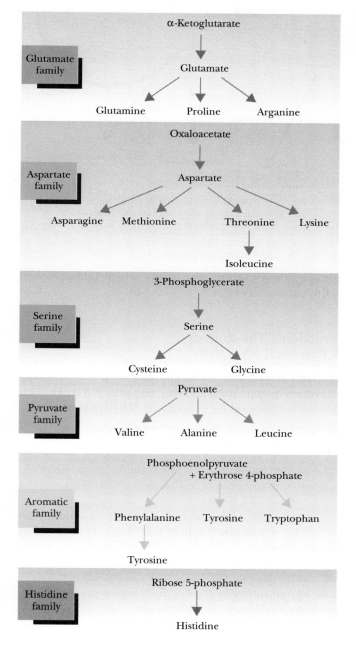

**FIGURE 18.5** Families of amino acids based on biosynthetic pathways. Each family has a common precursor.

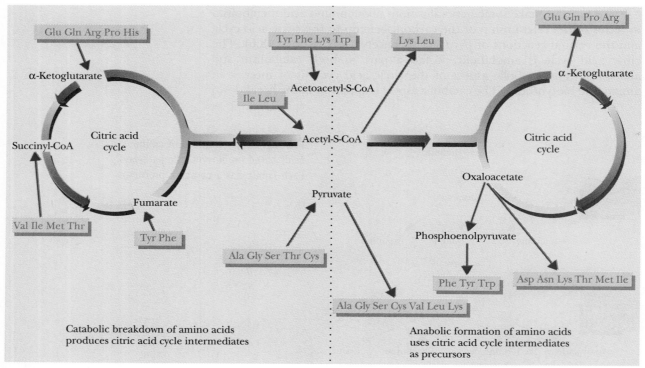

**FIGURE 18.6**  The relationship between amino acid metabolism and the citric acid cycle.

of amino acids, leading to their eventual excretion, which takes place in reactions related to the citric acid cycle.

### Transamination Reactions: The Role of Glutamate and of Pyridoxal Phosphate

Glutamate is formed from $NH_4^+$ and $\alpha$-ketoglutarate in a reaction that, strictly speaking, is a reductive amination and that requires NADPH. This reaction is reversible and is catalyzed by **glutamate dehydrogenase.**

$$NH_4^+ + \alpha\text{-ketoglutarate} + NADPH + H^+ \rightleftharpoons glutamate + NADP^+ + H_2O$$

Glutamate is a major donor of amino groups in transamination reactions, and $\alpha$-ketoglutarate is a major acceptor of amino groups (see Figure 18.4 (a)).

The conversion of glutamate to glutamine (strictly speaking, an amidation) is catalyzed by **glutamine synthetase** in a reaction that requires ATP (see Figure 18.4 (b)).

$$NH_4^+ + glutamate + ATP \longrightarrow glutamine + ADP + P_i + H_2O$$

These reactions fix inorganic nitrogen ($NH_3$), forming organic (carbon-containing) nitrogen compounds, such as amino acids.

Quite frequently, enzymes that catalyze transamination reactions require pyridoxal phosphate as a coenzyme (Figure 18.7). We discussed this

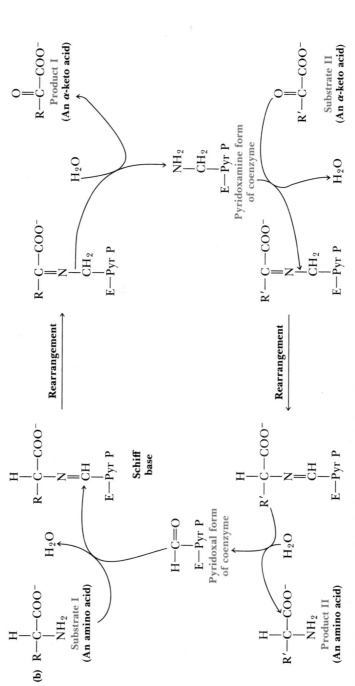

**FIGURE 18.7** The role of pyridoxal phosphate in transamination reactions. (a) The mode of binding of pyridoxal phosphate (PyrP) to the enzyme (E) and to the substrate amino acid. (b) The reaction itself. The original substrate, an amino acid, is deaminated, while an α-keto acid is aminated to form an amino acid. The net reaction is one of transamination. Note that the coenzyme is regenerated and that the original substrate and final product are both amino acids.

467

compound in Section 10.13 as a typical example of a coenzyme, and here we can see its mode of action in context.

Pyridoxal phosphate forms a Schiff base with the amino group of Substrate I (the amino group donor) (Figure 18.7 (a)). The next stage is a rearrangement followed by hydrolysis, which removes Product I (the α-keto acid corresponding to Substrate I). The coenzyme now carries the amino group (pyridoxamine). Substrate II (another α-keto acid) then forms a Schiff base with pyridoxamine. Again there is a rearrangement followed by a hydrolysis, which gives rise to Product II (an amino acid) and regenerates pyridoxal phosphate. The net reaction (Figure 18.7 (b)) is that an amino acid (Substrate I) reacts with an α-keto acid (Substrate II) to form an α-keto acid (Product I) and an amino acid (Product II). The amino group has been transferred from Substrate I to Substrate II, forming the amino acid, Product II.

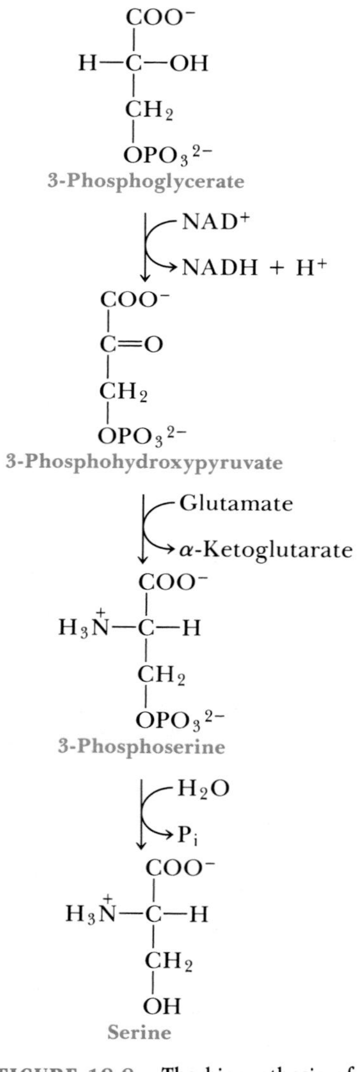

## One-Carbon Transfers and the Serine Family

In addition to transamination reactions, one-carbon transfer reactions occur frequently in amino acid biosynthesis. A good example of a one-carbon transfer can be found in the reactions that produce the amino acids of the serine family. This family also includes glycine and cysteine. Serine and glycine themselves are frequently precursors in other biosynthetic pathways. A discussion of the synthesis of cysteine will give us some insight into the metabolism of sulfur, as well as nitrogen.

The ultimate precursor of serine is 3-phosphoglycerate, which is obtainable from the glycolytic pathway. The hydroxyl group on carbon 2 is oxidized to a keto group, giving an α-keto acid. A transamination reaction in which glutamate is the nitrogen donor produces 3-phosphoserine. This reaction is another example of a transamination involving glutamate and α-ketoglutarate. Hydrolysis of the phosphate group then gives rise to serine (Figure 18.8).

The conversion of serine to glycine involves the transfer of a one-carbon unit from serine to an acceptor. This reaction is catalyzed by **serine hydroxymethylase,** with pyridoxal phosphate as a coenzyme. The acceptor in this reaction is **tetrahydrofolate,** a derivative of folic acid and a frequently encountered carrier of one-carbon units in metabolic pathways. Its structure has three parts: a substituted pteridine ring, $p$-aminobenzoic acid, and glutamic acid (Figure 18.9).

$$\text{Serine} + \text{tetrahydrofolate} \rightleftharpoons \text{glycine} + \text{methylenetetrahydrofolate} + \text{H}_2\text{O}$$

The one-carbon unit transferred in this reaction is bound to tetrahydrofolate, forming $N^5,N^{10}$-methylenetetrahydrofolate, in which the methylene (one-carbon) unit is bound to two of the nitrogens of the carrier (Figure 18.10). Tetrahydrofolate is not the only carrier of one-carbon units. We have already encountered biotin, a carrier of $CO_2$, and we have discussed the role that biotin plays in gluconeogenesis (Section 13.7) and in the anabolism of fatty acids (Section 16.6).

**FIGURE 18.8**   The biosynthesis of serine.

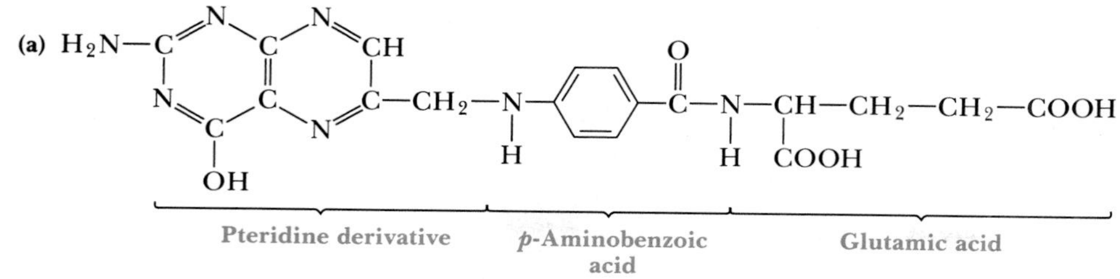

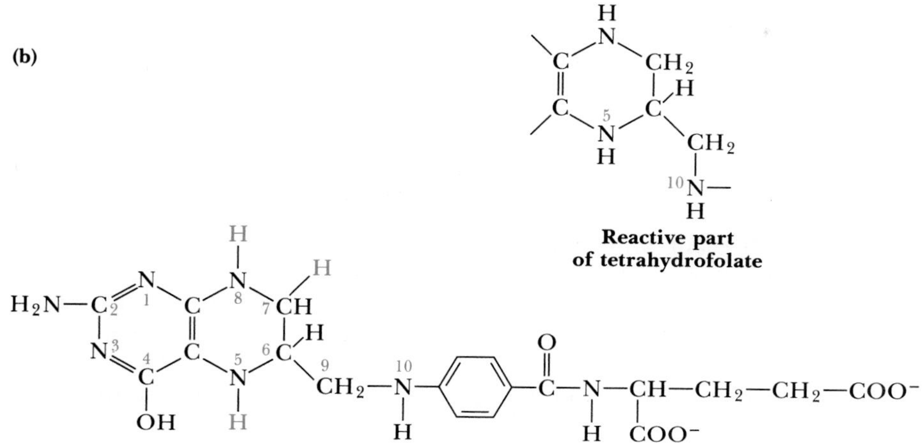

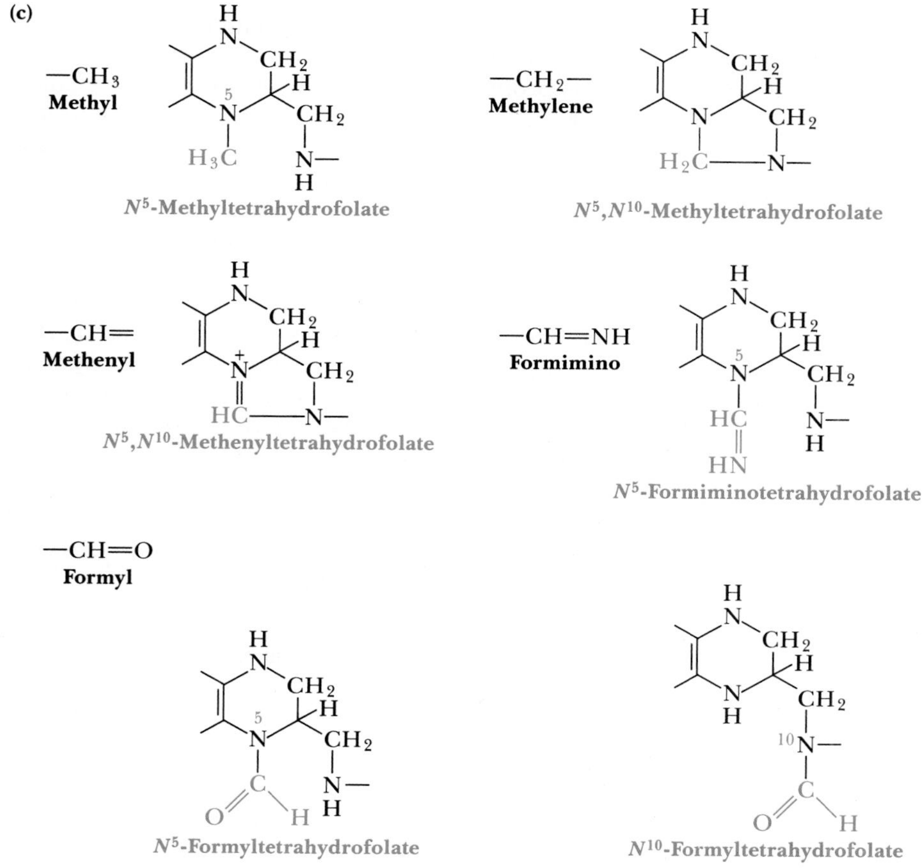

**FIGURE 18.9** (a) The structure of folic acid, shown in nonionized form. (b) The structure of tetrahydrofolate, shown in ionized form. Hydrogens are added at positions 5, 6, 7, and 8. The nitrogens at positions 5 and 10 are involved in the reactions of tetrahydrofolate. (c) The one-carbon groups transferred by tetrahydrofolate, showing the forms in which they are bound. Note that there are two positions, at nitrogen 5 and nitrogen 10, at which a formyl group is bound to tetrahydrofolate. There is only one binding site for the other one-carbon groups.

469

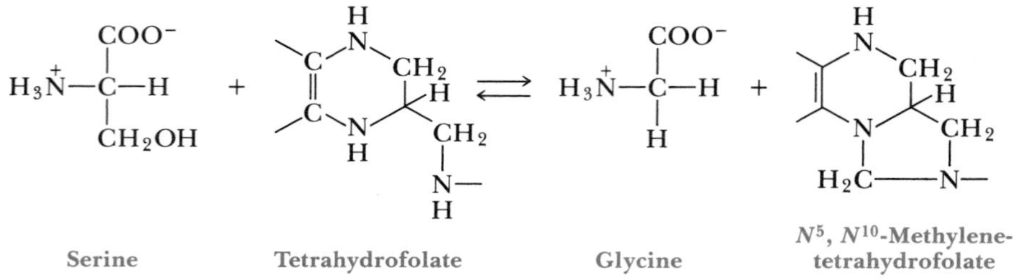

| Serine | Tetrahydrofolate | Glycine | $N^5, N^{10}$-Methylene-tetrahydrofolate |

FIGURE 18.10   The conversion of serine to glycine showing the role of tetrahydrofolate.

The conversion of serine to cysteine involves some interesting reactions. The source of the sulfur in animals differs from that in plants and bacteria. In plants and bacteria, serine is acetylated to form O-acetylserine. This reaction is catalyzed by **serine acyltransferase,** with acetyl-CoA as the acyl donor (Figure 18.11). Conversion of O-acetylserine to cysteine requires production of sulfide by a sulfur donor. The sulfur donor for plants and bacteria is 3'-phospho-5'-adenyl sulfate. The sulfate group is reduced first to sulfite and then to sulfide (Figure 18.12). The sulfide, in the conjugate acid form $HS^-$, displaces the acetyl group of the O-acetylserine to produce cysteine. Animals form cysteine from serine by a different pathway, since they do not have the enzymes to carry out the sulfate to sulfide conversion that we have just seen. The reaction sequence in animals involves the amino acid methionine.

Methionine, which is produced by reactions of the aspartate family (Section B.2 in Interchapter B) in bacteria and plants, cannot be produced by animals. It must be obtained from dietary sources. It is an **essential amino acid** because it cannot be synthesized by the body and must be obtained from the diet. The ingested methionine reacts with ATP to form **S-adenosylmethionine,** which has a highly reactive methyl group (Figure 18.13). This compound is a carrier of methyl groups in many reactions. The methyl group from S-adenosylmethionine can be transferred to any one of a number of acceptors, producing S-adenosylhomocysteine. Hydrolysis of S-adenosylhomocysteine in turn produces homocysteine. Cysteine can be synthesized from serine and homocysteine, and this pathway for cysteine biosynthesis is the only one available to animals (Figure 18.14). Serine and homocysteine react to produce cystathionine, which hydrolyzes to form cysteine, $NH_4^+$, and $\alpha$-ketobutyrate.

It is worth noting that we have now seen three important carriers of one-carbon units: biotin, a carrier of $CO_2$; tetrahydrofolate ($FH_4$), a

FIGURE 18.11   The biosynthesis of cysteine in plants and bacteria.

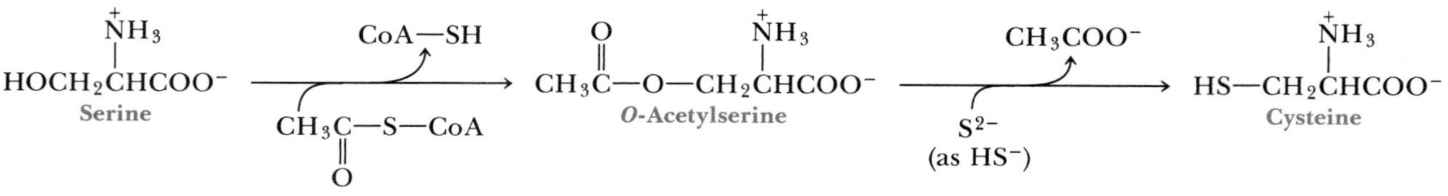

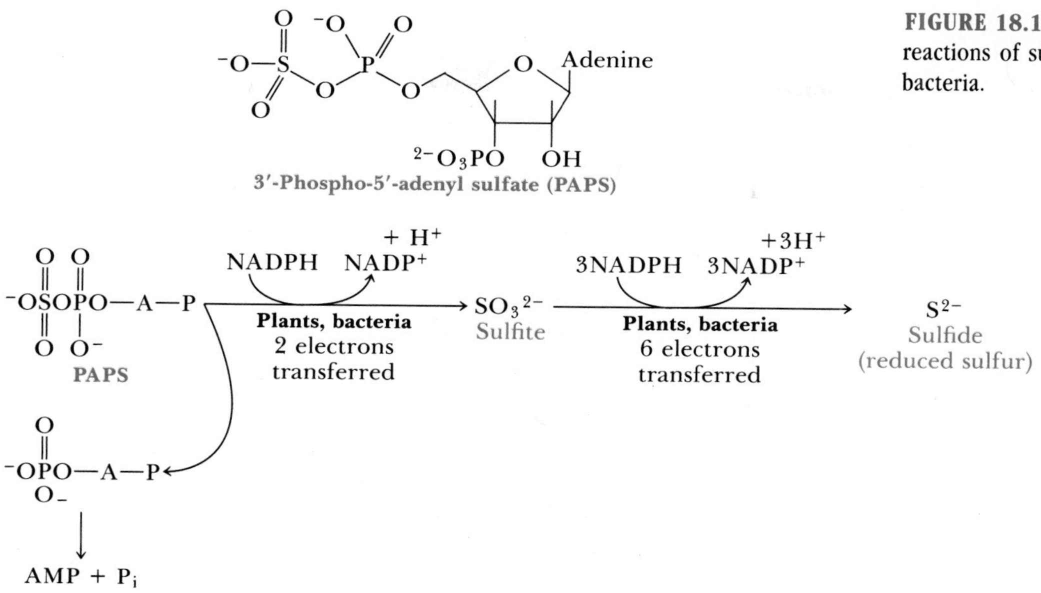

**FIGURE 18.12** Electron transfer reactions of sulfur in plants and bacteria.

3'-Phospho-5'-adenyl sulfate (PAPS)

carrier of methylene and formyl grups; and *S*-adenosylmethionine, a carrier of methyl groups.

## 18.5
## ESSENTIAL AMINO ACIDS

The biosynthesis of proteins requires the presence of all the constituent amino acids. If 1 of the 20 amino acids is missing or in short supply, protein biosynthesis is inhibited. Some organisms, such as *Escherichia coli*,

**FIGURE 18.13** The structure of *S*-adenosylmethionine, with the structure of methionine shown for comparison.

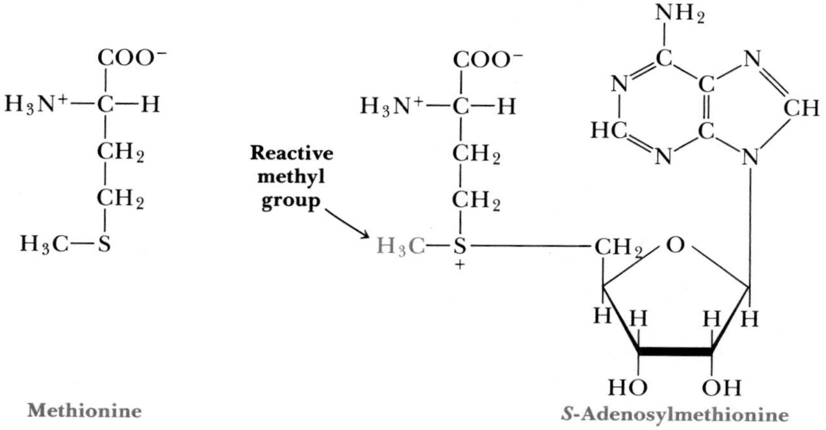

Methionine

*S*-Adenosylmethionine

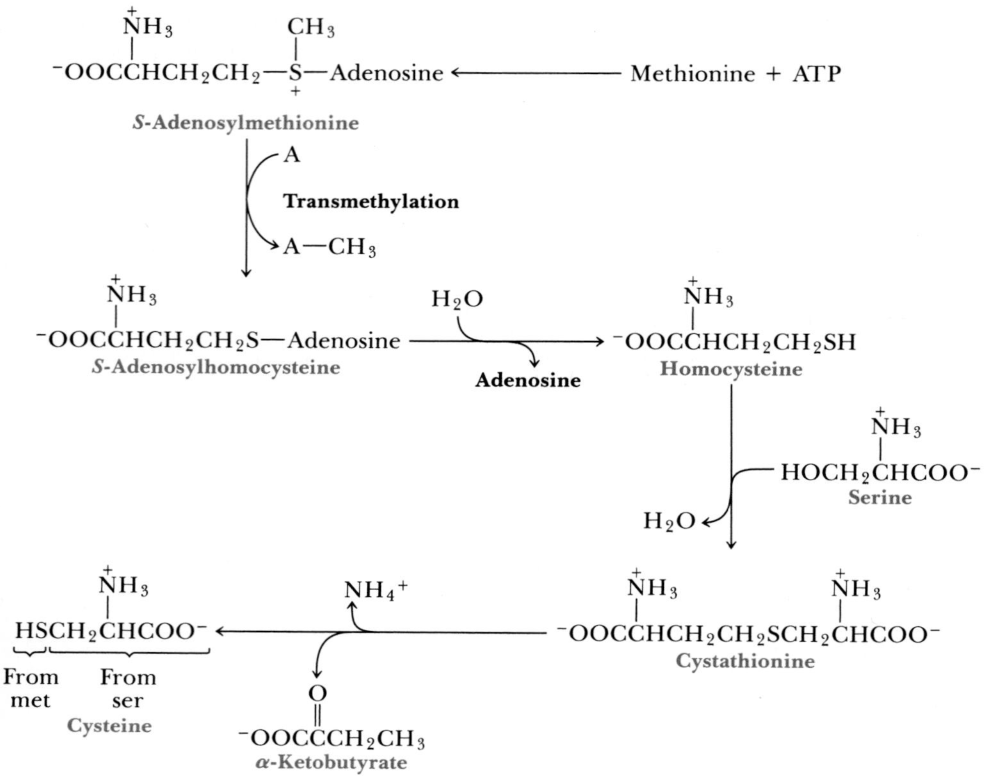

$$\overset{+}{N}H_3 \qquad\qquad CH_3$$
$$^-OOCCHCH_2CH_2\!-\!\overset{+}{S}\!-\!Adenosine \longleftarrow \text{———} Methionine + ATP$$

*S*-Adenosylmethionine

A

**Transmethylation**

A—CH$_3$

$$\overset{+}{N}H_3$$
$$^-OOCCHCH_2CH_2S\!-\!Adenosine \qquad H_2O \longrightarrow \overset{+}{N}H_3$$
*S*-Adenosylhomocysteine       Adenosine       $^-OOCCHCH_2CH_2SH$
Homocysteine

$$\overset{+}{N}H_3$$
$$HOCH_2CHCOO^-$$
Serine

$H_2O$

$$\overset{+}{N}H_3 \qquad\qquad NH_4^+ \qquad\qquad \overset{+}{N}H_3 \qquad\qquad \overset{+}{N}H_3$$
$$HSCH_2CHCOO^- \longleftarrow \qquad\qquad ^-OOCCHCH_2CH_2SCH_2CHCOO^-$$
Cystathionine

From   From
met    ser
Cysteine

$$\overset{O}{\underset{\parallel}{}}$$
$$^-OOCCCH_2CH_3$$
α-Ketobutyrate

**FIGURE 18.14** The biosynthesis of cysteine in animals. (A is acceptor.)

can synthesize all the amino acids that they need. Other species, including humans, must obtain some amino acids from dietary sources. The essential amino acids in human nutrition are listed in Table 18.1. Amino acids are not stored (except in proteins), and dietary sources of essential amino acids are needed at regular intervals. Protein deficiency, especially a prolonged deficiency in sources that contain essential amino acids, leads to the disease **kwashiorkor.** The problem in this disease, particularly severe in growing children, is not simply starvation, but the breakdown of the body's own proteins.

**TABLE 18.1**
**Amino Acid Requirements in Humans**

| ESSENTIAL | NONESSENTIAL |
|---|---|
| Arginine | Alanine |
| Histidine | Asparagine |
| Isoleucine | Aspartate |
| Leucine | Cysteine |
| Lysine | Glutamate |
| Methionine | Glutamine |
| Phenylalanine | Glycine |
| Threonine | Proline |
| Tryptophan | Serine |
| Valine | Tyrosine |

## 18.6
## THE ANABOLISM OF PORPHYRINS

The biosynthesis of porphyrins and heme is shown in Figure 18.15. Glycine and succinyl-CoA (an intermediate in the citric acid cycle) are the precursors of the porphyrin rings of hemes and chlorophylls. The first step of this pathway is a condensation reaction followed by the loss of $CO_2$, leading to the formation of δ-aminolevulinate. Two molecules of δ-aminolevulinate condense in turn, producing porphobilinogen, a molecule that contains a **pyrrole ring.** Four molecules of porphobilinogen condense further to form a **linear tetrapyrrole.** Three methylene bridges have been

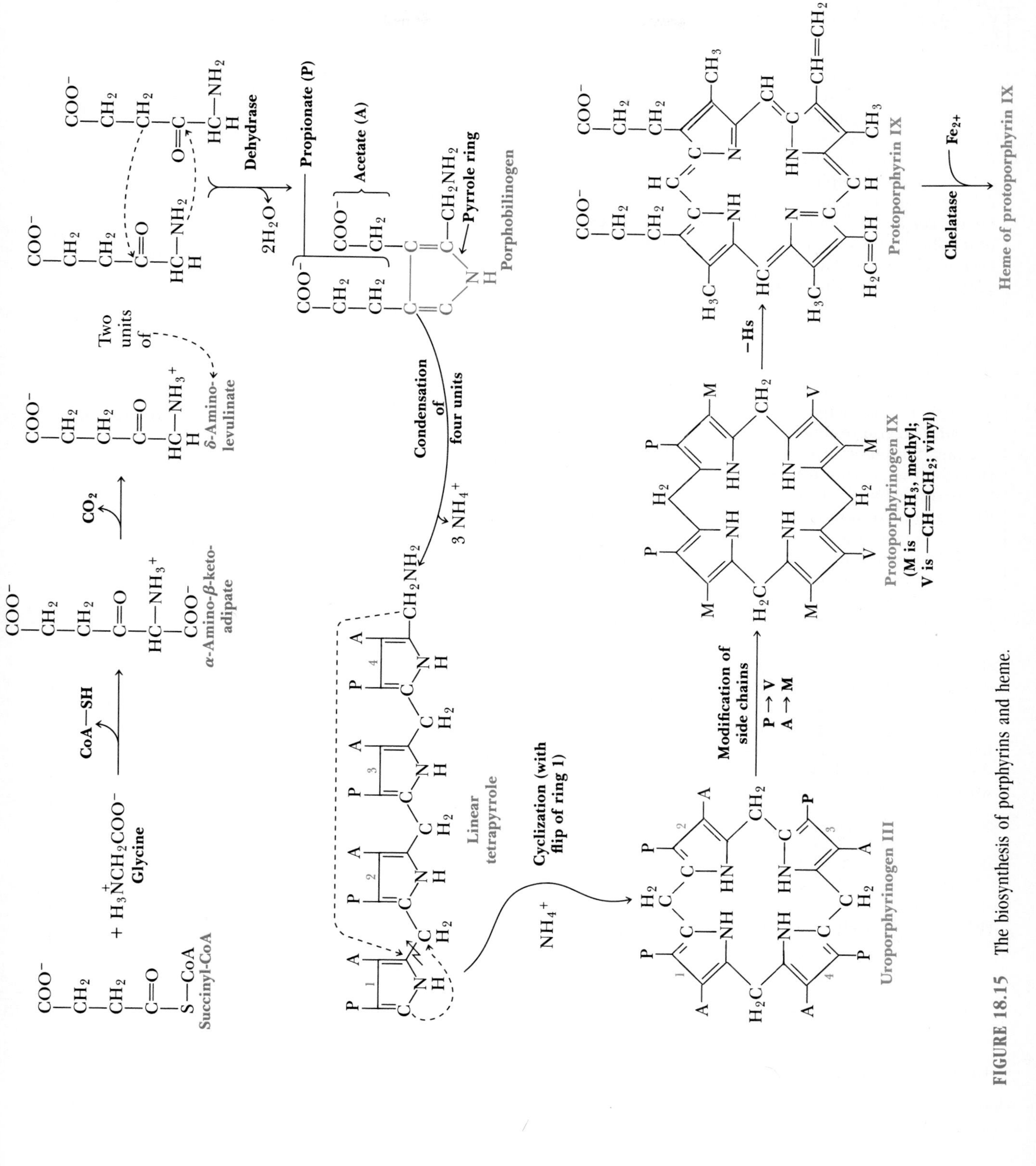

**FIGURE 18.15**   The biosynthesis of porphyrins and heme.

## BOX 18.1
## INBORN ERRORS OF METABOLISM INVOLVING AMINO ACIDS AND THEIR DERIVATIVES

Mutations leading to deficiencies in enzymes that catalyze reactions of amino acid metabolism frequently have drastic consequences, many of them leading to severe forms of mental retardation. **Phenylketonuria** (PKU) is a well-known example. The enzyme that is missing in this case is phenylalanine hydroxylase, which is responsible for converting phenylalanine to tyrosine. Phenylalanine, phenylpyruvate, phenyllactate, and phenylacetate accumulate in the blood and urine. Available evidence suggests that phenylpyruvate, which is a phenyl ketone, causes mental retardation by interfering with the conversion of pyruvate to acetyl-CoA in the brain. Metabolic activity in the brain is reduced as a result. Fortunately, this condition can be detected easily in newborns, and the consequences can be avoided by keeping the child on a diet that is restricted in phenylalanine.

**Albinism** also results because an enzyme required for the metabolism of an aromatic amino acid is missing. In this case the enzyme is **tyrosinase,** which is involved in the pathway by which tyrosine is converted to melanin, the material responsible for pigmentation of skin and hair. A temperature-sensitive form of tyrosinase is responsible for another easily visible effect on the pigmentation of an organism; this effect is the characteristic pattern of light and dark fur seen in Siamese cats. The tyrosinase found in these cats is inactive at their normal body temperature but is more active in parts of the body that have a lower temperature, such as the paws, tail, and ears. The more active enzyme leads to the production of melanin in these areas, and the fur is darker there.

Deficiencies in porphyrin metabolism can have severe neurological consequences. There are several classes of diseases, collectively known as **porphyrias,** associated with the lack of enzymes that catalyze reactions of porphyrins. In all cases the urine of the affected person is red; the neurological consequences can include a form of mania. King George III of England probably suffered from one type of porphyria, and his irrational behavior, which grew worse toward the end of his life, affected the course of history at the time of the American Revolution.

Reactions involved in the development of phenylketonuria (PKU). A deficiency in the enzyme that catalyzes the conversion of phenylalanine to tyrosine leads to the accumulation of phenylpyruvate, a phenyl ketone.

The production of melanin from tyrosine takes place in a multistep pathway.

formed, and three ammonium ions have been released in the process. Finally, the linear tetrapyrrole cyclizes to form cyclic tetrapyrrole, **uroporphyrinogen III.** Once again, an ammonium ion is lost in the process. The remaining steps are those of modifying the side chains of the porphyrin ring to produce the form that occurs in hemoglobin. In uroporphyrinogen III, the side chains are acetate or propionate groups. The conversion from uroporphyrinogen III to protoporphyrin IX requires the decarboxylation of the acetate side chains to methyl groups and the modification of two of the four propionate side chains to vinyl groups. The addition of $Fe^{2+}$ then produces heme.

Box 18.1 explains how the deficient metabolism of porphyrins, as well as other inborn errors of metabolism, can have serious consequences in affected persons.

## 18.7
## CATABOLISM OF AMINO ACIDS: THE UREA CYCLE

The first step to consider in the catabolism of amino acids is the removal of nitrogen by transamination. The amino nitrogen of the original amino acid is transferred to $\alpha$-ketoglutarate to produce glutamate, leaving behind the carbon skeleton. The fates of the carbon skeleton and of the nitrogen can be considered separately. Breakdown of the carbon skeleton of amino acids follows two general pathways, the difference between the two pathways depending on the type of end product. A **glucogenic** amino acid is one that yields pyruvate or oxaloacetate on degradation. Oxaloacetate is the starting point for the production of glucose by gluconeogenesis. A **ketogenic** amino acid is one that breaks down to acetyl-CoA or acetoacetyl-CoA, leading to the formation of ketone bodies (Table 18.2). (See Section 16.5.) The carbon skeletons of the amino acids give rise to

**TABLE 18.2    Glucogenic and Ketogenic Amino Acids**

| GLYCOGENIC | KETOGENIC | GLUCOGENIC AND KETOGENIC |
|---|---|---|
| Aspartate | Leucine | Isoleucine |
| Asparagine | | Lysine |
| Alanine | | Phenylalanine |
| Glycine | | Tryptophan |
| Serine | | Tyrosine |
| Threonine | | |
| Cysteine | | |
| Glutamate | | |
| Glutamine | | |
| Arginine | | |
| Proline | | |
| Histidine | | |
| Valine | | |
| Methionine | | |

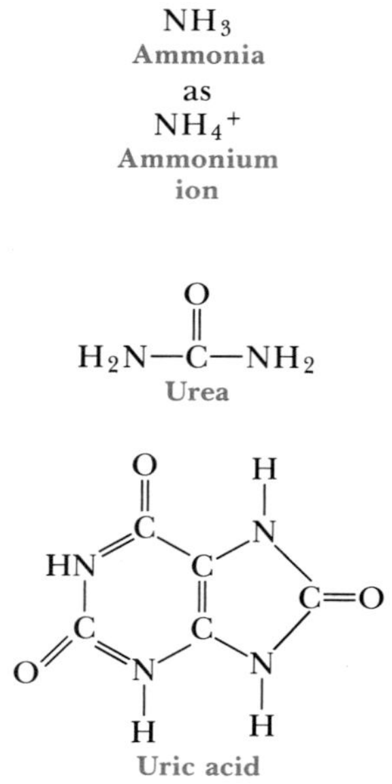

NH$_3$
Ammonia
as
NH$_4^+$
Ammonium
ion

$$\underset{\text{Urea}}{\text{H}_2\text{N}-\overset{\displaystyle\overset{\text{O}}{\|}}{\text{C}}-\text{NH}_2}$$

Uric acid

**FIGURE 18.16**   Products of amino acid catabolism.

metabolic intermediates such as pyruvate, acetyl-CoA, acetoacetyl-CoA, $\alpha$-ketoglutarate, succinyl-CoA, fumarate, and oxaloacetate (see Figure 18.6). Oxaloacetate is a key intermediate in the breakdown of the carbon skeleton of amino acids because of its dual role in the citric acid cycle and in gluconeogenesis. The amino acids degraded to acetyl-CoA and acetoacetyl-CoA are used in the citric acid cycle, but mammals cannot synthesize glucose from acetyl-CoA. This fact is the source of the distinction between glucogenic and ketogenic amino acids. Glucogenic amino acids can be converted to glucose, with oxaloacetate as an intermediate, but ketogenic amino acids cannot.

The nitrogen portion of amino acids is involved in transamination reactions in breakdown as well as in biosynthesis. Excess nitrogen is excreted in one of three forms: **ammonia** (as ammonium ion), **urea,** and **uric acid** (Figure 18.16). Animals, such as fish, that live in an aquatic environment excrete nitrogen as ammonia; they are protected from the toxic effects of high concentrations of ammonia not only by the removal of ammonia from their bodies but also by rapid dilution of the excreted ammonia by the water in the environment. The principal waste product of nitrogen metabolism in terrestrial animals is urea (a water-soluble compound); its reactions provide some interesting comparisons with the citric acid cycle. Birds excrete nitrogen in the form of uric acid, which is insoluble in water. They do not have to carry the excess weight of water, which could hamper flight, to rid themselves of waste products.

The fate of nitrogen on which we shall concentrate is the production of urea in the **urea cycle** (Figure 18.17). The nitrogens that enter the urea cycle do so first as ammonia in the form of ammonium ion. The immediate precursor is glutamate, but the ammonia nitrogens of glutamate have ultimately come from many sources as a result of transamination reactions. A condensation reaction between the ammonium ion and carbon dioxide produces **carbamoyl phosphate** in a reaction that requires two molecules of ATP for each molecule of carbamoyl phosphate. Carbamoyl phosphate reacts with **ornithine** to form **citrulline.** A second nitrogen enters the urea cycle when aspartate reacts with citrulline to form **argininosuccinate** in another reaction that requires ATP (AMP and PP$_i$ are produced in this reaction). The amino group of the aspartate is the source of the second nitrogen in the urea that will be formed in this series of reactions. Argininosuccinate is split to produce **arginine** and **fumarate.** Finally, arginine is hydrolyzed to give urea and to regenerate ornithine. The biosynthesis of arginine from ornithine is discussed in Interchapter B. Another way of looking at the urea cycle is to consider arginine the immediate precursor of urea, producing ornithine in the process. According to this point of view, the rest of the cycle is the regeneration of arginine from ornithine.

The synthesis of fumarate is a link between the urea cycle and the citric acid cycle. Fumarate is, of course, an intermediate of the citric acid cycle, and it can be converted to oxaloacetate. A transamination reaction can convert oxaloacetate to aspartate, providing another link between the two cycles (Figure 18.17). Four "high-energy" phosphate bonds are required because of the production of pyrophosphate in the conversion of aspartate to argininosuccinate.

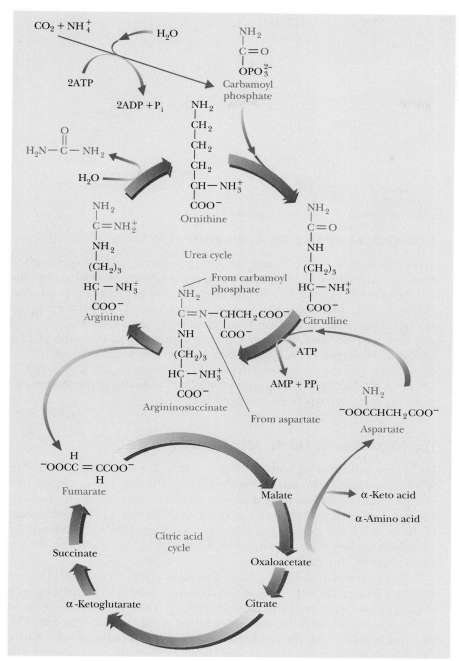

**FIGURE 18.17** The urea cycle and some of its links to the citric acid cycle.

## 18.8
## PURINE NUCLEOTIDE METABOLISM: THE ANABOLISM OF PURINE NUCLEOTIDES

We have already discussed the formation of ribose 5-phosphate as part of the pentose phosphate pathway (Section 13.6). The biosynthetic pathway for both purine and pyrimidine nucleotides makes use of preformed ribose 5-phosphate. Purines and pyrimidines are synthesized in different ways, and we shall consider them separately.

477

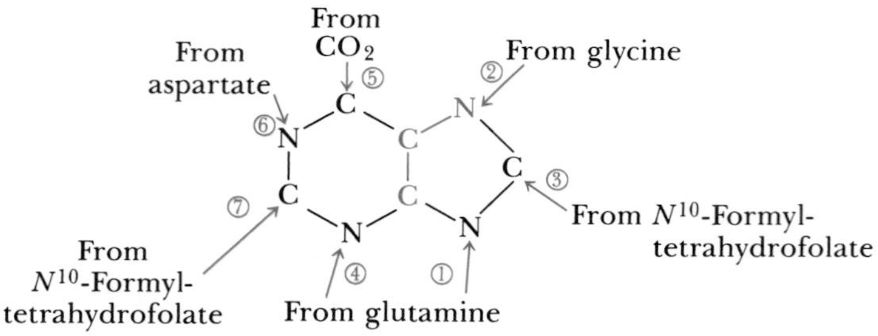

FIGURE 18.18    Sources of the atoms in the purine ring in purine nucleotide biosynthesis. The numbering system indicates the order in which each atom or group of atoms is added.

## Anabolism of Inosine Monophosphate (IMP)

In the synthesis of purine nucleotides, the growing ring system is bonded to the ribose phosphate while the purine skeleton is being assembled, first the five-membered ring and then the six-membered ring, eventually producing inosine 5'-monophosphate. All four nitrogen atoms of the purine ring are derived from amino acids: two from glutamine, one from aspartate, and one from glycine. Two of the five carbon atoms (adjacent to the glycine nitrogen) also come from glycine, two more come from tetrahydrofolate derivatives, and the fifth comes from $CO_2$ (Figure 18.18). The series of reactions producing IMP is long and complex; the details of the process can be found in Interchapter B, Section B.6.

## The Conversion of IMP to AMP and GMP

IMP is the precursor of both AMP and GMP. The conversion of IMP to AMP takes place in two stages (Figure 18.19). The first step is the reaction of aspartate with IMP to form adenylosuccinate. This reaction is catalyzed by adenylosuccinate synthetase and requires GTP, not ATP. The cleavage of fumarate from adenylosuccinate to produce AMP is catalyzed by adenylosuccinase. This enzyme also functions in the synthesis of the six-membered ring of IMP.

The conversion of IMP to GMP also takes place in two stages (Figure 18.20). The first of the two steps is an oxidation in which the C—H group at the C-2 position is converted to a keto group. The oxidizing agent in the reaction is $NAD^+$, and the enzyme involved is IMP dehydrogenase. The nucleotide formed by the oxidation reaction is xanthosine 5'-phosphate (XMP). An amino group from the side chain of glutamine replaces the C-2 keto group of XMP to produce GMP. This reaction is catalyzed by GMP synthetase; ATP is hydrolyzed to AMP and $PP_i$ in the process. Note that there is some control over the relative levels of purine nucleotides; GTP is needed for the synthesis of adenine nucleotides, while ATP is required for the synthesis of guanine nucleotides. Each of the purine nucleotides must occur at a reasonably high level for the other to be synthesized.

Subsequent phosphorylation reactions produce purine nucleoside diphosphates (ADP and GDP) and triphosphates (ATP and GTP). The purine nucleoside monophosphates, diphosphates, and triphosphates are all feedback inhibitors of the first stage of their own biosynthesis. Also, AMP, ADP, and ATP inhibit the conversion of IMP to adenine nucleo-

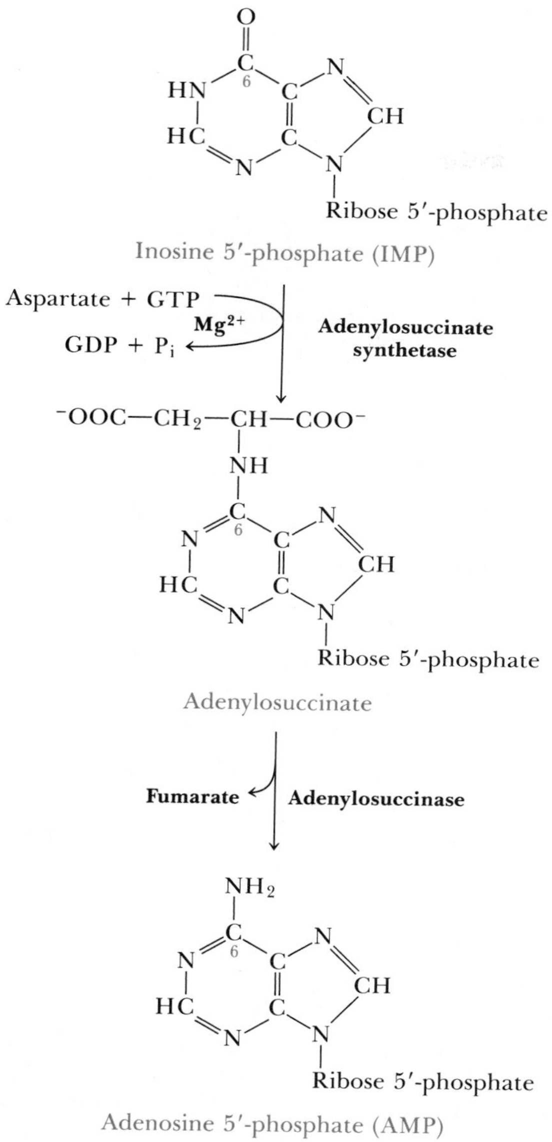

FIGURE 18.19   The synthesis of AMP from IMP.

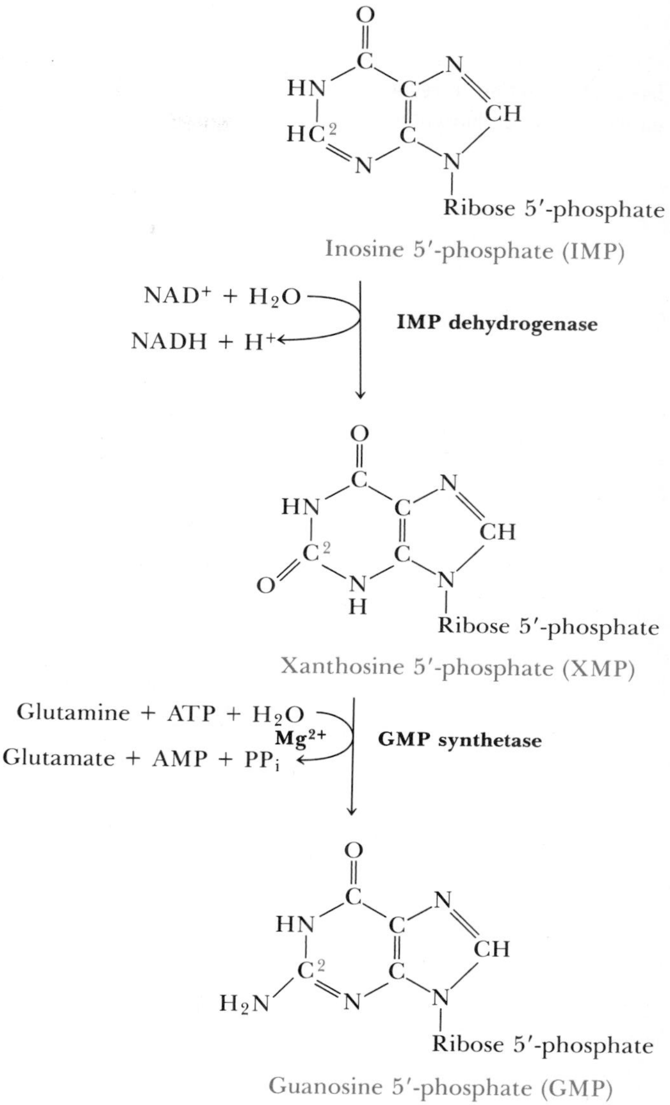

FIGURE 18.20   The synthesis of GMP from IMP.

tides, and GMP, GDP, and GTP inhibit the conversion of IMP to xanthylate and to guanine nucleotides (Figure 18.21).

## Energy Requirements for Production of AMP and GMP

The production of IMP starting with ribose 5-phosphate requires the equivalent of six ATP (see Section B.6 of Interchapter B). The conversion of IMP to AMP requires hydrolysis of an additional phosphate ester bond, in this case that of GTP. In the formation of AMP from ribose 5-phosphate, the equivalent of seven ATP is needed. The conversion of IMP to GMP requires two "high-energy" bonds, since a reaction occurs in

479

**FIGURE 18.21** The role of feedback inhibition in regulation of purine nucleotide biosynthesis.

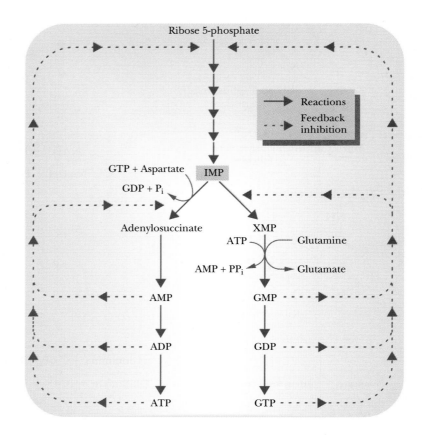

which ATP is hydrolyzed to AMP and $PP_i$. For the production of GMP from ribose 5-phosphate, the equivalent of eight ATP is necessary. The anaerobic oxidation of glucose produces only two ATP for each molecule of glucose (Section 13.1). Anaerobic organisms require four molecules of glucose (which produce eight ATP) for each AMP or GMP they form. The process is more efficient for aerobic organisms. Since 36 or 38 ATP result from each molecule of glucose, depending on the type of tissue, aerobic organisms can produce 5 AMP (requiring 35 ATP) or 4 GMP (requiring 32 ATP) for each molecule of glucose oxidized. A mechanism for reuse of purines, rather than complete turnover and new synthesis, saves energy for organisms.

## 18.9
## CATABOLIC REACTIONS OF PURINE NUCLEOTIDES

The catabolism of purine nucleotides proceeds by hydrolysis to the nucleoside and subsequently to the free base, which is further degraded. Deamination of guanine produces xanthine, and deamination of adenine produces hypoxanthine (the base corresponding to the nucleoside inosine) (Figure 18.22 (a)). Hypoxanthine can be oxidized to xanthine, so this base is a common degradation product of both adenine and guanine. Xanthine

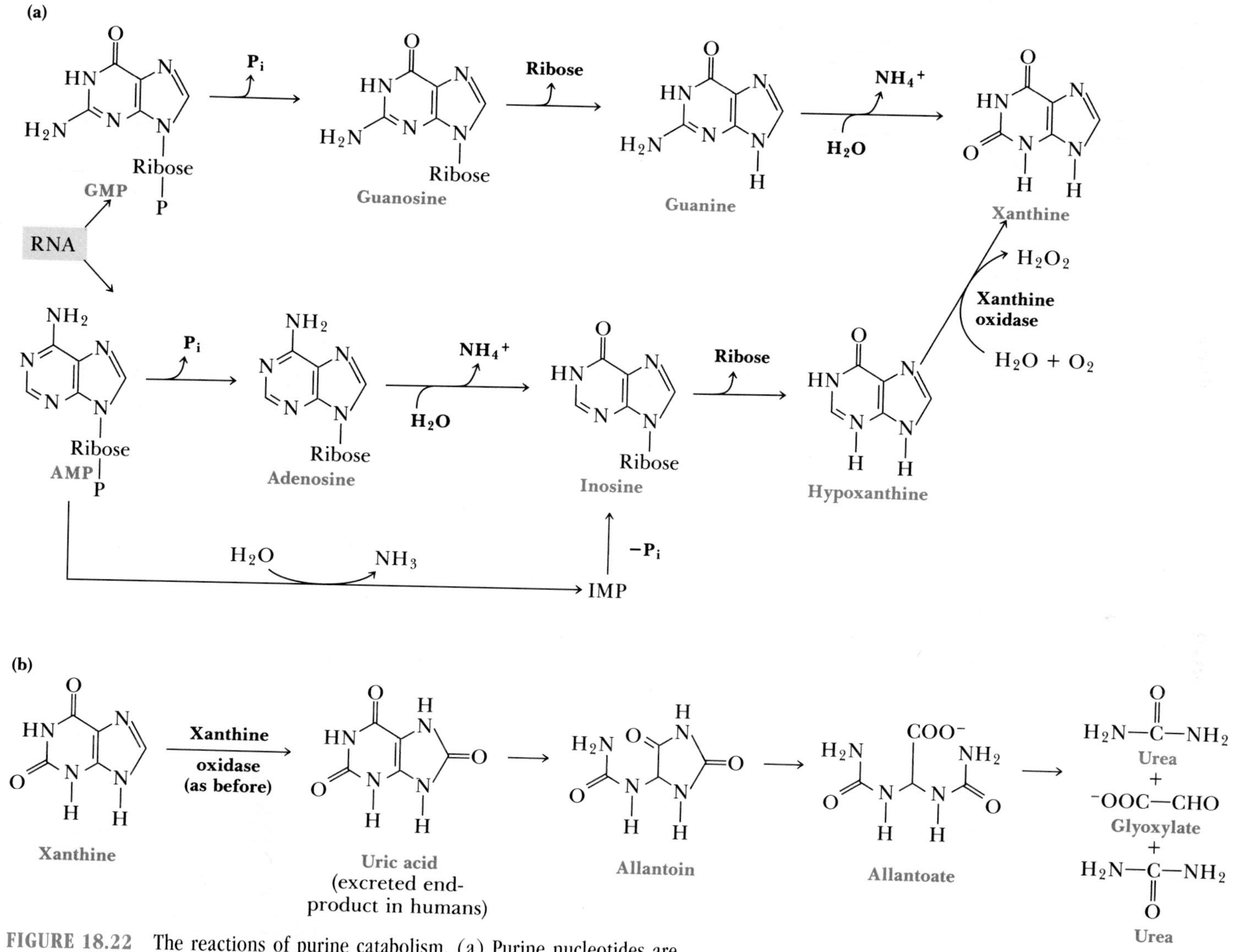

**FIGURE 18.22** The reactions of purine catabolism. (a) Purine nucleotides are converted to the free base and then to xanthine. (b) Catabolic reactions of xanthine.

481

Allopurinol, a substance used in the treatment of gout

is oxidized in turn to **uric acid.** In birds, some reptiles, and insects, and in dalmatian dogs and primates (including humans), uric acid is the end product of purine metabolism and is excreted. In all other terrestrial animals, including all other mammals, allantoin is the product excreted, while allantoate is the product in fish. Allantoate is further degraded to glyoxylate and urea by microorganisms and some amphibia (Figure 18.22 (b)). **Gout** is a disease in humans that is caused by overproduction of uric acid. Deposits of uric acid (which is not soluble in water) accumulate in the joints of the hands and feet. Benjamin Franklin suffered from gout, and he left eloquent testimony to the painful nature of this disease. **Allopurinol** is a compound used to treat gout; it inhibits the degradation of hypoxanthine to xanthine and of xanthine to uric acid, preventing the buildup of uric acid deposits.

**Salvage reactions** are important in the metabolism of purine nucleotides because of the amount of energy required for the synthesis of the purine bases. A free purine base that has been cleaved from a nucleotide can produce the corresponding nucleotide by reacting with the compound phosphoribosylpyrophosphate (PRPP), formed by a transfer of a pyrophosphate group from ATP to ribose 5-phosphate (Figure 18.23).

Two different enzymes with different specificity with respect to the purine base catalyze salvage reactions. The reaction

$$\text{Adenine} + \text{PRPP} \longrightarrow \text{AMP} + \text{PP}_i$$

is catalyzed by adenine phosphoribosyltransferase. The corresponding reactions of guanine and hypoxanthine

$$\text{Hypoxanthine} + \text{PRPP} \xrightarrow{\text{HPRT}} \text{IMP} + \text{PP}_i$$

$$\text{Guanine} + \text{PRPP} \xrightarrow{\text{HPRT}} \text{GMP} + \text{PP}_i$$

are catalyzed by hypoxanthine-guanine phosphoribosyltransferase (HPRT). A deficiency in HPRT can result in a serious disorder, the Lesch–Nyhan syndrome (Box 18.2).

## BOX 18.2
## LESCH–NYHAN SYNDROME

A deficiency of the HPRT enzyme is the cause of the Lesch–Nyhan syndrome, a genetic disease. The biochemical consequences include an elevated concentration of PRPP and increased production of purines and uric acid. The accumulation of uric acid leads to kidney stones and gout, but the most striking clinical manifestations are neurological in nature. There is a compulsive tendency to self-mutilation among patients with the Lesch–Nyhan syndrome; they tend to bite off fingertips and parts of the lips. The development of kidney stones and gouty symptoms can be prevented by the administration of allopurinol, but there is no real treatment for the self-destructive behavior and the mental retardation and spasticity that accompany it. The diverse manifestations of this disease show clearly that metabolism is extremely complex and that the failure of one enzyme to function can have consequences that reach far beyond the reaction that it catalyzes.

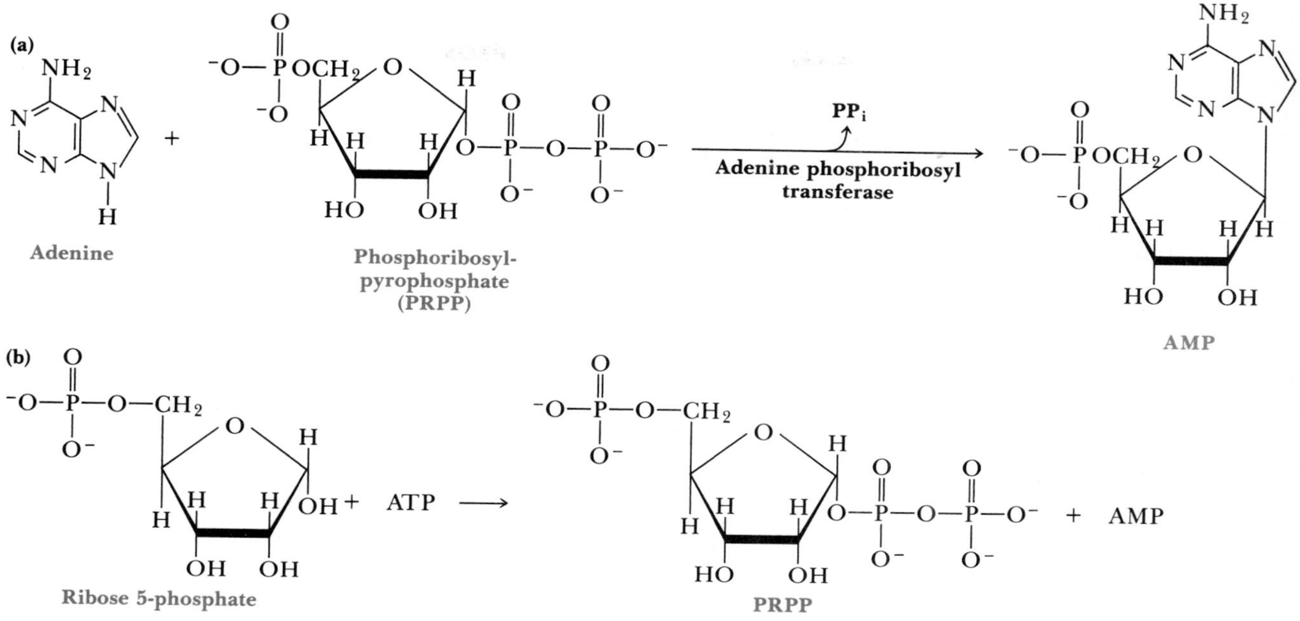

**FIGURE 18.23**   Purine salvage. (a) Adenine is the purine in this example. There are analogous reactions for salvage of guanine and hypoxanthine (see text). (b) The formation of PRPP.

## 18.10
## PYRIMIDINE NUCLEOTIDE METABOLISM: ANABOLISM AND CATABOLISM

### The Anabolism of Pyrimidine Nucleotides

The overall scheme of pyrimidine nucleotide biosynthesis differs from that of purine nucleotides in that the pyrimidine ring is assembled before it is attached to ribose 5-phosphate. The carbon and nitrogen atoms of the pyrimidine ring come from carbamoyl phosphate and aspartate. The production of carbamoyl phosphate for pyrimidine biosynthesis takes place in the cytosol, and the nitrogen donor is glutamine. (We have already seen a reaction for the production of carbamoyl phosphate when we discussed the urea cycle in Section 18.7. That reaction differs from this one because it takes place in mitochondria and the nitrogen donor is $NH_4^+$).

$$HCO_3^- + \text{glutamine} + 2 \text{ ATP} + H_2O \longrightarrow$$

$$\text{carbamoyl phosphate} + \text{glutamate} + 2 \text{ ADP} + P_i$$

The reaction of carbamoyl phosphate with aspartate to produce N-carbamoylaspartate is the committed step in pyrimidine biosynthesis. The compounds involved in reactions up to this point in the pathway can play other roles in metabolism; after this point, N-carbamoylaspartate can be used only to produce pyrimidines—thus the term "committed step."

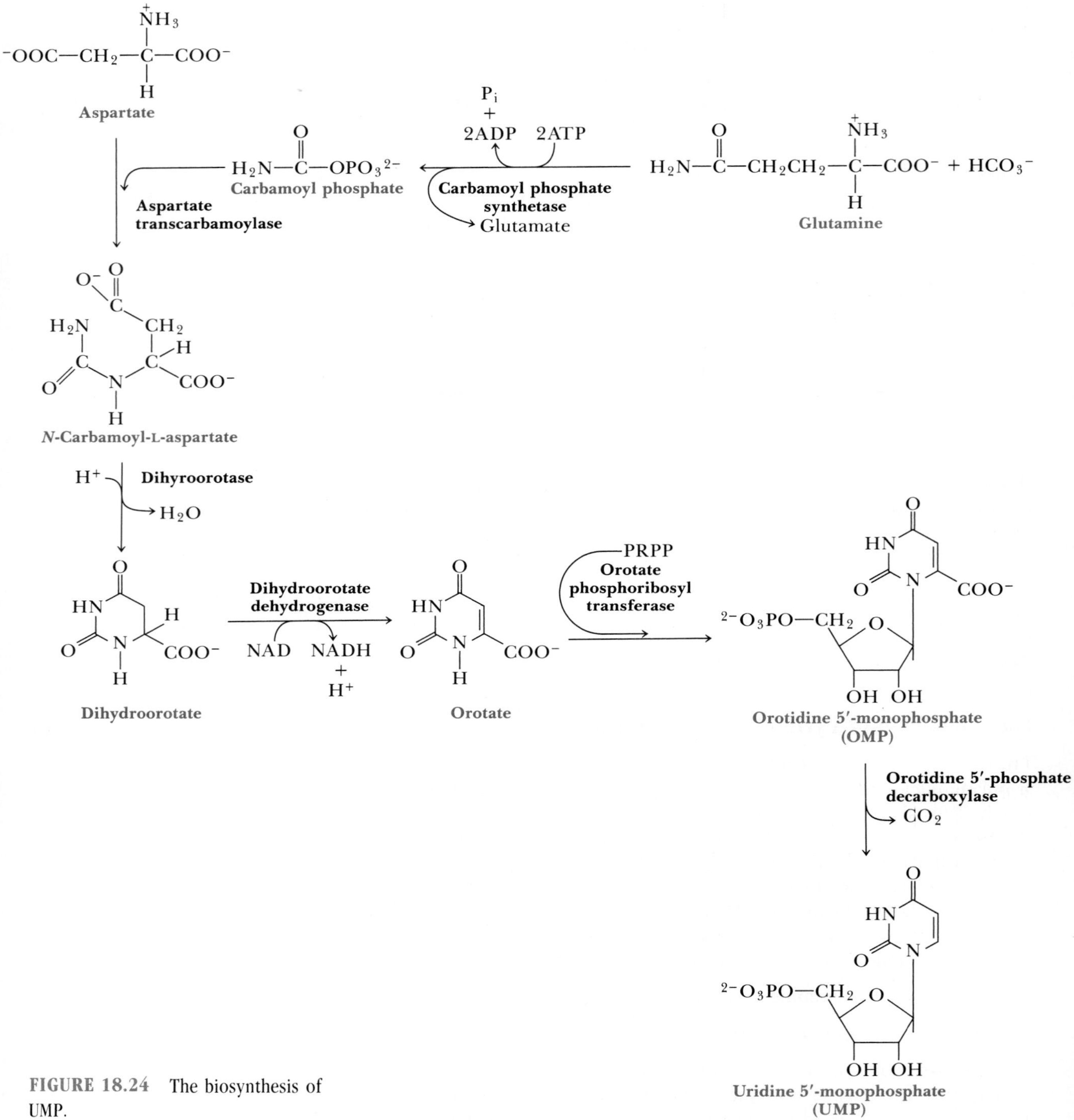

FIGURE 18.24    The biosynthesis of UMP.

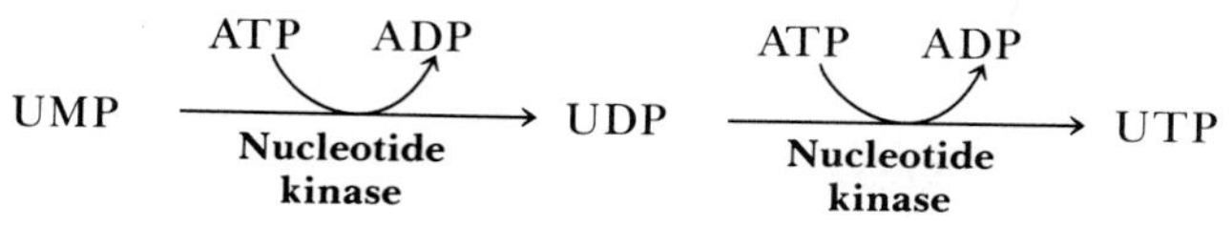

UMP $\xrightarrow[\substack{\textbf{Nucleotide} \\ \textbf{kinase}}]{\text{ATP \quad ADP}}$ UDP $\xrightarrow[\substack{\textbf{Nucleotide} \\ \textbf{kinase}}]{\text{ATP \quad ADP}}$ UTP

**FIGURE 18.25**   The conversion of UMP to UTP.

This reaction is catalyzed by aspartate transcarbamoylase, which we discussed in detail in Chapter 10 as a prime example of an allosteric enzyme subject to feedback regulation. The next step, which is the conversion of N-carbamoylaspartate to dihydroorotate, takes place in a reaction that involves an intramolecular dehydration (loss of water) as well as cyclization. This reaction is catalyzed by dihydroorotase. Dihydroorotate is converted to orotate by dihydroorotate dehydrogenase, with the concomitant conversion of $NAD^+$ to NADH. A pyrimidine nucleotide is now formed by the reaction of orotate with PRPP to give orotidine 5'-phosphate (OMP), which is a reaction similar to that which takes place in purine salvage (Section 18.9). Orotate phosphoribosyl transferase catalyzes this reaction. Finally, orotidine 5'-phosphate decarboxylase catalyzes the conversion of OMP to UMP (uridine 5'-phosphate), which is the precursor of the remaining pyrimidine nucleotides (Figure 18.24).

Two successive phosphorylation reactions convert UMP to UTP (Figure 18.25). The conversion of uracil to cytosine takes place in the triphosphate form, catalyzed by CTP synthetase (Figure 18.26). Glutamine is the nitrogen donor, and ATP is required, as we have seen earlier in similar reactions.

$$\text{UTP} + \text{glutamine} + \text{ATP} \longrightarrow \text{CTP} + \text{glutamate} + \text{ADP} + \text{P}_i$$

Feedback inhibition in pyrimidine nucleotide biosynthesis takes place in several ways. CTP is an inhibitor of aspartate transcarbamoylase and of CTP synthetase. UMP is an inhibitor of an even earlier step, the one catalyzed by carbamoyl phosphate synthetase (Figure 18.27).

## Pyrimidine Catabolism

Pyrimidine nucleotides are broken down first to the nucleoside and then to the base, as is the case with purine nucleotides. Cytosine can be deaminated to uracil, and the double bond of the uracil ring is reduced to produce

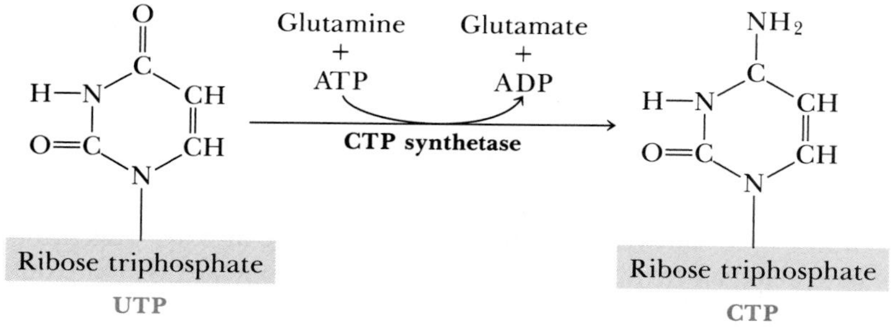

**FIGURE 18.26**   The conversion of UTP to CTP.

**FIGURE 18.27** The role of feedback inhibition in the regulation of pyrimidine nucleotide biosynthesis.

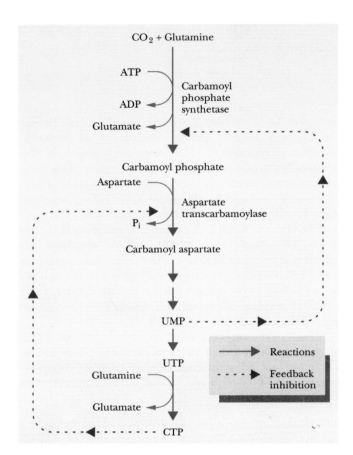

dihydrouracil. The ring opens to produce $N$-carbamoylpropionate, which in turn is broken down to $NH_4^+$, $CO_2$, and $\beta$-alanine (Figure 18.28).

## 18.11
## THE REDUCTION OF RIBONUCLEOTIDES TO DEOXYRIBONUCLEOTIDES

Ribonucleoside diphosphates are reduced to 2′-deoxyribonucleoside diphosphates in all organisms (Figure 18.29 (a)); NADPH is the reducing agent.

$$\text{Ribonucleoside diphosphate} + NADPH + H^+ \longrightarrow$$

$$\text{deoxyribonucleoside diphosphate} + NADP^+ + H_2O$$

The actual process, which is catalyzed by **ribonucleotide reductase,** is more complex than the equation above would indicate and involves some intermediate electron carriers. The ribonucleotide reductase system from *E. coli* has been extensively studied, and its mode of action gives some clues to the nature of the process. Two other proteins are required, **thioredoxin**

Cytosine $\xrightarrow[\text{H}_2\text{O} \quad \text{NH}_4^+]{}$ Uracil $\xrightarrow[\text{H}^+]{\text{NADPH} + \quad \text{NADP}^+}$ Dihydrouracil $\xrightarrow[\text{H}_2\text{O} \quad \text{H}^+]{}$

$$\underset{\text{N-Carbamoylpropionate}}{\text{H}_2\text{N} - \overset{\overset{\text{O}}{\|}}{\text{C}} - \text{NH} - \text{CH}_2 - \text{CH}_2 - \text{COO}^-} \xrightarrow{\text{H}_2\text{O}} \text{NH}_4^+ + \text{CO}_2 + \underset{\beta\text{-Alanine}}{\text{H}_3\overset{+}{\text{N}} - \text{CH}_2 - \text{CH}_2 - \text{COO}^-}$$

**FIGURE 18.28**    The catabolism of pyrimidines.

and **thioredoxin reductase.** Thioredoxin contains a disulfide (S—S) group in its oxidized form and two sulfhydryl (—SH) groups in its reduced form. NADPH reduces thioredoxin in a reaction catalyzed by thioredoxin reductase. The reduced thioredoxin in turn reduces a ribonucleoside diphosphate (NDP) to a deoxyribonucleoside diphosphate (dNDP) (Figure 18.29 (b)), and it is this reaction that is actually catalyzed by ribonucleotide reductase. Note that this reaction produces dADP, dGDP, dCDP, and dUDP. The first three are phosphorylated to give the corresponding triphosphates, which are substrates for the synthesis of DNA. Another required substrate for DNA synthesis is dTTP, and we shall now see how dTTP is produced from dUDP.

**FIGURE 18.29**    (a) The conversion of ribonucleotides to deoxyribonucleotides. TR(—S—S—) and TR (—SH)$_2$ refer to the oxidized (disulfide) and reduced (sulfhydryl) forms of thioredoxidin. (b) The structures of NDP and dNDP.

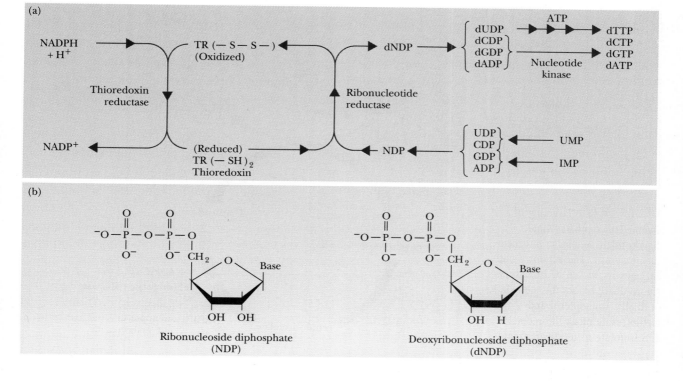

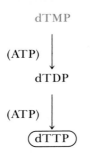

**FIGURE 18.30**    The conversion of dUDP to dTTP. ($FH_4$ is tetrahydrofolate; $FH_2$ is dihydrofolate.)

## 18.12
## THE CONVERSION OF dUDP TO dTTP

A one-carbon transfer is required for the conversion of uracil to thymine by attachment of the methyl group. The most important reaction in this conversion is that catalyzed by **thymidylate synthetase** (Figure 18.30). The source of the one-carbon unit is $N^5,N^{10}$-methylenetetrahydrofolate, which is converted to dihydrofolate in the process. The metabolically active form of the one-carbon carrier is tetrahydrofolate. Dihydrofolate must be reduced to tetrahydrofolate for this series of reactions to continue, and this process requires NADPH and **folate reductase.**

Since a supply of dTTP is necessary for DNA synthesis, inhibition of enzymes that catalyze the production of dTTP will inhibit the growth of rapidly dividing cells. Cancer cells, like all fast-growing cells, depend on continued DNA synthesis for growth. Inhibitors of thymidylate synthetase, such as fluorouracil (see Exercise 5), and inhibitors of folate reductase, such as aminopterin and methotrexate (structural analogues of folate), have been used in cancer chemotherapy (Figure 18.31). The intent of such therapy is to inhibit the formation of dTTP and thus of DNA in cancer cells, causing the death of the cancer cells with minimal effect on normal

**FIGURE 18.31**    Targeting of thymidylate synthetase and dihydrofolate reductase in cancer chemotherapy. The action of both enzymes is blocked by inhibitors. Fluoro-dUMP blocks the methylation of dUMP. Aminopterin and methotrexate block the reduction of dihydrofolate to tetrahydrofolate.

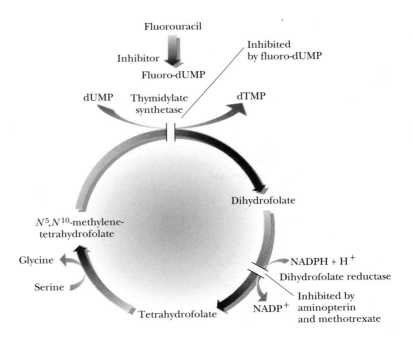

cells, which grow more slowly. There are adverse side effects to chemotherapy because of the highly toxic nature of most of the drugs involved, and enormous amounts of research are focused on finding safe and effective forms of treatment.

## SUMMARY

The metabolism of nitrogen encompasses a number of topics, including the anabolism and catabolism of amino acids, porphyrins, and nucleotides. Atmospheric nitrogen is the ultimate source of this element in biomolecules. Nitrogen fixation is the process by which molecular nitrogen from the atmosphere is made available to organisms, in the form of ammonia. Nitrification reactions convert $NO_3^-$ to $NH_3$ and provide another source of nitrogen. Feedback control mechanisms are a unifying factor in biosynthetic pathways involving nitrogen compounds.

In the anabolism of amino acids, transamination reactions play an important role. Glutamate and glutamine are frequently the amino group donors. The enzymes that catalyze transamination reactions frequently require pyridoxal phosphate as a coenzyme. One-carbon transfers also operate in the anabolism of amino acids. Carriers are required for the one-carbon groups transferred. Tetrahydrofolate is a carrier of methylene and formyl groups, while S-adenosylmethionine is a carrier of methyl groups. Some species, including humans, cannot synthesize all the amino acids required for protein synthesis and must obtain these essential amino acids from dietary sources. The biosynthesis of another important group of nitrogen-containing compounds, the porphyrins, begins with the amino acid glycine as its ultimate precursor.

The catabolism of amino acids has two parts: the fate of the nitrogen and the fate of the carbon skeleton. In the urea cycle, nitrogen released by the catabolism of amino acids is converted to urea. The carbon skeleton is converted to pyruvate or oxaloacetate, in the case of glucogenic amino acids, or to acetyl-CoA or acetoacetyl-CoA, in the case of ketogenic amino acids. Inborn errors of metabolism involving amino acids and their derivatives have serious consequences. Examples of such disorders include phenylketonuria and the porphyrias.

The anabolic pathway of nucleotide synthesis involving purines differs from that involving pyrimidines. Both pathways make use of preformed ribose 5-phosphate but differ with regard to the point in the pathway at which the sugar phosphate is attached to the base. In the case of purine nucleotides, the growing base is attached to the sugar phosphate during the synthesis. In pyrimidine biosynthesis, the base is first formed and then attached to the sugar phosphate. In catabolism, purine bases are frequently salvaged and reattached to sugar phosphates. Otherwise, purines are broken down to uric acid. Pyrimidines are degraded to β-alanine.

Deoxyribonucleotides for DNA synthesis are produced by the reduction of ribonucleoside diphosphates to deoxyribonucleoside diphosphates. Another reaction specifically needed to produce substrates for DNA synthesis is the conversion of uracil to thymine. This pathway, which requires a tetrahydrofolate derivative as the carrier for one-carbon transfer, is a target for cancer chemotherapy.

## EXERCISES

**1.** Lysine is frequently added to cereal products as a means of enhancing the quality of the product as a protein source. Suggest a reason for the choice.

**2.** Sketch the structure of folic acid; also show by a sketch how it serves as a carrier of one-carbon groups.

3. How many $\alpha$-amino acids participate directly in the urea cycle? Of these, how many can be used for protein synthesis?

4. Write an equation for the net reaction of the urea cycle; show how the urea cycle is linked to the citric acid cycle.

5. Suggest a mode of action for fluorouracil in cancer chemotherapy.

**5-Fluorouracil**

6. Why is there no net gain if homocysteine is converted to methionine with *S*-adenosylmethionine as the methyl donor?

7. Show, by the equation for a typical reaction, why glutamate plays a central role in the biosynthesis of amino acids.

8. List the essential amino acids for a phenyketonuric adult and compare them with the requirements for a normal adult.

9. What is an important difference between the biosynthesis of purine nucleotides and that of pyrimidine nucleotides?

10. List the intermediates in the flow of nitrogen from $N_2$ to the N-7 nitrogen of purines.

11. Sulfanilamide and related sulfa drugs were widely used to treat diseases of bacterial origin before penicillin and more advanced drugs were readily available. The inhibitory effect of sulfanilamide on bacterial growth can be reversed by *p*-aminobenzoate. Suggest a mode of action for sulfanilamide.

**Sulfanilamide**

12. By means of a structural formula, show how *S*-adenosylmethionine is a carrier of methyl groups.

13. Comment briefly on the usefulness to organisms of feedback control mechanisms in long biosynthetic pathways.

14. People on high-protein diets are advised to drink lots of water. Why?

15. One of the symptoms of the protein deficiency disease kwashiorkor is depigmentation of skin and hair. What is the biochemical basis for this observation?

16. Chemotherapy patients receiving cytotoxic (cell-killing) agents such as FdUMP (the UMP analogue that contains fluorouracil) and methotrexate bald temporarily. Why does this take place?

17. How many "high-energy" phosphate bonds must be hydrolyzed in the pathway that produces GMP from guanine and PRPP by the PRPP salvage reaction, compared with the number of such bonds hydrolyzed in the pathway leading to IMP and then to GMP?

## ANNOTATED BIBLIOGRAPHY

Bender, D.A. *Amino Acid Metabolism.* 2nd ed. New York: John Wiley, 1985. [A general treatment of the topic, with a particularly good section on tryptophan metabolism.]

Benkovic, S. On the Mechanism of Action of Folate- and Biopterin-Requiring Enzymes. *Ann. Rev. Biochem.* **49,** 227–254 (1980). [A review article on one-carbon transfers.]

Braunstein, A.E. Amino Group Transfer. In Boyer, P.D., ed. *The Enzymes.* 3rd ed. Vol. 9. New York: Academic Press, 1973. [Getting old, but a standard reference.]

Stadtman, E.R. Mechanisms of Enzyme Regulation in Metabolism. In Boyer, P.D., ed. *The Enzymes.* 3rd ed. Vol. 1. New York: Academic Press, 1970. [A review dealing with the importance of feedback control mechanisms.]

# The Anabolism of Nitrogen-Containing Compounds

**B**

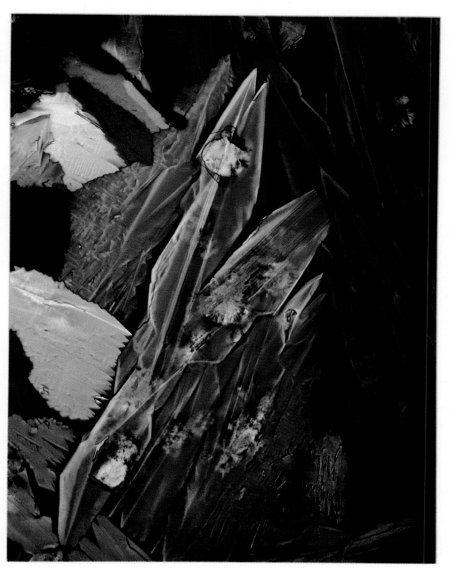

## OUTLINE

Crystals of L-tryptophan viewed under polarized light.

*Amino acids are more than the building blocks of protein molecules. They are the port of entry for organic nitrogen into a variety of nitrogen-containing compounds that includes the following: (1) The nitrogenous bases of DNA and RNA. (2) The nucleotide coenzymes, such as NAD$^+$ and NADP that serve as electron carriers in energy metabolism. (3) Heme, an oxygen binding component of the proteins myoglobin and hemoglobin. (4) Small physiologically active molecules like the hormone epinephrine (Adrenalin), and the neurotransmitter serotonin. Of the 20 amino acids, nine are synthesized by plants and microorganisms, but not by humans. These amino acids must be supplied from the diet and are called* essential amino acids. *Although each of the amino acids is synthesized by a different pathway, they may be grouped into families. For example, $\alpha$-ketoglutarate, a citric acid intermediate, is the precursor of the nonessential amino acid, glutamate; and glutamate is the precursor of glutamine, proline, and arginine. The pathways of the essential amino acids are more complex. In bacteria, for example, aspartate is the precursor of lysine, methionine, and threonine. Surplus amino acids are deaminated and nitrogen is excreted as urea, while their carbon skeletons are fed into the citric acid cycle to be used as fuel or recycled as components of other biomolecules.*

## B.1
## GLUTAMATE AS A PRECURSOR OF PROLINE AND ARGININE

Two amino acids, proline and arginine, can be classified as members of the glutamate family in addition to the ones we saw in Chapter 18. Proline is synthesized in several steps, involving the production of glutamate-$\gamma$-semialdehyde, which then cyclizes, ultimately forming proline (Figure B.1). The conversion of glutamate to the semialdehyde is inhibited by proline, an example of feedback inhibition in nitrogen metabolism. In addition to being a precursor of proline, glutamate-$\gamma$-semialdehyde undergoes other reactions, eventually forming ornithine, which gives rise to arginine in the urea cycle (Figure B.2). (See Section 18.7.) Arginine is an inhibitor of an early stage of this pathway.

## B.2
## THE ASPARTATE FAMILY

Aspartate, the precursor of the other amino acids in this family, is formed from oxaloacetate by a transamination reaction (Figure B.3). Aspartate in turn undergoes further reactions to produce isoleucine, threonine, lysine,

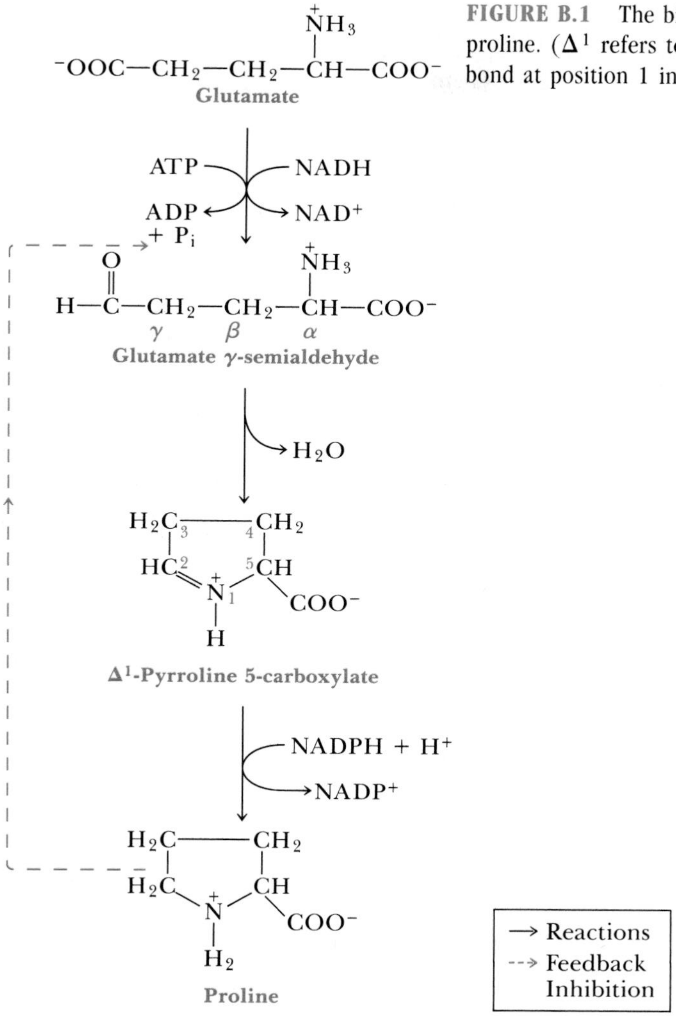

**FIGURE B.1** The biosynthesis of proline. ($\Delta^1$ refers to the double bond at position 1 in the ring.)

and methionine. The pathway that produces this group of amino acids comprises several branches. In methionine biosynthesis, the side-chain carboxylate of aspartate is phosphorylated in a reaction requiring ATP. The carboxyl group that has been phosphorylated is then reduced in two stages, first to the aldehyde, then to the alcohol, producing homoserine (Figure B.4). In bacteria the next step in methionine biosynthesis is the reaction of succinyl-CoA with homoserine to form $O$-succinylhomoserine. (There is an analogous reaction in plants, involving the reaction of homoserine with acetyl-CoA to form $O$-acetylhomoserine). The next step in bacterial methionine synthesis is the transfer of sulfur from cysteine to $O$-succinylhomoserine to form succinate and cystathionine. A hydrolysis reaction removes pyruvate and ammonia, producing homocysteine. Note

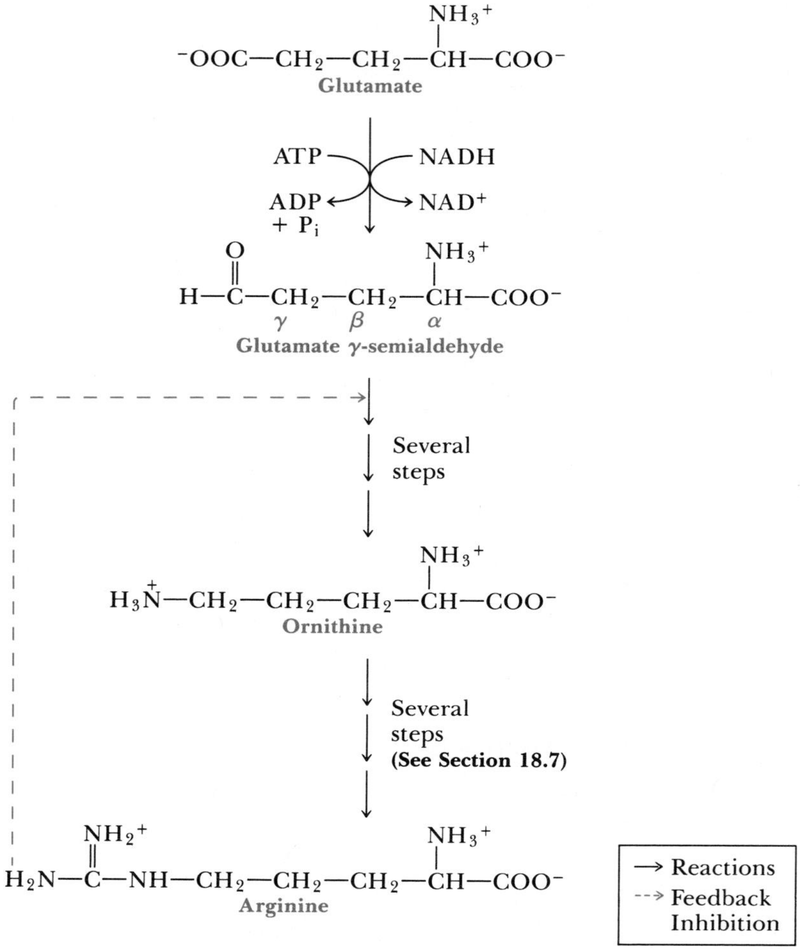

**FIGURE B.2**  The biosynthesis of arginine.

$$NH_3^+$$
$$^-OOC—CH_2—CH_2—\overset{|}{CH}—COO^-$$

**Glutamate**

ATP, NADH, ADP + $P_i$, NAD$^+$

$$\overset{O}{\overset{||}{H—C}}—CH_2—CH_2—\overset{\overset{NH_3^+}{|}}{CH}—COO^-$$
$$\gamma \qquad \beta \qquad \alpha$$

**Glutamate γ-semialdehyde**

Several steps

$$H_3\overset{+}{N}—CH_2—CH_2—CH_2—\overset{\overset{NH_3^+}{|}}{CH}—COO^-$$

**Ornithine**

Several steps
**(See Section 18.7)**

$$H_2N—\overset{\overset{NH_2^+}{||}}{C}—NH—CH_2—CH_2—CH_2—\overset{\overset{NH_3^+}{|}}{CH}—COO^-$$

**Arginine**

→ Reactions
--→ Feedback Inhibition

that homocysteine differs from homoserine only in the substitution of a thiol for a hydroxyl group in the side chain; homocysteine is also produced by plants in analogous reactions. The final step in methionine biosynthesis in both plants and bacteria is the transfer of a one-carbon unit from $N^5$-methyltetrahydrofolate to produce methionine from homocysteine. In animals no biosynthetic pathway for methionine exists, and thus it is an essential amino acid (See One-Carbon Transfers and the Serine Family in Section 18.4).

We shall not discuss in detail the biosynthesis of the other members of the aspartate family. Aspartyl-$\beta$-semialdehyde, which plays a role in the

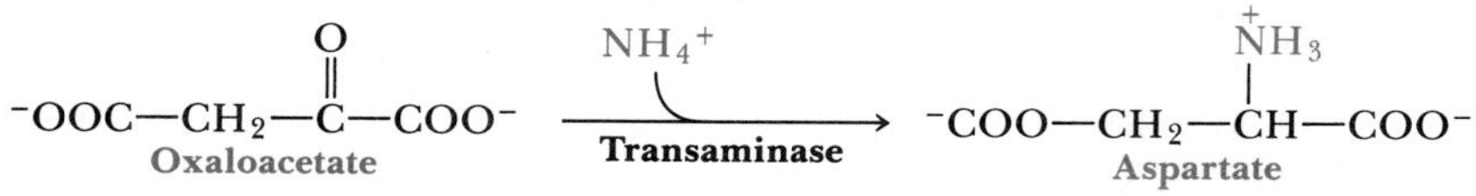

**FIGURE B.3**  Aspartate is produced by transamination of oxaloacetate.

$$^-OOC—CH_2—\overset{O}{\overset{||}{C}}—COO^- \xrightarrow[\textbf{Transaminase}]{NH_4^+} {}^-COO—CH_2—\overset{\overset{+}{NH_3}}{\overset{|}{CH}}—COO^-$$

**Oxaloacetate**           **Aspartate**

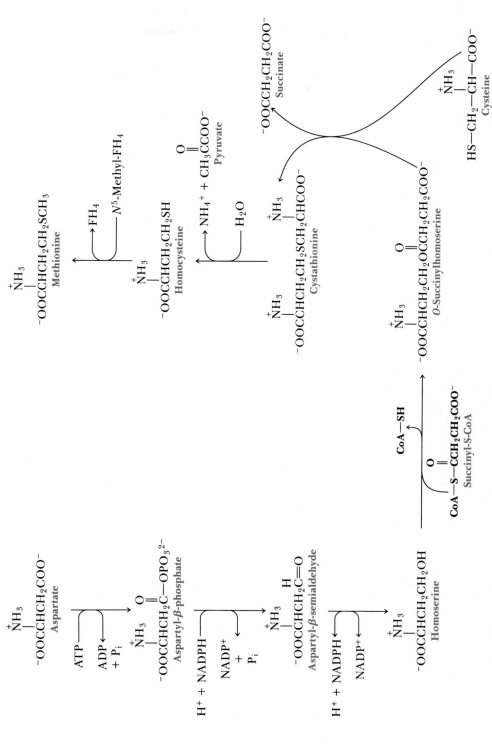

**FIGURE B.4**  The biosynthesis of methionine in bacteria. ($FH_4$ is tetrahydrofolate.)

biosynthesis of methionine, is also converted to lysine in a multistep process. Homoserine can be converted to threonine as well as to methionine. Threonine in turn is converted to isoleucine in a multistep process. Note in Figure B.5 that feedback inhibition occurs at nearly every step in this biosynthetic pathway.

FIGURE B.5    Feedback control in biosynthesis of amino acids of the aspartate family.

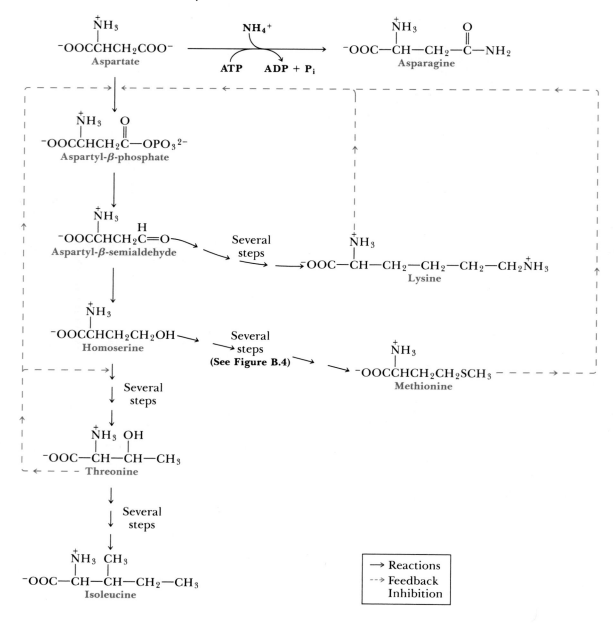

$$CH_3-\overset{\overset{\displaystyle O}{\|}}{C}-COO^- + {}^-OOC-\overset{\overset{\displaystyle \overset{+}{N}H_3}{|}}{CH}-CH_2-CH_2-COO^- \longrightarrow$$

Pyruvate           Glutamate

$$CH_3-\overset{\overset{\displaystyle \overset{+}{N}H_3}{|}}{CH}-COO^- + {}^-OOC-\overset{\overset{\displaystyle O}{\|}}{C}-CH_2-CH_2-COO^-$$

Alanine           α-Ketoglutarate

**FIGURE B.6**  Alanine is produced from pyruvate by transamination.

## B.3
## THE PYRUVATE FAMILY

The carbon skeletons of alanine, valine, and leucine are derived from pyruvate, the key metabolic intermediate that links glycolysis and the citric acid cycle. Alanine is obtained from pyruvate directly by a transamination reaction involving glutamate (Figure B.6). The formation of valine and leucine is a more complicated process. Several steps are necessary for the conversion of pyruvate to α-ketoisovalerate (Figure B.7). This intermediate

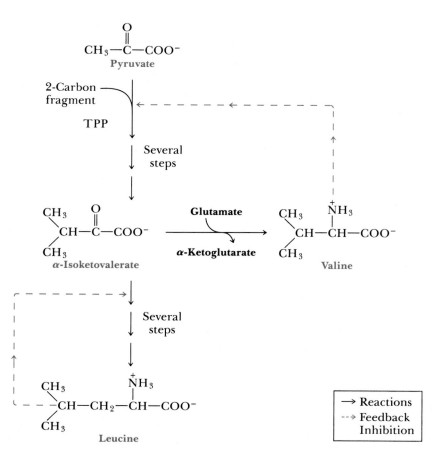

**FIGURE B.7**  The biosynthesis of leucine and valine from pyruvate showing feedback inhibition. (TPP is thiamine pyrophosphate.)

can be converted to valine by transamination, or it can undergo several other changes to produce $\alpha$-ketoisocaproate, which is converted to leucine by transamination. Both valine and leucine are inhibitors of early steps in their own biosynthesis.

## B.4
## THE AROMATIC AMINO ACIDS

The ultimate precursors of the carbon skeletons of the aromatic amino acids are phosphoenolpyruvate, which is a glycolytic intermediate, and erythrose 4-phosphate, which is an intermediate in the pentose phosphate pathway. The reaction of these two compounds eventually leads to the formation of shikimate, which in turn is converted to chorismate. The pathway for the synthesis of aromatic amino acids branches at chorismate; one branch leads to phenylalanine and tyrosine (Figure B.8), and the other branch leads to tryptophan. (Chorismate is also a precursor of folic acid, coenzyme Q (CoQ), and plastoquinone, but we shall not discuss those pathways.) Chorismate can be converted to phenylpyruvate (which leads to phenylalanine in a transamination reaction) or to $p$-hydroxyphenylpyruvate (which similarly leads to tyrosine by transamination). Phenylalanine can also be converted to tyrosine by a hydroxylation reaction.

The conversion of chorismate to tryptophan (Figure B.9) starts with the substitution of an amino group (which comes from the amide side chain of glutamine) on the ring system of chorismate, accompanied by the elimination of pyruvate, giving rise to anthranilate. Phosphoribosylpyrophosphate (PRPP) (see Section 18.9 and Figure 18.23) reacts with anthranilate, eventually giving rise to indole-3-glycerol phosphate. Finally, serine reacts with indole-3-glycerol phosphate to produce tryptophan and glyceraldehyde 3-phosphate. This reaction is catalyzed by tryptophan synthetase, an enzyme that requires pyridoxal phosphate. The most extensive studies of tryptophan synthetase have been done in *Neurospora crassa* and in bacterial systems such as *E. coli* and *Salmonella typhimurium*. Tryptophan is another example of an essential amino acid, since humans cannot synthesize it.

## B.5
## HISTIDINE BIOSYNTHESIS

Histidine is an essential amino acid for children, and it may be essential for adults as well; this last point has not been definitely established. The essential amino acids tend to have long and complex biosynthetic pathways, and histidine is no exception (Figure B.10). Most of the carbon skeleton of histidine comes from PRPP (Sections 18.9 and B.4). One carbon and one nitrogen of the imidazole ring come from ATP, and the second nitrogen of the imidazole ring comes from the side-chain amide of glutamine. The amino group comes from glutamate by the familiar transamination reaction. This pathway was established by studies in *Escherichia coli* and *Salmonella*. Histidine is a feedback inhibitor of the first stage of its own biosynthesis.

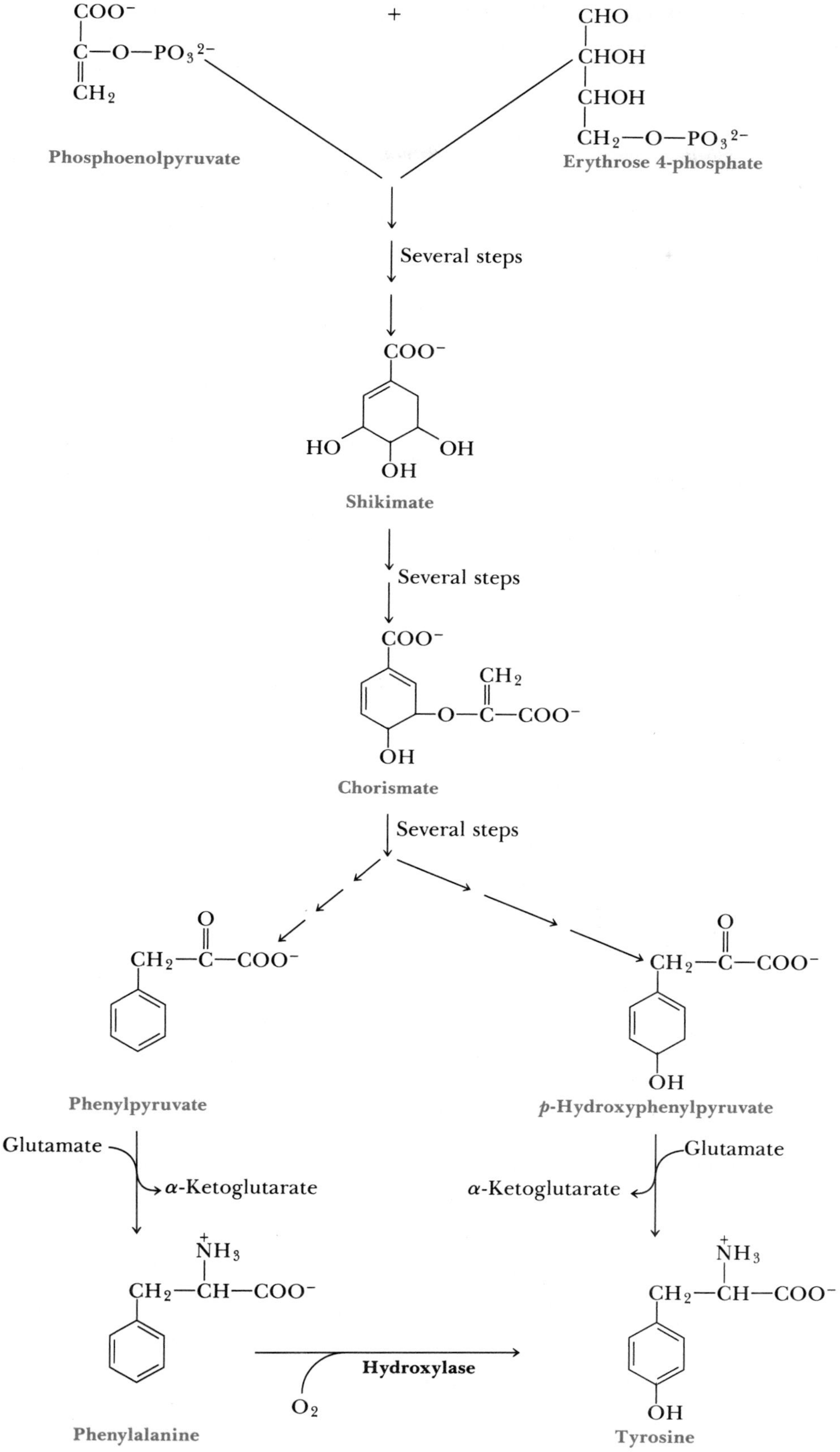

**FIGURE B.8** The biosynthesis of phenylalanine and tyrosine.

**FIGURE B.9**   The conversion of chorismate to tryptophan in *E. coli*. (P) is a phosphate group.

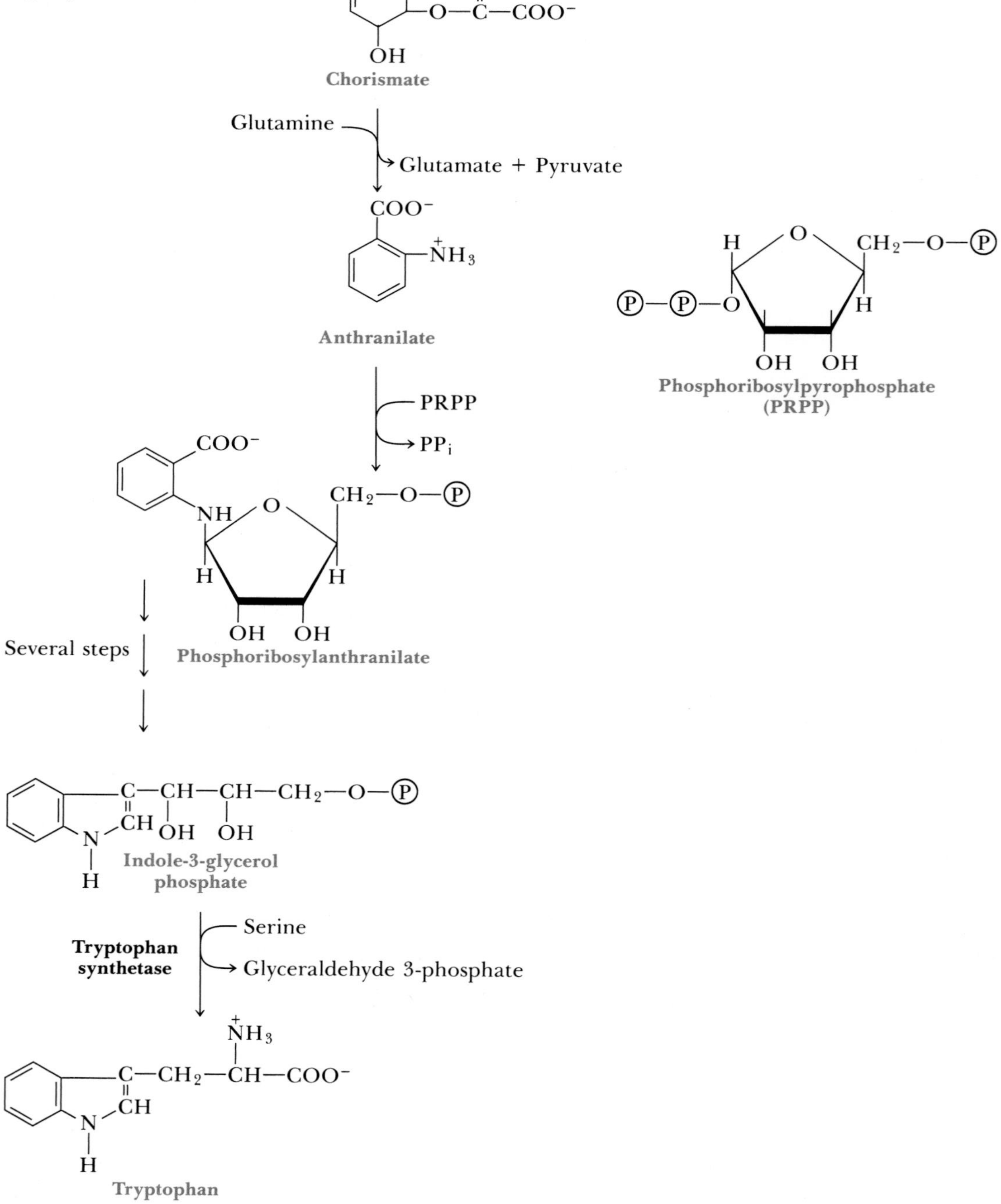

Chorismate

Glutamine

Glutamate + Pyruvate

Anthranilate

Phosphoribosylpyrophosphate
(PRPP)

PRPP

PP$_i$

Phosphoribosylanthranilate

Several steps

Indole-3-glycerol
phosphate

Serine

**Tryptophan
synthetase**

Glyceraldehyde 3-phosphate

Tryptophan

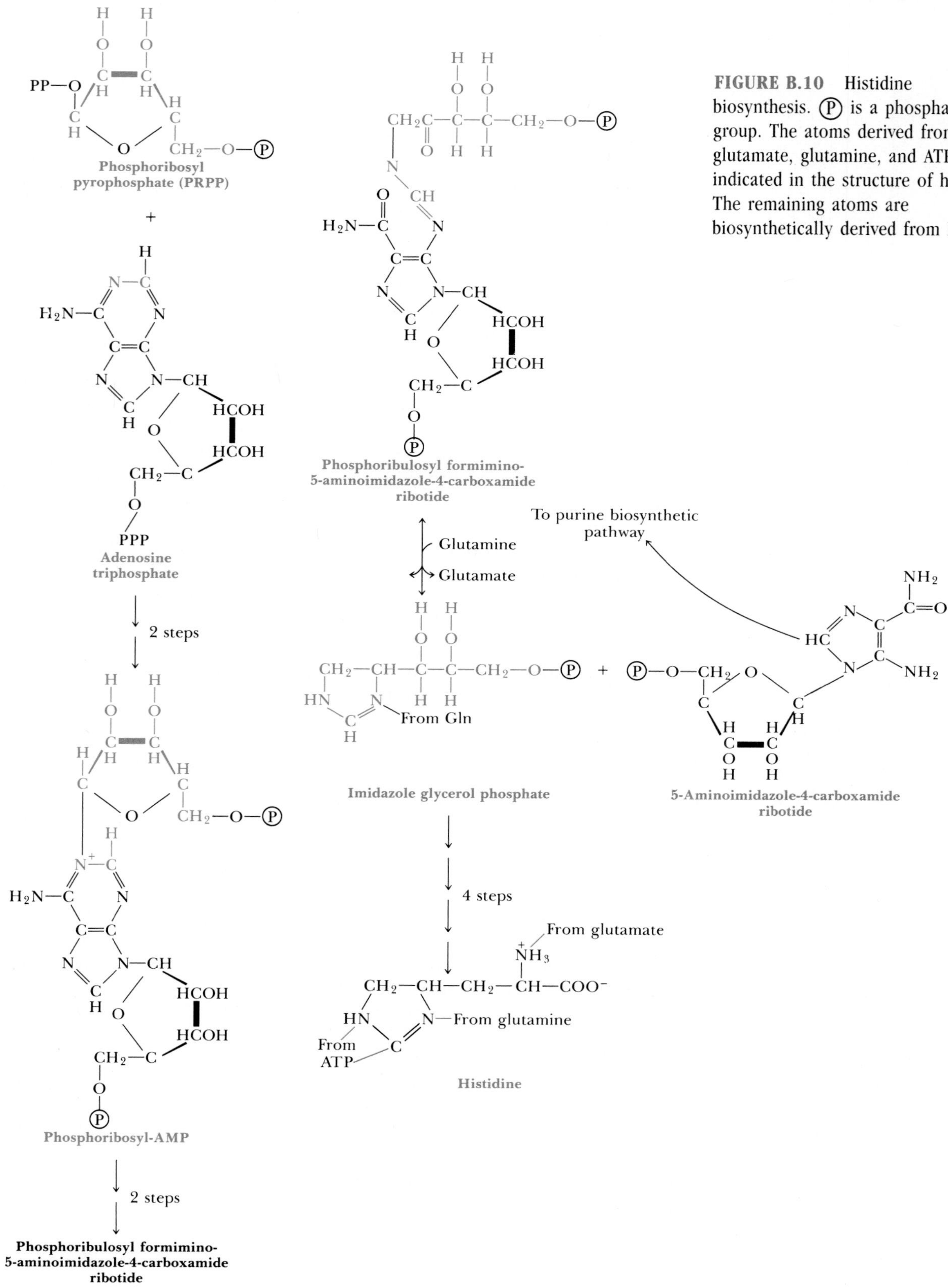

FIGURE B.10 Histidine biosynthesis. $\textcircled{P}$ is a phosphate group. The atoms derived from glutamate, glutamine, and ATP are indicated in the structure of histidine. The remaining atoms are biosynthetically derived from PRPP.

501

## B.6
## THE ANABOLIC PATHWAY FOR IMP

The biosynthetic pathway for purine nucleotides starts with ribose 5-phosphate; the reactions of the series attach the growing purine ring to the sugar phosphate. The form in which ribose 5-phosphate is used in the synthesis of purine nucleotides is PRPP. We shall use this pathway as a case study showing all the steps in a long and complex biosynthesis (Figure B.11).

PRPP reacts with glutamine as the nitrogen source to produce 5-phospho-$\beta$-D-ribosylamine. Glycine is then attached to the phosphoribosylamine by forming an amide linkage between the carboxyl group of the glycine and the amino group that came from glutamine. The product of this reaction is 5′-phosphoribosylglycinamide. This reaction also requires ATP. The next step is the addition of a one-carbon formyl group to the free amino group of the glycine. The source of the formyl group is a tetrahydrofolate derivative, $N^{10}$-formyltetrahydrofolate. The compound produced by the formylation reaction is 5′-phosphoribosyl-$N$-formylglycinamide. The keto group of an amide (the linkage between the carboxyl group contributed by glycine and the amino group contributed by the first glutamine) is next converted to an imino group, making use of nitrogen from a second glutamine. In the terminology of organic chemistry, the amide is converted to an amidine. The amidine produced is 5′-phosphoribosyl-$N$-formylamidine. Both ATP and $Mg^{2+}$ are required in this reaction. Finally, a ring closure takes place to produce the five-membered imidazole ring of the growing purine skeleton. The ring closure takes place between the carbon that came from the formyl group and the glycosidic nitrogen (the one bonded to the ribose). Once again ATP is required, and the product formed is 5′-phosphoribosyl-5-aminoimidazole. (The numbering used on the five-membered ring at this point is that for an imidazole, rather than that for a purine.)

Three of the six atoms of the six-membered ring, two carbons and a nitrogen, are already in place. Two carbon atoms and one nitrogen atom have yet to be added. The first of the two carbons comes from $CO_2$ in a carboxylation reaction. This reaction has the moderately unusual feature of requiring neither a cofactor, such as biotin, nor an energy source, such as ATP. The reason is that the substrate to which the $CO_2$ becomes attached, 5′-phosphoribosyl-5-aminoimidazole, is an eneamine, a highly reactive compound, which reacts directly with $CO_2$. The product of the carboxylation reaction is 5′-phosphoribosyl-5-aminoimidazole-4-carboxylate (Figure B.11). The nitrogen atom to be added to the six-membered ring comes from aspartate in a two-step process. The carboxyl group just added to the growing purine reacts with the amino group of aspartate to give 5′-phosphoribosyl-4-($N$-succinocarboxamide)-5-aminoimidazole. This reaction again requires ATP. Fumarate is then split off from the succinocarboxamide. The amino group is the only portion of the aspartate that will appear in the purine that is eventually formed. Addition of the final carbon atom is another one-carbon transfer involving $N^{10}$-formyltetrahydrofolate. The product of this reaction is 5′-phosphoribosyl-4-carboxamide-5-formamidoimidazole.

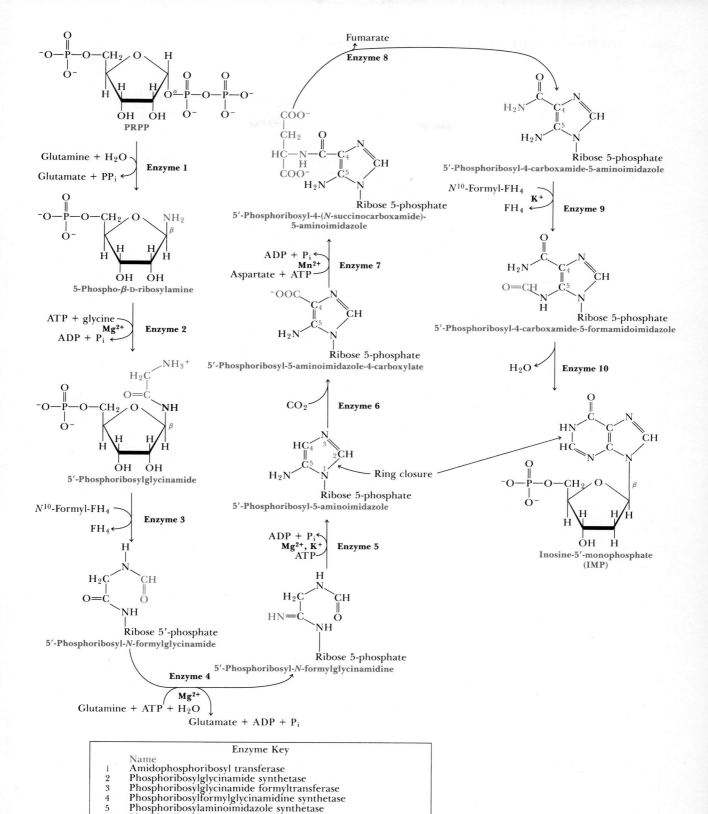

**FIGURE B.11** The formation of inosine 5'-monophosphate starting with PRPP.

Enzyme Key

| | Name |
|---|---|
| 1 | Amidophosphoribosyl transferase |
| 2 | Phosphoribosylglycinamide synthetase |
| 3 | Phosphoribosylglycinamide formyltransferase |
| 4 | Phosphoribosylformylglycinamidine synthetase |
| 5 | Phosphoribosylaminoimidazole synthetase |
| 6 | Phosphoribosylaminoimidazole carboxylase |
| 7 | Phosphoribosylaminoimidazole-succinocarboxamide synthetase |
| 8 | Adenylosuccinate lyase |
| 9 | Phosphoribosylaminoimidazolecarboxamide formyltransferase |
| 10 | IMP cyclohydrolase |

The last reaction is a ring closure that produces the six-membered ring of purines. The compound formed is inosine 5'monophosphate (IMP), which is a purine nucleotide. Note that four ATP were required for the synthesis of the purine ring, in addition to the two "high-energy" bonds (in the form of pyrophosphate) needed for the formation of PRPP, giving a total of six ATP required for the production of IMP starting with ribose 5-phosphate.

# Summary of Metabolism

## OUTLINE

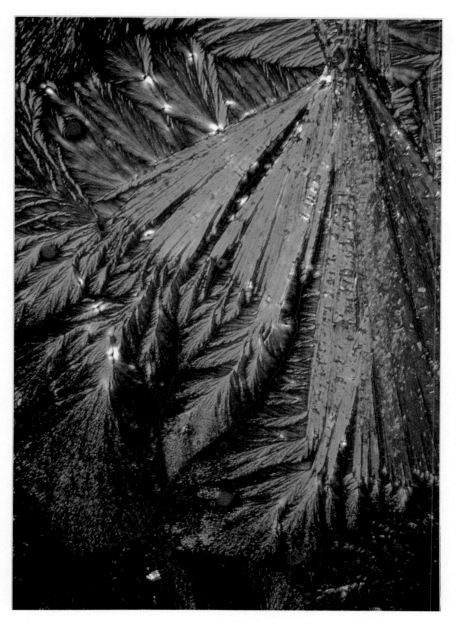

Crystals of ascorbic acid (vitamin C) viewed under polarized light.

*In the process of* catabolism, *large biopolymers are broken down to smaller molecules; while in* anabolism, *small precursors are built up into larger molecules. As a rule, catabolism provides the energy for building up large molecules. The citric acid cycle, as the hub of metabolic pathways, serves to connect the breakdown and synthesis of proteins, carbohydrates, and lipids. Pyruvate from glycolysis as well as fatty acids from lipids and catabolized protein molecules can be fed into the citric acid cycle as acetyl Co-A. Here, these metabolites can be oxidized to produce energy. Alternately, intermediates of the citric acid cycle can be converted to other biomolecules. Thus the citric acid cycle is* amphibolic, *meaning that its action is both catabolic and anabolic. The flow of molecules in the metabolic pathways can be regulated to either slow down or speed up the production of molecules. Overproduction can be controlled by regulatory enzymes that act by feedback inhibition to limit the creation of new molecules. On the other hand, hormones can send signals to speed up their production. Under stress, the hypothalamus in the brain sends hormone-releasing factors to the pituitary gland, which initiates a hormone cascade that increases the production of glucose to provide the extra energy needed to deal with a stressful situation.*

## 19.1
## ALL METABOLIC PATHWAYS ARE RELATED

In the preceding chapters we have learned about a number of individual metabolic pathways. Some metabolites, such as pyruvate, oxaloacetate, and acetyl-CoA, appear in more than one pathway. Furthermore, all the reactions of metabolism take place simultaneously. We shall now focus on some of the relationships among pathways by considering the related pathways themselves and the energetics of metabolism as a whole.

The **citric acid cycle** plays a central role in metabolism. Three main points can be considered in assigning a central role to the citric acid cycle. The first of these is its part in the **catabolism** of nutrients of the main types: carbohydrates, lipids, and proteins. The second is the function of the citric acid cycle in the **anabolism** of sugars, lipids, and amino acids. The third and final point is the relationship between individual metabolic pathways and the citric acid cycle, particularly in the storage and use of energy.

## 19.2
## THE CITRIC ACID CYCLE IN CATABOLISM

The nutrients taken in by an organism can include large molecules. This observation is especially true in the case of animals, which ingest polysaccharides and proteins, which are polymers, as well as lipids. Nucleic acids constitute a very small percentage of the nutrients present in foodstuffs, and we shall not consider their catabolism.

The first step in the breakdown of nutrients is the degradation of large molecules to smaller ones. Polysaccharides are hydrolyzed by specific enzymes to produce sugar monomers; an example is the breakdown of starch by amylases. Lipases hydrolyze triacylglycerols to give fatty acids and glycerol. Proteins are digested by proteases, with amino acids as the end products. Sugars, fatty acids, and amino acids then enter their specific catabolic pathways.

In Chapter 13 we discussed the glycolytic pathway, by which sugars are converted to pyruvate, which then enters the citric acid cycle. In Chapter 16 we saw how fatty acids are converted to acetyl-CoA; we learned about the fate of acetyl-CoA in the citric acid cycle in Chapter 14. Amino acids enter the cycle by various paths. We discussed specific catabolic reactions of amino acids in detail in Chapter 18.

Figure 19.1 shows schematically the various catabolic pathways that feed into the citric acid cycle. The catabolic reactions occur in the cytosol; the citric acid cycle takes place in mitochondria. Many of the end products of catabolism cross the mitochondrial membrane and then participate in

**FIGURE 19.1** Summary of catabolism showing the central role of the citric acid cycle. Note that the end products of the catabolism of carbohydrates, lipids, and amino acids all appear. (PEP is phosphoenolpyruvate; $\alpha$-KG is $\alpha$-ketoglutarate; TA is transamination; $\longrightarrow\!\!\longrightarrow\!\!\longrightarrow$ is a multistep pathway.)

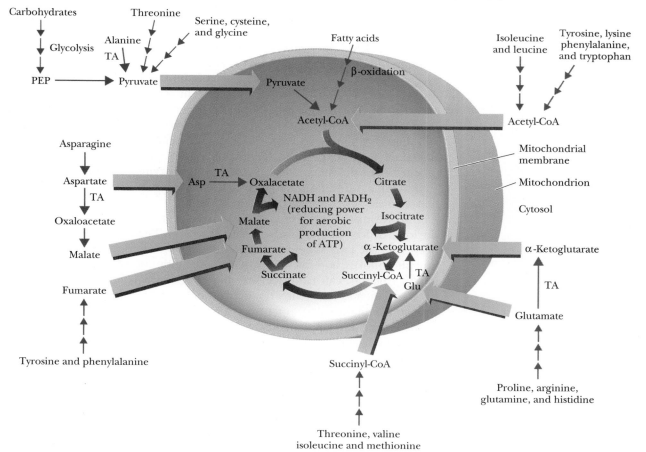

the citric acid cycle. This figure also shows the outline of pathways by which amino acids are converted to components of the citric acid cycle. Be sure to notice that sugars, fatty acids, and amino acids are all included in this overall catabolic scheme.

## Required Nutrients

In humans, the catabolism of **macronutrients** (carbohydrates, fats, and proteins) to supply energy is an important aspect of nutrition. In the United States, most diets provide a more than adequate number of nutritional calories. The typical American diet is high enough in fat that essential fatty acids (Section 16.6) are seldom, if ever, deficient. The only concern is that the diet contains an adequate supply of protein. If the intake of protein is sufficient, the supply of essential amino acids (Section 18.5) is normally sufficient as well. Packaging on food items frequently includes the protein content both in terms of the number of grams of protein and the percentage of the recommended daily allowance (RDA) suggested by the Food and Nutrition Board under the auspices of the National Research Council of the National Academy of Sciences (see Table 19.1).

    **Micronutrients** (vitamins and minerals) are also listed on food packaging. RDA are listed for the fat-soluble vitamins—vitamins A, D, and E (Section 7.3)—but care must be taken to avoid overdoses of these vitamins. Toxic effects are known to occur, especially with vitamin A, when excess amounts of fat-soluble vitamins accumulate in adipose tissue. With water-soluble vitamins, turnover is frequent enough that the danger of excess is not normally a problem. The water-soluble vitamins with listed RDA are vitamin C, necessary for the prevention of scurvy (Section 9.3); and the B vitamins—niacin, pantothenic acid, vitamin $B_6$, riboflavin, thiamine, folic acid, biotin, and vitamin $B_{12}$. The B vitamins are the precursors of the metabolically important coenzymes listed in Table 10.2. In that table, references are given to the reactions in which the coenzymes play a role.

    **Minerals** in the nutritional sense are inorganic substances required in the ionic or free element form for life processes. The macrominerals (those needed in the largest amounts) are sodium, potassium, chloride, magnesium, phosphorus, and calcium. The required amounts of all these minerals, except calcium, can easily be satisfied by a normal diet. Deficiencies of calcium can, and frequently do, occur. Such deficits lead to bone fragility, with concomitant risk of fracture, which is a problem especially for elderly women. Calcium supplements are indicated in such cases. Requirements for some microminerals (trace minerals) are not always clear. It is known, for example, from biochemical evidence that chromium is necessary for glucose metabolism and manganese for bone formation, but no deficiencies of these elements have been recorded. Requirements have been established for iron, copper, zinc, iodide, and fluoride; there is an RDA for all these minerals except fluoride. In the case of copper and zinc, needs are easily met by dietary sources. A deficiency of iodide, leading to an enlarged

**TABLE 19.1   U. S. Recommended Daily Allowances for the Average Man and Woman, ages 19 to 22**

| NUTRIENT | MAN | WOMAN |
| --- | --- | --- |
| Protein | 56 g | 44 g |
| Lipid-soluble vitamins | | |
|    Vitamin A | 1 mg RE* | 0.8 mg RE* |
|    Vitamin D | 7.5 $\mu$g† | 7.5 $\mu$g† |
|    Vitamin E | 10 mg $\alpha$-TE‡ | 8 mg $\alpha$-TE‡ |
| Water-soluble vitamins | | |
|    Vitamin C | 60 mg | 60 mg |
|    Thiamine (vitamin $B_1$) | 1.5 mg | 1.1 mg |
|    Riboflavin (vitamin $B_2$) | 1.7 mg | 1.3 mg |
|    Vitamin $B_6$ | 2.2 mg | 2 mg |
|    Vitamin $B_{12}$ | 3 $\mu$g | 3 $\mu$g |
|    Niacin | 19 mg | 14 mg |
|    Folic acid | 0.4 mg | 0.4 mg |
|    Pantothenic acid (estimate) | 10 mg | 10 mg |
|    Biotin (estimate) | 0.3 mg | 0.3 mg |
| Minerals | | |
|    Calcium | 800 mg | 800 mg |
|    Phosphorus | 800 mg | 800 mg |
|    Magnesium | 350 mg | 300 mg |
|    Zinc | 15 mg | 15 mg |
|    Iron | 10 mg | 18 mg |
|    Copper (estimate) | 3 mg | 3 mg |
|    Iodine | 150 $\mu$g | 150 $\mu$g |

*RE = retinol equivalent, where 1 retinol equivalent = 1 $\mu$g retinol or 6 $\mu$g $\beta$-carotene. See Section 7.5.
†As cholecalciferol. See Section 7.5.
‡$\alpha$-TE = $\alpha$-tocopherol equivalent, where 1 $\alpha$-TE = 1 mg $d$-$\alpha$-tocopherol. See Section 7.5. Data from the Food and Nutrition Board, National Academy of Sciences–National Research Council, Washington, D.C., 1988.

thyroid gland (Section 19.4), has been a problem in some parts of the United States for many years. Fluoride is necessary to prevent tooth decay in children and has been added with that end in mind to water supplies, sometimes causing considerable controversy. Iron is important because it is part of the structure of the ubiquitous heme proteins. Women of childbearing age are more susceptible to iron deficiencies than are other segments of the population, and in some cases supplements are advised.

# 19.3
## THE CITRIC ACID CYCLE IN ANABOLISM

### Carbohydrate Anabolism

The citric acid cycle is a source of starting materials for the biosynthesis of many important biomolecules, but the supply of the starting materials that are components of the cycle must be replenished if the cycle is to continue operating. In particular, the oxaloacetate in an organism must be main-

**FIGURE 19.2** The necessity of anaplerotic reactions in mammals. An anabolic reaction uses a citric acid cycle intermediate ($\alpha$-ketoglutarate is transaminated to glutamate in our example), competing with the rest of the cycle. The concentration of acetyl-CoA rises and signals the allosteric activation of pyruvate carboxylase to produce more oxaloacetate. (*Anaplerotic reaction; **part of glyoxylate pathway.)

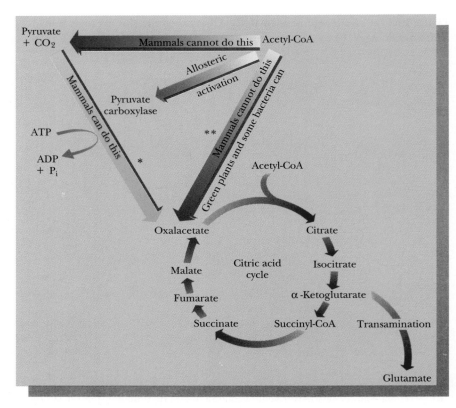

tained at a level sufficient to allow acetyl-CoA to enter the cycle. In some organisms acetyl-CoA can be converted to oxaloacetate and other citric acid cycle intermediates, but mammals cannot do this. In mammals oxaloacetate is produced from pyruvate by the enzyme **pyruvate carboxylase** (Figure 19.2). We have already encountered this enzyme and this reaction in the context of gluconeogenesis (see Oxaloacetate Is an Intermediate in the Production of Phosphoenolpyruvate in Gluconeogenesis, Section 13.7), and here we have another highly important role for this enzyme and the reaction it catalyzes. This type of reaction is called **anaplerotic,** a word derived from the Greek expression meaning "to fill up," since it maintains an adequate supply of a metabolic intermediate. The supply of oxaloacetate would soon be depleted if there were no means of producing it from a readily available precursor.

This reaction, which produces oxaloacetate from pyruvate, provides a connection between the amphibolic citric acid cycle and the anabolism of sugars by gluconeogenesis. On this same topic of carbohydrate anabolism, we should note again that pyruvate cannot be produced from acetyl-CoA in mammals. Since acetyl-CoA is the end product of catabolism of fatty acids, we can see that mammals cannot exist with fats or acetate as the sole carbon source. The intermediates of carbohydrate metabolism would soon be depleted. Carbohydrates are the principal energy and carbon source in animals (Figure 19.2). Plants can carry out the conversion of acetyl-CoA to

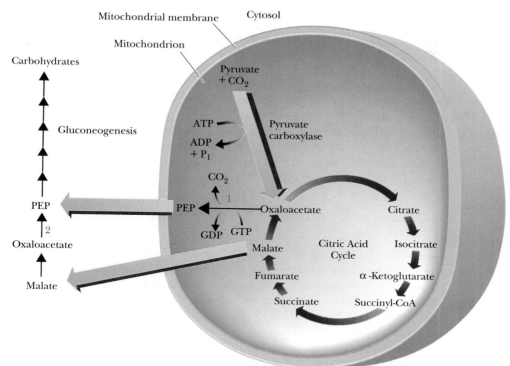

**FIGURE 19.3** Transfer of the starting materials of gluconeogenesis from the mitochondrion to the cytosol. Note that phosphoenolpyruvate (PEP) can be transferred from the mitochondrion to the cytosol, as can malate. Oxaloacetate is not transported across the mitochondrial membrane. (① is PEP carbokinase in mitochondria; ② is PEP carbokinase in cytosol; other symbols are as in Figure 19.1.)

pyruvate and oxaloacetate, so they can exist without carbohydrates as a carbon source. The conversion of pyruvate to acetyl-CoA does take place in both plants and animals (see Conversion of Pyruvate to Acetyl-CoA, Section 14.3).

The anabolic reactions of gluconeogenesis take place in the cytosol. Two mechanisms exist for the transfer of molecules needed for gluconeogenesis from mitochondria to the cytosol. One mechanism takes advantage of the fact that phosphoenolpyruvate can be formed from oxaloacetate in the mitochondrial matrix (this reaction is the next step in gluconeogenesis); phosphoenolpyruvate is then transferred to the cytosol, where the remaining reactions take place (Figure 19.3). Oxaloacetate is not transported across the mitochondrial membrane. The other mechanism relies on the fact that malate, which is another intermediate of the citric acid cycle, can be transferred to the cytosol. There is a **malate dehydrogenase** enzyme in the cytosol as well as in mitochondria, and malate can be converted to oxaloacetate in the cytosol.

$$\text{Malate} + \text{NAD}^+ + \text{H}^+ \longrightarrow \text{oxaloacetate} + \text{NADH}$$

Oxaloacetate is then converted to phosphoenolpyruvate, leading to the rest of the steps of gluconeogenesis (Figure 19.3).

Gluconeogenesis has many steps in common with the production of glucose in photosynthesis, but photosynthesis also has many reactions in common with the pentose phosphate pathway. Thus nature has evolved common strategies to deal with carbohydrate metabolism in all its aspects.

## Lipid Anabolism

The starting point of lipid anabolism is acetyl-CoA. The anabolic reactions of lipid metabolism, like those of carbohydrate metabolism, take place in the cytosol; these reactions are catalyzed by soluble enzymes that are not bound to membranes. Acetyl-CoA is mainly produced in mitochondria, whether from pyruvate or from breakdown of fatty acids. It is not clear whether acetyl-CoA is directly transferred to the cytosol (Figure 19.4), but an indirect transfer mechanism does exist in which citrate is transferred to the cytosol. Citrate reacts with CoA-SH to product citryl-CoA, which is then cleaved to yield oxaloacetate and acetyl-CoA. The enzyme that catalyzes this reaction requires ATP and is called ATP–citrate lyase. The overall reaction is

$$\text{Citrate} + \text{CoA-SH} + \text{ATP} \longrightarrow \text{acetyl CoA} + \text{oxaloacetate} + \text{ADP} + \text{P}_i$$

Acetyl-CoA is the starting point for lipid anabolism in both plants and animals. (An important source of acetyl-CoA is the catabolism of carbohydrates. We have just seen that animals cannot convert lipids to carbohydrates, but they can convert carbohydrates to lipids. The efficiency of the conversion of carbohydrates to lipids in animals is a source of considerable chagrin to many humans.)

Oxaloacetate can be reduced to malate by the reverse of a reaction we saw in the last section in the context of carbohydrate anabolism.

$$\text{Oxaloacetate} + \text{NADH} + \text{H}^+ \longrightarrow \text{malate} + \text{NAD}^+$$

Malate can move into and out of mitochondria by active transport processes, and the malate produced in this reaction can be used again in

**FIGURE 19.4**   Transfer of the starting materials of lipid anabolism from the mitochondrion to the cytosol. (① is ATP-citrate lyase; other symbols are as in Figure 19.1.) It is not definitely established whether acetyl-CoA is transported from the mitochondrion to the cytosol.

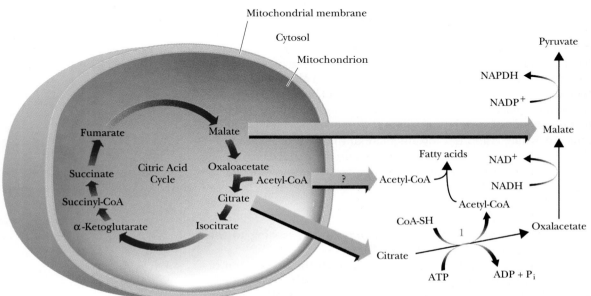

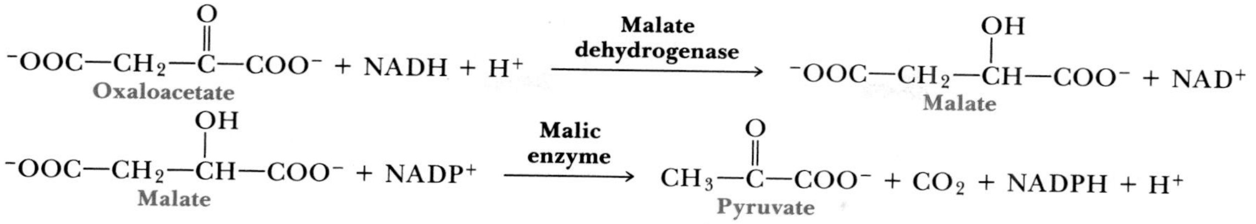

$$^{-}OOC-CH_2-\overset{\overset{O}{\|}}{C}-COO^- + NADH + H^+ \xrightarrow{\text{Malate dehydrogenase}} \ ^{-}OOC-CH_2-\overset{\overset{OH}{|}}{CH}-COO^- + NAD^+$$

Oxaloacetate                                                                 Malate

$$^{-}OOC-CH_2-\overset{\overset{OH}{|}}{CH}-COO^- + NADP^+ \xrightarrow{\text{Malic enzyme}} CH_3-\overset{\overset{O}{\|}}{C}-COO^- + CO_2 + NADPH + H^+$$

Malate                                                          Pyruvate

**FIGURE 19.5** Reactions involving citric acid cycle intermediates that produce NADPH for fatty acid anabolism. Note that these reactions take place in the cytosol.

the citric acid cycle. However, malate need not be transported back into mitochondria but can be oxidatively decarboxylated to pyruvate by **malic enzyme,** which requires $NADP^+$.

$$\text{Malate} + NADP^+ \longrightarrow \text{pyruvate} + CO_2 + NADPH + H^+$$

These last two reactions are a reduction reaction followed by an oxidation; there is *no net oxidation*. There is, however, a *substitution of NADPH for NADH*. This last point is an important one, since many of the enzymes of fatty acid synthesis require NADPH. The pentose phosphate pathway (Section 13.6) is the principal source of NADPH in most organisms, but here we have another source as well (Figure 19.5).

The two ways of producing NADPH clearly indicate that all metabolic pathways are related. The malate-citryl-CoA shuttle is a control mechanism in lipid anabolism, while the pentose phosphate pathway is part of carbohydrate metabolism. Both carbohydrates and lipids are important energy sources in many organisms, particularly animals.

## Anabolism of Amino Acids and Other Metabolites

The anabolic reactions that produce amino acids have, as a starting point, those intermediates of the citric acid cycle that can cross the mitochondrial membrane into the cytosol. We have already seen that malate can cross the mitochondrial membrane and give rise to oxaloacetate in the cytosol. Oxaloacetate can undergo a transamination reaction to produce aspartate, and aspartate in turn can undergo further reactions to form not only amino acids but also other nitrogen-containing metabolites, such as pyrimidines. Similarly, isocitrate can cross the mitochondrial membrane and produce $\alpha$-ketoglutarate in the cytosol. Glutamate arises from $\alpha$-ketoglutarate as a result of another transamination reaction, and glutamate undergoes further reactions to form still more amino acids. Succinyl-CoA is another citric acid cycle intermediate that can cross the mitochondrial membrane. It gives rise not to amino acids but to the porphyrin ring of the heme group.

The overall outline of anabolic reactions is shown in Figure 19.6. We used the same type of diagram in Figure 19.1 to show the overall outline of catabolism. The similarity of the two schematic diagrams points out that

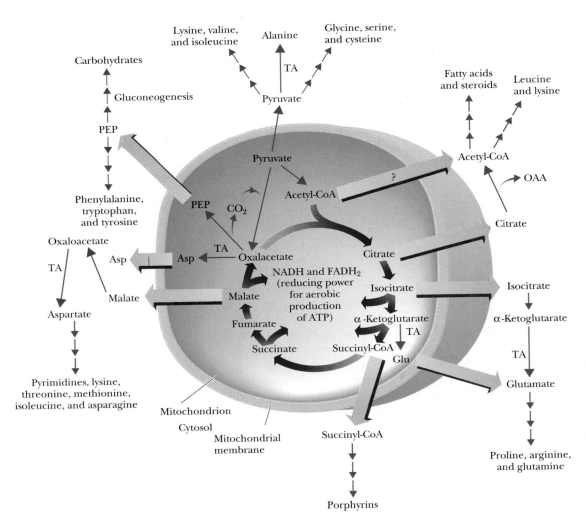

**FIGURE 19.6** Summary of anabolism showing the central role of the citric acid cycle. Note that there are pathways for the biosynthesis of carbohydrates, lipids, and amino acids. It is not definitely established whether acetyl-CoA is transported from the mitochondrion to the cytosol. OAA is oxaloacetate. Symbols are as in Figure 19.1.

catabolism and anabolism, while not exactly the same, are closely related. The operation of any metabolic pathway, anabolic or catabolic, can be "speeded up" or "slowed down" in response to the needs of an organism by control mechanisms such as feedback control. Regulation of metabolism takes place in similar ways in many different pathways.

## 19.4
## HORMONES AND SECOND MESSENGERS

The metabolic processes within a given cell can be, and frequently are, regulated by signals from outside the cell. A usual means of intercellular communication takes place through the workings of the **endocrine system,** in which the ductless glands produce **hormones** as intercellular

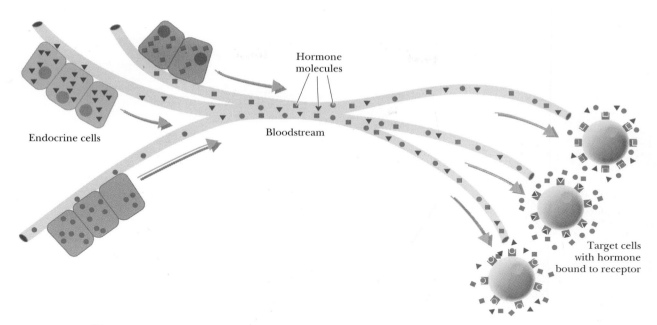

Endocrine cells

Hormone
molecules

Bloodstream

Target cells
with hormone
bound to receptor

messengers. Hormones are transported from the site of their synthesis to the site of action by the bloodstream (Figure 19.7). Typical hormones can be steroids, such as estrogens, androgens, and mineralocorticoids (Section 16.8); polypeptides, such as insulin and endorphins (Section 5.5); and amino acid derivatives, such as epinephrine and norepinephrine (Table 19.2).

Hormones have several important functions in the body. They help to maintain **homeostasis,** the balance of biological activities in the body. The effect of insulin in keeping the blood glucose level within narrow limits is an example of this function. The operation of epinephrine and norepinephrine in the "fight or flight" syndrome is an example of the way in which hormones mediate response to external stimuli. Finally, hormones play a role in growth and development, as seen in the roles of growth hormone and the sex hormones. The methods and insights of both biochemistry and physiology help illuminate the workings of the endocrine system.

The release of hormones exerts control on the cells of target organs, while other control mechanisms in turn determine the workings of the endocrine gland that releases the hormone in question. Simple feedback mechanisms can be postulated, in which the action of the hormone leads to feedback inhibition of the release of hormone (Figure 19.8). The workings of the endocrine system are in fact much less simple, with the added complexity allowing for a greater degree of control. To take a rather restricted example, insulin is released in response to a rapid rise in the level of blood glucose. In the absence of control mechanisms, an excess of insulin can produce **hypoglycemia,** the condition of low blood glucose. In addition to negative feedback control on the release of insulin, the action of the hormone glucagon tends to increase the level of glucose in the bloodstream. The two hormones together regulate blood glucose.

**FIGURE 19.7**   Endocrine cells secrete hormone into the bloodstream, which transports them to target cells.

**TABLE 19.2    Selected Human Hormones**

| HORMONE | SOURCE | MAJOR EFFECTS |
|---|---|---|
| **Polypeptides** | | |
| Corticotropin-releasing factor (CRF) | Hypothalamus | Stimulates release of ACTH |
| Gonadotropin-releasing factor (GnRF) | Hypothalamus | Stimulates FSH and LH release |
| Thyrotropin-releasing factor (TRF) | Hypothalamus | Stimulates TSH release |
| Growth hormone–releasing factor (GRF) | Hypothalamus | Stimulates growth hormone release |
| Adrenocorticotropic hormone (ACTH) | Anterior pituitary | Stimulates release of adrenocorticosteroids |
| Thyrotropin (TSH) | Anterior pituitary | Stimulates thyroxine release |
| Follicle-stimulating hormone (FSH) | Anterior pituitary | In ovaries: stimulates ovulation and estrogen synthesis; in testes: stimulates spermatogenesis |
| Luteinizing hormone (LH) | Anterior pituitary | In ovaries: stimulates estrogen and progesterone synthesis; in testes: stimulates androgen synthesis |
| Met-enkephalin | Anterior pituitary | Opioid effects on central nervous system |
| Leu-enkephalin | Anterior pituitary | Opioid effects on central nervous system |
| $\beta$-Endorphin | Anterior pituitary | Opioid effects on central nervous system |
| Vasopressin | Posterior pituitary | Stimulates water resorption by kidney and raises blood pressure |
| Oxytocin | Posterior pituitary | Stimulates uterine contractions and flow of milk |
| Insulin | Pancreas ($\beta$ cells of islets of Langerhans) | Stimulates uptake of glucose from bloodstream |
| Glucagon | Pancreas ($\alpha$ cells of islets of Langerhans) | Stimulates release of glucose to bloodstream |
| **Steroids** | | |
| Glucocorticoids | Adrenal cortex | Decrease inflammation, increase resistance to stress |
| Mineralocorticoids | Adrenal cortex | Maintain salt and water balance |
| Estrogens | Gonads and adrenal cortex | Development of secondary sex characteristics, particularly in females |
| Androgens | Gonads and adrenal cortex | Development of secondary sex characteristics, particularly in males |
| **Amino acid derivatives** | | |
| Epinephrine | Adrenal medulla | Increases heart rate and blood pressure |
| Norepinephrine | Adrenal medulla | Decreases peripheral circulation, stimulates lipolysis in adipose tissue |
| Thyroxine | Thyroid | General metabolic stimulation |

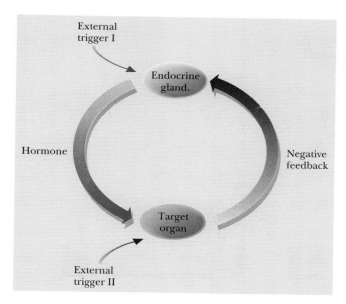

**FIGURE 19.8** A simple feedback control system involving an endocrine gland and a target organ.

A more sophisticated control system involves the action of the **hypothalamus,** the **pituitary,** and specific **endocrine** glands (Figure 19.9). The central nervous system sends a signal to the hypothalamus. The hypothalamus secretes a hormone-releasing factor, which in turn stimulates release of a **trophic hormone** by the anterior pituitary (Table 19.2). (The action of the hypothalamus on the posterior pituitary is mediated by nerve impulses.) Trophic hormones act on specific endocrine glands, which release the hormones to be transported to target organs. Note that feedback control is exerted at every stage of the process. Even more fine tuning is possible with zymogen activation mechanisms, which exist for many well-known hormones.

The releasing factors and trophic hormones listed in Table 19.2 tend to be polypeptides, but the chemical nature of the hormones released by specific endocrine glands shows greater variation. Thyroxine, for example, produced by the thyroid, is an iodinated derivative of the amino acid tyrosine (Section 5.2). Abnormally low levels of thyroxine lead to **hypothyroidism,** characterized by lethargy and obesity, while increased levels produce the opposite effect **(hyperthyroidism).** Low levels of iodine in the diet often lead to hypothyroidism and an enlarged thyroid gland **(goiter).** This condition has largely been eliminated by the addition of sodium iodide to commercial table salt ("iodized" salt).

Steroid hormones (Section 16.8) are produced by the adrenal cortex and the gonads (testes in males, ovaries in females). The **adrenocortical hormones** include **glucocorticoids,** which affect carbohydrate metabolism, modulate inflammatory reactions, and are involved in the reaction to stress. The **mineralocorticoids** control the level of excretion of water and salt by the kidney. If the adrenal cortex does not function adequately, one result is **Addison's disease,** characterized by hypoglycemia, weakness, and

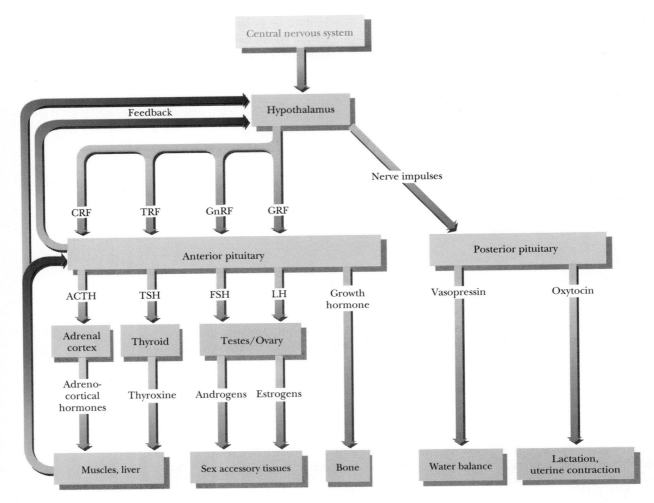

**FIGURE 19.9** Hormonal control systems showing the role of the hypothalamus, pituitary, and target tissues.

increased susceptibility to stress. This disease is eventually fatal unless treated by administration of mineralocorticoids and glucocorticoids to make up for what is missing. The opposite condition, hyperfunction, is frequently caused by a tumor of the adrenal cortex or of the pituitary. The characteristic clinical manifestation is **Cushing's syndrome,** marked by hyperglycemia, water retention, and the easily recognized "moon face."

The adrenal cortex produces some steroid sex hormones, the **androgens** and **estrogens,** but the main site of production is the gonads. Estrogens are required for female sexual maturation and function, but not for embryonic sexual development of female mammals. Information on sexual differentiation related to the action of steroid hormones has been obtained from studies of chromosome abnormalities in human patients of infertility clinics. While normal males have the XY genotype and normal females exhibit the XX genotype (have X and Y and 2 X chromosomes, respectively), there are XXY and XO (missing chromosome) individuals who are phenotypic (external-appearing) males and phenotypic females, respectively, but who are sterile. Extremely rare individuals are XX males and XY females. The XX males have a small segment of a normal Y chromosome attached to one of the X chromosomes; the XY females do not have this segment. A protein encoded by this part of the Y chromosome, the **testis-determining factor (TDF),** controls development of the undifferentiated embryonic gonads. In the presence of this factor, the gonads develop as testes; in its absence, they develop as ovaries. If a genotypic male animal has the gonads surgically removed, that individual will develop as a phenotypic female. Embryonic mammals develop as phenotypic females in the absence of male sex hormones.

As a final example, we shall discuss growth hormone (GH), which is a polypeptide. When overproduction of GH occurs, it is usually because of a pituitary tumor. If this condition occurs while the skeleton is still growing, the result is **giantism.** If the skeleton has stopped growing before the onset of GH overproduction, the result is **acromegaly,** characterized by enlarged hands, feet, and facial features. Underproduction of GH leads to **dwarfism,** but this condition can be treated by the injection of human GH before the skeleton reaches maturity. Animal GH is ineffective in treating dwarfism in humans. Supplies of human GH were very limited when it could be obtained only from cadavers, but it can now be synthesized by recombinant DNA techniques. (Another discussion of peptide hormones can be found in Box 5.1, which treats oxytocin and vasopressin.)

## Second Messengers

When a hormone binds to the specific receptor for it on a target cell, it sets off a chain of events in which the actual response within the cell is elicited. Several kinds of receptors are known. The receptors for steroid hormones tend to occur within the cell rather than as part of the membrane (steroids can pass the plasma membrane); steroid-receptor complexes affect the transcription of specific proteins. More frequently, the receptor proteins are a part of the plasma membrane. Binding of hormone to the receptor triggers the release of a **second messenger.** The second messenger brings about the changes within the cell as a result of a series of reactions.

Cyclic AMP (adenosine 3,'5'- phosphate, cAMP)

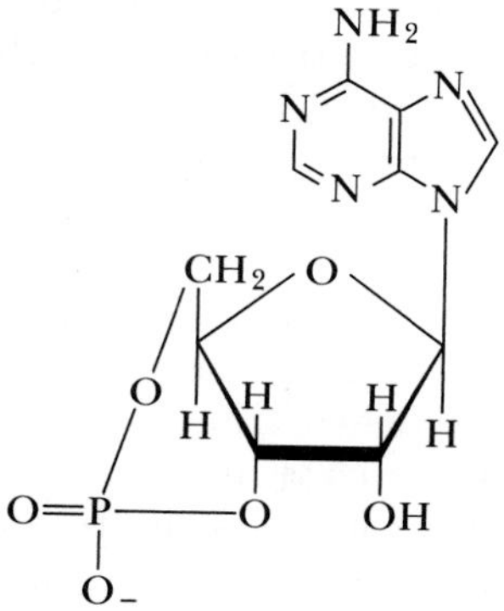

**Cyclic AMP**

**FIGURE 19.10** Activation of the G protein. The inactive G protein is a trimer, consisting of $\alpha$, $\beta$, and $\gamma$ subunits, with GDP bound to the $\alpha$ subunit. In the process of activation the $\beta\gamma$ dimer binds to a hormone-receptor complex. GTP binds to the $\beta$ subunit, replacing the GDP that is already there. The system returns to its inactive state when the GTPase activity of the $\alpha$ subunit hydrolyzes GTP to GDP.

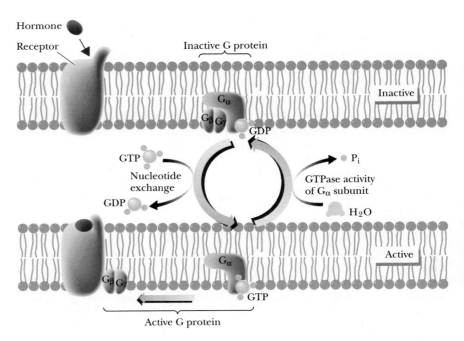

is a ubiquitous second messenger. The binding of hormone to its receptor triggers the production of cAMP from ATP, catalyzed by **adenylate cyclase.** This reaction is mediated by the G protein, a trimer consisting of three subunits—$\alpha$, $\beta$, and $\gamma$. Binding of a stimulatory hormone to the receptor activates the G protein; the $\beta$ and $\gamma$ subunits bind to the receptor, while the $\alpha$ subunit binds GTP, giving rise to the name of the protein. The active protein has GTPase activity and hydrolyzes GTP, returning the G protein to the inactive state. GDP remains bound to the $\alpha$ subunit and must be exchanged for GTP when the protein is activated the next time (Figure 19.10). The G protein and adenylate cyclase are bound to the plasma membrane, while cAMP is released into the interior of the cell to act as a second messenger.

In eukaryotic cells the usual mode of action of cAMP is to stimulate a protein kinase, a tetramer consisting of two regulatory subunits and two catalytic subunits. When cAMP binds to the dimer of regulatory subunits, the two active catalytic subunits are released. The active kinase catalyzes the phosphorylation of some target enzyme (Figure 19.11). The usual site of phosphorylation is the hydroxyl group of a serine or a threonine. ATP is the source of the phosphate group transferred to the enzyme. The target enzyme then elicits the cellular response.

The G protein is permanently activated by **cholera** toxin, leading to excessive stimulation of adenylate cyclase. The main danger in cholera, caused by the bacterium *Vibrio cholerae,* is severe dehydration as a result of diarrhea. The unregulated activity of adenylate cyclase in epithelial cells leads to diarrhea, since cAMP in epithelial cells stimulates active transport of $Na^+$. Excessive cAMP in epithelial cells produces a large flow of $Na^+$ and water from the epithelial cells to the intestines. If the lost fluid and salts

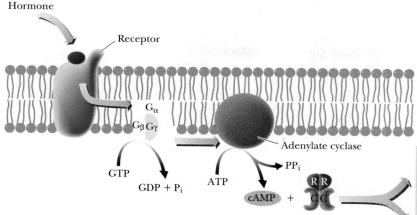

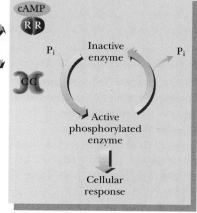

can be replaced in cholera victims, the immune system can deal with the actual infection within a few days.

Calcium ion ($Ca^{2+}$) is involved in another ubiquitous second messenger scheme. Much of the calcium-mediated response depends on release of $Ca^{2+}$ from intracellular reservoirs, similar to the release of $Ca^{2+}$ from the sarcoplasmic reticulum in the action of the neuromuscular junction (Section 11.5). A component of the inner layer of the phospholipid bilayer, **phosphatidylinositol 4,5-*bis*phosphate (PIP$_2$),** is also required in this scheme (Figure 19.12).

**FIGURE 19.11** The activation of adenylate cyclase (Ac) by the binding of hormone to the receptor and the mode of action of cAMP. The binding of hormone to the receptor leads to the production of cAMP from ATP, catalyzed by adenylate cyclase; this reaction is mediated by the G protein. Once cAMP is formed, it stimulates a protein kinase by binding to the regulatory subunits. The active catalytic subunits are released and catalyze the phosphorylation of a target enzyme. The target enzyme elicits the response of the cell to the hormonal signal.

R$_1$ R$_2$

Diacylglycerol (DAG) moiety

Inositol 1, 4, 5-triphosphate (IP$_3$) moiety

R$_1$ and R$_2$ = fatty acid residues
(P) = phosphate moiety

Phosphatidylinositol 4,5-*bis*phosphate
(PIP$_2$)

When the external trigger binds to its receptor on the cell membrane, it activates **phospholipase C,** which hydrolyzes PIP$_2$ to **inositol 1,4,5-triphosphate (IP$_3$)** and a **diacylglycerol (DAG),** in a process mediated by the G protein. The IP$_3$ is the actual second messenger. It diffuses through

521

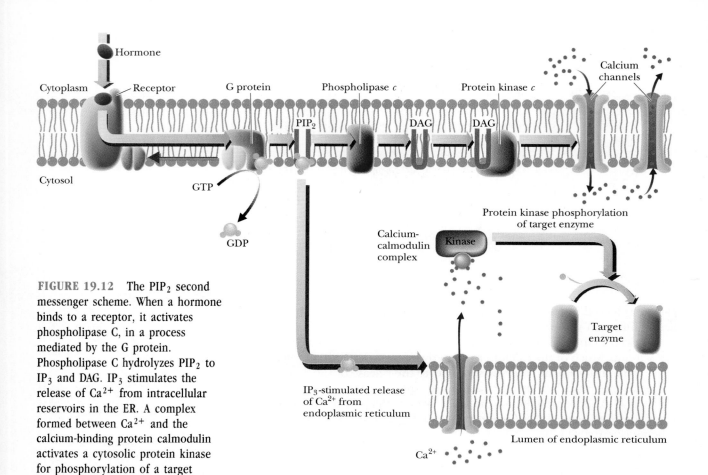

Cytoplasm  Receptor  G protein  Phospholipase *c*  Protein kinase *c*  Calcium channels

Hormone

PIP₂  DAG  DAG

Cytosol

GTP

GDP

Protein kinase phosphorylation of target enzyme

Calcium-calmodulin complex  Kinase

Target enzyme

IP₃-stimulated release of Ca²⁺ from endoplasmic reticulum

Lumen of endoplasmic reticulum

Ca²⁺

**FIGURE 19.12** The PIP₂ second messenger scheme. When a hormone binds to a receptor, it activates phospholipase C, in a process mediated by the G protein. Phospholipase C hydrolyzes PIP₂ to IP₃ and DAG. IP₃ stimulates the release of Ca²⁺ from intracellular reservoirs in the ER. A complex formed between Ca²⁺ and the calcium-binding protein calmodulin activates a cytosolic protein kinase for phosphorylation of a target enzyme. DAG remains bound to the plasma membrane, where it activates the membrane-bound protein kinase C (PKC). PKC phosphorylates channel proteins that control the flow of Ca²⁺ in and out of the cell. Ca²⁺ from extracellular sources can produce sustained responses even when the supply of Ca²⁺ in intracellular reservoirs is exhausted.

the cytosol to the endoplasmic reticulum (ER), where it stimulates the release of Ca²⁺. A complex is formed between the calcium-binding protein calmodulin and Ca²⁺. This complex activates a cytosolic protein kinase, which phosphorylates target enzymes in the same fashion as in the cAMP second messenger scheme. DAG also plays a role in this scheme; it is nonpolar and diffuses through the plasma membrane. When DAG encounters the membrane-bound **protein kinase C,** it too acts as a second messenger by activating this enzyme (actually a family of enzymes). Protein kinase C also phosphorylates target enzymes, including channel proteins that control the flow of Ca²⁺ into and out of the cell. By controlling the flow of Ca²⁺, this second messenger system can produce sustained responses even when the supply of Ca²⁺ in the intracellular reservoirs becomes exhausted. (For more information on this point, see the article by Rasmussen cited in the bibliography at the end of the chapter.)

## 19.5
## A FINAL LOOK AT USES AND SOURCES OF METABOLIC ENERGY

Carbohydrates and lipids are sources of metabolic energy. Degradation of lipids and carbohydrates releases energy and produces ATP. Biosynthesis, including that of lipids and carbohydrates, requires energy and uses ATP. Another important process that requires energy and uses ATP is muscle contraction. Athletes are concerned about using their energy in as efficient a manner as possible, while those who are interested in losing weight or controlling their weight frequently use exercise in conjunction with diet as a means to achieve this goal. Our last topic on the general subject of metabolism will be the use of metabolic energy for muscle contraction.

Muscle cells are made up of fibers called **myofibrils,** which in turn are made up of repeating units called **sarcomeres.** In the repeating unit, thick and thin filaments lie parallel to each other but do not overlap completely. The thick filament consists of the protein **myosin,** with an aggregate of several molecules forming the actual filament. The thin filament consists of an aggregate of the protein **actin.** Other proteins that are involved in muscle contraction occur in the thin filament, but actin and myosin are the most important contributors to the process; we shall concentrate on them.

The presence of thick and thin filaments is responsible for the appearance of the sarcomere, which is characterized by two alternating bands of varying degrees of darkness (Figure 19.13). The two bands are a

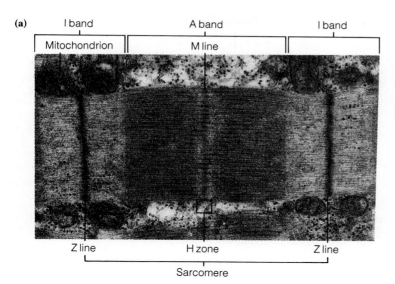

**(a)**

I band      A band      I band

Mitochondrion      M line

Z line      H zone      Z line

Sarcomere

**FIGURE 19.13** Arrangement of filaments in a myofibril. (a) The electron micrograph of a sarcomere. The overlapping thick and thin filaments give rise to the observed light and dark bands. (b) Schematic diagram of a sarcomere.

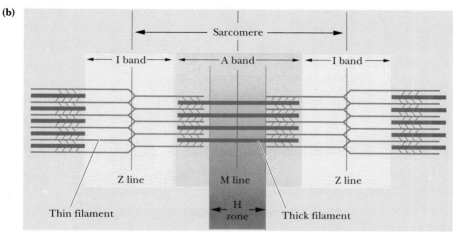

**(b)**

Sarcomere

I band    A band    I band

Z line    M line    Z line

Thin filament    H zone    Thick filament

light I band and a heavy A band. The thick filament occurs in the A band, by itself in the comparatively light H zone and parallel to the thin filament in the rest of the A band. Within the H zone is the M line, another dark portion representing the overlap of thick filaments. The I band consists of thin filaments only, but it is bisected by the dark Z line, an area rich in proteins other than actin.

In muscle contraction the thick and thin filaments slide past one another. The length of the filaments does not change, but the amount of overlap increases as the muscle contracts. The mode of interaction of actin and myosin changes in the process (Figure 19.14). Myosin is principally a fibrous protein, but it has globular portions that appear as projections from the thick filament. The globular portions of myosin interact with actin to

**FIGURE 19.14**   Muscle contraction. (a) In resting muscle, cross-bridges do not form between the thick and thin filaments. ATP becomes attached to the globular head of myosin and is split into ADP and $P_i$, which remain bound to myosin. (b) In the presence of $Ca^{2+}$, the ADP and $P_i$ are released. The myosin head binds to actin. (c) The power stroke consists of bending the myosin head at the neck, pulling the filaments past one another. (d) ATP is bound to the myosin head, which is no longer bound to actin, and ATP is split into ADP and $P_i$. The whole process repeats itself, with myosin binding to the next available binding site on actin.

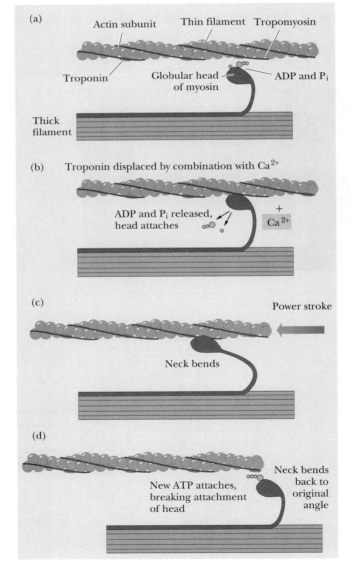

form an actomyosin complex, a process that requires ATP. When a muscle contracts, the globular regions of myosin first tilt and then become detached from the site to which they were bound on the actin complex and move to a new site. The tilting of the globular head of the myosin is considered to provide the **power stroke** for the contraction (Figure 19.14 (c)). The process requires $Ca^{2+}$ as well as ATP. The other protein components of the thin filament, troponin and tropomyosin, are involved in the role of calcium in muscle contraction. The presence of high levels of $Ca^{2+}$ triggers a conformational change in troponin, a change that is passed along to tropomyosin. The conformational change in tropomyosin in turn triggers the tilting and sliding motion of myosin, which is the main feature of muscle contraction. The increase in the level of $Ca^{2+}$, leading to muscle contraction, comes about as a result of the opening of the gated $Ca^{2+}$ channels in the sarcoplasmic reticulum, produced by the action of the neuromuscular junction (Section 11.5). This long series of steps points out clearly that a complex network of relationships exists in all biological processes.

In muscle contraction, myosin is an ATPase, that is, myosin exhibits enzymatic activity, specifically that of hydrolysis of ATP. Easily accessible sources of ATP are needed for prolonged periods of muscular activity. The amount of ATP usually found in a cell cannot sustain muscular activity for more than a fraction of a second. There is, however, an easily accessible reservoir of metabolic energy in resting muscle, in the form of "high-energy" phosphate compounds. Substances such as creatine phosphate have a larger negative free energy of hydrolysis than does ATP (Section 12.4). As a result, hydrolysis of these compounds can drive the phosphorylation of ADP (Figure 19.15). The reservoir of creatine phosphate is soon

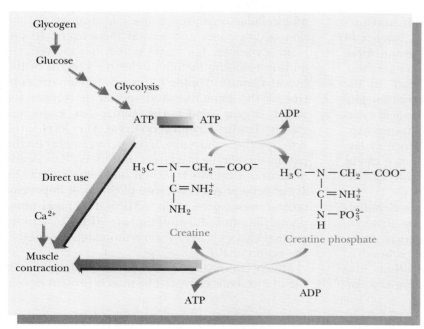

FIGURE 19.15   The role of creatine phosphate in the use of energy in muscle contraction. In resting muscle, energy is stored as the chemical energy of creatine phosphate. In active muscle, creatine phosphate is used as a source of "high-energy" phosphate. When creatine phosphate is depleted, ATP is used directly for contraction.

exhausted. Several minutes of strenuous activity are enough to deplete the supply, and metabolic processes must meet any further demand. In skeletal muscle the level of glycolysis is increased. Glycogen is hydrolyzed to provide a substrate for the glycolytic enzymes, and lactate is produced. The burning sensation in muscles after strenuous exercise comes from the increased level of lactate in the tissues. In muscle tissues such as those of the heart, oxidative phosphorylation is stimulated. As we have seen in Chapters 14 and 15, aerobic metabolism extracts far more ATP from substrates than does glycolysis. Long periods of activity (30 minutes or more) that involve aerobic metabolism can deplete available supplies of glycogen; then fatty acid oxidation is used as a source of ATP. Marathon runners make a point of eating a high-carbohydrate diet during the days before a race to build up their glycogen stores (carbohydrate loading); they are well known to have a very low percentage of body fat. Athletes in general want to make as efficient use of their energy as possible. Aerobic exercise, such as swimming and cycling, is also favored by those who want to use up as many "calories" as possible in the interests of weight control. Inefficient use of energy is, in a sense, what is desired here, since nonathletes who want to control their weight have to worry about an excess of stored energy in the form of fat.

## SUMMARY

All metabolic pathways are related. Some metabolites appear in several pathways. Furthermore, all the reactions of metabolism take place simultaneously. The citric acid cycle plays a central role in metabolism, both in catabolic and in anabolic pathways. The breakdown products of sugars, fatty acids, and amino acids all enter the citric acid cycle. A balance of nutrients in the diet is essential for proper nutrition.

While the citric acid cycle takes place in mitochondria, many anabolic reactions take place in the cytosol. Oxaloacetate, the starting material for gluconeogenesis, is a component of the citric acid cycle. Malate, but not oxaloacetate, can be transported across the mitochondrial membrane. Once malate from mitochondria is carried to the cytosol, it can be converted to oxaloacetate by malate dehydrogenase, an enzyme that requires $NAD^+$.

In the case of lipid anabolism, it is not definitely settled whether acetyl-CoA, the starting material, can be transported across the mitochondrial membrane. Malate, which does cross the mitochondrial membrane, plays a role in lipid anabolism, in a reaction in which malate is oxidatively decarboxylated to pyruvate by an enzyme that requires $NADP^+$, producing NADPH. This reaction is an important source of NADPH for lipid anabolism, with the pentose phosphate pathway the only other source.

Sophisticated fine tuning of metabolic processes in multicellular organisms is possible through the action of hormones and second messengers. In humans a complex hormonal system has evolved, requiring releasing factors (under the control of the hypothalamus), trophic hormones (under the control of the pituitary), and specific hormones for target organs (under the control of endocrine glands). Feedback control occurs at every level of the system. When a hormone binds to its receptor on the plasma membrane of a target cell, it sets off a cascade of reactions by which second messengers elicit the actual cellular response. Two of the most important second messengers, cyclic AMP (cAMP) and phosphatidylinositol 4,5-*bis*phosphate ($PIP_2$), activate protein kinases. Calcium ion is intimately involved in the action of $PIP_2$.

Muscle contraction is an important example of the use of metabolic energy. The muscle protein myosin acts as an ATPase during muscle contraction. Stored energy in the form of carbohydrates and lipids is converted to the chemical energy of ATP to allow the process to take place. As the thick and thin filaments of the muscle fiber move with respect to one another, the myosin of the thick filament detaches itself from the actin of the thin filament and slides to a new position. It is this process that requires ATP.

## EXERCISES

*Hint:* You may want to review material from other chapters in this section.

1. Immature rats are fed all the essential amino acids but one. Three hours later they are fed the missing amino acid. The rats fail to grow. Explain this observation.
2. NADH is an important coenzyme in catabolic processes, while NADPH appears in anabolic processes. Explain how an exchange of the two can be effected.
3. A cat named Lucullus is so spoiled that he will eat nothing but freshly opened canned tuna. Another cat, Griselda, is given only dry cat food by her far less indulgent owner. Canned tuna is essentially all protein, while dry cat food can be considered 70% carbohydrate and 30% protein. Assuming that these animals have no other sources of food, what can you say about the differences and similarities in their catabolic activities? (The pun is intended.)
4. Kwashiorkor is a protein-deficiency disease, particularly in small children, who characteristically have thin arms and legs and a bloated, distended abdomen due to fluid imbalance. When such children are placed on an adequate diet, they tend to lose weight at first. Explain this observation.
5. Recent recommendations on diet suggest that the sources of calories should be distributed as follows: 50 to 55% carbohydrate, 25 to 30% fats, and 20% protein. Suggest some reasons for these recommendations.
6. In contracting skeletal muscle, the concentration of phosphocreatine decreases, while that of ATP remains the same. Give a reason for this observation.
7. Suggest a reason why the oxygen consumption of a runner for a 10-second period is about 20 times that of a sedentary person.
8. When $PIP_2$ is hydrolyzed, why does $IP_3$ diffuse into the cytosol while DAG remains in the membrane?
9. Briefly describe the series of events that take place when cAMP acts as a second messenger.
10. How does the action of the hypothalamus and pituitary affect the workings of endocrine glands?
11. For each of three hormones discussed in this chapter, give its source and chemical nature; also discuss the mode of action of each hormone.

## ANNOTATED BIBLIOGRAPHY

Bagshaw, C.R. *Muscle Contraction.* New York: Chapman and Hall, 1982. [A short book about general aspects of muscle contraction.]

Eisenberg, E., and T.L. Hill. Muscle Contraction and Free-Energy Transduction in Biological Systems. *Science* **227**, 999–1006 (1985). [An account that covers many points discussed in this chapter.]

Katch, F.I. and W.D. McArdle. *Nutrition, Weight Control, and Exercise.* 3rd ed. Philadelphia: Lea and Febiger, 1988. [A book on nutrition, exercise, and health, intended to substitute facts for myth on the subject.]

Rasmussen, H. The Cycling of Calcium As an Intracellular Messenger. *Sci. Amer,* **261**(4), 66–73 (1989). [An article on the role of calcium as a second messenger.]

Uvnas-Moberg, K. The Gastrointestinal Tract in Growth and Reproduction. *Sci. Amer.* **261**(1), 78–83 (1989). [An article on how the endocrine and digestive systems cooperate to provide for the needs of the fetus and the newborn.]

See also the references for Chapters 13 to 18.

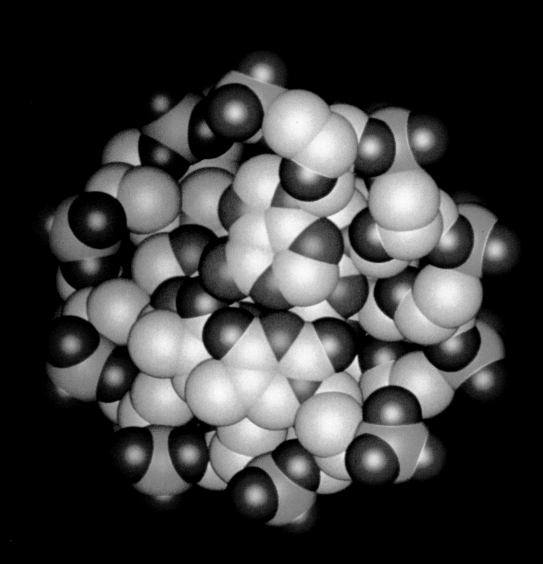

# The Workings of the Genetic Code

# INTERVIEW

# Jacqueline K. Barton

*Dr. Jacqueline K. Barton is a native New Yorker and was educated in that city. She received her B.A. degree from Barnard College in 1974 and her Ph.D. from Columbia University in 1979. While at Columbia she worked in an area of platinum chemistry closely related to the work done by Dr. Barnett Rosenberg. Following her Ph.D. work, Dr. Barton did further research at Yale University and Bell Laboratories and then joined the faculty of Hunter College. In 1983 she returned to Columbia University, where she rose rapidly to the rank of full professor, and in the fall of 1989 she assumed her present position as Professor of Chemistry at the California Institute of Technology.*

*Dr. Barton has done outstanding research in the field of biochemistry, particularly in the design of simple molecular probes to explore the variations in structure and conformation along the DNA helix. In spite of having been a research scientist for a relatively short time, she has done important new work and has received many honors. In 1985*

*she received the Alan T. Waterman Award of the National Science Foundation as the outstanding young scientist in the United States. In 1987 she was the recipient of the American Chemical Society's Eli Lilly Award in Biological Chemistry, and the following year she received the Society's Award in Pure Chemistry. That same year, 1988, she also received the Mayor of New York's Award of Honor in Science and Technology.*

*As with many chemists, she is interested in art and has a painting by the Spanish artist Miró on her office wall as well as a print by the French artist Vasareley, and this interest in form and color in art carries over into her research.*

**Dr. Barton, it is always of interest to learn how scientists came to their chosen field. Did you have a strong background in chemistry?**

I never took chemistry in high school. Maybe one shouldn't publicize that, but it's the truth.

However, I was always very interested in mathematics, so I took a lot of calculus when I was in high school. I also took a course in geometry, and that interest in geometry has carried over into my research, since the sort of science I do now is very much governed by structures and shapes.

When I went to college I thought that in addition to taking math I should take some science courses. I walked into the freshman chemistry class, and there were about 150 people there. However, there was also a small honors class with about 10 students. Even though I hadn't had chemistry before, I thought I would try it—and I loved it. What chemistry allowed me to do was to combine the abstract and the real. I was very excited by it.

But it was really the experience of the laboratory that got me interested in chemistry. Like many who are involved in chemistry, I was fascinated by color changes in reactions and the significance of these observations. However, I was also interested in trying to predict what would happen in a reaction and, if my prediction was not correct, to try to explain this and then to do more experiments that would solve the puzzle. That's really what got me started in science.

In addition, I also had an inspirational teacher and role model, Bernice Segal. She was an absolute inspiration to me. She gave a magnificent course, and was a tough lady who asked a lot of you—and you did it!

**Your Ph.D. thesis research focused on compounds known as "platinum blues." Could you tell us more about these compounds?**

Most platinum compounds are orange or red, and yet there are these magnificent blue complexes. What are they? What are their structures, and why are they blue? In fact, when Barney Rosenberg was looking at certain platinum compounds, he also found some 'platinum blues'; they were water soluble, unlike cisplatin, and so people had hopes that they might even be better with respect to chemotherapy. But it turned out that was not the case. Nonetheless, that finding and others got Steve Lippard, my advisor at Columbia, interested in trying to work on the problem. Therefore, I made the type of complex that Rosenberg had made, and it was indeed blue. It's an absolutely beautiful

molecule. We solved the structure of the molecule and found that it contains four platinum atoms in a line. We also found that it's mixed-valent, where the platinum has an average oxidation number of $2\frac{1}{4}$, a fact that helps to explain its blue color.[1]

**You are currently doing research in bioinorganic chemistry, which deals with the role of metals in biological systems. You have received numerous awards for your work, indicating that the scientific community places great importance on the field and your contributions to it. Could you explain why this work has such significance?**

The interest of my group is to exploit inorganic chemistry as a tool to ask questions of biological interest and to explore biological molecules. A lot of the work in bioinorganic chemistry thus far has been the exploration of metal centers in biology. Why is blood red? Why does the iron [in heme] do what it does? That's just one example, but there are hundreds of others. Many enzymes and proteins within the body, in fact, contain metals, and the reason we've looked at blood and then the heme center within it has been because it's colored. An obvious tool that transition metal chemistry provides is color, and so things change color when reactions occur. That is one of the things that fascinated me in the first place.

Another wonderful thing about transition metal chemistry is that it allows us to build molecules that

have interesting shapes and structures, depending upon the coordination geometry. In fact, you can create a wealth of different shapes, several of which are chiral, and that's something we take advantage of in particular. What we want to do is make a variety of molecules of different shapes, target these molecules to sites on a DNA strand, and then ask questions such as, 'Does DNA vary in its shape as a function of sequence?'[2] If we think about how proteins bind to DNA, do they also take advantage of shape recognition in binding to one site to activate one gene or turn off another gene? When scientists first wondered about these and other such problems, they would write down a one-dimensional sequence of DNA and would think about it in one-dimensional terms. How does the protein recognize a particular DNA sequence? DNA is clearly not one dimensional. It has a three-dimensional structure, and different sequences of bases will generate different shapes and different forms. Therefore, we think we can build transition metal complexes of particular shapes, target them to particular sequences of bases in DNA, and then use these complexes to plot out the topology of DNA.[3] We can then ask how nature takes advantage of this topology. We want to develop a true molecular understanding, a three-dimensional understanding, of the structure and the shapes of biologically important molecules such as DNA and RNA.

---

[1]Superconducting solids contain metal ions of different valences, that is, they are mixed-valent.

[2]Amino acids, proteins, nucleic acids such as DNA, and other aspects of biochemistry are discussed in the text.

[3]The results of such an experiment are seen in the structure of a complex formed between a ruthenium-based coordination compound and DNA.

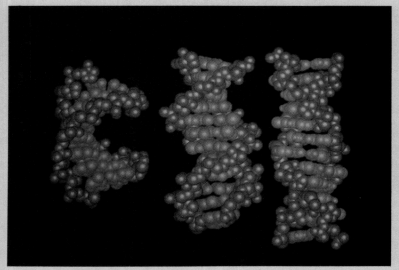

*Double helical conformations of DNA: (left) A-DNA, (center) B-DNA, (right) Z-DNA.*

**What are some recent developments in your field? Can you comment generally on the direction in which chemistry is moving?**

I think our work may be an example of where chemistry is going in general. I think there has been a revolution in chemistry in the past 10 years. The revolution is at the interface between chemistry and biology where we can now ask chemical questions about biological molecules. First of all, we can make biological molecules that are pure. I can now go to a machine called a DNA synthesizer, and I can type in a sequence of DNA; from that sequence I can synthesize a pure material, with full knowledge of where all of the bonds are. Then I can run it through an HPLC and get it 100% pure.[4] Therefore, I can now talk about these biopolymers in chemical terms as molecules rather than as impure cellular extracts. I couldn't do that before.

Not only do we have the ability to prepare biological molecules in pure form, but we also have the techniques to characterize them in ways that chemists think about molecules. The development of new techniques allows us to make a bridge between chemistry and biology and ask chemical questions with molecular detail. It's an exciting time to be doing chemistry, and that is why I see it as a new frontier area.

**Are you satisfied with the curriculum in science as it is, or would you make some changes?**

It is chemists who are making new materials and making and exploring biological systems. It's the chemist who looks at questions of molecular detail and asks about structure and its relationship to function. Since this involves so many areas of chemistry, I believe that we are going to have to stop making divisions between inorganic,

physical, analytical, and organic chemistry. We must all do a little bit of each. This is an attitude shared by many in chemical education today, and it means that we should perhaps rethink the curriculum in chemistry in particular and science in general.

No matter what the curricular structure, however, I believe what is important in the education of scientists is to get across the excitement that now we can know what biologically important molecules look like. And, from knowing what they look like, we can manipulate them and change them a little. Then we can ask how those changes affect the function, so we can relate the structure of the molecule and its macroscopic function.

A protein molecule of average size is so small you could put more than a billion billion of them on the head of a pin. We now know we can manipulate molecules that are of

---

[4]An HPLC is a "high-pressure liquid chromatograph," an instrument capable of separating one type of molecule from another.

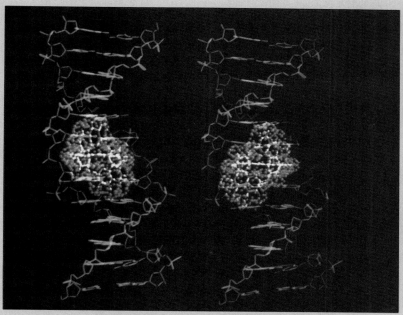

*Tris (phenanthroline) metal complexes (Λ, left and Δ, right) are shown intercalated into right-handed, double helical DNA.*

those dimensions and can know exactly what they look like. I can't imagine that we can't get people interested in chemistry if we can get across the excitement that comes from the realization that we are looking at things so small and yet can do surgery on them.

**Could you describe those special attributes of chemistry that first attracted you and have kept you interested? In other words, how would you "sell" chemistry?**

The bottom line is that chemistry is fun, it's addictive, and, if one has a sense of curiosity, it can be tremendously entertaining and appealing. And it is not so difficult. It's difficult when one thinks about it as rote memorization, which *is* difficult and boring. But that isn't what chemistry is. Chemistry is

trying to understand the world around us in some detail. For example, we are interested in knowing such things as what makes skin soft, what makes things different in color, why sugar is sweet,[5] or why a particular pharmaceutical agent makes us feel better.

**What is your perspective on the issue of women in science? Do you see special opportunities for women? Are there problems that women need to overcome or be aware of?**

Because I am a woman, and there are so few women currently in professional positions in chemistry, I'm asked those questions often. First of all, I am not an expert on the subject. What I like to think my best contribution to women in

chemistry can be is to do the best science I can, and to be recognized for my science, not for being a woman in science. I think that it is generally important when women go into science that they should appreciate that there are no special opportunities; that is, you will be treated like any other person doing science. But just as there should be no special opportunities in that respect, happily—maybe this is naive of me—I think there are also no special detriments or obstacles that one need consider in this day and age. One shouldn't think that 'because I am a woman I can't do it.' That's patently false. In fact, everyone is extremely supportive of women who do science. However, I remember talking to Bernice Segal, my former teacher at Barnard College, and having her explain to me that when she was a graduate student she had to do things behind a curtain, because the women weren't supposed to be doing chemistry. Mildred Cohn, another one of my role models, took over 20 years to have her own independent position as a professor, as opposed to being a laboratory assistant working for someone else. The bottom line is that I don't have a story like that to tell. That's the good news. In my generation there are few such stories of blatant discrimination. Now the world is a much better place for a woman to do science.

*Dr. Barton's enthusiasm for her work and for chemistry in general is obvious. It is evident that she will continue to do some of the most important work in science and that her infectious enthusiasm for chemistry will bring many more young people into the profession.*

[5]See "Aspartame, The Sweet Peptide," in Chapter 5.

# CHAPTER

# 20

# The Structure of Nucleic Acids

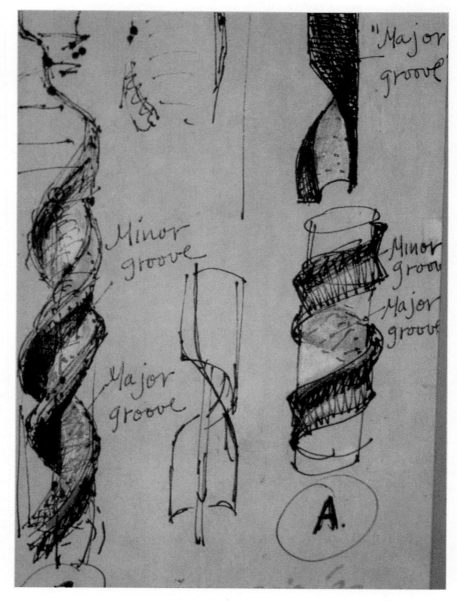

Three forms of DNA as they appear in an artist's sketchbook.

*Genes, the hereditary material of the chromosomes, are essentially long stretches of double helical DNA. Each gene specifies a single polypeptide (protein) chain. The sequence of DNA bases specifies the sequence of amino acids in a protein molecule; and the amino acid sequence, in turn, determines the protein's structure and function. Thus, the genetic code of DNA ultimately directs the activity of proteins, the essential machinery of life. Each cell carries in its DNA the instructions for making the complete organism. When the cell divides, each half carries a copy of the original DNA. Replication of the hereditary material is made possible by the complementary nature of the DNA bases. Adenine on one strand pairs with thymine on the opposite strand of the double helix. The same is true for the other two bases—guanine on one strand pairs with cytosine on an opposite strand. Thus, one strand of DNA is a template for the other strand. Another dimension was added to our understanding of DNA structure in the early 1980s when short sections of DNA were created from synthetic oligonucleotides. These sections were crystallized and studied by x-ray diffraction analysis. The results showed local variation in the three-dimensional geometry of base pairs, depending on the DNA sequence. This will provide valuable insight into the manner by which proteins interact with DNA molecules to control gene activation and repression.*

## 20.1
## LEVELS OF STRUCTURE IN NUCLEIC ACIDS

In Chapter 10 we saw that there are various levels of structure—primary, secondary, tertiary, and so on—in proteins. The same sort of situation applies in nucleic acids. By analogy to proteins, the primary structure of nucleic acids refers to the order of bases in the polynucleotide sequence, and the secondary structure refers to the three-dimensional conformation of the backbone. The tertiary structure refers specifically to supercoiling of the molecule. Important differences exist between DNA and RNA with regard to their secondary and tertiary structures. For this reason we shall describe these structural features separately for DNA and for RNA.

## 20.2
## THE COVALENT STRUCTURE OF POLYNUCLEOTIDES

The monomers of nucleic acids are **nucleotides.** An individual nucleotide consists in turn of three parts—a base, a sugar, and a phosphoric acid residue—all of which are covalently bonded together.

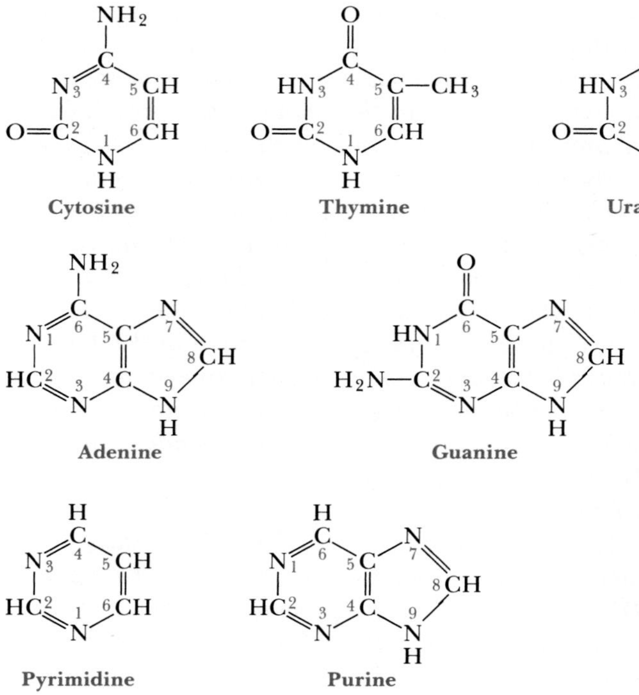

**FIGURE 20.1**   Structures of the common nucleobases. Structures of pyrimidine and purine are shown for comparison.

Cytosine          Thymine          Uracil

Adenine                    Guanine

Pyrimidine          Purine

The **nucleic acid bases** (also called **nucleobases**) are **pyrimidines** and **purines.** The genetic code is specified by the order of bases in the nucleic acids. Three pyrimidine bases commonly occur in nucleic acids: **cytosine, thymine,** and **uracil.** Cytosine is found in both RNA and DNA, while uracil occurs only in RNA. In DNA thymine is substituted for uracil; thymine is also found to a small extent in some forms of RNA (Figure 20.1).

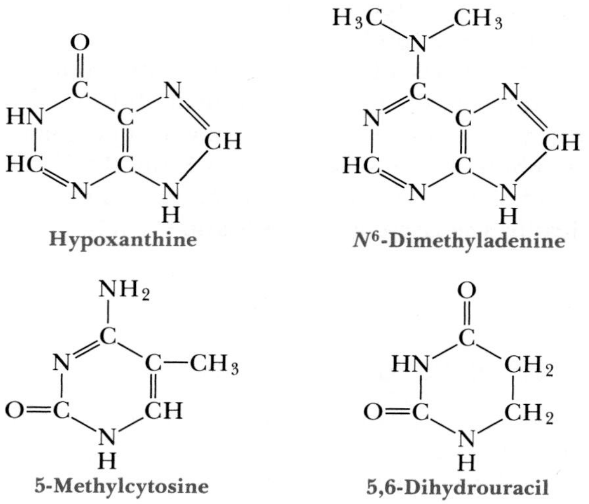

**FIGURE 20.2**   Structures of some of the less common nucleobases.

Hypoxanthine          $N^6$-Dimethyladenine

5-Methylcytosine          5,6-Dihydrouracil

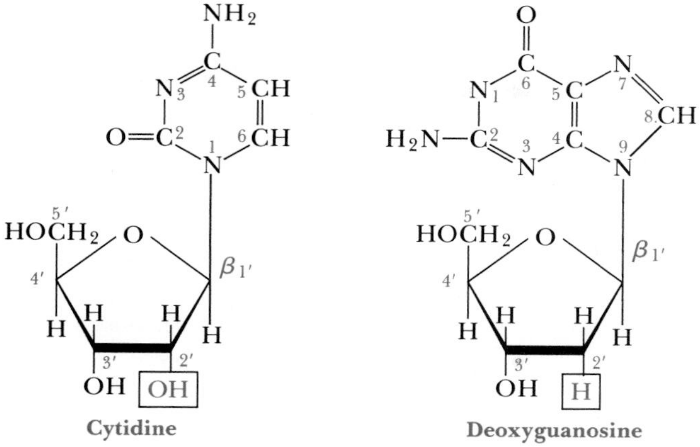

**FIGURE 20.3** Comparison of the structure of a ribonucleoside and a deoxyribonucleoside.

Cytidine

Deoxyguanosine

The common purine bases are **adenine** and **guanine,** both of which are found in both RNA and DNA (Figure 20.1). In addition to the five commonly occurring bases, there are other "unusual" bases with slightly different structures, found principally, but not exclusively, in transfer RNA (Figure 20.2). In most cases, but not all, the base is modified by methylation.

A **nucleoside** is a compound that consists of a base and a sugar covalently linked together. A nucleoside differs from a nucleotide by not having a phosphate group as part of its structure. When a base forms a glycosidic linkage with $\beta$-D-ribose, the resulting compound is a **ribonucleoside;** when the sugar is $\beta$-D-deoxyribose, the resulting compound is a **deoxyribonucleoside** (Figure 20.3). (Recall our discussion of glycosidic linkages and stereochemistry of sugars in The Formation of Glycosides in Section 6.2.) The glycosidic linkage is from the C-1 carbon of the sugar to the N-1 nitrogen of pyrimidines or to the N-9 nitrogen of purines. The ring atoms of the base and the carbon atoms of the sugar are numbered, with the numbers of the sugar atoms primed to prevent confusion. A nucleoside derivative that is very much in the news is 3'-azido-3'-deoxythymidine (AZT) (Figure 20.4). This compound has shown promise in the treatment of AIDS (acquired immune deficiency syndrome).

When phosphoric acid is esterified to one of the hydroxyl groups of the sugar portion of a nucleoside, a nucleotide is formed (Figure 20.5). A nucleotide is called by the name of the parent nucleoside with the suffix monophosphate; the position of the phosphate ester is specified by the number of the carbon atom at the hydroxyl group to which it is esterified, *e.g.,* adenosine 3'-monophosphate, deoxycytidine 5'-monophosphate. In nature the 5' nucleotides are the ones most commonly encountered.

The polymerization of nucleotides gives rise to nucleic acids. The linkage between monomers in nucleic acids involves formation of two ester bonds by phosphoric acid. The hydroxyl groups to which the phosphoric acid is esterified are those bonded to the 3' and 5' carbons on adjacent residues. The resulting repeated linkage is a **3', 5'-phosphodiester bond.**

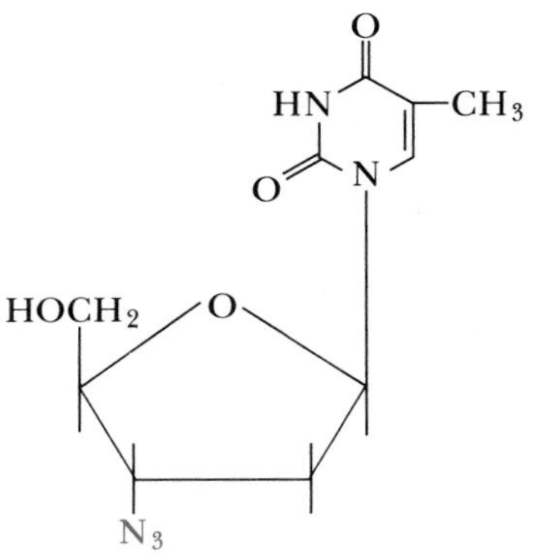

**FIGURE 20.4** The structure of 3'-azido-3'-deoxythymidine (AZT).

**FIGURE 20.5** The structures and names of the commonly occurring nucleotides. All structures are shown in the form that exists at pH 7. (a) Ribonucleotides. (b) Deoxyribonucleotides.

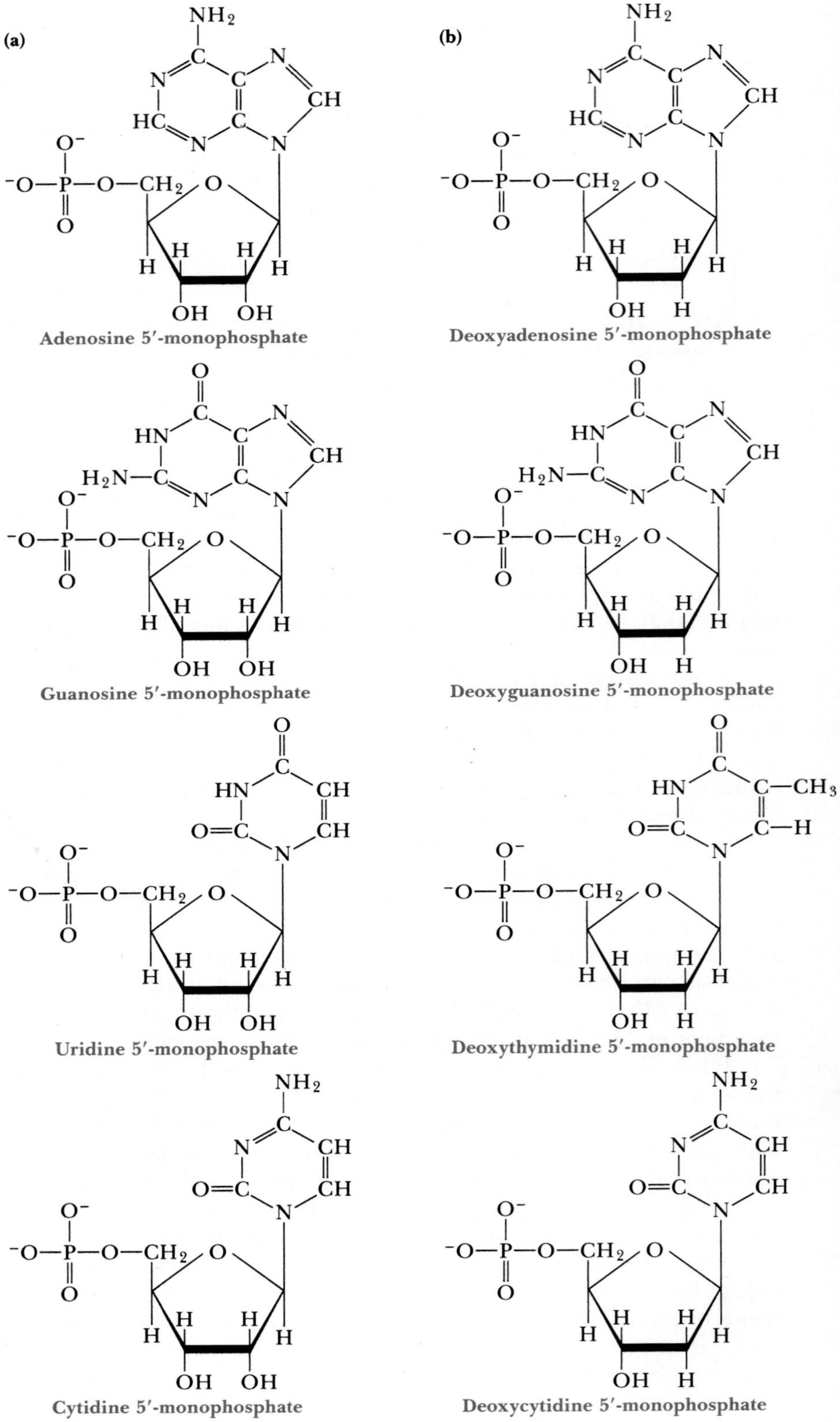

(a)

Adenosine 5'-monophosphate

Guanosine 5'-monophosphate

Uridine 5'-monophosphate

Cytidine 5'-monophosphate

(b)

Deoxyadenosine 5'-monophosphate

Deoxyguanosine 5'-monophosphate

Deoxythymidine 5'-monophosphate

Deoxycytidine 5'-monophosphate

The nucleotide residues of nucleic acids are numbered from the 5' end, which normally carries a phosphate group, to the 3' end, which normally has a free hydroxyl group.

The structure of a fragment of an RNA chain is shown in Figure 20.6. The **sugar-phosphate backbone** repeats itself down the length of the chain. The most important point about the structure of nucleic acids is the identity of the bases, and abbreviated forms of the structure can be written to convey the essential information. In one form of notation single letters, such as A, G, C, U, or T, represent the individual bases. A vertical line shows the position of the sugar moieties to which the individual bases are attached, and a diagonal line through the letter "p" represents a phosphodiester bond (Figure 20.6). A still more abbreviated form of notation uses only the single letters to show the order of the bases. When it is necessary to indicate the position on the sugar to which the phosphate group is

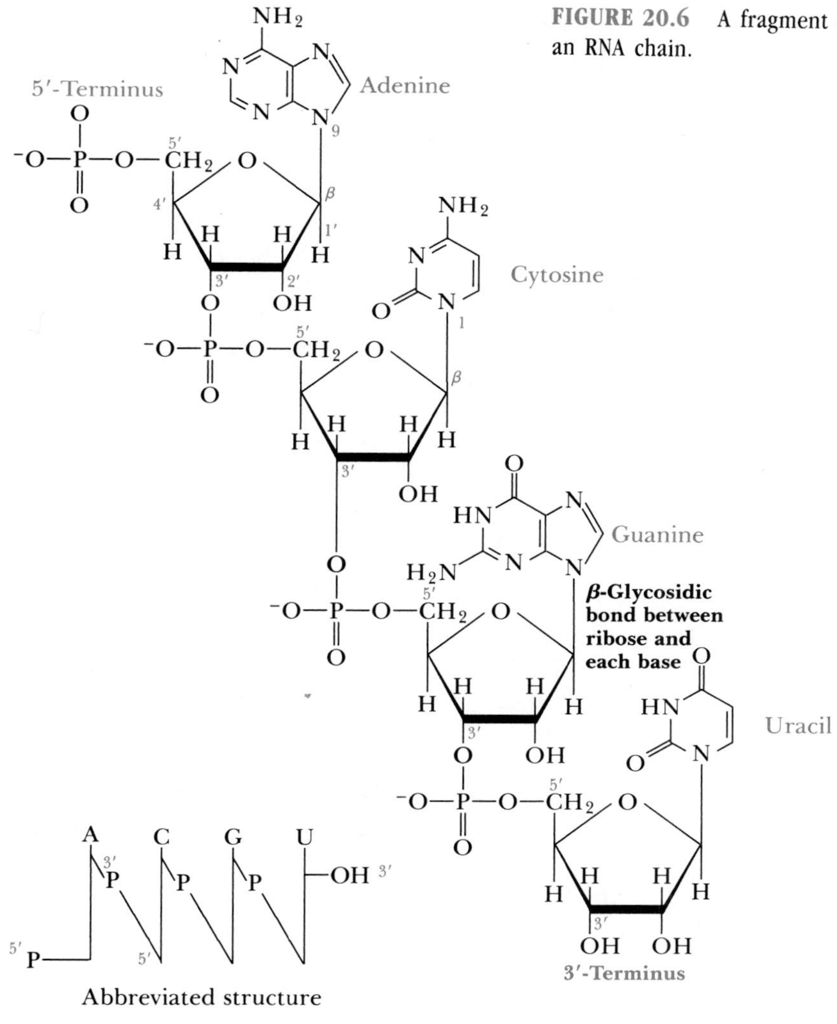

**FIGURE 20.6**  A fragment of an RNA chain.

bonded, the letter "p" is written to the left of the single-letter code for the base to represent a 5′ nucleotide and to the right to represent a 3′ nucleotide. For example, pA signifies 5′-AMP, and Ap signifies 3′-AMP. It is possible to represent the sequence of an oligonucleotide as pGpApCpApU, or even more simply as pGACAU, specifying only the phosphate group at the 5′ end but assuming the presence of the phosphates that link the rest of the nucleosides.

A portion of a DNA chain differs from the RNA chain we have just described only in the fact that the sugar is deoxyribose rather than ribose (Figure 20.7). In abbreviated notation the deoxyribonucleotide is specified

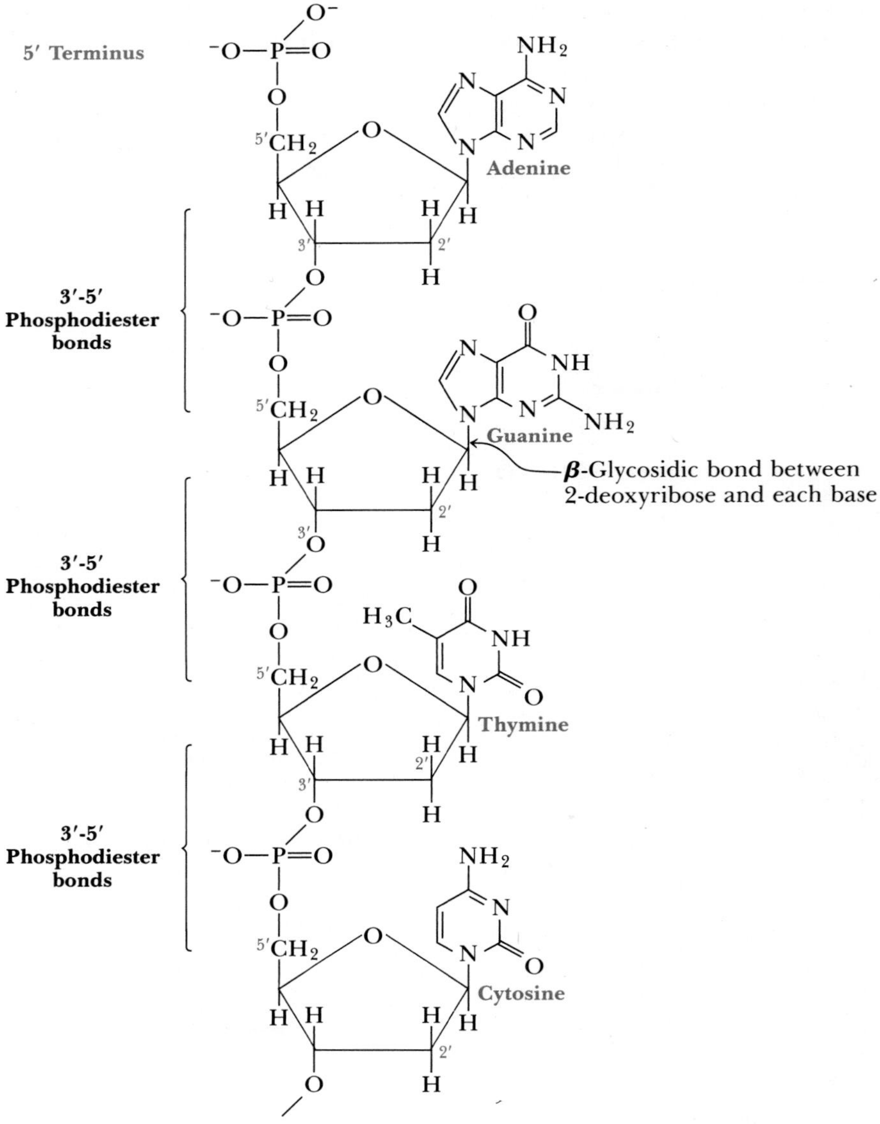

**FIGURE 20.7**   A portion of a DNA chain.

in the usual manner. Sometimes a "d" is added to indicate a deoxyribonu-cleotide residue, *e.g.*, dG is substituted for G, and the deoxy analogue of the ribooligonucleotide shown above would be pd(GACAT).

## 20.3
## THE STRUCTURE OF DNA

### Secondary Structure of DNA: The Double Helix

Representations of the double-helical structure of DNA have become common in the popular press as well as in the scientific literature. When this structure was proposed by Watson and Crick in 1953, it touched off a flood of research activity, leading to great advances in molecular biology. The determination of the double-helical structure was based on chemical analysis of DNA base composition and on x-ray diffraction patterns. Both these lines of evidence were necessary for the conclusion that DNA consists of two polynucleotide chains wrapped around each other to form a helix. The two chains are held together by hydrogen bonds between bases on opposite chains, with the paired bases lying in a plane perpendicular to the helix axis. The sugar-phosphate backbone is the outer part of the helix (Figure 20.8). The chains run in antiparallel directions, one 3' to 5' and one 5' to 3'.

The x-ray diffraction pattern of DNA demonstrated the helical structure. The combination of evidence from x-ray diffraction and chemical analysis led to the conclusion that the base pairing is **complementary,** a term that means that adenine pairs with thymine, and guanine with cytosine. (Since complementary base pairing occurs along the entire double helix, the two chains are also referred to as *complementary strands.*) Earlier studies of the base composition of DNA from many species had already shown by 1953 that, to within experimental error, the mole percentages (the number of moles of the substance in question out of the total number of moles present, expressed as a percentage) of adenine and thymine were equal; the same equality of mole percentages was found to be the case with guanine and cytosine. An adenine-thymine (A-T) base pair has two hydrogen bonds between the bases; a guanine-cytosine (G-C) base pair has three (Figure 20.9).

The inside diameter of the sugar-phosphate backbone of the double helix is about 11 Å (1.1 nm). The distance between the point of attachment of the bases to the two strands of the sugar-phosphate backbone is the same for the two base pairs (A-T and G-C), about 11 Å (1.1 nm) so these base pairs are the right size to fit into the helix. Base pairs other than A-T and G-C are possible, but they do not have the correct dimensions to fit the inside diameter of the double helix (Figure 20.9). The outside diameter of the helix is 20 Å (2 nm). The length of one complete turn of the helix along its axis is 34 Å (3.4 nm), containing ten base pairs. It is possible to draw a cylinder around the double helix. The atoms that make up the two polynucleotide chains of the double helix do not completely fill such a cylinder, leaving empty spaces known as grooves. There is a large major

**FIGURE 20.8**    The double helix. A complete turn of the helix spans 10 base pairs, covering a distance of 34 Å (3.4 nm). The individual base pairs are spaced 3.4 Å apart. The solid dots seen where strands cross represent base pairs that are aligned perpendicular to the viewer. The inside diameter is 11 Å, while the outside diameter is 20 Å. Within the cylindrical outline of the double helix there are two grooves, a small one and a large one. Both grooves are large enough to accommodate polypeptide chains. The raised minus signs represent the many negatively charged —$PO_2^-$— groupings along the entire length of each strand.

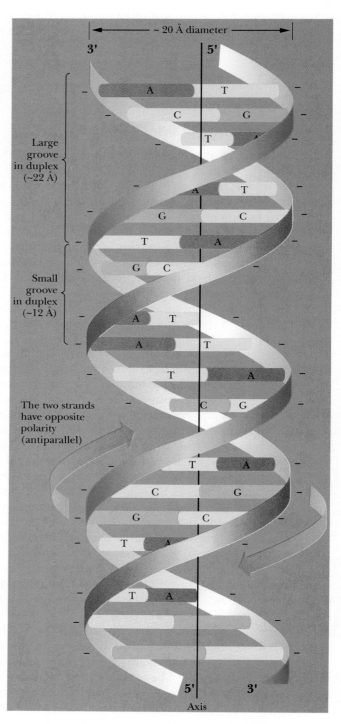

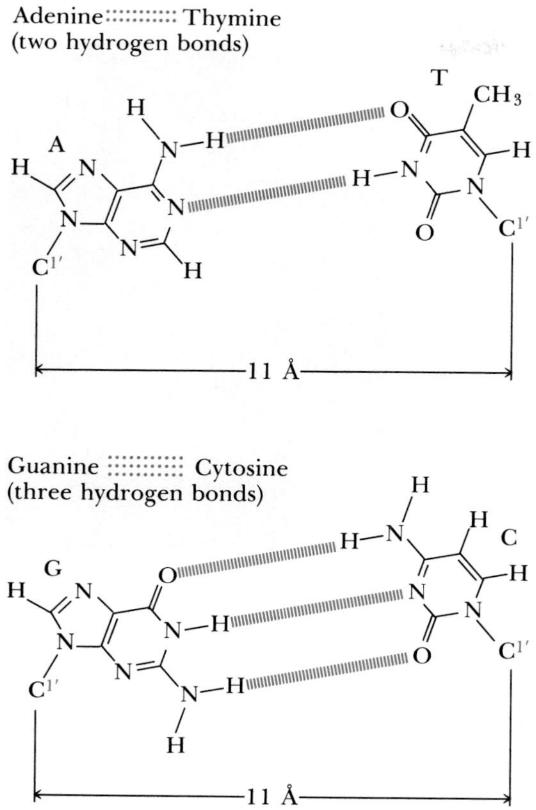

Adenine:::::::::::Thymine
(two hydrogen bonds)

**FIGURE 20.9**    Base pairing. The adenine-thymine (A-T) base pair has two hydrogen bonds, while the guanine-cytosine (G-C) base pair has three hydrogen bonds.

Guanine :::::::::: Cytosine
(three hydrogen bonds)

groove and a smaller minor groove in the double helix; both grooves can be sites at which drugs or polypeptides bind to DNA (see Figure 20.8). At neutral, physiological pH, each phosphate group of the backbone carries a negative charge; positively charged ions such as $Na^+$ or $Mg^{2+}$ or polypeptides with positively charged side chains are frequently associated with DNA as a result of electrostatic attraction (see Figure 20.8). The antiparallel direction of the two polynucleotide chains is an aspect of the complementarity of the two strands (see Figure 20.8).

The form of DNA that we have been discussing up to now is called B-DNA; it is thought to be the principal one that occurs in nature. However, it is not the only possible secondary structure that is observed for DNA. Different forms can exist, depending on conditions such as the nature of the positive ion associated with the DNA. One of these other forms is A-DNA, which has 11 base pairs for each turn of the helix. The base pairs in A-DNA are not perpendicular to the helix axis but lie at an angle of about 20° to the perpendicular, like the blades of a propeller (Figure 20.10). An important resemblance between A-DNA and B-DNA is that both are right-handed helices, that is, the helix winds upward in the direction in which the fingers of the right hand curl when the thumb is

**(a)**

**(b)**

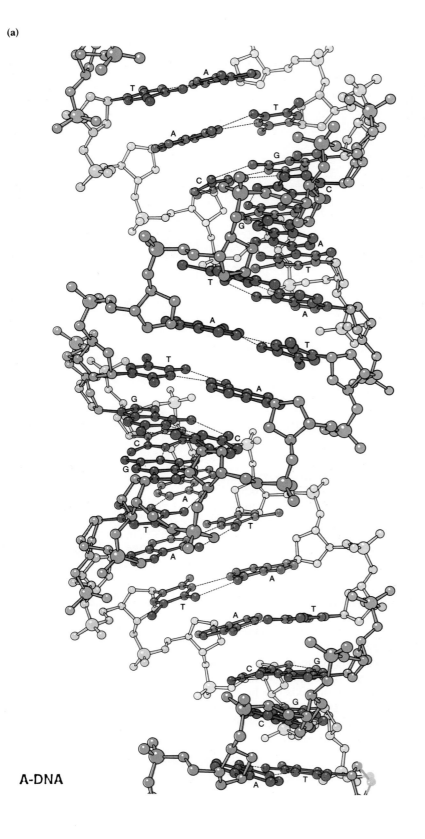

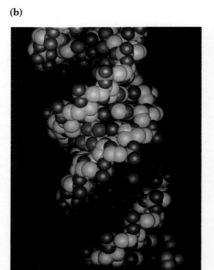

**A-DNA**

**FIGURE 20.10** A comparison between the A and B forms of DNA. In the A form, (a) ball and stick drawing, (b) computer-generated space-filling model, the base pairs have a marked propeller twist with respect to the helix axis. In the B form, (c) ball and stick drawing, (d) computer-generated space-filling model, the base pairs lie in a plane that is close to perpendicular to the helix axis.

(c)

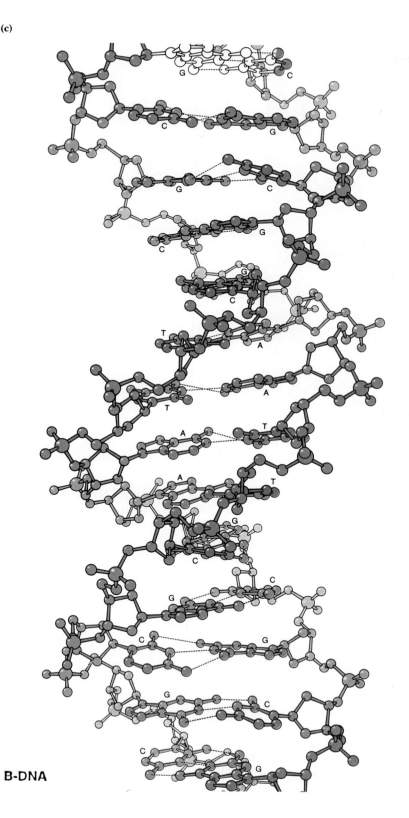

B-DNA

(d)

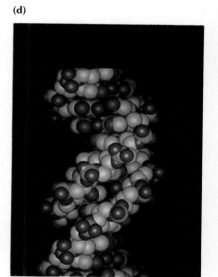

**FIGURE 20.11** Right- and left-handed helices are related to each other in the same way as the right and left hand.

pointing upward (Figure 20.11). There is variant form of the double helix in which the helix is left-handed; it winds in the direction of the fingers of the left hand. This left-handed form is called Z-DNA (Figure 20.12). It is known that Z-DNA occurs in nature, but its function has not been determined at this writing. It may play a role in the regulation of gene expression.

## Tertiary Structure of DNA: Supercoiling

The length of the DNA molecule is considerably greater than its diameter; it is not completely stiff. The double helix we have encountered so far is relaxed, which means that it has no twists in it other than the double helix itself. Further twisting and coiling of the double helix is possible. The first example of supercoiling we shall consider is the case in which DNA is not complexed to proteins. This "naked" DNA occurs in prokaryotes, but not in eukaryotes, where DNA is complexed to proteins of various types and the supercoiling pattern is different.

If the sugar-phosphate backbone of a prokaryotic DNA forms a covalently bonded circle, the structure is still relaxed. Some extra twists are

(a)

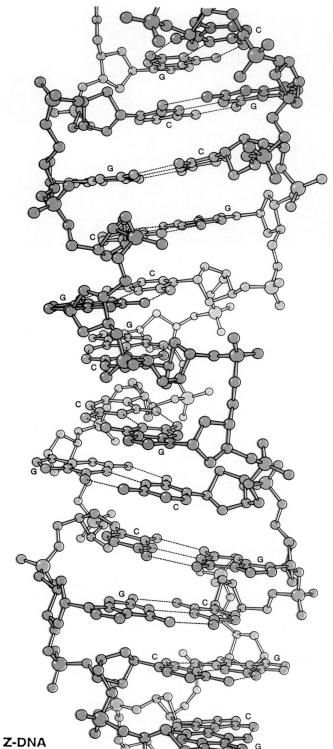

**Z-DNA**

(b)

**FIGURE 20.12** Z-DNA is a left-handed helix and in this respect differs from A-DNA and B-DNA, both of which are right-handed helixes: (a) ball and stick drawing, (b) computer-generated space-filling model.

**(a)**

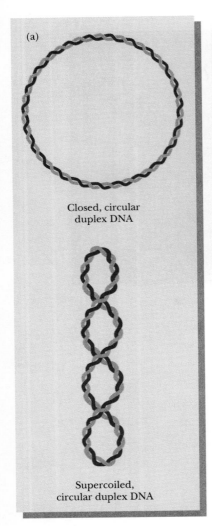

Closed, circular
duplex DNA

Supercoiled,
circular duplex DNA

**(b)**

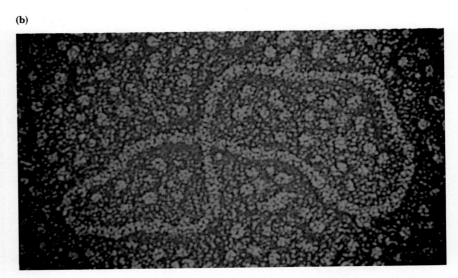

**FIGURE 20.13**    (a) A schematic representation of circular and supercoiled DNA.
(b) Electron micrograph of circular (top) and supercoiled DNA.

added if the DNA is unwound slightly before the ends are joined to form
the circle. A strain is introduced in the molecular structure, and the DNA
assumes a new conformation to compensate for the unwinding. If a
right-handed double helix acquires an extra left-handed helical twist
because of unwinding (a supercoil), the circular DNA is said to be
negatively supercoiled because of the new conformation added by the
unwinding (Figure 20.13(a)). Under different conditions it is possible to
form a right-handed supercoiled structure, in which there is overwinding
of the closed-circle double helix. The difference between the positively and
negatively supercoiled forms lies in the right- or left-handed nature of the
supercoil, which in turn depends on the overwinding or underwinding of
the double helix. Supercoiling has been observed experimentally in
naturally occurring DNA. Particularly strong evidence has come from
electron micrographs that clearly show coiled structures in circular DNA
from a number of different sources, including bacteria, viruses, mitochondria, and chloroplasts (Figure 20.13(b)). It has been known for some time
that prokaryotic DNA is normally circular, but the occurrence of supercoiling has been the subject of more recent research.

Enzymes that affect the supercoiling of DNA have been isolated from
various organisms. Naturally occurring circular DNA is negatively supercoiled, except during the process of DNA replication, when it becomes
positively supercoiled. It is critical for the cell to have control over the
process. Two classes of enzymes are involved in regulating the process, one
to relax the supercoil and one to rewind it. **Topoisomerases** are enzymes
that hydrolyze a phosphodiester linkage in one strand of the double helix,
relax the supercoiling by rotating one strand around the other, and then

reseal the break. **DNA gyrases** induce negative supercoiling in relaxed, closed-circular DNA.

The supercoiling of the nuclear DNA of eukaryotes such as plants and animals is more complicated than the supercoiling of the circular DNA from prokaryotes, which we have described so far. DNA from these sources is complexed with basic proteins that have an abundance of positively charged side chains at physiological (neutral) pH. Electrostatic attraction between the negatively charged phosphate groups on DNA and the positively charged groups on the proteins favors the formation of complexes of this sort. The resulting material is called **chromatin.** The supercoiling of chromatin must take into account the presence of the proteins.

The principal proteins in chromatin are the **histones.** There are five main types, called H1, H2A, H2B, H3, and H4. All these proteins contain a large number of basic amino acid residues, such as lysine and arginine. In the chromatin structure, the DNA is tightly bound to all the types of histone but H1. The H1 protein is comparatively easy to remove from chromatin, but dissociating the other histones from the complex is more difficult. Proteins other than histones are also complexed with the DNA of eukaryotes, but they are neither as abundant nor as well studied as histones.

In electron micrographs chromatin has the appearance of beads on a string. This appearance, shown in Figure 20.14, reflects the molecular composition of the protein-DNA complex. The "beads" are called nucleosomes, and the "string" portions are called spacer regions. **Nucleosomes** consist of DNA wrapped around a histone core. This protein core is an octamer, which includes two molecules of each type of histone but H1; the composition of the octamer is $(H2A)_2(H2B)_2(H3)_2(H4)_2$. **Spacer regions** consist of DNA complexed to some H1 histone and nonhistone proteins. As the DNA coils around the histones in the nucleosome, about 200 base pairs are in contact with the proteins; the spacer region is about 30 to 50 base pairs long.

## 20.4
## THE THREE KINDS OF RNA AND THEIR STRUCTURES

Three kinds of RNA, which are called **transfer RNA (tRNA), ribosomal RNA (rRNA),** and **messenger RNA (mRNA),** play an important role in the life processes of cells (Figure 20.15). All three are involved in the synthesis of proteins in a series of reactions ultimately directed by the sequence of bases in the DNA of the cell. The base sequence of all types of RNA is determined by that of DNA. The process by which the order of bases is passed from DNA to RNA is called **transcription** (Section 21.9).

Ribosomes, in which rRNA is associated with proteins, are the site for assembly of the growing polypeptide chain in protein synthesis. Amino acids are brought to the assembly site covalently bonded to tRNA as aminoacyl-tRNAs. The order of amino acids in the growing protein is specified by the order of bases in mRNA. This process is called **translation** of the genetic message. A sequence of three bases in mRNA directs the

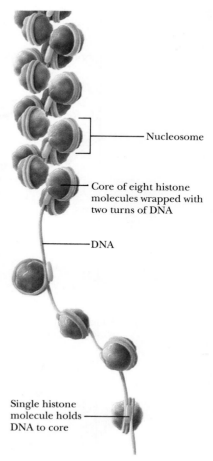

Nucleosome

Core of eight histone molecules wrapped with two turns of DNA

DNA

Single histone molecule holds DNA to core

**FIGURE 20.14** The structure of chromatin. DNA is associated with histones in an arrangement giving the appearance of beads on a string. The "string" is DNA, while the "beads" (nucleosomes) consist of DNA wrapped around a protein core of eight histone molecules. A single histone molecule holds the DNA to the core. Further coiling of the DNA spacer regions produces the compact form of chromatin found in the cell.

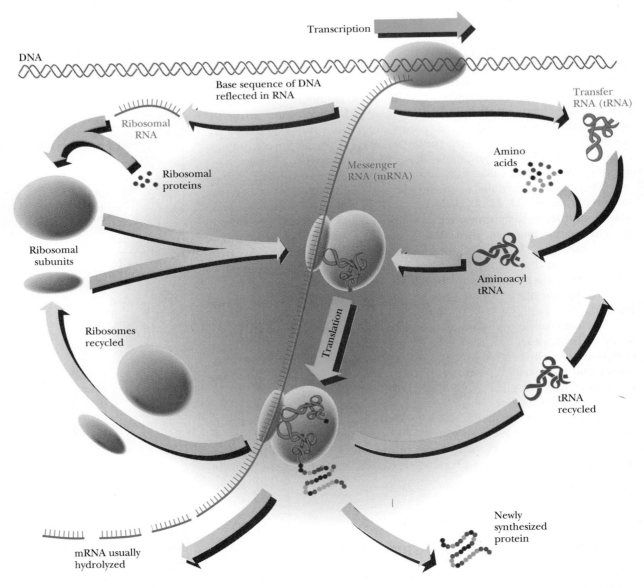

**FIGURE 20.15**   Flow chart showing the role of various types of RNA.

incorporation of a particular amino acid into the growing protein chain. We shall discuss the details of translation in Chapter 22, along with the **genetic code,** which specifies different amino acids during translation.

## Transfer RNA

The smallest of the three important kinds of RNA is tRNA. Dozens of different species of tRNA molecules can be found in all living cells, since there is at least one tRNA that bonds specifically to each of the amino acids that commonly occur in proteins. Frequently there are several tRNA

FIGURE 20.16   The cloverleaf structure of transfer RNA. Double-stranded regions (shown in red) are formed by folding of the molecule and stabilized by hydrogen bonds ( | | | | ) between complementary base pairs. Peripheral loops are shown in yellow. There are three major loops and one minor loop of variable size (dashed).

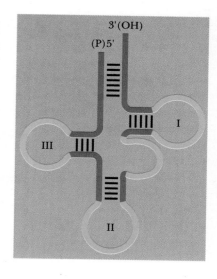

molecules for each amino acid. Transfer RNA molecules are single-stranded polynucleotide chains, usually about 80 nucleotide residues long, and their molecular weight is generally about 25,000 daltons.

**Intrachain hydrogen bonding** occurs in tRNA, forming A-U and G-C base pairs similar to those that occur in DNA except for the substitution of uracil for thymine. The molecule folds back on itself, represented by drawing a **cloverleaf structure,** which can be considered the secondary structure of tRNA (Figure 20.16). The hydrogen-bonded portions of the molecule are called stems, and the non–hydrogen-bonded portions are loops. Some of these loops contain modified bases (Figure 20.17). Both tRNA and mRNA are bound to the ribosome during protein synthesis in a definite spatial arrangement of ribosome, tRNA, and mRNA, which ultimately assures the correct order of the amino acids in the growing polypeptide chain.

The correct tertiary structure of tRNA is necessary for it to interact with the enzyme that covalently attaches the amino acid to the 3′ end. To produce this tertiary structure, the cloverleaf folds into an L-shaped conformation, the nature of which has been determined by x-ray diffraction (Figure 20.18).

## Ribosomal RNA

In contrast to tRNA, rRNA molecules tend to be quite large, and only a few types of rRNA exist in a cell. Because of the intimate association between rRNA and proteins, a useful approach to understanding the structure of rRNA is to look at ribosomes themselves.

The RNA portion of ribosomes accounts for 60 to 65% of the total weight, and the protein portion constitutes the remaining 35 to 40% of the weight. Dissociation of ribosomes into their components has proved to be a useful way of studying their structure and properties. A particularly important goal has been to determine both the number and the kind of RNA and protein molecules that make up ribosomes. This approach has helped elucidate the role of ribosomes in protein synthesis. In both prokaryotes and eukaryotes, ribosomes consist of two subunits, one of which is larger than the other. The smaller subunit in turn consists of 1 large RNA molecule and about 20 different proteins; the larger subunit consists of 2 RNA molecules in prokaryotes (there are 3 RNA molecules in eukaryotes) and about 35 different proteins in both prokaryotes and eukaryotes. The subunits are easily dissociated from one another in the laboratory by lowering the $Mg^{2+}$ concentration of the medium. Raising the $Mg^{2+}$ concentration to its original level reverses the process, and active ribosomes can be reconstituted by this method.

Pseudouridine ($\psi$)

4-Thiouridine

1-Methylguanosine (mG)

FIGURE 20.17   Structures of some modified bases found in transfer RNA. Note that in pseudouridine the pyrimidine is linked to ribose at C-5 rather than the usual N-1.

**FIGURE 20.18** The three-dimensional structure of transfer RNA. The amino acid is attached at the upper end, shown in blue. The site at which tRNA is bound to mRNA in the course of protein synthesis is shown in green at the lower end of the figure.

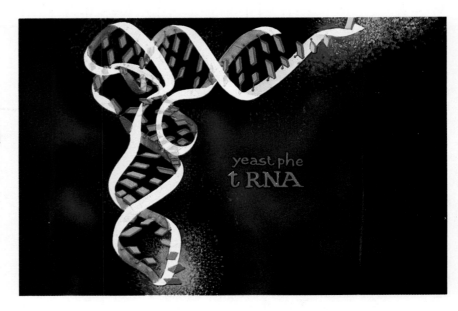

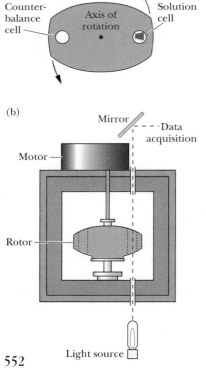

(a)

Counter-balance cell — Axis of rotation — Solution cell

(b)

Mirror ---·Data acquisition

Motor —

Rotor —

Light source

A technique called **analytical ultracentrifugation** has proved to be very useful in monitoring the dissociation and reassociation of ribosomes. An analytical ultracentrifuge is shown in Figure 20.19. We do not need to consider all the details of the method as long as it is clear that the basic principle of the experiment is observing the motion of ribosomes, RNA, or protein in a centrifuge. Both the size and the shape of a particle determine how fast it will move toward the bottom of the tube. The motion of the particle is characterized by a **sedimentation coefficient,** expressed in **Svedberg units (S).** This unit is named after The Svedberg, the Swedish scientist who invented the ultracentrifuge. The S value increases with the molecular weight of the sedimenting particle, but it is not directly proportional to it, because the shape of the particle also affects its sedimentation rate.

Ribosomes and ribosomal RNA have been studied extensively by determination of sedimentation coefficients. Most of the research on prokaryotic systems has been done in the bacterium *Escherichia coli,* which we shall use as an example here. An *E. coli* ribosome typically has a sedimentation coefficient of 70S. When an intact 70S bacterial ribosome dissociates, it produces a light 30S subunit and a heavy 50S subunit. Note that the values of sedimentation coefficients are not additive, an example of the dependence of the S value on the shape of the particle. The 30S subunit contains a 16S rRNA as well as 21 different proteins. The 50S subunit contains a 5S rRNA and a 23S rRNA as well as 34 different proteins (Figure 20.20). Eukaryotic ribosomes, for comparison, have a sedimentation coefficient of 80S, and the small and large subunits are 40S

**FIGURE 20.19** The analytical ultracentrifuge. (a) Top view of an ultracentrifuge rotor. The solution cell has optical windows; the cell passes through a light path once each revolution. (b) Side view of an ultracentrifuge rotor. The optical measurement done as the solution cell passes through the light path makes it possible to monitor the motion of sedimenting particles.

**FIGURE 20.20**   The subunit structure of ribosomes. The individual components can be mixed, producing functional subunits. Reassociation of subunits gives rise to an intact ribosome.

and 60S, respectively. The small subunit of eukaryotes contains an 18S rRNA, while the large subunit contains three types of rRNA molecules, 5S, 5.8S, and 28S.

The 5S rRNA has been isolated from many different types of bacteria, and the nucleotide sequences have been determined. A typical 5S rRNA is about 120 nucleotide residues long and has a molecular weight around 40,000 daltons. Some sequences have also been determined for the 16S and 23S rRNA molecules. These larger molecules are about 1500 and 2500 nucleotide residues long, respectively. The molecular weight of 16S rRNA is about 500,000 daltons, and that of 23S rRNA is about 1 million daltons. There appears to be a considerable degree of secondary and tertiary structure in the larger RNA molecules. A proposed secondary structure has been worked out for 16S rRNA (Figure 20.21), and suggestions have been made about the way in which the proteins associate

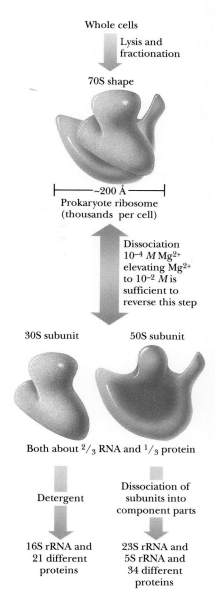

Whole cells

Lysis and fractionation

70S shape

├──~200 Å──┤
Prokaryote ribosome
(thousands per cell)

Dissociation
$10^{-4}$ M $Mg^{2+}$ elevating $Mg^{2+}$ to $10^{-2}$ M is sufficient to reverse this step

30S subunit          50S subunit

Both about $^2/_3$ RNA and $^1/_3$ protein

Detergent        Dissociation of subunits into component parts

16S rRNA and         23S rRNA and
21 different          5S rRNA and
proteins             34 different
                     proteins

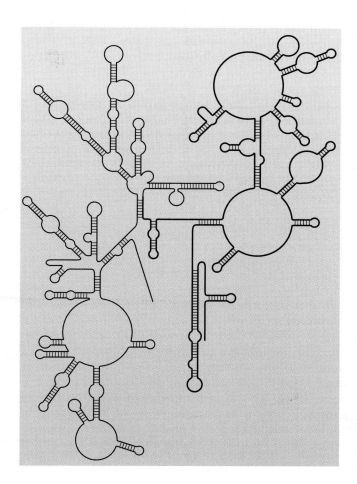

**FIGURE 20.21**   A proposed secondary structure for 16S rRNA. The intrachain folding pattern includes loops and double-stranded regions.

with the RNA to form the 30S subunit (see Figure 20.20). The **self-assembly of ribosomes** takes place in the living cell, and the process can be duplicated in the laboratory. Elucidating the structure of ribosomes is an active field of research. The binding of antibiotics to ribosomal subunits in such a manner as to prevent self-assembly of the ribosome is one such area of investigation. The structure of ribosomes is also one of the points used to compare and contrast eukaryotes, eubacteria, and the archaebacteria we discussed in Chapter 2. (See the articles by Lake, especially the review article, listed in the bibliography at the end of this chapter for more information on this subject.)

## Messenger RNA

The least abundant of the three types of RNA is mRNA. In most cells mRNA constitutes no more than 5 to 10% of the total cellular RNA, while tRNA accounts for 10 to 15% and the various types of rRNA compose 75 to 80% of the total. The sequence of bases in mRNA specifies the order of amino acids in proteins. In rapidly growing cells, many different proteins are needed in a short time. Fast turnover in protein synthesis becomes essential. Consequently, it is reasonable that mRNA is formed when it is needed, directs the synthesis of proteins, and then is degraded so that the nucleotides can be recycled. Of the three types of RNA—tRNA, rRNA, and mRNA—mRNA is the one that usually turns over most rapidly in the cell. Both tRNA and rRNA (as well as ribosomes themselves) can be recycled intact for many rounds of protein synthesis.

The sequence of bases in the mRNA that directs the synthesis of a protein reflects the sequence of DNA bases in the gene that codes for that protein. Messenger RNA molecules are heterogeneous in size, as are the proteins whose sequences they specify. There is probably no intrachain folding in mRNA; it is very likely an open chain. It is also likely that several ribosomes are associated with a single mRNA molecule at a time during the course of protein synthesis.

## 20.5
## DETERMINATION OF THE PRIMARY STRUCTURE OF NUCLEIC ACIDS

We have already seen that the primary structure of proteins determines their secondary and tertiary structure. The same is true of nucleic acids; the nature and order of monomer units determine the properties of the whole molecule. The base pairing in both RNA and DNA depends on a series of complementary bases, whether these bases are on different polynucleotide strands, as is the case in DNA, or on the same strand, as is frequently the case in RNA. The fact that the number of commonly occurring bases in nucleic acids is smaller than that of amino acids in proteins has made the determination of the primary structure of nucleic acids a challenging task. The situation is analogous to a jigsaw puzzle in which all the pieces are the same color compared with a jigsaw puzzle that has a marked pattern for identifying the pieces. Sequencing of nucleic acids is now fairly routine, and the relative ease with which it can be done would have been amazing to the scientists of the 1950s and 1960s.

## Hydrolysis of Nucleic Acids

The first step in determining the base sequence of nucleic acids is to hydrolyze the intact nucleic acid to produce fragments of convenient length. Like proteins, nucleic acids can be hydrolyzed by acids, bases, or enzymes. There are numerous hydrolytic enzymes that attack nucleic acids, with varying specificities regarding the point of attack. The hydrolysis of nucleic acids by acids and bases is less specific, and we shall concentrate on sequence determination in fragments produced by enzymatic cleavage.

The enzymes that hydrolyze nucleic acids are collectively called nucleases. Some nucleases are not specific with regard to the nature of the sugar moiety of the nucleic acid and hydrolyze either RNA or DNA. Some of them specifically hydrolyze DNA and are called **deoxyribonucleases (DNases);** others specifically hydrolyze RNA and are called **ribonucleases (RNases).** Another distinction is that some are **exonucleases,** which hydrolyze phosphodiester bonds from either the 3' or the 5' end of a nucleic acid. **Endonucleases** specifically cleave internal phosphodiester bonds in nucleic acids (Table 20.1).

Two other points of interest about the specificity of nucleases deserve mention before we go on to their role in the sequencing of nucleic acids. One point is the possibility of what is called $a$-type or $b$-type cleavage. In **$a$-type cleavage** a 5'-phosphate is the product; in **$b$-type cleavage** a 3'-phosphate is the product (Table 20.1). In $a$-type cleavage the products of hydrolysis of XpY are X (the nucleoside X) and pY (the 5'-phosphate of the nucleoside Y). In $b$-type cleavage the products of hydrolysis of XpY are Xp (the 3'-phosphate of nucleoside X) and Y (the nucleoside Y).

The other point is the question of **base specificity** or **sequence specificity.** A base-specific nuclease cleaves a nucleic acid molecule at a given base, *e.g.,* at every guanine site, or at every base of a given type, such as every pyrimidine site. A sequence-specific nuclease hydrolyzes a nucleic acid molecule at a given point in a strictly specified sequence, *e.g.,* at the G site in the sequence GAATTC (Table 20.1). The only currently known examples of this last type of nuclease are the **restriction endonucleases,** which act only on DNA, not on RNA, and they attack closely spaced sites on both strands. In the example given above, the site of attack is at both G residues in the double-stranded sequence GAATTC.
CTTAAG

The role of nucleases in the sequencing of nucleic acids is similar to that of the various types of proteases in the sequencing of proteins. Proteases are used to break up large protein molecules into fragments of manageable size for further analysis (Section 9.2). Similarly, the use of nucleases of different specificities on different samples of the same nucleic

**TABLE 20.1  Properties of Some Nucleases**

| ENZYME | SUBSTRATE | MODE OF ATTACK | SPECIFICITY | CLEAVAGE SITES |
|--------|-----------|----------------|-------------|----------------|
| 1. Pancreatic ribonuclease (RNase A) | RNA | Endo (b type) | Linkage between pyrimidine (Py) and a nonspecific base (X) on the 3′ side | Py p X p ↑ |
| 2. RNase T₁ (isolated from fungi) | RNA | Endo (b type) | Linkage between guanine (G) and nonspecific base (X) on 3′ side | G p X p ↑ |
| 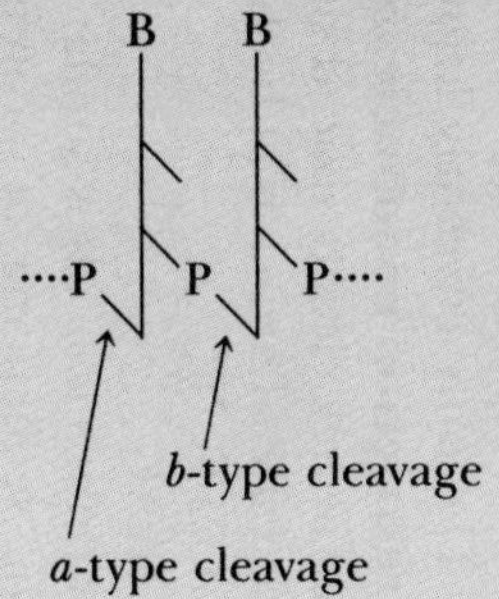 3. Micrococcal nuclease | RNA, DNA | Endo (b type) | Linkage between adenine (A) and nonspecific base (X) on 5′ side | X p A p ↑ |
| 4. Snake venom nuclease | RNA, DNA | Exo (a type) | Attack starts from 3′ hydroxyl end with no base specificity | X p Y p Z (3′-OH) ↑ ↑ |
| 5. Spleen nuclease | RNA, DNA | Exo (b type) | Attack starts from and requires a free 5′ hydroxyl terminus with no base specificity | [5′-OH]X p Y p Z ↑ ↑ |
| The next three are examples of restriction enzymes that cleave a specific bond in a specific DNA sequence. | | | | |
| 6. EcoRI (isolated from Escherichia coli) | DNA | Endo | | G A A T T C ↑ |
| 7. HaeIII (isolated from Hemophilus aegyptius) | DNA | Endo | | G G C C ↑ |
| 8. HindIII (isolated from Hemophilus influenzae) | DNA | Endo | | A A G C T T ↑ |

The phosphodiester bond can be cleaved on either of two sides

b-type cleavage

a-type cleavage

acid produces fragments with overlapping sequences, which are used to deduce the primary structure of the whole molecule, just as with proteins.

## Determination of the Primary Structure of RNA

The first RNA molecule for which the base sequence was determined was an alanine-specific RNA from yeast (Figure 20.22). This result was announced in 1965 by Robert W. Holley, who received the Nobel Prize in 1968 for this achievement. More than 200 RNA sequences have been determined since then. The RNA molecules that have been sequenced have been isolated from two main sources: yeast, which is eukaryotic, and *E. coli,* a bacterium; each of these RNA molecules is specific for a different amino acid, such as phenylalanine, isoleucine, or alanine. The general features of the secondary and tertiary structure of all RNA molecules are similar, even though some were isolated from a eukaryotic source and some from a prokaryotic source; the cloverleaf structure folded into an L-shaped conformation is a common structural theme.

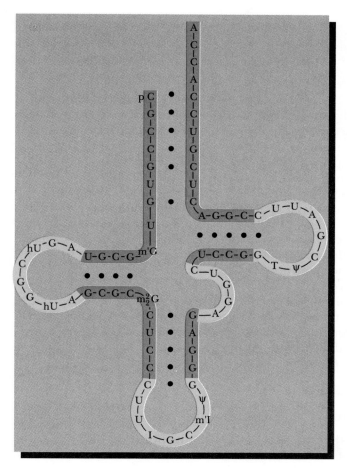

FIGURE 20.22   Structure of an alanyl-tRNA molecule from yeast. (From R.W. Holley, J. Apgar, G.A. Everett, et al., 1965. Structure of a Ribonucleic Acid. *Science* **147,** 1463. Copyright 1965 by the American Association for the Advancement of Science.)

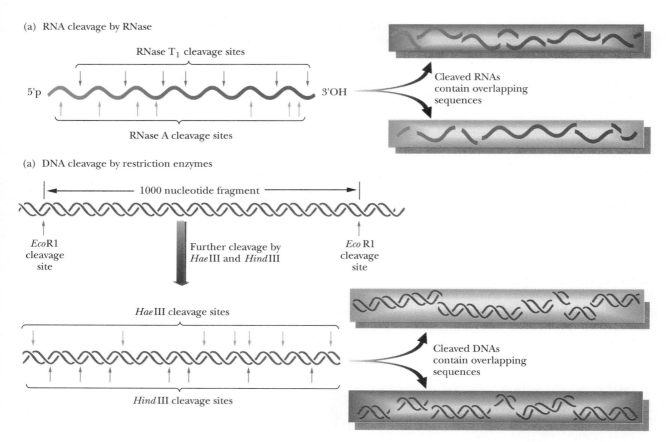

(a) RNA cleavage by RNase

RNase T₁ cleavage sites

5'p       3'OH

RNase A cleavage sites

Cleaved RNAs contain overlapping sequences

(a) DNA cleavage by restriction enzymes

1000 nucleotide fragment

EcoR1 cleavage site

Further cleavage by HaeIII and HindIII

Eco R1 cleavage site

HaeIII cleavage sites

HindIII cleavage sites

Cleaved DNAs contain overlapping sequences

**FIGURE 20.23** The role of enzymatic cleavage in nucleic acid sequencing. (a) Obtaining overlapping sequences in RNA by using two enzymes (RNase T₁ and RNase A) with different specificities. (b) Obtaining overlapping sequences in DNA. A 1000-nucleotide fragment has been obtained by using one restriction enzyme (*Eco*RI). Two other restriction enzymes (*Hae*III and *Hind*III) have been used for further hydrolysis.

The original strategy for sequence determination in tRNA has much in comon with methods used for the same purpose in proteins. The use of two specific ribonucleases of different specificities produced large fragments with overlapping sequences. Bovine pancreatic ribonuclease (RNase A) hydrolyzes phosphodiester linkages on the 3′ side of pyrimidine residues. Ribonuclease T₁ (RNase T₁) from the mold *Aspergillus oryzae* carries out the same type of reaction on guanine and inosine residues (Figure 20.23(a)).

The sequences of the fragments produced by the action of ribonucleases can then be determined by methods similar in principle to those used on peptide fragments of proteins; thus we shall not discuss them here. Once the sequence of the fragments produced by ribonuclease digestion has been determined, the overlapping sequences can be compared to arrive at the structure of the whole molecule. A more recent approach to nucleic acid sequencing relies on some of the unique properties of nucleic acid. This method was developed for DNA, and we shall discuss it in the context of DNA sequencing.

## Determining the Primary Structure of DNA

The first step in the sequencing of DNA is the hydrolysis of the molecule into fragments of a suitable size for further analysis, as is the case with other sequencing methods. This initial hydrolysis is carried out by restriction endonucleases. Several different enzymes of this type with different specificities are used to produce fragments 400 residues long or longer (Table 20.1 and Figure 20.23(b)). It is noteworthy that these fragments, which can be sequenced routinely and relatively easily, are more than twice as long as the 77-residue RNA that was the first nucleic acid for which a sequence was determined.

Several methods can be used for sequencing DNA fragments. All these methods depend on two factors. The first, the use of restriction endonucleases to obtain fragments of suitable size for sequencing, we have already mentioned. The second is the existence of convenient methods of **gel electrophoresis,** which can separate oligonucleotides on the basis of size, even when they differ in length by only one nucleotide monomer.

The actual sequencing is done using single-stranded oligonucleotides obtained from the DNA fragments by strand separation (Section 20.7), followed by further treatment. Oligonucleotides are separated by **poly-acrylamide gel electrophoresis (PAGE).** All electrophoretic methods use charge as the basis of separating molecules. The PAGE technique separates molecules on the basis of both charge and size; see Interchapter A, Section A.4, for a discussion of its usefulness in separating proteins and peptides. In nucleic acid fragments, each nucleotide residue contributes a negative charge from the phosphate to the overall charge of the fragment, but the mass of the oligonucleotide increases correspondingly. Thus the ratio of charge to mass remains very much the same regardless of the size of the oligonucleotide. As a result, the separation takes place simply on the basis of size and is due to the sieving action of the gel (Interchapter A, Section A.4). In a given amount of time, a smaller oligonucleotide will move farther than a larger one in an electrophoretic separation. The oligonucleotides move in the electric field because of their charge; the distance they move in a given time depends on their size.

The experimental design of the separation procedure takes advantage of the fact that the oligonucleotides differ in size. The separation is done with the gel in a vertical position. The negative electrode is at the top of the gel, and the positive electrode is at the bottom. There is room for several samples on each gel. Each sample is loaded at a given place (a separate track) at the top of the gel, and the current is applied until the separation is complete, with the smallest oligonucleotide products near the bottom of the gel (Figure 20.24).

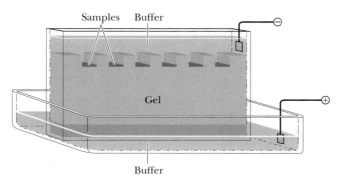

**FIGURE 20.24** The experimental setup for gel electrophoresis. The gel is placed in a vertical position. The samples are applied at the top of the gel. The negatively charged oligonucleotides migrate toward the positive electrode at the bottom when the current is applied.

**FIGURE 20.25** An example of an autoradiograph.

The most commonly used method for detecting the separated products is based on radioactive labeling of the sample. When the labeled oligonucleotides have been separated, the gel is placed in contact with a piece of x-ray film. The radioactively labeled oligonucleotides expose the portions of the film with which they are in contact. When the film is developed, the positions of the labeled substances show up as dark bands. This technique is called **autoradiography,** and the resulting film image is called an **autoradiograph** (Figure 20.25). Two widely used methods for sequencing DNA depend on these techniques. One is the direct chemical method of Maxam and Gilbert (see Box 20.1). The other is that of Sanger and Coulson, a method that makes use of some enzymes that play a role in DNA replication. We shall not discuss the second method now. We shall need some information from Chapter 21 to understand the Sanger–Coulson method, so we shall discuss it there (Box 21.2). Sanger and Gilbert were both recipients of the 1980 Nobel Prize in chemistry for their contributions to the techniques of determining nucleic acid sequences. Considerable research effort continues to be made to improve methods of DNA sequencing, particularly with regard to automating the process.

## 20.6
## RECOMBINANT DNA

Producing recombinant DNA requires cutting and splicing the two strands of DNA molecules in very specific ways. The cutting is done by restriction enzymes so as to have a convenient way to produce end sequences that are the same in DNA molecules from different sources. The end sequences match up, and the splicing is done by a group of enzymes called **DNA ligases.**

### Restriction Enzymes Produce "Sticky Ends"

Restriction endonucleases hydrolyze only a specific bond of a specific sequence in DNA. The sequences recognized by restriction enzymes read the same from left to right or from right to left (on the complementary strand). The term for such a sequence is a **palindrome.** "Able was I ere I saw Elba" and "Madam I'm Adam" are both well-known examples of palindromes. The sites of action of restriction enzymes are palindromic sequences in DNA. A typical restriction enzyme called *Eco*RI is isolated from *E. coli;* restriction endonucleases are designated by an abbreviation of the name of the organism in which they occur. The *Eco*RI site in DNA is 5′ GAATTC 3′, where the base sequence on the other strand is 3′ CTTAAG 5′. The sequence from left to right on one strand is the same as the sequence from right to left on the other strand. The phosphodiester bond between G and A is the one hydrolyzed. This same break is produced on both strands of the DNA. There are four nucleotide residues, two adenine and two thymine in each strand, between the two breaks on opposite strands, leaving the **"sticky ends,"** which can still be joined by hydrogen bonding between the complementary bases. The ends are held in place by the hydrogen bonds, and the two breaks can then be resealed covalently by the action of DNA ligases (Figure 20.26). If no ligase is present, the ends can remain separated. The hydrogen bonding at the "sticky ends" holds the molecule together until gentle warming or vigorous stirring effects a separation.

DNA from different sources can be treated with the same restriction enzyme, producing the same sequences of overlapping, sticky ends. The

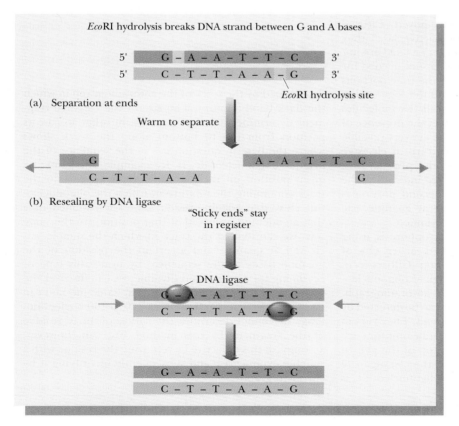

**FIGURE 20.26**   Hydrolysis of DNA by restriction endonucleases. (a) Separation of ends. (b) Resealing of ends by ligase.

## BOX 20.1
## THE MAXAM–GILBERT METHOD FOR DNA SEQUENCING

The sequencing approach of Maxam and Gilbert is called the **direct chemical method.** A restriction fragment of DNA is the starting point for chemical treatment. The DNA is radioactively labeled at the 5′ end with $^{32}P$, because the label will be used for detection of products later in the experiment. The sample is then divided into four portions for selective modification of each of the four bases. The chemical treatment causes cleavage of the DNA at the modified base. The four portions are then analyzed by gel electrophoresis.

As an example we shall consider the sequencing of an oligonucleotide 12 residues long, 5′ *GTACCAGCT-GCT 3′. The asterisk indicates the presence of the $^{32}P$ label, which has already been attached at the 5′ end.† The sample is divided into four portions for separate treatment of the four kinds of bases. When adenine is the base treated with subsequent cleavage, the following mixture of oligonucleotides is obtained:

| | |
|---|---|
| *GTACC | *GT |
| GCTGCT | CCAGCTGCT |

Note that the cleaved adenine is removed. Cleaved fragments are detected by the radioactive label they carry. Oligonucleotides that do not carry a label are not detected, but information about them is not needed for the sequence determination. Each of the four portions consists of many, many molecules, and the logic of the experiment depends on this fact. In all the molecules that make up the portion used to detect the presence of adenine, there should be a cleavage site at every possible adenine position. The result is a mixture of cleavage products of varying lengths. When the products are analyzed, the labeled molecules detected indicate all the sites at which adenine occurs. The second portion is similarly treated to produce cleavage at guanine, the third at cytosine, and the fourth at thymine.

The results are analyzed by gel electrophoresis, using autoradiographic detection as described earlier. One of the most striking features of the Maxam–Gilbert technique is that the sequence of the original oligonucleotide can be "read" directly from the autoradiograph. Recall that the cleavage products that can be detected by this technique are all labeled at the 5′ end. They vary in length, increasing in size by one nucleotide unit for each set of bands from the bottom to the top of the gel. In each band the identity of the base at the cleavage site (the cleavage site is the 3′ end of the oligonucleotide) is

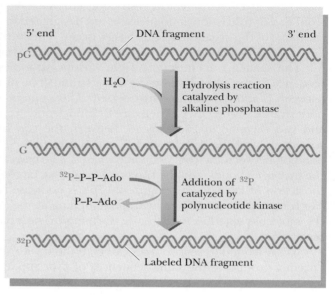

The labeling of the 5′ end of DNA with $^{32}P$.

determined by the track of the autoradiograph in which the dark band appears. The size of the cleavage fragment determines how far it moves; in other words, its position from top to bottom along the track shows where the base in question lies in the sequence of the original oligonucleotide. If, for example, the band at a given position on the autoradiograph appears in the track containing the mixture in which the oligonucleotide cleaved at adenine residues, we know immediately that an adenine occurred at that position. In the same way, we can look at the track in which the sample was cleaved at guanine and ascertain the position of the guanine bases in the sequence. By using this type of reasoning for cytosine and thymine, we can discover the sequence of the oligonucleotide. The same method can be used for restriction fragments 200 to 400 nucleotides long. We cannot determine the nature of the 5′ residue of the fragment by this method. We can, however, accomplish this objective by removing the residue with a suitable exonuclease and subsequently identifying it by chromatography.

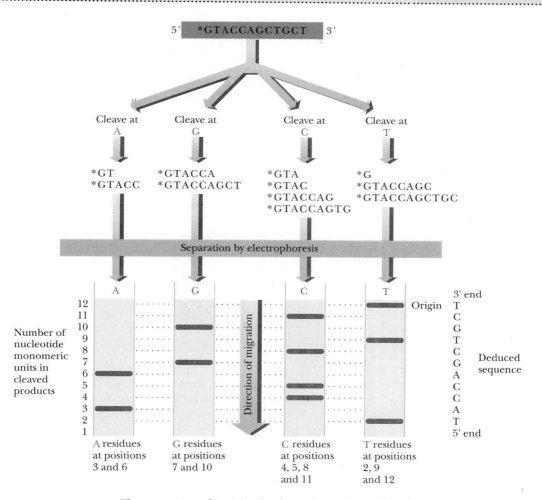

The sequencing of DNA by the direct chemical method. The asterisk indicates the $^{32}$P-labeled phosphate group at the 5' end. Note that the identity of the 5' residue is not determined by this method (see text). All other sequence information can be deduced from the electrophoretic pattern. Exonuclease digestion and identification of the product can be used to determine the nature of the 5' residue.

†The possibility of obtaining a CC oligonucleotide does exist; such an oligonucleotide would require cleavage at both sites. However, this would not affect the results of the experiment because the identity of the labeled oligonucleotide does not change.

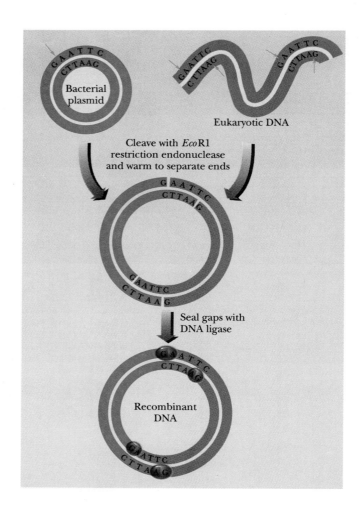

**FIGURE 20.27** The methodology for producing recombinant DNA.

overlapping sequences from the two different DNAs can hydrogen-bond with each other and then can be joined covalently by DNA ligases. More recently, techniques have been developed for the attachment of synthetic complementary "linkers" to the ends of DNAs that are to be recombined. This development is useful to researchers, who are no longer dependent on naturally occurring sites for the cut-and-splice process. Still other methods have been developed to cut DNA at chosen sites and only those sites. These methods allow researchers to obtain larger pieces of DNA than is possible using restriction enzymes. For more details, see the article by Roberts cited in the bibliography.

### Inserting Eukaryotic DNA into the Bacterial Genome

Quite frequently, the application of recombinant DNA technology involves insertion of a gene from some other organism into the DNA of a given bacterium. Since bacteria reproduce rapidly, it becomes possible to obtain a large quantity of a substance, *e.g.*, human insulin, that is normally produced by a slower-growing organism. Frequently, larger quantities of the gene product in question can be obtained from the bacteria than can normally be obtained from the original organism, a fact that can be useful if there is considerable need for that substance.

The exogenous gene (the "foreign" gene) is often inserted into a **bacterial plasmid,** a circular DNA molecule that is not a part of the main circular DNA chromosome of the bacterium. In addition to bacterial plasmids, other types of DNA molecules can be used as **vectors** (carrier molecules for recombinant genes). Another type of vector is DNA from bacteriophages (bacterial viruses), but we shall confine our discussion of vectors to bacterial plasmids.

*E. coli* is a common source of bacterial plasmids used as vectors. The DNA to be inserted can come from any source, including eukaryotic ones. Both isolated DNAs are treated with the same restriction enzyme, such as *Eco*RI. The end sequences of both types of DNA are the same, and the ends can overlap. The action of DNA ligases in resealing the ends produces **closed-circular recombinant DNA** (Figure 20.27). Another name for recombinanant DNA is **chimeric DNA,** from the chimera, a monster in Greek mythology that was a combination of several different animals.

The recombined plasmid is introduced into *E. coli* cells, and the cells that actually take up the plasmid are allowed to reproduce. In this way eukaryotic genes, such as the one for human insulin, can be introduced into bacteria. This process is known as **cloning of DNA.** Large amounts of human insulin can be produced by such a process, which takes advantage of the rapid growth of bacteria. It is necessary to test whether the recombinant plasmid has actually been taken up by the cell. In the example that we have used, the test is whether the cell now produces human insulin.

## 20.7
## DENATURATION OF DNA

We have already seen that the hydrogen bonds between base pairs are an important factor in holding the double helix together. Not a large amount of stabilizing energy is associated with hydrogen bond, but there are so many of them that the cumulative effect is considerable. In addition, the stacking of the bases in the native conformation of DNA contributes some stabilization energy. Energy must be added to a sample of DNA to break

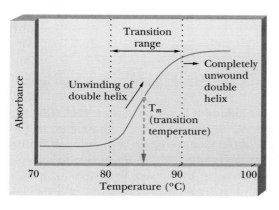

**FIGURE 20.28** The experimental determination of DNA denaturation. This is a typical melting curve profile of DNA, depicting the hyperchromic effect observed on heating. $T_m$ is displaced as the percentage of guanine and cytosine (the GC content) increases. (The entire curve would be shifted to the right for a DNA with higher GC content and to the left for a DNA with lower GC content.)

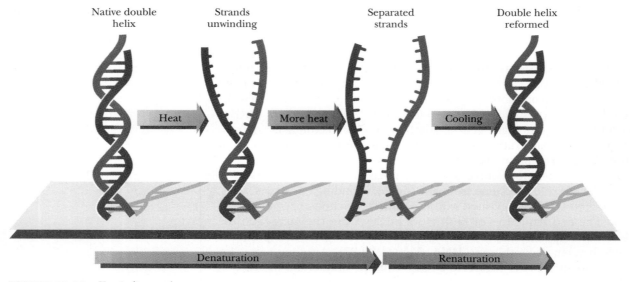

Native double helix · Strands unwinding · Separated strands · Double helix reformed

Heat · More heat · Cooling

Denaturation · Renaturation

**FIGURE 20.29** Unwinding and rewinding of the double helix in DNA denaturation and renaturation. The double helix unwinds when DNA is denatured, with eventual separation of strands. The double helix is reformed on renaturation with slow cooling and annealing.

the hydrogen bonds, a process that is usually carried out by heating a solution of DNA.

The heat denaturation of DNA can be monitored experimentally by observing the absorption of ultraviolet light. The bases absorb light in the 260-nm wavelength region. As the DNA is heated and the strands separated, the wavelength of absorption does not change, but the amount of light absorbed increases (Figure 20.28). This effect is called **hyperchromicity.** It is based on the fact that the bases, which are stacked on top of one another in native DNA, become unstacked as the DNA is denatured. Since the bases interact differently in the stacked and unstacked orientations, their absorbance changes. Heat denaturation of DNA is also called **melting** (Figure 20.29).

Under a given set of conditions, the midpoint of the melting curve (the transition temperature, or melting temperature, written as $T_m$) is characteristic of DNA from different sources. The underlying reason for this property is that each type of DNA has a given well-defined base composition. A G-C base pair has three hydrogen bonds, and an A-T base pair has only two. The higher the percentage of G-C base pairs, the higher will be the melting temperature of a DNA molecule.

Renaturation of denatured DNA is possible on slow cooling (Figure 20.29). The annealed, separated strands are able to recombine and to form the base pairs responsible for maintaining the double helix.

## SUMMARY

The double helix originally proposed by Watson and Crick is the most striking feature of DNA structure. The two coiled strands run in antiparallel directions and are held together by hydrogen bonds between complementary bases. Adenine pairs with thymine and guanine with cytosine. Supercoiling is a feature of DNA structure in both prokaryotes and eukaryotes. Eukaryotic DNA is complexed with histones and other basic proteins, while prokaryotic DNA occurs in "naked" form, not complexed to proteins.

The three kinds of RNA—transfer RNA (tRNA), ribosomal RNA (rRNA), and messenger RNA (mRNA)—differ somewhat in structure. Transfer RNA is relatively small, about 80 nucleotides long. There is extensive intrachain hydrogen bonding, represented by a cloverleaf structure. Ribosomal RNA molecules tend to be quite large and are complexed with proteins to form ribosomal subunits. Ribosomal RNA also exhibits extensive internal hydrogen bonding. The sequence of bases in a given mRNA determines the sequence of amino acids in a specified protein. The size of mRNA molecules varies with the size of the protein. No evidence exists for extensive hydrogen bonding in mRNA.

The base sequence of nucleic acids can be determined in a manner similar to that used for determining the amino acid sequence of proteins, but it is more efficient, particularly for DNA, to use methods that make use of specialized techniques. Restriction endonucleases can be used to cleave DNA molecules into fragments of suitable size. In a direct chemical method, four samples of a given restriction fragment can each be treated with a selective reagent, causing cleavage at a given base. The resulting mixtures can be analyzed by gel electrophoresis, which separates,

on the basis of size, the oligonucleotides produced by this treatment. The base sequence of the oligonucleotide can be "read" directly from the sequencing gel.

Restriction enzymes also have an important part in recombinant DNA technology. These endonucleases produce "sticky ends," short single-stranded stretches at each end. These "sticky ends" provide a way to link together DNAs from different sources, even to the point of inserting eukaryotic DNA into bacterial genomes.

When DNA is denatured, the double-helical structure breaks down, a phenomenon that can be followed by monitoring absorption of ultraviolet light. The temperature at which DNA becomes denatured by heat depends on its base composition; higher temperatures are needed to denature DNA rich in G-C base pairs.

## EXERCISES

1. Suggest a reason why AZT is useful in the treatment of AIDS. *Hint:* Compare the structure of AZT with the covalent structure of the DNA backbone.
2. The modified nucleoside 2′, 3′-dideoxycytidine is another compound that has been proposed for the treatment of AIDS. Draw the structure of this compound and suggest a reason for its usefulness.
3. Vidarabine (also called ara-A) is a modified nucleoside used as an antiviral agent. This compound differs from adenosine by the replacement of ribose by arabinose. Draw the structure of this compound and suggest a reason for its mode of action. (See Figure 6.4 for the structure of arabinose.)
4. Why does DNA with a high A-T content have a lower transition temperature ($T_m$) than DNA with a high G-C content?
5. Which of the following statements are true?
   (a) Bacterial ribosomes consist of 40S and 60S subunits.
   (b) Prokaryotic DNA is normally complexed with protein.
   (c) Prokaryotic DNA normally exists as a closed circle.
   (d) Circular DNA is supercoiled.
6. Binding sites for the interaction of polypeptides and drugs with DNA are found in the major and minor grooves. True or false?
7. Which of the following statements are true?
   (a) The two strands of DNA run in parallel directions from the 5′ to the 3′ end.

   (b) An adenine-thymine base pair contains three hydrogen bonds.
   (c) Positively charged counterions are associated with DNA.
   (d) DNA base pairs are always perpendicular to the helix axis.
8. Define the following terms:
   (a) supercoiling      (c) topoisomerase
   (b) positive supercoil    (d) negative supercoil
9. Briefly describe the structure of chromatin.
10. Sketch a typical cloverleaf structure for transfer RNA. Point out any similarities between the cloverleaf pattern and the proposed structures of ribosomal RNA.
11. How does the use of restriction endonucleases of different specificities aid in the sequencing of DNA?
12. What fragments, labeled and unlabeled, can be expected from selective modification and cleavage of *GAATCTACGT at cytosine?
13. Predict the electrophoretic pattern obtained when the oligonucleotide in Exercise 12 is subjected to the Maxam–Gilbert method.
14. Show by an example how a restriction endonuclease produces "sticky ends" in DNA.
15. Define cloning of DNA.
16. Give an example of a palindromic sequence in DNA.
17. Outline the methods you would use to produce human growth factor (a substance used in the treatment of dwarfism) in bacteria.

## ANNOTATED BIBLIOGRAPHY

Adams, R.L.P., J.T. Knowles, and D.P. Leader. *The Biochemistry of the Nucleic Acids*. 10th ed. New York: Chapman and Hall, 1986. [New authors have taken over a classic text originally written by J.N. Davidson.]

Anderson, W.F., and E.G. Diacumakos. Genetic Engineering in Mammalian Cells. *Sci. Amer.* **245**(1), 106–121 (1981). [A description of recombinant DNA techniques.]

Barton, J.K. Recognizing DNA Structures. *Chem. Eng. News* **66**, 30–42 (1988). [An account of the different conformations of DNA and how they change on drug binding.]

Bauer, W.R., F.H.C. Crick, and J.H. White. Supercoiled DNA. *Sci. Amer.* **243**(1), 118–133 (1980). [A description of the various types of circular DNA.]

Brimacombe, R. The Emerging Three-Dimensional Structure and Function of 16S Ribosomal RNA. *Biochemistry* **27**, 4208–4212 (1988). [A short review.]

Darnell, J.E. The Processing of RNA. *Sci. Amer.* **249**(2), 90–100 (1983). [A discussion of the ways in which RNA is modified after synthesis.]

Felsenfeld, G. DNA. *Sci. Amer.* **253**(4), 58–67 (1985). [A review of the main features of DNA structure.]

Holley, R.W. The Nucleotide Sequence of a Nucleic Acid. *Sci. Amer.* **214**(2), 30–46 (1966). [A description of the original methods used for sequencing transfer RNA. Of historical interest.]

Kornberg, R.D., and A. Klug. The Nucleosome. *Sci. Amer.* **244**(2), 52–64 (1981). [A discussion of the way in which DNA and histone proteins associate with each other.]

Lake, J.A. Evolving Ribosome Structure: Domains in Archaebacteria, Eubacteria, Eocytes and Eukaryotes. *Ann. Rev. Biochem.* **54**, 507–530 (1985). [A review article on the evolutionary implications of ribosome structure.]

Lake, J.A. The Ribosome. Sci. Amer. **245**(2), 84–97 (1981). [A look at some of the complexities of ribosome structure.]

Maxam, A.M., and W. Gilbert. A New Method for Sequencing DNA. *Proc. Natl. Acad. Sci. USA* **74**, 560–564 (1977). [The original article describing the direct chemical method for sequencing DNA. Mainly of historical interest.]

Nomura, M. The Control of Ribosome Synthesis. *Sci. Amer.* **250**(2), 102–114 (1984). [The assembly of ribosomes is the topic of discussion.]

Roberts, L. New Scissors for Cutting Chromosomes. *Science* **249**, 127 (1990). [A Research News article about a method for choosing exactly the right place to cut DNA into large pieces.]

Ross, J. The Turnover of Messenger RNA. *Sci. Amer.* **260** (4), 48–55 (1989). [An article describing the regulation of the rate at which messenger RNA is degraded in the cell.]

Saenger, W. *Principles of Nucleic Acid Structure*. New York: Springer-Verlag, 1984. [A fairly advanced treatment of the subject.]

Scovell, W.M. Supercoiled DNA. *J. Chem. Ed.* **63**, 562–565 (1986). [An article that deals mainly with the topology of circular DNA.]

Sinden, R.A. Supercoiled DNA: Biological Significance. *J. Chem. Ed.* **64**, 294–301 (1987). [An article that concentrates on the enzymes involved in DNA supercoiling.]

Smith, L.M., J.Z. Sanders, R.J. Kaiser, et al. Fluorescence Detection in Automated DNA Sequence Analysis. *Nature* **321**, 674–679 (1986). [The first report of an automated method for sequencing of DNA.]

Wasserman, S.A., and N.R. Cozzarelli. Biochemical Topology: Applications to DNA Recombination and Replication. *Science* **232**, 951–960 (1986). [A discussion of the ways in which DNA recombination and replication can produce knotted DNA structures.]

Watson, J.D., and F.H.C. Crick. Molecular Structure of Nucleic Acid. A Structure for Deoxyribose Nucleic Acid. *Nature* **171**, 737–738 (1953). [The original article describing the double helix. Of historical interest.]

Westheimer, F.H. Why Nature Chose Phosphates. *Science* **235**, 1173–1178 (1987). [A discussion of the importance of phosphate groups in biochemistry, particularly in the backbone of nucleic acids. The author of this article is an eminent organic chemist.]

Most textbooks of organic chemistry have a chapter on nucleic acids.

# Biosynthesis of Nucleic Acids

*Fibers of purified DNA viewed under polarized light.*

569

*Before double helical DNA can be replicated, helical sections of DNA must be unwound so that the two parental strands can serve as templates for the synthesis of new daughter strands, thus making a precise copy of the original double helix. DNA polymerases synthesize DNA by joining nucleotides to the exposed single stranded DNA template, with each new base bound to its complementary partner. The fidelity of DNA synthesis is of utmost importance, since errors of replication will be passed to future generations. One of the polymerases has "proofreading" powers capable of self-correcting. In the next stage of DNA processing, the sequence of DNA bases is* transcribed *into a complementary sequence of RNA bases called messenger RNA. The RNA message differs from DNA in one respect: the DNA base thymine (T) is replaced by the RNA base uracil (U). In eukaryotes, messenger RNA carries the genetic code from the nucleus to the ribosomes in the cytosol where the sequence of RNA bases is* translated *into the amino acid sequence of proteins. Copying the genetic message is a powerful way to amplify the production of protein molecules, as seen from one vivid example. A single fibroin gene from the silkworm can, in four days, be copied in a hundred thousand RNA messenger molecules, which, in turn, can code for ten billion polypeptide chains of silk.*

## 21.1
## THE FLOW OF GENETIC INFORMATION IN THE CELL

The sequence of bases in DNA contains the **genetic code.** The duplication of DNA, giving rise to a new DNA molecule with the same base sequence as the original, is necessary whenever a cell divides to produce daughter cells. This duplication process is called **replication.** The actual formation of gene products requires RNA; the production of RNA on a DNA template is called **transcription** of the genetic message. The base sequence of DNA is reflected in the base sequence of RNA.

Three kinds of RNA are involved in the biosynthesis of proteins; of the three, messenger RNA (mRNA) is of particular importance. A sequence of three bases in mRNA specifies the identity of one amino acid in a manner directed by the genetic code. The process by which the base sequence directs amino acid sequence is called the **translation** of the genetic message. In nearly all organisms, the flow of genetic information is DNA $\longrightarrow$ RNA $\longrightarrow$ protein. The only major exceptions are viruses (called retroviruses) in which RNA, rather than DNA, is the genetic material. In those viruses RNA can direct its own synthesis as well as that of DNA (see the article by Varmus cited in the bibliography at the end of this chapter). Figure 21.1 shows ways in which information is transferred in the cell.

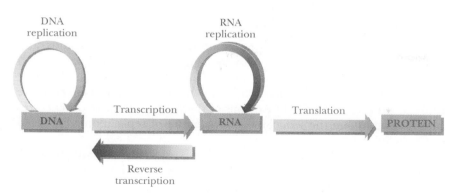

FIGURE 21.1 **FIGURE 21.1**  Mechanisms for transfer of information in the cell. The yellow arrows represent general cases, while the blue arrows represent special cases (mostly in RNA viruses).

The nucleic acids are the focus of some of the most exciting current research in molecular biology. Study in this field advances so quickly that some discoveries are out of date within months of the time they appear in print. The material in this chapter is well-established basic information on the subject. Considerably more needs to be learned, however; perhaps some of the students now reading this chapter will do research on nucleic acids in the future.

## 21.2
## THE REPLICATION OF DNA: SOME GENERAL CONSIDERATIONS

Naturally occurring DNA exists in many forms. Single- and double-stranded DNAs are known, and both can exist in linear and circular forms. As a result, it is difficult to generalize about all possible cases of DNA replication. Since many DNAs are double-stranded, we can present some general features of the replication of double-stranded DNA, features that apply to both linear and circular DNA. Most of the details of the process that we shall discuss here were first investigated in prokaryotes, particularly *Escherichia coli*. We shall use information obtained by experiments on this organism throughout our discussion of the topic.

The process by which one double-helical DNA molecule is duplicated to produce two such double-stranded molecules is a complex one. The very complexity allows for a high degree of fine tuning, which in turn ensures considerable fidelity in replication. The cell faces three important challenges in carrying out the necessary steps. The first challenge is how to *separate the two DNA strands*. In addition to achieving continuous unwinding of the double helix, the cell also has to protect the unwound portions of DNA from the action of nucleases that preferentially attack single-stranded DNA. The second task involves the *synthesis of DNA from the 5' to the 3' end.* Two antiparallel strands must be synthesized in the same direction on antiparallel templates. The third task is how to *guard against errors in replication,* ensuring that the correct base is added to the growing polynucleotide chain. The answers to these questions require the material in this section and in the three following sections.

## Semiconservative Replication

DNA replication involves separation of the two original strands and production of two new strands with the original ones as templates. Each new DNA molecule contains one strand from the original DNA and one newly synthesized strand. This situation is what is called **semiconservative replication** (Figure 21.2). The details of the process differ in prokaryotes and eukaryotes, but the semiconservative nature of replication is observed in all organisms.

Semiconservative replication of DNA was established unequivocally in the late 1950s by experiments performed by Meselson and Stahl. *E. coli* bacteria were grown with $^{15}NH_4Cl$ as the sole nitrogen source, where $^{15}N$ is a heavy isotope of nitrogen. The usual form of nitrogen is $^{14}N$. In such a medium, all newly formed nitrogen compounds, including purine and pyrimidine nucleobases, become labeled with $^{15}N$. The $^{15}N$-labeled DNA has a higher density than unlabeled DNA, which contains the usual isotope, $^{14}N$. In this experiment, the $^{15}N$-labeled cells were then transferred to a medium that contained only $^{14}N$. The cells continued to grow in the new medium. With every new generation of growth, a sample of DNA was extracted and analyzed by the technique of **density-gradient centrifugation** (Figure 21.3). This technique depends on the fact that heavy $^{15}N$ DNA (DNA that contains $^{15}N$ alone) will form a band at the bottom of the tube; light $^{14}N$ DNA ($^{14}N$ alone) will appear at the top of the

**FIGURE 21.2**   The labeling pattern of $^{15}N$ strands in semiconservative replication. ($G_0$ is original strands; $G_1$ is new strands after first generation; $G_2$ is new strands after second generation.)

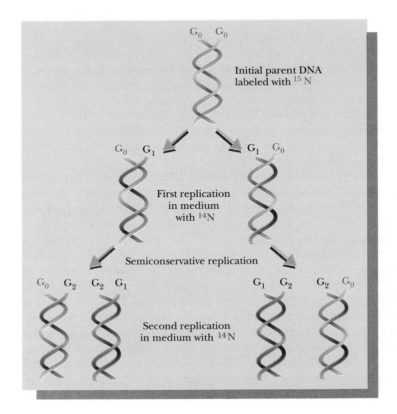

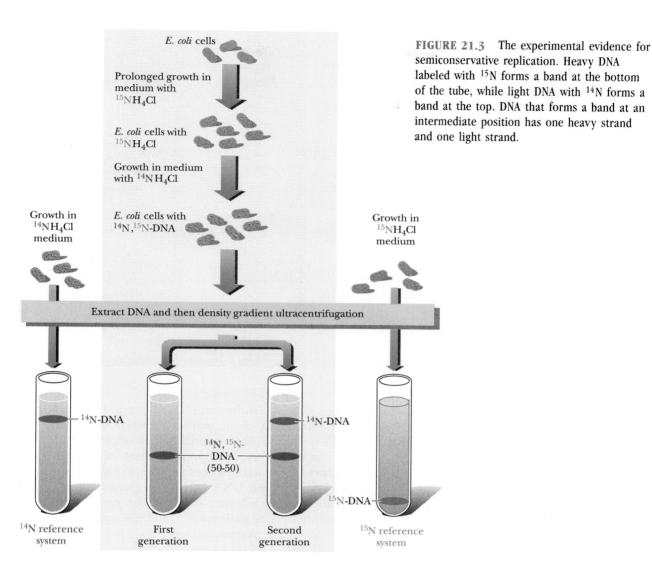

**FIGURE 21.3** The experimental evidence for semiconservative replication. Heavy DNA labeled with $^{15}N$ forms a band at the bottom of the tube, while light DNA with $^{14}N$ forms a band at the top. DNA that forms a band at an intermediate position has one heavy strand and one light strand.

tube. DNA containing a 50-50 mixture of $^{14}N$ and $^{15}N$ DNA will appear at a position that lies halfway between the two bands. In the actual experiment this 50-50 hybrid DNA was observed after one generation, a result to be expected with semiconservative replication. After two generations in the lighter medium, half the DNA in the cells should be the 50-50 hybrid and half should be the lighter $^{14}N$ DNA. This prediction of the kind and amount of DNA observed is confirmed by the experiment.

## Bidirectional Replication

During replication the DNA double helix unwinds at a specific point called the **origin of replication** (Ori). New polynucleotide chains are synthesized using each of the exposed strands as templates. Two possibilities exist for

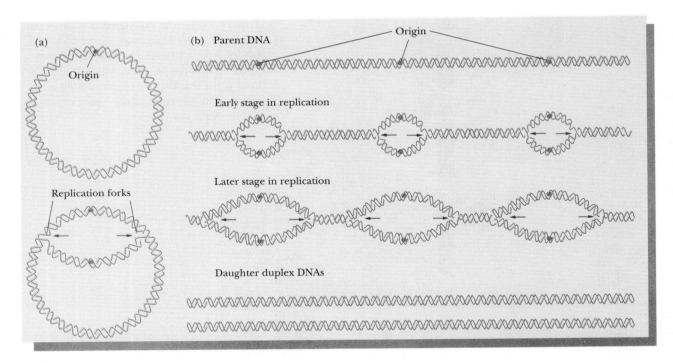

**FIGURE 21.4**   Bidirectional replication of DNA in prokaryotes (one origin of replication) and in eukaryotes (several origins). Bidirectional replication refers to overall synthesis (compare this figure with Figure 21.5). (a) Replication of the chromosome of *E. coli*, a typical prokaryote. There is one origin of replication and two replication forks. (b) Replication of a eukaryotic chromosome. There are several origins of replication and two replication forks for each origin. The "bubbles" that arise from each origin eventually coalesce.

the growth of the new strands; synthesis can take place in both directions from the origin of replication or in only one direction. It has been established that DNA synthesis is bidirectional in most organisms, with the exception of a few viruses. For each origin of replication there are two points **(replication forks)** at which new polynucleotide chains are formed. A "bubble" of newly synthesized DNA between regions of the original DNA is a manifestation of the advance of the two replication forks in opposite directions. There is one such bubble (and one origin of replication) in the circular DNA of prokaryotes (Figure 21.4 (a)). In eukaryotes several origins of replication and thus several bubbles exist (Figure 21.4 (b)). The bubbles grow larger and eventually coalesce, giving rise to two complete daughter DNAs. This bidirectional growth of both new polynucleotide chains refers to **net chain growth.** Both new polynucleotide chains are synthesized in the 5' to 3' direction.

## 21.3
## THE DNA POLYMERASE REACTION

### One Strand of DNA Is Synthesized Discontinuously

A major challenge for the cell in DNA replication is how to achieve 5'⟶3' polymerization in the opposite direction from the template strand, which is itself exposed from the 5'⟶3' direction. (There is no

**FIGURE 21.5**  Discontinuous and continuous synthesis of DNA strands at a replication fork. The leading strand of the DNA is synthesized continuously in the $5' \longrightarrow 3'$ direction. In the lagging strand, short fragments (Okazaki fragments) are synthesized, also in the $5' \longrightarrow 3'$ direction.

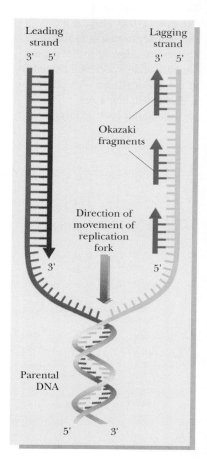

problem with the other strand, which is exposed by unwinding from the 3' to the 5' end.)

The problem is solved by different modes of polymerization for the two growing strands. One newly formed strand (the leading strand) is formed continuously from its 5' end to its 3' end at the replication fork on the exposed 3' to 5' template strand. The other strand (the lagging strand) is formed discontinuously in small fragments (typically 1000–2000 nucleotides long), sometimes called **Okazaki fragments** after the scientist who first studied them (Figure 21.5). The 5' end of each of these fragments is closer to the replication fork than the 3' end. The fragments of the lagging strand are then linked enzymatically by an enzyme called **DNA ligase.**

## DNA Polymerase from *E. Coli*

The first DNA polymerase discovered was found in *E. coli.* A universal feature of DNA replication is that the nascent chain grows from the 5' to the 3' end; there is a 5'-phosphate on the sugar at one end and a free 3'-hydroxyl on the sugar at the other end. **DNA polymerase** catalyzes the successive addition of each new nucleotide to the growing chain. The 3'-hydroxyl group at the end of the growing chain is a nucleophile. It attacks the phosphorus adjacent to the sugar in the nucleotide to be added to the growing chain, leading to the elimination of the pyrophosphate and the formation of a new phosphodiester bond (Figure 21.6). We discussed nucleophilic attack by a hydroxyl group at length in the case of serine proteases (Section 10.12); here we see another instance of this kind of mechanism.

There are actually three DNA polymerases in *E. coli;* some of their properties are listed in Table 21.1. DNA polymerase I (Pol I) was discovered first, with the subsequent discovery of polymerases II (Pol II) and III (Pol III). Polymerases I and II consist of a single polypeptide chain, but polymerase III is a multisubunit protein (Figure 21.7). All these enzymes add nucleotides to a growing polynucleotide chain but have different roles in the overall replication process. All three enzymes require the presence of a **primer,** a short strand of RNA to which the growing polynucleotide chain is covalently attached in the early stages of replication.

The DNA polymerase reaction requires all four deoxyribonucleoside triphosphates: dTTP, dATP, dGTP, and dCTP (Figure 21.8). $Mg^{2+}$ and DNA itself are also necessary. Because of the requirement for an RNA primer, all four ribonucleoside triphosphates—ATP, UTP, GTP, and

**FIGURE 21.6** The addition of a nucleotide to a growing DNA chain. The 3'-hydroxyl group at the end of the growing DNA chain is a nucleophile. It attacks at the phosphorus adjacent to the sugar in the nucleotide, which will be added to the growing chain. Pyrophosphate is eliminated, and a new phosphodiester bond is formed.

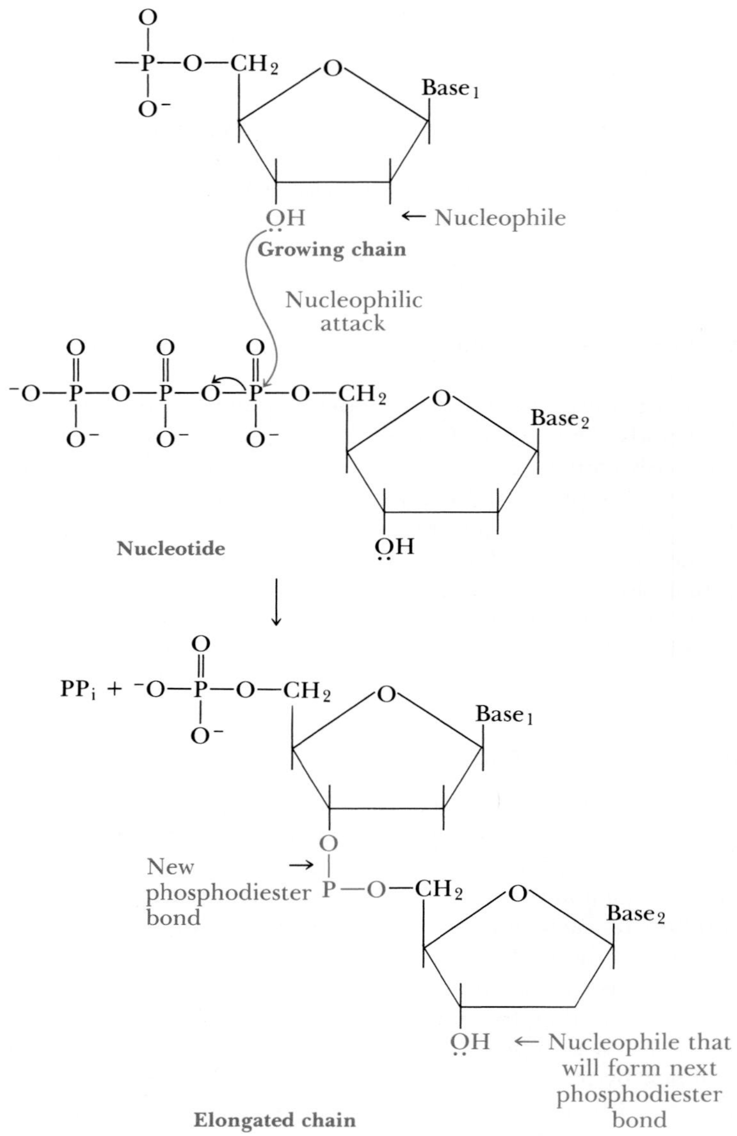

**TABLE 21.1  Properties of DNA Polymerases of *E. coli***

|  | POLYMERASE I | POLYMERASE II | POLYMERASE III |
|---|---|---|---|
| **Functions** |  |  |  |
| Polymerization: 5' ⟶ 3' | + | + | + |
| Exonuclease: 3' ⟶ 5' | + | + | + |
| Exonuclease: 5' ⟶ 3' | + | − | + |
| Molecular weight | 109,000 | 120,000 | 140,000 |
| Molecules/cell | 400 |  | 10–20 |
| Number of nucleotides polymerized/min at 37° by one enzyme molecule | 600 | 30 | 9000 |

Data from A. Kornberg, 1980. *DNA Replication.* New York: W.H. Freeman, p. 169.

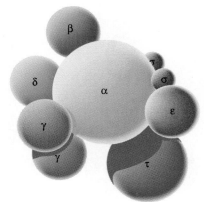

FIGURE 21.7   The subunit structure of DNA polymerase III of *E. coli*. The $\alpha$ subunit is the polymerase itself (the protein of molecular weight 140,000 listed in Table 21.1). The whole complex, called the DNA polymerase holoenzyme, has a molecular weight of 550,000. The $\beta$ subunit is required for recognition of the parental DNA. Once the complex is bound to the DNA to be copied, the $\beta$ subunit dissociates, leaving an active polymerase of molecular weight 400,000.

$$\text{dTTP, dATP, dGTP, dCTP} \xrightarrow[\text{DNApolymerase}]{\text{DNA-dependent}} \text{DNA (polymer)} + PP_i$$
(All four required)

FIGURE 21.8   The requirements for the DNA polymerase reaction. Template DNA, $Mg^{2+}$, and an RNA primer are also required. Because of the need for an RNA primer, there is also an implicit requirement for all four ribonucleoside triphosphates (ATP, UTP, GTP, and CTP) for formation of the primer.

CTP—are needed as well; they are incorporated into the primer. The primer (RNA) is hydrogen-bonded to the template (DNA); the primer provides a stable framework on which the nascent chain can start to grow. The newly synthesized DNA strand begins to grow by forming a covalent linkage to the free 3'-hydroxyl group of the primer.

It is now known that DNA polymerase I has a specialized function in replication, that of repair and "patching" of DNA, and that DNA polymerase III is the enzyme primarily responsible for the polymerization of the newly formed DNA strand. The major function of DNA polymerase II is not known. The exonuclease activity listed in Table 21.1 is part of the **proofreading** and **repair** functions of DNA polymerases; it is a process by which incorrect nucleotides are removed from the polynucleotide so that the correct nucleotides can be incorporated. The 3'⟶5' exonuclease activity, which all three polymerases possess, is part of the proofreading function; incorrect nucleotides are removed in the course of replication and replaced by the correct ones. Proofreading is done one nucleotide at a time. The 5'⟶3' exonuclease activity clears away short stretches of nucleotides during repair, usually involving several nucleotides at a time.

## 21.4
## DNA REPLICATION REQUIRES THE COMBINED ACTION OF SEVERAL ENZYMES

### Unwinding the Double Helix

Two questions arise in separating the two strands of the original DNA so that it can be replicated. The first is how to achieve continuous unwinding of the double helix. This question is complicated by the fact that prokaryotic DNA exists in supercoiled, closed-circular form (see Tertiary Structure of DNA: Supercoiling, in Section 20.3). The second, related

**FIGURE 21.9**    DNA gyrase introduces supertwisitng in circular DNA.

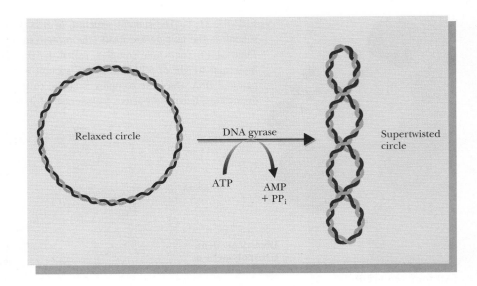

Relaxed circle

DNA gyrase

ATP      AMP + PP$_i$

Supertwisted circle

question is how to protect single-stranded stretches of DNA exposed to intracellular nucleases as a result of the unwinding.

An enzyme called **DNA gyrase** catalyzes the conversion of relaxed, circular DNA with a nick in one strand to the supercoiled form with the nick sealed (Figure 21.9). A slight unwinding of the helix before the nick is sealed introduces the supercoiling. The energy required for the process is supplied by the hydrolysis of ATP. Some evidence exists that DNA gyrase causes a double-strand break in DNA in the process of converting the relaxed, circular form to the supercoiled form (a mode of action typical of type II topoisomerases, the class of enzymes to which DNA gyrase belongs). In replication, the role of the gyrase is somewhat different. It introduces a nick in supercoiled DNA; the reverse of the reaction that produces supercoiling gives rise to a **swivel point** in the DNA at the site of the nick. The gyrase opens and reseals the swivel point in advance of the replication fork (Figure 21.10). (The newly formed DNA automatically assumes the supercoiled form, since it does not have the nick at the swivel point.) A helix-destabilizing protein, also called a **helicase,** promotes unwinding by binding at the replication fork. This protein is also called the **rep protein.** Another protein called **SSB** (single-strand binding) protein stabilizes the single-stranded regions by binding tightly to these portions of the molecule. The presence of the DNA-binding protein protects the single-stranded regions from hydrolysis.

## The Primase Reaction

One of the great surprises in early studies of DNA replication was the discovery that *RNA serves as a primer in DNA replication.* In retrospect, it is not surprising at all, since RNA can be formed *de novo* without a primer, while DNA synthesis requires a primer. This finding lends support to

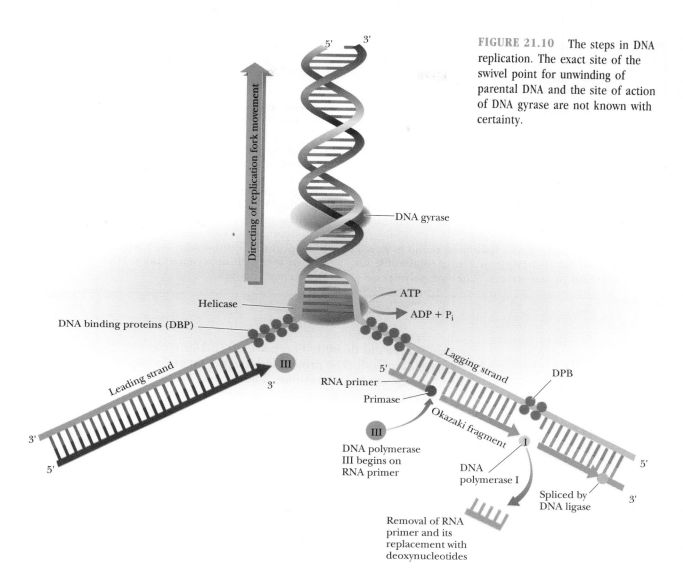

**FIGURE 21.10** The steps in DNA replication. The exact site of the swivel point for unwinding of parental DNA and the site of action of DNA gyrase are not known with certainty.

theories of the origin of life in which RNA, rather than DNA, was the original genetic material. A primer in DNA replication should have a free 3′-hydroxyl to which the growing chain can attach, and both RNA and DNA can provide this group. The primer activity of RNA was first observed *in vivo*. In some of the original *in vitro* experiments, DNA was used as a primer, since a primer consisting of DNA was expected. Living organisms are, of course, far more complex than isolated molecular systems and, as a result, can be full of surprises for researchers. It has subsequently been found that a separate enzyme, called **primase,** is responsible for copying a short stretch of the DNA template strand to produce the RNA primer sequence. The first primase was discovered in *E. coli;* the enzyme consists of a single polypeptide chain, with a molecular

---

## BOX 21.1
## A SUMMARY OF DNA REPLICATION IN PROKARYOTES

1. DNA synthesis is bidirectional. Two replication forks advance in opposite directions from an origin of replication.
2. The direction of DNA synthesis is from the 5′ to the 3′ end of the newly formed strand. One strand (the leading strand) is formed continuously, while the other strand (the lagging strand) is formed discontinuously. In the lagging strand, small fragments of DNA (Okazaki fragments) are subsequently linked together.
3. Three DNA polymerases have been found in *E. coli*. Polymerase III (Pol III) is primarily responsible for the synthesis of new strands. The first polymerase enzyme discovered, polymerase I (Pol I), is a repair enzyme. The function of polymerase II is unknown.
4. DNA gyrase introduces a swivel point in advance of the movement of the replication fork. A helix-destabilizing protein, helicase, binds at the replication fork and promotes unwinding. The exposed single-stranded regions of the template DNA are stabilized by a DNA-binding protein.
5. Primase catalyzes the synthesis of an RNA primer.
6. The synthesis of new strands is catalyzed by Pol III. The primer is removed by Pol I, which also replaces the primer with deoxynucleotides. DNA ligase seals any remaining nicks.

---

weight areound 60,000. There are 50 to 100 molecules of primase in a typical *E. coli* cell. The primer and the protein molecules at the replication fork comprise the **primosome.** The general features of DNA replication, including the use of an RNA primer, appear to be common to all prokaryotes (Figure 21.10).

### Synthesis and Linking of New DNA Strands

The synthesis of two new strands of DNA is begun by DNA polymerase III. The newly formed DNA is linked to the 3′-hydroxyl of the RNA primer, and synthesis proceeds from the 5′ to 3′ end on both the leading and the lagging strands. Two molecules of Pol III, one for the leading strand, one for the lagging strand, are physically linked to the primosome; the resulting multiprotein complex is called the **replisome.** As the replication fork moves away, the RNA primer is removed by polymerase I, using its exonuclease activity. The primer is replaced by deoxynucleotides, also by DNA polymerase I, using its polymerase activity. (The removal of the RNA primer and its replacement with the missing portions of the newly formed DNA strand by polymerase I is the repair function we mentioned earlier.) None of the DNA polymerases can seal the nicks that remain; DNA ligase is the enzyme responsible for the final linking of the new strand.

### 21.5
### PROOFREADING AND REPAIR

On the average, DNA replication takes place only once each generation in each cell, unlike other processes, such as RNA and protein synthesis, which occur many times. It is essential that the fidelity of the replication process

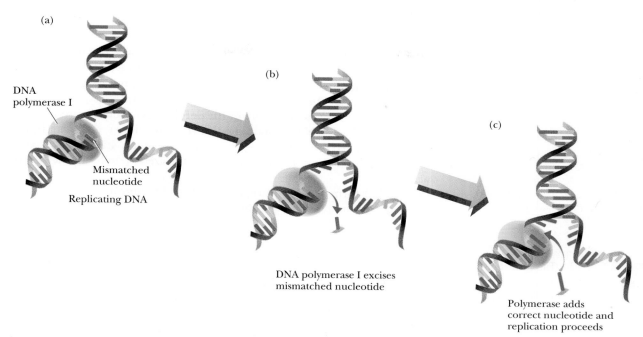

(a)

DNA
polymerase I

Mismatched
nucleotide

Replicating DNA

(b)

DNA polymerase I excises
mismatched nucleotide

(c)

Polymerase adds
correct nucleotide and
replication proceeds

be as high as possible to prevent **mutations,** which are errors in replication. Most mutations are harmful, even lethal, to organisms. Nature has devised several ways to ensure that the base sequence of DNA is copied faithfully. Errors in replication occur spontaneously only once in every $10^9$ to $10^{10}$ base pairs.

Proofreading removes incorrect nucleotides immediately after they are added to the growing DNA during the replication process. Errors in hydrogen bonding lead to the incorporation of an incorrect nucleotide into a growing DNA chain once in every $10^4$ to $10^5$ base pairs (Figure 21.11 (a)). DNA polymerase III uses its 3′ exonuclease activity to remove the incorrect nucleotide (Figure 21.11 (b)). Replication resumes when the correct nucleotide is added, also by DNA polymerase III (Figure 21.11 (c)). The specificity of hydrogen-bonded base pairing accounts for one error in every $10^4$ to $10^5$ base pairs; the proofreading function of DNA polymerase improves the fidelity of replication to one error in every $10^9$ to $10^{10}$ base pairs.

During replication, a **cut-and-patch** process catalyzed by polymerase I takes place. The cutting is the removal of the RNA primer by the 5′ exonuclease function of the polymerase, while the patching is the incorporation of the required deoxynucleotides, done by the polymerase function of the same enzyme. Existing DNA can also be repaired by polymerase I, using the cut-and-patch method, if one or more bases have been damaged by an external agent. In addition to experiencing those spontaneous mutations caused by misreading of the genetic code, organisms are frequently exposed to **mutagens** (agents that produce mutations). Common mutagens include ultraviolet light, ionizing radiation (radioactivity), and various chemical agents, all of which lead to changes in DNA over and above those produced by spontaneous mutation. In the repair of existing DNA, the process of cutting and patching is called **excision-repair;** the

**FIGURE 21.11** Proofreading in DNA replication. (a) When an incorrect nucleotide has been incorporated into a growing polynucleotide chain, (b) DNA polymerase uses its exonuclease activity to remove the mismatched nucleotide. (c) The polymerase then adds the correct nucleotide, and replication proceeds. (From M. Radman and R. Wagner, 1988. The High Fidelity of DNA Replication. *Sci. Amer.* **259** (2), 40–46.)

damaged portion of the DNA is removed, and the new, correct portion is substituted.

A type of damage that frequently occurs in DNA is the formation of **thymine dimers,** which result from the effect of ultraviolet light (Figure 21.12). This dimer cannot fit into the double helix, and it cannot form the proper base pairs. The repair process occurs in several steps. An endonuclease nicks the DNA near the site of the defect. DNA polymerase I excises a short stretch of DNA containing the thymine dimer and inserts the corresponding stretch with the correct base sequence. In the last step of the process, DNA ligase joins the newly synthesized "patch" to the rest of the DNA.

Another form of excision-repair involves the removal of cytosines that have been deaminated to produce uracil, a reaction that can happen

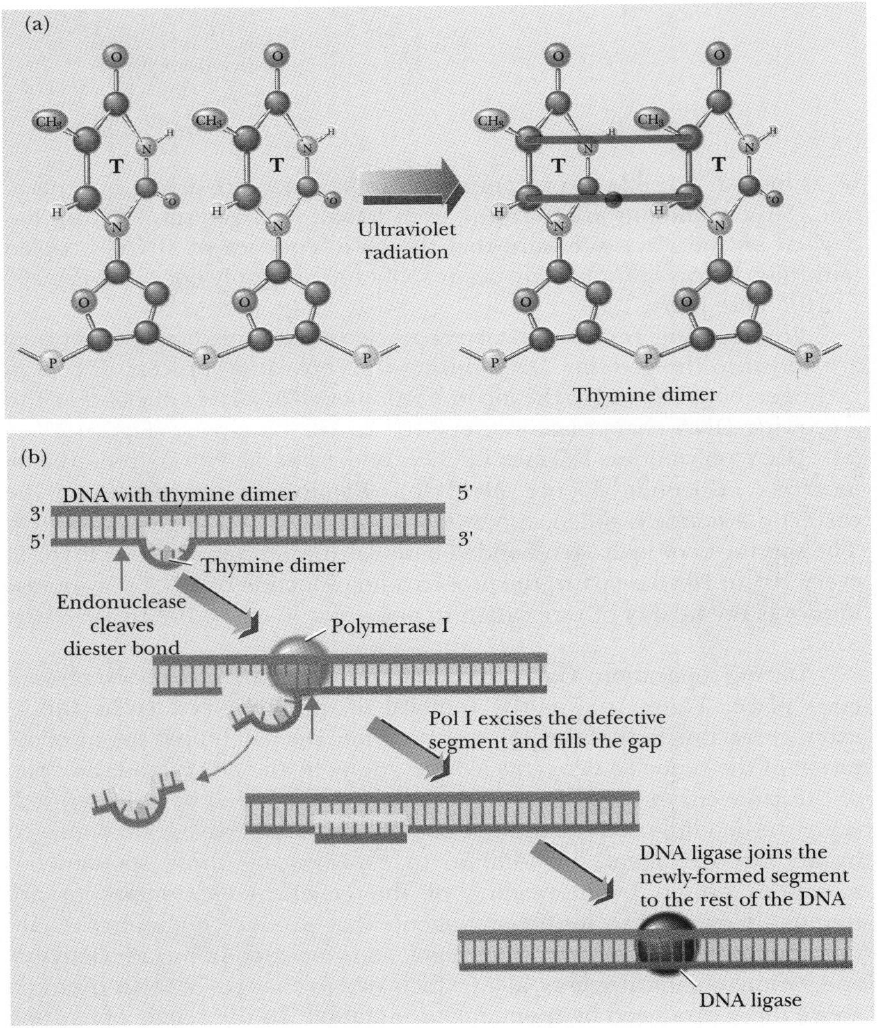


**FIGURE 21.12** The formation and repair of thymine dimers. (a) Thymine dimers can be formed by the action of ultraviolet light on DNA. (b) In the excision-repair scheme, an endonuclease nicks the DNA near the site of the dimer. Polymerase I removes a short stretch containing the thymine dimer and fills the gap with a short stretch of DNA. DNA ligase makes the final link joining the newly formed portion to the rest of the DNA.

**FIGURE 21.13** Removal of uracil and replacement by cytosine. Uracil produced from cytosine by deamination can be excised and replaced by cytosine by the successive action of several enzymes.

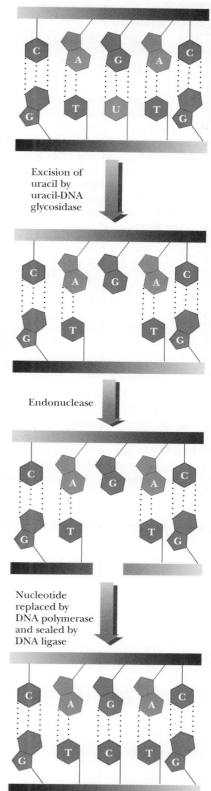

spontaneously or be induced by chemical agents. The uracil produced by this change forms hydrogen bonds with adenine rather than with guanine, the complementary base for cytosine, in subsequent rounds of replication (Figure 21.13). As with the repair of thymine dimers, the repair takes place in several steps. The enzyme **uracil-DNA glycosidase** cleaves the glycosidic bond between uracil and deoxyribose, leaving a DNA with an intact backbone but with one missing base. An endonuclease nicks the DNA, the damaged portion is removed and replaced with a cytosine nucleotide by DNA polymerase I, and the nick is sealed by DNA ligase. (In some cases, more than one nucleotide may be removed, but the general repair scheme remains the same.) It is noteworthy that if uracil, rather than thymine, normally occurred in DNA as the base that pairs with adenine, the repair system would not be able to distinguish uracils that should be present from uracils produced from cytosine by deamination. The methyl group of thymine serves as a "tag" to distinguish it from uracil. The distinction among the various pyrimidines makes the repair system more efficient, since there is no mistaking the fact that any uracil must have come from cytosine.

A segment of DNA polymerase I (the Klenow fragment) has no exonuclease activity, which is why it is used in the Sanger method of DNA sequencing; see Box 21.2.

Defects in DNA repair mechanisms can have drastic consequences. One of the most remarkable examples is the disease **xeroderma pigmentosum.** Affected individuals develop numerous skin cancers at an early age because they do not have the repair system to correct damage caused by ultraviolet light. The endonuclease that nicks the damaged portion of the DNA is probably the missing enzyme. The cancerous lesions eventually spread throughout the body, causing death.

## 21.6
## MODIFICATION OF DNA AFTER REPLICATION

Once DNA is formed by the replication mechanism of the cell, modifications can take place for specific purposes. The most important of these modifications is **methylation.** Specific adenine and thymine residues are altered in prokaryotes, but only cytosine residues are affected in eukaryotes. The products include $N^6$-methyladenine and 5-methylcytosine. *S*-Adenosylmethionine (see One-Carbon Transfers and the Serine Family, in Section 18.4) is the methyl group donor. The residues modified are frequently found in sequences that would otherwise be attacked by restriction endonucleases. These enzymes (see Restriction Enzymes Produce "Sticky Ends," in Section 20.6), which occur mainly in bacteria, produce breaks in specific places in both strands of DNA. The methylation

BOX 21.2

# THE SANGER METHOD FOR SEQUENCING DNA

In several ways, the Sanger method is similar to the Maxam–Gilbert method (see Section 20.5 and Box 20.1) for sequencing DNA. In both cases, gel electrophoresis is used to separate labeled oligonucleotides that differ in length by only one nucleotide. The important difference between the two is in the way the labeled oligonucleotides are obtained. In the Maxam–Gilbert method, selective chemical cleavage is used to obtain the oligonucleotides, while in the Sanger method, interrupted synthesis of DNA on a template yields the oligonucleotide mixture.

The interruption of synthesis depends on the presence of 2',3'-dideoxyribonucleoside triphosphates (ddNTP).

The 3'-hydroxyl group of deoxyribonucleoside triphosphates (the usual monomer unit for DNA synthesis) has been replaced by a hydrogen. These ddNTPs can be incorporated in a growing DNA chain, but they lack a 3'-hydroxyl group to form a bond to another nucleoside triphosphate. The incorporation of a ddNTP into the growing chain causes termination at that point. The presence of ddNTPs in a replicating mixture will cause random termination of chain growth.

The DNA to be sequenced is mixed with a short oligonucleotide that serves as a primer for synthesis of the complementary strand. The primer is hydrogen-bonded to the 3' end of the DNA to be sequenced. The DNA with primer is divided into four separate reaction mixtures. Each of the reaction mixtures contains all four deoxyribonucleoside triphosphates (dNTPs), one of which is labeled to allow the newly synthesized fragments to be visualized by autoradiography. In addition, each of the reaction mixtures contains one of the four ddNTPs. The polymerase enzyme used is the Klenow fragment of DNA polymerase I; this fragment has the polymerase activity without the 5',3'-exonuclease activity. (Exonuclease activity could confuse the results by degrading some of the products.) In each reaction mixture, chain termination will occur at all possible sites for that nucleotide.

When gel electrophoresis is done on each reaction mixture, there will be a band corresponding to each position of chain termination. The sequence of the newly formed strand, which is complementary to that of the template DNA, can be "read" directly from the sequencing gel. A variation on this method is to use a single reaction mixture with a different fluorescent label on each of the four ddNTPs. Each of the four fluorescent labels can be detected by its characteristic spectrum, requiring only a single gel electrophoresis experiment. The use of fluorescent labels makes it possible to automate DNA sequencing, with the whole process under computer control. The Sanger approach is now more frequently used than the Maxam–Gilbert method.

protects the DNA of the bacterium from the restriction endonuclease (Figure 21.14), but DNA of an infecting bacteriophage does not have the methyl groups and is cleaved. The cleavage of the bacteriophage DNA restricts the growth of the bacteriophage in the infected cell, giving rise to the name "restriction endonuclease."

## 21.7
## THE REPLICATION OF DNA IN EUKARYOTES

Three different DNA polymerases have been isolated from animal systems. The use of animals rather than plants for study avoids the complication of any DNA synthesis in chloroplasts. The three different polymerases are called $\alpha$, $\beta$, and $\gamma$. The $\alpha$ and $\beta$ enzymes are found in the nucleus, while the $\gamma$ form occurs in mitochondria. DNA polymerase $\alpha$ plays a role similar to that of DNA polymerase III in prokaryotes, in the sense that the $\alpha$ enzyme

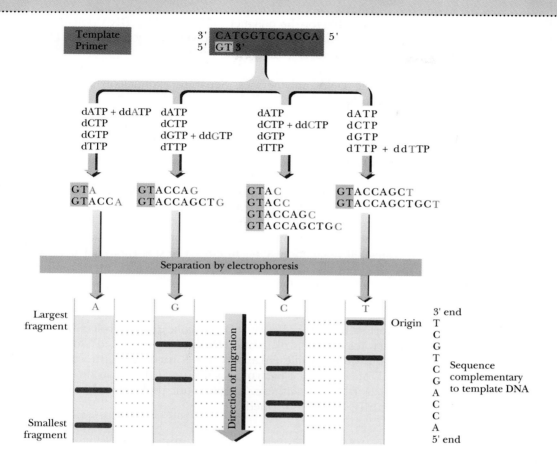

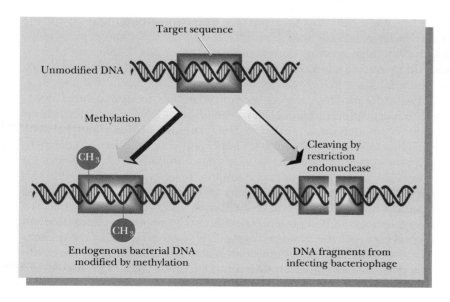

**FIGURE 21.14** Methylation of endogenous DNA protects it from cleavage by its own restriction endonucleases.

is primarily responsible for the polymerization of DNA. DNA polymerase $\beta$ appears to be a repair enzyme. DNA polymerase $\gamma$ carries out DNA replication in mitochondria. None of the DNA polymerases isolated from animals acts as an exonuclease, and in this regard the animal enzymes differ from prokaryotic DNA polymerases. Separate exonucleolytic enzymes exist in animal cells.

The general features of DNA replication in eukaryotes are similar to those in prokaryotes but are not as extensively studied. Many replication origins are found in eukaryotes, rather than a single one, as in prokaryotes, but the steps in the replication process are basically the same. As with prokaryotes, DNA replication in eukaryotes is semiconservative. There is a leading strand with continuous synthesis in the $5' \longrightarrow 3'$ direction and a lagging strand with discontinuous synthesis in the $5' \longrightarrow 3'$ direction. An RNA primer is formed by a specific enzyme in eukaryotic DNA replication, as is the case with prokaryotes. The formation of Okazaki fragments (typically 150–200 nucleotides long in eukaryotes) is catalyzed by DNA polymerase $\alpha$. The RNA primer is hydrolyzed, and the gaps left by removal of the primer are filled in by a reaction that is also catalyzed by DNA polymerase $\alpha$. Finally, DNA ligase seals the nicks that separate the fragments.

An important difference between DNA replication in prokaryotes and that in eukaryotes is that prokaryotic DNA exists in "naked" form, not complexed to proteins, while eukaryotic DNA is complexed to proteins, principally histones. Histone biosynthesis occurs at the same time and at the same rate as DNA biosynthesis. In eukaryotic replication, histones are associated with DNA as it is formed.

## 21.8
## SOME ASPECTS OF DNA RECOMBINATION

We have already discussed restriction endonucleases and recombinant DNA in Chapter 20, but since we now know something more about the nature of DNA and the way in which it is replicated, we can explore more deeply the subject of recombination. An especially important topic is **genetic recombination.** Mechanisms for DNA recombination are shown in Figure 21.15.

Two different DNA molecules can recombine to form a third DNA, which is itself different from each of the original DNA molecules. This is genetic recombination, a process in which **regions of homology** between the original DNA molecules play an important role. In other words, there has to be a stretch of matching base sequence in both DNA molecules for the process to take place with maximum efficiency. The reason for this requirement will become apparent as we discuss the process. Both the original DNA molecules are nicked; the nicks will eventually be sealed in a different arrangement. The actual details of the process can be carried out in various ways, but the example we shall give here is typical of the manner in which genetic recombination takes place. An endonuclease nicks two homologous chromosomes on different DNA molecules. The two nicked

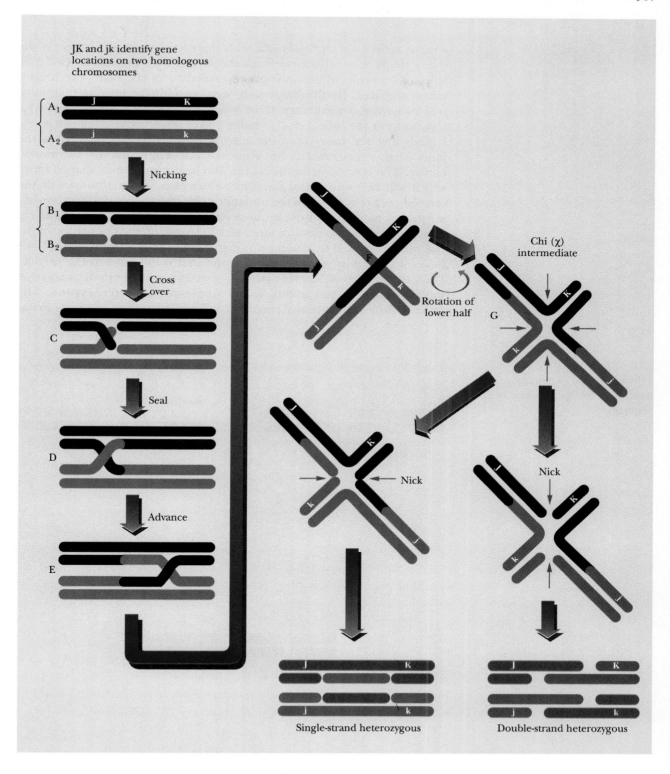

**FIGURE 21.15**   Mechanisms for DNA recombination.

strands cross over and base-pair with the complementary strand on the other DNA. A suitable ligase seals the nicked ends in their crossed-over form. As a result of unwinding and rewinding of the helix, the branch point migrates. In this migration step, two single strands that were originally parts of different DNA molecules form a double helix. The formation of the hybrid double helices takes place because the two single strands involved have complementary base sequences. This point is the basis of the requirement for sequence homology that we mentioned earlier. The recombined product can also be drawn in an X-shaped form, which will be more useful for future discussion. A rotation of half the crossed-over product gives the untangled **chi** intermediate; this intermediate gets its name from its resemblance to the Greek letter chi ($\chi$). The chi intermediate has two possible fates, which depend on the orientation of the two nicks, followed by sealing. In one orientation the result is two DNA molecules, each of which has one strand that is the same as the original DNA strand and one strand that incorporates parts of both original DNA strands. This product is thus called **single-strand heterozygous.** The other possible orientation of two nicks, followed by sealing, gives rise to two

**FIGURE 21.16**  A possible mechanism for transposition of DNA segments.

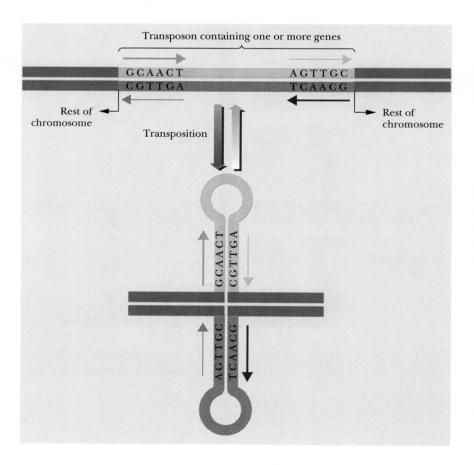

DNA molecules in which both strands incorporate parts of both original DNA strands. This second product is **double-strand heterozygous.**

**Transposition of DNA** within a chromosome or from one chromosome to another ("jumping genes") was an idea greeted with general skepticism when it was first proposed, but it is now well established that genes and clusters of genes can change position. Both prokaryotes and eukaryotes are subject to gene transposition. This phenomenon certainly plays a part in the rapid development of mutations in bacteria, including the appearance of antibiotic resistance. The movable segments of DNA are called **transposable elements,** and they begin and end with an inverted-repeat sequence. The existence of the inverted-repeat sequence could allow the transposon to loop out from the rest of the DNA, but this point is not established (Figure 21.16). Many questions remain about DNA transposition, and research is under way to elucidate the details of the process.

## 21.9
## RNA BIOSYNTHESIS: TRANSCRIPTION OF THE GENETIC MESSAGE

The details of RNA transcription differ somewhat in prokaryotes and eukaryotes. Most of the research on the subject has been done in prokaryotes, especially *E. coli,* but some general features are found in all organisms except RNA viruses.

### General Features of RNA Synthesis

1. All RNAs are synthesized on a DNA template; the enzyme that catalyzes the process is **DNA-dependent RNA polymerase.**
2. All four ribonucleoside triphosphates (ATP, GTP, CTP, and UTP) are required, as is $Mg^{2+}$.
3. A primer is not needed in RNA synthesis, but a DNA template is required.
4. As is the case with DNA biosynthesis, the RNA chain grows from the 5′ to the 3′ end (Figure 21.17). The nucleotide at the 5′ end of the chain retains its triphosphate group (abbreviated ppp).
5. The enzyme uses one strand of the DNA (the sense strand) as the template for RNA synthesis. The base sequence of the DNA contains signals for initiation and termination of RNA synthesis. The enzyme binds to the sense strand and moves along it in the 3′ to 5′ direction.
6. The template is unchanged (Figure 21.17).

### RNA Polymerase from *Escherichia coli*

The most extensively studied RNA polymerase is that isolated from *E. coli.* The molecular weight of this enzyme is about 500,000, and it has a multisubunit structure. Four different types of subunit, designated $\alpha, \beta, \beta'$,

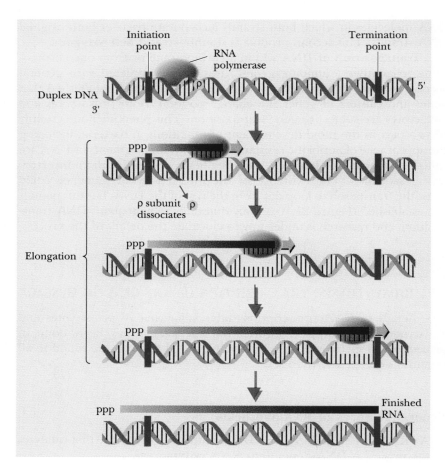

**FIGURE 21.17**    The transcription of RNA on a DNA template. In *E. coli*, transcription begins when the RNA polymerase holoenzyme ($\alpha_2\beta\beta'\sigma$) binds to DNA. The $\sigma$ subunit is required for binding but dissociates from the complex after initiation of transcription. Synthesis of RNA proceeds from the 5' to the 3' end. Termination of chain growth requires the $\rho$ subunit.

and $\sigma$, have been identified. The actual composition of the enzyme is $\alpha_2\beta\beta'\sigma$. The $\sigma$ subunit is rather loosely bound to the rest of the enzyme (the $\alpha_2\beta\beta'$ portion), which is called the **core enzyme.** The **holoenzyme** consists of all the subunits, including the $\sigma$ subunit. The essential role of the $\sigma$ subunit is recognition of the **promoter locus** (a DNA sequence that signals the start of RNA transcription). The loosely bound $\sigma$ subunit is released after transcription begins.

The promoter region to which RNA polymerase binds is closer to the 3' end of the DNA than is the actual gene for the RNA to be synthesized. (The RNA is formed from the 5' to the 3' end, so the polymerase moves along the DNA from the 3' to the 5' end.) The binding site for the polymerase is said to lie **upstream** of the start of transcription. The base sequence of promoter regions has been determined for a number of prokaryotic genes, and a striking feature is that they contain many sequences in common **(consensus sequences).** Promoter regions are A-T–rich, with two hydrogen bonds per base pair, and thus more easily unwound than G-C–rich regions, with three hydrogen bonds per base pair. In prokaryotes these consensus regions frequently occur 35 base pairs (bp) and 10 bp upstream from the start of transcription. These two regions are calleld the **−35 box** and the **−10 box** (also called the **Pribnow box**). The consensus sequence for a prokaryotic promoter is

$$-35 \qquad\qquad\qquad -10$$
$$\text{TTGACA} \qquad 17 \text{ bp} \qquad \text{TATAAT}$$

In eukaryotes a similar sequence called the **TATA box** lies about 25 bp upstream from the start of transcription.

A good deal more can be said about the operation of the promoter region of genes, but we are going to need some information about protein synthesis for an effective discussion of the subject. The control of genetic expression is complex and important enough that we shall return to the topic in Section 22.9 to look at it in light of new information.

Termination of RNA transcription also involves specific sequences **downstream** of the actual gene for the RNA to be transcribed, and in prokaryotes a protein designated $\rho$ (rho) is involved in the termination process. This $\rho$ **protein,** which is a tetramer with a molecular weight of approximately 200,000, binds to DNA, RNA, and RNA polymerase. The $\rho$ factor triggers the release of the completed RNA from the DNA-RNA polymerase complex (Figure 21.17). A $\rho$ protein has not been detected in eukaryotic systems.

## 21.10
## POSTTRANSCRIPTIONAL MODIFICATION OF RNA

The three kinds of RNA—transfer RNA (tRNA), ribosomal RNA (rRNA), and messenger RNA (mRNA)—are all modified enzymatically after transcription to give rise to the functional form of the RNA in question. The type of processing in prokaryotes can differ greatly from that in eukaryotes, especially in the case of mRNA. The initial size of the RNA transcripts is larger than the final size because of leader sequences at the 5′ end and trailer sequences at the 3′ end. The leader and trailer sequences must be removed, and other forms of **trimming** are possible as well. **Terminal sequences** can be added after transcription, and **base modification** is frequently observed, especially in tRNA.

### Transfer RNA and Ribosomal RNA

The precursor of several tRNA molecules is frequently transcribed in one long polynucleotide sequence. All three types of modification—trimming, addition of terminal sequences, and base modification—take place in the transformation of the initial transcript to the mature tRNAs (Figure

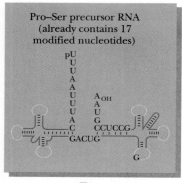

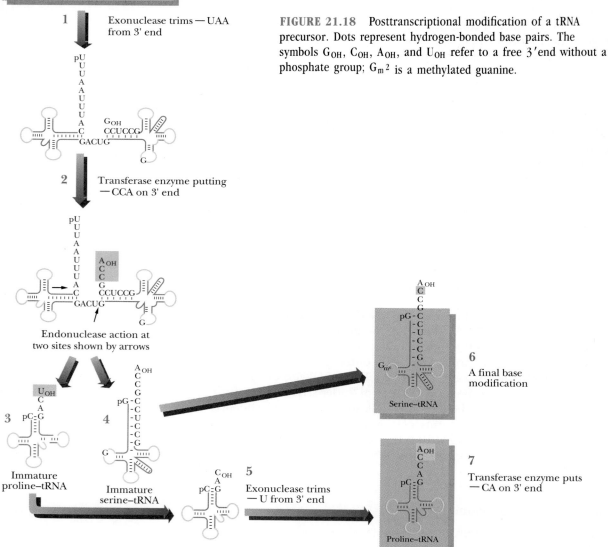

**FIGURE 21.18** Posttranscriptional modification of a tRNA precursor. Dots represent hydrogen-bonded base pairs. The symbols $G_{OH}$, $C_{OH}$, $A_{OH}$, and $U_{OH}$ refer to a free 3′ end without a phosphate group; $G_{m^2}$ is a methylated guanine.

21.18). Some base modifications take place before trimming, and some occur after. Methylation and substitution of sulfur for oxygen are two of the more usual types of base modification. (See Section 20.2 and Transfer RNA in Section 20.4 for the structures of some of the modified bases.) One type of methylated nucleotide found only in eukaryotes contains a 2'-O-methylribosyl group (Figure 21.19).

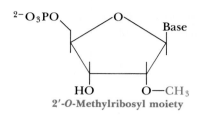

**FIGURE 21.19**  The structure of a nucleotide containing a 2'-O-methylribosyl group.

The trimming and addition of terminal nucleotides produce tRNAs with the proper size and base sequence. All tRNAs contain a CCA sequence at the 3' end. The presence of this portion of the molecule is of great importance in protein synthesis, since the 3' end is the acceptor for amino acids to be added to a growing protein chain. Trimming of large precursors of eukaryotic tRNAs takes place in the nucleus, but most methylating enzymes occur in the cytosol.

The processing of rRNAs is primarily a matter of methylation and of trimming to the proper size. In prokaryotes there are three rRNAs in an intact ribosome, which has a sedimentation coefficient of 70S. (Recall that we discussed sedimentation coefficients and some aspects of ribosomal structure in Ribosomal RNA, in Section 20.4.) In the smaller subunit, which has a sedimentation coefficient of 30S, there is one RNA molecule that has a sedimentation coefficient of 16S. The 50S subunit contains two kinds of RNA, with sedimentation coefficients of 5S and 23S. The ribosomes of eukaryotes have a sedimentation coefficient of 80S, with 40S and 60S subunits. The 40S subunit contains an 18S RNA, while the 60S subunit contains a 5S RNA, a 5.8S RNA, and a 28S RNA. Base modificaitons in both prokaryotic and eukaryotic rRNA are accomplished primarily by methylation.

## Messenger RNA

Extensive processing takes place in eukaryotic mRNA. Modifications include **capping** of the 5' end, **polyadenylation** (addition of a poly A sequence) of the 3' end, and **splicing of coding sequences.** Such processing is not a feature of the synthesis of prokaryotic mRNA.

The cap at the 5' end of eukaryotic mRNA is a guanylate residue that is methylated at the N-7 position. This modified guanylate residue is attached to the neighboring residue by a 5'⟶5' triphosphate linkage (Figure 21.20). The 2'-hydroxyl group of the ribosyl portion of the neighboring residue is frequently methylated and sometimes that of the next nearest neighbor as well. The polyadenylate "tail" at the 3' end of a messenger (typically 100–200 nucleotides long) is added before the mRNA leaves the nucleus. It is thought that the presence of the tail serves to protect the mRNA from nucleases and phosphatases, which would degrade it. According to this point of view, the adenylate residues would be cleaved off before the portion of the molecule that contains the actual message is attacked.

The genes of prokaryotes are continuous; every base pair in a continuous prokaryotic gene is reflected in the base sequence of mRNA. The genes of eukaryotes are not necessarily continuous; eukaryotic genes frequently contain intervening sequences that do not appear in the final

**FIGURE 21.20**   The structures of some typical mRNA caps.

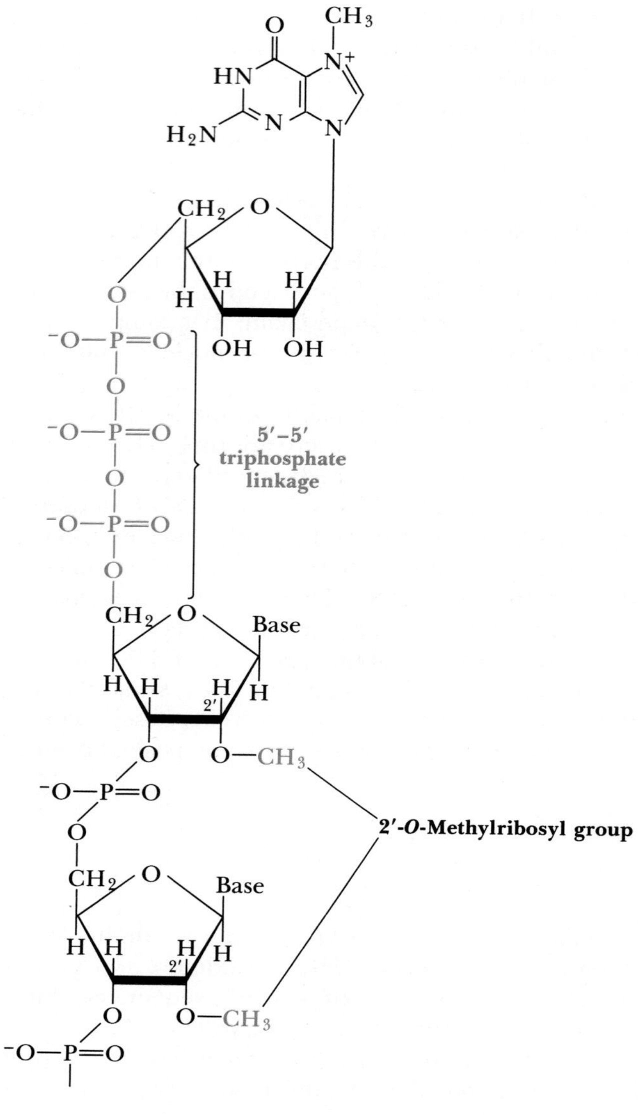

5'–5' triphosphate linkage

2'-*O*-Methylribosyl group

base sequence of the mRNA for that gene product. The DNA sequences that are **expressed** (the ones actually retained in the final product) are called **exons.** The **intervening** sequences, which are not expressed, are called **introns.** The β-globin gene of the mouse, which codes for the β-chain of hemoglobin, is a well-known example. The actual gene is split into three parts, with two intervening sequences (Figure 21.21).

In eukaryotes the entire DNA sequence, both introns and exons, is transcribed to produce a precursor of the mature mRNA. In the processing of mRNA, the noncoding sequences, the introns, must be excised, and the coding sequences, the exons, must be spliced together. This process is done by the sequential action of suitable nucleases and ligases. The number and size of the introns vary in different genes. There is 1 intron in the gene

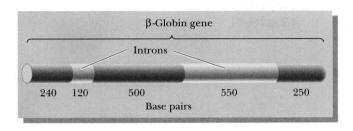

FIGURE 21.21 Intervening sequences (introns) in the β-globin gene.

for the muscle protein actin, 2 for both the α- and β-chains of hemoglobin, 3 for lysozyme, and so on, up to as many as 50 introns in a single gene (the gene for one of the subunits of collagen).

The removal of intervening sequences takes place in the nucleus, where small nuclear ribonucleoproteins, or snRNPs (pronounced "snurps"), mediate the process. The snRNPs, as their name implies, contain both RNA and proteins, such as the nucleases and ligases involved in the splicing process. It is now widely recognized that some RNAs can catalyze their own self-splicing; the present process involving ribonucleoproteins may well have evolved from the self-splicing of RNAs. An important similarity between the two processes is that both proceed by a lariat mechanism by which the splice sites are brought together (Figure 21.22). In both cases, the splicing of mRNA by snRNPs and the self-

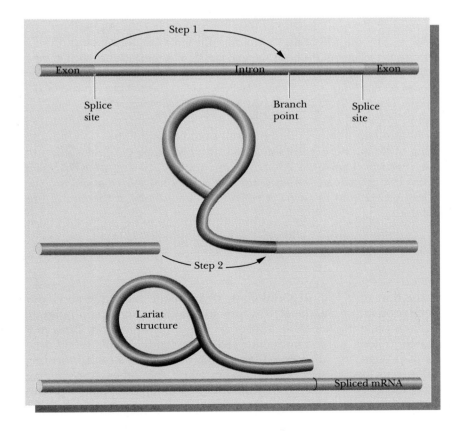

FIGURE 21.22 The formation of a lariat structure as an intermediate in the splicing of mRNA. In the first step of the splicing process, one splice site is cut and attached to a binding site near the second splice site, forming a lariat. In the course of splicing, mRNA is bound to small nuclear ribonucleoproteins (snRNPs) (not shown here), which hold the splice sites in position. In the second step, the intron lariat is excised, and the mRNA exons are spliced together. A similar lariat mechanism is involved in the self-splicing of RNA. (From J. Steitz, 1988. "Snurps." *Sci. Amer.* **258** (6), 56–63.)

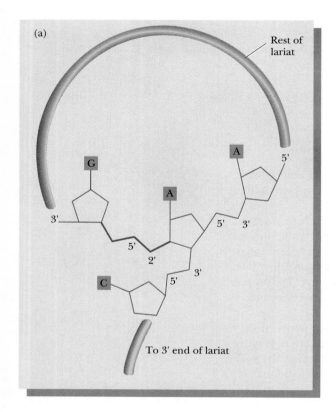

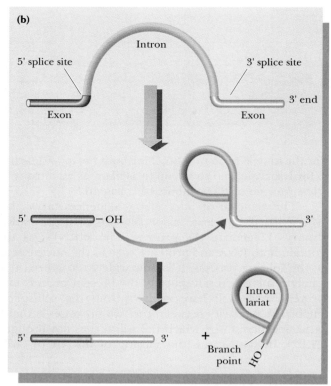

**FIGURE 21.23** The self-splicing of RNA from the protozoan *Tetrahymena*. (a) The backbone (yellow) of the lariat showing the 2′, 5′ phosphodiester bond at the branch point (red). (b) The lariat is formed when the 2′-OH group of a guanine nucleotide residue forms a 2′, 5′ phosphodiester bond. The RNA chain branches at that point. (From T. Cech, 1986. RNA As an Enzyme. *Sci. Amer.* **255** (5), 64–75.)

splicing of RNA, there is still much to learn; both topics are the subject of active research. Figure 21.23 shows how the self-splicing of RNA takes place in the protozoan *Tetrahymena*. (For additional information, see the articles by Cech and by Steitz cited in the bibliography at the end of this chapter.)

## SUMMARY

In all organisms except RNA viruses, the flow of genetic information is DNA ⟶ RNA ⟶ protein. The duplication of DNA is called replication, while the production of RNA on a DNA template is called transcription. Translation is the process of protein synthesis, in which the sequence of amino acids is directed by the sequence of bases in the genetic material.

Replication of DNA is semiconservative and bidirectional. Two replication forks advance in opposite directions from an origin of replication. Both new polynucleotide chains are synthesized in the 5′ to 3′ direction. One strand (the leading strand) is synthesized continuously, while the other (the lagging strand) is synthesized discontinuously, in fragments that are subsequently linked together. Two DNA polymerases play an important role in replication. Polymerase III (Pol III) is primarily responsible for the synthesis of new strands. The first polymerase enzyme discovered, polymerase I (Pol I), is mainly a repair enzyme. DNA gyrase introduces a swivel point in advance of the movement of the replication fork. A helix-destabilizing protein, helicase, binds at the replication fork and promotes unwinding. The exposed single-stranded regions of the template DNA are protected from nuclease digestion by a DNA-binding protein. Primase catalyzes the synthesis of an RNA primer. The synthesis of new strands linked to the primer is catalyzed by Pol III. The primer is removed by Pol I, which also replaces the primer with deoxynucleotides. DNA ligase seals any remaininng nicks.

DNA replication takes place only once each generation in each cell. It is essential that the fidelity of the replication process be as high as possible to prevent mutations, which are errors in replication. Pol III does proofreading in the course of replication. In addition, Pol I carries out a cut-and-patch process, removing the RNA primer and replacing it with deoxyribonucleotides during replication. Pol I uses the same cut-and-patch process to repair existing DNA. Replication has been most extensively studied in prokaryotes. Replication in eukaryotes follows the same general outline, with the most important difference being the presence of proteins complexed to eukaryotic DNA.

RNA synthesis is the transcription of the base sequence of DNA to that of RNA. All RNAs are synthesized on a DNA template; the enzyme that catalyzes the process is DNA-dependent RNA polymerase. All four ribonucleoside triphosphates (ATP, GTP, CTP, and UTP) are required, as is $Mg^{2+}$. There is no need for a primer in RNA synthesis. As is the case with DNA biosynthesis, the RNA chain grows from the 5' to the 3' end. The enzyme uses one strand of the DNA (the sense strand) as the template for RNA synthesis. Posttranscriptional processing takes place in RNA. Base modification frequently occurs, as does trimming of long polynucleotide chains. In eukaryotes, the removal of intervening sequences, which reflect the base sequence of a portion of the DNA not expressed in the mature RNA, is an important step in the processing of RNA.

## EXERCISES

1. Is the following statement true or false? Why? The flow of genetic information in the cell is always DNA ⟶ RNA ⟶ protein.
2. Define the following terms: replication, transcription, and translation.
3. Why is the replication of DNA referred to as a semiconservative process? What is the experimental evidence for the semiconservative nature of the process?
4. What is a replication fork and why is it important in replication?
5. Compare and contrast the properties of the enzymes DNA polymerase I and polymerase III from *E. coli*.
6. List the substances required for replication of DNA catalyzed by DNA polymerase.
7. Describe the discontinuous synthesis of the lagging strand in DNA replication.
8. What are the functions of the gyrase, primase, and ligase enzymes in DNA replication?
9. How does proofreading take place in the process of DNA replication?
10. Describe the excision-repair process in DNA using the excision of thymine dimers as an example.
11. How does DNA replication in eukaryotes differ from the process in prokaryotes?
12. What is the role of the chi intermediate in genetic recombination?
13. Define the term "transposon." Outline the role of transposons in gene rearrangement.
14. List three important properties of RNA polymerase from *E. coli*.
15. Define the term "promoter region" and list three of its properties.
16. Why is a trimming process important in converting precursors of tRNA and rRNA into the active forms?
17. List three molecular changes that take place in the processing of eukaryotic mRNA.
18. Define the terms "exon" and "intron."
19. What are snRNPs and what is their role in the processing of eukaryotic mRNAs?

## ANNOTATED BIBLIOGRAPHY

Adams, R.L.P., J.T. Knowles, and D.P. Leader. *The Biochemistry of the Nucleic Acids.* 10th ed. New York: Chapman and Hall, 1986. [New authors have taken over a classic text originally written by J.N. Davidson.]

Brenner, S., F. Jacob, and M. Meselson. An Unstable Intermediate Carrying Information from Genes to Ribosomes for Protein Synthesis. *Nature* **190,** 576–581 (1961). [One of the first descriptions of the concept of mRNA. Of historical interest.]

Cech, T.R. RNA As an Enzyme. *Sci. Amer.* **255** (5), 64–75 (1986). [A description of the discovery that some RNAs can catalyze their own self-splicing. The author was a recipient of the 1989 Nobel Prize in chemistry for this work.]

Cohen, S.N., and J.A. Shapiro. Transposable Genetic Elements. *Sci. Amer.* **242** (2), 40–49 (1980). [A discussion of the nature and significance of the movement of genetic elements.]

Darnell, J.E. The Processing of RNA. *Sci. Amer.* **249** (2), 90–100 (1983). [A discussion of the ways in which RNA is modified after synthesis.]

Darnell, J.E. RNA. *Sci. Amer.* **253** (4), 68–78 (1985). [An article on the function and processing of RNA.]

Echols, H. Multiple DNA-Protein Interactions Governing High-Precision DNA Transaction. *Science* **233,** 1050–1056 (1986). [A discussion of factors needed for faithful replication.]

Kolata, G. Fitting Methylation into Development. *Science* **228,** 1183–1184 (1985). [A *Research News* article on the role of methylated bases in the structure of DNA.]

Kornberg, A. *DNA Replication.* San Francisco: W.H. Freeman, 1980. [Most aspects of DNA biosynthesis are covered. The author received a Nobel Prize for his work in this field.]

Lewin, R. Making mRNA from a Pair of Precursors.

*Science* **230,** 55 (1985). [A report on some unusual forms of splicing.]

McCorkle, G.M., and S. Altman. RNA's As Catalysts. *J. Chem. Ed.* **64,** 221–226 (1987). [An article on the mechanism of action of catalytic RNA. The second author shared the 1989 Nobel Prize in chemistry with T. Cech for work on the role of RNA as a catalyst.]

Ogawa, T., and T. Okazaki. Discontinuous DNA Replication. *Ann. Rev. Biochem.* **49,** 421–458 (1980). [A review article on replication and joining of DNA fragments.]

Radman, M., and R. Wagner. The High Fidelity of DNA Duplication. *Sci. Amer.* **259** (1), 40–46 (1988). [A description of replication concentrating on the mechanisms for minimizing errors.]

Smith, L.M., J.Z. Sanders, R.J. Kaiser, et al. Fluorescence Detection in Automated DNA Sequence Analysis. *Nature* **321,** 674–679 (1986). [The first report of an automated method for sequencing of DNA.]

Stahl, F.W. Genetic Recombination. *Sci. Amer.* **256** (2), 90–101 (1987). [A discussion of how DNA recombination is related to the rearrangement of chromosomes.]

Steitz, J.A. Snurps. *Sci. Amer.* **258** (6), 56–63 (1988). [A discussion of the role of small nuclear ribonucleoproteins, or snRNPs (pronounced "snurps"), in the removal of introns from mRNA.]

Varmus, H. Reverse Transcription. *Sci. Amer.* **257** (3), 56–64 (1987). [An article that describes RNA-directed DNA synthesis. The author was one of the recipients of the 1989 Nobel Prize in medicine or physiology for his work on the role of reverse transcription in cancer.]

Watson, J.D. *Molecular Biology of the Gene.* 4th ed. Menlo Park, CA: Benjamin/Cummings, 1987. [A highly readable book on molecular biology.]

# Biosynthesis of Proteins

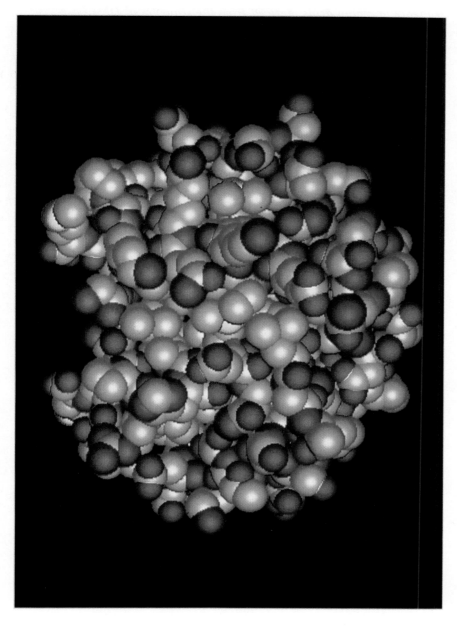

*Computer-generated model of the three-dimensional structure of a completed protein: uteroglobin.*

599

*After the DNA base sequence has been transcribed to RNA, the genetic code is needed to* translate *the RNA sequence into the amino acid sequence of proteins. In eukaryotic cells, DNA is transcribed in the nucleus, but translated in the cytosol. The transcript is exported from the nucleus in the form of* messenger RNA, *which is read and translated at the ribosome. Molecules of transfer RNA (tRNA), one for each amino acid, are required to collect activated amino acids and deliver them, one at a time, to the ribosome. Here they are sequentially joined to synthesize the polypeptide chain of a protein molecule. The sequence of amino acids, derived from the sequence of DNA bases, is specified by the genetic code using the four RNA bases A,U,G,C taken three at a time. With this triplet code, there are 64 possible "code words," called codons. Three of the codons are stop signals and 61 codons specify the 20 amino acids with considerable redundancy. In the actual mechanism of translation, a triplet codon of messenger RNA is temporarily bonded to an anti-codon of transfer RNA. As a succession of tRNA molecules deliver their amino acids to the ribosome, peptide bonds covalently join the amino acids to form the growing polypeptide chain. A stop signal on the messenger RNA terminates the protein chain, which is then released from the ribosome.*

## 22.1
## THE TRANSLATION OF THE GENETIC MESSAGE: PROTEIN BIOSYNTHESIS

Protein biosynthesis is a complex process requiring ribosomes, messenger RNA (mRNA), transfer RNA (tRNA), and a number of protein factors. The ribosome is the site of protein synthesis. The mRNA and tRNA, which are bound to the ribosome in the course of protein synthesis, are responsible for the correct order of amino acids in the growing protein chain.

Before an amino acid can be incorporated into a growing protein chain, it must first be **activated,** a process involving both tRNA and a specific enzyme of the class known as **aminoacyl-tRNA synthetases.** The amino acid is covalently bonded to the tRNA in the process, forming an aminoacyl-tRNA. The actual formation of the polypeptide chain occurs in three steps. In the **initiation** step the first aminoacyl-tRNA is bound to the mRNA at the site that encodes the start of polypeptide synthesis. In this complex the mRNA and the ribosome are bound to each other. The next aminoacyl-tRNA forms a complex with the ribosome and with mRNA. The binding site for the second aminoacyl-tRNA is close to that for the first aminoacyl-tRNA. A peptide bond is formed between the amino acids (chain **elongation**). The chain elongation process repeats itself until the polypeptide chain is complete. Finally, chain **termination** takes place. Each of these steps has many distinguishing features (Figure 22.1), and we shall look at each of them in detail.

## 22.2
## AMINO ACID ACTIVATION

The activation of the amino acid and the formation of the aminoacyl-tRNA take place in two separate steps, both of which are catalyzed by the aminoacyl-tRNA synthetase (Figure 22.2). First, the amino acid forms a covalent bond to an adenine nucleotide, producing an aminoacyl-AMP. The free energy of hydrolysis of ATP provides energy for bond formation. The aminoacyl moiety is then transferred to tRNA, forming an aminoacyl-tRNA.

$$\text{Amino acid} + \text{ATP} \longrightarrow \text{aminoacyl-AMP} + \text{PP}_i$$

$$\underline{\text{Aminoacyl-AMP} + \text{tRNA} \longrightarrow \text{aminoacyl-tRNA} + \text{AMP}}$$
$$\text{Amino acid} + \text{ATP} + \text{tRNA} \longrightarrow \text{aminoacyl-tRNA} + \text{AMP} + \text{PP}_i$$

Aminoacyl-AMP is a mixed anhydride of a carboxylic and a phosphoric acid. Since anhydrides are reactive compounds, the free energy change for the hydrolysis of aminoacyl-AMP favors the second step of the overall

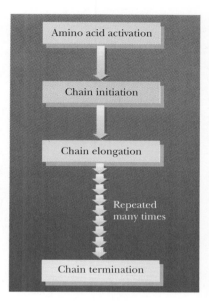

**FIGURE 22.1** A flow chart showing the steps in protein biosynthesis.

**FIGURE 22.2** The two steps of amino acid activation. (a) Formation of aminoacyl-AMP intermediate. (b) Formation of aminoacyl-tRNA from aminoacyl-AMP.

reaction. Another point that favors the process is the energy released when pyrophosphate ($PP_i$) is hydrolyzed to orthophosphate ($P_i$) to replenish the phosphate pool in the cell. The synthetase enzyme requires $Mg^{2+}$ and is highly specific both for the amino acid and for the tRNA. A separate synthetase exists for each amino acid.

In the second part of the reaction, an ester linkage is formed between the amino acid and the 3'-hydroxyl end of the tRNA. Several tRNAs can exist for each amino acid, but a given tRNA will not bond to more than one amino acid. Each synthetase has a high degree of specificity for the correct tRNA and for the correct amino acid. The specificity of the enzyme contributes to the accuracy of the translation process.

## 22.3
## CHAIN INITIATION

The details of chain initiation differ somewhat in prokaryotes and eukaryotes. Like DNA and RNA synthesis, this process has been more thoroughly investigated in prokaryotes. We shall use *Escherichia coli* as our principal example, since all aspects of protein synthesis have been most extensively studied in this bacterium. In all organisms the synthesis of polypeptide chains starts at the N-terminal end; the chain grows from the N-terminal to the C-terminal end.

In prokaryotes the initial N-terminal amino acid residue of all proteins is *N*-formylmethionine (fmet) (Figure 22.3). However, this residue can be, and often is, removed by posttranslational processing after the polypeptide chain is synthesized. There are two different tRNAs for methionine in *E. coli*, one for unmodified methionine and one for *N*-formylmethionine. These two tRNAs are called tRNA$^{met}$ and tRNA$^{fmet}$, respectively (the superscript identifies the tRNA). The aminoacyl-tRNAs that they form with methionine are called met-tRNA$^{met}$ and met-

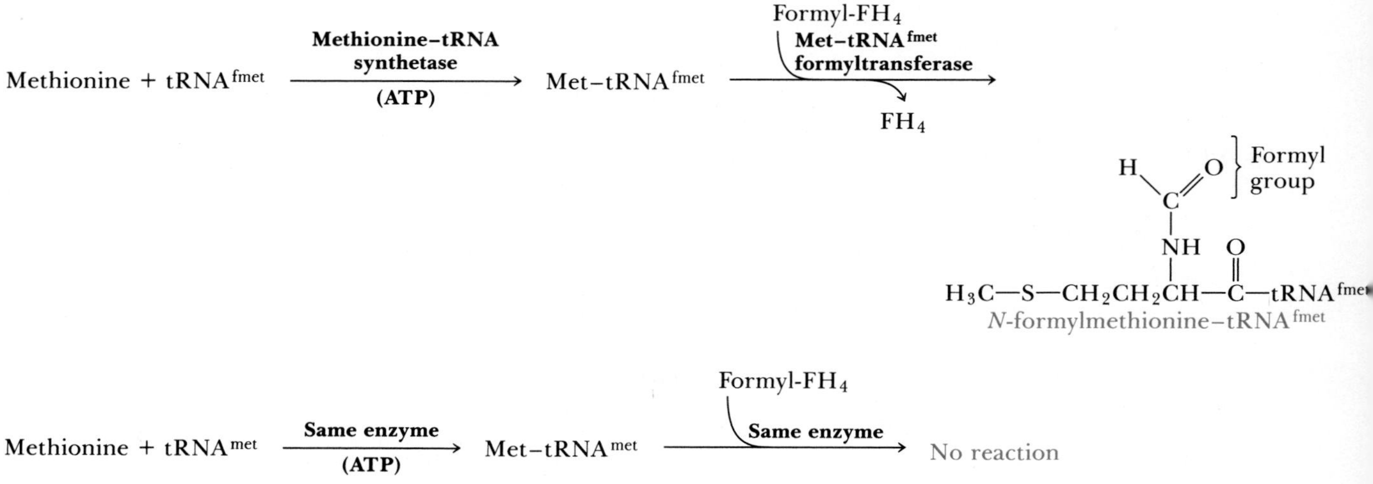

**FIGURE 22.3**  The formation of *N*-formylmethionine–tRNA$^{fmet}$ (first reaction). Methionine must be bound to tRNA$^{fmet}$ to be formylated. (FH$_4$ is tetrahydrofolate.) Methionine bound to tRNA$^{met}$ is not formylated (second equation).

tRNA$^{fmet}$, respectively (the prefix identifies the bound amino acid). In the case of met-tRNA$^{fmet}$, a formylation reaction takes place after methionine is bonded to the tRNA, producing **N-formylmethionine–tRNA$^{fmet}$ (fmet-tRNA$^{fmet}$).** The source of the formyl group is N$^{10}$-formyltetrahydrofolate (see One-Carbon Transfers and Serine Family, in Section 18.4). Methionine bound to tRNA$^{met}$ is not formylated.

Both tRNAs (tRNA$^{met}$ and tRNA$^{fmet}$) contain a specific sequence of three bases (a triplet), 3′-UAC-5′, which base-pairs with the triplet sequence 5′-AUG-3′ in the mRNA sequence. The tRNA$^{fmet}$ triplet in question, UAC, recognizes the AUG triplet, which is the start signal when it occurs at the beginning of the mRNA sequence that directs the synthesis of the polypeptide. The same UAC triplet in tRNA$^{met}$ recognizes the AUG triplet when it is found in an internal position in the mRNA sequence. The mRNA triplet, called the **codon,** specifies the nature of the amino acid to be added to the growing polypeptide chain. The tRNA triplet that hydrogen-bonds to the codon is called the **anticodon.** Note that the anticodon is antiparallel and complementary to the mRNA codon. (The list of amino acids specified by each triplet codon is given in Section 22.8.) The start signal is preceded by a leader segment of mRNA, the Shine–Dalgarno sequence (5′-GGAGGU-3′), which usually lies about ten nucleotides upstream of (nearer the 5′ end of the RNA) the AUG start signal. The genetic message of mRNA is read from the 5′ to the 3′ end. A portion of the mRNA leader segment binds to the 30S subunit of the ribosome by forming base pairs with the 3′ portion of the 16S rRNA of the subunit.

The start of polypeptide synthesis requires formation of an **initiation complex.** At least eight components enter into the formation of the initiation complex, including mRNA, the 30S ribosomal subunit, fmet-tRNA$^{fmet}$, GTP, and three protein initiation factors, called IF-1, IF-2, and IF-3. The IF-3 protein facilitates the binding of mRNA to the 30S ribosomal subunit. The other two factors, IF-1 and IF-2, are involved in the binding of fmet-tRNA$^{fmet}$ to the mRNA-30S subunit complex. The resulting combination is the **30S initiation complex** (Figure 22.4). A 50S ribosomal subunit binds to the 30S initiation complex to produce the **70S initiation complex.** The hydrolysis of GTP to GDP and P$_i$ favors the process by providing energy; the initiation factors are released at the same time.

The process of chain initiation in eukaryotes has not been studied to the same extent as it has in bacteria, but it is known that several differences exist. Methionine, not N-formylmethionine, is the amino acid used for initiation in eukaryotes. (Questions have been raised regarding whether eukaryotes use amino acids other than methionine for initiation. N-acetylated amino acids have been found in some proteins.) Two tRNAs for methionine are found in eukaryotes, and with one of them, the formylase enzyme from *E. coli* can produce fmet-tRNA. Apparently eukaryotes have lost the ability to carry out the reaction because of lack of the proper enzyme rather than because of the properties of the tRNA. There are at least eight initiation factors in eukaryotes; they are designated eIF-1, eIF-2, and so on, with the "e" (for eukaryotic) used to distinguish them from prokaryotic initiation factors. Considerably less is known about eukaryotic

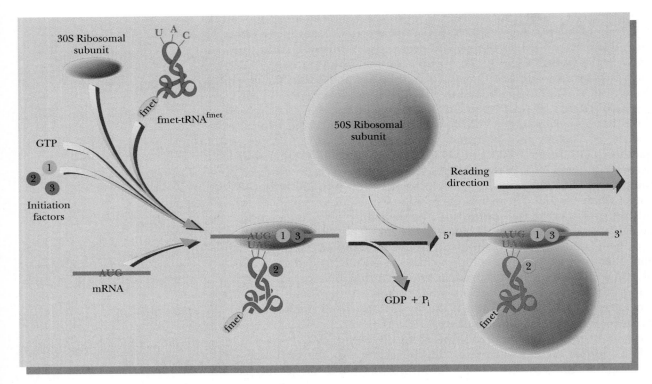

**FIGURE 22.4**    The formation of an initiation complex. The 30S ribosomal subunit binds to mRNA and fmet–tRNA$^{fmet}$ in the presence of GTP and the three initiation factors, IF 1, 2, and 3, forming the 30S initiation complex. The 50S ribosomal subunit is added, forming the 70S initiation complex.

initiation factors than about those from prokaryotes, but the eukaryotic proteins are the subject of current research. It is thought, for example, that the mode of action of the antiviral agent interferon may be to reduce the level of activity of initiation factors in eukaryotic cells infected by a virus; this topic is the focus of particularly active investigation.

## 22.4
## CHAIN ELONGATION

The elongation phase of protein synthesis (Figure 22.5) makes use of the fact that two binding sites for tRNA are present on the 50S subunit of the 70S ribosome. (Note that we shall continue to use the extensively studied prokaryotic system from *E. coli* as our example.) The two tRNA binding sites are called the P (peptidyl) site and the A (aminoacyl) site. The P site binds a tRNA that carries a peptide chain, and the A site binds an aminoacyl-tRNA. (More advanced discussions of chain elongation include a third site on the ribosome, the E (exit) site for deacylated tRNA. For more details, see the article by Nierhaus cited in the bibliography at the end of this chapter.) Chain elongation begins with the addition of the second amino acid specified by the mRNA to the 70S initiation complex. The P site on the ribosome is the one occupied by the fmet-tRNA$^{fmet}$ in the 70S initiation complex. The second aminoacyl-tRNA binds at the A site. A

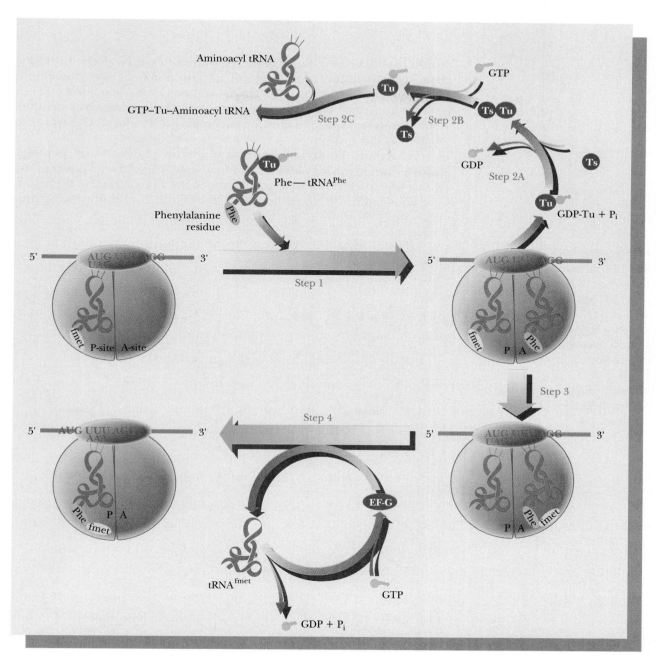

**FIGURE 22.5**  Summary of the steps in chain elongation. *Step 1*. An aminoacyl-tRNA is bound to the A site on the ribosome. Elongation factor EF-Tu (Tu) and GTP are required. The P site on the ribosome is already occupied. *Step 2*. Elongation factor EF-Tu is released from the ribosome and regenerated in a process requiring elongation factor EF-Ts (Ts) and GTP. *Step 3*. The peptide bond is formed, leaving an uncharged tRNA at the P site. *Step 4*. In the translocation step, the uncharged tRNA is released. The peptidyl-tRNA is translocated to the P site, leaving an empty A site. Elongation factor EF-G and GTP are required. (UUU is the codon for phenylalanine; AAA is the anticodon for phenylalanine.)

triplet of tRNA bases (the anticodon AAA in our example) forms hydrogen bonds with a triplet of mRNA bases (UUU, the codon for phenylalanine, in this example). In addition, GTP and two protein elongation factors, EF-Tu and EF-Ts (temperature-unstable and temperature-stable elongation factors, respectively), are required for the binding of aminoacyl-tRNAs to the A site. GTP is hydrolyzed to GDP and $P_i$ in the process.

A **peptide bond** is formed in a reaction catalyzed by **peptidyl transferase,** which is a part of the 50S subunit. The carboxyl group of the $N$-formylmethionyl residue is transferred to the amino group of the amino acid bound to the tRNA at the A site. (See Figure 22.6 for the mechanism

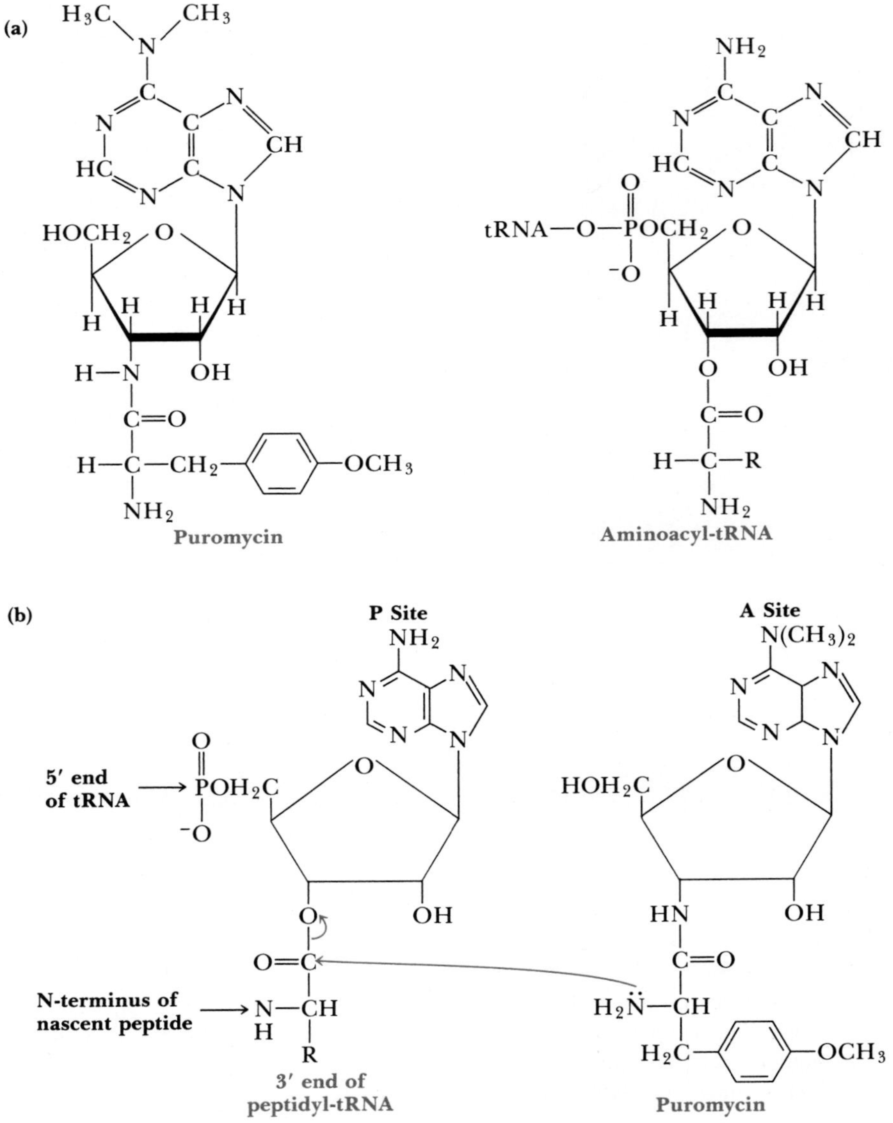

**FIGURE 22.6** The mode of action of puromycin. (a) A comparison of the structures of puromycin and the 3′ end of an aminoacyl-tRNA. (b) Formation of a peptide bond between a peptidyl-tRNA bound at the P site of a ribosome and puromycin bound at the A site. Protein synthesis cannot continue, and the product dissociates from the ribosome.

of this reaction.) There is now a dipeptidyl-tRNA at the A site and a tRNA with no amino acid attached (an uncharged tRNA) at the P site.

A **translocation** step then takes place before another amino acid can be added to the growing chain. In the process the uncharged tRNA is released from the P site, and the peptidyl-tRNA moves from the A site to the vacated P site. In addition, the mRNA moves with respect to the ribosome. Another elongation factor, EF-G, also a protein, is required at this point, and once again GTP is hydrolyzed to GDP and $P_i$.

The three steps of the chain elongation process are aminoacyl-tRNA binding, peptide bond formation, and translocation. They are repeated for each amino acid specified by the genetic message of the mRNA until the stop signal is reached.

Much of the information about this phase of protein synthesis has been gained from the use of inhibitors. Puromycin is a structural analogue of the 3′ end of an aminoacyl-tRNA, making it a useful probe to study chain elongation (Figure 22.6). In an experiment of this sort, puromycin binds to the A site, and a peptide bond is formed between the C-terminal of the growing polypeptide and the puromycin. The peptidyl puromycin dissociates from the ribosome, resulting in premature termination and a defective protein. Puromycin also binds to the P site and blocks the translocation process, although it does not react with peptidyl-tRNA in this case. The existence of A and P sites was determined by these experiments with puromycin.

The main features of the elongation process are the same in prokaryotes and eukaryotes, but the details differ. These differences can be seen in the response to inhibitors of protein synthesis and to toxins. The antibiotic chloramphenicol (a trade name is Chloromycetin) binds to the A site and inhibits peptidyl transferase activity in prokaryotes, but not in eukaryotes. This property has made chloramphenicol useful in treating bacterial infections. In eukaryotes diphtheria toxin is a protein that interferes with protein synthesis by decreasing the activity of the eukaryotic elongation factor eEF-2.

## 22.5
## CHAIN TERMINATION

A stop signal is required for the termination of protein synthesis. The codons UAA, UAG, and UGA are the stop signals. One of two protein release factors (RF-1 or RF-2) is also required, as is GTP, which is bound to a third release factor, RF-3. RF-1 binds to UAA and UAG, while RF-2 binds to UAA and UGA. RF-3 does not bind to any codon, but it does facilitate the activity of the other two release factors. Either RF-1 or RF-2 is bound near the A site of the ribosome when one of the termination codons is reached. The release factor not only blocks the binding of a new aminoacyl-tRNA but also affects the activity of the peptidyl transferase, so that the bond between the carboxyl end of the peptide and the tRNA is hydrolyzed. GTP is hydrolyzed in the process. The whole complex dissociates, setting free release factors, tRNA, mRNA, and the 30S and 50S

**TABLE 22.1    Components Required for Each Step of Protein Synthesis in** *Escherichia coli*

| STEP | COMPONENTS |
|---|---|
| Amino acid activation | Amino acids<br>tRNAs<br>Aminoacyl-tRNA synthetases<br>ATP, $Mg^{2+}$ |
| Chain initiation | fmet-tRNA$^{fmet}$<br>Initiation codon (AUG) of mRNA<br>30S ribosomal subunit<br>50S ribosomal subunit<br>Initiation factors (IF-1, IF-2, and IF-3)<br>GTP, $Mg^{2+}$ |
| Chain elongation | 70S ribosome<br>Codons of mRNA<br>Aminoacyl-tRNAs<br>Elongation factors<br>    (EF-Tu, EF-Ts, and EF-G)<br>GTP, $Mg^{2+}$ |
| Chain termination | 70S ribosome<br>Termination codons (UAA, UAG, and UGA)<br>    of mRNA<br>Release factors (RF-1, RF-2, and RF-3)<br>GTP, $Mg^{2+}$ |

ribosomal subunits. All these components can be reused in further protein synthesis. Table 22.1 summarizes the steps in protein synthesis and the components required for each step.

## 22.6
## POSTTRANSLATIONAL MODIFICATION OF PROTEINS

Newly synthesized polypeptides are frequently processed before they reach the form in which they have biological activity. We have already mentioned that *N*-formylmethionine in prokaryotes is cleaved off. Specific bonds in precursors can be hydrolyzed, exemplified by the cleavage of chymotrypsinogen to chymotrypsin. Proteins destined for export to specific parts of the cell or from the cell have leader sequences at their N-terminal ends. These leader sequences, which direct the proteins to their proper destination, are recognized and removed by specific proteases associated with the endoplasmic reticulum. The finished protein then enters the Golgi apparatus, which directs it to its final destination.

In addition to the processing of proteins by breaking bonds, other substances can be linked to the newly formed polypeptide. Various cofactors such as heme groups are added, and disulfide bonds are formed. Some amino acid residues are also covalently modified, such as the conversion of proline to hydroxyproline. Other covalent modifications can take place, an example being the addition of carbohydrates to yield glycoproteins. These processes are quite common, and the result is to produce an active final form of the protein in question.

## 22.7
## POLYSOMES AND THE SIMULTANEOUS PRODUCTION OF SEVERAL COPIES OF THE SAME POLYPEPTIDE

In our description of protein synthesis, we have considered, up to now, the reactions that take place at one ribosome. It is, however, not only possible, but quite usual, for several ribosomes to be attached to the same mRNA. Each of these ribosomes will bear a polypeptide in various stages of completion, depending on the position of the ribosome as it moves along the mRNA (Figure 22.7). This complex of mRNA with several ribosomes is called a **polysome;** an alternate name is polyribosome.

In prokaryotes, translation begins very soon after mRNA transcription. It is possible for a molecule of mRNA that is still in the process of being transcribed to have a number of ribosomes attached to it that are in

**FIGURE 22.7**   Simultaneous protein synthesis on polysomes. A single mRNA molecule is translated by several ribosomes simultaneously. Each ribosome produces one copy of the polypeptide chain specified by the mRNA. When the protein has been completed, the ribosome dissociates into subunits that are used in further rounds of protein synthesis.

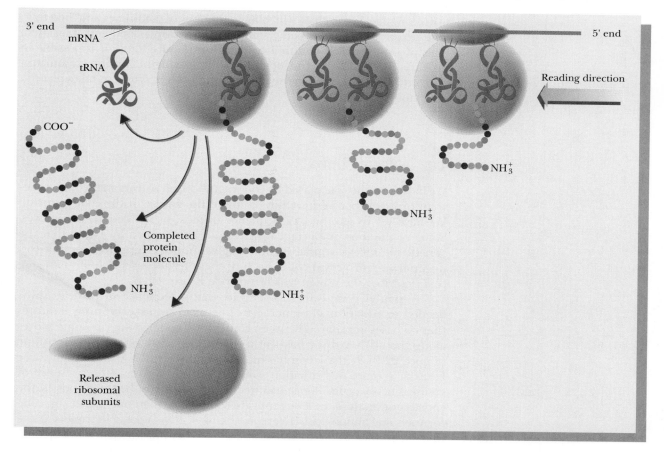

FIGURE 22.8   Electron micrograph showing coupled translation. The dark spots are ribosomes, arranged in clusters on a strand of mRNA. Several mRNAs have been transcribed from one strand of DNA (diagonal line from center to upper right).

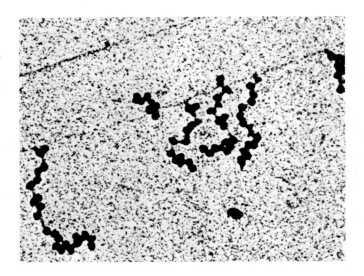

various stages of translating that mRNA. It is also possible for DNA to be in various stages of being transcribed. In this situation several molecules of RNA polymerase are attached to a single gene, giving rise to several mRNA molecules, each of which has a number of ribosomes attached to it. The prokaryotic gene is being simultaneously transcribed and translated. This process is called **coupled translation** (Figure 22.8); it is possible in prokaryotes because of the lack of cell compartmentalization. In eukaryotes, mRNA is produced in the nucleus, and protein synthesis takes place in the cytosol.

## 22.8
## THE GENETIC CODE

We have seen the role played by codon–anticodon pairing in the biosynthesis of proteins, but we have not discussed the nature of the code in detail. We have yet to see the base sequence of most of the codons themselves. Some of the most important features can be specified by saying that the genetic message is contained in a **triplet, nonoverlapping, commaless, degenerate, universal code.** Each of these terms has a definite meaning that describes the way in which the code is translated.

A **triplet** code refers to the fact that a sequence of three bases is needed to specify one amino acid. The term **nonoverlapping** indicates that no bases are shared between consecutive codons; the ribosome moves along the mRNA three bases at a time rather than one or two at a time (Figure 22.9). If the ribosome moved along the mRNA more than three bases at a time, this situation would be referred to as "the presence of commas in the code." Since no intervening bases exist between codons, the code is **commaless.** In a **degenerate** code, more than one triplet can code for the same amino acid. There are 64 (4 × 4 × 4) possible triplets of the four bases that occur in RNA, and all of them are used to code for 20 amino acids or for one of the three stop signals. A **universal** code is one

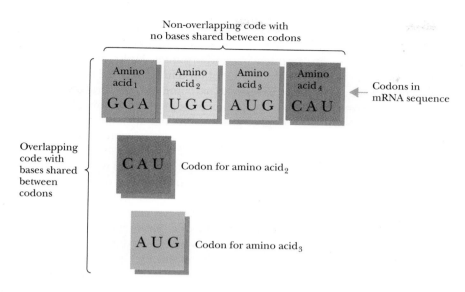

Non-overlapping code with no bases shared between codons

| Amino acid₁ | Amino acid₂ | Amino acid₃ | Amino acid₄ |
| G C A | U G C | A U G | C A U |

← Codons in mRNA sequence

Overlapping code with bases shared between codons

**C A U**   Codon for amino acid₂

**A U G**   Codon for amino acid₃

**FIGURE 22.9**   A comparison of nonoverlapping and overlapping codes. In a nonoverlapping code, adjacent codons do not share any nucleotides in the mRNA sequence. In an overlapping code, adjacent codons have one or more nucleotides in common.

that is the same in all organisms. The universality of the code has been observed in viruses, prokaryotes, and eukaryotes; the only exceptions are the differences in some codons seen in mitochondria. The evolutionary origin of these differences is not known at this writing.

All 64 codons have been assigned a meaning, with 61 of them coding for an amino acid and the remaining 3 serving as the termination signals (Table 22.2).

**TABLE 22.2    The Assignment of the 64 Triplet Codons in the 5′——→3′ Sequence of mRNA**

| BASE AT 5′ END OF CODON ↓ | MIDDLE BASE OF CODON ——→ | | | | BASE AT 3′ END OF CODON ↓ |
| | U | C | A | G | |
| U | phe (UUU) | ser | tyr | cys | U |
|  | phe | ser | tyr | cys | C |
|  | leu | ser | termination | termination | A |
|  | leu | ser | termination | trp | G |
| C | leu | pro | his | arg | U |
|  | leu | pro | his | arg | C |
|  | leu | pro | gln | arg | A |
|  | leu | pro | gln | arg | G |
| A | ile | thr | asn | ser | U |
|  | ile | thr | asn | ser | C |
|  | ile | thr | lys | arg | A |
|  | met (and initiation) | thr | lys | arg | G |
| G | val | ala | asp | gly | U |
|  | val | ala | asp | gly | C |
|  | val | ala | glu | gly | A |
|  | val | ala | glu | gly | G |

Two amino acids, tryptophan and methionine, have only one codon, but the rest have more than one. A single amino acid can have as many as six codons, as is the case with leucine and arginine. Multiple codons for a single amino acid are not randomly distributed in Table 22.2 but have one or two bases in common. The bases that are common to several codons are usually the first and second base, with more room for variation in the third base. Because of the greater degree of variation, the third base is called the "wobble" base. The variations in the genetic code of mitochondria have also been determined (Table 22.3).

The assignment of triplets in the genetic code is based on several types of experiment. One of the most significant involves the use of synthetic polyribonucleotides as messengers. This approach can give some information about the nature of the code, but ambiguities remain. Unambiguous assignments require another important type of experiment that is based on binding studies involving tRNAs and triplets synthesized in the laboratory.

When homopolynucleotides (polyribonucleotides that contain only one type of base) are used as a **synthetic mRNA** in laboratory systems for polypeptide synthesis, homopolypeptides (polypeptides that contain only one kind of amino acid) are produced. When poly U is the messenger, the product is polyphenylalanine. With poly A as the messenger, polylysine is formed. The product for poly C is polyproline, and with poly G, polyglycine results. When an alternating copolymer (a polymer with an alternating sequence of two bases) is the messenger, the product is an alternating polypeptide (a polypeptide with an alternating sequence of two amino acids). For example, when the sequence of the polynucleotide is -ACACACACACACACACACAC-, the polypeptide produced is poly-(thr-his), with the alternating sequence threonine-histidine. There are two types of coding triplets in this polynucleotide, ACA and CAC, but this experiment cannot establish which one codes for threonine and which one for histidine. More information is needed for an unambiguous assignment, but it is interesting that this result proves that the code is a triplet code. If it were a doublet code, the product would be a mixture of two homopolymers, one specified by the codon AC and the other by the codon CA. (The terminology for the different ways of reading this message as a doublet is to say that they have different **reading frames,** /AC/AC/ and /CA/CA/. In a triplet code only one reading frame is possible, namely, /ACA/CAC/ ACA/CAC/, which gives rise to an alternating polypeptide.) Using other synthetic polynucleotides can yield other coding assignments, but, as in our example here, many questions remain.

Other methods are needed to answer the remaining questions about codon assignment. One of the most useful methods is the **binding assay.**

**TABLE 22.3   Differences Between the Genetic Code in Mitochondria and the Universal Code**

| CODONS | UGA | AUA | AGU | AGG | AUU |
|---|---|---|---|---|---|
| Universal code | termination | ile | arg | arg | ile |
| Mitochondrial code | trp | met and initiation | termination | termination | ile and possibly initiation |

This technique depends on the fact that aminoacyl-tRNAs bind strongly to ribosomes in the presence of trinucleotides. In this situation, the trinucleotide plays the role of an mRNA codon. The possible trinucleotides are synthesized by chemical methods, and binding assays are carried out with each type of trinucleotide. Aminoacyl-tRNAs are tested for their ability to bind in the presence of a given trinucleotide. For example, if the aminoacyl-tRNA for isoleucine binds to the ribosome in the presence of the trinucleotide AUC, the sequence AUC is established as a codon for isoleucine. About 50 of the 64 codons have been identified by this method.

## Codon–Anticodon Pairing and "Wobble"

A codon forms base pairs with a complementary anticodon of a tRNA when an amino acid is incorporated during protein synthesis. Some tRNAs bond to one codon exclusively, but many of them can recognize more than one codon because of variations in the allowed pattern of hydrogen bonding. This variation is called "wobble" (Figure 22.10), and it applies to the first base of an anticodon, the one at the 5' end, but not to the second or the third base. Recall that mRNA is read from the 5' to the 3' end. The first (wobble) base of the anticodon hydrogen-bonds to the third base of the codon, the one at the 3' end. The base in the wobble position of the anticodon can base-pair with several different bases in the codon, not just the base specified by Watson–Crick base pairing (Table 22.4).

When the wobble base of the anticodon is uracil, it can base-pair not only with adenine, as expected, but also with guanine, the other purine base. When the wobble base is guanine, it can base-pair with cytosine, as expected, and also with uracil, the other pyrimidine base. The purine base hypoxanthine frequently occurs in the wobble position in many tRNAs, and it can base-pair with adenine, cytosine, and uracil in the codon (Figure 22.11). Adenine and cytosine do not form any base pairs other than the expected ones with uracil and guanine, respectively (Table 22.4). To summarize, when the wobble position is occupied by I (from inosine, the nucleoside made up of ribose and hypoxanthine), G, or U, variations in hydrogen bonding are allowed; when the wobble position is occupied by A or C, these variations do not occur.

The wobble hypothesis provides insight into some aspects of the degeneracy of the code. In many, but not all, cases the degenerate codons

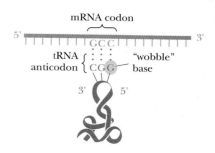

FIGURE 22.10   "Wobble" base pairing. (a) The "wobble" base of the anticodon is the one at the 5' end; it forms hydrogen bonds with the last base of the mRNA codon, the one at the 3' end of the codon.

**TABLE 22.4   Base-Pairing Combinations in the "Wobble" Scheme**

| BASE AT 5' END OF ANTICODON | BASE AT 3' END OF CODON |
|---|---|
| I* | A, C, or U |
| G | C or U |
| U | A or G |
| A | U |
| C | G |

*I = hypoxanthine.

Note that there are no variations in base pairing when the wobble position is occupied by A or C.

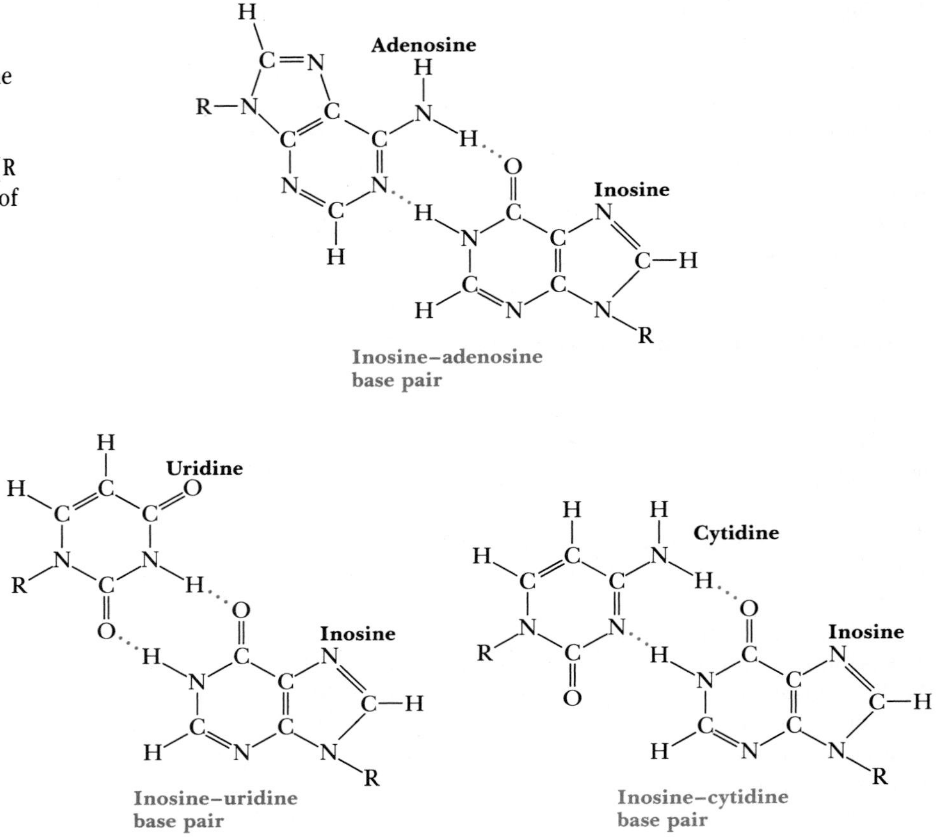

**FIGURE 22.11** Base pairing involving inosine. When inosine occupies the wobble position of the anticodon, it can form hydrogen-bonded base pairs with adenosine, uridine, and cytidine. (R indicates the point of attachment of the base to the ribose.)

Inosine–adenosine base pair

Inosine–uridine base pair

Inosine–cytidine base pair

for a given amino acid differ in the third base, the one that pairs with the wobble base of the anticodon. Fewer different tRNAs are needed, since a given tRNA can base-pair with several codons. As a result, a cell would have to invest less energy in the synthesis of needed tRNAs. The existence of wobble also minimizes the damage that can be caused by misreading of the code. If, for example, a leucine codon CUU were to be misread as CUC, CUA, or CUG during transcription of mRNA, this codon would still be translated as leucine during protein synthesis; no damage to the organism would occur. We have seen earlier that drastic consequences can result from misreading of the genetic code in other codon positions, but here we see that such effects are not inevitable.

## 22.9
## GENETIC REGULATION OF TRANSCRIPTION AND TRANSLATION

Some proteins are not synthesized by cells at all times. Rather, the production of these proteins can be triggered by the presence of a suitable substance, called the **inducer.** This phenomenon is called **induction;** the process is under genetic control. A particularly well studied example of an inducible protein is the enzyme *β*-**galactosidase** in *E. coli,* which we shall use as a case study.

The disaccharide **lactose** (a *β*-galactoside) (Section 6.3) is the substrate of *β*-galactosidase. The enzyme hydrolyzes the glycosidic linkage between galactose and glucose, the monosaccharides that are the component parts of lactose. *E. coli* can survive with lactose as its sole carbon source. To do so, the bacterium needs *β*-galactosidase to catalyze the first step in lactose degradation. The production of *β*-galactosidase takes place only in the presence of lactose, not in the presence of other carbon sources, such as glucose. Lactose is the inducer, and *β*-galactosidase is an **inducible enzyme.**

In 1961, Jacob and Monod proposed a theory to account for the experimental facts given in the last paragraph; the main features of the theory have been supported by further experimental results. According to this point of view, the actual production of an inducible protein such as *β*-galactosidase is under the control of a **structural gene** (*Z*). The base sequence of the structural gene specifies the amino acid sequence of the protein. The expression of one or more structural genes is in turn under the control of a **regulatory gene** (*I*), and the mode of operation of the regulatory gene is the most important feature of the theory. The regulatory gene is responsible for the production of a protein, the **repressor.** As the name indicates, the repressor inhibits the expression of the structural gene. In the presence of an inducer, this inhibition is removed.

The repressor operates by binding to a portion of the DNA known as the **operator** (*O*). When a repressor is bound to the operator, RNA polymerase cannot bind to the adjacent **promoter gene** (*P*), which facilitates the expression of the structural gene. The operator and promoter together constitute the **control sites.**

In induction, the inducer binds to the repressor, producing an inactive repressor that cannot bind to the operator (Figure 22.12). Since the operator is no longer bound to the repressor, RNA polymerase can now bind to the promoter, and transcription and eventual translation of the structural gene (or genes) can take place. The whole assemblage of promoter, operator, and structural genes is called an **operon.** The control sites, the promoter and operator, are physically adjacent to the structural genes in the DNA sequence, but the regulatory gene can be quite far removed from the operon. (In *E. coli,* the regulatory gene for this operon is adjacent to the promoter).

When *E. coli* is presented with lactose as a carbon source, *β*-galactosidase is not the only protein induced. In other words, several structural genes are found in this operon, which is called the ***lac* operon.** In addition to *β*-galactosidase, the *lac* operon is responsible for the production of two other enzymes, a permease and an acetyltransferase. The permease mediates active transport of lactose into the cell, but the function of the acetyltransferase is not well understood. The structural genes of the *lac* operon are the *Z* gene, which codes for *β*-galactosidase; the *Y* gene, which codes for the permease; and the *A* gene, which codes for the acetyltransferase.

The *lac* operon is induced when *E. coli* has lactose, and no glucose, available to it as a carbon source. When both glucose and lactose are

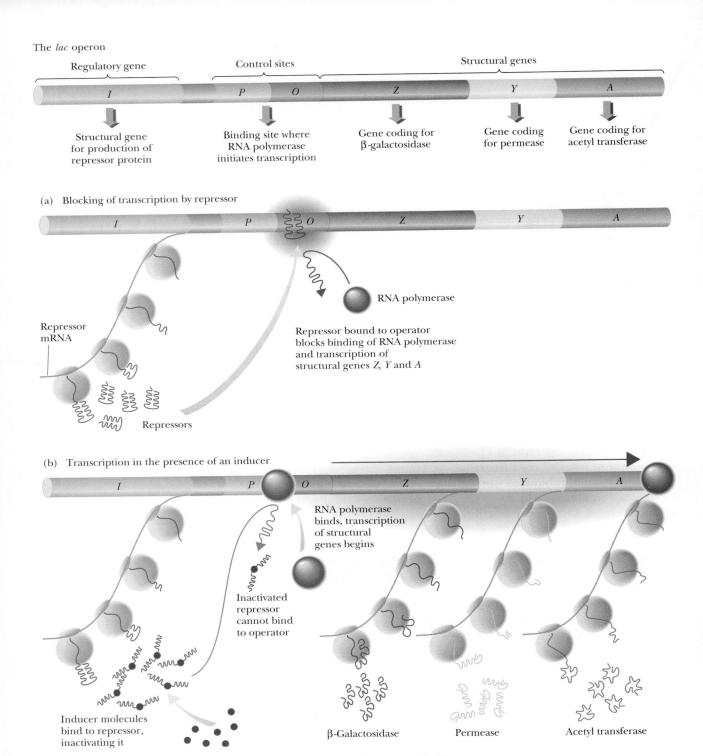

The *lac* operon

Regulatory gene | Control sites | Structural genes

I | P | O | Z | Y | A

Structural gene for production of repressor protein

Binding site where RNA polymerase initiates transcription

Gene coding for β-galactosidase

Gene coding for permease

Gene coding for acetyl transferase

(a) Blocking of transcription by repressor

I | P | O | Z | Y | A

RNA polymerase

Repressor bound to operator blocks binding of RNA polymerase and transcription of structural genes Z, Y and A

Repressor mRNA

Repressors

(b) Transcription in the presence of an inducer

I | P | O | Z | Y | A

RNA polymerase binds, transcription of structural genes begins

Inactivated repressor cannot bind to operator

Inducer molecules bind to repressor, inactivating it

β-Galactosidase

Permease

Acetyl transferase

**FIGURE 22.12** The *lac* operon in the absence and presence of inducer. (a) When no inducer is present, the regulatory gene (*I*) is transcribed, leading to the production of repressor. The repressor in turn binds to the operator gene (*O*), blocking transcription of the structural genes that code for β-galactosidase and the other proteins produced under the control of the *lac* operon. (b) In the presence of inducer, the repressor is still produced but becomes inactive when the inducer is bound to it. The inactive repressor does not bind to the operator, and the structural genes of the *lac* operon are expressed.

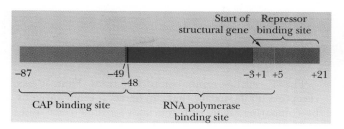

**FIGURE 22.13** Binding sites in the *lac* operon. Numbering refers to base pairs. Negative numbers are assigned to base pairs in the regulatory sites. Positive numbers indicate the structural gene, starting with base pair + 1. (There is some overlap between promoter and operator regions.)

present, the cell does not make the *lac* proteins. The repression of the synthesis of *lac* proteins by glucose is called **catabolite repression.** The mechanism by which *E. coli* recognizes the presence of glucose involves the promoter gene. The promoter has two regions. One is the entry site for RNA polymerase, and the other is the binding site for another regulatory protein, the **catabolite activator protein** (CAP) (Figure 22.13). The binding site for RNA polymerase also overlaps the binding site for repressor in the operator region (the promoter and operator overlap).

The binding of the CAP protein to the promoter depends on the presence or absence of 3′,5′-cyclic AMP (cAMP). When glucose is not present, cAMP is formed, serving as a "hunger signal" for the cell. CAP forms a complex with cAMP. The complex binds to the CAP site in the promoter region. When the complex is bound to the CAP site on the promoter, the RNA polymerase binds strongly to the entry site for it on the promoter. If, at the same time, lactose is present, the repressor-inducer complex forms, so the RNA polymerase can bind at the entry site available to it and proceed with transcription (Figure 22.14).

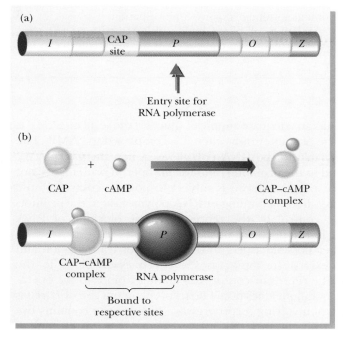

**FIGURE 22.14** Catabolite repression. (a) The control sites of the *lac* operon. The CAP-cAMP complex, not CAP alone, binds to the CAP site of the *lac* promoter. When the CAP site on the promoter is not occupied, RNA polymerase cannot bind to the entry site for it on the promoter. (b) In the absence of glucose, cAMP forms a complex with CAP. The complex binds to the CAP site, allowing RNA polymerase to bind to the entry site on the promoter and transcribe the structural genes.

## BOX 22.1
## THE BODY'S DEFENSES: *THE IMMUNE SYSTEM*

If harmful pathogens, such as viruses and bacteria, get past the outer defenses of the body and invade its interior, they are met by a powerful surveillance and attack force called the immune system. The chief talent of the specialized cells and molecules that make up the system resides in their ability to distinguish self from non-self. If the invading substance is different in composition from a similar substance in the body being invaded, then it is marked for destruction. Organ or cellular transplants from a donor who is not an identical twin (or closely matched genetically) are attacked and destroyed.

A major component of the immune system in vertebrates is the white blood cell called the lymphocyte. These cells circulate through lymphatic vessels and also enter the bloodstream. Swollen lymph nodes (where lymphocytes are generated) indicate that the body is making weapons to fight off the hostile invaders.

T and B cells are lymphocytes that have their origin in the bone marrow; but T cells also mature in the thymus gland. The function of B cells is to manufacture antibodies—protein molecules that bind to specific portions of foreign invaders. The precise binding (molecule to molecule) of antibody to invader is called the antibody–antigen complex. The B cell is genetically programmed to make an antibody to fit every conceivable antigen. Thus there is the possibility of millions of different antibodies, each with a different (and highly specific) binding site. Only recently has it been learned that this great diversity is made possible by genetic recombinations of DNA within the cell.

The T cell has a number of functions. First, it has receptors on its surface that can detect foreign antigens by comparing them with the body's own surface antigens. If the foreign invader does not have the proper "password," it is marked for destruction by a killer T cell. The second function of the T cell is that of "helper" to the B cell, which manufactures antibodies. In the *immune response,* foreign antigen is detected by scavenger cells called macrophages. When the macrophage ingests an antigen, it is processed and displayed on its surface to be recognized by a T cell. When the T cell contacts this antigen, it is activated to become a helper cell to the B lymphocyte. Binding of the same antigen to the B cell causes it to become activated. The B cells proliferate, each cell making precisely the same antibody against the same antigen by which it was activated.

An even more startling effect of the immune system is that some of the activated B and T cells of the initial immune response become *memory* cells and can respond strongly to a second antigen attack. In fact, this is the very basis of vaccination (immunization). The polio vaccine, for example, contained killed polio viruses that, when injected, raised an initial antibody response. Some time later, when a live virus of the same kind strikes, the body responds with a massive outpouring of antibodies, derived from surviving memory cells left over from the initial antibody response to inoculation.

When a cell of the body is infected with a virus, the viral proteins still embedded in the infected cell can be

When the cell has an adequate supply of glucose, the level of cAMP is low. CAP binds to the promoter only when it is complexed to cAMP. The entry site is not available to the RNA polymerase when the CAP-cAMP complex is not bound to the promoter, and the *lac* operon proteins are not produced. Note that the *lac* operon is subject to positive control by the cAMP-CAP complex (allowing binding of RNA polymerase to the promoter) and to negative control by the repressor (steric blocking of RNA polymerase binding in the presence of the repressor).

There is another possible aspect to the control of transcription and translation by repressors. In some cases a **corepressor** binds to the repressor protein. The repressor-corepressor complex binds to the operator, while the repressor alone does not. The tryptophan operon of *E. coli* is an example of control involving a corepressor. This operon contains five structural genes for the enzymes that convert chorismate to tryptophan detected by killer T cells that will then eliminate the infected cell.

Although the immune system is wonderfully efficient in repelling enemies, a key cell of the immune system, the T cell, is itself vulnerable to attack by the HIV (human immunodeficiency virus) or AIDS virus. The T cell is infected when the AIDS virus binds to the T cell and injects its core of the genetic material RNA. A single strand of RNA is converted into single-stranded DNA by the enzyme reverse transcriptase. (This mechanism gives viruses like the AIDS virus the name *retroviruses*.) A second copy makes another DNA strand. Combined into double-stranded DNA, the transcribed viral genome can migrate to the DNA of the infected cell. It can remain dormant, or the DNA can be used to make more AIDS viruses, which in turn can further infect other T cells. Binding of the virus takes place between a molecule on the surface of the virus and a surface molecule on the T cell called CD 4. Depletion of T cells by the AIDS virus naturally weakens the immune system and leaves the patient subject to infections such as pneumonia and Kaposi's sarcoma (a rare cancer), which ultimately are fatal.

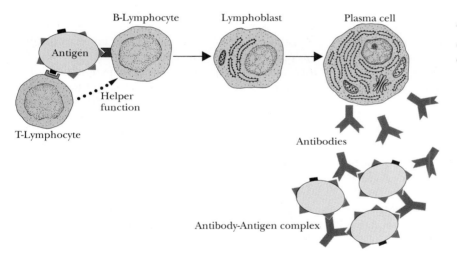

The immune response. Helper T cells facilitate activation of the B cells, producing antibodies.

(see Interchapter B, Section B.4). The corepressor of this operon is tryptophan itself. When tryptophan is abundant, the cell does not need to invest the energy needed to produce the mRNA for these five enzymes or to produce the enzymes themselves. When tryptophan is required, the cell forms it as well as the necessary enzymes and mRNAs.

Regulation of the production of inducible enzymes by the induction-repression mechanism allows for a high degree of fine tuning of metabolism by organisms, even apparently simple ones, such as bacteria. Parenthetically, there are no known operons in eukaryotes. The simplest living cell is a highly complex and well-organized entity, far more than just a "bag of enzymes"; it is a great challenge for present and future biochemists to discover the details of metabolic processes and their control mechanisms.

Box 22.1 describes the way in which some cells respond to a challenge to their capacity to synthesize proteins.

## SUMMARY

Protein biosynthesis requires ribosomes, messenger RNA (mRNA), transfer RNA (tRNA), and a number of protein factors. The ribosome is the site of protein synthesis. The mRNA and tRNA, which are bound to the ribosome in the course of protein synthesis, are responsible for the correct order of amino acids in the growing protein chain.

Before an amino acid can be incorporated into a growing protein chain, it must first be activated. A covalent bond is formed between the amino acid and a tRNA, yielding an aminoacyl-tRNA. The actual formation of the polypeptide chain takes place in three steps. In the initiation step, the first aminoacyl-tRNA is bound to the ribosome and to mRNA. A second aminoacyl-tRNA forms a complex with the ribosome and with mRNA. The binding site for the second aminoacyl-tRNA is close to that for the first aminoacyl-tRNA. A peptide bond is formed between the amino acids (chain elongation). The chain elongation process—which involves translocation of the ribosome along the mRNA, in addition to peptide bond formation—repeats itself until the polypeptide chain is complete. Finally, chain termination takes place. Newly synthesized polypeptides frequently undergo posttranslational modification to produce the final active form of the protein. In the actual translation process, it is usual for several ribosomes to be bound to the same mRNA. Such a complex is called a polysome. Each of the ribosomes in the polysome has a polypeptide in various stages of completion, depending on the position of the ribosome as it moves along the mRNA and transcribes the genetic message.

The genetic message is contained in a triplet, nonoverlapping, commaless, degenerate, universal code. A codon—in other words, a series of three bases adjacent to one another in sequence (nonoverlapping and commaless)—specifies a given amino acid. Several codons can and usually do specify the same amino acid (degeneracy of the code). The same code has been observed in viruses, prokaryotes, and eukaryotes (universality of the code); the only exceptions are the differences in some codons observed in mitochondria. All 64 possible codons have been assigned a meaning, with 61 of them coding for an amino acid and the remaining 3 serving as the termination signals.

Some proteins are not synthesized by all cells at all times. The process of induction, which triggers the production of such proteins in the presence of a suitable substance called the inducer, is under genetic control. The actual synthesis of an inducible protein is regulated by a structural gene. The expression of one or more structural genes is in turn under the control of a regulatory gene. The regulatory gene is responsible for the production of a protein, the repressor, which binds to the operator site. When a repressor is bound to the operator, RNA polymerase cannot bind to the adjacent promoter site. The operator and promoter together constitute the control sites. The whole assemblage of promoter, operator, and structural genes is called an operon. In induction, the inducer binds to the repressor, producing an inactive repressor that cannot bind to the operator. Since the operator is no longer bound to the repressor, RNA polymerase can now bind to the promoter, and transcription and eventual translation of the structural gene (or genes) can take place. Other variations in operon activity, such as catabolite repression and the action of corepressors, are possible, allowing for several different ways to control protein synthesis. The working of the immune system presents an enormous challenge to the protein-synthesizing capacity of the cells involved.

## EXERCISES

1. Prepare a flow chart showing the stages of protein synthesis.
2. What is the role of ATP in amino acid activation?
3. A friend tells you that she is starting a research project on aminoacyl esters. She asks you to describe the biological role of this class of compounds. What do you tell her?
4. *E. coli* has two tRNAs for methionine. What is the basis for the distinction between the two?
5. Describe the recognition process by which the

tRNA for *N*-formylmethionine interacts with the portion of mRNA that specifies the start of transcription.

6. What are the components of the initiation complex in protein synthesis and how do they interact with one another?

7. What are the A site and the P site? How are their roles in protein synthesis similar and how do they differ?

8. Identify the following by describing their function: EF-G, EF-Tu, EF-Ts, and peptidyl transferase.

9. How does puromycin function as an inhibitor of protein synthesis?

10. Describe the role of the stop signals in protein synthesis.

11. The amino acid hydroxyproline is found in collagen. There is no codon for hydroxyproline. Give an explanation for the occurrence of this amino acid in a common protein.

12. In the early days of research on protein synthesis, some scientists observed that their most highly purified ribosome preparations, containing single ribosomes almost exclusively, were less active than preparations that were less highly purified. Suggest an explanation for this observation.

13. A genetic code in which two bases code for a single amino acid is not adequate for protein synthesis. Give a reason for this statement.

14. It is possible for the codons for a single amino acid to have the first two bases in common and to differ in the third base. Why is this experimental observation consistent with the concept of "wobble"?

15. The base hypoxanthine (the corresponding nucleoside is inosine) frequently occurs as the third base in codons. What role does hypoxanthine play in wobble base pairing?

16. Define the terms "inducer" and "inducible enzyme."

17. What are the component parts of an operon and what roles do they play in inducing enzyme synthesis?

18. What is the distinction between a repressor and a corepressor?

## ANNOTATED BIBLIOGRAPHY

Abraham, A.K., T.S. Eikhon, and I.F. Pryme, eds. *Protein Synthesis: Translational and Post-Translational Events.* Clifton, NJ: Humana Press, 1983. [A collection of articles on all aspects of protein synthesis, with extensive coverage of modification of proteins after synthesis.]

Adams, R.L.P., J.T. Knowles, and D.P. Leader. *The Biochemistry of the Nucleic Acids.* 10th ed. New York: Chapman and Hall, 1986. [New authors have taken over a classic text originally written by J.N. Davidson.]

Brenner, S., F. Jacob, and M. Meselson. An Unstable Intermediate Carrying Information from Genes to Ribosomes for Protein Synthesis. *Nature* **190**, 576–581 (1961). [One of the first descriptions of the concept of mRNA. Of historical interest.]

Crick, F.H.C. Codon-Anticodon Pairing: The Wobble Hypothesis. *J. Mol. Biol.* **19**, 548–555 (1966). [The first statement of the wobble hypothesis and still one of the best.]

Gualerzi, C. and C. Pon. Initiation of mRNA Translation in Prokaryotes. *Biochemistry* **29**, 588–589 (1990). [A review on the initiation step in protein synthesis.]

Jacob, F., and J. Monod. Genetic Regulatory Mechanisms in the Synthesis of Proteins. *J. Mol. Biol.* **3**, 318–356 (1961). [The original article in which the concept of repression was postulated. Also one of the first descriptions of mRNA. Mostly of historical interest but quite well written.]

Marrack, P. and J. Kappler. The T Cell and Its Receptor. *Sci. Amer.* **254** (2), 36–45 (1986). [A description of the workings of the immune system.]

Marx, J. Taming Rogue Immune Reactions. *Science* **249**, 246–248 (1990). [A Research News article about promising treatments for autoimmune diseases.]

Moldave, K., ed. *RNA and Protein Synthesis.* New York: Academic Press, 1981. [A collection of articles on all aspects of protein synthesis, with emphasis on the role of tRNA.]

Nierhaus, K. The Allosteric Three-Site Model for the Ribosomal Elongation Cycle: Features and Future. *Biochemistry* **29**, 4997–5008 (1990). [A review on the role of ribosomal binding sites for tRNA in protein chain elongation.]

Ptashne, M. *A Genetic Switch: Gene Control and Phage λ.* Palo Alto, CA: Cell Press/Blackwell Scientific, 1986. [An extensive discussion of induction and repression. The life cycle of a bacteriophage that infects *E. coli* is used as an example.]

Ptashne, M. How Gene Activators Work. *Sci. Amer.* **260** (1), 41–47 (1989). [An article on the mode of action of repressors in eukaryotic and prokaryotic systems.]

Ross, J. The Turnover of Messenger RNA. *Sci. Amer.* **260**

(4), 48–55 (1989). [An article describing the regulation of the rate at which mRNA is degraded in the cell.]

Tonegawa, S. The Molecules of the Immune System. *Sci. Amer.* **253** (4), 122–131 (1985). [A description of the immune system by a Nobel laureate.]

Watson, J.D. *Molecular Biology of the Gene.* 4th ed. Menlo Park, CA: Benjamin/Cummings, 1987. [A highly readable book on molecular biology.]

# ANSWERS TO EXERCISES

## CHAPTER 1

1. A polymer is a very large molecule formed by the linking together of smaller units (monomers). A protein is a polymer formed by the linking together of amino acids. A nucleic acid is a polymer formed by the linking together of nucleotides. Catalysis is the process which increases the rate of chemical reactions compared to the uncatalyzed reaction. Biological catalysts are proteins in almost all cases; the only exceptions are a few types of RNA which can catalyze some of the reactions of their own metabolism. The genetic code is the means by which the information for the structure and function of all living things is passed from one generation to the next. The sequence of purines and pyrimidines in DNA carries the genetic code (RNA is the coding material in some viruses). Anabolism is the process of synthesis of important biomolecules from simpler compounds. Catabolism is the breakdown of biomolecules to release energy.

2. The theory that proteins arose first in the origins of life gives a good explanation of catalysis and metabolic pathways but is vague about the origin of coding. The theory that nucleic acids arose first gives prime importance to coding but does not address the problem of lack of stability of unprotected nucleic acids. The double-origin theory that life arose on the surface of clay particles suggests a stable coding system (the clay surface), which also served as a site for catalysis. Later, more efficient biomolecules replaced clay particles in life coding and catalysis for life processes.

3. With respect to coding, RNA has been produced from monomers in the absence of either a preexisting RNA to be copied or an enzyme to catalyze the process. The observations that some existing RNA molecules can catalyze their own processing suggests a role for RNA in catalysis. With this dual role, RNA may have been the original informational macromolecule in the origin of life.

## CHAPTER 2

1. Five differences between prokaryotes and eukaryotes are: (1) Prokaryotes do not have a well-defined nucleus, but eukaryotes have a nucleus marked off from the rest of the cell by a double membrane. (2) Prokaryotes have only a plasma (cell) membrane; eukaryotes have an extensive internal membrane system. (3) Eukaryotic cells contain membrane-bounded organelles, while prokaryotic cells do not. (4) Eukaryotic cells are normally larger than those of prokaryotes. (5) Prokaryotes are single-celled organisms, while eukaryotes can be multicellular as well as single-celled.

2. See Section 2.3 for the functions of the parts of an animal cell, which are shown in Figure 2.2(a).

3. See Section 2.3 for the functions of the parts of a plant cell, which are shown in Figure 2.2(b).

4. In green plants photosynthesis takes place in the membrane system of chloroplasts, which are large membrane-bounded organelles. In photosynthetic bacteria there are extensions of the plasma membrane into the interior of the cell called chromatophores, which are the sites of photosynthesis.

5. Nuclei, mitochondria and chloroplasts are bounded by a double membrane.

6. Nuclei, mitochondria and chloroplasts all contain DNA. The DNA found in mitichondria and in chloroplasts differs from that in the nucleus.

7. Mitochondria are the primary sites of ATP synthesis, since they carry out a high percentage of the oxidation reactions which release energy for the cell.

8. Protein synthesis takes place on ribosomes in both prokaryotes and eukaryotes. In eukaryotes ribosomes may be bound to the endoplasmic reticulum or found free in the cytoplasm; in prokaryotes ribosomes are only found free in the cytoplasm.

9. Microtubules form the mitotic spindle, allowing for the separation of chromosomes to be included in the two daughter cells.

10. The Golgi apparatus is involved in carbohydrate metabolism and in the export of substances from the cell. Lysosomes contain hydrolytic enzymes, peroxisomes contain catalase (needed for the metabolism of peroxides), and glyoxysomes contain enzymes needed by plants for the glyoxylate cycle. All these organelles have the appearance of flattened sacs bounded by a single membrane.

# CHAPTER 3

1. The C—H bond is not sufficiently polar for greatly unequal distribution of electrons at its two ends. Also, there are no unshared pairs of electrons to serve as hydrogen bond acceptors.

2. In a hydrogen-bonded dimer of acetic acid the —OH portion of the carboxyl group on molecule 1 is hydrogen-bonded to the —C=O portion of the carboxyl group on molecule 2, and vice versa.

$$CH_3-\underset{OH\text{---}O}{\overset{O\text{---}HO}{\diagdown\diagup}}-CH_3$$

3. $(CH_3)_3NH^+$ (conjugate acid)
   $\overline{(CH_3)_3N}$ (conjugate base)

   $^+H_3N-CH_2-COOH$ (conjugate acid)
   $\overline{^+H_3N-CH_2-COO^-}$ (conjugate base)

   $^+H_3N-CH_2-COO^-$ (conjugate acid)
   $\overline{H_2N-CH_2-COO^-}$ (conjugate base)

   $^-OOC-CH_2-COOH$ (conjugate acid)
   $\overline{^-OOC-CH_2-COO^-}$ (conjugate base)

   $^-OOC-CH_2-COOH$ (conjugate base)
   $\overline{HOOC-CH_2-COOH}$ (conjugate acid)

4. | | |
   |---|---|
   | Blood plasma, pH 7.4 | $[H^+] = 4.0 \times 10^{-8}$ M |
   | Orange juice, pH 3.5 | $[H^+] = 3.2 \times 10^{-4}$ M |
   | Human urine, pH 6.2 | $[H^+] = 6.3 \times 10^{-7}$ M |
   | Household ammonia, pH 11.5 | $[H^+] = 3.2 \times 10^{-12}$ M |
   | Gastric juice, pH 1.8 | $[H^+] = 1.6 \times 10^{-2}$ M |

5. In all cases the suitable buffer range covers a pH range of $pK_a +/- 1$ pH units.
   Lactic acid ($pK_a = 3.86$) and its sodium salt, pH 2.86–4.86
   Acetic acid ($pK_a = 4.76$) and its sodium salt, pH 3.76–5.76
   TRIS (see Table 3.4, $pK_a = 8.3$) in its protonated form and its free amine form, pH 7.3–9.3
   HEPES (see Table 3.4, $pK_a = 7.55$) in its zwitterionic and its anionic form, pH 6.55–8.55

6. Use the Henderson-Hasselbalch equation

$$pH = pK_a + \log\left(\frac{[CH_3COO^-]}{[CH_3COOH]}\right)$$

$$5.00 = 4.76 + \log\left(\frac{[CH_3COO^-]}{[CH_3COOH]}\right)$$

$$0.24 = \log\left(\frac{[CH_3COO^-]}{[CH_3COOH]}\right)$$

$$\left(\frac{[CH_3COO^-]}{[CH_3COOH]}\right) = \frac{1.7}{1}$$

7. At pH 7.5, the ratio of $[HPO_4{}^{2-}]/[H_2PO_4{}^-]$ is 2/1 ($pK_a$ of $H_2PO_4{}^- = 7.2$), as calculated using the Henderson-Hasselbalch equation. $K_2HPO_4$ is a source of the base form, and HCl must be added to convert one-third of it to the acid form, according to the 2/1 base/acid ratio. Weigh out 8.7 grams of $K_2HPO_4$ (0.05 moles, based on a formula weight of 174 grams/mole), dissolve in a small quantity of distilled water, add 16.7 mL of 1 M HCl (gives 1/3 of 0.05 moles of hydrogen ion, which converts 1/3 of the 0.05 moles of $HPO_4{}^{2-}$ to $H_2PO_4{}^-$) and dilute the resulting mixture to one liter.

8. A 2/1 ratio of the base form to acid form is still needed, because the pH of the buffer is the same in both problems. $NaH_2PO_4$ is a source of the acid form, and NaOH must be added to convert two thirds of it to the base form. Weigh out 6.0 grams of $NaH_2PO_4$ (0.05 moles, based on a formula weight of 120 grams/mole), dissolve in a small quantity of distilled water, add 33.3 mL of 1 M NaOH (gives 2/3 of 0.05 moles of hydroxide ion, which converts 2/3 of the 0.05 moles of $H_2PO_4{}^-$ to $HPO_4{}^{2-}$) and dilute the resulting mixture to one liter.

9. At the equivalence point of the titration a small amount of acetic acid remains because of the equilibrium $CH_3COOH \rightleftarrows H^+ + CH_3COO^-$. There is a small, but non-zero, amount of acetic acid left.

10. Buffering capacity refers to the amounts of the acid and base forms present in the buffer solution. A solution with a high buffering capacity can react with large amount of added acid or base without drastic changes in pH. A solution with a low buffer-

ing capacity can react with only comparatively small amounts of acid or base before showing changes in pH. The more concentrated the buffer, the higher is its buffering capacity. The first buffer listed here has 10 times less buffering capacity than the second, which in turn has 10 times less buffering capacity than the third. All three buffers have the same pH, since they all have the same relative amounts of the acid and base form.

11. The only zwitterion is $^+H_3N$—$CH_2$—$COO^-$.

12. The solution is a buffer because it contains equal concentrations of TRIS in the acid and free amine forms. When the two solutions are mixed, the concentrations of the resulting solution (in the absence of reaction) are 0.05 M HCl and 0.1 M TRIS because of dilution. The HCl reacts with half the TRIS present, giving 0.05 M TRIS (protonated form) and 0.05 M TRIS (free amine form).

13. $[H^+] = 7.9 \times 10^{-3}$ M

14. Use the Henderson-Hasselbalch equation. [Acetate ion]/[acetic acid] = 2.3/1

15. A substance with a $pK_a'$ of 3.9 has a buffer range of 2.9 to 4.9. It will not buffer effectively at pH 7.5.

16. Hypoventilation decreases the pH of blood.

17. Aspirin is electrically neutral at the pH of the stomach and can pass the membrane more easily there than in the small intestine.

# CHAPTER 4

1. The correct match of functional groups and compounds containing that functional group is given in the following list.

| Amino group | $CH_3CH_2NH_2$ |
|---|---|
| Carbonyl group (ketone) | $CH_3COCH_3$ |
| Hydroxyl group | $CH_3OH$ |
| Carboxyl group | $CH_3COOH$ |
| Carbonyl group (aldehyde) | $CH_3CH_2CHO$ |
| Thiol group | $CH_3SH$ |
| Ester linkage | $CH_3COOCH_2CH_3$ |
| Double bond | $CH_3CH{=}CHCH_3$ |
| Amide linkage | $CH_3CON(CH_3)_2$ |
| Ether | $CH_3CH_2OCH_2CH_3$ |

2. Identify the functional groups in the compounds shown below.

**glucose**

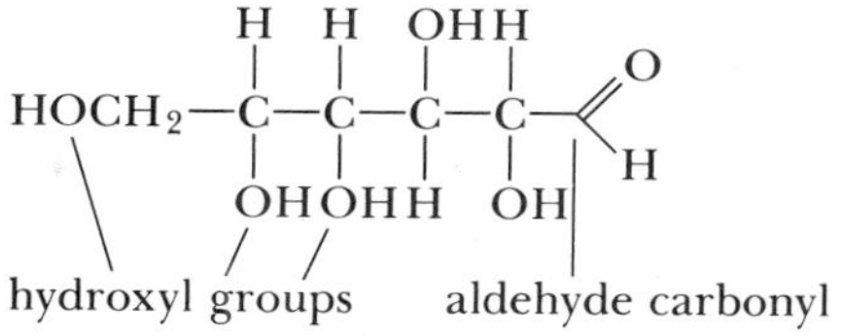

hydroxyl groups    aldehyde carbonyl

**a triglyceride**

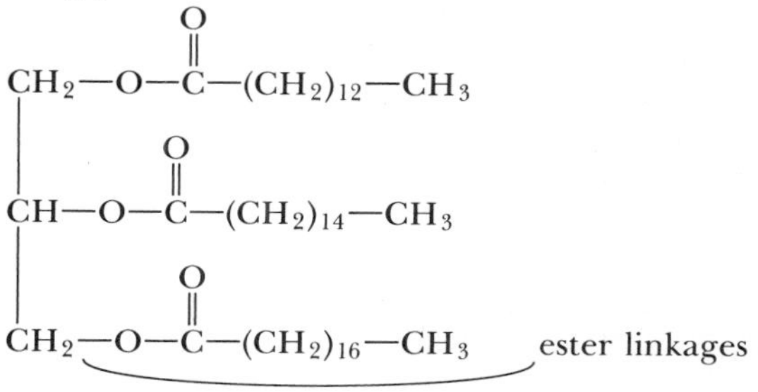

ester linkages

**a peptide**

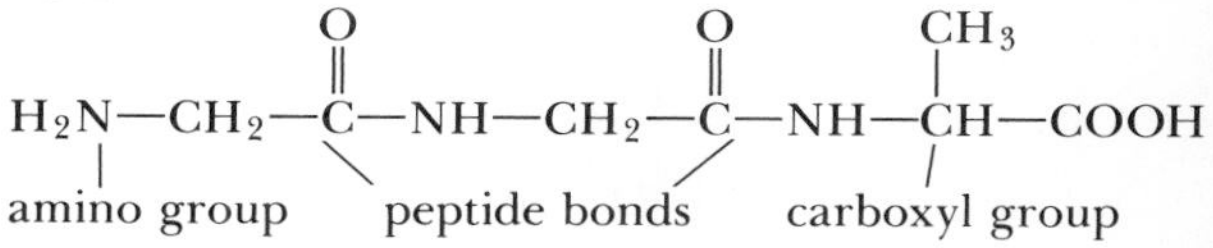

amino group    peptide bonds    carboxyl group

**vitamin A**

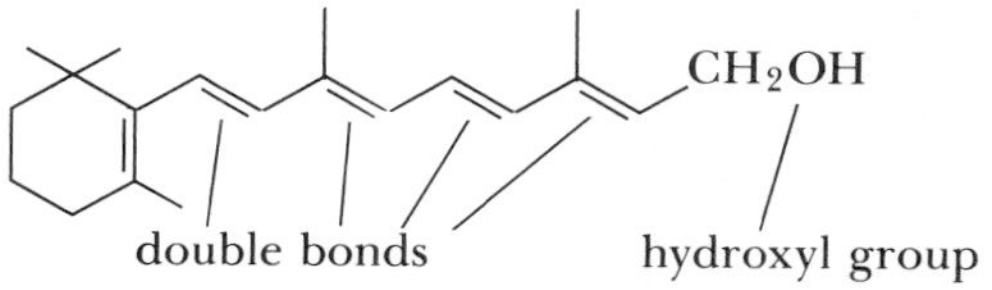

double bonds    hydroxyl group

3.    $RCONR_2' + H_2O \longrightarrow RCOOH + R_2'N$

Functional group: carboxyl group    amino group
Class of compound: carboxylic acid    amine

$RCOOR' + H_2O \longrightarrow RCOOH + R'OH$

Functional group: carboxyl group    hydroxyl group
Class of compound: carboxylic acid    alcohol

4. Urea, like all organic compounds, has the same molecular structure whether it is produced by a living organism or not.

**5.** Ester linkages are important in the formation of lipids.

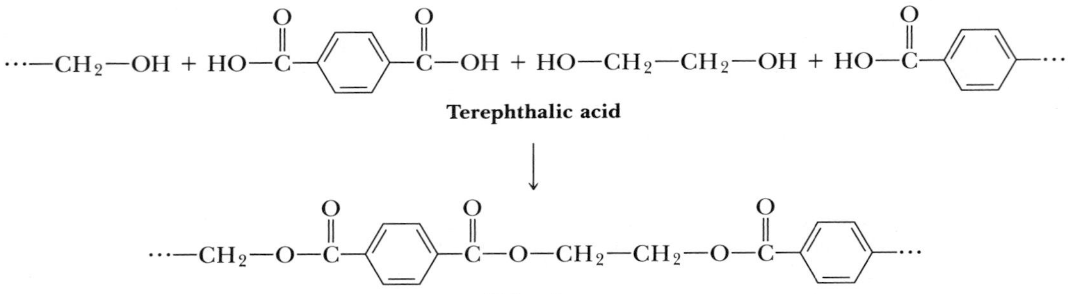

$$\cdots-CH_2-OH \;+\; HO-\overset{\overset{\displaystyle O}{\|}}{C}-\!\!\!\raisebox{-1ex}{⬡}\!\!\!-\overset{\overset{\displaystyle O}{\|}}{C}-OH \;+\; HO-CH_2-CH_2-OH \;+\; HO-\overset{\overset{\displaystyle O}{\|}}{C}-\!\!\!\raisebox{-1ex}{⬡}\!\!\!-\cdots$$

**Terephthalic acid**

↓

$$\cdots-CH_2-O-\overset{\overset{\displaystyle O}{\|}}{C}-\!\!\!\raisebox{-1ex}{⬡}\!\!\!-\overset{\overset{\displaystyle O}{\|}}{C}-O-CH_2-CH_2-O-\overset{\overset{\displaystyle O}{\|}}{C}-\!\!\!\raisebox{-1ex}{⬡}\!\!\!-\cdots$$

**Polyester**

**6.** Amide linkages are important in the formation of proteins.

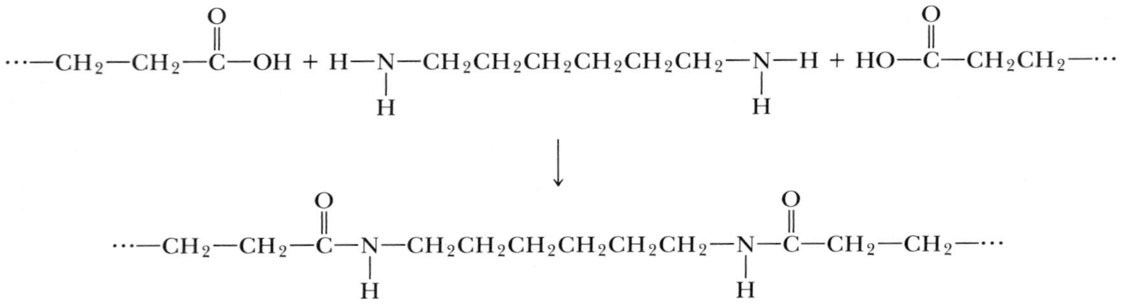

$$\cdots-CH_2-CH_2-\overset{\overset{\displaystyle O}{\|}}{C}-OH \;+\; H-\underset{\underset{\displaystyle H}{|}}{N}-CH_2CH_2CH_2CH_2CH_2CH_2-\underset{\underset{\displaystyle H}{|}}{N}-H \;+\; HO-\overset{\overset{\displaystyle O}{\|}}{C}-CH_2CH_2-\cdots$$

↓

$$\cdots-CH_2-CH_2-\overset{\overset{\displaystyle O}{\|}}{C}-\underset{\underset{\displaystyle H}{|}}{N}-CH_2CH_2CH_2CH_2CH_2CH_2-\underset{\underset{\displaystyle H}{|}}{N}-\overset{\overset{\displaystyle O}{\|}}{C}-CH_2-CH_2-\cdots$$

**7.** Ethyl acetate is an ester with the typical ester linkage

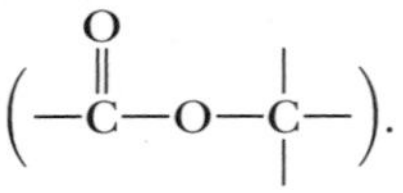

$$\left(-\overset{\overset{\displaystyle O}{\|}}{C}-O-\overset{|}{\underset{|}{C}}-\right).$$

**8.** Caffeine is a purine, as are two of the nucleobases, adenine and guanine.

**9.**

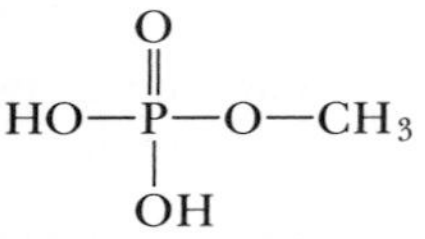

$$HO-\overset{\overset{\displaystyle O}{\|}}{\underset{\underset{\displaystyle OH}{|}}{P}}-O-CH_3 \qquad\qquad CH_3-\overset{\overset{\displaystyle O}{\|}}{C}-O-CH_3$$

**Phosphoric acid ester** 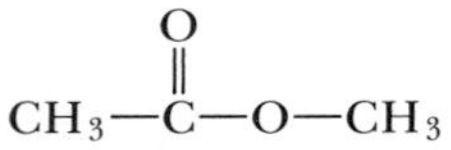        **Carboxylic acid ester**

**10.** $CH_3-CH_2OH \longrightarrow CH_3-\overset{\overset{\displaystyle O}{\|}}{C}-OH$

**Ethyl alcohol (in wine)  Acetic acid (in vinegar)**

## CHAPTER 5

**1.** For the ionic dissociation reactions of the following amino acids: aspartic acid, valine, histidine, serine and lysine.

**Aspartic acid**

$$\underset{\text{+1 net charge}}{\overset{\displaystyle \text{COOH}}{\underset{\displaystyle \text{COOH}}{\overset{\displaystyle |}{\underset{\displaystyle |}{H_3\overset{\oplus}{N}-\underset{\displaystyle CH_2}{C}-H}}}}} \underset{\substack{pK'_a \\ 2.09}}{\rightleftharpoons} \underset{\text{0 net charge}}{\overset{\displaystyle \text{COO}^{\ominus}}{\underset{\displaystyle \text{COOH}}{H_3\overset{\oplus}{N}-\underset{\displaystyle CH_2}{C}-H}}} \underset{\substack{pK'_a \\ 3.86}}{\rightleftharpoons} \underset{\text{−1 net charge}}{\overset{\displaystyle \text{COO}^{\ominus}}{\underset{\displaystyle \text{COO}^{\ominus}}{H_3\overset{\oplus}{N}-\underset{\displaystyle CH_2}{C}-H}}} \underset{\substack{pK'_a \\ 9.82}}{\rightleftharpoons} \underset{\text{−2 net charge}}{\overset{\displaystyle \text{COO}^{\ominus}}{\underset{\displaystyle \text{COO}^{\ominus}}{H_2N-\underset{\displaystyle CH_2}{C}-H}}}$$

**Valine**

$$\underset{\text{+1 net charge}}{\overset{\displaystyle \text{COOH}}{\underset{\displaystyle CH_3}{H_3\overset{\oplus}{N}-\underset{\displaystyle H_3C-C-H}{C}-H}}} \underset{\substack{pK'_a \\ 2.32}}{\rightleftharpoons} \underset{\text{0 net charge}}{\overset{\displaystyle \text{COO}^{\ominus}}{\underset{\displaystyle CH_3}{H_3\overset{\oplus}{N}-\underset{\displaystyle H_3C-C-H}{C}-H}}} \underset{\substack{pK'_a \\ 9.62}}{\rightleftharpoons} \underset{\text{−1 net charge}}{\overset{\displaystyle \text{COO}^{\ominus}}{\underset{\displaystyle CH_3}{H_2N-\underset{\displaystyle H_3C-C-H}{C}-H}}}$$

**Histidine**

+2 net charge $\underset{\substack{pK'_a \\ 1.83}}{\rightleftharpoons}$ +1 net charge $\underset{\substack{pK'_a \\ 6.0}}{\rightleftharpoons}$ 0 net charge $\underset{\substack{pK'_a \\ 9.2}}{\rightleftharpoons}$ −1 net charge

**Serine**

$$\underset{\text{+1 net charge}}{\overset{\displaystyle \text{COOH}}{\underset{\displaystyle CH_2OH}{H_3\overset{\oplus}{N}-C-H}}} \underset{\substack{pK'_a \\ 2.21}}{\rightleftharpoons} \underset{\text{0 net charge}}{\overset{\displaystyle \text{COO}^{\ominus}}{\underset{\displaystyle CH_2OH}{H_3\overset{\oplus}{N}-C-H}}} \underset{\substack{pK'_a \\ 9.15}}{\rightleftharpoons} \underset{\text{−1 net charge}}{\overset{\displaystyle \text{COO}^{\ominus}}{\underset{\displaystyle CH_2OH}{H_2N-C-H}}}$$

**Lysine**

+2 net charge $\underset{\substack{pK'_a \\ 2.18}}{\rightleftharpoons}$ +1 net charge $\underset{\substack{pK'_a \\ 8.95}}{\rightleftharpoons}$ 0 net charge $\underset{\substack{pK'_a \\ 10.53}}{\rightleftharpoons}$ −1 net charge

2. For the ionized form of the following amino acids at pH 7: glutamic acid, leucine, threonine, histidine and arginine.

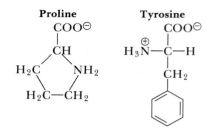

**Glutamic acid**

$COO^\ominus$
$H_3\overset{\oplus}{N}-C-H$
$CH_2$
$CH_2$
$COO^\ominus$

**Leucine**

$COO^\ominus$
$H_3\overset{\oplus}{N}-C-H$
$CH_2$
$CH$
$CH_3 \quad CH_3$

**Threonine**

$COO^\ominus$
$H_3\overset{\oplus}{N}-C-H$
$CHOH$
$CH_3$

pH7

**Proline**

$COO^\ominus$
$CH$
$H_2C \quad NH_2$
$H_2C-CH_2$

**Tyrosine**

$COO^\ominus$
$H_3\overset{\oplus}{N}-C-H$
$CH_2$

**Histidine**

$COO^\ominus$
$H_3\overset{\oplus}{N}-C-H$
$CH_2$
$N{=}\!\!{\diagup}\,NH$

**Arginine**

$COO^\ominus$
$H_3\overset{\oplus}{N}-C-H$
$(CH_2)_3$
$NH$
$C{=}\overset{\oplus}{N}H_2$
$NH_2$

**3.** The $pK_a'$ for the ionization of the thiol group of cysteine is 8.33, so this amino acid could serve as a buffer in the —SH and —S— forms over the pH range 7.33–9.33. The $\alpha$-amino groups of asparagine and lysine have $pK_a'$ values of 8.80 and 8.95 respectively; these are also possible buffers, but they are both near the end of their buffer ranges.

**4.** In the peptide, Val-Met-Ser-Ile-Phe-Arg-Cys-Tyr-Leu, the polar amino acids are Ser, Arg, Cys, and Tyr; the aromatic amino acids are Phe and Tyr; and the sulfur-containing amino acids are Met and Cys.

**5.** At pH 1 the charged groups are the N-terminal $NH_3^+$ on valine and the protonated guanidino group on arginine. The charged groups at pH 7 are the same as pH 1 with the addition of the carboxylate group on the C-terminal leucine.

**6.**
| | | |
|---|---|---|
| Ser-Leu-Phe | Leu-Ser-Phe | Phe-Ser-Leu |
| Ser-Phe-Leu | Leu-Phe-Ser | Phe-Leu-Ser |

**7.**

**Histidine**

$COO^\ominus$
$H_3\overset{\oplus}{N}-C-H$
$CH_2$
$\diagup NH$

**Asparagine**

$COO^\ominus$
$H_3\overset{\oplus}{N}-C-H$
$CH_2$
$O{=}C$

**Tryptophan**

$COO^\ominus$
$H_3\overset{\oplus}{N}-C-H$
$CH_2$
$NH$

pH4

**8.** Both peptides, Phe-Glu-Ser-Met and Val-Trp-Cys-Leu have a charge of +1 at pH 1 because of the protonated N-terminal amino group. At pH 7, the peptide on the right has no net charge because of the protonated N-terminal amino group and the ionized C-terminal carboxylate negative charge. The peptide on the left has a net charge of −1 at pH 7 because of the sidechain carboxyate group on the glutamate in addition to the charges on the N-terminal and C-terminal groups.

**9.** Cysteine will have: no net charge at pH 5.02 = $\dfrac{(1.71 + 8.33)}{2}$

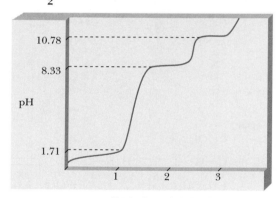

Equivalents OH⁻ added

**10.** The conjugate acid-base pair will act as a buffer in the pH range 1.09 to 3.09.

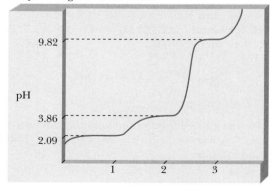

Equivalents OH⁻ added

11. The two peptides differ in amino acid sequence but not in composition. Consequently, they will have titration curves of the same shape. The p$K_a'$ values of the $\alpha$-amino and $\alpha$-carboxyl groups will differ.

12. Oxytocin has an isoleucine at position 3 and a leucine at position 8; it stimulates smooth muscle contraction in the uterus during labor and in the mammary glands during lactation. Vasopressin has a phenylalanine at position 3 and an arginine at position 8; it stimulates resorption of water by the kidneys, thus raising blood pressure.

13. The reduced form of glutathione consists of three amino acids with a sulfhydryl group; the oxidized form consists of six amino acids and can be considered the result of linking two molecules of reduced glutathione by a disulfide bridge.

14. Gramicidin S is an antibiotic; its sequence is L-Val-L-Orn-L-Leu-D-Phe-L-Pro-L-Val-D-Orn-L-Leu-L-Phe-L-Pro.

15. The different stereochemistry of the two peptides leads to different binding with taste receptors and to the sweet taste for one and the bitter taste for the other.

16. See Figure 5.5 for the structures of modified amino acids. Hydroxyproline and hydroxylysine are found in collagen; and thyroxine is found in thyroglobulin.

17. See Figure 5.9. The resonance structures contribute to the planar arrangement by giving the C—N bond partial double bond character.

## CHAPTER 6

1. Ester linkage

   Repeating disaccharide of pectin (see next column)

4. This polymer would be expected to have a structural role. The presence of the $\beta$-glycosidic linkage makes it useful as food only to animals such as termites or to ruminants such as cows and horses; these animals harbor bacteria capable of attacking the $\beta$-linkage in their digestive tracts.

**galacturonic acid ($\alpha$ form)**

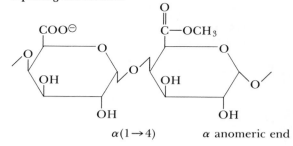

unmethylated          methylated

**repeating disaccharide**

$\alpha(1\rightarrow4)$          $\alpha$ anomeric end

2. **structure of gentibiose**

$\beta(1\rightarrow6)$

3. To 2500. One place 0.02% To 1000 Four places 0.08% To 200 24 places 0.48%.

$\alpha(1\rightarrow4)$          $\beta(1\rightarrow4)$          $\alpha(1\rightarrow4)$          $\beta(1\rightarrow4)$          $\alpha(1\rightarrow4)$

**5.** A polysaccharide is a polymer of simple sugars, which are compounds that contain a single carbonyl group and several hydroxyl groups. A furanose is a cyclic sugar that contains a five-membered ring similar to that in furan.

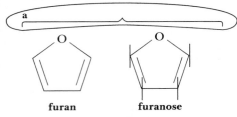

A pyranose is a cyclic sugar that contains a six-membered ring similar to that in pyran.

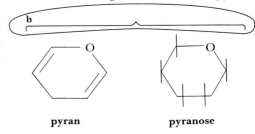

An aldose is a sugar that contains an aldehyde group; a ketose is a sugar that contains a ketone group. A glycosidic bond is the acetal linkage that joins two sugars. An oligosaccharide is a compound formed by the linking of several simple sugars (monosaccharides) by glycosidic bonds. A glycoprotein is formed by the covalent bonding of sugars to a protein.

**6.** D-mannose and D-galactose are both epimers of D-glucose with inversion of configuration around carbon atoms 2 and 4, respectively; D-ribose has only five carbons, while the rest of the sugars named in this question have six.

**7.** All groups are aldose-ketose pairs.

$$CH_2OH$$
$$|$$
$$C=O$$
$$|$$
$$H-C-OH \quad \textbf{D-ribulose}$$
$$|$$
$$H-C-OH$$
$$|$$
$$CH_2OH$$

**8.** In some cases the enzyme that degrades lactose (milk sugar) to its components, glucose and galactose, is missing. In other cases, the enzyme is the one which isomerizes galactose to glucose for further metabolic breakdown.

**9.** Enantiomers: a and f, b and d. Epimers: a and c, a and d, a and e, b and f.

Five carbon sugars

(a)
$$CHO$$
$$|$$
$$H-C-OH$$
$$|$$
$$H-C-OH$$
$$|$$
$$H-C-OH$$
$$|$$
$$CH_2OH$$

(b)
$$CHO$$
$$|$$
$$H-C-OH$$
$$|$$
$$HO-C-H$$
$$|$$
$$HO-C-H$$
$$|$$
$$CH_2OH$$

(c)
$$CHO$$
$$|$$
$$H-C-OH$$
$$|$$
$$H-C-OH$$
$$|$$
$$HO-C-H$$
$$|$$
$$CH_2OH$$

(d)
$$CHO$$
$$|$$
$$HO-C-H$$
$$|$$
$$H-C-OH$$
$$|$$
$$H-C-OH$$
$$|$$
$$CH_2OH$$

(e)
$$CHO$$
$$|$$
$$H-C-OH$$
$$|$$
$$HO-C-H$$
$$|$$
$$H-C-OH$$
$$|$$
$$CH_2OH$$

(f)
$$CHO$$
$$|$$
$$HO-C-H$$
$$|$$
$$HO-C-H$$
$$|$$
$$HO-C-H$$
$$|$$
$$CH_2OH$$

**10.**

(a)
$$\beta(1\rightarrow4)$$

**(b)**

$\alpha,\alpha(1\rightarrow)$ =

CHAPTER 7

1. In both types of lipids glycerol is esterified to carboxylic acids, with three such ester linkages formed in triacylglycerols and two in phosphatidyl ethanolamines. The structural difference comes in the nature of the third ester linkage to glycerol. In phosphatidyl ethanolamines, the third hydroxyl group of glycerol is esterified not to a carboxylic acid, but to phosphoric acid. The phosphoric acid moiety is esterified in turn to ethanolamine. (See Figures 7.2 and 7.5.)

**(c)**

$\beta(1\rightarrow 6)$

2. Both sphingomyelins and phosphatidylcholines contain phosphoric acid esterified to an amino alcohol, which must be choline in the case of a phosphatidyl choline and may be choline in the case of a sphingomyelin. They differ in the second alcohol to which phosphoric acid is esterified. In phosphatidylcholines the second alcohol is glycerol, which has also formed ester bonds to two carboxylic acids. In sphingomyelins the second alcohol is another amino alcohol, sphingosine, which has formed an amide bond to a fatty acid. (See Figure 7.6.)

3. Triacylglycerols are not found in animal membranes.

11. The cell walls of plants consist mainly of cellulose, while those of bacteria consist mainly of polysaccharides with peptide crosslinks.

12. Chitin is a polymer of $N$-acetyl-$\beta$-D-glucosamine, while cellulose is a polymer of D-glucose.

13. Glycogen and starch differ mainly in the degree of chain branching.

14. The enzyme $\beta$-amylase is an exoglycosidase, degrading polysaccharides from the ends. The enzyme $\alpha$-amylase is an endoglycosidase, cleaving internal glycosidic bonds.

15. The sugar portions of the blood group glycoproteins are the source of the antigenic difference.

4.

$$CH_2-O-\overset{\overset{\displaystyle O}{\|}}{C}-(CH_2)_7CH=CH-(CH_2)_7CH_3 \quad \textbf{oleic acid moiety}$$

**glycerol moiety**

$$CH_2-O-\overset{\overset{\displaystyle O}{\|}}{C}-(CH_2)_{16}CH_3 \quad \textbf{stearic acid moiety}$$

$$CH_2-O-\overset{\overset{\displaystyle O}{\|}}{\underset{\underset{\displaystyle O^{\ominus}}{|}}{P}}-O-(CH_2)_2-\overset{\oplus}{N}\overset{\diagup CH_3}{\underset{\diagdown CH_3}{\big|_{CH_3}}} \quad \textbf{choline moiety}$$

5. This lipid is a ceramide, which is one kind of sphingolipid.

**6.**

$$CH_2-O-\overset{\overset{\displaystyle O}{\|}}{C}-(CH_2)_{14}CH_3 \quad \textbf{palmitic acid moiety}$$

glycerol moiety $\quad CH-O-\overset{\overset{\displaystyle O}{\|}}{C}-(CH_2)_7CH=CH-CH_2-CH=CH(CH_2)_4CH_3 \quad \textbf{linoleic acid moiety}$

$$CH_2-O-\overset{\overset{\displaystyle O}{\|}}{C}-(CH_2)_7(CH=CHCH_2)_3CH_3 \quad \textbf{linolenic acid moiety}$$

Any combination of fatty acids is possible.

**7.**

$$CH_2-O-\overset{\overset{\displaystyle O}{\|}}{C}-(CH_2)_{14}CH_3$$

$$CH-O-\overset{\overset{\displaystyle O}{\|}}{C}-(CH_2)_7CH=CH-CH_2-CH=CH-(CH_2)_4CH_3$$

$$CH_2-O-\overset{\overset{\displaystyle O}{\|}}{C}-(CH_2)_7-(CH=CH-CH_2)_3CH_3$$

$$\downarrow \quad \begin{array}{l}\textbf{aqueous}\\ \textbf{NaOH}\end{array}$$

$$CH_2OH \qquad CH_3-(CH_2)_{14}-\overset{\overset{\displaystyle O}{\|}}{C}-O^{\ominus}Na^{\oplus}$$

$$CHOH \quad + \quad CH_3-(CH_2)_4-CH=CH-CH_2-CH=CH-(CH_2)_7-\overset{\overset{\displaystyle O}{\|}}{C}-O^{\ominus}Na^{\oplus}$$

$$CH_2OH \qquad CH_3(CH_2-CH=CH)-(CH_2)_7-\overset{\overset{\displaystyle O}{\|}}{C}-O^{\ominus}Na^{\oplus}$$

8. Myelin is a multilayer sheath consisting mainly of lipids (with some proteins) that insulates the axons of nerve cells, facilitating transmission of nerve impulses.

9. Steroids contain a characteristic fused-ring structure, which other lipids do not.

10. The *cis-trans* isomerization of retinal in rhodopsin triggers the transmission of an impulse to the optic nerve and is the primary photochemical event in vision.

11. Lipid-soluble vitamins accumulate in fatty tissue, leading to toxic effects. Water-soluble vitamins are excreted, drastically reducing the chances of an overdose.

12. Cholesterol is a precursor of Vitamin $D_3$; the conversion reaction involves ring opening.

13. Vitamin E is an antioxidant.

14. Prostglandins and leukotrienes are derived from arachidonic acid. They play a role in inflammation and in allergy and asthma attacks.

# CHAPTER 8

1. The system is the nonpolar solute and water, which become more disordered when a solution is formed; $\Delta S_{sys}$ is positive but comparatively small. The $\Delta S_{surr}$ is negative and comparatively large, since it is a reflection of the unfavorable enthalpy change for forming the solution ($\Delta H_{sys}$). Since $\Delta S_{univ} = \Delta S_{sys} + \Delta S_{surr}$, $\Delta S_{univ}$ is negative and does not favor the dissolving of nonpolar solutes in water.

2. Processes (a) and (b) are spontaneous, while processes (c) and (d) are not. The spontaneous processes represent an increase in disorder (increase in the entropy of the universe) and have a negative $\Delta G°$ at constant temperature and pressure, while the opposite is true of the nonspontaneous processes.

3. In all cases there is an increase in entropy. In all cases the final state has more possible random arrangements than the initial state.

4. The first statement is true, but the second is not.

The standard state of solutes is normally defined as unit activity (1 M for all but the most careful work). In biological systems the pH is frequently in the neutral range (i.e. $H^+$ is close to $10^{-7}$ M); the modification is a matter of convenience. Water is the solvent, not a solute, and its standard state is the pure liquid.

**5.** $\Delta G^{\circ\prime} = -2.303\ RT\ \log K'_{eq}$

$K'_{eq} = 10^{-(\Delta G^{\circ\prime}/2.303\ RT)}$

$K'_{eq} = 10$ $-(-43.0\ \text{kJ mol}^{-1}\ (1000\ \text{J/kJ})/((2.303)$

$(8.31\ \text{J mol}^{-1}\ K^{-1})(310\ \text{K})))$

$K'_{eq} = 1.77 \times 10^7$

**6.** $\Delta G^{\circ\prime} = \Delta H^{\circ\prime} - T\Delta S^{\circ\prime}$

$\Delta S^{\circ\prime} = 34.9\ \text{J mol}^{-1}\ K^{-1} = 8.39\ \text{cal mol}^{-1}\ K^{-1}$

There are two particles on the reactant side of the equation and three on the product side, representing an increase in disorder.

**7.** Statements (a), (c) and (d) are correct. (a) The unfavorable entropy change for the water is reflected in the unfavorable heat of solution for nonpolar solutes in water. (c) This statement is a way of defining the enthalpy change. (d) Heat is a less useful form of energy than others, resulting from the degradation of ordered molecular motion into disordered motion. Statements (b) and (e) are incorrect. (b) The entropy of the universe reflects its randomness, which increases in spontaneous processes. (e) An endergonic reaction is one in which energy is taken up rather than given off, as is the case in spontaneous processes.

**8.** Nonpolar residues tend to be found in the interior of proteins because of hydrophobic interactions. Polar residues tend to be found on the exterior because of dipolar interactions with solvent water.

**9.** $K_{eq} = 0.90$

**10.** The local decrease in entropy associated with living organisms is balanced by the larger increase in the entropy of the surroundings caused by their presence.

**11.** Hydrophobic interactions stabilize the nonpolar interior of membranes, with the exterior charged and polar groups in contact with water.

**12.** The biosynthesis of proteins is endergonic and is accompanied by a large decrease in entropy.

**13.** The ATP constantly generated by living organisms is used as a source of chemical energy for endergonic processes. There is a good deal of turnover of molecules, but no net change.

**14.** There is a large increase in entropy accompanying the hydrolysis of one molecule to five separate molecules.

## CHAPTER 9

**1.** (A)(2); (B)(4); (C)(1); (D)(3)

**2.**

$$\text{(phenyl)}-\text{NH}-\overset{\displaystyle N}{\underset{\displaystyle S-C}{C}}\!\!\text{CH}-\text{CH}_2-\text{CH}\overset{\text{CH}_3}{\underset{\text{CH}_3}{}} + H_3\overset{\oplus}{N}\text{—rest of peptide}$$

thiazoline derivative
of leucine

aqueous
acid

$$S=C\cdots N-C\cdots CH-CH_2-CH\overset{\text{CH}_3}{\underset{\text{CH}_3}{}}$$

phenylthiohydantoin
derivative of leucine

3. Val-Leu-Gly-Met-Ser-Arg-Asn-Thr-Trp-Met-Ile-Lys-Gly-Tyr-Met-Gln-Phe

4. The sequence is Met-Val-Ser-Thr-Lys-Leu-Phe-Asn-Glu-Ser-Arg-Val-Ile-Trp-Thr-Leu-Met-Ile

5. (1) Backbone H-bonds, involving the CO and NH groups of the peptide chain (2) Side-chain H-bonds, involving any possible hydrogen bond donor or acceptors on the side chains (3) Hydrophobic interactions, involving the nonpolar groups on the protein (4) Electrostatic interactions, involving any charged groups on the protein (5) Metal ligation, involving coordination bonds between side chains and a metal ion.

6. When a protein is denatured, the interactions that determine secondary, tertiary and any quaternary structure are overcome by the presence of the denaturing agent. Only the primary structure remains intact.

7. Similarities—both contain heme group, oxygen binding, secondary structure primarily $\alpha$-helix. Differences—hemoglobin is a tetramer, while myoglobin is a monomer—oxygen binding to hemoglobin is cooperative, non-cooperative to myoglobin.

8. The function of hemoglobin is oxygen transport; its sigmoidal binding curve reflects the fact that it can bind easily to oxygen at comparatively high pressures and release oxygen at lower pressures. The function of myoglobin is oxygen storage, and, as a result it is easily saturated with oxygen at low pressures, as shown by its hyperbolic binding curve.

9. Deoxygenated hemoglobin is a weaker acid (has a higher $pK_a'$) than oxygenated hemoglobin. In other words, deoxygenated hemoglobin binds more strongly to $H^+$ than does oxygenated hemoglobin. The binding of $H^+$ (and of $CO_2$) to hemoglobin favors the change in quaternary structure to the deoxygenated form of hemoglobin.

10. When a protein is covalently modified, its primary structure is changed. The primary structure determines the final three-dimensional structure of the protein. The modification disrupts the folding process.

11. The $\alpha$-helix is not fully extended, and its hydrogen bonds are parallel to the protein fiber. The $\beta$-pleated sheet structure is almost fully extended, and its hydrogen bonds are perpendicular to the protein fiber.

12. The $\alpha\alpha$ unit, the $\beta\alpha\beta$ unit, the $\beta$ meander, the Greek key, the $\beta$ barrel.

13. (a) Serine has a small side chain that can fit in any relatively polar environment. (b) Tryptophan has the largest side chain of any of the common amino acids, and it tends to require a nonpolar environment. (c) Lysine and arginine are both basic amino acids; exchanging one for the other would not affect the side chain $pK_a$ in a significant way. Similar reasoning applies to the substitution of a nonpolar isoleucine for a nonpolar leucine.

14. Persons with sickle-cell trait have some abnormal hemoglobin, impairing their capacity to transport oxygen in the bloodstream. At high altitudes, there is less oxygen, and the decreased efficiency becomes more apparent.

15. In the presence of $H^+$ and $CO_2$, both of which bind

to hemoglobin, the oxygen-binding capacity of hemoglobin decreases.

16. The geometry of the proline residue is such that it does not fit into the $\alpha$-helix, but it does fit exactly for a reverse turn. See Figure 9.12, part c.

17. In the absence of 2,3-*bis*phosphoglycerate, the binding of oxygen by hemoglobin resembles that of myoglobin, characterized by lack of cooperativity. 2,3-*Bis*phosphoglycerate binds at the center of the hemoglobin molecule, increases cooperativity, and modulates the binding of oxygen so that it can easily be released in the capillaries.

18. Fetal hemoglobin binds oxygen more strongly than adult hemoglobin.

19. Hb S (sickle-cell anemia), Glu A3(6)$\beta$ $\longrightarrow$ Val; Hb E, Glu B8(26)$\beta$ $\longrightarrow$ Lys; Hb Savannah, Gly B6(24)$\beta$ $\longrightarrow$ Val; Hb Bibba, Leu H19(136)$\alpha$ $\longrightarrow$ Pro; HbM Iwate, His F8(87)$\alpha$ $\longrightarrow$ Tyr; HbM Milwaukee, Val E11(67)$\beta$ $\longrightarrow$ Glu.

20. In fetal hemoglobin, the subunit composition is $\alpha_2\gamma_2$, with replacement of the $\beta$ chains by the $\gamma$ chains. The sickle-cell mutation affects the $\beta$ chain, so the fetus homozygous for HbS has normal fetal hemoglobin.

21. Blood changes color from red to brown when the iron ions in it are oxidized from Fe(II) to Fe(III) in air. In methemoglobin (an abnormal hemoglobin) the iron is already Fe(III).

# INTERCHAPTER A

1. Glutamic acid will be eluted from the column first. It will be necessary to raise the pH to elute lysine from the column.

2. Phenylalanine will have the largest $R_f$ value; glutamic acid will have the smallest.

3. Glutamic acid will move fastest, while phenylalanine will move slowest, the reverse of the situation in Exercise 2.

4. Small particles, such as the ammonium and sulfate ions, enter the pores in the molecular sieve material, while large molecules, such as the protein, do not. The protein is eluted from the molecular sieve column before the salts.

5. The protein consists of two polypeptide chains, with alanine as the N-terminal amino acid of one and methionine as the N-terminal amino acid of the other.

6. The fingerprinting technique, consisting of a combination of paper chromatography and electrophoresis was used to detect the difference between normal hemoglobin and sickle-cell anemia hemoglobin.

7. Molecular weights of newly isolated proteins can be estimated on a molecular sieve column by comparing elution volumes of known proteins with those of the unknown.

# CHAPTER 10

1. The reaction of glucose with oxygen is thermodynamically favored, as shown by the negative free energy change. The fact that glucose can be maintained in an oxygen atmosphere is a reflection of the kinetic aspects of the reaction, requiring overcoming an activation-energy barrier.

2. The reaction is first order with respect to A, first order with respect to B, and second order overall. The detailed mechanism of the reaction is likely to involve one molecule each of A and B.

3. In the lock-and-key model the substrate fits into a comparatively rigid protein that has an active site with a well-defined shape. In the induced-fit model the enzyme undergoes a conformational change on binding to substrate. The active site takes shape around the substrate.

4. See Figures 10.4 and 10.5.

5. The steady state assumption is that the concentration of the enzyme-substrate complex does not change appreciably with time. The rate of appearance of the complex is equal to its rate of disappearance, simplifying the equations for enzyme kinetics.

6. Use equation 10.18. (a) $V = 0.5\ V_{max}$ (b) $V = 0.33\ V_{max}$ (c) $V = 0.09\ V_{max}$ (d) $V = 0.67\ V_{max}$ (e) $V = 0.91\ V_{max}$.

7. Turnover number $= V_{max}/[E]_0$

8. In the case of competitive inhibition, the value of $K_M$ increases, while the value of $K_M$ remains unchanged in noncompetitive inhibition.

9.

$K_M = 7.42$ mM; $V_{max} = 15.9$ mmol min$^{-1}$; noncompetitive inhibition

10. A competitive inhibitor binds to the active site of an enzyme, preventing binding of the substrate. A noncompetitive inhibitor binds at a site different from the active site, causing a conformational change which renders the active site less able to bind substrate and convert it to product.

11. The graph of rate against substrate concentration is sigmoidal for an allosteric enzyme but hyperbolic for an enzyme that obeys the Michaelis-Menten equation. Allosteric enzymes have multi-subunit structures. Michaelis-Menten enzymes may be multi-subunit or a single polypeptide chain.

12. In the concerted model, a conformational change caused by binding of an allosteric effector takes place simultaneously in all subunits. In the sequential model, a conformational change takes place in one subunit and is subsequently passed on to the other subunits.

13. Trypsin, chymotrypsin, fibrin.

14. See Table 10.2.

15. False. The mechanisms of enzymic catalysis are the same as those encountered in organic chemistry, operating in a complex environment.

16. The results do not prove that the mechanism is correct, since results from different experiments could contradict the proposed mechanism. In that case, the mechanism would have to be modified to accommodate the new experimental results.

17. Metal ions can provide a "steering effect" by forming coordination bonds of specified geometry both to the substrate and to enzyme side chains. Metal ions can also be an aid to catalysis by acting as Lewis acids.

18. In the first step of the reaction the serine hydroxyl is the nucleophile that attacks the substrate peptide bond. In the second step, water is the nucleophile that attacks the acyl-enzyme intermediate.

19. As a result of cleavage of the peptide bond, isoleucine 16 now has a free protonated amino group that forms an ionic bond with the carboxylate of aspartate 194. This linkage stabilizes the active form of the enzyme.

20. Instead of a phenylalanine moiety (similar to the usual substrates of chymotrypsin), use a nitrogen-containing basic group similar to the usual substrates of trypsin.

21. The easiest way to follow the rate of this reaction is to monitor the decrease in absorbance at 340 nm, reflecting the disappearance of NADH.

22. In this case methanol is acting as a competitive inhibitor. The administration of ethanol (the usual substrate of this enzyme) takes advantage of the fact that competitive inhibition can be overcome by a high enough substrate concentration.

## CHAPTER 11

1. The transition temperature is lower in a lipid bilayer with mostly unsaturated fatty acids compared to one with a high percentage of saturated fatty acids. The bilayer with the unsaturated fatty acids is already more disordered than the one with a high percentage of saturated fatty acids.

2. In a 100-gram sample of membrane, there are 50 grams of protein and 50 of phosphoglycerides.

$$50 \text{ g lipid} \times \frac{1 \text{ mole lipid}}{800 \text{ g lipid}} = 0.0625 \text{ mole lipid}$$

$$50 \text{ g protein} \times \frac{1 \text{ mole protein}}{50,000 \text{ g protein}} = 0.001 \text{ mole protein}$$

The molar ratio of lipid to protein is 0.0625/0.001 or 62.5/1.

3. Statements (c) and (d) are consistent with what is known about membranes. If there is any covalent bonding between lipids and proteins (statement (e)), it is rare. Proteins "float" in the lipid bilayers rather than being sandwiched between them (statement (a)). Bulkier molecules tend to be found in the outer lipid layer (statement (b)).

4. Biological membranes are highly nonpolar environments. Charged ions tend to be excluded from such environments rather than dissolving in them, as they would have to do to pass the membrane by simple diffusion.

5. Statements (c) and (d) are correct. Transverse diffusion is normally not observed (statement (b)), and the term "mosaic" refers to the pattern of distribution of proteins in the lipid bilayer (statement (e)). Peripheral proteins are also considered part of the membrane (statement (a)).

6. Statements (a) and (c) are correct; statement (b) is not correct since ions and larger molecules, especially polar ones, require channel proteins.

7. At the lower temperature, the membrane would tend to be less fluid. The presence of more unsaturated fatty acids would tend to compensate by increasing the fluidity of the membrane compared to one at the same temperature with a higher proportion of saturated fatty acids.

8. The binding site of the LDL receptor recognizes the protein portion of the LDL particle, specifically a protein exposed to the aqueous environment of the bloodstream. The LDL protein is likely to contain polar amino acids on its surface, as is the active site of the receptor.

9. Hydrophobic interactions among the hydrocarbon tails are the main thermodynamic driving force in the formation of lipid bilayers.

10. The higher percentage of unsaturated fatty acids in membranes in cold climates is an aid to membrane fluidity.
11. The waxy surface coating is a barrier that prevents loss of water.
12. The lecithin in the egg yolks serves as an emulsifying agent by forming closed vesicles. The lipids in the butter (frequently triacylglycerols) are retained in the vesicles and do not form a separate phase.

## CHAPTER 12

1. The second half reaction (the one involving NADH) is that of oxidation, while the first half reaction (the one involving $O_2$) is that of reduction. The overall reaction is

   $$1/2\ O_2 + NADH + H^+ \longrightarrow H_2O + NAD^+$$

   $O_2$ is the oxidizing agent and NADH is the reducing reagent.
2. Reaction (a) will not proceed as written; $\Delta G^{\circ\prime} = +12.6$ kJ. Reaction (b) will proceed as written; $\Delta G^{\circ\prime} = -20.8$ kJ. Reaction (c) will not proceed as written; $\Delta G^{\circ\prime} = +31.4$ kJ. Reaction (d) will proceed as written; $\Delta G^{\circ\prime} = -18.0$ kJ.
3. Sprints and similar short periods of exercise rely on anaerobic metabolism as a source of energy, producing lactic acid. Longer periods of exercise draw on aerobic metabolism as well.
4. Creatine phosphate + ADP $\longrightarrow$ creatine + ATP; $\Delta G^{\circ\prime} = -12.6$ kJ

   ATP + glycerol $\longrightarrow$ ADP + glycerol 3-phosphate; $\Delta G^{\circ\prime} = -20.8$ kJ

   Creatine phosphate + glycerol $\longrightarrow$
   creatine + glycerol 3-phosphate $\Delta G^{\circ\prime}$ overall $= -33.4$ kJ
5. Glucose 1-phosphate $\longrightarrow$ glucose + $P_i$; $\Delta G^{\circ\prime} = -20.9$ kJ mol$^{-1}$

   Glucose + $P_i$ $\longrightarrow$ glucose 6-phosphate; $\Delta G^{\circ\prime} = +12.5$ kJ mol$^{-1}$

   Glucose 1-phosphate $\longrightarrow$
   glucose 6-phosphate; $\Delta G^{\circ\prime} = -8.4$ kJ mol$^{-1}$
6. In both pathways the overall reaction is

   ATP + 2 $H_2O$ $\longrightarrow$ AMP + 2 $P_i$.

   Thermodynamic parameters such as energy are additive. The overall energy is the same, since the overall pathway is the same.

7. Phosphoarginine + ADP $\longrightarrow$ arginine + ATP; $\Delta G^{\circ\prime} = -1.7$ kJ

   ATP + $H_2O$ $\longrightarrow$ ADP + $P_i$; $\Delta G^{\circ\prime} = -30.5$ kJ

   Phosphoarginine + $H_2O$ $\longrightarrow$ arginine + $P_i$; $\Delta G^{\circ\prime} = -32.2$ kJ
8. Glucose 6-phosphate is oxidized, and $NADP^+$ is reduced. $NADP^+$ is the oxidizing agent, and glucose 6-phosphate is the reducing agent.
9. FAD is reduced, and succinate is oxidized. FAD is the oxidizing agent, and succinate is the reducing agent.
10. $NAD^+$, $NADP^+$, and FAD all contain an ADP moiety.

## CHAPTER 13

1. The bubbles in beer are $CO_2$, produced by alcoholic fermentation. Tired and aching muscles are caused by a buildup of lactic acid, a product of anaerobic glycolysis.
2. Reactions that require ATP: Phosphorylation of glucose to give glucose 6-phosphate, phosphorylation of fructose 6-phosphate to give fructose 1,6-diphosphate, carboxylation of pyruvate to give oxaloacetate.
   Reactions that produce ATP: Transfer of phosphate group from 1,3-diphosphoglycerate to ADP, transfer of phosphate group from phosphoenolpyruvate to ADP.
   Enzymes that catalyze reactions requiring ATP: Hexokinase, glucokinase, phosphofructokinase, pyruvate carboxylase.
   Enzymes that catalyze reactions producing ATP: Phosphoglycerate kinase, pyruvate kinase.
3. Reactions that require NADH: Reduction of pyruvate to lactate, reduction of acetaldehyde to ethanol.
   Reactions that require $NAD^+$: Oxidation of glyceraldehyde 3-phosphate to give 1,3-diphosphoglycerate.
   Enzymes that catalyze reactions requiring NADH: Lactate dehydrogenase, alcohol dehydrogenase.
   Enzymes that catalyze reactions requiring $NAD^+$: Glyceraldehyde 3-phosphate dehydrogenase.
4. NADPH has one more phosphate group than NADH (at the 2′ position of the ribose ring of the adenine nucleotide portion of the molecule). NADH is produced in oxidative reactions that give rise to ATP. NADPH is a reducing agent in biosynthesis. Different enzymes use NADH as a coenzyme compared to those that require NADPH.

5. The rate of a reaction depends on the concentration of reactant. The hexokinase reaction controls the concentration of glucose 6-phosphate, the starting material for glycolysis, and thus controls the rate.

6. The rate-determining step of glycolysis refers to the kinetics of the overall reaction. The term "committed step" refers to the fact that, once fructose 1,6-diphosphate is formed, it must undergo the reactions of glycolysis. The earlier components can play a part in other pathways, but not fructose 1,6-diphosphate.

7. Lactate dehydrogenase, alcohol dehydrogenase, glyceraldehyde 3-phosphate dehydrogenase.

8. Acetyl-CoA is an allosteric effector (activator) of pyruvate carboxylase. This enzyme catalyzes the conversion of pyruvate to oxaloacetate, the first step in gluconeogenesis. Biotin is the carrier of carbon dioxide in the carboxylation of pyruvate to oxaloacetate.

9. Aldolase catalyzes the reverse aldol condensation of fructose 1,6-diphosphate to glyceraldehyde 3-phosphate and dihydroxyacetone phosphate.

10. Hemolytic anemia is caused by defective working of the pentose phosphate pathway. There is a deficiency of NADPH, which indirectly contributes to the integrity of the red blood cells. The pentose phosphate pathway is the only source of NADPH in red blood cells.

11. The energy released by all the reactions of glycolysis is 184.5 kJ mol glucose$^{-1}$. The energy released by glycolysis drives the phosphorylation of two ADP to ATP for each molecule of glucose, trapping 61.0 kJ mol glucose$^{-1}$. The estimate of 33 percent efficiency comes from the calculation $(61.0/184.5) \times 100 = 33$ percent.

12. Three reactions of glycolysis are irreversible under physiological conditions. They are the production of pyruvate and ATP from phosphoenolpyruvate, the production of fructose 1,6-diphosphate from fructose 6-phosphate and the production of glucose 6-phosphate from glucose. These are the reactions that are bypassed in gluconeogenesis; the reactions of gluconeogenesis differ from those of glycolysis at these points and are catalyzed by different enzymes.

13. Add the $\Delta G^{\circ\prime}$ mol$^{-1}$ values for the reactions from glucose to glyceraldehyde 3-phosphate. The result is 2.5 kJ mol$^{-1}$ = 0.6 kcal mol$^{-1}$.

14. Phosphoenolpyruvate $\longrightarrow$ pyruvate + P$_i$
$$\Delta G^{\circ\prime} = -61.9 \text{ kJ mol}^{-1} = -14.8 \text{ kcal mol}^{-1}$$

ADP + P$_i$ $\longrightarrow$ ATP
$$\Delta G^{\circ\prime} = 30.5 \text{ kJ mol}^{-1} = 7.3 \text{ kcal mol}^{-1}$$

Phosphoenolpyruvate + ADP $\longrightarrow$ pyruvate + ATP
$$\Delta G^{\circ\prime} = -31.4 \text{ kJ mol}^{-1} = -7.5 \text{ kcal mol}^{-1}$$

15.

(a) $CH_3-\overset{\text{:C:}}{\underset{}{C}}-COO^-$ + NADH + H$^+$ $\longrightarrow$ lactate + NAD$^+$

(b) glucose 6-phosphate + NADP$^+$ $\longrightarrow$

A-16

**6-phosphoglucono lactone** + **NADPH** + $H^+$

(c) $CH_3-C::\ddot{O}:$ + NADH + $H^+$ $\longrightarrow$ $CH_3-C:\ddot{O}:H$ + $NAD^+$

**acetaldehyde** (see Exercise 13.15a) **ethanol**

(d) $H:C::\ddot{O}:$
CHOH
$CH_2OPO_3^{2-}$ + $NAD^+$ + $P_i$ $\longrightarrow$ $:\ddot{O}:$ C:phosphate + NADH + $H^+$
CHOH
$CH_2-OPO_3^{2-}$

**glyceraldehyde 3-phosphate** (see $NADP^+$ in Exercise 13.15b) **3-phosphoglycerate**

16. There is a net gain of two ATP molecules per glucose molecule consumed in glycolysis.

17. Pyruvate can be converted to lactate, ethanol, or acetyl-CoA.

18. The reaction of 2-phosphoglycerate to phosphoenolpyruvate is a dehydration (loss of water) rather than a redox reaction.

19. The hexokinase molecule changes shape drastically on binding to substrate, consistent with the induced fit theory of an enzyme adapting itself to its substrate.

20. ATP inhibits phosphofructokinase, consistent with the fact that ATP is produced by later reactions of glycolysis.

21. Phosphate ion, rather than ATP, is the source of phosphorus in substrate-level phosphorylation. An example is the conversion of glyceraldehyde-3-phosphate to 1,3-*bis*phosphoglycerate.

22. Isozymes are oligomeric enzymes that have slightly different amino acid compositions in different organs. Lactate dehydrogenase is an example, as is phosphofructokinase.

23. There is a net gain of three, rather than two, ATP when glycogen, not glucose, is the starting material of glycolysis.

24. Transketolase catalyzes the transfer of a two-carbon unit, while transaldolase catalyzes the transfer of a three-carbon unit.

25. Thiamine pyrophosphate is a coenzyme in the transfer of two-carbon units. It is required for catalysis by pyruvate carboxylase in alcoholic fermentation and by transketolase in the pentose phosphate pathway.

# CHAPTER 14

1.

2.

**pyruvate decarboxylase and pyruvate dehydrogenase**

**transketolase**

3. The NADH and FADH$_2$ produced by the citric acid cycle are the electron donors in the electron-transport chain linked to oxygen. Because of this connection the citric acid cycle is considered part of aerobic metabolism.

4. There is an adenine nucleotide portion in the structure of NADH, with a specific binding site on NADH-linked dehydrogenases for this portion of NADH.

5. If the amount of ADP in a cell increases relative to the amount of ATP, the cell needs energy (ATP). This situation favors the reactions of the citric acid cycle, which release energy, activating isocitrate dehydrogenase. Also, stimulates the formation of NADH and FADH$_2$ for ATP production by electron transport and oxidative phosphorylation.

6. If the amount of NADH in a cell increases relative to the amount of NAD$^+$, the cell has completed a number of energy-releasing reactions. There is less need for the citric acid cycle to be active, and, as a result, the activity of pyruvate dehydrogenase is decreased.

7. The citric acid cycle is less active when a cell has a high ATP/ADP and a high NADH/NAD$^+$ ratio. Both ratios indicate a high "energy charge" in the cell, indicating less of a need for the energy-releasing reactions of the citric acid cycle.

8.

(a) Pyruvate + CoA-SH $\longrightarrow$ Acetyl CoA + H$^+$ + Carbon dioxide + 2e$^-$

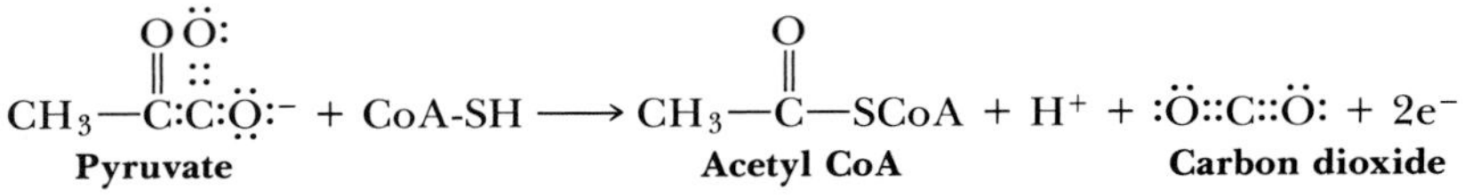

(b) isocitrate $\longrightarrow$ $\alpha$-ketoglutarate + CO$_2$ + 2e$^-$

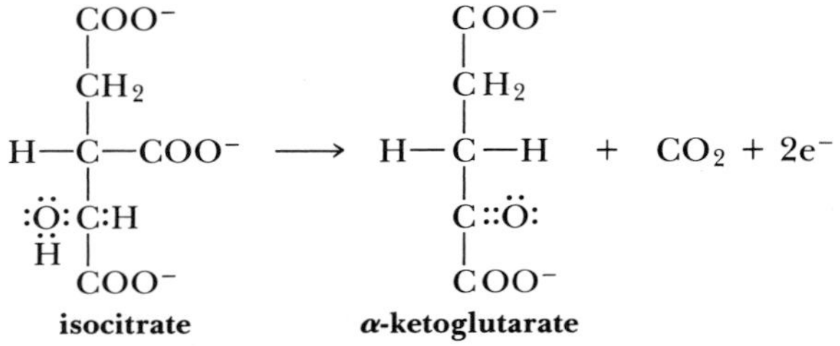

(c) $\alpha$-ketoglutarate + CoA-SH $\longrightarrow$ succinyl-CoA + :Ö::C::Ö: + 2e$^-$

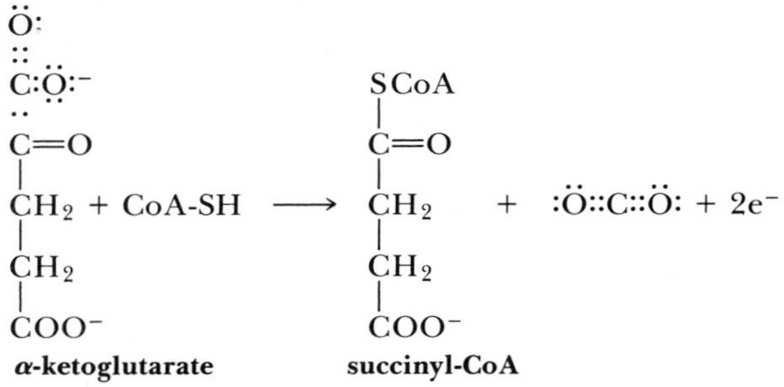

(d) succinate $\longrightarrow$ fumarate + 2e$^-$

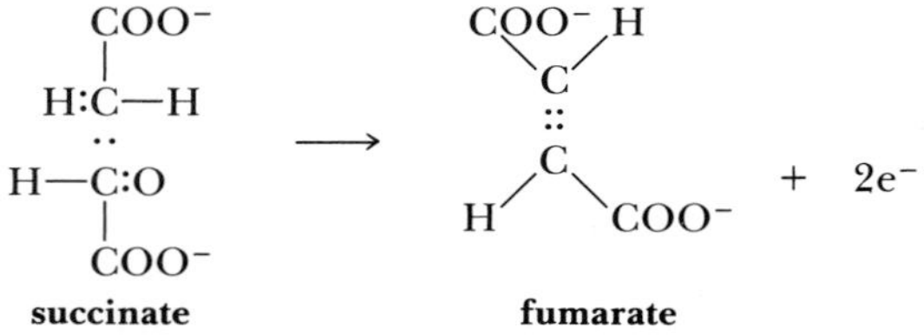

(e) malate $\longrightarrow$ oxaloacetate + 2e$^-$

A-18

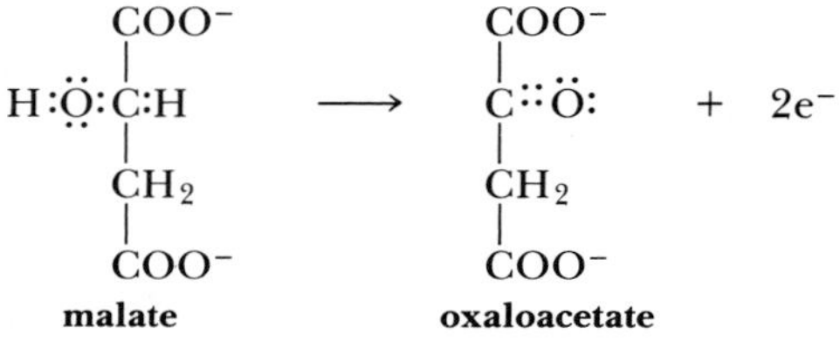

9. The conversion of fumarate to malate is a hydration reaction, not a redox reaction.
10. See Equation 14.3 on page 341.
11. A condensation reaction is one in which a new carbon-carbon bond is formed. The reaction of acetyl-CoA and oxaloacetate to produce citrate involves formation of such a carbon-carbon bond.
12. Thioesters are "high-energy" compounds that play a role in group-transfer reactions; consequently their $\Delta G^{\circ\prime}$ of hydrolysis is large and negative to provide energy for the reaction.
13. The citric acid cycle takes place in the mitochondrial matrix, while glycolysis occurs in the cytosol.
14. $NAD^+$ and FAD are the primary electron acceptors of the citric acid cycle.
15. Lipoic acid plays a role both in redox and acetyl-transfer reactions.
16. In oxidative decarboxylation the molecule that is oxidized loses a carboxyl group as carbon dioxide. Examples of oxidative decarboxylation include the conversion of pyruvate to acetyl-CoA, isocitrate to $\alpha$-ketoglutarate, and $\alpha$-ketoglutarate to succinyl-CoA.
17. Table 14.1 shows that the sum of the energies of the individual reactions is $-44.3$ kJ ($-10.6$ kcal) for each mole of acetyl-CoA that enters the cycle.
18. The glyoxylate cycle bypasses the two oxidative decarboxylations of the citric acid cycle by splitting isocitrate to glyoxylate and succinate. Glyoxylate reacts with acetyl-CoA to give malate, which is converted to oxaloacetate by reactions of the citric acid cycle.

# CHAPTER 15

1. Electrons are passed from NADH to a flavin-containing protein to coenzyme Q. From coenzyme Q the electrons pass to cytochrome $b$, then to cytochrome $c$, followed by cytochromes $a$ and $a_3$. From the cytochrome $a/a_3$ complex the electrons are finally passed to oxygen.
2. (a) 40; (b) 38; (c) 15; (d) 20; (e) 3; (f) 15
3. The half reaction of oxidation

$$NADH + H^+ \longrightarrow NAD^+ + 2H^+ + 2e^-$$

is strongly exergonic ($E^{\circ\prime} = +0.32$ v), as is the overall reaction

$$Pyruvate + NADH + H^+ \longrightarrow lactate + NAD^+$$

($E^{\circ\prime} = +0.13$ v)
4. The maximum yield of ATP, to the nearest whole number, is 3.

$$102.3 \text{ kJ released} \times \frac{1 \text{ ATP}}{30.5 \text{ kJ}} = 3.35 \text{ ATP}$$

One ATP is actually produced, so the efficiency of the process is

$$\frac{1 \text{ ATP}}{3 \text{ ATP}} \times 100 = 33.3\%$$

5. (a) Azide inhibits the transfer of electrons from cytochrome $a/a_3$ to oxygen. (b) Antimycin A inhibits the transfer of electrons from cytochrome $b$ to cytochrome $c_1$. (c) Amytal inhibits the transfer of electrons from NADH reductase to coenzyme Q. (d) Rotenone inhibits the transfer of electrons from NADH reductase to coenzyme Q. (e) Dinitrophenol is an uncoupler of oxidative phosphorylation. (f) Gramicidin A is an uncoupler of oxidative phosphorylation. (g) Carbon monoxide inhibits the transfer of electrons from cytochrome $a/a_3$ to oxygen.
6. A P/O ratio of 2 can be expected because oxidation of succinate passes electrons to coenzyme Q $via$ a flavoprotein intermediate, bypassing the first respiratory complex.
7. Succinate $+ \frac{1}{2}O_2 \longrightarrow$ fumarate $+ H_2O$
8. Cytochrome $c$ is not tightly bound to the mitochondrial membrane and can easily be lost in the course of cell fractionation. This protein is so similar in most aerobic organisms that cytochrome $c$ from one source can easily be substituted for that from another source.
9. The chemiosmotic coupling mechanism is based on the difference in hydrogen ion concentration between the intermembrane space and the matrix of actively respiring mitochondria. The hydrogen ion gradient is created by the proton pumping that accompanies the transfer of electrons. The flow of hydrogen ions back into the matrix through a channel in the mitochondrial ATPase (ATP synthetase) is directly coupled to the phosphorylation of ADP.
10. The $F_1$ portion of the mitochondrial ATP synthetase, which projects into the matrix, is the site of ATP synthesis.
11. The complete oxidation of glucose produces 36 molecules of ATP in muscle and brain and 38 ATP in liver, heart, and kidney. The underlying reason is the different shuttle mechanisms for transfer to mitochondria of electrons from the NADH produced in the cytosol by glycolysis.
12. (a) $E^{\circ\prime} = +0.13$ v, $\Delta G^{\circ\prime} = -25$ kJ/mol
    (b) $E^{\circ\prime} = +0.31$ v, $\Delta G^{\circ\prime} = -60$ kJ/mol
    (c) $E^{\circ\prime} = +0.12$ v, $\Delta G^{\circ\prime} = -23$ kJ/mol

**13.** In all reactions electrons are passed from the reduced form of one reactant to the oxidized form of the next reactant in the chain. The notation [Fe-S] refers to any one of a number of iron-sulfur proteins.

Reactions of Complex I

$$\left.\begin{array}{l} NADH + E\text{-}FMN \longrightarrow NAD^+ + E\text{-}FMNH_2 \\[6pt] E\text{-}FMNH_2 + [Fe\text{-}S]_{ox} \longrightarrow E\text{-}FMN + [Fe\text{-}S]_{red} \end{array}\right\} 1 \text{ ATP produced}$$

Transfer to Coenzyme Q

$$[Fe\text{-}S]_{red} + CoQ \longrightarrow [Fe\text{-}S]_{ox} + CoQH_2$$

Reactions of Complex II

$$\left.\begin{array}{l} CoQH_2 + cyt\ b_{ox} \longrightarrow CoQ + cyt\ b_{red} \\[6pt] cyt\ b_{red} + cyt\ c_{1\ ox} \longrightarrow cyt\ b_{ox} + cyt\ c_{1\ red} \\[6pt] cyt\ c_{1\ red} + [Fe\text{-}s]_{ox} \longrightarrow cyt\ c_{1\ ox} + [Fe\text{-}S]_{red} \end{array}\right\} 1 \text{ ATP produced}$$

Transfer to Cytochrome $c$

$$[Fe\text{-}S]_{red} + cyt\ c_{ox} \longrightarrow [Fe\text{-}S]_{ox} + cyt\ c_{red}$$

Reactions of Complex III

$$\left.\begin{array}{l} cyt\ c_{red} + cyt\ a,a_{3\ ox} \longrightarrow cyt\ c_{ox} + cyt\ a,a_{3\ red} \\[6pt] cyt\ a,a_{3\ red} + 1/2\ O_2 \longrightarrow cyt\ a,a_{3\ ox} + H_2O \end{array}\right\} 1 \text{ ATP produced}$$

There are three sites of oxidative phosphorylation in the electron transport chain, which results in a P/O ratio of 3/1.

**14.** When $FADH_2$ is the starting point for electron transport, electrons are passed from $FADH_2$ to Coenzyme Q in a reaction that bypasses Complex 1.

$$FADH_2 + [Fe\text{-}S]_{ox} \longrightarrow FAD + [Fe\text{-}S]_{red}$$
$$[Fe\text{-}S]_{red} + CoQ \longrightarrow [Fe\text{-}S]_{ox} + CoQH_2$$

There are only two sites of oxidative phosphorylation in the rest of the electron transport chain, giving rise to a P/O ratio of 2.

# CHAPTER 16

**1.** Features in common: involvement of acetyl-CoA and thioesters, each round of breakdown or synthesis involves two-carbon units.

Differences: involvement of malonyl-CoA in biosynthesis, not in breakdown; thioesters involve CoA in breakdown, acyl carrier proteins in biosynthesis; biosynthesis occurs in the cytosol, breakdown in the mitochondrial matrix; breakdown is an oxidative process that requires $NAD^+$ and FAD and produces ATP by electron transport and oxidative phosphorylation, while biosynthesis is a reductive process that requires NADPH and ATP.

**2.** From 7 cycles of $\beta$-oxidation: 8 acetyl-CoA, 7 $FADH_2$, 7NADH

From the processing of 8 acetyl-CoA in the citric acid cycle: 8 $FADH_2$, 24 NADH, 8 GTP

From reoxidation of all $FADH_2$ and NADH: 30 ATP from 15 $FADH_2$, 93 ATP from 31 NADH

From 8 GTP: 8 ATP

Subtotal: 131 ATP

2 ATP equivalent used in activation step

Grand total: 129 ATP

**3.** Seven carbon-carbon bonds are broken in the course of $\beta$-oxidation (see Figure 16.5).

**4.** Acyl groups are esterified to carnitine to cross the inner mitochondrial matrix. There are transesterification reactions from the acyl-CoA to carnitine and from acyl-carnitine to CoA (see Figure 16.3).

**5.** False. The oxidation of unsaturated fatty acids to acetyl-CoA requires a *cis-trans* isomerization and an epimerization (see Figure 16.7), reactions which are not found in the oxidation of saturated fatty acids.

**6.** The humps of camels contain lipids that can be degraded as a source of metabolic water, rather than water as such.

**7.** Acetone is one of the "ketone bodies" produced by

breakdown of lipids. Diabetics have impaired ability to metabolize carbohydrates and degrade an excessive amount of lipids as a result.

8. *Step 1* Biotin is carboxylated using bicarbonate ion ($HCO_3^-$) as the source of the carboxyl group.
*Step 2* The carboxylated biotin is brought into proximity with enzyme-bound acetyl-CoA by a biotin carrier protein.
*Step 3* The carboxyl group is transferred to acetyl-CoA, forming malonyl-CoA.

9. Linoleate and linolenate cannot be synthesized by the body and must be obtained from dietary sources.

10. Acyl-CoA intermediates are essential in the conversion of fatty acids to other lipids.

11. In steroid biosynthesis, three acetyl-CoA molecules condense to form the six-carbon mevalonate, which then gives rise to a five-carbon isoprenoid unit. A second and then a third isoprenoid unit condenses, giving rise to a 10-carbon and then a 15-carbon unit. Two of the 15-carbon units condense, forming a 30-carbon precursor of cholesterol.

12. See Figure 16.19.

13. Acetyl groups condense with oxaloacetate to form citrate, which can cross the mitochondrial membrane. Acetyl groups are regenerated in the cytosol by the reverse action.

14. All steroids have a characteristic fused-ring structure, implying a common biosynthetic origin.

15. Cholesterol is nonpolar and cannot dissolve in blood, which is an aqueous medium.

## CHAPTER 17

1. In cyclic photophosphorylation, the excited chlorophyll of Photosystem I passes electrons directly to the electron transport chain that normally links Photosystem II to Photosystem I. This electron transport chain is coupled to ATP production (see Figure 17.8).

2. Electron transport and ATP production are coupled to one another by the same mechanism in mitochondria and choroplasts. In both cases the coupling depends on the generation of a proton gradient across the inner mitochondrial membrane or across the thylakoid membrane as the case may be.

3. From the standpoint of thermodynamics, the production of sugars in photosynthesis is the reverse of the complete oxidation of a sugar such as glucose to $CO_2$ and water. The complete oxidation reaction produces six moles of $CO_2$ for each mole of glucose oxidized. To get the energy change for the fixation of one mole of $CO_2$, change the sign of the energy for the complete oxidation of glucose and divide by six.

4. In the fall the chlorophyll in leaves is lost, and the red and yellow colors of the accessory pigments become visible, accounting for fall foliage colors.

5. The proton gradient is created by the operation of the electron transport chain that links the two photosystems in noncyclic photophosphorylation (see Exercise 1).

6. A prokaryotic organism that contains both chlorophyll *a* and chlorophyll *b* could be a relict of an evolutionary way-station in the development of chloroplasts.

7. When light impinges on the reaction center of *Rhodopseudomonas*, the special pair of chlorophylls there is raised to an excited energy level. An electron is passed from the special pair to accessory pigments, first pheophytin, then plastoquinone. The electron lost by the special pair of chlorophylls is replaced by a soluble cytochrome, which diffuses away. The separation of charge represents stored energy (see Figure 17.5).

8. In Photosystem I and in Photosystem II light energy is needed to raise the reaction-center chlorophylls to a higher energy level. Energy is needed to generate strong enough reducing agents to pass electrons to the next of the series of components in the pathway.

9. Glucose synthesized by photosynthesis is not uniformly labeled because only one molecule of $CO_2$ is incorporated into each molecule of ribulose 1,5-*bis*phosphate, which then goes on to give rise to sugars.

10. In tropical plants the $C_4$ pathway is operative in addition to the Calvin cycle.

## CHAPTER 18

1. Lysine is an essential amino acid that is frequently lacking in cereals.

2. See Figure 18.9.

3. Four $\alpha$-amino acids—ornithine, citrulline, argininosuccinate, and arginine—participate in the urea cycle; of these, only arginine can be used for protein synthesis.

4. $H^+ + HCO_3^- + 2\ NH_3 + 3\ ATP \longrightarrow$
$NH_2CONH_2 + 2\ ADP + 2\ P_i + AMP + PP_i + 2\ H_2O$
The urea cycle is linked to the citric acid cycle by fumarate and by aspartate, which can be converted to malate by transamination (see Figure 18.17).

5. Fluorouracil substitutes for thymine in DNA synthesis. In rapidly dividing cells, such as cancer cells, the result is the production of defective DNA.

6. Conversion of homocysteine to methionine using S-adenosylmethionine as the methyl donor gives no net gain; one methionine is needed to produce another methionine.

7. Glutamate + $\alpha$-keto acid $\longrightarrow$
$$\alpha\text{-ketoglutarate} + \alpha\text{-amino acid}$$

8. In both cases the requirements are those given in Table 18.1.

9. In purine nucleotide biosynthesis the growing purine ring is covalently bonded to ribose, while in pyrimidine nucleotide biosynthesis the ribose is added after the ring is synthesized.

10. $N_2 \longrightarrow NH_4^+ \longrightarrow$ 3-phosphoserine $\longrightarrow$
$$\text{serine} \longrightarrow \text{glycine} \longrightarrow \text{N-7}$$

11. Sulfanilamide inhibits folic acid biosynthesis.

12. See the S-adenosylmethionine structure in Figure 18.13. The reactive methyl group is indicated.

13. Feedback control mechanisms slow down long biosynthetic pathways at or near their beginnings, saving energy for the organism.

14. A high-protein diet leads to increased production of urea. Drinking more water increases the volume of urine, ensuring elimination of the urea from the body with less strain on the kidneys than if urea were at higher concentration.

15. The pigmentation of skin and hair is due to the presence of melanin. The amino acid tyrosine is the precursor of melanin. In a situation of protein deficiency the essential amino acid, tyrosine, is lacking.

16. The DNA of fast-growing cells, such as those of the hair follicles, is damaged by chemotherapeutic agents.

17. The purine salvage reaction that produces GMP requires the equivalent of 2 ATP. The pathway to IMP and then to GMP requires the equivalent of 8 ATP.

## CHAPTER 19

1. All amino acids must be present at the same time for protein synthesis to occur. Newly synthesized proteins are necessary for growth in the immature rats.

2. The following series of reactions exchanges NADH for NADPH.

Oxaloacetate + NADH + $H^+$ $\longrightarrow$ malate + $NAD^+$

Malate + $NADP^+$ $\longrightarrow$ pyruvate + $CO_2$ + NADPH + $H^+$

3. Lucullus breaks down the protein in the tuna to amino acids, which in turn undergo the urea cycle and the breakdown of the carbon skeleton described in Chapter 18, eventually leading to the citric acid cycle and electron transport. In addition to protein catabolism, Griselda breaks down the carbohydrates to sugars, which then undergo glycolysis and enter the citric acid cycle. (Gratuitous information: Lucullus was a notorious Roman gourmand. In medieval literature, Griselda was the name usually given to a forbearing, long-suffering woman.)

4. The weight loss is due to correction of the bloating caused by retention of liquids.

5. Carbohydrates are the main energy source. Excess fat consumption can lead to the formation of "ketone bodies" and to atherosclerosis. Diets extremely high in protein can put a strain on the kidneys.

6. Phosphocreatine has a higher free energy of hydrolysis than ATP. When ATP is used in muscle contraction, the supply is restored by using phosphocreatine reserves to phosphorylate ADP.

7. The sprinter is using much more energy than the sedentary person under conditions that lead to lactic acid buildup.

8. $IP_3$ is a polar compound and can dissolve in the aqueous environment of the cytosol, while DAG is nonpolar and interacts with the side chains of the membrane phospholipids.

9. When a stimulatory hormone binds to its receptor on a cell surface, it stimulates the action of adenylate cyclase, mediated by the G protein. The cAMP produced elicits the desired effect on the cell by stimulating a kinase that phosphorylates a target enzyme.

10. The hypothalamus secretes hormone-releasing factors. Under the influence of these factors, the pituitary secretes trophic hormones, which act on specific endocrine glands. Individual hormones are then released by the specific endocrine glands.

11. See Table 19.2.

## CHAPTER 20

1. AZT can be bonded into DNA at its 5' end, but its 3' end cannot form a phosphodiester bond; it causes a break in DNA.

**2.**

NH₂ ... (chemical structure)

This compound cannot form a phosphodiester bond at its 3' end and causes a break in DNA.

**3.**

NH₂ ... (chemical structure)

The inversion of configuration at carbon-2 causes a change in the conformation of the sugar moiety so that this nucleoside disrupts the structure of viral RNA.

**4.** A-T base pairs have two hydrogen bonds, while G-C base pairs have three. It takes more energy and higher temperature to disrupt the structure of DNA rich in G-C base pairs.

**5.** Statements (c) and (d) are true; statements (a) and (b) are not.

**6.** True. There is room for binding and access to the base pairs in both the major and minor grooves of DNA.

**7.** Statement (c) is true. Statements (a) and (b) are false. Statement (d) is true for the B form of DNA but not for the A and Z forms.

**8.** Supercoiling refers to twists in DNA over and above those of the double helix. Positive supercoiling refers to an extra twist in DNA caused by overwinding of the helix before sealing the ends to produce circular DNA. A topoisomerase is an enzyme that induces a single-strand break in supercoiled DNA, relaxes the supercoiling, and reseals the break. Negative supercoiling refers to unwinding of the double helix before sealing the ends to produce circular DNA.

**9.** Chromatin is the complex consisting of DNA and basic proteins found in eukaryotic nuclei (see Figure 20.14).

**10.** See Figures 20.16 and 20.21.

**11.** The use of restriction endonucleases with different specificities gives overlapping sequences that can be combined to give an overall sequence.

**12.** *GAAT      TACGT
    *GAATCTA  GT
(TA will also occur if the molecule is cleaved in two places.)

**13.** Cleavage at

| Length of fragment (bases) | | A | C | G | T | | |
|---|---|---|---|---|---|---|---|
| | 9 | | | | — | T | 3' |
| | 8 | | | — | | C | |
| | 7 | | — | | | G | |
| | 6 | — | | | | A | |
| | 5 | | | | — | T | |
| | 4 | | — | | | C | |
| | 3 | | | | | T | |
| | 2 | — | | | | A | |
| | 1 | — | | | | A | |
| | | | | | | ? | |

The 5'-terminal G is destroyed by the G cleavage reaction.

**14.** In the cleavage by *Eco*RI there are single strand breaks between the G and A, but the two strands are held together by the A-T hydrogen bonds.
G A-A-T-T-C
C-T-T-A-A G

**15.** A portion of exogenous DNA is introduced into a suitable vector, frequently a bacterial plasmid, and many copies of the DNA are produced when the bacteria grow.

**16.** The sequence in Exercise 14 is an example of a palindrome. See Table 20.1 for others.

**17.** Isolate the DNA that codes for the growth factor by means of suitable probes. Introduce the DNA into a bacterial genome. Allow the bacteria to grow and to produce human growth factor.

## CHAPTER 21

**1.** False. In retroviruses, the flow of information is RNA ⟶ DNA.

**2.** Replication is the production of new DNA on a DNA template. Transcription is the production of RNA on a DNA template. Translation is the synthesis of proteins directed by mRNA, which reflects the base sequence of DNA.

**3.** The semiconservative replication of DNA means that a newly formed DNA molecule has one new strand and one strand from the original DNA. The experimental evidence for semiconservative replication comes from density-gradient centrifugation (Figure 21.3).

4. A replication fork is the site of formation of new DNA. The two strands of the original DNA separate, and a new strand is formed on each original strand.

5. DNA polymerase I is primarily a repair enzyme. DNA polymerase III is mainly responsible for the synthesis of new DNA. See Table 21.1.

6. All four deoxyribonucleoside triphosphates, template DNA, DNA polymerase, all four ribonucleoside triphosphates, primase, helicase, single-strand binding protein, DNA gyrase, DNA ligase.

7. DNA is synthesized from the 5' to the 3' end, and the new strand is antiparallel to the template strand. One of the strands is exposed from the 5' to the 3' end as a result of unwinding. Small stretches of new DNA are synthesized, still in an antiparallel direction from the 5' to the 3' end and are linked by DNA ligase. See Figure 21.5.

8. DNA gyrase introduces a swivel point in advance of the replication fork. Primase synthesizes the RNA primer. DNA ligase links small newly formed strands to produce longer ones.

9. When an incorrect nucleotide is introduced into a growing DNA chain as a result of mismatched base pairing, DNA polymerase acts as a 3'-exonuclease, removing the incorrect nucleotide. The same enzyme then incorporates the correct nucleotide.

10. An exonuclease nicks the DNA near the site of the thymine dimers. Polymerase I acts as a nuclease and excises the incorrect nucleotides, then acts as a polymerase to incorporate the correct ones. DNA ligase seals the nick.

11. The general features of DNA replication are similar in prokaryotes and eukaryotes. The main differences are that eukaryotic DNA polymerases do not have exonuclease activity. After synthesis, eukaryotic DNA is complexed with proteins, while prokaryotic DNA is not.

12. The chi intermediate can give rise to either of the possibilities for recombinant DNA—single-strand or double-strand heterozygous—depending on how it is nicked with subsequent resealing of the nicks.

13. A transposon is the portion of DNA that is moved from one place on a chromosome to another in the course of gene rearrangement. Homologous sequences in DNA may play a role in the looping out of transposons from the rest of the DNA.

14. 1. RNA polymerase from *E. coli* has a molecular weight of about 500,000 and four different kinds of subunit. 2. It uses one strand of the DNA template to direct RNA synthesis. 3. It catalyzes polymerization from the 5' to the 3' end.

15. The promoter region is the portion of DNA to which RNA polymerase binds at the start of transcription. This region lies upstream (nearer the 3' end of the DNA) of the actual gene for the RNA. The promoter regions of DNA from many organisms have sequences in common (consensus sequences.) The consensus sequences frequently lie 10 base pairs and 35 base pairs upstream of the start of transcription.

16. Trimming is necessary to obtain RNA transcripts of the proper size. Frequently several tRNAs are transcribed in one long RNA molecule and must be trimmed to obtain active tRNAs.

17. Capping, polyadenylation, and splicing of coding sequences take place in the processing of eukaryotic mRNA.

18. Exons are the portions of DNA that are expressed, which means that they are reflected in the base sequence of the final mRNA product. Introns are the intervening sequences that do not appear in the final product but are removed during the splicing of mRNA.

19. The snRNPs are small nuclear ribonucleoprotein particles. They are the site of mRNA splicing.

## CHAPTER 22

1. See Figure 22.1

2. The hydrolysis of ATP to AMP and $PP_i$ provides the energy to drive the activation step.

3. The linkage of amino acids to tRNA is as an aminoacyl ester.

4. Methionine bound to tRNA$^{fmet}$ can be formylated, but methionine bonded to tRNA$^{met}$ cannot.

5. The methionine anticodon (UAC) on the tRNA base pairs with the methionine codon AUG in the mRNA sequence that signals the start of protein synthesis.

6. The initiation complex in *E. coli* requires mRNA, the 30S ribosomal subunit, fmet-tRNA$^{fmet}$, GTP, and three protein initiation factors, called IF-1, IF-2, and IF-3. The IF-3 protein is needed for the binding of mRNA to the ribosomal subunit. The other two protein factors are required for the binding of fmet-tRNA$^{fmet}$ to the mRNA-30S complex.

7. The A site and the P site on the ribosome are both binding sites for charged tRNAs taking part in protein synthesis. The P (peptidyl) site binds a tRNA to which the growing polypeptide chain is bonded. The A (aminoacyl) site binds to an aminoacyl tRNA.

The amino acid moiety will be the next added to the nascent protein.

8. Peptidyl transferase catalyzes the formation of a new peptide bond in protein synthesis. The elongation factors, EF-Tu and EF-Ts, are required for binding of aminoacyl tRNA to the A site. The third elongation factor, EF-G, is needed for the translocation step in which the mRNA moves with respect to the ribosome, exposing the codon for the next amino acid.

9. Puromycin terminates the growing polypeptide chain, by forming a peptide bond with its C-terminus, which prevents the formation of new peptide bonds (see Figure 22.6).

10. The stop codons bind to release factors, proteins that block binding of aminoacyl tRNAs to the ribosome and release the newly formed protein.

11. Hydroxyproline is formed from proline, an amino acid for which there are four codons, by posttranslational modification of the collagen precursor.

12. The less highly purified ribosome preparations contained polysomes, which are more active in protein synthesis than single ribosomes.

13. A code in which two bases code for a single amino acid allows for only 16 (4 × 4) possible codons, not adequate to code for 20 amino acids.

14. The concept of "wobble" specifies that the first two bases of a codon remain the same, while there is room for variation in the third base. This is precisely what is observed experimentally.

15. Hypoxanthine is the most versatile of the "wobble" bases; it can base pair with adenine, cytosine, or uracil.

16. An inducible enzyme is one that is not produced at all times by a cell. The synthesis of such an enzyme is triggered by the presence of a specific substance, the inducer.

17. An operon consists of an operator gene, a promoter gene, and structural genes. When a repressor is bound to the operator, RNA polymerase cannot bind to the promoter to start transcription of the structural genes. When an inducer is present, it binds to the repressor, rendering it inactive. The inactive repressor can no longer bind to the operator. As a result, RNA polymerase can bind to the promoter, leading to the eventual transcription of the structural genes.

18. A repressor is a protein that binds to the operator gene. A corepressor is a substance that binds to the repressor. Where there is a corepressor, the repressor-corepressor complex binds to the operator and the repressor alone does not.

# PHOTOGRAPH AND ILLUSTRATION CREDITS

Frontmatter, "About the Authors": Mary Campbell (Michael Zide); Irving Geis (Sandy Geis).

**Part I**
Part opener, p. 1: © Phillip A. Harrington/Fran Heyl Associates.
**Chapter 1**, p. 10 (bottom): Courtesy NASA/Marshall Space Flight Center; p. 14 (margin): © 1986 Wetmore, Science Source/Photo Researchers.
**Chapter 2**: Chapter opener, p. 25: © Dan McCoy/Rainbow. 2.1, 2.3, 2.4, 2.5, 2.6, 2.7: Courtesy of Dr. Sue Ellen Gruber, Mt. Holyoke College.
**Chapter 3**: Chapter opener, p. 42: © Frans Lanting/Minden Pictures.

**Part II**
**Part opener**, p. 62: Altman interview, pp. 64, 66, 67: Courtesy of T. Charles Erickson and Michael Marsland, Yale University, Office of Public Information. © Leonard Lessin/Waldo Feng/Mt. Sinai CORE.
**Chapter 4**: Chapter opener, p. 68: © Herb Charles Ohlmeyer/Fran Heyl Associates; 4.1(a.2), (a.3), (b.2); 4.3(a.2), (a.3), (b.2), (b.3); 4.4(a.2), (a.3), (a.4), (a.5), (a.6); 4.5(a.2), (a.3), (a.4) © Leonard Lessin/Waldo Feng/Mt. Sinai CORE.
**Chapter 5**: Chapter opener, p. 79: © Professors Hadley and Hruby, University of Arizona; 5.1(b), 5.2(b), 5.3(b), 5.8(c), Box art (b), 5.12(b): © Leonard Lessin/Waldo Feng/Mt. Sinai CORE.
**Chapter 6**: Chapter opener, p. 96: © Irving Geis; 6.1(b.2), (b.4), © Leonard Lessin/Waldo Feng/Mt. Sinai CORE. 6.4(b) Courtesy of Tito Simboli, photographer. 6.7(c.3) Leonard Lessin/Waldo

Feng/Mt. Sinai CORE. Bottom figure, p. 113: © Irving Geis. Top figure, p. 114: © Irving Geis.
**Chapter 7**: Chapter opener, p. 126: © J. J. Sullivan/Rainbow

**Part III**
Part opener, p. 144: © Leonard Lessin/Waldo Feng/Mt. Sinai CORE. Stubbe interview, p. 146, (Barry Hethesington)
**Chapter 8**: Chapter opener, p. 150: © David Muench; p. 151(a): © Elizabeth Weiland/Photo Researchers; p. 151(b): © Mike Neumann/Photo Researchers; p. 151(c): © Dick Rowan/Photo Researchers. p. 152, margin: The Bettmann Archive; p. 155, margin: The Bettmann Archive. p. 161(a), (b): © Irving Geis.
**Chapter 9**: Chapter opener, p. 164: © Irving Geis; p. 167: © Irving Geis. 9.9: © Irving Geis, 9.10(a): © Irving Geis, (b): © Leonard Lessin/Waldo Feng/Mt. Sinai CORE. p. 179 (margin): © Irving Geis. p. 180 (bottom): © Irving Geis. 9.16(a) © Irving Geis; (b) © Leonard Lessin/Waldo Feng/Mt. Sinai CORE. 9.18(b) © Leonard Lessin/Waldo Feng/Mt. Sinai CORE. 9.19(a) © Irving Geis; (b) © Leonard Lessin/Waldo Feng/Mt. Sinai CORE. 9.21: © Irving Geis. 9.22 (a), (b), © Irving Geis. 9.25: © Irving Geis.
**Interchapter A**: Chapter opener, p. 197: © Professor T. L. Blundell, Dept. of Crystallography, Berkbeck College/Science Photo Library/Photo Researchers; A.12: © Gelteach, Inc/Lynn Kraft-Sibley.
**Chapter 10**: Chapter opener, p. 213: © Irving Geis.
**Chapter 11**: Chapter opener, p. 254: © Herb Charles Ohlmeyer/Fran Heyl Associates; 11.9: © Dr. L. Andrew Staehelin, University of Colorado.

# GLOSSARY

**absolute specificity** the property of an enzyme such that it acts on one, and only one, substrate

**accessory pigments** plant pigments other than chlorophyll that play a role in photosynthesis

**acid dissociation constant** a number that characterizes the strength of an acid

**acid strength** the tendency of an acid to dissociate to a hydrogen ion and its conjugate base

**acromegaly** a disease caused by an excess of growth hormone after the skeleton has stopped growing, characterized by enlarged hands, feet, and facial features

**activation energy** the energy required to start a reaction

**activation step** the beginning of a multistep process, in which a substrate is converted to a more reactive compound

**active site** the part of an enzyme to which the substrate binds and at which the reaction takes place

**active transport** the energy-requiring process of moving substances into a cell against a concentration gradient

**acyl-CoA synthetase** the enzyme that catalyzes the activation step in lipid catabolism

**Addison's disease** a disease caused by deficiency in steroid hormones of the adrenal cortex, characterized by hypoglycemia, weakness, and increased susceptibility to stress

**adenine** one of the purine bases found in nucleic acids

**adenylate cyclase** the enzyme that catalyzes the production of cyclic AMP

**ADP** (adenosine diphosphate) a compound that can serve as an energy carrier when it is phosphorylated to form ATP

**albinism** the condition of depigmentation in an organism, caused by a lack of an enzyme responsible for the conversion of the amino acid tyrosine to melanin

**alcohol dehydrogenase** an NADH-linked redox enzyme; catalyzes the conversion of acetaldehyde to ethanol

**alcoholic fermentation** the anaerobic pathway that converts glucose to ethanol

**aldolase** in glycolysis, the enzyme that catalyzes the reverse aldol condensation of fructose 1,6-*bis*phosphate

**aldose** a sugar that contains an aldehyde group as part of its structure

**allosteric** the property of multisubunit proteins such that a conformational change in one subunit induces a change in another subunit

**allosteric effector** a substance—substrate, inhibitor, or activator—that binds to an allosteric enzyme and affects its activity

**amino acid activation** the formation of an ester bond between an amino acid and its specific tRNA, catalyzed by a suitable synthetase

**amino acid analyzer** an instrument that gives information on the number and kind of amino acids in a protein

**aminoacyl-tRNA synthetases** enzymes that catalyze the formation of an ester linkage between an amino acid and tRNA

**aminopeptidase** an enzyme that removes amino acids from the N-terminal end of a protein

**amphibolic** able to be a part of both anabolism and catabolism

**amphiphilic** the property of a molecule such that it has one end that dissolves in water and another end that dissolves in nonpolar solvents

**α-amylase** an enzyme that hydrolyzes glycosidic linkages anywhere along a polysaccharide chain

**β-amylase** an enzyme that hydrolyzes glycosidic linkages starting at the end of a polysaccharide chain

**anabolism** the synthesis of biomolecules from simpler compounds

**anaerobic glycolysis** the pathway of conversion of glucose to lactate, distinguished from glycolysis, which is the conversion of glucose to pyruvate

**analytical ultracentrifugation** the technique for observing the motion of particles as they sediment in a centrifuge

**anaplerotic** a reaction that ensures an adequate supply of an important metabolite

**androgens** male sex hormones, steroids in chemical nature

**anomer** one of the possible stereoisomers formed when a sugar assumes the cyclic form

**anomeric carbon** the chiral center created when a sugar cyclizes

**antibody** a glycoprotein that binds to and immobilizes a substance the cell recognizes as foreign

**anticodon** the sequence of three bases (triplet) in tRNA that hydrogen bonds with the mRNA triplet that specifies a given amino acid

**antigenic determinant** the portion of a molecule that antibodies recognize as foreign and to which they bind

**antioxidant** a strong reducing agent, which is easily oxidized and thus prevents the oxidation of other substances

**arachidonic acid** a fatty acid that contains 20 carbon atoms and four double bonds, the precursor of prostaglandins and leukotrienes

**aspartate transcarbamoylase (ATCase)** a classic example of an allosteric enzyme, catalyzes an early reaction in pyrimidine biosynthesis

**ATP** (adenosine triphosphate) a universal energy carrier

**ATP synthetase** the enzyme responsible for production of ATP in mitochondria

**atherosclerosis** the blockage of arteries by cholesterol deposits

**autoradiography** the technique of locating radioactively labeled substances by allowing them to expose photographic film

**bacterial plasmid** a portion of circular DNA separate from the main genome of the bacterium

**bacteriochlorophyll** the form of chlorophyll that occurs in bacteria that carry out photosynthesis not linked to oxygen

**β-barrel** a β-pleated sheet extensive enough to fold back on itself

**bidentate ligand** a substance that forms two bonds when it binds to another molecule

**binding assay** an experimental method for selecting one molecule out of a number of possibilities by specific binding; used to determine the nature of many triplets of the genetic code

**biotin** a carbon dioxide carrier molecule

**Bohr effect** the decrease in oxygen binding by hemoglobin caused by binding of carbon dioxide and hydrogen ion

**buffer solution** a solution that resists a change in pH on addition of moderate amounts of strong acid or strong base

**buffering capacity** a measure of the amount of acid or base that reacts with a given buffer solution

**C-terminal** the end of a protein or peptide with a carboxyl group not bonded to another amino acid

**Calvin cycle** the pathway of carbon dioxide fixation in photosynthesis

**carboxypeptidase** an enzyme that removes amino acids from the C-terminal end of a protein

**β-carotene** an unsaturated hydrocarbon, the precursor of vitamin A

**carrier protein** a membrane protein to which a substrate binds in passive transport into the cell

**catabolism** the breakdown of nutrients to provide energy

**catabolite activator protein (CAP)** a protein that can bind to a promoter when complexed with cAMP, allowing RNA polymerase to bind to its entry site on the same promoter

**catabolite repression** repression of the synthesis of *lac* proteins by glucose

**catalysis** the process of increasing the rate of chemical reactions

**cell membrane** the outer membrane of the cell that separates it from the outside world

**cell wall** the outer coating of bacterial and plant cells

**cellulose** a polymer of glucose, an important structural material in plants

**cerebroside** a glycolipid that contains sphingosine and a fatty acid in addition to the sugar moiety

**channel protein** a membrane protein that has a channel through which a substance passes without binding in passive transport into the cell

**chemiosmotic coupling** the mechanism for coupling electron transport to oxidative phosphorylation; requires a proton gradient across the inner mitochondrial membrane

**chimeric DNA** DNA from more than one species covalently linked together

**chiral** refers to an object that is not superimposable on its mirror image

**chlorophyll** the principal photosynthetic pigment, responsible for trapping light energy from the sun

**chloroplast** the organelle that is the site of photosynthesis in green plants

**cholera** a disease caused by the bacterium *Vibrio cholerae*, characterized by dehydration due to excessive $Na^+$ transport in epithelial cells

**cholesterol** a steroid that occurs in cell membranes, the precursor of other steroids

**chromatin** a complex of DNA and protein found in eukaryotic nuclei

**chromatography** an experimental method for separating substances

**chromosome** a linear structure that contains the genetic material

**chymotrypsin** a proteolytic enzyme that preferentially hydrolyzes amide bonds adjacent to aromatic amino acid residues

**chymotrypsinogen** the precursor of chymotrypsin, converted to it by proteolytic cleavage of specific peptide bonds

**cilia** short projections from eukaryotic cells

***cis-trans* isomerase** an enzyme that catalyzes a *cis-trans* isomerization in the catabolism of unsaturated fatty acids

**citric acid cycle** a central metabolic pathway, part of aerobic metabolism

**cloning of DNA** the introduction of a section of DNA into a genome, in which it can be reproduced many times

**clotting factors** proteins that play a role in blood clotting

**cloverleaf structure** the characteristic secondary structure of tRNA; includes hydrogen-bonded stems and loops without hydrogen bonds

**codon** sequence of three bases on mRNA that specifies a given amino acid

**coenzyme** a nonprotein substance that takes part in an enzymatic reaction and is regenerated at the end of the reaction

**coenzyme A** a carrier of carboxylic acids bound to its thiol group by a thioester linkage

**coenzyme Q** an oxidation–reduction coenzyme in mitochondrial electron transport

**column chromatography** a form of chromatography in which the stationary phase is packed in a column

**committed step** in a metabolic pathway, the formation of a substance that can play no other role in metabolism but to undergo the rest of the reactions of the pathway

**competitive inhibition** a decrease in enzymatic activity caused by binding of a substrate analogue to the active site

**complementary** the specific hydrogen bonding of adenine with thymine (or uracil) and guanine with cytosine in nucleic acids

**concerted model** a description of allosteric activity in which the conformations of all subunits change simultaneously

**configuration** the three-dimensional arrangement of groups around a chiral carbon atom

**conformational coupling** a mechanism for coupling of electron transport to oxidative phosphorylation that depends on a conformational change in the ATP synthetase

**consensus sequences** DNA sequences to which RNA polymerase binds, identical in many organisms

**contact inhibition** the process by which cell growth stops when cells touch each other, absent in cancer cells

**control sites** the operator and promoter genes that modulate the production of proteins whose amino acid sequence is specified by the structural genes under their control

**cooperative binding** binding to several sites such that when the first ligand is bound the binding of subsequent ones is easier

**cooperative transition** one that takes place in an all-or-nothing fashion, such as the melting of a crystal

**corepressor** a substance that binds to a repressor protein, making it active and able to bind to an operator gene

**coupled translation** in prokaryotes, the situation in which a gene is simultaneously transcribed and translated

**coupling** the process by which an exergonic reaction provides energy for an endergonic one

**cut and patch** a mechanism for repair of DNA by enzymatically removing incorrect nucleotides and substituting correct ones

**cyanogen bromide** a reagent that cleaves proteins at internal methionine residues

**cyanosis** a short supply of oxygen in the bloodstream, characterized by bluish skin

**cyclic AMP** a nucleotide in which the same phosphate group is esterified to the 3' and 5' hydroxyl groups of a single adenosine, an important second messenger

**cytochrome** any one of a group of heme-containing proteins in the electron transport chain

**cytosine** one of the pyrimidine bases found in nucleic acids

**dansyl chloride** a reagent used for labeling the N-terminal amino acid residue in proteins

**debranching enzyme** one that hydrolyzes the linkages in a branched-chain polymer such as amylopectin

**degenerate code** a code in which more than one triplet of bases can code for the same amino acid

**denaturation** the unraveling of the three-dimensional structure of a macromolecule caused by the breakdown of noncovalent interactions

**denitrification** the process by which nitrates are broken down to molecular nitrogen

**density-gradient centrifugation** the technique of separating substances in an ultracentrifuge by applying the sample to the top of a tube that contains a solution of varying densities

**deoxyribonucleoside** a compound formed when a nucleobase and deoxyribose form a glycosidic bond

**deoxyribose** a sugar that is part of the structure of DNA

**deoxy sugar** one in which one of the hydroxyl groups has been reduced to a hydrogen

**diastereomers** nonsuperimposable, non–mirror-image stereoisomers

**disaccharide** two monosaccharides (monomeric sugars) linked by a glycosidic bond

**DNA** deoxyribonucleic acid, the molecule that contains the genetic code

**DNase** (deoxyribonuclease) an enzyme that specifically hydrolyzes DNA

**DNA gyrase** an enzyme that introduces supercoiling into closed circular DNA

**DNA ligase** the enzyme that links separate stretches of DNA

**DNA polymerase** the enzyme that forms DNA from deoxyribonucleotides on a DNA template

**domain** a portion of a polypeptide chain that folds independently of other portions of the chain

**double-strand heterozygous** a recombinant DNA in which both strands incorporate portions of the two original molecules

**downstream** in transcription, a portion of the DNA sequence nearer the 5′ end than the gene to be transcribed, where the DNA is read from the 3′ to the 5′ end and the RNA is formed from the 5′ to the 3′ end; in translation, nearer the 3′ end of mRNA

**dwarfism** a disease caused by a deficiency of growth hormone

**Edman degradation** a method for determining the amino acid sequence of peptides and proteins

**electron transport to oxygen** a series of oxidation–reduction reactions by which the electrons derived from oxidation of nutrients are passed to oxygen

**electrophile** an electron-poor substance that tends to react with centers of negative charge or polarization

**electrophoresis** a method for separating molecules on the basis of the ratio of charge to size

**elongation step** in protein synthesis, the succession of reactions in which the peptide bonds are formed

**enantiomers** mirror-image, nonsuperimposable stereoisomers

**endergonic** energy-absorbing

**endocrine system** the series of ductless glands that release hormones into the bloodstream

**endocytosis** the process by which portions of a cell membrane are pinched off into the cell

**endoglycosidase** an enzyme that hydrolyzes glycosidic linkages anywhere along a polysaccharide chain

**endonuclease** an enzyme that hydrolyzes nucleic acids, attacking linkages in the middle of the polynucleotide chain

**enthalpy** a thermodynamic quantity, measured as the heat of reaction at constant pressure

**entropy** a thermodynamic quantity, a measure of the disorder of the universe

**enzyme** a biological catalyst, usually a globular protein, with self-splicing RNA as the only exception

**epimerase** an enzyme that catalyzes the inversion of configuration around a single carbon atom

**epimers** stereoisomers that differ only in configuration around one of several chiral carbon atoms

**equilibrium** the state in which a forward process and reverse process occur at the same rate

**essential amino acids** ones that cannot be synthesized by the body and must be obtained in the diet

**essential fatty acids** the polyunsaturated fatty acids (such as linoleic acid) that the body cannot synthesize; must be obtained from dietary sources

**estradiol** a steroid sex hormone

**estrogens** female sex hormones, steroids in chemical nature

**excision repair** repair of DNA by enzymatic removal of incorrect nucleotides and replacement by the correct ones

**exergonic** energy-releasing

**exoglycosidase** an enzyme that hydrolyzes glycosidic linkages starting at the end of a polysaccharide chain

**exon** a DNA sequence that is expressed in the sequence of mRNA

**exonuclease** an enzyme that hydrolyzes nucleic acids, starting at the end of the polynucleotide chain

**facilitated diffusion** a process by which substances enter a cell by binding to a carrier protein; does not require energy

**familial hypercholesterolemia** a disease characterized by high cholesterol levels in the bloodstream and early heart attacks

**farnesyl pyrophosphate** a 15-carbon intermediate in cholesterol biosynthesis

**fatty acid** a compound with a carboxyl group at one end and a long, normally unbranched hydrocarbon tail at the other; the hydrocarbon tail may be saturated or unsaturated

**feedback inhibition** the process by which the final product of a series of reactions inhibits the first reaction in the series

**fibrin** an insoluble protein formed as the final stage in the blood-clotting process

**fibrinogen** a soluble protein transformed into fibrin by selective hydrolysis of peptide bonds

**fingerprinting** the use of two methods in sequence for separating peptides

**first order** refers to a reaction whose rate depends on the first power of the concentration of a single reactant

**Fischer projection** a two-dimensional representation of the stereochemistry of three-dimensional molecules

**fluid mosaic model** the model for membrane structure in which proteins and a lipid bilayer exist side by side without covalent bonds between the proteins and lipids

**folate reductase** the enzyme that reduces dihydrofolate to tetrahydrofolate, a target for cancer chemotherapy

**N-formylmethionine-tRNA**$^{fmet}$ an essential factor for the start of protein synthesis, the amino acid methionine formylated at the amino group and covalently bonded to its specific tRNA

**free energy** a thermodynamic quantity, diagnostic for the spontaneity of a reaction at constant temperature and pressure

**free radical** a molecule that contains one or more unpaired electrons

**functional group** one of the groups of atoms that give rise to the characteristic reactions of organic compounds

**furanose** a cyclic sugar with a six-membered ring, named for its resemblance to the ring system in furan

**furanoside** a glycoside involving a furanose

**G-protein** a membrane-bound protein that mediates the action of adenylate cyclase

**β-galactosidase** the enzyme that hydrolyzes lactose to galactose and glucose, the classic example of an inducible enzyme

**gated channels** proteins that permit transient passage of external substances into a cell when suitably triggered

**gel electrophoresis** a method for separating molecules on the basis of charge-to-size ratio using a gel as a support and sieving material

**gene** the individual unit of inheritance

**general acid–base catalysis** a form of catalysis that depends on transfer of protons

**genetic code** the information for the structure and function of all living organisms

**genetic recombination** the combining of two different DNA molecules to produce a third molecule that is different from either of the original ones

**genome** the total DNA of the cell

**geranyl pyrophosphate** a ten-carbon intermediate in cholesterol biosynthesis

**giantism** a disease caused by overproduction of growth hormone before the skeleton has stopped growing

**glucocorticoid** a kind of steroid hormone involved in the metabolism of sugars

**glucogenic amino acid** one that has pyruvate or oxaloacetate as a catabolic breakdown product

**glucose** a monosaccharide, a ubiquitous metabolite

**glyceraldehyde 3-phosphate** a key intermediate in the reactions of sugars

**glycerol phosphate shuttle** a mechanism for transferring electrons from NADH in the cytosol to $FADH_2$ in the mitochondrion

**glycogen** a polymer of glucose, an important energy storage molecule in animals

**glycolipid** a lipid to which a sugar moiety is bonded

**glycolysis** the anaerobic breakdown of glucose to three-carbon compounds

**glycoside** a compound in which one or more sugars are involved in a linkage to another molecule

**glyoxysomes** membrane-bounded organelles that contain the enzymes of the glyoxylate cycle

**Golgi apparatus** a system of flattened membranous sacs, usually involved in secretion of proteins

**gout** a disease characterized by painful deposits of uric acid in the joints of the fingers and toes

**grana** bodies within the chloroplast that contain the thylakoid disks, the site of photosynthesis

**Greek key** a form of polypeptide chain folding that resembles a motif found in ancient pottery

**guanine** one of the purine bases found in nucleic acids

**half reaction** an equation that shows either the oxidative or the reductive part of an oxidation–reduction reaction

**Haworth projection formulas** a perspective representation of the cyclic forms of sugars

**helicase (rep protein)** unwinds the double helix of DNA in the process of replication

**α-helix** one of the most frequently encountered folding patterns in the protein backbone

**heme** an iron-containing cyclic compound, found in cytochromes, hemoglobin, and myoglobin

**hemiacetal** a compound formed by reaction of an aldehyde with an alcohol, found in the cyclic structure of sugars

**hemiketal**  a compound formed by reaction of a ketone with an alcohol, found in the cyclic structure of sugars

**hemolytic anemia**  a disease characterized by destruction of red blood cells

**hemophilia**  a molecular disease characterized by uncontrollable bleeding

**Henderson–Hasselbalch equation**  a mathematical relationship between the pK′a of an acid and the pH of a solution containing the acid and its conjugate base

**heparin**  a polysaccharide that has anticoagulant properties

**heteropolysaccharide**  one that contains more than one kind of sugar monomer

**hexose monophosphate shunt**  a synonym for the pentose phosphate pathway, in which glucose is converted to five-carbon sugars with concomitant production of NADPH

**histones**  basic proteins found complexed to eukaryotic DNA

**holoenzyme**  an enzyme that has all component parts, including coenzymes and all subunits

**homeostasis**  the balance of biological activities in the body

**homologous sequences**  regions of different macromolecules that have the same order of monomers

**homopolysaccharide**  one that contains only one kind of sugar monomer

**hormone**  a substance produced by endocrine glands and delivered by the bloodstream to target cells, producing a desired effect

**HPLC (high-performance liquid chromatography)**  a form of column chromatography

**hyaluronic acid**  a polysaccharide that occurs in the lubricating fluid of joints

**hydrazine**  a reagent for labeling the C-terminal end of proteins

**hydride ion transfer**  the transfer of a proton ($H^+$) and two electrons that occurs in many biological redox reactions

**hydrogen bonding**  a noncovalent association formed between a hydrogen atom covalently bonded to one electronegative atom and a lone pair of electrons on another electronegative atom

**hydrophilic**  tending to dissolve in water

**hydrophobic**  tending not to dissolve in water

**β-hydroxy-β-methylglutaryl-CoA**  an intermediate in the biosynthesis of cholesterol

**hyperbolic**  a characteristic of a curve on a graph such that it rises quickly and then levels off

**hyperchromicity**  the increase in absorption of ultraviolet light that accompanies the denaturation of DNA

**hypothalamus**  the portion of the brain that controls much of the workings of the endocrine system

**induced-fit model**  a description of substrate binding to an enzyme such that the conformation of the enzyme changes to accommodate the shape of the substrate

**inducible enzyme**  one whose synthesis can be triggered by the presence of some substance, the inducer

**induction of enzyme synthesis**  the triggering of the production of an enzyme by the presence of a specific inducer

**initial rate**  the rate of a reaction immediately after it starts, before any significant accumulation of product

**initiation complex**  the aggregate of mRNA, $N$-formylmethionine-tRNA$^{fmet}$, ribosomal subunits, and initiation factors needed at the start of protein synthesis

**initiation step**  the start of protein synthesis, the formation of the initiation complex

**integral protein**  one that is embedded in a membrane

**intermembrane space**  the region between the inner and outer mitochondrial membranes

**intron**  an intervening sequence in DNA that does not appear in the final sequence of mRNA

**ion exchange chromatography**  a method for separating substances on the basis of charge

**ion product constant for water**  a measure of the tendency of water to dissociate to give hydrogen ion and hydroxide ion

**ionophore**  a peptide or protein that serves as a channel through membranes for ions

**irreversible inhibition**  covalent binding of an inhibitor to an enzyme, causing permanent inactivation

**isoelectric focusing**  a method for separating substances on the basis of their isoelectric points

**isoelectric point**  the pH at which a molecule has no net charge

**isoprene**  a five-carbon unsaturated group, part of the structure of many lipids

**ketogenic amino acid**  one that has acetyl-CoA or acetoacetyl-CoA as a catabolic breakdown product

**α-ketoglutarate dehydrogenase complex**  one of the enzymes of the citric acid cycle; catalyzes the conversion of α-ketoglutarate to succinyl-CoA

**ketose**  a sugar that contains a ketone group as part of its structure

**Krebs cycle**  an alternate name for the citric acid cycle

**kwashiorkor**  a disease caused by serious protein deficiency

**L and D amino acids**  ones whose stereochemistry is the same as the stereochemical standards L and D glyceraldehyde, respectively

**labeling**  covalent modification of a specific residue on an enzyme

***lac* operon**  the promoter, operator, and structural

genes involved in the induction of $\beta$-galactosidase and related proteins

**lactate dehydrogenase** an NADH-linked dehydrogenase; catalyzes the conversion of pyruvate to lactate

**lactone** a cyclic ester

**lagging strand** in DNA replication, the strand that is formed in small fragments subsequently joined by DNA ligase

**lanosterol** a precursor of cholesterol

**leading strand** in DNA replication, the strand that is continuously formed in one long stretch

**Lesch–Nyhan syndrome** a metabolic disease characterized by severe retardation and a tendency to self-mutilation, caused by a deficiency in an enzyme of the purine salvage pathway

**leukotriene** a substance derived from leukocytes (white blood cells) that has three double bonds, of pharmaceutical importance

**Lewis acid–base catalysis** a form of catalysis that depends on the Lewis definition of an acid as an electron pair acceptor and a base as an electron pair donor

**ligand-gated** a channel protein that opens transiently on binding a specific molecule

**lignin** a polymer of coniferyl alcohol, a structural material found in woody plants

**Lineweaver–Burk double-reciprocal plot** a graphical method for analyzing the kinetics of enzyme-catalyzed reactions

**lipase** an enzyme that hydrolyzes lipids

**lipid** a compound insoluble in water and soluble in organic solvents

**lipid bilayers** aggregates in which the polar head groups are in contact with water and the hydrophobic parts are not

**lipoic acid** a coenzyme that can function either in redox reactions or as an acyl transfer agent

**liposome** an aggregate of lipids arranged so that the polar ends are in contact with water and the nonpolar tails are sequestered from water

**lock-and-key model** a description of binding of substrate to an enzyme such that the active site and the substrate exactly match each other in shape

**lysosomes** membrane-bounded organelles that contain hydrolytic enzymes

**macronutrients** ones needed in large amounts, such as proteins, carbohydrates, or fats

**malate-aspartate shuttle** a mechanism for transferring electrons from NADH in the cytosol to NADH in the mitochondrion

**malonyl-CoA** a three-carbon intermediate important in the biosynthesis of fatty acids

**matrix (mitochondrial)** the part of a mitochondrion enclosed within the inner mitochondrial membrane

**melting** the denaturation of DNA by heat

**mercaptoethanol** a reagent for reducing the disulfide bonds in proteins to sulfhydryl groups

**metabolic water** the water produced as a result of complete oxidation of nutrients; sometimes the only water source of desert-dwelling organisms

**methemoglobin** an abnormal form of hemoglobin in which the iron is Fe(III) rather than the normal Fe(II)

**micelle** an aggregate formed by amphiphilic molecules such that their polar ends are in contact with water and their nonpolar portions are on the interior

**Michaelis constant** a numerical value for the strength of binding of a substrate to an enzyme, an important parameter in enzyme kinetics

**micronutrients** vitamins and minerals, needed in small amounts

**microspheres** spherical aggregates of artificially synthesized proteinoids

**microtrabecular lattice (cytoskeleton)** the network of microtubules that pervades the cell

**microtubules** filaments made up of the protein tubulin, the material of the cytoskeleton, the mitotic spindle, and undulipodia

**mineralocorticoid** a kind of steroid hormone involved in the regulation of levels of inorganic ions, or "minerals"

**minerals** in nutrition, inorganic substances required as the ion or free element

**mitochondrion** an organelle that contains the apparatus responsible for aerobic oxidation of nutrients

**mobile phase (eluent)** in chromatography, the portion of the system in which the mixture to be separated moves

**molecular sieve chromatography** a method for separating molecules on the basis of size

**monomer** a small molecule that bonds to many others to form a macromolecule

**monosaccharide** a compound that contains a single carbonyl group and two or more hydroxyl groups

**mucopolysaccharide** a polysaccharide that has a gelatinous consistency

**multifunctional enzyme** one in which a single protein catalyzes several reactions

**multiple sclerosis** a disease in which the lipid sheath of nerve cells is progressively destroyed

**muscular dystrophy** a disease characterized by muscle weakness

**mutagen** an agent that brings about a mutation; includes radiation and chemical substances that alter DNA

**mutation** a change in DNA, causing subsequent changes in the organism that can be transmitted genetically

**myelin** the lipid-rich sheath of nerve cells

**N-terminal** the end of a protein or polypeptide with its amino group not linked to another amino acid by a peptide bond

**native conformation** a three-dimensional shape of a protein with biological activity

**nitrification** the conversion of nitrates to ammonia

**nitrogen fixation** the conversion of molecular nitrogen to ammonia

**nitrogenase** the enzyme complex that catalyzes nitrogen fixation

**noncompetitive inhibition** a form of enzyme inactivation in which a substance binds to a place other than the active site but distorts the active site so that the reaction is inhibited

**nonheme (iron–sulfur) protein** one that contains iron and sulfur but no heme group

**nonoverlapping, commaless code** a code in which no bases are shared between the sequences of three bases (triplets) that specify an amino acid, with no intervening bases

**nonpolar bond** a bond in which two atoms share electrons evenly

**nuclear region** the portion of a prokaryotic cell that contains the DNA

**nucleic acid** a macromolecule formed by polymerization of nucleotides; carries the genetic message

**nucleic acid base (nucleobase)** one of the nitrogen-containing aromatic compounds that make up the coding portion of a nucleic acid

**nucleophile** an electron-rich substance that tends to react with sites of positive charge or polarization

**nucleophilic substitution reaction** one in which one functional group is replaced by another as the result of nucleophilic attack

**nucleoside** a purine or pyrimidine base bonded to a sugar (ribose or deoxyribose)

**nucleosome** a structure in which DNA is wrapped around an aggregate of histones

**nucleolus** a portion of the nucleus rich in RNA

**nucleotide** a purine or pyrimidine base bonded to a sugar (ribose or deoxyribose), which in turn is bonded to a phosphate group

**nucleus** the organelle that contains the main genetic apparatus in eukaryotes

**Okazaki fragments** short stretches of DNA formed in the lagging strand in replication, subsequently linked by DNA ligase

**oligomer** an aggregate of several smaller units (monomers), for which bonding may be covalent or noncovalent

**oligosaccharide** a few sugars linked by glycosidic bonds

**one-carbon transfers** reactions in which the transfer usually involves carbon dioxide, a methyl group, or a formyl group

**operator** the gene to which a repressor of protein synthesis binds

**operon** a group of operator, promoter, and structural genes

**opsin** a protein in the rod and cone cells of the retina; plays a crucial role in vision

**optical isomers** (see **stereoisomers**)

**order of a reaction** the experimentally determined dependence of the rate on substrate concentrations

**organelle** a membrane-bounded portion of a cell with a specific function

**organic chemistry** the study of compounds of carbon, especially of carbon and hydrogen and their derivatives

**origin** in chromatography, the place to which the mixture to be separated is applied

**origin of replication** the point at which the DNA double helix begins to unwind at the start of replication

**oxidation** the loss of electrons

**β-oxidation** the main pathway of catabolism of fatty acids

**oxidative decarboxylation** loss of carbon dioxide accompanied by oxidation

**oxidative phosphorylation** a process for generating ATP; depends on the creation of a pH gradient within the mitochondrion as a result of electron transport

**oxidizing agent** a substance that accepts electrons from other substances

**palindrome** a message that reads the same from left to right or right to left

**palmitate** a 16-carbon saturated fatty acid, the end product of fatty acid biosynthesis in mammals

**paper chromatography** a method for separating substances on the basis of polarity, using paper as an inert support

**partial double-bond character** the characteristic of a bond normally written as a single bond that can also be written as a double bond by a simple shift of electrons

**passive transport** the process by which a substance enters a cell without expenditure of energy by the cell

**pectin** a polymer of galacturonic acid that occurs in plant cell walls

**pentose phosphate pathway** a pathway in sugar metabolism that gives rise to five-carbon sugars and NADPH

**peptide bond** an amide bond between amino acids in a protein

**peptides** molecules formed by linking two to several dozen amino acids by amide bonds

**peptidoglycan** a polysaccharide that contains peptide cross-links; found in bacterial cell walls

**peptidyl transferase** in protein synthesis, the enzyme that catalyzes formation of the peptide bond; part of the 50S ribosomal subunit

**peripheral proteins** ones loosely bound to the outside of a membrane

**peroxisomes** membrane-bounded sacs that contain enzymes involved in the metabolism of hydrogen peroxide ($H_2O_2$)

**pH** a measure of the acidity of a solution

**phenylketonuria** a disease characterized by mental retardation in developing children, caused by a lack of the enzyme that converts phenylalanine to tyrosine

**phosphatidic acid** a compound in which two fatty acids and phosphoric acid are esterified to the three hydroxyl groups of glycerol

**phosphatidylinositol *bis*phosphate (PIP$_2$)** a membrane-bound substance that mediates the action of $Ca^{2+}$ as a second messenger

**phosphoacylglycerol (phosphoglyceride)** a phosphatidic acid (*vide supra*) with another alcohol esterified to the phosphoric acid moiety

**3',5'-phosphodiester bond** a covalent linkage in which phosphoric acid is esterified to the 3' hydroxyl of one nucleoside and the 5' hydroxyl of another nucleoside; forms the backbone of nucleic acids

**phosphofructokinase** the key allosteric control enzyme in glycolysis; catalyzes the phosphorylation of fructose 6-phosphate

**phospholipase** an enzyme that hydrolyzes phospholipids

**photophosphorylation** the synthesis of ATP coupled to photosynthesis

**photorespiration** the process by which plants oxidize carbohydrates aerobically in the light

**photosynthesis** the process of using light energy from the sun to drive the synthesis of carbohydrates

**photosynthetic unit** the assemblage of chlorophylls that includes light-harvesting molecules and the special pair that actually carry out the reaction

**Photosystem I** the portion of the photosynthetic apparatus responsible for the production of NADPH

**Photosystem II** the portion of the photosynthetic apparatus responsible for the splitting of water to oxygen

**pili** short projections from the cell wall of bacteria; involved in sexual conjugation

**pituitary** the gland that releases trophic hormones to specific endocrine glands, under the control of the hypothalamus

**plasma membrane** another name for the cell membrane, the outer boundary of the cell

**plastocyanin** a copper-containing protein, part of the electron transport chain that links the two photosystems in photosynthesis

**plastoquinone** a substance similar to coenzyme Q; part of the electron transport chain that links the two photosystems in photosynthesis

**β-pleated sheet** one of the most important types of secondary structure, in which the protein backbone is almost fully extended

**P/O ratio** the ratio of ATP produced by oxidative phosphorylation to oxygen atoms consumed in electron transport

**polar bond** a bond in which two atoms have an unequal share in the bonding electrons

**polyacrylamide gel electrophoresis (PAGE)** a form of electrophoresis in which a polyacrylamide gel serves as both a sieve and a supporting medium

**polymer** a macromolecule formed by bonding of smaller units

**polypeptide chain** the backbone of a protein, formed by linking of amino acids by peptide (amide) bonds

**polysaccharide** a polymer of sugars

**polysome** the assemblage of several ribosomes bound to one mRNA

**porphyrins** large-ring compounds formed by linking four pyrrole rings; combine with iron ions to form the heme group

**power stroke** in muscle contraction, the stage in the process that causes the muscle fibers to slide past one another

**pregnenolone** a steroid sex hormone

**primary structure** the order in which the amino acids in a protein are linked by peptide bonds

**primosome** the complex at the replication fork in DNA synthesis; consists of the RNA primer, primase, and helicase

**progesterone** a steroid sex hormone

**prokaryote** an organism without a well-defined nucleus

**promoter** the portion of DNA to which RNA polymerase binds at the start of transcription

**proofreading** the process of removing incorrect nucleotides as DNA replication is in progress

**prostaglandin** one of a group of derivatives of arachidonic acid; contains a five-membered ring; of pharmaceutical importance

**prosthetic group** a portion of a protein that does not consist of amino acids

**protease** an enzyme that hydrolyzes proteins

**protein** a macromolecule formed by polymerization of amino acids

**proteolysis** the hydrolysis of proteins

**prothrombin** a protein that plays a part in blood clotting after activation by selective hydrolysis to form thrombin

**proton gradient** the difference between the hydrogen ion concentrations in the mitochondrial matrix and the intermembrane space, the basis of coupling between oxidation and phosphorylation

**purine** a nitrogen-containing aromatic compound that contains a six-membered ring fused to a five-membered ring, the parent compound of two nucleobases

**pyranose** a cyclic form of a sugar containing a five-membered ring, named for its resemblance to pyran

**pyranoside** a glycoside involving a pyranose

**pyrimidine** a nitrogen-containing aromatic compound that contains a six-membered ring; the parent compound of several nucleobases

**pyrrole ring** a five-membered ring that contains one nitrogen atom; part of the structure of porphyrins and heme

**pyruvate carboxylase** the enzyme that catalyzes the conversion of pyruvate to oxaloacetate

**pyruvate dehydrogenase complex** the enzyme that catalyzes the conversion of pyruvate to acetyl-CoA and carbon dioxide

**quaternary structure** the interaction of several polypeptide chains in a multisubunit protein

**R group** the side chain of an amino acid that determines its identity

**rate constant** a proportionality constant in the equation that describes the rate of a reaction

**rate-limiting step** the slowest step in a reaction mechanism; determines the maximum velocity of the reaction

**reaction center** the site of the special pair of chlorophylls responsible for trapping light energy from the sun

**reading frame** the starting point for reading a genetic message

**receptor protein** a protein on a cell membrane with specific binding site for extracellular substances

**reducing agent** a substance that gives up electrons to other substances

**reducing sugar** a sugar that has a free carbonyl group, one that can react with an oxidizing agent

**reduction** the loss of electrons

**regulatory gene** one that directs the synthesis of a repressor protein

**relative specificity** the property of an enzyme such that it acts on several related substrates

**repair** the enzymatic removal of incorrect nucleotides from DNA and their replacement by correct ones

**replication** the process of duplication of DNA

**replication fork** in DNA replication, the point at which new DNA strands are formed

**replisome** a complex of DNA polymerase, the RNA primer, primase, and helicase at the replication fork

**repressor** a protein that binds to an operator gene blocking the transcription and eventual translation of structural genes under the control of that operator

**residues** the portion of a monomer unit included in a polymer after splitting out of water between the monomers

**resonance structures** structural formulas that differ from one another only by the position of electrons

**respiratory complexes** the multienzyme systems in the inner mitochondrial membrane that carry out the reactions of electron transport

**restriction endonuclease** an enzyme that catalyzes a double-strand hydrolysis of DNA at a defined point in a specific sequence

**retinal** the aldehyde form of vitamin A

**retinol** the alcohol form of vitamin A

**retrovirus** one in which the base sequence of RNA directs the synthesis of DNA

**reverse turn** a part of a protein where the polypeptide chain folds back on itself

**reversible inhibitor** one that is not covalently bound to an enzyme; can be removed with restoration of activity

**rhodopsin** a molecule crucial in vision, formed by the reaction of retinal and opsin

**ribonucleoside** a compound formed when a nucleobase forms a glycosidic bond with ribose

**ribose** a sugar that is part of the structure of RNA

**ribosome** the site of protein synthesis in all organisms, consisting of RNA and protein

**ribulose 1,5-*bis*phosphate** a key intermediate in the production of sugars in photosynthesis

**rickets** a disease characterized by skeletal deformities, caused by a deficiency of vitamin D

**RNA** ribonucleic acid

**RNA polymerase** the enzyme that catalyzes the production of RNA on a DNA template

**mRNA** the kind of RNA that specifies the order of amino acids in a protein

**rRNA** the kind of RNA found in ribosomes

**tRNA** the kind of RNA to which amino acids are

bonded as a preliminary step to being incorporated into a growing polypeptide chain

**S-adenosylmethionine** a carrier of methyl groups

**salvage reactions** ones that reuse compounds, such as purines, that require a large amount of energy to produce

**saponification** the reaction of a triacylglycerol with base to produce glycerol and three molecules of fatty acid

**saturated** having all carbon–carbon bonds as single bonds

**Schiff base** a linkage of a carbonyl-containing substrate to an amino group on an enzyme

**sclerotic plaque** the damage to the myelin sheath of nerve cells in multiple sclerosis

**second messenger** a substance produced or released by a cell in response to hormone binding to a receptor on the cell surface; elicits the actual response in the cell

**secondary structure** the arrangement in space of the backbone atoms in a polypeptide chain

**sedimentation coefficient** the number describing the rate at which particles move to the bottom of the tube during centrifugation

**self-assembly of ribosomes** the reversible aggregation of ribosomal subunits in the presence of $Mg^{2+}$

**semiconservative replication** the mode in which DNA reproduces itself such that one strand comes from parent DNA and the other strand is newly formed

**sequential model** a description of the action of allosteric proteins in which a conformational change in one subunit is passed along to the other subunits

**serine protease** a proteolytic enzyme in which a serine hydroxyl plays an essential role in catalysis

**sickle-cell anemia** a disease caused by the change of one amino acid in two of the four polypeptide chains of hemoglobin; characterized by blockage of blood vessels

**sigmoidal** an S-shaped curve on a graph, characteristic of cooperative interactions

**simple diffusion** the process of passing through a pore or opening in a membrane without a requirement for a carrier or for energy

**single-strand–binding protein (SSB)** in DNA replication, a protein that protects exposed single-strand sections of DNA from nucleases

**single-strand heterozygous** a form of recombinant DNA in which one strand comes entirely from one of the original DNA molecules and the other contains sequences from both the original molecules

**sodium–potassium ion pump** the export of sodium ion from the cell with simultaneous inflow of potassium ion, both against concentration gradients

**solvent front** in chromatography, the leading edge of the mobile phase

**sphingolipid** a lipid whose structure is based on sphingosine

**sphingosine** a long-chain amino alcohol, the basis of the structure of a number of lipids

**spin labeling** the tagging of a part of a macromolecule with a group that has an unpaired electron that can be detected by magnetic measurements

**standard potential** the voltage, under standard conditions, of a half reaction of reduction compared with the hydrogen half reaction

**standard state** the standard set of conditions used for comparison of chemical reactions

**starch** a polymer of glucose that plays an energy storage role in plants

**stationary phase** in chromatography, the substance that selectively retards the flow of the sample, effecting the separation

**steady state** the condition in which the concentration of an enzyme–substrate complex remains constant in spite of continuous turnover

**stereochemistry** the three-dimensional shape of molecules

**stereoisomers (optical isomers)** molecules that differ from one another only in their configuration (three-dimensional shape)

**stereospecific** able to distinguish between stereoisomers

**steroid** a lipid with a characteristic fused-ring structure

**"sticky ends"** short, single-stranded stretches at the ends of double-stranded DNA; can overlap and provide sites at which DNA molecules can be linked

**stroma** a portion of the chloroplast equivalent to the mitochondrial matrix, the site of production of sugars in photosynthesis

**structural gene** one that directs the synthesis of a protein under the control of some regulatory gene

**substrate-level phosphorylation** a reaction in which the source of phosphorus is inorganic phosphate ion, not ATP

**sugar–phosphate backbone** the series of ester bonds between phosphoric acid and deoxyribose (in DNA) or ribose (in RNA)

**supercoiling** the presence of extra twists (over and above those of the double helix) in closed circular DNA

**supersecondary structure** specific clusters of secondary structure motifs in proteins

**Svedberg unit (S)** the unit of the sedimentation coefficient, a number that characterizes the motion of a particle in an ultracentrifuge tube

**swivel point** in DNA replication, the nick at which DNA unwinds before new strands can be formed

**Tay–Sachs disease** a disease caused by the lack of an enzyme that breaks down gangliosides, characterized by retardation and early death

**termination step** in protein synthesis, the point at which the stop signal is reached, releasing the newly formed protein from the ribosome

**tertiary structure** the arrangement in space of all the atoms in a protein

**testosterone** a steroid sex hormone

**tetrahydrofolate** the metabolically active form of the vitamin folic acid; a carrier of one-carbon groups

**thiamine pyrophosphate** a coenzyme involved in the transfer of two-carbon units

**thioester** a sulfur-containing analogue of an ester

**thrombin** a protein that catalyzes the last stage in blood clotting

**thylakoid disks** the site of the light-trapping reaction in chloroplasts

**thylakoid space** the portion of the chloroplast between the thylakoid disks

**thymidylate synthetase** the enzyme that catalyzes the production of thymine nucleotides needed for DNA synthesis; a target for cancer chemotherapy

**thymine** one of the pyrimidine bases found in nucleic acids

**thymine dimers** a defect in DNA structure caused by the action of ultraviolet light

**titration** an experiment in which a measured amount of base is added to an acid

**α-tocopherol** the most active form of vitamin E

**topoisomerase** an enzyme that relaxes supercoiling in closed circular DNA

**transaldolase** an enzyme that transfers a three-carbon unit in reactions of sugars

**transamination** the transfer of amino groups from one molecule to another, an important process in the anabolism and catabolism of amino acids

**transcription** the process of formation of RNA on a DNA template

**transition state** the intermediate stage in a reaction in which old bonds break and new bonds are formed

**transketolase** an enzyme that transfers a two-carbon unit in reactions of sugars

**translation** the process of protein synthesis in which the amino acid sequence of the protein reflects the sequence of bases in the gene that code for that protein

**translocation** in protein synthesis, the motion of the ribosome along the mRNA as the genetic message is being read

**transport protein** a component of a membrane that mediates the entry of specific substances into a cell

**transposable elements** genes that can change their position on a chromosome

**triacylglycerol (triglyceride)** a lipid formed by esterification of three fatty acids to glycerol

**tricarboxylic acid cycle** another name for the citric acid cycle

**triosephosphate isomerase** the enzyme that catalyzes the conversion of dihydroxyacetone phosphate to glyceraldehyde 3-phosphate

**triplet code** a sequence of three bases (a triplet) in mRNA specifies one amino acid in a protein

**trophic hormones** ones produced by the pituitary gland under the direction of the hypothalamus; in turn cause the release of specific hormones by individual endocrine glands

**trypsin** a proteolytic enzyme specific for basic amino acid residues at the site of hydrolysis

**turnover number** the number of moles of substrate that react per second per mole of enzyme

**uncoupler** a substance that overcomes the proton gradient in mitochondria, allowing electron transport to proceed in the absence of phosphorylation

**undulipodia** long strands projecting from eukaryotic cells, an aid to locomotion

**universal code** the genetic code, which is the same in all organisms

**unsaturated** having some carbon–carbon double bonds

**upstream** in transcription, a portion of the sequence nearer the 3′ end than the gene to be transcribed, where the DNA is read from the 3′ to the 5′ end and the RNA is formed from the 5′ to the 3′ end; in translation, nearer the 5′ end of the mRNA

**uracil** one of the pyrimidine bases found in nucleic acids

**urea cycle** a pathway that leads to excretion of waste products of nitrogen metabolism, especially that of amino acids

**uric acid** a product of catabolism of nitrogen-containing compounds, especially purines; accumulation in joints causes gout in humans

**vacuoles** membrane-bounded sacs that tend to be found in plant rather than animal cells, and that function as a storage location for toxic waste materials

**van der Waals bond** a noncovalent association based on the attraction of transient dipoles for one another

**vector**  a carrier molecule for transfer of genes in DNA recombination

**voltage-gated**  refers to a kind of channel protein transiently opened by changes in the membrane voltage

**"wobble"**  the possible variation in the third base of a codon allowed by several acceptable forms of base pairing between mRNA and tRNA

**x-ray crystallography**  an experimental method for determining the tertiary structure of proteins

**zero order**  refers to a reaction that proceeds at a constant rate, independent of the concentration of reactant

**zwitterion**  a molecule that has both a positive and a negative charge

**zymogen**  an inactive protein that can be activated by specific hydrolysis of peptide bonds

# INDEX

Page numbers in **boldface** refer to a major discussion of the entry. F after a page number refers to a figure or a structural formula. T after a page number refers to a table. Positional and configurational designations in chemical names (e.g., 3-, N-, $\alpha$-) are ignored in alphabetizing.

Abiosis, 12
Acetaldehyde, reduction of, 316
Acetic acid, dissociation constant of, 52T
  NaOH titration of, 54, 54F
Acetoacetyl-CoA, in ketone body production, 400, 401F
Acetone ($CH_3COCH_3$), in ketone body production, 400, 401F
  water dissolution of, 44
N-Acetyl amino sugars, 109–110, 110F
N-Acetyl-$\beta$-D-glucosamine, in chitin, 115F
  structure of, 110, 110F
Acetyl-CoA, 289, 395
  in fatty acid biosynthesis, 401–403, 402F, 403F
  in lipid anabolism, 512, 512F
  pyruvate conversion to, in citric acid cycle, **340–344**, 343F, 345F
Acetyl-CoA carboxylase, in fatty acid biosynthesis, 403, 403F
Acetyl phosphate, hydrolysis of, free energy of, 286T
Acetylcholine, in neuromuscular junction activation, 268, 268F
Acetylcholinesterase, turnover number of, 226T, 227
N-Acetylgalactosamine-6-sulfate, of chondroitin sulfate, 117F
N-Acetylglucosamine, structure of, 118, 119F
N-Acetylmuramic acid, structure of, 110, 110F, 118, 119F
Acid(s), **50–51**
  amino, 71–72, 72F, 80F, **80–82**, 81F. See also Amino acid(s).
  bile, from cholesterol, 420F, 421

carboxylic, 70
  definition of, 50
  dissociation of, 50–51
  strength of, 50
Acid-base catalysis, 246
Acid dissociation constant ($K_a'$), 50
Aconitase, fluorocitrate inhibition of, 347
  in citric acid cycle, 347
cis-Aconitate, in citric acid cycle, 348
Acromegaly, 519
Actin, 523, 523F
Activation, metabolic stage of, 288F, 288–290, 289F
Acyl carrier protein, in palmitate synthesis, 404F, 404–406, 405F
Acyl-CoA cholesterol acyltransferase, 422, 424F
Acyl-CoA synthetase, in fatty acid oxidation, 393, 393F
Acylglycerols, anabolism of, **408–414**, 409F–412F
N-Acylsphingosine, 412F, 413
  structure of, 133F
Addison's disease, 517
Adenine, 75, 537
  structure of, 74F, 536F
Adenosine diphosphate (ADP), 17, 18
  phosphorylation of, 152, 284–287, 285F, 286F, 286T
  glucose oxidation in, 286–287
Adenosine monophosphate (AMP), cyclic (cAMP), 519
  inosine monophosphate conversion to, **478–479**, 479F
  energy requirements for, **479–480**, 480F
  structure of, 74F, 538F

Adenosine triphosphate (ATP), 17, 18
  aspartate transcarbamoylase catalysis and, 233
  citrate synthetase inhibition by, 346
  from glucose oxidation, 386, 387T
  from stearic acid oxidation, 396–397
  hydrolysis of, 154–155, 285–286, 286T
  in energy utilization, 286F
  in metabolism, 289–290, 290F
  production of, 284–287, 285F, 286F, 286T
    in citric acid cycle, 350
    in photosynthesis, **439–441**, 440F, 441F
  structure of, 71F, 74F
S-Adenosylhomocysteine, 470
S-Adenosylmethionine, 470
  structure of, vs. methionine structure, 471F
Adenylate cyclase, in hormone receptor binding, 520
Adenylosuccinate lyase, in IMP anabolism, 503F
Adipose tissue, brown, 379
ADP. See Adenosine diphosphate.
Adrenocortical hormones, 517
Adrenocorticotropic hormone (ACTH), 516T
Alanine, abbreviation for, 84T
  biosynthesis of, 497F
  ionization of, 86, 86F
  p$K_a'$ of, 87T
  structure of, 83F
  titration curve of, 86, 86F
L-Alanine, 81, 81F
D-Alanine, 81, 81F
Alanyl-tRNA, structure of, 557, 557F

I-1

(Coupling)
conformational, in oxidative phosphorylation, 379F, **379–380**
in metabolic reactions, 284, 287
Creatine phosphate, hydrolysis of, free energy of, 286T
in muscle contraction, 525, 525F
Cristae, membrane, eukaryotic, 30, 31F
Crystallography, x-ray, 183
Cushing's syndrome, 518
Cyanobacteria, 20F, 21, 26, 38, 443
*Cyanophora paradoxa*, 38
Cyanosis, 193
Cysteine, abbreviation for, 84T
biosynthesis of, 470, 472F
p$K_a'$ of, 87T
structure of, 83F
Cytidine, structure of, 537F
Cytidine monophosphate (CMP), structure of, 538F
Cytidine triphosphate (CTP), aspartate transcarbamoylase inhibition by, 231–233, 232F
in phosphoacylglycerol anabolism, 410–411
production of, 231F, 231–233, 232F
uridine triphosphate conversion to, 485, 485F
Cytochrome(s), absorption peaks of, 380, 382F
heme groups of, 380–381, 381F
in electron transport, 366–371, 367F, 371T, **380–382**, 381F, 382F
Cytochrome c, three-dimensional structure of, 161, 161F
Cytochrome oxidase, 367F
in electron transport, 370–371
Cytoplasm, 27
Cytosine, 75, 536
deamination of, uracil production and, 582–583, 583F
structure of, 74F, 536F
Cytoskeleton, 33, 34F
Cytosol, 27
eukaryotic, 33
prokaryotic, 28

Dansyl chloride, in N-terminal amino acid identification, 210, 210F
Decarboxylation, oxidative, in citric acid cycle, 337, 338F
7-Dehydrocholesterol, in vitamin D reactions, 137F
Dehydrogenase, NADH-linked, binding sites of, 307F
Denaturation, of proteins, **169–171**, 170F, 171F
Denitrification, 460
β-D-Deoxyribose (2-deoxy-β-D-ribose), structure of, 106F
2-Deoxy-α-D-ribose, structure of, 73F
2-Deoxy-D-ribose, structure of, 73F

Deoxyadenosine 5′-monophosphate, structure of, 74F, 538F
Deoxycytidine 5′-monophosphate, structure of, 538F
Deoxyguanosine, structure of, 537F
Deoxyguanosine 5′-monophosphate, structure of, 538F
Deoxyhemoglobin, structure of, 190F
Deoxyribonucleases (DNases), 555
Deoxyribonucleic acid (DNA), 535, **541–549**
A, 543, 544F
B, 543, 545F
chain fragment of, 540, 540F
chimeric, 565
denaturation of, 565F, **565–566**, 566F
excision-repair of, 581–583, 582F, 583F
hydrogen bonds of, 50
methylation of, 583, 585F
modification of, after replication, **583–584**, 585F
of chloroplasts, 38
of mitochondria, 39
recombinant, **560–565**, 561F, 564F, **586–589**, 587F, 588F
chi intermediate in, 587F, 588
closed-circular, 565
heterozygous, double-strand, 587F, 588–589
single-strand, 587F, 588
homologous regions in, 586, 587F, 588–589
in DNA cloning, 564–565
restriction enzymes in, **560–564**, 561F, 564F
"sticky ends" in, 560–561, 561F
renaturation of, 566
replication of, 570, 571F–574F, **571–574**
bidirectional, 571F, **573–574**
cut-and-patch process in, 581
discontinuous, **574–575**, 575F
DNA ligase in, 575
Okazaki fragments in, 575, 575F
DNA gyrase in, 578, 578F
DNA polymerase in, **575–577**, 576F, 576T, 577F
DNA strand separation in, 571
double helix unwinding in, **577–578**, 578F, 579F
enzymes in, **577–580**, 578F, 579F
helicase in, 578, 579F
in eukaryotes, **584**, **586**
net chain growth in, 574
origin of replication in, 573–574, 574F
primase reaction in, **578–580**
primosome in, 580
proofreading in, 577, **580–583**, 581F–583F
repair in, 577, **580–583**, 583F
replication forks in, 574, 574F
replisome in, 580

semiconservative, 572F, **572–573**, 573F
experimental evidence for, 572–573, 573F
SSB protein in, 578, 579F
swivel point in, 578, 579F
sequencing of, direct chemical method for, 562–563
Maxam-Gilbert method for, 562–563
Sanger method for, 584–585
structure of, **541–549**, 542F–549F
primary, **558–560**, 559F, 560F
secondary, **541–546**, 542F–546F
complementary base pairing of, 541, 543F
complementary strands of, 541, 543F
sugar-phosphate backbone of, 541
supercoiling of, **546–549**, 547F–549F
tertiary, **546–549**, 547F–549F
circular, 548, 548F
thymine dimer formation and, 582, 582F
transposition of, 588
Z, 546, 547F
A-Deoxyribonucleic acid (DNA), 543, 544F
B-Deoxyribonucleic acid (DNA), 543, 545F
Z-Deoxyribonucleic acid (DNA), 546, 547F
Deoxyribonucleic acid (DNA) gyrase, in DNA replication, 578, 578F
Deoxyribonucleic acid (DNA) ligase, in DNA replication, 575
Deoxyribonucleic acid (DNA) polymerase, in DNA replication, **575–577**, 576F, 576T, 577F
of E. coli, 575, 576T
Deoxyribonucleic acid (DNA) polymerase I, 576T, 577
in Sanger DNA sequencing, 583
turnover number of, 226T, 227
Deoxyribonucleic acid (DNA) polymerase II, 576T, 577
Deoxyribonucleic acid (DNA) polymerase III, 576T, 577
Deoxyribonucleoside, 537, 537F
Deoxyribonucleotides, production of, **486–487**, 487F
Deoxythymidine 5′-monophosphate, structure of, 538F
Detergents, protein denaturation with, 170F, 171
Diacylglycerol (DAG), in calcium ion second messenger scheme, 520
in triacylglycerol synthesis, 408, 409F
Diacylglycerol acyltransferase, in triacylglycerol synthesis, 408, 409F
Diastereomers, 99